Lecture Notes in Computer Science   6617

Commenced Publication in 1973
Founding and Former Series Editors:
Gerhard Goos, Juris Hartmanis, and Jan van Leeuwen

Mihaela Bobaru Klaus Havelund
Gerard J. Holzmann Rajeev Joshi (Eds.)

NASA
Formal Methods

Third International Symposium, NFM 2011
Pasadena, CA, USA, April 18-20, 2011
Proceedings

 Springer

Volume Editors

Mihaela Bobaru
Klaus Havelund
Gerard J. Holzmann
Rajeev Joshi
NASA Jet Propulsion Laboratory
4800 Oak Grove Drive, M/S 301-285, Pasadena, CA 91109, USA
E-mail: {mihaela.bobaru, klaus.havelund, gh, rajeev.joshi}@jpl.nasa.gov

ISSN 0302-9743   e-ISSN 1611-3349
ISBN 978-3-642-20397-8   e-ISBN 978-3-642-20398-5
DOI 10.1007/978-3-642-20398-5
Springer Heidelberg Dordrecht London New York

Library of Congress Control Number: 2011924552

CR Subject Classification (1998): D.2.4, D.2, D.3, F.3, D.1

LNCS Sublibrary: SL 2 – Programming and Software Engineering

Typesetting: Camera-ready by author, data conversion by Scientific Publishing Services, Chennai, India

Printed on acid-free paper

Springer is part of Springer Science+Business Media (www.springer.com)

Preface

This publication contains the proceedings of the Third NASA Formal Methods Symposium (NFM 2011), which was held April 18–20, 2011, in Pasadena, CA, USA. The NASA Formal Methods Symposium is a forum for theoreticians and practitioners from academia, industry, and government, with the goal of identifying challenges and providing solutions to achieving assurance in safety-critical systems.

Within NASA, such systems include manned and unmanned spacecraft, orbiting satellites, and aircraft. Rapidly increasing code size, as well as the adoption of new software development paradigms, e.g., code generation and code synthesis, static source code analysis techniques and tool-based code review methods, bring new challenges and opportunities for significant improvement. Also gaining increasing importance in NASA applications is the use of more rigorous software test methods, founded in theory.

The focus of the symposium is understandably on formal methods, their foundation, current capabilities, as well as their current limitations. The NASA Formal Methods Symposium is an annual event that was created to highlight the state of the art in formal methods, both in theory and in practice. The series was originally started as the Langley Formal Methods Workshop, and was held under that name in 1990, 1992, 1995, 1997, 2000, and 2008. In 2009 the first NASA Formal Methods Symposium was organized by NASA Ames Research Center, and took place at Moffett Field, CA. In 2010 the symposium was organized by NASA Langley Research Center and NASA Goddard Space Flight Center, and held at NASA Headquarters, in Washington DC. This year's symposium was organized by the Laboratory for Reliable Software at the Jet Propulsion Laboratory / California Institute of Technology, and held in Pasadena CA.

The topics covered by NFM 2011 included but were not limited to: theorem proving, logic model checking, automated testing and simulation, model-based engineering, real-time and stochastic systems, SAT and SMT solvers, symbolic execution, abstraction and abstraction refinement, compositional verification techniques, static and dynamic analysis techniques, fault protection, cyber security, specification formalisms, requirements analysis, and applications of formal techniques.

Two types of papers were considered: regular papers describing fully developed work and complete results or case studies, and tool papers describing an operational tool, with examples of its application. The symposium received 141 submissions (112 regular papers and 29 tool papers) out of which 38 were accepted (26 regular papers and 12 tool papers), giving an acceptance rate of 27%. All submissions went through a rigorous reviewing process, where each paper was read by at least three reviewers.

In addition to the refereed papers, the symposium featured three invited talks and three invited tutorials. The invited talks were presented by Rustan Leino from Microsoft Research, on "From Retrospective Verification to Forward-Looking Development," Oege de Moor from the University of Oxford in England, and CEO of Semmle/Inc., on "Do Coding Standards Improve Software Quality?," and Andreas Zeller from Saarland University in Germany, on "Specifications for Free." The invited tutorials were presented by Andreas Bauer from the Australian National University in Australia, and Martin Leucker from Institut für Softwaretechnik und Programmiersprache, Universität zu Lübeck in Germany, on "The Theory and Practice of SALT—Structured Assertion Language for Temporal Logic," Bart Jacobs from the Katholieke Universiteit Leuven in Belgium on "VeriFast: A Powerful, Sound, Predictable, Fast Verifier for C and Java," and Michał Moskal from Microsoft Research, on "Verifying Functional Correctness of C Programs with VCC."

The organizers are grateful to the authors for submitting their work to NFM 2011 and to the invited speakers for sharing their insights. NFM 2011 would not have been possible without the collaboration of the outstanding Steering Committee, Program Committee, and external reviewers, and the general support of the NASA Formal Methods community. The NFM 2011 website can be found at http://lars-lab.jpl.nasa.gov/nfm2011.

Support for the preparation of these proceedings was provided by the Jet Propulsion Laboratory, California Institute of Technology, under a contract with the National Aeronautics and Space Administration.

February 2011   Mihaela Bobaru
                                                          Klaus Havelund
                                                          Gerard Holzmann
                                                          Rajeev Joshi

Organization

Program Chairs

Mihaela Bobaru	NASA Jet Propulsion Laboratory, USA
Klaus Havelund	NASA Jet Propulsion Laboratory, USA
Gerard Holzmann	NASA Jet Propulsion Laboratory, USA
Rajeev Joshi	NASA Jet Propulsion Laboratory, USA

Program Committee

Rajeev Alur	University of Pennsylvania, USA
Tom Ball	Microsoft Research, USA
Howard Barringer	University of Manchester, UK
Saddek Bensalem	Verimag Laboratory, France
Nikolaj Bjørner	Microsoft Research, USA
Eric Bodden	Technical University Darmstadt, Germany
Marsha Chechik	University of Toronto, Canada
Rance Cleaveland	University of Maryland, USA
Dennis Dams	Bell Labs/Alcatel-Lucent, Belgium
Ewen Denney	NASA Ames Research Center, USA
Ben Di Vito	NASA Langley, USA
Matt Dwyer	University of Nebraska at Lincoln, USA
Cormac Flanagan	University of California at Santa Cruz, USA
Dimitra Giannakopoulou	NASA Ames Research Center, USA
Patrice Godefroid	Microsoft Research, USA
Alex Groce	Oregon State University, USA
Radu Grosu	Stony Brook University, USA
John Hatcliff	Kansas State University, USA
Mats Heimdahl	University of Minnesota, USA
Mike Hinchey	Lero, the Irish SE Research Centre, Ireland
Sarfraz Khurshid	University of Texas at Austin, USA
Orna Kupferman	Jerusalem Hebrew University, Israel
Kim Larsen	Aalborg University, Denmark
Rupak Majumdar	Max Planck Institute, Germany
Kenneth McMillan	Microsoft Research, USA
César Muñoz	NASA Langley, USA
Madan Musuvathi	Microsoft Research, USA
Kedar Namjoshi	Bell Labs/Alcatel-Lucent, USA
Corina Păsăreanu	NASA Ames Research Center, USA
Shaz Qadeer	Microsoft Research, USA
Grigore Roşu	University of Illinois at Urbana-Champaign, USA

Nicolas Rouquette	NASA Jet Propulsion Laboratory, USA
Kristin Rozier	NASA Ames Research Center, USA
John Rushby	SRI International, USA
Wolfram Schulte	Microsoft Research, USA
Koushik Sen	University of California at Berkeley, USA
Sanjit Seshia	University of California at Berkeley, USA
Natarajan Shankar	SRI International, USA
Willem Visser	University of Stellenbosch, South Africa
Mahesh Viswanathan	University of Illinois at Urbana-Champaign, USA
Mike Whalen	University of Minnesota, USA

Steering Committee

Ewen Denney	NASA Ames Research Center, USA
Ben Di Vito	NASA Langley, USA
Dimitra Giannakopoulou	NASA Ames Research Center, USA
Klaus Havelund	NASA Jet Propulsion Laboratory, USA
Gerard Holzmann	NASA Jet Propulsion Laboratory, USA
César Muñoz	NASA Langley, USA
Corina Păsăreanu	NASA Ames Research Center, USA
James Rash	NASA Goddard Space Flight Center, USA
Kristin Rozier	NASA Ames Research Center, USA

External Reviewers

Ki Yung Ahn	Jaco Geldenhuys
Mihail Asavoae	Shalini Ghosh
Richard Banach	Alwyn Goodloe
Ezio Bartocci	Divya Gopinath
Ananda Basu	Andreas Griesmayer
Bert van Beek	Arie Gurfinkel
Shoham Ben-David	George Hagen
Simon Bliudze	Ian J. Hayes
Dragan Bosnacki	Daniel Holcomb
Marius Bozga	Cornelia Inggs
Bryan Brady	Ethan Jackson
Sebastian Burckhardt	Alan Jeffrey
Jacob Burnim	Susmit Jha
Katherine Coons	Dongyun Jin
Pierpaolo Degano	Barbara Jobstmann
Xianghua Deng	Taylor Johnson
Parasara Sridhar Duggirala	Anjali Joshi
Bruno Dutertre	Panagiotis Katsaros
Tayfun Elmas	Shadi Khalek
Vaidas Gasiunas	Robert Koenighofer

Ruurd Kuiper
Benoit Lahaye
Axel Legay
Wenchao Li
Karthik Manamcheri
Daniel Marino
Patrick Meredith
Michał Moskal
MohammadReza Mousavi
Anthony Narkawicz
Than Hung Nguyen
Sam Owre
Ganesh Pai
Chang-Seo Park
Fiona Polack
Pavithra Prabhakar
Vishwanath Raman
Giles Reger
Elaine Render
Robby
Neha Rungta
David Rydeheard
Mehrdad Sabetzadeh

Rick Salay
Ralf Sasse
Lucas Satabin
Traian Serbanuta
Peter Sestoft
Andreas Sewe
Shalini Shamasunder
K.C. Shashidhar
Elena Sherman
Junaid Siddiqui
Natalia Sidorova
Élodie-Jane Sims
Andrei Stefanescu
Christos Stergiou
Nikolai Tillmann
Ashish Tiwari
Arnaud Venet
Ou Wei
Tim Willemse
Guowei Yang
Razieh Zaeem
Hans Zantema
Chaoqiang Zhang

Table of Contents

IV. Tool Papers

From Retrospective Verification
to Forward-Looking Development

K. Rustan M. Leino

Microsoft Research, Redmond, WA, USA
leino@microsoft.com

Abstract. One obstacle in applying program verification is coming up with specifications. That is, if you want to verify a program, you need to write down what it means for the program to be correct. But doesn't that seem terribly wrong? Why don't we see it as "one obstacle in program design is coming up with code"? That is, if you want to realize a specification, you need to write down how the machine is supposed do it. Phrased this way, we may want to change our efforts of verification into efforts of what is known as correct-by-construction or stepwise-refinement. But the choice is not so clear and there are plenty of obstacles on both sides. For example, many programs are developed from specifications, but the specifications are not in a form suitable for refinement tools. For other programs, the clearest specifications may be given by pseudo-code, but such specification may not be suitable for some verification tools. In this talk, I will discuss verification tools and refinement-based tools, considering how they may be combined.

M. Bobaru et al. (Eds.): NFM 2011, LNCS 6617, p. 1, 2011.
© Springer-Verlag Berlin Heidelberg 2011

Specifications for Free

Andreas Zeller

Saarland University, Saarbrücken, Germany
zeller@cs.uni-saarland.de
http://www.st.cs.uni-saarland.de/

Abstract. Recent advances in software validation and verification make
it possible to widely automate the check whether a specification is satisfied. This progress is hampered, though, by the persistent difficulty of
writing specifications. Are we facing a "specification crisis"? By mining
specifications from existing systems, we can alleviate this burden, reusing
and extending the knowledge of 60 years of programming, and bridging
the gap between formal methods and real-world software. In this NFM
2011 invited keynote, I present the state of the art in specification mining, its challenges, and its potential, up to a vision of seamless integration
of specification and programming.

1 Introduction

Automated validation of software systems has finally come of age. In software
testing, test case generation now routinely automatically explores the entire program structure. In formal verification, the *Coq* prover has been used to produce
a fully verified C compiler; the *Verisoft* project has formally proven correctness
for a 10,000-line operating system kernels[1].

Dependable software, however, requires *specifications* of the intended behavior. Writing such specifications has always been hard; and while research has
made tremendous advances in systematically verifying and validating program
behavior, there has been little to no progress in actually *specifying* this behavior. This lack of specifications effectively hinders widespread adoption of rigorous
methods in industry—not only formal methods, but also systematic automated
testing. On top, developing new, dependable systems from scratch is risky as one
may get the specification wrong—the software would be correct with respect to
its specification, but still may be full of unpleasant surprises.

At his keynote at the SIGSOFT/FSE 2010 conference, Ralph Johnson identified specifications as the single missing piece for formal verification: "Everybody
agrees it is crucial but nobody works on it." All the pieces for rigorous, high-
quality software development are in place—except for one: *Where are we going
to get good specifications from?*

---

[1] Verisoft project, http://www.verisoft.de/, and Verisoft XT project,
http://www.verisoftxt.de/

M. Bobaru et al. (Eds.): NFM 2011, LNCS 6617, pp. 2–12, 2011.

2   Specification Mining

In this talk, I give an overview of *specification mining*. The idea is to extract *specifications from existing systems*, effectively leveraging the knowledge encoded into billions of code lines. These specifications are *models of software behavior*, but models so precise and concise so that they can act as specifications for building, verifying, and synthesizing new or revised systems. Rather than writing specifications from scratch, developers would thus rely on this existing knowledge base, overcoming specification inertia.

Let us use a simple example to illustrate the challenges and perspectives of specification mining. *NanoXML* is a popular open source Java package for parsing XML files. Its central data structure is a tree of *XMLElement* objects, defined in Figure 1. Each *XMLElement* node can have a number of children, publicly accessible via the *enumerateChildren()* method. Methods like *removeChild()* manipulate the set of children. Note that the individual methods are well-documented—for humans, that is. If we were to validate *removeChild()* (or any program using it), we need a formal specification denoting its pre- and postconditions. How can we obtain such a specification?

```
1    public class XMLElement implements IXMLElement, Serializable
2    {
3        // The name.
4        private String name;
5
6        // The child elements.
7        private Vector children;
8
9        // Returns an enumeration of all child elements.
10       public Enumeration enumerateChildren() { ... }
11
12       // Returns the number of children.
13       public int getChildrenCount() { ... }
14
15       // Returns true iff children exist
16       public bool hasChildren() { ... }
17
18       // Removes a child element.
19       public void removeChild(IXMLElement child) { ... }
20
21       // more methods and attributes...
22   }
```

Fig. 1. The XMLElement class from the NanoXML parser

3   Static Analysis

One way to infer properties from software systems is *static analysis* of the program code.

```
1    public void removeChild(IXMLElement child) {
2        if (child == null)
3            throw new IllegalArgumentException
4                ("child must not be null");
5        this.children.removeElement(child);
6    }
```

Fig. 2. *XMLElement.removeChild()* source

From the *removeChild()* code (Figure 2), any static analysis can easily deduce the precondition *child ≠ null*. But how would it know the postcondition, namely that *child* has been removed? The code of called methods such as *Vector.removeElement()* might be dynamically dispatched, remote, inaccessible, or come in another language.

The central problem with static analysis, though, is that it determines *all that could happen,* without differentiating normal usage from exceptions. For instance, static analysis cannot determine that the *child* parameter normally denotes a child node of the target (*"child? ∈ enumerateChildren"*). Differentiating "normal" from "abnormal" usage can only be deduced from actual call sites—and thus, our reasoning must be complemented *from actual usage.*

4   Dynamic Invariants

The concept of *dynamic invariants* addresses the issues of static analysis by analyzing actual executions. The key idea is to observe variable values at the beginning and the end of each call, check them against a library of *fixed invariant patterns,* and retain those patterns that match. This concept was pioneered by Ernst's DAIKON dynamic invariant detector [6]. From the NanoXML test suite execution, DAIKON indeed infers that no element of *children* is ever *null* and that all children again are of class *XMLElement* (Figure 3).

Dynamic invariants can be extremely helpful to get first insights into the behavior of a program. As specifications, however, dynamic invariants are still much too limited. They might refer to *internal details* (such as the private attribute *"this.children"*) or simply be *irrelevant* (*"this.lineNr != size(this.children[])"*).

```
1    this.children[] elements != null
2    this.children[].getClass() elements == net.n3.nanoxml.XMLElement.class
3    this.name.toString one of { "BAR", "FOO" }
4    this.lineNr != size(this.children[])
5    // 26 more...
```

Fig. 3. DAIKON invariants for *XMLElement*

The worst problem of all existing dynamic approaches to specification mining, though, is that their conclusions heavily depend on the set of *observed executions.*

In the NanoXML test suite, for instance, *this.name* always is "BAR" or "FOO", resulting in a misleading overgeneralization. Furthermore, *removeChild()* is never executed; hence, we obtain no pre- or postconditions.

5 Exploring Behavior

The problem of not having enough executions to observe can be alleviated by *generating* appropriate executions. Today's test case generation has made tremendous advances; successful test generation techniques include random selection of test inputs [13], applying meta-heuristic search techniques to find test data [11], or representing the test generation problem as a constraint solving problem, leveraging the power of modern constraint solvers to derive test cases [9].

Interestingly enough, the power of test case generation has rarely been combined with specification mining so far. In recent work [5], we have explored the combination of test case generation and specification mining to systematically *enrich* specifications. Our TAUTOKO prototype is poised towards extracting *typestate automata*—finite state machines that represent legal sequences of method calls on individual objects. After initial bootstrapping via a small set of (possibly random) executions, the key challenge is to explore (and test) the full behavior of a system. For this purpose, TAUTOKO *generates additional executions* as needed, systematically exploring the full behavior.

Let us assume we want to obtain a specification for *XMLElement* as defined in Figure 1, characterizing the correct usage of the *addChild()* and *removeChild()* methods. Let us also assume that we have one given run, consisting of a constructor call and adding one child. From this single run, TAUTOKO constructs the initial model shown in Figure 4.

How does this model come to be? To characterize states, TAUTOKO leverages *inspector methods* (i.e., methods without parameters or side effects) as detected in the object interface. In the case of *XMLElement*, the two inspector methods would be *hasChildren()* and *enumerateChildren()*. Before and after each call to a public method, TAUTOKO retrieves the values of these inspectors, thus characterizing the current object state. For numerical values, each sign of the value returned becomes a state; boolean and enumeration values each map to an individual state.

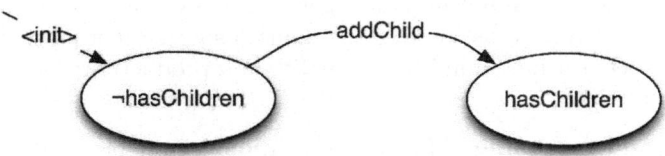

Fig. 4. *XMLElement* initial model

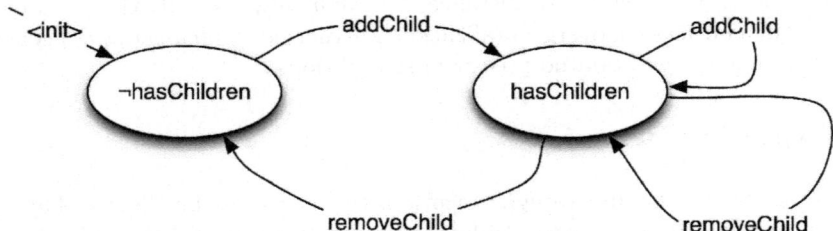

Fig. 5. *XMLElement* state transitions

In Figure 4, we can already see the pre- and postcondition of *addChild()* as observed so far. But what happens if we invoke *addChild()* when we are in the *hasChildren()* state? And what do methods like *removeChild()* do? TAUTOKO generates *test cases* that explore this behavior to *enrich* the original model. (Among others, this also ensures that *removeChild()* will actually be executed.) From these test cases, it then obtains *state transitions* as shown in Figure 5.

The specification now shows the full postcondition *hasChildren()* for *add-Child()*, which is also a precondition for *removeChild()*. Invoking *removeChild()* from the *¬hasChildren()* state results in an exception; hence this is not part of the resulting model. Besides being useful as documentation, such a mined typestate model can be immediately used for verification purposes. In our experiments, we fed them into static typestate verifiers and were able to discover bugs with very high precision [5].

6   Pre- and Postconditions

TAUTOKO is a very dedicated system with a limited, yet powerful model of program behavior. However, the general principle, combining test case generation with specification mining, can easily be extended to the much more expressive specifications as mined by DAIKON, for instance. Already right now, a typestate model as in Figure 5 directly translates into a full-fledged specification with pre- and postconditions; expressed as a parametric test case, this reads like the code shown in Figure 6.

Now assume that we could extract specifications like these for every single method in our system—*complete* as the generated test cases cover all aspects of behavior and *expressive* as we leverage domain-specific methods. For a program *P*, such specifications could boost quality and productivity in all software development activities:

Verification and modeling. Mined specifications could be abstract enough to be used for formal verification of *P* as well as its clients. In particular, *P*'s mined specifications are ideal starting points for modeling, building, or synthesizing *software that is similar to P or interacts with P*. If one were to rebuild current banking software in a rigorous, dependable fashion, mined axioms can ensure that while legacy code goes, its behavior prevails.

```
1   public void testRemoveChild(child c)
2   {
3       // Precondition
4       assume element.enumerateChildren().contains(c);
5       assume c != null;
6       assume element.hasChildren();
7       old_getChildrenCount = element.getChildrenCount();
8
9       element.removeChild(c);
10
11      // Postcondition
12      assert !element.enumerateChildren().contains(c);
13      assert element.getChildrenCount() == old_getChildrenCount - 1;
14  }
```

Fig. 6. A parametric *removeChild()* test

Software testing. Expressed as parametric test cases (Figure 6), one obtains *a test suite for free*—a test suite that covers all behavior of *P* previously mined. The test suite includes *oracles* (assertions) that validate pre- and postconditions as well as invariants such as "$doors\_open() \Rightarrow current\_speed() = 0$" for a bus on-board system. By systematically exploring corner cases ($\neg doors\_open()$, $current\_speed() > 0, \dots$), we could cover all predicate variations. Such fully automated testing will have a profound impact on development, as the programmer can assess the system at a much higher abstraction level; all she has to do is to validate the mined specification against the (implicitly) intended behavior.

Defect detection. Validating mined specifications can reveal *undesired* properties. A mined predicate like "$2010 \le year < 2016 \Rightarrow balance' = balance$" explicitly shows that the ATM card will not provide cash in 2010–2015[2]. Each such property would come with a test case demonstrating it.

Maintenance. Understanding what a piece of software does is an essential prerequisite for all maintenance activities; mining specifications could very much ease these tasks. With specifications, one can also determine the *impact of changes:* If a new revision violates the previous invariant $vertical\_sensor() < 32767$, the failing test case will automatically indicate a problem[3].

Seamless integration. Specifications need not be extracted from legacy systems alone. They could also be *derived from code as it is being written,* informing programmers immediately of the consequences of their actions: "This change causes this assertion to fail. Do you want to revise your code, or use this new assertion instead?"

By combining test case generation with dynamic state observation, we can apply these techniques to arbitrary legacy code; all we need is the ability to execute individual functions.

---

[2] "German banks hope to repair faulty cards", Wall Street Journal, January 6, 2010.
[3] "Space dreams of Europeans crash, burn with Ariane 5", Herald Journal, June 5, 1996.

7   Some Challenges

After these promises, let us now get back to earth; indeed, there is still loads of work to do and issues to face. Here's a non-exhaustive list of the challenges extracting the *XMLElement* behavior.

7.1   Domain-Specific Vocabularies

Current approaches to extracting dynamic invariants do not differentiate between levels of abstraction, as shown in Figure 3. Invariants are typically expressed over internal variables, rendering them irrelevant for clients—and making them useless as a basis for extracted specifications.

However, as we have demonstrated with TAUTOKO, modern design mandates the usage of specific *inspector methods* to access an object's state. The *XMLElement* class in Figure 1, for instance, provides methods like *hasChildren()* or *getChildrenCount()* for this purpose; Figure 5 demonstrates the usage of such inspectors in mined specifications.

7.2   Exploring Behavior

By systematically covering program structure (Figure 2), test cases would explore yet uncovered conditions such as *child = null* and, for instance, enrich a state model with a forbidden transition towards an *IllegalArgument* exception state.

However, if the *Vector* state is not accessible, we have little chance to systematically cover all of the *Vector*'s behavior. For instance, the vector will only remove the argument if it is an element of the Vector. Hence, *XMLElement.removeChild()* will do nothing if the argument is not a child; but how would we discover this? The challenge here will be to leverage given ground specifications for base classes like *Vector* as it comes to deriving high-level specifications.

7.3   Expressive Specifications

Approaches to extracting dynamic invariants take an "all or nothing" approach: A potential invariant is retained only if it matches *all* observed runs. This brings a problem with *conditional behavior:* If a postcondition holds only under certain circumstances, it will not be retained unless these circumstances are especially identified.

As an example, consider Figure 5. If we ever invoke *removeChild* with a non-child, *removeChild* leaves everything unchanged. Therefore, the pre- and postconditions observed so far no longer hold for all observed runs and are not retained—*removeChild* loses its specification. What we need is a *splitting condition* that differentiates between $child \in enumerateChildren$ (in which case the given predicates hold) and $child \notin enumerateChildren$ (in which case the object stays unchanged). Identifying such splitting conditions is hard due to the large number of potential premises; possible solutions to this problem include leveraging internal code structure or client usage.

7.4   Ranking Specifications

Looking back at *removeChild()*, we now have covered three situations:

1. A child is passed and removed:

$$child \in enumerateChildren \Rightarrow child \notin enumerateChildren'$$

2. A non-child is passed; everything stays as is:

$$child \in enumerateChildren \Rightarrow \Xi XMLElement$$

3. The argument is *null*, raising an exception

$$child? = null \Rightarrow IllegalArgumentException$$

Each of these three preconditions results in a different behavior, and together, they make up the complete *removeChild()* specification. However, some of these behaviors are more relevant than others—because they are relevant for the client, or because they are helpful in detecting errors. How can we *rank* the individual predicates that make up a specification such that the most relevant come out first? One of the most promising solutions is *client usage*—focusing on those specifications that are most *relevant* for program behavior.

7.5   Integration with Symbolic Approaches

Despite test generation attempting to cover all corner cases, mined specifications can never claim completeness, as they are derived from a finite set of observed executions. This becomes evident in *corner cases,* such as this one: If the same child is added *twice* to an *XMLElement*, the *removeChild()* method will only remove the first instance (because *Vector.removeElement()* does so). We either have to state that all *XMLElement*s span a tree (leveraging, say, an *XMLElement.isTree()* helper method), or extend the *removeChild()* specification.

While such corner cases may be ranked down as such (see above), one may wonder whether there is a way we can ensure the specifications are complete descriptions of behavior, making them universally valid. The challenge here is to explore missing aspects and resolve inconsistencies using state-of-the-art SMT solvers and model checkers.

7.6   Scalability and Efficiency

Finally, any approach for specification mining suffers from scalability issues. Tools like DAIKON or TAUTOKO require a significant run time overhead for their dynamic instrumentation; likewise, multiple executions of long-running programs consume a large amount of time.

8   Related Work

The term "mining specifications" was first applied to mining *finite state automata* from programs. Originally proposed by Ammons et al. [1], several researchers have extracted such automata statically or dynamically; the most recent work [8] focuses on inferring regular behavior from observed traces.

Mining specifications as pre- and postconditions from executions was first explored in the DAIKON tool by Michael Ernst et al. [6]. DAIKON works by instantiating *invariant patterns* such as "$1 = $2" against all possible variable values, retaining only those patterns that universally match. The most important usage of this work is the integration with theorem provers [2]; however, a recent comparison of DAIKON-inferred and programmer-written specifications [14] finds two major issues, shared by all dynamic approaches: *lack of detail* as it is hard to identify conditional behavior, and *incompleteness* as one is restricted to observed executions.

Static specification mining aims at discovering specifications from program code alone. For programs already annotated with pre- and postconditions, one can infer additional candidates [7]. However, static mining is frequently seen as a complement to dynamic mining [15,4].

Hybrid approaches combine specification mining with testing. The approach by Henkel and Diwan [10] systematically generates test cases to extract algebraic specifications, yet is limited to functional methods without side effects. Approaches integrating model checking and testing to refine abstractions [3] are promising, but limited to finite-state behavior. Combinations of DAIKON and random testing [13,12] so far focus on improving testing rather than the mined specifications.

9   Conclusion and Consequences

In the past decade, automated validation of software systems has made spectacular progresses. On the testing side, it is now possible to automatically generate test cases that effectively explore the entire program structure; on the verification side, we can now formally prove properties for software as complex as operating systems. To push validation further, however, we need *specifications* of what the software actually should do. But writing such specifications has always been hard—and so far significantly inhibited the deployment of rigorous development methods.

By combining specification mining (extracting specifications from executions) and *test case generation* (generating additional runs to explore execution space), we have the chance to obtain specifications that are both *complete* and *useful*. The proposed techniques all easily adapt to legacy programs; all we need is the ability to execute individual functions.

Before we get there, many challenges remain, as listed in section 7. As researchers, we should not see these challenges as obstacles, but rather as research opportunities. Formal methods just need this single piece—let's go and solve the puzzle!

Disclaimer. This paper was an invited paper and was not assessed by any member of the NFM 2011 organization. No-one but the author is to blame for weird ideas, inconsistencies, incompleteness, or plain nonsense.

Acknowledgments. Special thanks go to Valentin Dallmeier and Gordon Fraser for their hard work and ongoing inspiration.

Information on our recent work can be found at

http://www.st.cs.uni-saarland.de/

References

1. Ammons, G., Bodík, R., Larus, J.R.: Mining specifications. In: Proc. POPL 2002, pp. 4–16. ACM, New York (2002)
2. Burdy, L., Cheon, Y., Cok, D.R., Ernst, M.D., Kiniry, J.R., Leavens, G.T., Leino, K.R.M., Poll, E.: An overview of JML tools and applications. STTT 7(3), 212–232 (2005)
3. Clarke, E., Grumberg, O., Jha, S., Lu, Y., Veith, H.: Counterexample-guided abstraction refinement for symbolic model checking. J. ACM 50, 752–794 (2003)
4. Csallner, C., Tillmann, N., Smaragdakis, Y.: DySy: dynamic symbolic execution for invariant inference. In: Proc. ICSE 2008, pp. 281–290. ACM, New York (2008)
5. Dallmeier, V., Knopp, N., Mallon, C., Hack, S., Zeller, A.: Generating test cases for specification mining. In: Proceedings of the 19th International Symposium on Software Testing and Analysis, ISSTA 2010, pp. 85–96. ACM, New York (2010)
6. Ernst, M.D., Cockrell, J., Griswold, W.G., Notkin, D.: Dynamically discovering likely program invariants to support program evolution. IEEE TSE 27(2), 99–123 (2002)
7. Flanagan, C., Leino, K.R.M.: Houdini, an Annotation Assistant for ESC/Java. In: Oliveira, J.N., Zave, P. (eds.) FME 2001. LNCS, vol. 2021, pp. 500–517. Springer, Heidelberg (2001)
8. Ghezzi, C., Mocci, A., Monga, M.: Synthesizing intensional behavior models by graph transformation. In: Proc. ICSE 2009, pp. 430–440. IEEE Computer Society, Washington, DC (2009)
9. Godefroid, P., Klarlund, N., Sen, K.: DART: directed automated random testing. In: Proc. PLDI 2005, pp. 213–223. ACM, New York (2005)
10. Henkel, J., Diwan, A.: Discovering algebraic specifications from Java classes. In: Cardelli, L. (ed.) ECOOP 2003. LNCS, vol. 2743, pp. 431–456. Springer, Heidelberg (2003)
11. McMinn, P.: Search-based software test data generation: a survey. Software Testing, Verification & Reliability 14(2), 105–156 (2004)
12. Pacheco, C., Ernst, M.D.: Eclat: Automatic generation and classification of test inputs. In: Gao, X.-X. (ed.) ECOOP 2005. LNCS, vol. 3586, pp. 504–527. Springer, Heidelberg (2005)
13. Pacheco, C., Lahiri, S.K., Ernst, M.D., Ball, T.: Feedback-directed random test generation. In: Proc. ICSE 2007, pp. 75–84. IEEE Computer Society, Washington, DC (2007)

14. Polikarpova, N., Ciupa, I., Meyer, B.: A comparative study of programmer-written and automatically inferred contracts. In: Proc. ISSTA 2009, pp. 93–104. ACM, New York (2009)
15. Shoham, S., Yahav, E., Fink, S., Pistoia, M.: Static specification mining using automata-based abstractions. In: Proc. ISSTA 2007, pp. 174–184. ACM, New York (2007)

The Theory and Practice of SALT

Andreas Bauer[1] and Martin Leucker[2]

[1]NICTA Canberra Research Lab and The Australian National University
[2]Institut für Softwaretechnik und Programmiersprachen,
University of Lübeck, Germany

Abstract. SALT is a general purpose specification and assertion language developed for creating concise temporal specifications to be used in industrial verification environments. It incorporates ideas of existing approaches, such as PSL or Specification Patterns, in that it provides operators to express scopes and exceptions, as well as support for a subset of regular expressions. On the one hand side, SALT exceeds specific features of these approaches, for example, in that it allows the nesting of scopes and supports the specification of real-time properties. On the other hand, SALT is fully translatable to LTL, if no real-time operators are used, and to TLTL (also known as state-clock logic), if real-time operators appear in a specification. The latter is needed in particular for verification tasks to do with reactive systems imposing strict execution times and deadlines. SALT's semantics is defined in terms of a translation to temporal (real-time) logic, and a compiler is freely available from the project web site, including an interactive web interface to test drive the compiler. This tutorial paper details on the theoretical foundations of SALT as well as its practical use in applications such as model checking and runtime verification.

1 Introduction

When considering specification language formalisms, we have at least three different characteristics for their classification, which are (i) *expressiveness*, (ii) *conciseness*, and (iii) *readability*. In simple words, expressiveness means, which kind of languages can be defined at all within the considered formalism, while conciseness studies the question, how long the shortest spefications for a given family of languages is. Readability, on the other hand, deals with the question, how *easy* it is, to specify a certain language within the given formalism for a typical human beeing—and is thus a vague, not formal notion. SALT, which is an acronym for *structured assertion language for temporal logic*, aims to be a general purpose specification and assertion language, and was first introduced in [1]. It has been designed especially with *readability* in mind. Thus, one of the main goals of SALT is to offer users, who are not necessarily experts in formal specification and verification, a versatile tool that allows them to express system properties in a formal and concise, yet intelligible manner. In that respect, SALT has the look and feel of a general purpose programming language (e.g., it uses if-then-else constructs, supports (a subset of) regular expressions, and allows the definition

M. Bobaru et al. (Eds.): NFM 2011, LNCS 6617, pp. 13–40, 2011.

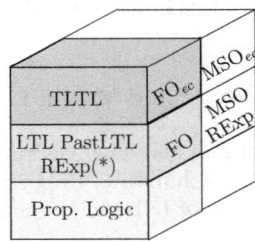

Fig. 1. Relationships between propositional, first-order, and temporal logics

of macros), yet it is fully translatable into standard temporal logics, such as *linear time temporal logic* (LTL [2]) or, if dedicated real-time operators appear inside a specification, to TLTL (also known as *state-clock logic* [3,4]). In other words, the untimed fragment of SALT is equally expressive as LTL, whereas the timed fragment is equally expressive as TLTL. D'Souza has shown in [4] that TLTL corresponds exactly to the first-order fragment of monadic second order logic interpreted over timed words. This resembles the correspondence of LTL and first-order logic over words as shown by Kamp [5]. However, LTL is strictly less expressive than second-order logic over words which, in turn, is expressively equivalent to ω-regular expressions. This also explains why full support of regular expressions is not possible when only LTL-expressible properties are in question (see Figure 1 for an overview).

As such it is possible to employ SALT as a higher level specification front-end to be used with standard model checking or runtime verification tools that otherwise would accept plain LTL formulae with a minimal set of operators as input. As a matter of fact, the freely available SALT compiler[1], which takes as input a SALT specification and returns a temporal logic formula, already supports the syntax of two powerful and commonly used model checking tools, namely SMV [6] and SPIN [7], such that deployment should be relatively straightforward, irrespective of the choice of verification tool.

The emphasis of this paper, however, is less on motivating the overall approach (which was already done in [1]), but rather to give an overview of the main language features, and to demonstrate how it can be used to specify complex temporal properties in a concise and accurate manner. Another important objective, which did not play an important role in [1], is to deepen our understanding of the similarities between SALT and some closely related approaches, in particular

- Dwyer et al.'s frequently cited *specification patterns* [8] which have become part of the Bandera system used to model checking Java programs [9],
- the *Property Specification Language* (PSL [10,11]), which is also a high-level temporal specification language, predominantly used for the specification of integrated circuits, and recently standardised by the IEEE (IEEE-Std 1850[TM]–2005),

---

[1] For details, see http://salt.in.tum.de/ and/or the authors' homepages.

As a matter of fact, parts of the design of SALT were directly influenced by the features existent in these approaches, but the goal was not to create yet another domain-specific tool, but a generic one which like LTL is more or less application-agnostic.

For instance, SALT offers operators to express complex temporal scopes (e.g., by means of from, upto, between, etc.), which is one of the main features underlying the specification patterns. On the other hand, SALT also offers operators to express so called exceptions (by means of accepton and rejecton), which similarly appear in PSL. While SALT's use of scopes exceeds the possibilities of the specification patterns, in that they allow the nesting of scopes, Sec. 4 will show that the exception operators in SALT are basically equivalent to those of PSL.

The rest of the paper is structured as follows. The next section is a guided tour of the language itself and highlights its main features; as such this section intersects most with work already presented in [1], except for the extended list of practical examples. Sec. 3 provides details on the translation of SALT expressions into LTL, i.e., semantics and implementation. In Sec. 4, we discuss in a more detailed manner the differences and similarities of SALT and other languages, in particular PSL and the Bandera input language, whereas in Sec. 5, we discuss complexity and experimental results of using SALT as a general purpose specification language. Finally, Sec. 6 concludes the paper.

2   Feature Overview

A SALT specification contains one or many assertions that together formulate the requirements associated with a system under scrutiny. Each assertion is translated into a separate LTL/TLTL formula, which can then be used in, say, a model checker or a runtime verification framework. SALT uses mainly textual operators, so that the frequently used LTL formula $\Box(p \rightarrow \Diamond q)$ would be written as

```
assert always (p implies eventually q).
```

Note that the assert keyword precedes all SALT specifications, except metadefinitions such as macros.

The SALT language itself consists of the following three layers, each covering different aspects of a specification:

- *The propositional layer* provides the atomic, Boolean propositions as well as the well-known Boolean operators.
- *The temporal layer* encapsulates the main features of the SALT language for specifying temporal system properties. The layer is divided into a future fragment and a symmetrical past fragment.
- *The timed layer* adds real-time constraints to the language. It is equally divided into a future and a past fragment, similar to the temporal layer.

Within each layer, macros and parameterised expressions can be defined and instantiated by iteration operators, enlarging the expressiveness of each layer into the orthogonal dimension of functions.

As pointed out in the introduction, depending on which layers are used for specification, the SALT compiler generates either LTL or TLTL formulae (resp. with or without past operators). For instance, if only operators from the propositional layer are used, the resulting formulae are purely propositional formulae. If only operators from the temporal and the propositional layer are used, the resulting formulae are LTL formulae, whereas if the timed layer is used, the resulting formulae are TLTL formulae.

2.1   Propositional Layer

Atomic propositions. Boolean propositions are the atomic elements from which SALT expressions are built. They usually resemble variables, signals, or complete expressions of the system under scrutiny. SALT is parameterised with respect to the propositional layer: any term that evaluates to either *true* or *false* can be used as atomic proposition. This allows, for example, propositions to be Java expressions when used for runtime verification of Java programs, or, simple bit-vectors when SALT is used as front end to verification tools like SMV [6].

Usually, every identifier that is used in the specification and that was not defined as a macro or a formal parameter is treated as an atomic proposition, which means that it appears in the output as it has been written in the specification. Additionally, arbitrary strings can be used as atomic propositions. For example,

```
assert always "state!=ERROR"
```

is a valid SALT specification and results in the output (here, in SMV syntax)

```
LTLSPEC G state!=ERROR.
```

However, the SALT compiler can also be called with a customised parser provided as a command line parameter, which is then used to perform additional checks on the syntactic structure of the propositions thus, making the use of structured propositions more reliable.

Boolean operators. The well-known set of Boolean operators $\neg$, $\wedge$, $\vee$, $\rightarrow$ and $\leftrightarrow$ can be used in SALT both as symbols (!, &, |, ->, <->), or as textual operators (not, and, or, implies, equals). Additionally, the conditional operators if-then and if-then-else, which have been already mentioned in the introduction, can be used. They tend to make specifications easier to read, because if-then-else constructs are familiar to programmers in almost any language. With the help of conditional operators, the introductory example could be reformulated as

```
assert always (if p then eventually q).
```

More so, any such formula can be arbitrarily combined using the Boolean connectives.

2.2 Temporal Layer

The temporal layer consists of a future and a past fragment. Although past operators do not add expressiveness [12], they can help to write formulae that are easier to understand and more efficient for processing [13].

In the following, we concentrate on the future fragment of SALT. The past fragment is, however, completely symmetrical. SALT's future operators are translated using only LTL future operators, and past operators are translated using only LTL past operators. This leaves users the complete freedom as to whether they do or do not want to have past operators in the result. This is useful as not all verification frameworks support both fragments. That said, it would be likewise possible to extend the current compilation process of SALT: The plain SALT compiler translates past operators into LTL with past operators. If the specification should be used say for a verification tool that does not support past operators, a further translation process may be started compiling past formulas to equivalent, possibly non-elementary longer future formulas. If, on the other hand, both future and past operators are supported, either output might be used, depending on how efficiently past operators are supported.

Standard LTL operators. Naturally, SALT provides the common LTL operators U, W, R, $\Box$, $\Diamond$ and $\circ$, written as `until`, `until weak`, `releases`, `always`, `eventually`, and `next`.

Extended operators. SALT provides a number of extended operators that help express frequently used requirements.

- `never`. The `never` operator is dual to `always` and requires that a formula never holds. While this could of course be easily expressed with the standard LTL operators, using `never` can, again, help to make specifications easier to understand.
- Extended `until`. SALT provides an extended version of the LTL U operator. The users can specify whether they want it to be *exclusive* (i. e., in φ U ψ, φ has to hold until the moment ψ occurs) or *inclusive* (i. e., φ has to hold until and during the moment ψ occurs)[2]

  They can also choose whether the end condition is *required* (i. e., must eventually occur), *weak* (i. e., may or may not occur), or *optional* (i. e., the expression is only considered if the end condition eventually occurs).
- Extended `next`. Instead of writing long chains of `next` operators, SALT users can specify directly that they want a formula to hold at a certain step in the future. It is also possible to use the extended `next` operator with an interval, e. g., specifying that a formula has to hold at some time between 3 and 6 steps in the future. Note that this operator refers only to states at certain positions in the sequence, not to real-time constraints.

---

[2] This has nothing to do with strict or non-strict U: strictness refers to whether the present state (i. e., the left end of the interval where φ is required to hold) is included or not in the evaluation, while inclusive/exclusive defines whether φ has to hold in the state where ψ occurs (i. e., the right end of the interval). Strict SALT operators can be created by adding a preceding `next`-operator.

Counting quantifiers. SALT provides two operators, `occurring` and `holding`, that allow to specify that an event has to occur a certain number of times. `occurring` deals with events that may last more than one step and are separated by one or more steps in which the condition does not hold. `holding` considers single steps in which a condition holds. Both operators can also be used with an interval, e. g., expressing the fact that an event has to occur *at most* 2 times in the future. To express this requirement manually in LTL, one would have to write

$$\neg p \ W \ (p \ W \ (\neg p \ W \ (p \ W \ \Box \neg p))).$$

The corresponding SALT specification is written concisely as

```
assert occurring[<=2] p.
```

Exceptions. SALT also includes exception operators, named `rejecton` and `accepton`, which interrupt the evaluation of a formula upon occurrence of an abort condition. `rejecton` evaluates a formula to false if the abort condition occurs and the formula has not been accepted before. For example, monitoring a formula $\Diamond \varphi$ when there has been no occurrence of φ yet would evaluate to false. The dual operator, `accepton`, evaluates a formula to true if it has not been rejected before.

While exceptions do not add expressiveness to the language (i.e., untimed SALT using exceptions is fully translatable to standard LTL), they can be very useful, for example, when specifying a communication protocol that requires certain messages to be sent, but allows to abort the communication at any time by sending a reset message. This would be expressed in SALT as

```
assert (con_open and next (data until con_close))
    accepton reset.
```

Exceptions also play an important role in the specification and verification of hardware systems. This is why languages such as PSL or ForSpec, which are used in this domain, both include this feature (see Sec. 4).

Scope operators. Many temporal specifications use requirements restricted to a certain *scope*, i. e., they state that the requirement has to hold only before, after, or between some events, and not on the whole sequence [8]. This can be expressed in SALT using the operators `upto` (or `before`), `from` (or `after`) and `between`.

Figure 2 illustrates scopes. It should be clear from the figure that it is mandatory in SALT to specify whether the delimiting events are part of the interval (i. e., *inclusive*) or not (i. e., *exclusive*).

Furthermore, for scope operators it has to be stated whether the occurrence of the delimiting events is strictly required. For example, the following specification

```
assert p
    between inclusive optional call,
            inclusive optional answer
```

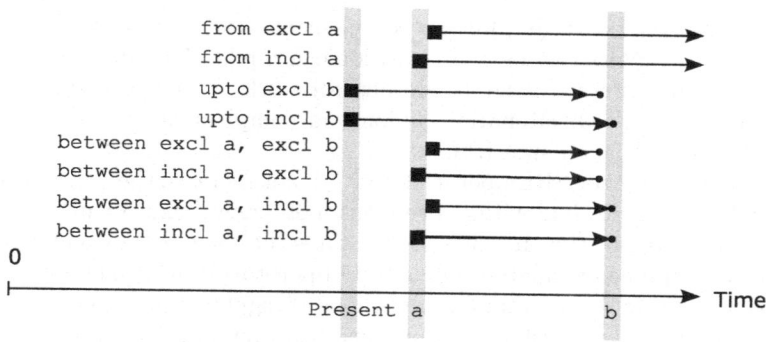

Fig. 2. Scopes of `upto`, `from` and `between`

means that p has to hold within the interval delimited by *call* and *answer*, provided such an interval exists. Without the keyword *optional*, such an interval would be required and within this interval, p must occur.

While it is possible to implement a translation of the `from` operator into LTL relatively straightforward (see Sec. 3), the `upto` operator proves to be more difficult, as can be seen in the following example.

A specification `always` φ `upto` b expresses that φ must always hold until the occurrence of the end condition b. A naïve translation into LTL would be φ W b. This is in order for a purely propositional φ, but might be wrong when temporal operators are used: Consider for example $\varphi := $ `p -> (eventually s)` yielding the formula $(p \rightarrow \Diamond s)Wb$, intending to say "p should be followed by s before b". The sequence `pbs` is a model for the latter formula, although `s` occurs after the end condition `b`, which clearly violates our intuitions. To meet our intuition, the negated end condition b has to be inserted into the U and $\bigcirc$ statements of φ in various places, e. g., like this: $(p \rightarrow (\neg b$ U $(\neg b \wedge s)))$ W b. Dwyer et al. describe this procedure in the notes of their specification pattern system [8]. It is however a tedious and highly error-prone task if undertaken manually.

SALT supports automatic translation by internally defining a stop operator. Using stop, the above example can be formulated as $((p \rightarrow \Diamond s)$ stop $b)Wb$ with stop b expressing that $(p \rightarrow \Diamond s)$ shall not take into account states after the occurrence of b. It is then transformed into an LTL expression in a similar way as the `rejecton` and `accepton` operators. Details can be found in Sec. 3.

Regular expressions. Regular expressions are well-known to many programmers. They provide a convenient way to express complex patterns of events, and appear also in many specification languages (see Sec. 4). However, arbitrary regular languages can be defined using regular expressions, while LTL only allows to define so-called *star-free* languages (cf. [14]). Thus, regular expressions have to be restricted in SALT in order to stay translatable to standard LTL. The main operators of SALT regular expressions (SREs) are concatenation (`;`), union (`|`)

and Kleene-star operators (*), but no complement. The argument of a Kleene-star operator is required to be a propositional formula. The advantage of this operator set—in contrast to the usual operator set for star-free regular expressions, which contains concatenation, union and complement—is that it can be translated more efficiently into LTL.

SALT provides further SRE operators that do not increase the expressiveness, but, arguably, make dealing with expressions more convenient for users. For example, the overlapping sequence operator : states that one expression follows another one, overlapping in one step. The ? and + operators (optional expression and repetition at least once) are common extensions of regular expressions. Moreover, there are a number of variations of the Kleene-star operator such as *[=n] to express how many steps from now the argument has to consecutively hold, *[>n] (resp. *[>n]) to express a minimum (resp. maximum) bound on the consecutive occurrence of the argument, *[n..m] to express an exact bound, etc. All these operators, however, have to adhere to the same restriction as the standard Kleene-star operator; that is, their argument needs to be a propositional formula.

While traditional regular expressions match only finite sequences, a SALT regular expression holds on an (infinite) sequence if it matches a finite prefix of the sequence.

Finally, with the help of regular expressions, we can rewrite the example using exception operators as

```
assert /con_open; data*; con_close/ accepton reset.
```

2.3   Timed Layer

SALT contains a timed extension that allows the specification of real-time constraints. Timed operators are translated into TLTL [3,4], a timed variant of LTL.

Timing constraints in SALT are expressed using the modifier timed[$\sim$], which can be used together with several untimed SALT operators in order to turn them into timed operators. $\sim$ is one of <, <=, =, >=, > for next timed and either < or <= for all other timed operators.

- next timed[$\sim c$] φ
  states that the next occurrence of φ is within the time bounds $\sim c$. This corresponds to the operator $\triangleright_{\sim c}\varphi$ in TLTL.
- φ until timed[$\sim c$] ψ
  states that φ is true until the next occurrence of ψ, and that this occurrence of ψ is within the time bounds $\sim c$. The extended variants of until can be used as timed operators as well.
- always timed[$\sim c$] φ
  states that φ must always be true within the time bounds $\sim c$.
- never timed[$\sim c$] φ
  states that φ must never be true within the time bounds $\sim c$.
- eventually timed[$\sim c$] φ
  states that φ must be true at some point within the time bounds $\sim c$.

2.4   Macros and Parameterised Expressions

SALT allows user-defined sub-expressions as *macros* and to parameterise macros and sub-expressions. Macro definitions do not begin with the `assert` keyword. They can be called in the same way as built-in SALT operators. Within certain limits, this allows the user to extend the SALT language using their own operators. For example, the following macro is called in infix notation:

```
define respondsto(x, y) := y implies eventually x
assert always (reply respondsto request)
```

Iteration operators allow to instantiate a parameterised sub-expression or macro with a list of values provided by the user. For example, the following specification states that either a or !b or c must hold forever.

```
assert someof list [a, !b, c] as i in always i
```

Parameters defined in a macro or an iteration expression can also be used to parameterise Boolean variables, as in the following example, which states that exactly one of the four variables, `state_1`, `state_2`, `state_3` and `state_4`, must be true.

```
assert exactlyoneof enumerate[1..4] as i in state_$i$
```

Macros can help to make a specification easier to understand, because complicated sub-expressions can be transparently hidden from the user, and accessed via an intuitive name that explains what the expression actually stands for. Sub-expressions that are used several times have to be written down only once.

2.5   Further Examples

In this section, a concluding look at some more SALT specifications is taken, and their corresponding LTL versions examined. The examples are mostly borrowed from the survey presented in [8], except where indicated otherwise. Note that propositions appearing in the specifications are not necessarily marked as such and are denoted in plain text only, indicating their intuitive meaning wrt. the respective application.

1. The requirement that a system should operate until a queue of jobs is either empty, or an abort signal issued can be formulated in LTL as

$$\neg((\neg(queuelength == 0 \lor abort))\ \mathrm{U}$$
$$(\neg working \land (\neg(queuelength == 0 \lor abort)))).$$

   The accompanying SALT specification would be:

   ```
   assert working until weak
       ("queuelength == 0" | abort),
   ```

   where abort is a proposition, and not the abort operator.

2. To specify idle behaviour, the following LTL specification could be used:

$$\Box(\neg return\_Execute \lor (return\_Execute \land ((\Diamond call\_Execute) \Rightarrow$$
$$(\neg(\neg call\_Execute \ \mathrm{U} \ (call\_doWork \land \neg call\_Execute)))))).$$

It asserts that between the moment in which an execution completes, and before a new one begins, there is no work done. In SALT, this example would be written as:

```
assert always
  (never call_doWork
    between inclusive optional return_Execute,
    exclusive optional call_Execute).
```

3. Coming back to an example from the area of protocol specification, one might assert that an answer was immediately preceded by a request. In LTL this would be written as:

$$\Box(answer \Rightarrow (\bigcirc request)).$$

Using a macro, in SALT, precedes can be expressed as follows:

```
define precedes(x, y) := if y then once x
assert always (request precedes answer).
```

4. A system with n input channels, may be using at most one at a time. Given that $n = 4$, this simple requirement would require

$$\Box((((in\_0 \land (\neg(in\_1 \lor (in\_2 \lor in\_3)))) \lor$$
$$((in\_1 \land (\neg(in\_0 \lor (in\_2 \lor in\_3)))) \lor$$
$$((in\_2 \land (\neg(in\_0 \lor (in\_1 \lor in\_3)))) \lor$$
$$(in\_3 \land (\neg(in\_0 \lor (in\_1 \lor in\_2)))))))) \lor$$
$$(\neg(in\_0 \lor (in\_1 \lor (in\_2 \lor in\_3)))))$$

if specified in LTL. The shorter SALT specification appears to be less error-prone and more readable:

```
assert always
  (exactlyoneof enumerate [0..3] as i in in_i) |
  (noneof enumerate [0..3] as i in in_i).
```

5. To show that regular expressions can be very useful for specification purposes, in the following it is expressed that a connection signal is eventually answered by an acknowledgement, followed by at least four data packets and a close signal. Again, this is first examined in LTL:

$$\Box(connection \Rightarrow$$
$$(\Diamond(answer \land (\bigcirc(data \ \mathrm{U} \ (data \land$$
$$(\bigcirc(data \land (\bigcirc(data \land (\bigcirc(data \land (\bigcirc close))))))))))))).$$

Now, consider the SALT counterpart using a regular expression:

```
assert always
    (if connection then
        eventually /answer; data*[>=4]; close/)
```

6. Consider an elevator: A possible requirement could be that between the time an elevator is called at a floor and the time it opens its doors at that floor, the elevator can arrive at that floor at most twice. In SALT, this can be specified as:

```
assert always
    (occurring[<=2] atfloor
      between inclusive optional call, exclusive
        optional open)
```

7. This section is now concluded by extending this example further and thus, showing most of SALT's features in one use-case. The following specification describes the following behaviour: On all three floors in a building, calling the elevator at floor i implies that it may pass at most two times at that floor without opening its doors, and that it must finally open its doors at that floor within 60 seconds.

```
define max_twice_at_floor_before_open(i) :=
    always (occurring[<=2] atfloor_$i$
                between inclusive optional call_$i$,
                    exclusive optional open_$i$)

define max_60s_before_open(i) :=
    always (call_$i$ implies
                eventually timed[<=60.0] open_$i$)

assert allof enumerate[1..3] as floor in
                max_twice_at_floor_before_open(floor)
            and max_60s_before_open(floor)
```

The modifiers optional in the between-statement make sure that atfloor_i is only checked provided call_i occurs.

Note that the equality between the LTL specifications in the above examples and their SALT counterparts, was established using the model checker SMV. For this purpose the SALT specifications were first compiled into plain LTL using the SALT compiler and then compared with the manually written requirements.

3   Semantics

SALT comes with a precisely defined semantics. It can be translated into either LTL or TLTL; the latter only when timed operators are used in a specification. Therefore, we define the semantics of SALT's operators by means of their corresponding LTL or respectively TLTL formulae.

More precisely, we define a translation function $\mathcal{T}$ to translate a valid SALT specification ψ into a temporal logic formula $\mathcal{T}(\psi)$, and define that an infinite word w over a finite alphabet of actions satisfies ψ iff $w \models \mathcal{T}(\psi)$ (using the standard satisfaction relation $\models$ defined for LTL/TLTL [15]).

For brevity, we exemplify the translation on a few selected operators only and refer to the extensive language reference and manual available from SALT's homepage at http://salt.in.tum.de/ for the remaining cases.

In what follows, let ψ, φ, and φ' denote SALT specifications. Many of SALT's operators can be considered as simple syntactic sugaring and are easily translated to LTL. For example, $\mathcal{T}(\varphi \text{ or } \varphi'))$ is translated inductively to $\mathcal{T}(\varphi) \vee \mathcal{T}(\varphi')$. The aforementioned accepton operator, which adds an exception to a specification is inductively defined as follows:

$$\mathcal{T}(b \text{ accepton } a) = b \vee a$$

$$\mathcal{T}((\neg\varphi) \text{ accepton } a) = \neg\mathcal{T}(\varphi \text{ rejecton } a)$$

$$\mathcal{T}((\varphi \wedge \psi) \text{ accepton } a) = \mathcal{T}(\varphi \text{ accepton } a) \wedge \mathcal{T}(\psi \text{ accepton } a)$$

$$\mathcal{T}((\varphi \vee \psi) \text{ accepton } a) = \mathcal{T}(\varphi \text{ accepton } a) \vee \mathcal{T}(\psi \text{ accepton } a)$$

$$\mathcal{T}((\varphi \text{ U } \psi) \text{ accepton } a) = \mathcal{T}(\varphi \text{ accepton } a) \text{ U } \mathcal{T}(\psi \text{ accepton } a)$$

$$\mathcal{T}((\bigcirc\varphi) \text{ accepton } a) = (\bigcirc\mathcal{T}(\varphi \text{ accepton } a)) \vee a$$

$$\mathcal{T}((\square\varphi) \text{ accepton } a) = \neg(\neg a \text{ U } \neg\mathcal{T}(\varphi \text{ accepton } a))$$

$$\mathcal{T}((\lozenge\varphi) \text{ accepton } a) = \lozenge\mathcal{T}(\varphi \text{ accepton } a),$$

Whereas the rejecton operator, which is used in the above definition, is given in terms of:

$$\mathcal{T}(b \text{ rejecton } r) = b \wedge \neg r$$

$$\mathcal{T}((\neg\varphi) \text{ rejecton } r) = \neg\mathcal{T}(\varphi \text{ accepton } r)$$

$$\mathcal{T}((\varphi \wedge \psi) \text{ rejecton } r) = \mathcal{T}(\varphi \text{ rejecton } r) \wedge \mathcal{T}(\psi \text{ rejecton } r)$$

$$\mathcal{T}((\varphi \vee \psi) \text{ rejecton } r) = \mathcal{T}(\varphi \text{ rejecton } r) \vee \mathcal{T}(\psi \text{ rejecton } r)$$

$$\mathcal{T}((\varphi \text{ U } \psi) \text{ rejecton } r) = \mathcal{T}(\varphi \text{ rejecton } r) \text{ U } \mathcal{T}(\psi \text{ rejecton } r)$$

$$\mathcal{T}((\bigcirc\varphi) \text{ rejecton } r) = (\bigcirc\mathcal{T}(\varphi \text{ rejecton } r)) \wedge \neg r$$

$$\mathcal{T}((\square\varphi) \text{ rejecton } r) = \square\mathcal{T}(\varphi \text{ rejecton } r)$$

$$\mathcal{T}((\lozenge\varphi) \text{ rejecton } a) = \neg r \text{ U } \mathcal{T}(\varphi \text{ rejecton } r).$$

However, not all SALT operators translate in such a straightforward inductive manner, since their translation depends on what is defined by the according sub-formulae occurring in a given expression. To guide the translation process for such operators, we have introduced an artificial or helper operator, stop, which is inductively defined as follows:

$$\mathcal{T}(b \ \mathrm{stop}_{\mathrm{excl}} \ s) = b$$

$$\mathcal{T}((\neg\varphi) \ \mathrm{stop}_{\mathrm{excl}} \ s) = \neg\mathcal{T}(\varphi \ \mathrm{stop}_{\mathrm{excl}} \ s)$$

$$\mathcal{T}((\varphi \wedge \psi) \ \mathrm{stop}_{\mathrm{excl}} \ s) = \mathcal{T}(\varphi \ \mathrm{stop}_{\mathrm{excl}} \ s) \wedge \mathcal{T}(\psi \ \mathrm{stop}_{\mathrm{excl}} \ s)$$

$$\mathcal{T}((\varphi \vee \psi) \ \mathrm{stop}_{\mathrm{excl}} \ s) = \mathcal{T}(\varphi \ \mathrm{stop}_{\mathrm{excl}} \ s) \vee \mathcal{T}(\psi \ \mathrm{stop}_{\mathrm{excl}} \ s)$$

$$\mathcal{T}((\varphi \ \mathrm{U} \ \psi) \ \mathrm{stop}_{\mathrm{excl}} \ s) = (\neg s \wedge \mathcal{T}(\varphi \ \mathrm{stop}_{\mathrm{excl}} \ s)) \ \mathrm{U} \ (\neg s \wedge \mathcal{T}(\psi \ \mathrm{stop}_{\mathrm{excl}} \ s))$$

$$\mathcal{T}((\varphi \ \mathrm{W} \ \psi) \ \mathrm{stop}_{\mathrm{excl}} \ s) = \mathcal{T}(\varphi \ \mathrm{stop}_{\mathrm{excl}} \ s) \ \mathrm{W} \ (s \vee \mathcal{T}(\psi \ \mathrm{stop}_{\mathrm{excl}} \ s))$$

$$\mathcal{T}((\bigcirc\varphi) \ \mathrm{stop}_{\mathrm{excl}} \ s) = \bigcirc(\neg s \wedge \mathcal{T}(\varphi \ \mathrm{stop}_{\mathrm{excl}} \ s))$$

$$\mathcal{T}((\bigcirc_W\varphi) \ \mathrm{stop}_{\mathrm{excl}} \ s) = \bigcirc(s \vee \mathcal{T}(\varphi \ \mathrm{stop}_{\mathrm{excl}} \ s))$$

$$\mathcal{T}((\Box\varphi) \ \mathrm{stop}_{\mathrm{excl}} \ s) = \mathcal{T}(\varphi \ \mathrm{stop}_{\mathrm{excl}} \ s) \ \mathrm{W} \ s$$

$$\mathcal{T}((\Diamond\varphi) \ \mathrm{stop}_{\mathrm{excl}} \ s) = (\neg s) \ \mathrm{U} \ (\neg s \wedge \mathcal{T}(\varphi \ \mathrm{stop}_{\mathrm{excl}} \ s))$$

where b denotes an atomic proposition from the action alphabet and s an arbitrary formula, possibly atomic also.

Thus, stop selects certain aspects of a formula, and in $\psi \equiv \varphi_1$ stop φ_2, intuitively asserts that the validity of ψ does not depend on events occurring after φ_2 has occurred. Again, for brevity, we consider only the exclusive variant of stop and only for the future fragment of SALT. The past fragment and inclusive semantics, however, are each symmetrical.

The more complicated scope operator upto, which was discussed earlier in Sec. 2.2, and whose translation depends on stop, is then defined as:

$$\mathcal{T}(\varphi \ \texttt{upto excl req} \ b) =$$
$$\text{if } \mathcal{T}(\varphi) = \Box\psi: \quad (\psi \ \mathrm{stop}_{\mathrm{excl}} \ b) \ \mathrm{U} \ b$$
$$\text{if } \mathcal{T}(\varphi) = \neg\Diamond\psi: \quad (\neg\psi \ \mathrm{stop}_{\mathrm{excl}} \ b) \ \mathrm{U} \ b$$
$$\text{else}: \quad (\Diamond b) \wedge (\mathcal{T}(\varphi) \ \mathrm{stop}_{\mathrm{excl}} \ b)$$

$$\mathcal{T}(\varphi \ \texttt{upto excl opt} \ b) =$$
$$\text{if } \mathcal{T}(\varphi) = \Diamond\psi: \quad \neg((\neg\psi \ \mathrm{stop}_{\mathrm{excl}} \ b) \ \mathrm{U} \ b)$$
$$\text{else}: \quad (\Diamond b) \rightarrow (\mathcal{T}(\varphi) \ \mathrm{stop}_{\mathrm{excl}} \ b)$$

$$\mathcal{T}(\varphi \ \texttt{upto excl weak} \ b) = (\mathcal{T}(\varphi) \ \mathrm{stop}_{\mathrm{excl}} \ b)$$

$$\mathcal{T}(\texttt{req} \ \varphi \ \texttt{upto excl req} \ b) =$$
$$\text{if } \mathcal{T}(\varphi) = \Box\psi: \quad \neg b \wedge ((\psi \ \mathrm{stop}_{\mathrm{excl}} \ b) \ \mathrm{U} \ b)$$
$$\text{if } \mathcal{T}(\varphi) = \neg\Diamond\psi: \quad \neg b \wedge ((\neg\psi \ \mathrm{stop}_{\mathrm{excl}} \ b) \ \mathrm{U} \ b)$$
$$\text{else}: \quad (\Diamond b) \wedge \neg b \wedge (\mathcal{T}(\varphi) \ \mathrm{stop}_{\mathrm{excl}} \ b)$$

$$\mathcal{T}(\texttt{req} \ \varphi \ \texttt{upto excl opt} \ b) =$$
$$\text{if } \mathcal{T}(\varphi) = \Diamond\psi: \quad \neg((\neg\psi \ \mathrm{stop}_{\mathrm{excl}} \ b) \ \mathrm{U} \ b)$$
$$\text{else}: \quad (\Diamond b) \rightarrow (\neg b \wedge (\mathcal{T}(\varphi) \ \mathrm{stop}_{\mathrm{excl}} \ b))$$

$$\mathcal{T}(\texttt{req}\ \varphi\ \texttt{upto excl weak}\ b) \quad = \neg b \wedge (\mathcal{T}(\varphi)\ \text{stop}_{\text{excl}}\ b)$$

$\mathcal{T}(\texttt{weak}\ \varphi\ \texttt{upto excl req}\ b) \quad =$
        if $\mathcal{T}(\varphi) = \Box\psi$: $(\psi\ \text{stop}_{\text{excl}}\ b)\ \text{U}\ b$
        if $\mathcal{T}(\varphi) = \neg\Diamond\psi$: $(\neg\psi\ \text{stop}_{\text{excl}}\ b)\ \text{U}\ b$
        else: $(\Diamond b) \wedge (b \vee (\mathcal{T}(\varphi)\ \text{stop}_{\text{excl}}\ b))$

$\mathcal{T}(\texttt{weak}\ \varphi\ \texttt{upto excl opt}\ b) \quad =$
        if $\mathcal{T}(\varphi) = \Diamond\psi$: $b \vee \neg((\neg\psi\ \text{stop}_{\text{excl}}\ b)\ \text{U}\ b)$
        else: $(\Diamond b) \rightarrow (b \vee (\mathcal{T}(\varphi)\ \text{stop}_{\text{excl}}\ b))$

$$\mathcal{T}(\texttt{weak}\ \varphi\ \texttt{upto excl weak}\ b) = b \vee (\mathcal{T}(\varphi)\ \text{stop}_{\text{excl}}\ b)$$

$$\mathcal{T}(\varphi\ \texttt{upto incl req}\ b) \quad = (\Diamond b) \wedge (\mathcal{T}(\varphi)\ \text{stop}_{\text{incl}}\ b)$$

$$\mathcal{T}(\varphi\ \texttt{upto incl opt}\ b) \quad = (\Diamond b) \rightarrow (\mathcal{T}(\varphi)\ \text{stop}_{\text{incl}}\ b)$$

$\mathcal{T}(\varphi\ \texttt{upto incl weak}\ b) \quad =$
        if $\mathcal{T}(\varphi) = \Box\psi$: $\neg(\neg b\ \text{U}\ \neg(\psi\ \text{stop}_{\text{incl}}\ b))$
        if $\mathcal{T}(\varphi) = \neg\Diamond\psi$: $\neg(\neg b\ \text{U}\ (\psi\ \text{stop}_{\text{incl}}\ b))$
        else: $(\mathcal{T}(\varphi)\ \text{stop}_{\text{incl}}\ b)$

where, of course, $\text{stop}_{\text{excl}}$ and $\text{stop}_{\text{incl}}$ are references to the exclusive and inclusive variants of stop, respectively.

Similar translation schemes are defined for the remaining operators' semantics, which are detailed in the SALT language reference and manual.

4   Comparison with Existing Approaches

As already mentioned in the introduction, the design of SALT is influenced by a number of existing (domain-specific) specification languages, in particular PSL and specification patterns. In order to see the differences between these approaches and SALT, besides their domain-specifitivity, we go again through the main list of SALT features and discuss similarities and differences between the approaches.

4.1   Overview

Like SALT, the Property Specification Language PSL [11] is a high level specification language, but predominantly used in the area of integrated circuit design. The initial version of PSL, which underwent standardisation by the IEEE, became available in March 2003, whereas the latest version, version 1.1, became available in April 2004. The PSL standard incorporates concepts and ideas of various other specification languages, such as ForSpec, which has been developed at Intel and been donated to Accellera in order to facilitate the then ongoing standardisation efforts (cf. [16]). PSL also comprises different layers, (i) a Boolean, (ii) temporal, (iii) verification, and (iv) modelling layer. The Boolean layer is much like SALT's propositional layer, and available in different flavours, depending on the concrete application domain; for instance, there exists a VHDL flavour which means that Boolean connectives follow roughly the same syntax used in

VHDL, which is a standard hardware specification language and simulation environment (cf. [17]). The temporal layer, in turn, is divided into two separate languages: the *foundation language* and the *optional branching extension*, with the main difference being that the former employs a linear model of time, and the latter a branching one. Therefore, in the following comparison with SALT, we focus mainly on the foundation language, although some features are the same for both languages. The foundation language, basically, consists of

- the usual Boolean connectives,
- future-time LTL operators,
- clocking operators,
- Sequential Extended Regular Expressions (SEREs), as well as
- an `abort` operator to model exceptions.

The verification layer consists of directives which describe how the temporal properties should be used by verification tools. Unlike SALT, the `assert` keyword is part of the verification layer, and not inherent to all specifications alike. Instead of `assert` it is also possible to `assume` that a specification holds, or to check if certain parts of a trace are `covered` by an SERE. As such, the verification layer instructs an employed verification tool how to treat the specification. Note that SALT specifications, basically, always use `assert`, except to define macros. Finally, the modelling layer of PSL is used to introduce domain-specific modelling constructs, e.g., in a VHDL-flavour or other hardware description language, to model directly the behaviour of hardware designs, and therefore augment what is possible using PSL alone. For example, it can be used to calculate an expected value of an output or use custom data structures, but then typically exceeding what is expressible using ω-regular languages alone. In fact, there exists currently no accepted formal semantics for PSL's modelling layer (cf. [18]). However, when augmented models are merely used for simulation, e.g., to compare the observed behaviour with a specified one (runtime verification), then such formal properties of the system are of no concern as a runtime verification toolkit would not care *how* a trace has been generated. SALT, on the other hand, aiming to be used as a general specification front-end, rather than a system modelling tool does not currently offer the integration of third-party languages as a means of extension.

Like PSL, the Bandera system [19,20] is also a domain-specific tool, in that it targets the Java language as platform to perform software model checking on. Dwyer et al.'s specification patterns [8] have been adopted by the Bandera Specification Language (BSL), which has a compiler to translate high-level specifications to LTL. Basically, Dwyer et al. analysed ca. 600 real-world specifications in order to identify common patterns among them [8]. These patterns were then formalised, and formed the foundation of their well-known *specification patterns*. Conceptually, specification patterns are similar to design patterns in software engineering [21]; that is, a pattern provides a solution to a recurring problem, often including notes about its advantages, drawbacks, and alternatives. As such it enables inexperienced users to reuse expert knowledge. The specification patterns themselves consist of *requirements*, such as "absence" (i.e., a condition is false) or "response" (i.e., an event triggers another one), that can be expressed under different *scopes*, like "globally", "before an event r", "after an event q", or "between two events r and q". Similarly to PSL and SALT, BSL consists of layers: the *assertion property specification layer* allows developers to define constraints on program contexts, whereas the *temporal property specification layer* provides support for temporal properties.

4.2 A Comparison of Features

In what follows, we go through a list of core features of the SALT language, as presented in Sec. 2, and discuss how the other specification languages mentioned above realise them, if at all. Since this discussion is guided by the features existing in SALT, it is not meant to distill a single best approach, but rather to show where the similarities and the differences are between all three languages. An objective comparison is also difficult because BSL and PSL each are optimised for a different purpose, namely hardware design/verification and software model checking.

Extended operators. SALT's extended operators aim at providing a richer set of LTL-like primitives, e.g., such as **never** as opposed to the frequently used LTL-operator **always**. PSL also has the **never** operator and its own equivalent operators to SALT's different versions of the **next** and **until** operators. BSL, on the other hand, discourages the use of low-level LTL primitives in favour of high-level patterns and scopes. Hence it does not provide these operators although it is easy to express them in terms of the standard LTL operators.

Scopes. Scopes have been identified by Dwyer et al. as an important issue in the specification pattern system. However, their pattern system is restricted to predefined requirements. That is, it does not allow nested scopes, and by default only certain combinations of inclusive/exclusive and required/optional delimiters. Some—but by far not all—scopes can also be expressed in PSL using the **next_event** and different variants of **before** operators. SALT's distinguishing feature here is that scope operators can be used with arbitrary formulae, even with nested scope operators as in the following example:

```
assert weak e between inclusive optional
  (eventually (required a before exclusive required b)
    from exclusive optional c),
  exclusive required d
```

Here, the outer-most scope is a **between**, which uses a **from** scope which, in turn, uses a **before** scope as one of its arguments. Admittedly, the example is hard to read and rather artificial, but it does highlight this particular feature of SALT in a very obvious manner.

Exceptions. Interestingly one of the main changes between versions 1.0 and 1.1 of PSL, besides precedence ordering, is the treatment of PSL's **abort** operator, and whose SALT counterpart are the operators **accepton** and **rejecton**.

The reason for this change is described in [22]. In this paper, Armoni et al. describe their discovery that the original definition of the **abort** operator would cause, in the worst-case, a non-elementary blow-up when translating a specification into an alternating Büchi automaton. In essence, this meant that PSL, as it was defined in version 1.0, could render subsequent formal verification an unnecessarily difficult if not impossible task, as the performance of most such tools, which use temporal specifications, directly depends on the size of the resulting automata representations. This problem has been addressed by basically adopting the semantics of a similar language wrt. this operator, called ForSpec [23], and which has been mainly developed at Intel and donated to Accellera in 2003. ForSpec also offers exception operators, called **accept** and **reject**, and specifications are translatable into a logic termed Reset-LTL in [22]. Although Reset-LTL contains two additional operators when compared to Pnueli's LTL, the two languages are actually equally expressive [22]. At this stage we only give an intuitive semantics of Reset-LTL, and refer the reader to the Appendix for a formal account.

An infinite word w at position i over some alphabet is said to satisfy a Reset-LTL formula, φ, if $\langle w^i, \text{false}, \text{false} \rangle \models \varphi$ holds. But unlike in standard LTL, this satisfaction relation is not only defined between an infinite word and a formula, but also between two additional Boolean formulae, which capture the exception conditions for the **accept** and **reject** operators, respectively. Let us refer to the former by a and the latter by r. Initially, when evaluating a formula, false and false are used for a and r, and the definition of the semantics (see Appendix) ensures that during evaluation, it is not possible for a and r to be true at the same time. That is, in some relation $\langle w^i, a, r \rangle \models \varphi$, if a is satisfied in state w_i then the entire word w^i is a model, irrespective of whether or not φ is satisfied by w^i using the standard LTL semantics. On the other hand, if r is satisfied, then w^i is not a model, irrespective of whether or not φ is satisfied using standard LTL semantics. As such a and r are, indeed, exception conditions, and set to a value other than false by the definition of the **accept** and **reject** operators above. What is interesting to note is that SALT's exception operators **accepton** and **rejecton** are, in fact, compatible with Reset-LTL's exception operators in the following sense.

Theorem 1. *The following relationship holds between the Reset-LTL operators* **abort** *and* **reject** *and the* SALT *operators* **accepton** *and* **rejecton***:*

$$\langle w^i, \text{false}, \text{false} \rangle \models \text{accept } e \text{ in } \phi \text{ if and only if } w^i \models \phi \text{ accepton } e,$$

and

$$\langle w^i, \text{false}, \text{false} \rangle \models \text{reject } e \text{ in } \phi \text{ if and only if } w^i \models \phi \text{ rejecton } e.$$

Again, for a formal proof of this statement, see the Appendix.

Note also that although PSL adheres to Reset-LTL semantics with respect to the two exception operators, it has adopted its own keyword (**abort**) and, unlike SALT's exception operators which are defined in a mutually recursive manner in

Sec. 3, uses a direct definition, expressed in terms of two "helper" symbols, ⊤ and ⊥, instead of the two Boolean context formulae as in Reset-LTL. These helper symbols are not part of the underlying alphabet. Basically, the symbol ⊤ is such that everything holds on it, including false, and ⊥ is such that nothing holds on it, including true. As the two semantic definitions for the exception operators are expressively equivalent, we abstain from giving further details at this point, but the interested reader may refer to [22,24] and [10, §B2.1.1.2].

Regular expressions. As pointed out in Sec. 2, SALT supports a subset of regular expressions, which is translatable to LTL. Note that as is the case with PSL, SALT regular expressions (SREs) do not offer complementation as an operator. The reason being not to restrict expressiveness, but the fact that arbitrary use of complementation in a specification can lead to exponentially larger LTL formulae in the translation. It is, however, possible to negate the language an expression defines by using **not** as can be seen in the example already employed in Sec. 2:

```
assert not /con_open; data*; con_close/
    accepton reset
```

As SEREs form a superset of SREs (modulo a different semantics, e.g., SEREs are typically enclosed by brackets instead of slashes), the above is also a valid PSL expression. SALT basically supports the same repetition operators as SEREs do, but with further restrictions on their arguments to not allow specifications that would otherwise exceed the expressiveness of star-free languages, and thus LTL:

- The argument of *, *[>n], *[>=n] and + has to be a propositional formula.
- All expressions except for the last in an SRE must be either Boolean propositions, or they must be other SRE combined by |. No other Boolean connectives are allowed for the combination of SRE (although they can be used to form propositional expressions).
- The last element in an SRE may be any SALT expression, however, because of operator precedences it may be necessary to surround it with parentheses.

Other operators, like the overlapping sequence operator (":") are also inspired from SEREs, and its semantics defined accordingly.

To the best of our knowledge, BSL does not currently offer any kind of support for regular expressions.

Real-time support. As also pointed out in Sec. 2, SALT has dedicated support for real-time specifications, in that temporal operators can be enriched with discrete timeouts as is shown in the last example in Sec. 2.5. Recall that all specifications employing real-time directives are translated into TLTL, and although the timing constraints that appear in a SALT specification can only be discrete, TLTL's underlying model of time is continuous [4]. TLTL basically enriches standard LTL with two operators, each accepting a discrete value as argument: one operator is used to express when a proposition was true in the past, and the other one to express when it will be true in the future. Based on these operators, it is easy to derive time-bounded variants of the typical LTL modalities.

Neither PSL nor BSL currently offer real-time support in the above sense. However, PSL supports clocked expressions using the "@" operator, which can be appended to unclocked expressions, similarly as time-bounds in SALT can be appended to untimed expressions. Clocks, however, are not used to model real-time, but to match (parts of) expressions with different parts of the clock cycles of the hardware system under scrutiny. For example, @rose defines that something has to hold on a rising edge, @negedge on a negative edge, and so on. As such, PSL adopts a hardware designer's point of view. SALT on the other hand adopts, more or less, a purely behavioural point of view, in that the intention is not to let users model the actual implementation of an event-driven real-time system, but its abstract behaviour. Arguably, a continuous model of time, as is offered by TLTL and discrete time-outs, are an adequate language to achieve this goal.

Macros and parameterised expressions. In comparison to SALT, PSL's macro definition capabilities are more akin to C or C++'s preprocessor. PSL defines the well-known directives for #define, #ifdef, #undef, etc. which behave in the expected way. In addition it offers two less common directives, %for and %if, whose semantics can be explained as follows. The %for directive replicates something a number of times. The syntax is as follows:

```
// using a range
%for var in expr1 .. expr2 do
...
%end
// using a list
%for var in { item1 , item2, ... , itemN } do
...
%end
```

where var is an identifier, expr1 and expr2 are statically computable expressions, and item1, item2 etc. are either a number or a simple identifier. In the first case the text inside the %for...%end pairs will be replicated $expr2 - expr1 + 1$ times (assuming that $expr2 \geq expr1$). In the second case the text will be replicated according to the number of items in the list (cf. [11, §8.5]). The following PSL macro definition using %for

```
%for ii in 0..3 do
assign aa[ii] = ii > 2;
%end
```

is therefore equivalent to this slightly longer piece of PSL code:

```
assign aa[0] =  0 > 2;
assign aa[1] =  1 > 2;
assign aa[2] =  2 > 2;
assign aa[3] =  3 > 2;
```

As such, the %for directive is PSL's counterpart to SALT's enumeration operator, whereas %if is similar to the #ifdef construct known from C/C++. However

Table 1. Comparison of SALT language features with those of other specification languages

	Ext. ops	Scopes	Exceptions	Reg. exp.	Real-time	Macros	Iterators
SALT	●	●	●	◐	●	●	●
PSL	●	◐	●	●	○	●	●
BSL	○	●	○	○	○	◐	●

`%if` must be preferred over `#ifdef` when the condition refers to variables defined in an encapsulating `%for`. For further details, refer to [11,10].

While BSL doesn't directly support macros in the above sense, it has a rich and powerful *assertion language* as well as *predicate definition sublanguage*. While neither offers an if-then-else construct, the assertion language lets users define assertions of the form of C's conditional operator ":", which is also part of C++, Java, and other languages, e.g., as in a? b : c, which is equivalent to if a then b else c. Also, assertions in BSL define static properties, in that they are Boolean conditions which can be checked at certain control-flow points throughout the execution of a Java program, such as method entry and return. However, SALT's parameterised expressions (and as such PSL's `%for` operator) have a match in BSL. Consider, for example, the following excerpt from a specification given in [19]:

```
FullToNonFull:  forall[b:BoundedBuffer].
{Full(b)} leads to {!Full(b)} globally
```

which is translated into a parameterised specification, which during verification is instantiated accordingly by all objects of type `BoundedBuffer`:

$$\Box(\texttt{Full(b)} \rightarrow \bigcirc(\neg\texttt{Full(b)})).$$

Obviously, this form of parameterisation using type information is geared towards the verification of Java programs.

Summary. As an overview, we present a brief summary of our findings in the form of a table in Table 1.

4.3   Further Related Work

EAGLE [25], is a temporal logic with a small but flexible set of primitives. The logic is based on recursive parameterised equations with fix-point semantics and merely three temporal operators: next-time, previous-time, and concatenation. Using these primitives, one can construct the operators known from various other formalisms, such as LTL or regular expressions. While EAGLE allows the specification of real-time constraints, it lacks most high level constructs such as nested scopes, exceptions, counting quantifiers currently present in SALT.

Duration calculus [26] and similar interval temporal logics overcome some of the limitations of LTL that we mentioned. These logics can naturally encode past operators, scoping, regular expressions, and counting. However, it is unclear how to translate specifications in these frameworks to LTL such that standard model checking and runtime verification tools based on LTL can be employed.

Notably, [27] describes a symmetric approach by providing a more low-level and formal framework in which the various different aspects of different temporal logics can be expressed. The observational mu-calculus is introduced as an "assembly language" for various extensions of temporal logic. In a follow-up paper [28], first results from an integration of the observational mu-calculus into the *Object Constraint Language* (OCL), which also forms part of the UML are described. However, the goal of this work was not to provide a more rich and natural syntax, but rather a sufficient set of temporal operators.

5   Realisation and Results

Specification languages like SALT, PSL, BSL, etc. aim at offering as many convenience operators to users as possible, in order to make the specifications more concise, thus readable, and the task of specification ultimately less error-prone. However, increased conciseness often comes at a price, namely that the complexity of these formalisms increases. Although, to the best of our knowledge, there does not exist a complexity result for PSL's satisfiability problem, there exist results for LTL and specific SERE features: While LTL is known to be PSpace-complete, it turns out that adding even just a single operator of the ones offered by SEREs makes the satisfiability problem at least ExpSpace hard [29]. On the other hand in [22] it is noted, that Reset-LTL, which we have used to express PSL's exception operators in Sec. 4 is only PSpace-complete. As, due to Theorem 1 we can easily create a Reset-LTL formula for every untimed SALT specification that uses only the LTL operators, extended operators, and exceptions, it follows that this fragment is also in PSpace. In fact, due to the PSpace-completeness of LTL, it follows that this fragment is PSpace-complete.

The situation is different when we consider the complete untimed fragment of SALT. In particular, it contains a variant of the $\bigcirc$-operator, as in $\mathtt{nextn}[n]\varphi$, which states that φ is required to hold n steps from now in the future. It was pointed out in [30], that the succinctness gains of this operator alone push the complexity of a logic up by one exponent as the formula $\mathtt{nextn}[2^n]$ is only of length $O(n)$. This is the same argument used in [23] to explain ExpSpace-hardness of FTL, the logic underlying ForSpec. We thus get:

Theorem 2. *The untimed* SALT *fragment consisting of LTL-, extended-, and abort-operators is PSpace-complete. By adding the* $\mathtt{nextn}$ *operator, one obtains an ExpSpace-complete fragment.*

Note that we currently have no similar result for full SALT as it would require analysing many more features than the ones above. In fact, to the best of our knowledge, there does not exist a similar result for PSL, despite the fragments considered in [29], presumably for the same reason.

5.1 Experimental Results

We have implemented our concepts in terms of a compiler for the SALT language. The compiler front end is currently implemented in Java, while its back end, which also optimises specifications for size, is realised via the functional programming language Haskell. Basically, the compiler's input is a SALT specification and its output a temporal logic formula. Like with programming languages, compilation of SALT is done in several stages. First, user-defined macros, counting quantifiers and iteration operators are expanded to expressions using only a core set of SALT operators. Then, the SALT operators are replaced by expressions in the subset SALT--, which contains the full expressiveness of LTL/TLTL as well as exception handling and stop operators. The translation from SALT-- into LTL/TLTL is treated as a separate step since it requires weaving the abort conditions into the whole subexpression. The result is an LTL/TLTL formula in form of an abstract syntax tree that is transformed easily into concrete syntax via a so-called *printing function*. Currently, we provide printing functions for SMV [6] and SPIN [7] syntax, but the users can easily provide additional printing functions to support their tool of choice. The use of optimised, context-dependent translation patterns as well as a final optimisation step performing local changes also help reducing the size of the generated formulae.

As the time required for model checking depends exponentially on the size of the formula to check, efficiency was an important issue for the development of SALT and its compiler. Because of the arguments presented in the discussion above, one might suspect that generated formulae are necessarily bigger and less efficient to check than handwritten ones. But our experiments show that the compiler is doing a good job of avoiding this worst-case scenario in practice.

In order to quantify the efficiency of the SALT compiler, existing LTL formulae were compared to the formulae generated by the compiler from a corresponding SALT specification. This was done for two data sets: the specification pattern system [8] (50 specifications) and a collection of real-world example specifications, mostly from the Dwyer's et al.'s survey data [8] (26 specifications). The increase or decrease of the formula was measured using the following parameters:

BA [Fri]: Number of states of the Büchi automaton (BA) generated using the algorithm proposed by Fritz [31], which is one of the best currently known. This is probably the most significant parameter, as a BA is usually used for model checking, and the duration of the verification process depends highly on the size of this automaton.

BA [Odd]: Number of states of the BA generated using the algorithm proposed by Oddoux [32].

U: Number of U, R, $\square$ and $\lozenge$ in the formula.

X: Number of $\bigcirc$ in the formula.

Boolean: Number of Boolean leafs, i. e., variable references and constants. This is a good parameter for estimating the length of the formula.

The results can be seen in Figure 3. The formulae generated by the SALT compiler contain a greater number of Boolean leafs, but use *less temporal operators* and,

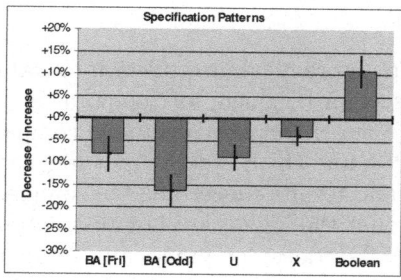 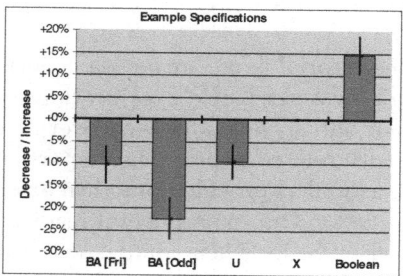

Fig. 3. Size of generated formulae

therefore, also yield a smaller BA. The error markers in the figure indicate the simple standard error of the mean.

Discussion. As it turned out, using SALT for writing specifications does not deprave model checking efficiency in practice. On the contrary, one can observe that it often leads to more succinct formulae. The reason for this result is that SALT performs a number of optimisations. For instance, when translating a formula of the form $\varphi W \psi$, the compiler can choose between the two equivalent expressions

$$\neg(\neg\psi \; U \; (\neg\varphi \wedge \neg\psi)) \quad \text{and} \quad (\varphi \; U \; \psi) \vee \Box\varphi.$$

While the first expression duplicates ψ in the resulting formula, the second expression duplicates φ, and introduces a new temporal operator. In most cases, the first expression, which is less intuitive for humans, yields better technical results.

Another equivalence utilised by the compiler is: $\Box(\varphi \; W \; \psi) \iff \Box(\varphi \vee \psi)$. With $\varphi \; W \; \psi$ being equivalent to $(\varphi \; U \; \psi) \vee \Box\varphi$, the left hand side reads as $\Box((\varphi \; U \; \psi) \vee \Box\varphi)$. When φ and ψ are propositions, this expression results in a BA with four states (using the algorithm proposed by Fritz [31]). $\Box(\varphi \vee \psi)$, however, is translated into a BA with only a single state.

Of course, the benefit obtained from using the SALT approach is of no principle nature: The rewriting of LTL formulae could be done without having SALT as a high-level language. What is more, given an LTL-to-BA translator that produces a minimal BA for the language defined by a given formula, no optimisations on the formula level would be required, and such a translation function exists—at least theoretically[3]. Nevertheless, the high abstraction level realised by SALT makes the mentioned optimisations *easily* possible, and produces BAs that are smaller than without such optimisations—despite the fact that today's LTL-to-BA translators already perform many optimisations.

---

[3] As the class of BAs is enumerable and language equivalence of two BAs decidable, it is possible to enumerate the class of BAs ordered by size and take the first one that is equivalent to the one to be minimised. Clearly, such an approach is not feasible in practice—and feasible minimisation procedures are hard to achieve.

6 Conclusions

In this tutorial paper we gave an overview and a practical introduction to SALT, a high-level extensible specification and assertion language for temporal logic. We not only gave an overview over its core features, but also a detailed comparison with related approaches, in particular PSL and the Bandera input language BSL, as well as provided practical examples and results concerning the complexity of SALT. Our experimental results show that the higher level of abstraction, offered by SALT when compared to normal LTL, does not practically result in an efficiency penalty, as compiled specifications are often considerably smaller than manually written ones. This is somewhat in contrast with our more theoretical considerations, in that the satisfiability problem of SALT specifications, depending on which features are used, can be exponentially harder than that of LTL. However, the experiments show that this exponential gap does not show up in many practical examples, and that our compiler, on the contrary, is able to optimise formulae to result in smaller automata.

Our feature comparison between SALT, PSL, and the Bandera input language BSL shows that SALT incorporates many of the features present in these domain-specific languages, while still being fully translatable to standard temporal logic. However, one could argue that this is also a shortcoming of SALT, in that it is not possible to express the full fragment of ω-regular languages as can be done in other approaches, but then of course not being able to map all specifications to LTL formulae any longer. This fact could be compensated for by adding a direct translation of SALT into automata as is suggested, for example, in [33], which introduces a regular form of LTL, i.e., expressively complete wrt. ω-regular languages. Moreover, the feature comparison does not show a clear "winner" among specification languages, since they have been designed for different purposes. In fact, SALT could be used in combination with other approaches, such as BSL where it would be possible to use the output of the SALT compiler, i.e., standard LTL formulae, as input to BSL's temporal property specification layer, which offers support for LTL specifications.

SALT as presented in this tutorial paper is ready to use and we invite the reader to explore it via an interactive web interface at `http://salt.in.tum.de/`, or to download the compiler from the same location.

References

1. Bauer, A., Leucker, M., Streit, J.: SALT—Structured Assertion Language for Temporal Logic. In: Liu, Z., He, J. (eds.) ICFEM 2006. LNCS, vol. 4260, pp. 757–775. Springer, Heidelberg (2006)
2. Pnueli, A.: The temporal logic of programs. In: Proc. 18th IEEE Symposium on the Foundations of Computer Science (FOCS), Providence, Rhode Island, pp. 46–57. IEEE, Los Alamitos (1977)
3. Raskin, J.-F., Schobbens, P.-Y.: State clock logic: A decidable real-time logic. In: Maler, O. (ed.) HART 1997. LNCS, vol. 1201, pp. 33–47. Springer, Heidelberg (1997)

4. D'Souza, D.: A logical characterisation of event clock automata. International Journal of Foundations of Computer Science 14(4), 625–639 (2003)
5. Kamp, J.A.W.: Tense Logic and the Theory of Linear Order. PhD thesis, University of California, Los Angeles (1968)
6. McMillan, K.L.: The SMV system, symbolic model checking - an approach. Technical Report CMU-CS-92-131, Carnegie Mellon University (1992)
7. Holzmann, G.J.: The model checker Spin. IEEE Trans. on Software Engineering 23, 279–295 (1997)
8. Dwyer, M.B., Avrunin, G.S., Corbett, J.C.: Patterns in property specifications for finite-state verification. In: Proc. 21st Int. Conf. on Software Engineering (ICSE), pp. 411–420. IEEE, Los Alamitos (1999)
9. Corbett, J.C., Dwyer, M.B., Hatcliff, J., Laubach, S., Pasareanu, C.S., Robby, Zheng, H.: Bandera: Extracting finite-state models from Java source code. In: Proc. 22nd Int. Conf. on Software Engineering (ICSE), IEEE, Los Alamitos (2000)
10. Accellera Property Specification Language. Reference Manual 1.1 (April 2004)
11. Eisner, C., Fisman, D.: A Practical Introduction to PSL (Series on Integrated Circuits and Systems). Springer, Heidelberg (2006)
12. Gabbay, D., Pnueli, A., Shelah, S., Stavi, J.: On the temporal analysis of fairness. In: Proc. 7th ACM SIGPLAN-SIGACT Symposium on Principles of Programming Languages (POPL), pp. 163–173. ACM, New York (1980)
13. Markey, N.: Temporal logic with past is exponentially more succinct, concurrency column. Bulletin of the EATCS 79, 122–128 (2003)
14. Lichtenstein, O., Pnueli, A., Zuck, L.D.: The glory of the past. In: Proc. Conference on Logic of Programs, pp. 196–218. Springer, Heidelberg (1985)
15. Manna, Z., Pnueli, A.: Temporal Verification of Reactive Systems. Springer, Heidelberg (1995)
16. Fix, L.: Fifteen years of formal property verification in intel. In: Grumberg, O., Veith, H. (eds.) 25 Years of Model Checking, pp. 139–144. Springer, Heidelberg (2008)
17. Ashenden, P.J.: The Designer's Guide to VHDL, 2nd edn. Morgan Kaufmann Publishers Inc., San Francisco (2001)
18. Ferro, L., Pierre, L.: Formal semantics for PSL modeling layer and application to the verification of transactional models. In: Proc. Conference on Design, Automation and Test in Europe (DATE), pp. 1207–1212. European Design and Automation Association (2010)
19. Corbett, J.C., Dwyer, M., Hatcliff, J., Robby: A language framework for expressing checkable properties of dynamic software. In: Havelund, K., Penix, J., Visser, W. (eds.) SPIN 2000. LNCS, vol. 1885. Springer, Heidelberg (2000)
20. Corbett, J., Dwyer, M., Hatcliff, J., Robby: Expressing checkable properties of dynamic systems: The Bandera specification language. Technical Report 04, Kansas State University, Department of Computing and Information Sciences (2001)
21. Gamma, E., Helm, R., Johnson, R., Vlissides, J.: Design Patterns: Elements of Reusable Object-Oriented Software. Addison-Wesley, Reading (1994)
22. Armoni, R., Bustan, D., Kupferman, O., Vardi, M.Y.: Resets vs. Aborts in linear temporal logic. In: Garavel, H., Hatcliff, J. (eds.) TACAS 2003. LNCS, vol. 2619, pp. 65–80. Springer, Heidelberg (2003)
23. Armoni, R., Fix, L., Flaisher, A., Gerth, R., Ginsburg, B., Kanza, T., Landver, A., Mador-Haim, S., Singerman, E., Tiemeyer, A., Vardi, M.Y., Zbar, Y.: The forSpec temporal logic: A new temporal property-specification language. In: Katoen, J.-P., Stevens, P. (eds.) TACAS 2002. LNCS, vol. 2280, pp. 211–296. Springer, Heidelberg (2002)

24. Eisner, C.: PSL for Runtime Verification: Theory and Practice. In: Sokolsky, O., Taşıran, S. (eds.) RV 2007. LNCS, vol. 4839, pp. 1–8. Springer, Heidelberg (2007)

25. Barringer, H., Goldberg, A., Havelund, K., Sen, K.: Rule-based runtime verification. In: Fifth International Conference on Verification, Model Checking and Abstract Interpretation (2004)

26. ChaoChen, Z., Hoare, T., Ravn, A.P.: A calculus of durations. Information Processing Letters 40(5), 269–276 (1991)

27. Bradfield, J., Stevens, P.: Observational mu calculus. In: Proc. Workshop on Fixed Points in Computer Science (FICS), pp. 25–27 (1998); An extended version is available as BRICS-RS-99-5

28. Bradfield, J.C., Filipe, J.K., Stevens, P.: Enriching OCL using observational mu-calculus. In: Kutsche, R.-D., Weber, H. (eds.) FASE 2002. LNCS, vol. 2306, pp. 203–217. Springer, Heidelberg (2002)

29. Lange, M.: Linear time logics around psl: Complexity, expressiveness, and a little bit of succinctness. In: Caires, L., Li, L. (eds.) CONCUR 2007. LNCS, vol. 4703, pp. 90–104. Springer, Heidelberg (2007)

30. Alur, R., Henzinger, T.A.: Real-time logics: complexity and expressiveness. Technical report, Stanford, CA, USA (1990)

31. Fritz, C.: Constructing Büchi Automata from Linear Temporal Logic Using Simulation Relations for Alternating Büchi Automata. In: Ibarra, O.H., Dang, Z. (eds.) CIAA 2003. LNCS, vol. 2759, pp. 35–48. Springer, Heidelberg (2003)

32. Gastin, P., Oddoux, D.: Fast LTL to büchi automata translation. In: Berry, G., Comon, H., Finkel, A. (eds.) CAV 2001. LNCS, vol. 2102, pp. 53–65. Springer, Heidelberg (2001)

33. Leucker, M., Sánchez, C.: Regular Linear Temporal Logic. In: Jones, C.B., Liu, Z., Woodcock, J. (eds.) ICTAC 2007. LNCS, vol. 4711, pp. 291–305. Springer, Heidelberg (2007)

A   Proofs

In this appendix, we summarise the formal semantics of Reset-LTL [22] and give a detailed proof of Theorem 1.

Definition 1 (Reset-LTL). *Let* $\Sigma := 2^{AP}$ *be a finite alphabet made up of propositions in the set* AP, $w \in \Sigma^\omega$ *an infinite word, and* a, r *be two Boolean formulae over* AP, *then*

- $\langle w^i, a, r \rangle \models p$ *if* $w^i \models a \vee (p \wedge \neg r)$,
- $\langle w^i, a, r \rangle \models \neg\varphi$ *if* $\langle w^i, r, a \rangle \not\models \neg\varphi$
- $\langle w^i, a, r \rangle \models \varphi \vee \psi$ *if* $\langle w^i, a, r \rangle \models \varphi$ *or* $\langle w^i, a, r \rangle \models \psi$
- $\langle w^i, a, r \rangle \models \bigcirc\varphi$ *if* $w^i \models a$ *or* $\langle w^{i+1}, a, r \rangle \models \varphi$ *and* $w^i \not\models r$,
- $\langle w^i, a, r \rangle \models \varphi U \psi$ *if* $\exists k \geq i.\ \langle w^k, a, r \rangle \models \psi \wedge \forall i \leq l < k.\ \langle w^l, a, r \rangle \models \varphi$,
- $\langle w^i, a, r \rangle \models$ accept e in φ *if* $\langle w^i, a \vee (e \wedge \neg r), r \rangle \models \varphi$,
- $\langle w^i, a, r \rangle \models$ reject e in φ *if* $\langle w^i, a, r \vee (e \wedge \neg a) \rangle \models \varphi$,

where w^i *denotes* w*'s infinite suffix after the* i-*th position, i.e.,* $w^i = w_i w_{i+1} \ldots$

Theorem 1. The following relationship holds between the Reset-LTL operators abort and reject and the SALT operators accepton and rejecton:

$$\langle w^i, \text{false}, \text{false} \rangle \models \texttt{accept } e \texttt{ in } \phi \text{ if and only if } w^i \models \phi \texttt{ accepton } e,$$

and

$$\langle w^i, \text{false}, \text{false} \rangle \models \texttt{reject } e \texttt{ in } \phi \text{ if and only if } w^i \models \phi \texttt{ rejecton } e.$$

Proof. By structural induction. Note that the relevant semantic definitions for SALT's exception operators are given in Sec. 3. Let $p \in AP$ and $\phi := p$.

$$\begin{aligned}
&\langle w^i, \text{false}, \text{false} \rangle \models \texttt{accept } e \texttt{ in } p \\
&\Leftrightarrow \langle w^i, e, \text{false} \rangle \models p \\
&\Leftrightarrow w^i \models e \vee p \\
&\Leftrightarrow w^i \models p \texttt{ accepton } e.
\end{aligned}$$

$$\begin{aligned}
&\langle w^i, \text{false}, \text{false} \rangle \models \texttt{reject } e \texttt{ in } p \\
&\Leftrightarrow \langle w^i, \text{false}, e \rangle \models p \\
&\Leftrightarrow w^i \models p \wedge \neg e \\
&\Leftrightarrow w^i \models p \texttt{ rejecton } e.
\end{aligned}$$

$\phi := \varphi \vee \psi$:

$$\begin{aligned}
&\langle w^i, \text{false}, \text{false} \rangle \models \texttt{accept } e \texttt{ in } \varphi \vee \psi \\
&\Leftrightarrow \langle w^i, e, \text{false} \rangle \models \varphi \vee \psi \\
&\Leftrightarrow \langle w^i, e, \text{false} \rangle \models \varphi \vee \langle w^i, e, \text{false} \rangle \models \psi \\
&\Leftrightarrow w^i \models \varphi \texttt{ accepton } e \vee w^i \models \psi \texttt{ accepton } e \\
&\Leftrightarrow w^i \models \varphi \vee \psi \texttt{ accepton } e
\end{aligned}$$

The case for reject e in $\varphi \vee \psi$ is analogous.

$\phi := \bigcirc \varphi$:

$$\begin{aligned}
&\langle w^i, \text{false}, \text{false} \rangle \models \texttt{accept } e \texttt{ in } \bigcirc \varphi \\
&\Leftrightarrow w^i \models e \vee \langle w^{i+1}, e, \text{false} \rangle \models \varphi \\
&\Leftrightarrow w^i \models e \vee w^{i+1} \models \varphi \texttt{ accepton } e \\
&\Leftrightarrow w^i \models e \vee w^i \models \bigcirc(\varphi \texttt{ accepton } e) \\
&\Leftrightarrow w^i \models (\bigcirc \varphi) \texttt{ accepton } e.
\end{aligned}$$

The case for reject e in $\bigcirc \varphi$ is analogous.

$\phi := \varphi \ \mathrm{U} \ \psi$:

$$\begin{aligned}
&\langle w^i, \text{false}, \text{false} \rangle \models \texttt{accept } e \texttt{ in } \varphi \ \mathrm{U} \ \psi \\
&\Leftrightarrow w^i \models e \vee \langle w^i, e, \text{false} \rangle \models \varphi \ \mathrm{U} \ \psi \\
&\Leftrightarrow \exists k \geq i. \ \langle w^k, e, \text{false} \rangle \models \psi \wedge \forall i \leq l < k. \ \langle w^l, e, \text{false} \rangle \models \varphi \\
&\Leftrightarrow \exists k \geq i. \ w^k \models \psi \texttt{ accepton } e \wedge \forall i \leq l < k. \ w^l \models \varphi \texttt{ accepton } e \\
&\Leftrightarrow w^i \models \varphi \texttt{ accepton } e \ \mathrm{U} \ \psi \texttt{ accepton } e \\
&\Leftrightarrow w^i \models \varphi \ \mathrm{U} \ \psi \texttt{ accepton } e.
\end{aligned}$$

The case for reject e in $\varphi \ \mathrm{U} \ \psi$ is analogous.

$\phi := \neg \varphi$: Negation is somewhat a special case due to the mutual recursive definition of the semantics. Here, we treat the Reset-LTL side first by itself, and use the duality between the accept and reject operators as follows.

$$\begin{aligned}
&\langle w^i, \text{false}, \text{false} \rangle \models \texttt{accept } e \texttt{ in } \neg \varphi \\
&\Leftrightarrow \langle w^i, \text{false}, \text{false} \rangle \models \neg(\texttt{reject } e \texttt{ in } \varphi) \\
&\Leftrightarrow \langle w^i, \text{false}, \text{false} \rangle \not\models (\texttt{reject } e \texttt{ in } \varphi).
\end{aligned}$$

Next, we observe that the following holds in SALT:

$$w^i \models \neg\varphi \text{ accepton } e \Leftrightarrow w^i \not\models \varphi \text{ rejecton } e.$$

Now, equivalence follows from case two of the induction hypothesis, i.e.,

$$w^i \models \varphi \text{ rejecton } e \Leftrightarrow \langle w^i, \text{false}, \text{false}\rangle \models (\text{reject } e \text{ in } \varphi).$$

The case for reject e in $\neg\varphi$ is dual.   □

VeriFast: A Powerful, Sound, Predictable, Fast Verifier for C and Java

Bart Jacobs, Jan Smans*, Pieter Philippaerts, Frédéric Vogels,
Willem Penninckx, and Frank Piessens

Department of Computer Science, Leuven, Belgium
firstname.lastname@cs.kuleuven.be

Abstract. VeriFast is a prototype verification tool for single-threaded and multithreaded C and Java programs. In this paper, we first describe the basic symbolic execution approach in some formal detail. Then we zoom in on two technical aspects: the approach to permission accounting, including fractional permissions, precise predicates, and counting permissions; and the approach to lemma function termination in the presence of dynamically-bound lemma function calls. Finally, we describe three ongoing efforts: application to JavaCard programs, integration of shape analysis, and application to Linux device drivers.

1 Introduction

VeriFast is a prototype verification tool for single-threaded and multithreaded C and Java programs annotated with preconditions and postconditions written in separation logic. To enable rich specifications, the programmer may define inductive datatypes, primitive recursive pure functions over these datatypes, and abstract separation logic predicates. To enable verification of these rich specifications, the programmer may write *lemma functions*, i.e., functions that serve only as proofs that their precondition implies their postcondition. The verifier checks that lemma functions terminate and do not have side-effects. Since neither VeriFast itself nor the underlying SMT solver need to do any significant search, verification time is predictable and low. VeriFast comes with an IDE that enables interactive annotation insertion and symbolic debugging and is available for download at http://www.cs.kuleuven.be/~bartj/verifast/.

For an introduction to VeriFast, we refer to earlier work [1]; furthermore, a tutorial text is available on the web site. In this invited paper, we take the opportunity to zoom in on three aspects of VeriFast that have not yet been covered in the same level of detail in earlier published work: in Section 2 we present in some formal detail the essence of VeriFast's symbolic execution algorithm; in Section 3 we present VeriFast's support for permission accounting; and in Section 4 we present our approach for ensuring termination of lemma functions that perform dynamically bound calls. Additionally, in Section 5, we briefly discuss some of the projects currently in progress at our group.

---

* Jan Smans is a Postdoctoral Fellow of the Research Foundation - Flanders (FWO).

M. Bobaru et al. (Eds.): NFM 2011, LNCS 6617, pp. 41–55, 2011.

2 Symbolic Execution

In this section, we present the essence of VeriFast's verification algorithm in some formal detail.

2.1 Symbolic Execution States

VeriFast modularly verifies a C or Java program by symbolically executing each routine (function or method) in turn, using other routines' contracts to verify calls. A symbolic execution state is much like a concrete execution state, except that terms of an SMT solver, containing logical symbols, are used instead of concrete values. For example, at the start of the symbolic execution of a routine, each routine parameter's value is represented using a fresh logical symbol.

Specifically, a symbolic state $\sigma = (\Sigma, h, s)$ consists of a *path condition* Σ, a *symbolic heap* h, and a *symbolic store* s. The path condition is a set of formulae of first-order logic that constrain the values of the logical symbols that appear in the symbolic heap and the symbolic store. The symbolic heap is a multiset of *heap chunks*. Each heap chunk is of the form $[f]p\langle\overline{\tau}\rangle(\overline{t})$, where f is the *coefficient*, p the *predicate name*, $\overline{\tau}$ the *type arguments*, and $\overline{t}$ the *arguments* of the chunk. The coefficient f is a term representing a real number; if it is different from 1, the chunk represents a fractional permission (see Section 3). The predicate name is a term that denotes the predicate of which the chunk is an instance; it is either the symbol associated with a built-in predicate, such as a struct or class field predicate, or a user-defined predicate, or it is a *predicate constructor application*, which is essentially a partially applied predicate (see [2] for more information), or it is some other term, which typically means the predicate name was passed into the function as a value. VeriFast supports type parameters on user-defined predicates; hence the type arguments, which are VeriFast types. Finally, each chunk specifies argument terms for the predicate's parameters. The symbolic store maps local variable names to terms that represent their value.

2.2 Algorithm: Preliminary Definitions

To describe the essence of the symbolic execution algorithm formally, we define a highly stylized syntax of assertions a, commands c, and routines r (given an unspecified syntax for arithmetic expressions e and boolean expressions b, and given a set of variables x):

$$a ::= [e]e(e, ?x) \mid b \mid a * a \mid \textbf{if } b \textbf{ then } a \textbf{ else } a$$
$$c ::= x := r(\overline{e}) \mid (c; c) \mid \textbf{if } b \textbf{ then } c \textbf{ else } c$$
$$rdef ::= \textbf{routine } r(\overline{x}) \textbf{ req } a \textbf{ ens } a \textbf{ do } c$$

We assume all predicates have exactly two parameters, and we consider only predicate assertions where the first argument is an expression and the second argument is a pattern.

We will give the semantics of a symbolic execution step by means of symbolic transition relations, which are relations from initial symbolic states σ to outcomes o. An outcome is either a final symbolic state or the special outcome **abort**, which signifies that an error was found. A given initial state may be related to multiple outcomes due to case splitting, and it may be related to no outcomes if all symbolic execution paths are found to be infeasible.

VeriFast sometimes makes arbitrary choices. Specifically, it arbitrarily chooses a matching chunk when consuming a predicate assertion that has multiple matching chunks (a situation we call an *ambiguous match*). To model this, we define the semantics of a symbolic execution step not as a single transition relation, but as a set of transition relations. Each element of this set makes different choices in the event of ambiguous matches. The soundness theorem states that if, for a given initial symbolic state, there is any transition relation where this initial state does not lead to an **abort**, then all concrete states represented by this initial symbolic state are safe. It is possible that some choices cause VeriFast to fail (i.e., lead to an **abort**), while others do not. It is up to the user to avoid such unfortunate matches, for example by wrapping chunks inside predicates defined just for that purpose to temporarily hide them.

A note about picking fresh logical symbols. We will use the function

$$\mathsf{nextFresh}(\Sigma) = (u, \Sigma')$$

which given a path condition Σ returns a symbol u that does not appear free in Σ, and a new path condition $\Sigma' = \Sigma \cup \{u = u\}$, which is equivalent to Σ but in which u appears free. Since path conditions are finite sets of finite formulae, and there are infinitely many logical symbols, this function is well-defined. We will also use this function to generate sequences of fresh symbols.

We use the following operations on sets of transition relations. Conjunction $W \wedge W'$ denotes the pairwise union of relations from W and W':

$$W \wedge W' = \{R \cup R' \mid R \in W \wedge R' \in W'\}$$

Similarly, generalized conjunction $\bigwedge i \in I.\ W(i)$ denotes the set where each element is obtained by taking the union of one element of each $W(i)$:

$$\left(\bigwedge i \in I.\ W(i)\right) = \{\bigcup i \in I.\ \psi(i) \mid \forall i \in I.\ \psi(i) \in W(i)\}$$

We omit the range I if it is clear from the context. Sequential composition $W; W'$ denotes the pairwise sequential composition of relations from W and W':

$$W; W' = \{R; R' \mid R \in W \wedge R' \in W'\}$$

where the sequential composition of transition relations $R; R'$ is defined as

$$R; R' = \{(\sigma, \mathbf{abort}) \mid (\sigma, \mathbf{abort}) \in R\} \cup \{(\sigma, o) \mid (\sigma, \sigma') \in R \wedge (\sigma', o) \in R'\}$$

We denote the term or formula resulting from evaluation of an arithmetic expression e or boolean expression b under a symbolic store s as $s(e)$ or $s(b)$, respectively. We abuse this notation for sequences of expressions as well.

2.3   The Algorithm

A basic symbolic execution step is an *assumption step* assume(b), defined as follows:

$$\mathsf{assume}(b) = \{\{((\Sigma, h, s), (\Sigma \cup \{s(b)\}, h, s)) \mid \Sigma \not\vdash_{\mathrm{SMT}} \neg s(b)\}\}$$

It consists of a single transition relation, which adds b to the path condition, unless doing so would lead to an inconsistency, in which case the symbolic execution path ends (i.e., the initial state does not map to any outcome).

Symbolic execution of a routine starts by *producing* the precondition, then verifying the body, and finally *consuming* the postcondition. Producing an assertion means adding the chunks and assumptions described by the assertion to the symbolic state:

$\mathsf{produce}([e]e'(e'', ?x)) =$
$\quad \{\{((\Sigma, h, s), (\Sigma', h \uplus \{[s(e)]s(e')(s(e''), u)\}, s[x := u])) \mid$
$\quad\quad (u, \Sigma') = \mathsf{nextFresh}(\Sigma)\}\}$
$\mathsf{produce}(b) = \mathsf{assume}(b)$
$\mathsf{produce}(a * a') = \mathsf{produce}(a); \mathsf{produce}(a')$
$\mathsf{produce}(\mathbf{if}\ b\ \mathbf{then}\ a\ \mathbf{else}\ a') = \mathsf{assume}(b); \mathsf{produce}(a) \wedge \mathsf{assume}(\neg b); \mathsf{produce}(a')$

Conversely, consuming an assertion means removing the chunks described by the assertion from the symbolic heap, and checking the assumptions described by the assertion against the path condition.

$\mathsf{consume}([e]e'(e'', ?x)) =$
$\quad \mathsf{choice}(\{\mathsf{matches}(\Sigma, h, s) \mid \mathsf{matches}(\Sigma, h, s) \neq \emptyset\})$
$\quad \wedge \{\{((\Sigma, h, s), \mathbf{abort}) \mid \mathsf{matches}(\Sigma, h, s) = \emptyset\}\}$
$\quad \mathrm{where}\ \mathsf{choice}(C) = \{\{\psi(c) \mid c \in C\} \mid \forall c \in C.\ \psi(c) \in c\}$
$\quad \mathrm{and}\ \mathsf{matches}(\Sigma, h, s) =$
$\quad\quad \{((\Sigma, h, s), (\Sigma, h', s[x := t'])) \mid$
$\quad\quad\quad h = h' \uplus \{[f]p(t, t')\} \wedge \Sigma \vdash_{\mathrm{SMT}} s(e, e', e'') = f, p, t\}$
$\mathsf{consume}(b) =$
$\quad \{\{((\Sigma, h, s), (\Sigma, h, s)) \mid \Sigma \vdash_{\mathrm{SMT}} s(b)\} \cup \{((\Sigma, h, s), \mathbf{abort}) \mid \Sigma \not\vdash_{\mathrm{SMT}} s(b)\}\}$
$\mathsf{consume}(a * a') = \mathsf{consume}(a); \mathsf{consume}(a')$
$\mathsf{consume}(\mathbf{if}\ b\ \mathbf{then}\ a\ \mathbf{else}\ a') =$
$\quad \mathsf{assume}(b); \mathsf{consume}(a) \wedge \mathsf{assume}(\neg b); \mathsf{consume}(a')$

Notice that consuming a predicate assertion generates one transition relation for each choice function ψ that picks one match for each initial state that has matches.

Verifying a routine call means consuming the precondition (under the symbolic store obtained by binding the arguments), followed by picking a fresh symbol to represent the return value, followed by producing the postcondition, followed by binding the return value into the caller's symbolic store. The other commands are straightforward.

$\text{verify}(x := r(\overline{e})) =$
$\bigwedge s. \{\{((\Sigma, h, s), (\Sigma, h, [\overline{x} := s(\overline{e})]))\}\}; \text{consume}(a);$
  $\bigwedge r.\{\{((\Sigma, h, s'), (\Sigma', h, s'[\text{result} := r])) \mid (r, \Sigma') = \text{nextFresh}(\Sigma)\}\};$
  $\text{produce}(a'); \{\{((\Sigma, h, s''), (\Sigma, h, s[x := r]))\}\}$
  where **routine** $r(\overline{x})$ **req** a **ens** a'
$\text{verify}((c; c')) = \text{verify}(c); \text{verify}(c')$
$\text{verify}(\textbf{if } b \textbf{ then } c \textbf{ else } c') = \text{assume}(b); \text{verify}(c) \wedge \text{assume}(\neg b); \text{verify}(c')$

Verifying a routine means binding the parameters to fresh symbols, then producing the precondition, then saving the resulting symbolic store s', then verifying the body under the original symbolic store, then restoring the symbolic store s' and binding the result value, and then finally consuming the postcondition. The routine is valid if in at least one transition relation, the initial state does not lead to **abort**.

$\text{valid}(\textbf{routine } r(\overline{x}) \textbf{ req } a \textbf{ ens } a' \textbf{ doc}) =$
$\exists R \in W. ((\Sigma_0, \emptyset, [\overline{x} := \overline{u}]), \textbf{abort}) \notin R$
  where $(\overline{u}, \Sigma_0) = \text{nextFresh}(\emptyset)$
  and $W = \bigwedge s. \{\{((\Sigma, h, s), (\Sigma, h, s))\}\}; \text{produce}(a);$
    $\bigwedge s'. \{\{((\Sigma, h, s'), (\Sigma, h, s))\}\}; \text{verify}(c);$
    $\bigwedge s''. \{\{((\Sigma, h, s''), (\Sigma, h, s'[\text{result} := s''(\text{result})]))\}\}; \text{consume}(a')$

A program is valid if all routines are valid.

2.4   Soundness Proof Sketch

We now sketch an approach for proving the soundness of this algorithm. First, we define *abstracted execution* operations aproduce, aconsume, and averify, that differ from the corresponding symbolic execution operations only in that they use concrete values instead of logical terms in heap chunks and store bindings. We then prove that the relation between an abstracted state and a symbolic state that represents it (through some interpretation of the logical symbols) is a simulation relation: if some symbolic state represents some abstracted state, then for every transition relation in the symbolic execution, there is a transition relation in the abstracted execution such that if the abstracted state aborts, then the symbolic state aborts, and if the abstracted state leads to some other abstracted state, then the symbolic state either aborts or leads to some other symbolic state that represents this abstracted state. It follows that if a program is valid under symbolic execution, it is valid under abstracted execution.

We then prove two lemmas about abstracted execution. Firstly, we prove that all abstracted execution operations are *local*, in the sense that heap contraction is a simulation relation: for state $(s, h \uplus h_0)$ and contracted state (s, h), for every transition relation R_2 there is a transition relation R_1 such that if $(s, h \uplus h_0)$ aborts in R_1, then (s, h) aborts in R_2, and otherwise if $(s, h \uplus h_0)$ leads to a state (s', h') in R_1, then either (s, h) aborts in R_2 or $h_0 \subseteq h'$ and (s, h) leads to state $(s', h' - h_0)$ in R_2.

Secondly, we prove the soundness of abstracted assertion production and consumption. Specifically, we prove that if we consume an assertion a in a state (h, s), then this either aborts or we obtain some state (h', s'), and for every such final state it holds that producing a in some state (h'', s) leads to state $(h'' \uplus (h - h'), s')$.

Finally, given a big-step operational semantics of the programming language, we prove that if all routines are valid, then concrete execution is simulated by abstracted execution: for every initial state, if concrete execution leads to some outcome, then in each transition relation either abstracted execution aborts or leads to the same or a contracted outcome[1]. We detail the case of routine call.

Consider a routine call $x := r(\bar{e})$ started in a state (s, h). Now, consider an arbitrary transition relation of consumption of r's precondition in state $([\bar{x} := s(\bar{e})], h)$. Either this aborts, in which case abstracted execution of the routine call aborts and we are done. Otherwise, it leads to a state (s', h'). Then, by the second lemma, production of the precondition in state $([\bar{x} := s(\bar{e})], h')$ leads to state (s', h). Now, consider the execution of the body of r in state $([\bar{x} := s(\bar{e})], h)$. If this aborts, then by the induction hypothesis, we have that abstracted execution of the body aborts in all transition relations when started in the same state. By locality, it follows that production of the precondition in state $([\bar{x} := s(\bar{e})], \emptyset)$ leads to state $(s', h - h')$ and abstracted execution of the body in state $(s', h - h')$ aborts. This contradicts the assumption that the routine is valid.

Now consider the case where execution of the body of the routine, when started in state $([\bar{x} := s(\bar{e})], h)$, leads to some state (s'', h''). Consider an arbitrary transition relation of consumption of r's postcondition in state $(s'[\text{result} := s''(\text{result})], h'')$. Either consumption aborts, in which case, by locality, the routine is invalid and we obtain a contradiction. Or it leads to some state (s''', h'''). Then, by the second lemma, production of r's postcondition in state $(s'[\text{result} := s''(\text{result})], h')$ leads to state $(s''', h' \uplus (h'' - h'''))$. By locality, we have $h' \subseteq h'''$; as a result, we have $h' \uplus (h'' - h''') \subseteq h''' \uplus (h'' - h''') = h''$, so the abstracted execution leads to a contraction of the final concrete execution state $(s[x := s''(\text{result})], h'')$.

3   Permission Accounting

This section presents VeriFast's support for permission accounting. Specifically, to enable convenient sharing of heap locations, mutexes, and other resources among multiple threads, and for other purposes, VeriFast has built-in support for fractional permissions (Section 3.1), and library support for counting permissions (Section 3.4). To facilitate the application of fractional permissions to user-defined predicates, VeriFast has special support for *precise predicates* (Section 3.2). Finally, to facilitate unrestricted sharing of resources in case reassembly is not required, VeriFast supports *dummy fractions* (Section 3.3).

---

[1] A contracted outcome (i.e., with a smaller heap) occurs in the case of routine calls if heap chunks remain after the routine's postcondition is consumed. When verifying a C program, VeriFast signals a leak error in this case; for a Java program, however, this is allowed.

3.1   Fractional Permissions

VeriFast has fairly convenient built-in support for the fractional permissions system proposed by Bornat et al. [3]. The basics of this support consist of the following elements: a coefficient term in each heap chunk, a relaxed proof rule for read-only memory accesses, fractional assertions, opening and closing of fractional user-defined predicate chunks, and autosplitting. Some more advanced features are explained in later subsections.

Fractional Heap Chunks and Memory Reads. As mentioned in Section 2, in VeriFast's symbolic heap data structure, each heap chunk specifies a term known as its *coefficient*. This term belongs to the SMT solver's sort of real numbers. If the real number represented by this term is different from 1, we say the chunk is a *fraction*. On any feasible symbolic execution path, the coefficient of any chunk that represents a memory location lies between 0, exclusive, and 1, inclusive, where 1 represents exclusive write access, and a smaller value represents shared read access. However, coefficients of user-defined predicates may feasibly lie outside this range.

Fractional Assertions. Both points-to assertions and predicate assertions may mention a coefficient f, which is a pattern of type real, using the syntax $[f]\ell$ |-> v or $[f]p(\overline{v})$. Just like other patterns, the coefficient pattern may be an expression, such as 1/2 or x, where x is a previously declared variable of type real. It may also be a question mark pattern ?x, which existentially quantifies over the coefficient and binds it to x. Finally, it may be a dummy pattern _, which also existentially quantifies over the coefficient but does not bind it to any variable. Dummy coefficient patterns are treated specially; see Section 3.3. If a points-to assertion or predicate assertion does not mention a coefficient, it defaults to 1.

The following simple example illustrates a common pattern:

```
int read_cell(int *cell)
    requires [?f]integer(cell, ?v);
    ensures [f]integer(cell, v) &*& result == v;
{ return *cell; }
```

The above function requires an arbitrary fraction of the `integer` chunk that permits access to the `int` object at location `cell`, and returns the same fraction.

Fractions and User-Defined Predicates. The syntax of open and close ghost statements allows mentioning a coefficient: open $[f]p(\overline{v})$, close $[f]p(\overline{v})$. By definition, applying a coefficient f to a user-defined predicate is equivalent to multiplying the coefficient of each chunk mentioned in the predicate's body by f. There is no restriction on the value of f. If no coefficient is mentioned, a close operation defaults to coefficient 1, and an open operation defaults to the coefficient found in the symbolic heap.

Autosplitting. When consuming a predicate assertion, the nano-VeriFast algorithm presented in Section 2 requires a precise match between the coefficient expression specified in the predicate assertion and the coefficient term in a heap chunk. Full VeriFast is more relaxed: for assertion coefficient f_a and chunk coefficient f_c, it requires either $f_a = f_c$ or $0 < f_a < f_c$. In the latter case, consumption does not remove the chunk, but simply reduces the chunk's coefficient to $f_c - f_a$.

3.2   Precise Predicates

Autosplitting is sound both for built-in memory location predicates and for arbitrary user-defined predicates. For built-in memory location predicates, we also have a merge law:

$$[f_1]\ell \mapsto v_1 * [f_2]\ell \mapsto v_2 \Rightarrow [f_1 + f_2]\ell \mapsto v_1 \wedge v_2 = v_1$$

This law states not only that two fractions whose first arguments are equal can be merged into one, but also that their second arguments are equal. VeriFast automatically performs this merge operation and adds this equality to the path condition when producing a built-in predicate chunk if a matching chunk is already present in the symbolic heap. Merging of fractional permissions is important because it enables modifying or deallocating memory locations once they are no longer shared between multiple threads.

However, VeriFast does not automatically merge arbitrary predicate chunks, even if they have identical argument lists. Doing so would be unsound, as illustrated by the following pathological user-defined predicate:

```
predicate foo() = integer(_, _);
lemma void evil()
    requires integer(_, _) &*& integer(_, _);
    ensures [2]integer(_, _);
{ close foo(); close foo(); open [2]foo(); }
```

Specifically, this would violate the invariant that on feasible paths, memory location chunks never appear with a coefficient greater than 1.

Therefore, VeriFast automerges only *precise predicates*. A user-defined predicate may be declared as precise by using a semicolon in the parameter list to separate the *input parameters* from the *output parameters*. If a predicate is declared as precise, VeriFast performs a static analysis on the predicate body to check that the merge law holds for this predicate. The merge law for a predicate p with input parameters $\overline{x}$ and output parameters $\overline{y}$ states:

$$[f_1]p(\overline{x}, \overline{y}_1) * [f_2]p(\overline{x}, \overline{y}_2) \Rightarrow [f_1 + f_2]p(\overline{x}, \overline{y}_1) \wedge \overline{y}_2 = \overline{y}_1$$

For example, the static analysis accepts the following definition of the classic list segment predicate:

```
struct node { struct node *next, int value };
predicate lseg(struct node *f, struct node *l; list<int> vs) =
    f == l ? vs == nil :
    f->next |-> ?n &*& f->value |-> ?v &*& malloc_block_node(f) &*&
    lseg(n, l, ?vs0) &*& vs == cons(v, vs0);
```

As a result, the following lemma is verified automatically:

```
lemma void lseg_merge(struct node *f, struct node *l)
    requires [?f1]lseg(f, l, ?vs1) &*& [?f2]lseg(f, l, ?vs2);
    ensures [f1+f2]lseg(f, l, vs1) &*& vs2 == vs1;
{}
```

The static analysis for a predicate definition **predicate** $p(\overline{x}; \overline{y}) = a$; checks that given fixed variables $\overline{x}$, assertion a is precise and fixes variables $\overline{y}$; formally: $\overline{x} \vdash a \rightsquigarrow \overline{y}$. The meaning of this judgment is given by a merge law for assertions:

$$[f_1]a_1 * [f_2]a_2[\overline{x}_1/\overline{x}_2] \Rightarrow [f_1 + f_2]a_1 \wedge \overline{y}_2 = \overline{y}_1$$

where a_1 is a with all free variables subscripted by 1 and a_2 is a with all free variables subscripted by 2. The static analysis proceeds according to the inference rules shown in Figure 1. Notice that the analysis allows both expressions and

$$\frac{\textbf{predicate } q(\overline{x}; \overline{y}) \quad |\overline{e}| = |\overline{x}| \quad \mathsf{FreeVars}(\overline{e}) \subseteq X}{X \vdash q(\overline{e}, \overline{pat}) \rightsquigarrow X \cup \mathsf{FixedVars}(pat)} \qquad \frac{\mathsf{FreeVars}(e) \subseteq X}{X \vdash x = e \rightsquigarrow X \cup \{x\}}$$

$$X \vdash e \rightsquigarrow X \qquad \frac{\mathsf{FreeVars}(e) \subseteq X \quad X \vdash a \rightsquigarrow Y}{X \vdash [e]a \rightsquigarrow Y} \qquad \frac{X \vdash a \rightsquigarrow Y}{X \vdash [\_]a \rightsquigarrow Y} \qquad \frac{X \vdash a_1 \rightsquigarrow Y \quad Y \vdash a_2 \rightsquigarrow Z}{X \vdash a_1 * a_2 \rightsquigarrow Z}$$

$$\frac{\mathsf{FreeVars}(b) \subseteq X \quad X \vdash a_1 \rightsquigarrow Y \quad X \vdash a_2 \rightsquigarrow Y}{X \vdash b ? a_1 : a_2 \rightsquigarrow Y} \qquad \frac{X \vdash a \rightsquigarrow Y \quad Y' \subseteq Y}{X \vdash a \rightsquigarrow Y'}$$

where

$$\mathsf{FixedVars}(x) = \{x\} \quad \mathsf{FixedVars}(e) = \emptyset \quad \mathsf{FixedVars}(?x) = \{x\} \quad \mathsf{FixedVars}(\_) = \emptyset$$

Fig. 1. The static analysis for preciseness of assertions

dummy patterns as coefficients (but not question mark patterns). In allowing dummy patterns, VeriFast's notion of preciseness deviates from the separation logic literature, where an assertion is precise if for any heap, there is at most one subheap that satisfies the assertion. Indeed, in the presence of dummy fractions, there may be infinitely many fractional subheaps that satisfy the assertion; however, the merge law still holds.

3.3 Dummy Fractions for Leakable Resources

VeriFast treats dummy coefficients in predicate assertions specially, to facilitate
scenarios where reassembly of fractions of a given resource is not required, and
as a result the resource can be shared arbitrarily. Specifically, when consuming
a predicate assertion with a dummy coefficient, VeriFast always performs an
autosplit; that is, it does not remove the matched chunk but merely replaces its
coefficient by a fresh symbol.

Furthermore, when verifying a C program, dummy fractions affect leak checking. In general, when verifying a C function, if after consuming the postcondition
the symbolic heap is not empty, VeriFast signals a leak error. However, VeriFast
does not signal an error if for all remaining resources, the user has indicated
explicitly that leaking this resource is acceptable. The user can do so using a
leak a; command. This command consumes the assertion a, and then reinserts
all consumed chunks into the symbolic heap, after replacing their coefficients
with fresh symbols and registering these symbols as *dummy coefficient symbols*.
Leaking a chunk whose coefficient is a dummy coefficient symbol is allowed.

To allow this leakability information to be carried across function boundaries,
dummy coefficients in assertions are considered to match only dummy coefficient
symbols. That is, consuming a dummy fraction assertion matches only chunks
whose coefficients are dummy coefficient symbols, and producing a dummy fraction assertion produces a chunk whose coefficient is a dummy coefficient symbol.

To understand the combined benefit of these features, consider the common
type of program where the main method creates a mutex and then starts an
unbounded number of threads, passing a fraction of the mutex to each thread.
Each thread leaks its mutex fraction when it dies. If the user performs a leak operation on the mutex directly after it is created, VeriFast automatically splits the
mutex chunk when a thread is started, and silently leaks each thread's fraction
when the thread finishes.

VeriFast also uses dummy fractions to represent C's string literals.

3.4 Counting Permissions

Fractional permissions are sufficient in many sharing scenarios; however, an important example of a scenario where they are not applicable is when verifying
a program that uses reference counting for resource management. For this scenario, another permission accounting scheme known as counting permissions [3]
is appropriate.

VeriFast does not have built-in support for counting permissions. However,
using VeriFast's support for higher-order predicates, VeriFast offers counting
permissions support in the form of a trusted library, specified by header file
counting.h, reproduced in Figure 2. This library allows any precise predicate
of one input parameter and one output parameter to be shared by means of
counting permissions. VeriFast's built-in memory location predicates satisfy this
constraint, so they can be used directly. Precise predicates that are of a different
shape can be wrapped in a helper predicate that bundles the input and output
arguments into tuples.

```
predicate counting<a, b>(predicate(a; b) p, a a, int count; b b);
predicate ticket<a, b>(predicate(a; b) p, a a, real frac;);

lemma void start_counting<a, b>(predicate(a; b) p, a a);
    requires p(a, ?b);
    ensures counting(p, a, 0, b);

lemma void counting_match_fraction<a, b>(predicate(a; b) p, a a);
    requires counting(p, a, ?count, ?b1) &*& [?f]p(a, ?b2);
    ensures counting(p, a, count, b1) &*& [f]p(a, b2) &*& b2 == b1;

lemma real create_ticket<a, b>(predicate(a; b) p, a a);
    requires counting(p, a, ?count, ?b);
    ensures counting(p, a, count + 1, b)
        &*& ticket(p, a, result) &*& [result]p(a, b) &*& 0 < result;

lemma void destroy_ticket<a, b>(predicate(a; b) p, a a);
    requires counting(p, a, ?count, ?b1)
        &*& ticket(p, a, ?f) &*& [f]p(a, ?b2) &*& 0 != count;
    ensures counting(p, a, count - 1, b1) &*& b2 == b1;

lemma void stop_counting<a, b>(predicate(a; b) p, a a);
    requires counting(p, a, 0, ?b);
    ensures p(a, b);
```

Fig. 2. The specification of VeriFast's counting permissions library

Once a chunk is wrapped into a `counting` chunk using the `start_counting` lemma, tickets can be created from it using the `create_ticket` lemma. This lemma not only increments the `counting` chunk's counter and produces a `ticket` chunk; it also produces an unspecified fraction of the wrapped chunk. In case of built-in memory location chunks, this allows the memory location to be read immediately. The `ticket` chunk remembers the coefficient of the produced fraction. The same fraction is consumed again when the ticket is destroyed using lemma `destroy_ticket`. When the counter reaches zero, the original chunk can be unwrapped using lemma `stop_counting`.

Notice that this library is sound even when applied to predicates that are not *unique*, i.e., predicates that can appear with a fraction greater than one. However, the existence of non-unique precise predicates does mean that we cannot assume that the counter of a `counting` chunk remains nonnegative, as illustrated in Figure 3.

4   Lemma Function Termination and Dynamic Binding

VeriFast supports *lemma functions*, which are like ordinary C functions, except that lemma functions and calls of lemma functions are written inside annotations, and VeriFast checks that they have no side-effects on non-ghost memory and that they terminate. Lemma functions serve mainly to encode inductive

```
predicate foo(int *n; int v) = [1/2]integer(n, v);
predicate hide(int *n; int v) = counting(foo, n, 1, v);
lemma void test(int *n)
    requires integer(n, ?v);
    ensures counting(foo, n, -1, v) &*& hide(n, v);
{
    close [2]foo(n, v);
    start_counting(foo, n); create_ticket(foo, n);
    close hide(n, v);
    start_counting(foo, n); destroy_ticket(foo, n);
}
```

Fig. 3. Example where a counter decreases below zero

proofs of lemmas about inductive datatypes, such as the associativity of appending two mathematical lists, or inductive proofs of lemmas about recursive predicates, such as a lemma stating that a linked list segment from node n_1 to node n_2 separately conjoined with a linked list segment from node n_2 to 0 implies a linked list segment from node n_1 to 0.

To enable such inductive proofs, lemma functions are allowed to be recursive. Specifically, to ensure termination, VeriFast allows a statically bound lemma function call if either the callee is defined before the caller in the program text, or the callee equals the caller and one of the following hold: 1) after consuming the precondition, at least one full (i.e., non-fractional) memory location predicate remains, or 2) the body of the lemma function is a switch statement over one of the function's parameters whose type is an inductive datatype, and the callee's argument for this parameter is a component of the caller's argument for this parameter, or 3) the body of the lemma function is not a switch statement and the first chunk consumed by the callee's precondition was obtained from the first chunk produced by the caller's precondition through one or more **open** operations. These three cases constitute induction on the size of the concrete heap, induction on the size of an argument, and induction on the derivation of the first conjunct of the precondition.

However, VeriFast supports not just statically bound lemma function calls, but dynamically bound calls as well. Specifically, VeriFast supports *lemma function pointers* and *lemma function pointer calls*. The purpose of these is as follows.

VeriFast supports the modular specification and verification of fine-grained concurrent data structures. It does so by modularizing Owicki and Gries's approach based on auxiliary variables. The problem with their approach is that it requires application-specific auxiliary variable updates to be inserted inside critical sections. If the critical sections are inside a library that is to be reused by many applications, this is a problem. In earlier work [4], we propose to solve this problem by allowing applications to pass auxiliary variable updates into the library in a simple form of higher-order programming. In VeriFast, this can be realized through lemma function pointers.

A simple approach to ensure termination of lemma functions in the presence of lemma function pointers would be to allow lemma function pointer calls only in non-lemma functions. However, when building fine-grained concurrent data structures on top of other fine-grained concurrent data structures, layer N needs to be able to call lemma function pointers it receives from layer $N + 1$ inside of its own lemma function, which it passes to layer $N - 1$.

To support this, we introduced a new kind of built-in heap chunks, called *lemma function pointer chunks*. A call of a lemma function pointer p is allowed only if the symbolic heap contains a lemma function pointer chunk for p, and this chunk becomes unavailable for the duration of the call. Non-lemma functions may produce lemma function pointer chunks arbitrarily. A lemma function may only produce lemma function pointer chunks for lemma functions that appear before itself in the program text, and furthermore, these chunks are consumed again before the producing lemma function terminates, so the pointer calls must occur within the dynamic scope of the producing lemma function.

We prove that this approach guarantees lemma function termination, by contradiction. Consider an infinite chain of nested lemma function calls. Since we have termination of statically bound calls, the chain must contain infinitely many pointer calls. Of all functions that appear infinitely often, consider the one that appears latest in the program text. It must be called infinitely often through a pointer. Therefore, infinitely many pointer chunks must be generated during the chain. However, these can only be generated by functions that appear later, which is a contradiction.

5 Ongoing Efforts

In this section, we briefly describe three projects currently proceeding in our group.

5.1 JavaCard Programs

JavaCard is a trimmed-down version of the Java Platform for smart cards such as cell phone subscriber cards, payment cards, identity cards, etc. We are applying VeriFast to a number of JavaCard programs, to prove absence of runtime exceptions and functional correctness properties.

An interesting aspect of the JavaCard execution environment is the fact that by default, objects allocated by a JavaCard program (called an *applet*) are *persistent*. That is, once a JavaCard applet is installed on a card, the applet object and objects reachable from its fields persist for the entire lifetime of the smart card. This interacts in interesting ways with the phenomenon of *card tearing*, which occurs when the user removes the smart card from the card reader while a method call on the applet object is in progress. To allow the programmer to preserve the consistency of the applet object, JavaCard offers a transaction mechanism, that ensures that modifications to objects during a transaction are rolled back if a card tear occurs before the transaction is committed.

We developed a specification of the JavaCard API that is sound in the presence of card tearing. We did not need to modify the VeriFast tool itself. In our specification, when a newly installed applet is registered with the virtual machine, the virtual machine takes ownership of the applet's state as defined by its `valid` predicate. When an applet receives a method call, the method receives a 1/2 fraction of the applet's `valid` predicate. This allows the method to inspect but not modify the applet's state. As a result, the method is forced to call the `beginTransaction` method before modifying the state. This API method produces the other half of the `valid` chunk. Conversely, API method `commitTransaction` consumes the entire `valid` chunk and produces a 1/2 fraction.

The soundness argument for this approach goes as follows. We need to show that in every execution, even one where card tears occur, at the start of each toplevel method call on the applet, the `valid` predicate is fully owned by the VM. We do so by showing that at every point during a method call, either we are in a transaction, or the VM owns half of `valid` and the method call owns the other half. When a method call terminates, either normally or due to a card tear, the method call's fraction is simply transfered to the VM. This proof explains the contract of `commitTransaction`: if `commitTransaction` merely consumed 1/2 of `valid`, it would not guarantee that the thread owned the other half.

5.2   Integrating Shape Analysis

We are in the process of integrating separation logic-based shape analysis algorithms from the literature [5,6] into VeriFast. The goal is to enable a scenario where annotations are inserted into a program using a mixed manual-automatic process: the user writes some manual annotations; then they invoke the shape analysis algorithm, which, given the existing annotations, infers additional ones; then, the user adds further annotations where the algorithm failed; etc. We are not yet at the point where we can report if this approach works or not.

We are currently targeting the scenario where the code is not evolving, i.e., the scenario where an existing, unannotated program is annotated for the first time. In a later stage, we intend to consider the question whether shape analysis can help to adapt existing annotations to code evolution. The latter problem seems much more difficult, especially if the generated shape annotations have been extended manually with functional information.

5.3   Linux Device Drivers

We are applying VeriFast to the verification of device drivers for the Linux operating system. These programs seem particularly suited for formal verification, because they are at the same time tricky to write, critical to the safety of the system, written by many different people with varying backgrounds and priorities, and yet relatively small and written against a relatively small API.

A significant part of the effort consists in writing specifications for the Linux kernel facilities used by the driver being verified. Part of the challenge here is that these facilities are often documented poorly or not at all, so we often find ourselves inventing a specification based on inspection of the kernel source code. Another part of the challenge is that VeriFast does not yet support all C language features required to interface with these facilities. As a temporary measure, in these cases we write a thin intermediate library that implements a VeriFast-friendly interface on top of the actual kernel interface.

We are only in the early stages of this endeavor. We have so far verified a small "Hello, world" driver that exposes a simple /proc file with an incrementing counter. This example, and the preliminary version of the VeriFast Linux Kernel Module Verification Kit that enables its verification, are included in the current VeriFast distribution. We are currently looking at a small USB keyboard driver.

Acknowledgements

This research is partially funded by the Interuniversity Attraction Poles Programme Belgian State, Belgian Science Policy, and by the Research Fund K.U. Leuven.

References

1. Jacobs, B., Smans, J., Piessens, F.: A quick tour of the veriFast program verifier. In: Ueda, K. (ed.) APLAS 2010. LNCS, vol. 6461, pp. 304–311. Springer, Heidelberg (2010)
2. Jacobs, B., Smans, J., Piessens, F.: The VeriFast program verifier: A tutorial (2010), http://www.cs.kuleuven.be/~bartj/verifast/
3. Bornat, R., Calcagno, C., O'Hearn, P., Parkinson, M.: Permission accounting in separation logic. In: POPL (2005)
4. Jacobs, B., Piessens, F.: Expressive modular fine-grained concurrency specification. In: POPL (2011)
5. Distefano, D., O'Hearn, P.W., Yang, H.: A local shape analysis based on separation logic. In: Hermanns, H. (ed.) TACAS 2006. LNCS, vol. 3920, pp. 287–302. Springer, Heidelberg (2006)
6. Calcagno, C., Distefano, D., O'Hearn, P., Yang, H.: Compositional shape analysis by means of bi-abduction. In: POPL (2009)

Verifying Functional Correctness
of C Programs with VCC

Michał Moskal

Microsoft Research Redmond
michal.moskal@microsoft.com

Abstract. VCC [2] is an industrial-strength verification environment for low-level concurrent systems code written in C. VCC takes a program (annotated with function contracts, state assertions, and type invariants) and attempts to prove the correctness of these annotations. VCC's verification methodology [4] allows global two-state invariants that restrict update of shared state and enforces simple, semantic conditions sufficient for checking those global invariants modularly. VCC works by translating C, via Boogie [1] intermediate verification language, to verification conditions handled by the Z3 [5] SMT solver.

The environment includes tools for monitoring proof attempts and constructing partial counterexample executions for failed proofs and has been used to verify functional correctness of tens of thousands of lines of Microsoft's Hyper-V virtualization platform and of SYSGOs embedded real-time operating system PikeOS.

In this talk, I am going to showcase various tools that come with VCC: the verifier itself, VCC Visual Studio plugin, and Boogie Verification Debugger. I am going to cover the basics of VCC's verification methodology on various examples: concurrency primitives, lock-free data-structures, and recursive data-structures.

The sources and binaries of VCC are available for non-commercial use at http://vcc.codeplex.com/. A tutorial [3] is also provided. VCC can be also tried online at http://rise4fun.com/Vcc.

References

1. Barnett, M., Chang, B.-Y.E., DeLine, R., Jacobs, B., Leino, K.R.M.: Boogie: A modular reusable verifier for object-oriented programs. In: de Boer, F.S., Bonsangue, M.M., Graf, S., de Roever, W.-P. (eds.) FMCO 2005. LNCS, vol. 4111, pp. 364–387. Springer, Heidelberg (2006)
2. Cohen, E., Dahlweid, M., Hillebrand, M.A., Leinenbach, D., Moskal, M., Santen, T., Schulte, W., Tobies, S.: VCC: A practical system for verifying concurrent C. In: Berghofer, S., Nipkow, T., Urban, C., Wenzel, M. (eds.) TPHOLs 2009. LNCS, vol. 5674, pp. 23–42. Springer, Heidelberg (2009)
3. Cohen, E., Hillebrand, M.A., Moskal, M., Schulte, W., Tobies, S.: Verifying C programs: A VCC tutorial. Working draft, http://vcc.codeplex.com/

M. Bobaru et al. (Eds.): NFM 2011, LNCS 6617, pp. 56–57, 2011.

4. Cohen, E., Moskal, M., Schulte, W., Tobies, S.: Local verification of global invariants in concurrent programs. In: Touili, T., Cook, B., Jackson, P. (eds.) CAV 2010. LNCS, vol. 6174, pp. 480–494. Springer, Heidelberg (2010)
5. de Moura, L.M., Bjørner, N.: Z3: An efficient SMT solver. In: Ramakrishnan, C.R., Rehof, J. (eds.) TACAS 2008. LNCS, vol. 4963, pp. 337–340. Springer, Heidelberg (2008)

Bakar Kiasan: Flexible Contract Checking for Critical Systems Using Symbolic Execution[*]

Jason Belt[1], John Hatcliff[1], Robby[1], Patrice Chalin[2], David Hardin[3], and Xianghua Deng[4,**]

[1] Kansas State University
{belt,hatcliff,robby}@ksu.edu
[2] Concordia University
chalin@encs.concordia.ca
[3] Rockwell Collins Advanced Technology Center
dshardin@rockwellcollins.com
[4] Penn State University Harrisburg
wdeng@google.com

Abstract. SPARK, a subset of Ada for engineering safety and security-critical systems, is designed for verification and includes a software contract language for specifying functional properties of procedures. Even though SPARK and its static analysis components are beneficial and easy to use, its contract language is almost never used due to the burdens the associated tool support imposes on developers. In this paper, we present: (a) SymExe techniques for checking software contracts in embedded critical systems, and (b) Bakar Kiasan, a tool that implements these techniques in an integrated development environment for SPARK. We describe a methodology for using Bakar Kiasan that provides significant increases in automation, usability, and functionality over existing SPARK tools, and we present results from experiments on its application to industrial examples.

1 Introduction

Though first proposed by King [17] over three and a half decades ago, symbolic execution (SymExe) has experienced a renaissance in recent years as researchers have looked for techniques that automatically discover wide-ranging properties of a program's behavior with little or no developer intervention. Research has centered around using symbolic execution for detection of common faults such as null-pointer de-referencing, buffer overflows, array bounds violations, assertion checking, and test case generation [16,22,23,5]. Much of the work has been carried out in the context of object-oriented languages such as Java [16,8,22], C++, and C#.

While SymExe can be applied in many contexts, we are exploring the effectiveness of SymExe in developing and assuring critical systems. Thus, in addition to an emphasis on bug-finding and test-case generation, we also aim to support checking of formal

---

[*] Work supported in part by the US National Science Foundation (NSF) CAREER award 0644288, the US Air Force Office of Scientific Research (AFOSR), Rockwell Collins, and the Natural Sciences and Engineering Research Council (NSERC) of Canada grant 261573.
[**] Current affiliation: Google Inc.

code contracts written in rich specification languages capable of capturing complex functional correctness properties. In this paper, we investigate how SymExe can add value to a commercial framework for developing and verifying critical systems that is based on the Spark/Ada (Spark for short) language [1].

Spark is a subset of Ada designed for programming and verifying high assurance applications such as avionics applications certified to DO-178B Level A. It deliberately omits constructs that are difficult to reason about such as dynamically created data structures, pointers, exceptions, and recursion. Spark annotations allow developers to capture pre/post-conditions and embedded assertions as well as information flow relationships between procedure parameters and global variables accessed in the procedure. Spark tooling can be used to perform static checking of such annotations, mainly through the generation and proof of verification conditions (VCs).

Our experience with Spark is derived from its use in security critical projects at Rockwell Collins including the Janus high-speed cryptography engine and several other embedded information assurance devices. Even though Spark and its static analysis components are beneficial and easy to use, its contract language is almost never used due to the burdens the associated tool support imposes on developers. In fact, we are not aware of any industrial development effort that makes significant use of the Spark pre/post-condition notation. We believe there are several reasons for this.

– Many uses of pre/post-conditions (e.g., those that include quantification for reasoning about arrays) will produce VCs that cannot be discharged automatically by Spark tooling, thus developers must fall back on traditional code inspection or (laborious) manual proof using an interactive checker.
– Verification conditions and proof rules necessary for discharging contracts are represented using Functional Description Language (FDL) that is very different from the Spark programming language. Shifting from source code to a specialized proof language that requires additional training is disruptive to developer workflows.
– Specifying desired functionality as logical expressions in contracts is difficult for complex properties; it is often necessary to introduce "helper" specification functions. In Spark, the only mechanism to achieve this is to introduce functions without implementations whose behavior is subsequently axiomatized in FDL. Reasoning about these functions usually requires manual proof.
– Although they capture a variety of useful semantic properties, Spark contracts are not leveraged by other quality assurance (QA) techniques in a manner that would increase the "value proposition" of the framework to developers.

In practice, the above problems often cause projects to avoid using Spark contracts.

We believe that the foundational approach to symbolic execution that we have been pursuing can significantly improve the usability and effectiveness of the Spark contract language by providing a completely automated bounded verification technology that scales to complex Spark contracts for industrial code bases. Our aim is not to replace the VCGen framework of Spark but to complement it by offering highly automated developer-friendly techniques that be used directly in the code(specify)-test(check)-debug(understanding feedback) loop of typical developer workflows. Specifically, the main contributions of this paper are:

- Description of how SPARK contracts can be represented to enable SymExe.
- Presentation of a SymExe algorithm and associated bounding strategies used to check SPARK code against our contract representation.
- Presentation of our SymExe tool for SPARK, Bakar Kiasan[1], which in addition to leveraging contracts, includes behavior visualization and test case generation.
- A methodology, that includes use of Bakar Kiasan, that we believe will be effective in critical system development.
- Illustrations of how SymExe provides greater flexibility including the ability to: (a) specify complex behaviors working directly at the source code level as opposed to a separate proof language, (b) forgo conventional compositional checking (as required by SPARK's existing VCGen approach) when methods are not fully specified, and (c) to freely mix logical and executable specifications in SPARK contracts.
- Evaluations that demonstrate significant improvements in the degree of automation required for checking SPARK contracts.

2   Example

Figure 1 shows excerpts of a SPARK package LinkedIntegerSet that provides a representation of a set of Item_Type records. The intention is that the ID field uniquely identifies the record within the set, while the Value holds data (the details of which are irrelevant for this example—though, for sake of concreteness, we have made it Integer, hence the module name). This code (minus the contracts) is taken directly from the code base of an embedded security device developed at Rockwell Collins (only variables have been renamed to avoid revealing the nature of the application), and was provided as a challenge problem to the academic authors for demonstrating contract specification/checking capabilities. Academic authors worked with Rockwell Collins engineers to develop what the engineers considered a reasonable approach to contract specification for this example.

The set representation is based on a single-linked list. Since SPARK does not include heap-allocated data, the linked list is implemented using two arrays: (1) Item_List that holds the current elements of the set as well as free slots for elements to be added, and (2) Next_List implements "links" from a set element to another. Used_Head gives the index position in Item_List of the first set item. Similarly, Free_Head marks the index position of the first of the free array elements.

SPARK includes both *procedures* (which may have side-effects) and (side-effect-free) *functions*; we refer to these collectively as *methods*. The parameter passing mechanism is call-by-value-result. Each parameter and global variable referenced by a procedure must be be classified as **in**, **out**, or **in out**. Each method can have a behavioral contract, embedded in Ada comments beginning with a special delimiter # recognized by the SPARK tools. Procedures can have both **pre** and **post** conditions. The symbol ~ is used (e.g., Elem_Array~ in the post-condition of Add) to denote the pre-state value of the variable. Instead of post-conditions, functions can have *return constraints*. Method implementations can include *in-line assertions*, which may be used to state loop invariants. The SPARK contract language includes quantifiers and the usual boolean operators.

---

[1] "Bakar" is the word for "spark" in Indonesian, while "Kiasan" is a word meaning "symbolic".

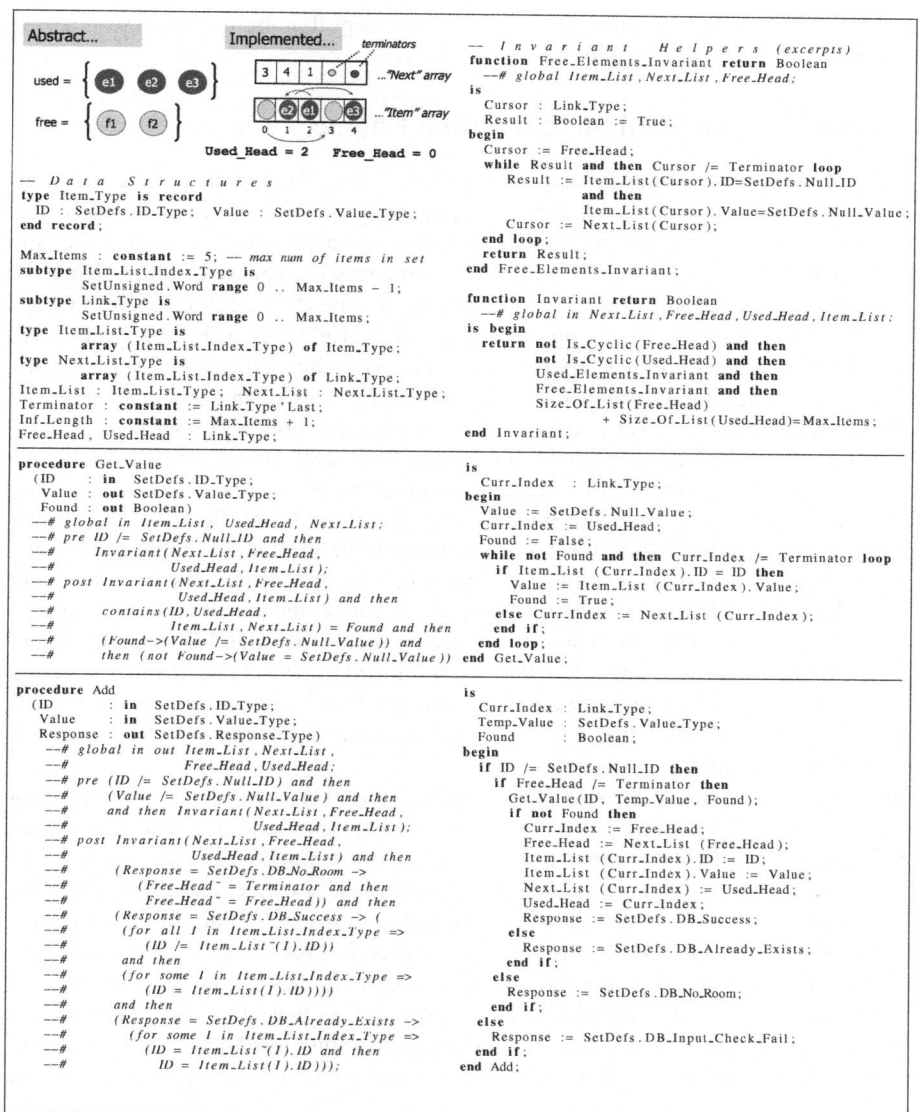

Fig. 1. LinkedIntegerSet Example (excerpts)

Since SPARK excludes dynamically-allocated data, all SPARK arrays are allocated statically and must have statically determined bounds (whose sizes are known at compile time). SPARK arrays are values; passing arrays as parameters and assignment between variables of an array type results in an array copy. Arrays can be compared for value equality (structurally) in both method implementations and contracts. In addition, contracts can utilize the array update notation A[I => V] which denotes an array value that is identical to that currently held by array A except that the position given by the expression I maps to the value of the expression V.

3   Bakar Kiasan Symbolic Execution Engine

SymExe characterizes values flowing through a program using logical constraints. Consider the Min example in Figure 2 that computes the minimum of two values. In this case, we are interested in proving that the assertion in line 9 is never executed (i.e., the true-branch in line 8 is infeasible) without knowing specific concrete values. Thus, we introduce special symbolic values a and b to act as placeholders for concrete values of A and B, respectively. The computation tree on the right side of Figure 2 illustrates SymExe on the procedure by keeping track of the symbolic values bound to each variable as well as logical constraints (*i.e.*, the *path condition* given in curly brackets {..}).

Initially, the constraint set is empty because we know nothing about how a and b are related. After executing line 3, we know that Z=a, thus, (A=a and B=b and Z=a). At line 5, both the condition $(a < b)$ and its negation $(b >= a)$ are satisfiable (i.e., there are integer values for a and b that satisfy these conditions because a and b are currently unconstrained), thus, we have to consider both program executions following the conditional's true-branch and its false-branch; thus, the initial path in on the right side of the figure splits into two possible cases. At line 8, the program state is characterized by either (A=a and B=b and Z=a and $\{a < b\}$) or (A=a and B=b and Z=b and $\{b >= a\}$). The constraints imply that the if-condition at line 8 is false in either situation (as indicated by the F for the path condition for the "true" cases) – there is no *feasible path* (no possible assignment of concrete values to inputs) that lead to line 9—and thus exploration along these paths is ignored.

Bakar Kiasan includes an interface to underlying decision procedures (including CVC3 [2], Yikes [10], and Z3 [7]). These are used to determine if constraints in path conditions are satisfiable in order to make decisions at branching points such as the ones at lines 5 and 8, and more generally, to determine if boolean conditions in method contracts are satisfiable. Constraints to be passed to conventional decision procedures are first passed through Kiasan's Lightweight Decision Procedure (LDP) module [4]. LDP contains a collection of rules for rapid solving of common constraint shapes and for implementing various forms of constant propagation that allow many constraints to be solved without the overhead of pushing constraints all the way out to an external decision procedure. It has been demonstrated that LDP can give a significant

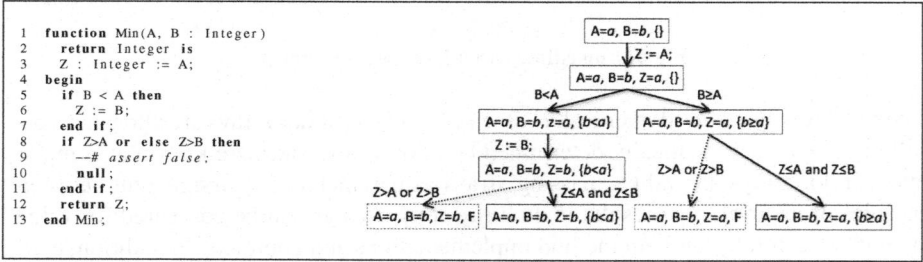

Fig. 2. Illustration of Symbolic Execution for a simple example

reduction in analysis time (by an order of magnitude) [4]. For SPARK scalar types, Bakar Kiasan depends on LDP and the underlying decision procedures for constraint solving. Developing SymExe to better support dynamically-allocated objects [16,23,8] has been a significant focus of previous research. Since SPARK uses only statically-allocated and value-based composite structures (array/records), previous approaches need to be adapted to this setting.

In contrast to other approaches, however, Bakar Kiasan supports *both* logical and graph-based symbolic representations of complex structures. In the logical representation, Bakar Kiasan uses supported theories in underlying decision procedures for representing values of arrays and records. In the graph-based representation, Kiasan uses an adaptation of [8] in which an explicit-state representation (similar to what would be used in explicit-state model checking) is used to model composite structures, and decision procedure support is used only to handle constraints on scalar values. Our approach for Java is adapted for SPARK by optimizing away aliasing cases, and it is enhanced to handle value-based structures instead of reference-based structures by using an optimized form of copy-on-write state tree structures.

The symbolic value manipulation above is incorporated in a depth-first exploration. Since SymExe does not merge state information at program joint points after branches and loops, the analysis may not terminate when the program being checked contains loops or recursion, unless inductive predicates such as loop invariants are provided at these loops and recursion points, as shown by [13]. However, in the context of programs manipulating complex structures, precise loop invariants are difficult to obtain.

A key goal of Bakar Kiasan is to offer developers an approach that provides meaningful checking without requiring the effort of writing loop invariants. The usual approach to address the termination issue is to employ some form of bounding. There are a variety of bounding mechanisms that have been used in the literature, such as loop bounding, depth bounding (*i.e.*, limiting the number of execution steps), bounding on the length of method call chains, etc. The use of these bounding mechanisms leads to an under-approximation of program behaviors. Technically, this means that the analysis is unsound in general, and care must be taken when interpreting analysis reports that indicate no bugs are found (errors may exist in the portion of the program's state space that was not explored). To compensate, Bakar Kiasan notifies users when a bound is exhausted with the program point (and state) where it occurs, thus, users are warned of potential behaviors that are not analyzed.

These are trade-offs that we are certainly willing to accept. The under-approximation and path splitting means that the analysis yields no false positives. Moreover, the analysis will provide complete verification when the procedure includes no loops (as often occurs in embedded programs). Most importantly, as we will explore in the following section, it allows developers to easily check sophisticated properties that, in practice, they would never check using totally automatic non-bounded methods.

4   Checking SPARK Contracts in Kiasan

Most contract checking tools work compositionally and require that every method be given a contract. Bakar Kiasan provides greater flexibility by providing the ability to check SPARK program behaviors either compositionally, non-compositionally, or mixed. Intuitively, when analyzing a procedure P, Kiasan starts by *assuming* P's pre-condition (*i.e.*, adding the pre-condition to the path condition as a conjunct). The analysis proceeds by symbolically executing P's body, and then *asserting* P's post-condition (*i.e.*, branching into two paths; one assumes the post-condition, and the other assumes the negation of the post-condition that leads to an error state). If P calls another procedure Q, Kiasan can symbolically execute Q directly (non-compositional), or substitute Q by its contract (compositional, via translation of contracts to executable form described below), as instructed by the user.

When applying non-compositional checking to Q, Kiasan *asserts* Q's pre-condition, performs appropriate parameter passing mechanics, and continues on with its depth-first exploration. Kiasan *asserts* Q's post-condition when it is encountered along each explored path. When applying compositional checking to Q, Kiasan *asserts* Q's pre-condition, havocs (i.e., assigns fresh unconstrained symbolic values to) all Q's **out** variables, and then *assumes* Q's post-condition. These steps ensure that Q's effect on **out** variables as specified by its contract is captured by: (a) starting with no knowledge about the variables' values, and then (b) applying the constraints in Q's post-condition which would typically constrain the values of the **out** variables. If no contract is supplied for a procedure, the procedure is treated as if its pre/post-condition expressions are both **true** (*i.e.*, checking always succeeds but the values of **out** variables are unconstrained).

Kiasan processes contracts by (automatically) translating each contract to an executable representation in the SPARK programming language that can be processed using the same interpretive engine used to process SPARK procedure implementations. Since SPARK's contract language is a super set of the expression language of its programming language, many aspects of this translation process are achieved rather directly. In the following paragraphs, we describe how we obtain an executable representation for additional elements of the contract language.

An "old" expression (i.e., $e^\sim$, for an expression e) in the post-condition of procedure P is handled via transformation. Intuitively, e's value is saved to a (fresh) variable x before P is executed, and upon post-condition checking, $e^\sim$ is replaced with x. This strategy applies both to SPARK scalar and complex structure (array and record) values (recall that SPARK has a value semantics for complex structures). In addition, since SPARK's contract expression language is side-effect free, several occurrences of $e^\sim$ can be substituted with the same variable. Old expression processing is illustrated for the Inc procedure below; the code on the left hand side is transformed by Kiasan into the one in the right hand side:

```
procedure Inc(I: in out Integer)              ...
    --# pre I > 0;                             begin
    --# post I = I~ + 1;                           assume I > 0;
is                                                 oldI := I;
begin                                              I := I+1;
    I := I+1;                                      assert I = oldI + 1;
end Inc;                                        end Inc;
```

Since SPARK records and arrays are value-based, the contract language provides an equality operation (=) that can be used to test entire records and arrays for equality regardless of their level of complexity/nesting. In Kiasan's logical representation of arrays/records, there is a fairly direct translation to the equality operators and function update notation used by underlying decision procedures. In the graph-based arrays/ records symbolic mode, Kiasan uses its own optimized algorithm tailored specifically for checking SPARK value-based arrays and records adapted from [8].

Kiasan transforms quantifications into loops. Universal and existential quantification of the forms for all x in τ => $\phi(x)$ and for some x in τ => $\phi(x)$, respectively, are transformed as illustrated below (using a pseudo-code notation to capture the intuition):

```
-- universal
Result := True;
S := KiasanValues(τ);
for x in S loop
  if not φ(x) then
    Result := False;
    exit;
  end if;
end loop;
```

```
-- existential
Result := False;
S := KiasanValues(τ);
for x in S loop
  if φ(x) then
    Result := true;
    exit;
  end if;
end loop;
```

Using this transformation scheme, nested quantifications become nested loops. Use of the Kiasan function KiasanValues allows the analysis to be configured to implement different exploration strategies for the array elements. In its simplest form, the function simply returns all elements of the range specified by τ. Executing the loop body implementing the quantification may lead to Kiasan's loop bound being exhausted if the number of array elements exceeds the loop bound. To work around this issue, users can configure Kiasan to return a bounded number of distinct and ordered (*i.e.*, strictly ordered) fresh symbolic (or concrete) values in τ, with the hope that these values would act as witnesses to uncover inconsistency between the program and its specification. Regardless, Kiasan warns the users if there are potential behaviors that are not analyzed (*i.e.*, potentially unsound). We adopt this pragmatic approach for the sake of giving users some helpful feedback due to the inherent limitation (i.e., incompleteness) of decision procedures on general quantifications.

5 Bakar Kiasan Methodology and Tools

Developers can interact with Bakar Kiasan in two ways: (1) via the command line interface for which a comprehensive HTML report is generated, and (2) via a GUI built as an Eclipse plug-in. The GUI, which is integrated with both AdaCore's GNATBench and the Hibachi Eclipse plug-ins, provides the ability to invoke the AdaCore GNAT compiler and SPARK tools, and visualize typical Eclipse error mark-ups corresponding to errors reported by the compiler and SPARK tools. Space constraints do not allow us to give screen-shots or details of the HTML report or GUI, but these can be found in the extended version of this paper [3]. Below, we give a brief overview of Bakar Kiasan capabilities that directly impact developer workflows and the methodology of the tool.

Visualizing procedure inputs/outputs and constraints: Gaining intuition and a proper understanding of a procedure's input and output behavior is an important element of writing and debugging contracts. To assist in this, for each path explored in a contract/procedure, Kiasan generates a *use case visualization* that provides an example of a concrete pre/post-state for that path. These use cases are constructed by calling model

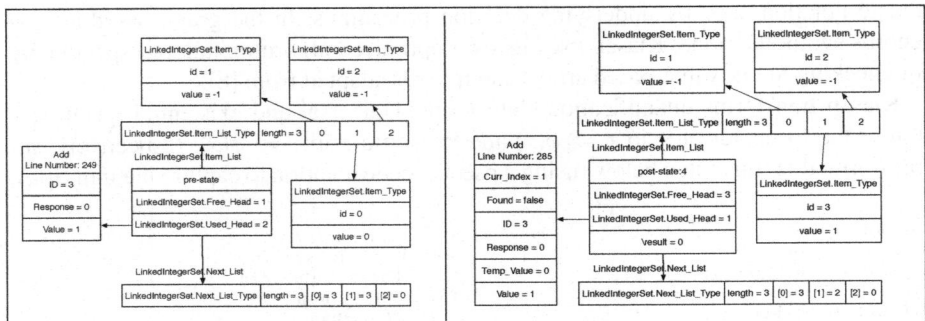

Fig. 3. Bakar Kiasan Pre/Post Use Case

finders that are usually part of decision procedures such as Yices and Z3 to find a so-lution to the symbolic pre/post-state constraints for the path. Figure 3 shows a use case for the LinkedIntegerSet.Add method of Figure 1 corresponding to a path through the method where an entry is successfully added. The pre-state shows the element to be added has ID=3 and Value=1. Index position 1 of the Item_List holds a null entry, and the post-state shows the newly added item replacing this entry. The index 3 is the ter-minator for this example, thus the post-state with Free_Head=3 indicates that the "free list" is now empty.

Although this example is taken from a procedure with a contract, the use case vi-sualizations are also useful when applied to procedures before contracts are written to help understand input/output relationships and guide the developer in writing contracts. In the GUI, developers can also step through the statements along each path in the use case and see both concrete values and symbolic constraints at each step.

Test case generation: SymExe is widely used for test case generation, and the same techniques can be applied here to generate, *e.g.*, a test in the AUnit format using the concrete values provided in the use cases[2]. Although the application of Kiasan already "checks the code" and no additional bugs would be uncovered by running the generated tests, we have found that generation of test cases can provide additional confidence to people unfamiliar with formal methods because they provide evidence external to the tool using quality assurance concepts that they are well acquainted with (*i.e.*, testing) that Kiasan is correctly exploring the program's state-space.

Coverage information: Gaining an appropriate understanding of what portions of a program's behavior have been explored or omitted is an important methodological as-pect of applying bounded verification. To aid in this, both the Kiasan HTML report and GUI provide extensive branch and statement coverage information that allows devel-opers to see portions of the code that may have been omitted during analysis due to bounding. This information typically drives an iterative process where contracts/code are debugged using smaller bounds and then bounds are increased to obtain desired levels of coverage. It is important to note that Kiasan gives an exhaustive (*i.e.*, com-plete in a technical sense) exploration of program behavior within bounds (relative to

---

[2] This capability is not yet provided but will be implemented in the near future.

limitations of decision procedures such as non-linear arithmetic). The factor limiting coverage is not precision of the analysis (*e.g.*, within analysis bounds, Kiasan always gives 100% MCDC coverage of reachable code), it is rather the required analysis time.

We now turn to key novel features of Bakar Kiasan that dramatically improve the usefulness of the SPARK contract language.

No loop invariants required: The existing SPARK tool chain and other VCGen techniques that only aim for complete verification require loop invariants. While progress has been made on research for inferring invariants [19,6], these techniques typically do not perform very well when complex data structures are involved. For example, consider the loop in Get_Value of Figure 1. An invariant for this loop would be very difficult for a developer to write using the logical expressions of SPARK's contract language because the loop is not iterating directly over the array (sequential progression through indices), but rather over the logical structure of the "used list" which jumps back and forth among index positions of the Item_List array via indirection realized by index values held in the Next_List. We believe that it is unlikely that a typical developer would *ever* use the existing SPARK tools to check this contract/procedure. Insisting on the presence of loop invariants before providing any sort of conclusive information is a serious impediment to the practical use of contracts. In contrast, Kiasan checks this contract without loop invariants automatically. Even though Kiasan requires relatively small bounds to be tractable, this gives a very thorough analysis of the structure of the code/contract.

Blended logical/executable contracts: Some constraints are quite simple and natural to express directly in SPARK's contract language. For example, the constraints on DB.Success and DB.Already_Exists in the bottom of Figure 1, even though non-trivial, can be coded using universal/existential quantification. However, the invariant properties and the Contains predicate used in Get_Value would be very difficult for a developer to write in the contract language. It is possible in the current SPARK tools to introduce *calls* to helper functions such as Invariant and Contains. However, VCGen in SPARK makes no connection between the use of such helper functions and their semantics as provided by their implementations. In SPARK, the semantics of such functions must be specified as axioms (rewrite rules) in the FDL proof language. This adds a considerable level of complexity to developer effort that almost always will result in developers not using contracts. In contrast, the SymExe foundation of Bakar Kiasan enables the semantics of such functions to be directly specified in the SPARK programming language as we illustrate in Figure 1—a task that would be straightforward for developers. Since all functions are guaranteed to be side-effect free in SPARK, this feature allows one to express complex properties (e.g., the contains function of Figure 1) in a form that is simpler and familiar to developers. We have found this capability to be extremely useful in practice. Ultimately, we envision a methodology which would gracefully move toward full functional verification by enabling the executable semantics of the helper functions to be incrementally switched out and replaced by corresponding definitions in an interactive theorem prover.

Compositional/Non-compositional checking: Kiasan's ability to support both compositional and non-compositional checking (as described in Section 4) also makes it easier for developers to start benefiting from contracts. Contracts can be specified for the most important methods and omitted for the rest.

Bounded checking matching well with SPARK applications: The style of bounded checking, while technically not providing the complete checking of VCGen/theorem-prover frameworks, in practice, it often completes the verification of procedures in embedded applications because such applications often have procedures without loops. In addition, embedded applications usually statically bound the size of data structures— which is a requirement in SPARK. Compared to conventional applications, this increases the likelihood that significant portions of a program's state-space can be covered within the bounding employed by Kiasan.

6   Evaluation

In this section, we report on the effectiveness of Bakar Kiasan when applied to a collection of examples representative of code found in embedded information assurance applications. Information about individual methods from these examples is displayed in Table 1. The sorting examples are a collection of library methods that manipulate array-based data structures as might be used to maintain configurable rules for managing message processing. **IntegerSet** and **LinkedIntegerSet** are representative of data structures used to maintain data packet filtering and transformation. **IntegerSet** provides an array-based implementation of an integer set data structure that adds an element by inserting it at the end of the occupied slots in the array and deletes an element by sliding the contents of occupied slots at higher index positions down one slot to reclaim the slot at which the element was deleted. **LinkedIntegerSet**, described earlier in Section 2, comes directly from a Rockwell Collins code base and uses two arrays to provide a set implementation with more efficient additions/deletions. The MMR (MILS Message Router) is an idealized version of a MILS infrastructure component (first proposed by researchers at the University of Idaho [20]) designed to mediate communication between partitions in a *separation kernel* [21]—the foundation of specialized real-time platforms used in security contexts to provide strong data and temporal separation. The MMR example is especially challenging to reason about because messages flow through a shared pool of memory slots (represented as one large array) where the partition "ownership" of slot contents changes dynamically and is maintained indirectly via two other two-dimensional arrays that hold indices into the memory array.

For each of these examples, **C-LoC** and **I-LoC** in Table 1 gives the number of lines of code in the method contract and implementation, respectively, broken down as X/Y where X is the LoC appearing directly in the contract or implementation and Y is the LoC appearing in helper functions. For **Helper** X/Y, X is the number of helper functions used in the contract; Y is the number of methods called in the implementation.

We seek to answer two primary questions with this evaluation: **Question (I):** can Bakar Kiasan provide a significant increase over the existing SPARK tool chain VCGen approach in the level of automation of contract checking? and **Question (II):** is the time required for Kiasan contract checking short enough to allow the tool to be employed as part of the developer code/test/debug cycles?

Table 1. Experiment Data (excerpts)

Package.Procedure Name	C-LoC	I-LoC	Helper	Loop	VC	k=3	k=4	k=5	k=6	k=7	k=8
Sort.Bubble	1/23	14/4	3/1	2	13/18	0.17	0.96	2.09	8.43	71.72	890.18
Sort.Insertion	1/21	11/0	3/0	2	10/14	0.15	0.98	2.06	8.24	70.72	892.17
Sort.Selection	1/21	15/0	3/0	2	28/30	0.16	1.06	2.28	9.95	90.14	1356.18
Sort.Shell	1/21	15/0	3/0	3	17/18	0.15	0.98	2.12	8.47	74.09	941.99
IntegerSet.Get_Element_Index	7/0	8/0	0/0	1	8/11	0.04	0.05	0.06	0.07	0.08	0.10
IntegerSet.Add	8/29	4/2	4/3	0	3/5	0.24	0.44	0.62	0.79	0.80	1.04
IntegerSet.Remove	8/27	6/0	4/1	0	5/6	0.16	0.30	0.56	0.96	1.21	1.36
IntegerSet.Empty	1/0	2/0	0/0	0	3/3	0.02	0.02	0.02	0.02	0.02	0.02
LinkedIntegerSet.Get_Value	6/45	12/0	6/0	1	9/10	0.64	0.88	1.13	1.51	2.19	2.85
LinkedIntegerSet.Add	15/51	23/12	6/1	0	14/16	0.43	0.73	1.66	5.26	34.96	379.34
LinkedIntegerSet.Delete	14/45	22/0	6/0	1	18/21	0.52	0.72	1.03	1.56	2.10	2.75
LinkedIntegerSet.Init	1/37	10/0	5/0	2	16/17	0.05	0.04	0.04	0.05	0.05	0.05
MMR.Fill_Mem_Row	3/1	6/1	0/1	1	8/10	0.18					
MMR.Zero_Mem_Row	5/1	3/1	0/1	1	6/7	0.19					
MMR.Zero_Flags	4/0	3/0	0/0	1	6/7	0.05					
MMR.Read_Msgs	15/63	5/13	6/5	0	3/4	1.71					
MMR.Send_Msg	10/24	6/1	3/3	0	4/5	0.50					
MMR.Route	22/82	22/1	9/2	2	62/67	13.90					

Regarding **Question (I)**, there are at least three forms of manual activity required to use the SPARK contract checking framework that go beyond what is required by Bakar Kiasan: (1) the need to supply loop invariants, (2) the need to add axioms to provide the semantics for uninterpreted functions used in contracts, and (3) the need to manually discharge VCs that are left unproven by the SPARK tools (we refer to these as "undischarged VCs").

Loop invariants: The **Loop** column records the number of loops in the implementation and indicates that well over 50% of the methods require a loop invariant describing properties of arrays when using SPARK's VCGen approach. For example, LinkedIntegerSet.Get_Value requires a complex loop invariant that depends on logic encoded in the helper function contains—which would either need to be coded in logical form or axiomatized in the FDL proof language (either approach would be very difficult and would likely fall outside of the scope of effort that a typical developer would be expected to expend). Kiasan allows developers to obtain effective bounded contract checking without having to add these loop invariants.

Verification Conditions: A **VC** column entry of X/Y indicates that X VCs were automatically discharged by SPARK out of Y generated VCs. Our experiments show that almost any contract that requires quantification (often required by functions that manipulates arrays) will have undischarged VCs. To give an indication of the amount of effort required to manually discharge VCs in the SPARK Proof Checker[3], a faculty member of our team with extensive experience in automated proof checking used the Proof Checker to prove the 3 undischarged VCs from one of our simplest examples—Value_Present which looks for an occurrence of a specified value in an array; 4 out of 7 of the VCs (those dealing with simple range checks on integer subtypes and array bounds) were automatically discharged by SPARK. Of the 3 remaining, 2 of the VCs required two proof steps to discharge while one required ten steps. Our best estimate is that it would take a Proof Checker *expert* user approximately 15 minutes to proof these three VCs. Given that the more realistic examples that we considered are much more complicated,

---

[3] Proof Checker is SPARK's interactive prover.

we can conclude that it is extremely unlikely that the SPARK tool chain would be used in its present form by typical developers to check contracts other than those that capture simple numeric constraints (though it is possible that the Proof Checker could be used by verification engineers in an extensive verification period at the conclusion of development). In contrast, Kiasan provides effective bounded contract checking automatically for *all* the methods in all of our examples. Thus, it significantly improves the accessibility and usefulness of SPARK contracts.

Regarding **Question (II)**, as discussed in the previous section, checking must be bounded for Kiasan checking to be tractable. The bounding philosophy used here is similar to that of Alloy [14]—bounded verification with relatively small bounds can be very useful in uncovering program flaws (in design and implementation). Moreover, due to the bounded nature of SPARK, we believe our bounded approach fits well with how developers use SPARK. Table 1 shows timing data (in seconds) with array sizes from $k = 3 \ldots 8$ elements (an exception is the MMR, which uses two-dimensional arrays of size 3 and a single dimensional array of size 9). The data shows that contract checking for even an entire package (except the MMR.Route) can be completed in 1-2 seconds for bounds of $k = 3, 4$ – indicating that Kiasan is clearly viable for incorporation in the code/test/debug loop of the developers. As an indication of how the performance scales, when we increased the array size for the **LinkedIntegerSet** example to 8, **Delete** and **Get_Value** completed in under 3 seconds each while **Add** required just over 6 minutes. This suggests that Kiasan could be deployed to check within small bounds during typical development activity, and then applied to check within larger bounds over night.

7   Related Work

Our long term research plan seeks to demonstrate that SymExe can serve as a true verification technique (albeit bounded at this point) that can provide high confidence in the domain of embedded safety/security-critical systems in a highly automated fashion. As part of our effort to provide a rigorous foundation for SymExe, in previous work we have justified SymExe execution algorithms by providing proofs of correctness for complex optimizations [4,8] and by providing mathematical approaches to calculate minimum number of test-cases and execution paths needed to achieve exhaustive exploration of program's data state [9].

There has been a lot of work on SymExe for programs that manipulate dynamically-allocated structures (*e.g.*, [16,12,22,11,23,5]). Bakar Kiasan directly leverages existing decision procedures on complex structures (i.e., records and arrays) when it is in logical representation mode [18]. This is similar to symbolic execution approaches that use a logical approach such as XRT [12] (and many others), however, without the complication of modeling program heap and pointer aliasing due to SPARK characteristics. When graph-based symbolic representation is used, the underlying algorithm in Bakar Kiasan is an adaptation of lazy initialization algorithms that were designed for Java [8], but optimized for inherent properties of SPARK programs. In addition, we focus on bounded verification of program behavioral contracts, as opposed to mainly finding bugs.

Carrying out work that is crucial for moving the SPARK infrastructure forward, Jackson and Passmore [15] aim to improve the usability of SPARK by building the

Victor [25] tool that translates SPARK VCs to the various SMT solvers including CVC3 [2], Yices [10], and Z3 [7]. They note that concerns over the cost of handling non-automatically proven VCs cause "most SPARK users [to] settle for verifying little more than the absence of run-time exceptions caused by arithmetic overflow, divide by zero, or array bounds violations" and that "the number of non-automatically-proved VCs is usually significant." Using examples that primarily consist of VCs from run-time checks instead of full contracts, they show that using SMT solvers instead of the SPARK tooling discharges roughly the same number of VCs, but provides better performance and better error explanations. Their conclusions substantiate our arguments that, even though better support for VC proving can improve the SPARK tools, substantially increasing the automation of VC proving is difficult even with state-of-the-art solvers. In our opinion, this provides addition justification for considering a tool like Bakar Kiasan to complement VC proving by trading off complete verification for highly-automated bounded checking of expressive contracts.

8   Conclusion and Future Work

We have illustrated how symbolic execution techniques can increase the practicality of contract-based specification and checking in development of safety and security critical embedded systems. These techniques are complementary and can be used in conjunction with other contract verification techniques such as VCGen that more directly target full functional verification. SymExe hits a "sweet spot" between trade-offs of: the completeness of full functional verification to obtain a much greater degree of automation, the ability to more naturally blend processing of declarative and executable specifications, better support for error trace explanation and visualization, and stronger connections to other quality assurance methods such as testing. Although we have illustrated these techniques in the context of SPARK, they can be adapted easily to other contexts as well, e.g., for safety critical subsets of C with contract languages [24]. In the case of the current SPARK tool chain, SymExe can play a key role in moving the SPARK contract framework from a method that is rarely used into one that is quite usable and quite effective in development of critical systems.

Our experience in using the SPARK contract language has exposed the need for several extensions including first class support for specifying package and invariants and data refinement. We are also investigating how contract extensions supporting rich secure information flow specifications can be integrated with the work presented here.

References

1. Barnes, J.: High Integrity Software – the SPARK Approach to Safety and Security. Addison-Wesley, Reading (2003)
2. Barrett, C., Tinelli, C.: CVC3. In: Damm, W., Hermanns, H. (eds.) CAV 2007. LNCS, vol. 4590, pp. 298–302. Springer, Heidelberg (2007)
3. Belt, J., Hatcliff, J., Robby, Chalin, P., Hardin, D., Deng, X.: Bakar Kiasan: Flexible contract checking for critical systems using symbolic execution. Technical Report SAnToS-TR2011-01-03, Kansas State University (2011),
http://people.cis.ksu.edu/~belt/SAnToS-TR2011-01-03.pdf

4. Belt, J., Robby, Deng, X.: Sireum/Topi LDP: A lightweight semi-decision procedure for optimizing symbolic execution-based analyses. In: Proceedings of the ACM SIGSOFT Symposium on the Foundations of Software Engineering (ESEC/FSE), pp. 355–364 (2009)
5. Cadar, C., Dunbar, D., Engler, D.R.: Klee: Unassisted and automatic generation of high-coverage tests for complex systems programs. In: 8th USENIX Symposium on Operating Systems Design and Implementation (OSDI), pp. 209–224. USENIX Association (2008)
6. Chang, B.-Y.E., Leino, K.R.M.: Inferring object invariants: Extended abstract. Electr. Notes Theor. Comput. Sci. 131, 63–74 (2005)
7. de Moura, L.M., Bjørner, N.: Z3: An efficient SMT solver. In: Ramakrishnan, C.R., Rehof, J. (eds.) TACAS 2008. LNCS, vol. 4963, pp. 337–340. Springer, Heidelberg (2008)
8. Deng, X., Lee, J., Robby: Efficient symbolic execution algorithms for programs manipulating dynamic heap objects. Technical Report SAnToS-TR2009-09-25, Kansas State University (September 2009)
9. Deng, X., Walker, R., Robby: Program behavioral benchmarks for evaluating path-sensitive bounded verification techniques. Technical Report SAnToS-TR2010-08-20, Kansas State University (2010)
10. Dutertre, B., de Moura, L.: The Yices SMT solver (August 2006), Tool paper at http://yices.csl.sri.com/-tool-paper.pdf
11. Godefroid, P., Klarlund, N., Sen, K.: DART: Directed automated random testing. In: ACM SIGPLAN 2005 Conference on Programming Language Design and Implementation (PLDI), pp. 213–223. ACM Press, New York (2005)
12. Grieskamp, W., Tillmann, N., Schulte, W.: XRT–exploring runtime for .NET: Architecture and applications. In: Workshop on Software Model Checking (2005)
13. Hantler, S.L., King, J.C.: An introduction to proving the correctness of programs. ACM Computing Surveys (CSUR) 8(3), 331–353 (1976)
14. Jackson, D.: Alloy: a lightweight object modelling notation. ACM Transactions on Software Engineering and Methodology (TOSEM) 11(2), 256–290 (2002)
15. Jackson, P., Passmore, G.: Proving SPARK verification conditions with SMT solvers (2009), Draft journal article, http://homepages.inf.ed.ac.uk/pbj/papers/vct-dec09-draft.pdf
16. Khurshid, S., Păsăreanu, C.S., Visser, W.: Generalized symbolic execution for model checking and testing. In: Garavel, H., Hatcliff, J. (eds.) TACAS 2003. LNCS, vol. 2619, pp. 553–568. Springer, Heidelberg (2003)
17. King, J.C.: Symbolic execution and program testing. Communications of the ACM 19(7), 385–394 (1976)
18. Kroening, D., Strichman, O.: Decision Procedures – An Algorithmic Point of View. Springer, Heidelberg (2008)
19. Leino, K.R.M., Logozzo, F.: Loop invariants on demand. In: Yi, K. (ed.) APLAS 2005. LNCS, vol. 3780, pp. 119–134. Springer, Heidelberg (2005)
20. Rossebo, B., Oman, P., Alves-Foss, J., Blue, R., Jaszkowiak, P.: Using SPARK-Ada to model and verify a MILS message router. In: Proceedings of the International Symposium on Secure Software Engineering (2006)
21. Rushby, J.: The design and verification of secure systems. In: 8th ACM Symposium on Operating Systems Principles, vol. 15(5), pp. 12–21 (1981)
22. Sen, K., Agha, G.: CUTE: A concolic unit testing engine for C. In: ACM SIGSOFT Symposium on the Foundations of Software Engineering (FSE), pp. 263–272 (2005)
23. Tillmann, N., de Halleux, J.: Pex–white box test generation for.NET. In: Beckert, B., Hähnle, R. (eds.) TAP 2008. LNCS, vol. 4966, pp. 134–153. Springer, Heidelberg (2008)
24. Frama-C website, http://frama-c.com/
25. Victor website, http://homepages.inf.ed.ac.uk/pbj/spark/victor.html

Approximate Quantifier Elimination for Propositional Boolean Formulae

Jörg Brauer[1] and Andy King[2]

[1] Embedded Software Laboratory, RWTH Aachen University, Germany
[2] Portcullis Computer Security, Pinner, UK

Abstract. This paper describes an approximate quantifier elimination procedure for propositional Boolean formulae. The method is based on computing prime implicants using SAT and successively refining over-approximations of a given formula. This construction naturally leads to an *anytime* algorithm, that is, it can be interrupted at anytime without compromising soundness. This contrasts with classical monolithic (all or nothing) approaches based on resolution or model enumeration.

1 Introduction

Model checking and abstract interpretation are sub-disciples of formal methods that, for many years, have been diametrically opposed. In model checking a programmer prescribes a so-called model that formally specifies the behaviour of the system or program. All paths through the program are then exhaustively checked against this requirement. Either the requirement is discharged or a counterexample is found that illustrates how the program is faulty. The detailed nature of the requirements entails that the program is simulated in a fine-grained way, sometimes down to the level of individual bits. Enumerating all these combinations is computationally infeasible. Thus, there has been much interest in representing all the states of a program symbolically, which enables states that share commonality to be represented without duplicating their commonality.

In abstract interpretation, the key idea is to abstract away from the detailed nature of states. Then the program checker operates over classes of related states — collections of states that are equivalent in some sense — rather than individual states. If the number of classes is small, then all the paths through the program can be enumerated one-by-one without incurring the problems of state-space explosion. When carefully constructed, the classes of states can preserve sufficient information to prove the correctness requirements.

Despite their philosophical differences, the fields of model checking and abstract interpretation are converging, partly because they draw on similar computational techniques. A case in point is given by Boolean formulae that are typically either represented with BDDs [5] or manipulated using SAT [22]. BDDs have been widely applied, both in symbolic model checking [7], and as an abstract domain for tracking dependences [1]. Although some niche problems remain difficult for SAT [6], clever ideas and careful engineering have advanced DPLL-based

M. Bobaru et al. (Eds.): NFM 2011, LNCS 6617, pp. 73–88, 2011.

SAT solvers [22] to the point they can rapidly decide the satisfiability of structured problems that involve thousands of variables. Conseqently SAT has been almost universally adopted within symbolic model checking [9].

1.1 Quantifier Elimination and Abstract Interpretation

Yet SAT remains a comparative novelty in abstract interpretation where it is more often than not relegated to solving auxiliary problems such as that of synthesising best transformers [4,20,30] rather then being integrated into the heart of the analysis itself [18]. This is not because there is no interest in using Boolean functions as an abstract domain [1,16,18] but rather because projection operations, namely existential and universal quantifier elimination, fit less comfortably with SAT than with BDDs. Eliminating a single variable from a BDD, either existentially or universally, is worst-case quadratic in size of the input BDD [5, Sect. 3.3]. By way of contrast, the natural way to existentially quantify using a SAT solver is to systematically enumerate the models of a formula using blocking clauses. Even when the blocking clauses only constrain the variables in the projection space, such methods are inefficient when compared to BDD-based techniques because of the large number of models that may need to be enumerated [6]. This would be less of a problem if projection was an infrequent operation in abstract interpretation; the guiding principle in domain design is that the commonly arising operations should be fast whereas the speed of the infrequent operations is less critical. However, in dependency analysis, elimination is applied whenever a call is encountered. This is because the dependencies at the call site need to be restricted to those variables that occur as the arguments of a call so as to propagate dependency information across the body of the callee. Existential quantification is applied to flow information in the direction of the control-flow [1] whereas universal quantification is needed to propagate requirements against the control-flow [13]. The frequency of call handling and the inefficiency of SAT-based elimination methods have tended to bias abstract interpretation towards BDDs [1], though new algorithms for elimination would break this dependency.

1.2 Quantifier Elimination by Resolution and Striking Out Literals

For formulae presented in CNF, existential and universal quantifiers can alternatively be eliminated by resolution and striking out literals [22]. To illustrate, let $f = (\wedge_{i=0}^{n_1} x \vee C_i) \wedge (\wedge_{j=0}^{n_2} \neg x \vee D_j) \wedge (\wedge_{k=0}^{n} E_k)$ and consider $\exists x : f$ and $\forall x : f$ where C_i, D_j and E_k are clauses that involve neither x nor $\neg x$. A quantifier-free version of $\exists x : f$ can be obtained by resolving each $x \vee C_i$ with $\neg x \vee D_j$ to give $\exists x : f = (\wedge_{i=0}^{n_1} \wedge_{j=0}^{n_2} C_i \vee D_j) \wedge (\wedge_{k=0}^{n} E_k)$, *increasing* the representation size by as many as $n_1 n_2 - n_1 - n_2$ clauses. By way of contrast, $\forall x : f$ can be found by removing the x and $\neg x$ literals to give $\forall x : f = (\wedge_{i=0}^{n_1} C_i) \wedge (\wedge_{j=0}^{n_2} D_j) \wedge (\wedge_{k=0}^{n} E_k)$, *reducing* the size of the representation.

One might be forgiven for thinking that calculating a quantifier-free version of $\forall \boldsymbol{y} : f$ is straightforward when f is propositional and $\boldsymbol{y}$ is a vector of variables.

For such an f, an equisatisfiable CNF formula g can be found [28] by introducing fresh variables z to give $f = \exists z : g$ [33] . But then $\forall y : f$ amounts to solving $\forall y : \exists z : g$ and the quadratic nature of resolution compromises the tractability of this approach as the size of z increases.

1.3   Contributions to Approximate Quantifier Elimination

In this paper, we show how upper-approximation can be applied to eliminate z from $\exists z : g$ where g is presented in CNF. We show how a SAT solver can be repeatedly called to compute a sequence of CNF formulae $h_0, h_1, \ldots$ that converge onto $\exists z : g$ from above in the sense that $\exists z : g$ entails h_i (each model of $\exists z : g$ is also a model of h_i). Each h_{i+1} strictly entails h_i so the sequence is ultimately stationary. However, each h_i is free from all variables in z, hence this approach has the attractive property that generation of the sequence $h_0, h_1, \ldots, h_t$ can be stopped prematurely, at any time t, without compromising soundness since each h_i is an upper-approximation of $\exists z : g$.

This approach leads to a so-called *anytime* (or *interruptible* [2, Sect. 2.6]) formulation of projection that compares favourably against resolution and model enumeration techniques, which lead to all or nothing, monolithic approaches. Specifically, if $g_0 = g$ and g_{i+1} is obtained from g_i by applying resolution to remove another variable of z, then it is only the final formula $g_{|z|}$ that is free from z. Moreover, the number of clauses in g_i do not necessarily decrease as i increases, and the size of intermediate g_i can be significantly larger than both g and its projection $g_{|z|}$. By way of contrast, the size of the h_i increases monotonically as the sequence converges. We also show how to construct a sequence $h_0, h_1, \ldots, h_t$ which rapidly converges onto $\exists z : g$ based on the enumeration of prime implicants, that is, small conjunctions of literals which entail $\exists z : g$. As a final contribution, we show how this scheme can be implemented with incremental SAT [35] and sorting networks [14,21].

Our paper makes a specific contribution to a specific problem, yet that problem appears in various guises in model checking and abstract interpretation. As already stated, projection arises in dependency analysis which is itself finding new applications in, for example, information flow analysis [16]. Projection arises when computing transfer functions [4] and, very recently, in the synthesis of ranking functions from template constraints for low-level code [10]. The existence of a ranking function on a path π with a transition $r_\pi(x, x')$ amounts to solving the formula $\exists c : \forall x : \forall x' : r_\pi(x, x') \rightarrow p(c, x) < p(c, x')$ where $p(c, x)$ is a polynomial over the bit-vector x whose coefficients constitute the vector c. However, if intermediate variables are needed to express $r_\pi(x, x')$, the polynomials $p(c, x)$ and $p(c, x')$ or the size relation $<$ in CNF, then the quantifiers take the form $\exists c : \forall x : \forall x' : \exists z$ where z is the vector of intermediate variables. The authors proceed by instantiating elements of the c vector to values drawn from the set $\{-1, 0, 1\}$, then testing the formula $\neg \exists x : \exists x' : r_\pi(x, x') \wedge \neg(p(c, x) < p(c, x'))$ for *unsatisfiability*. The method advocated in this paper suggests a more direct approach, which avoids enumerating combinations of coefficients, and restricts the coefficients to a small set of allowable values.

2   Existential Quantification in Five Steps

The idea behind our approach is to converge onto the set of solutions of a formula φ by adding constraints formed from the prime implicants of $\neg\varphi$ that are derived using SAT solving. This approach contrasts with existing techniques in that it is based on successive refinement and thereby provides an anytime approach to existential quantifier elimination. We build towards the technique in five steps.

2.1   Under-Approximation Using Implicants

We first show how to under-approximate an existentially quantified formula by deriving an implicant ν of $\exists z : \varphi$, that is, $\nu \models \exists z : \varphi$. To illustrate, let:

$$\varphi = (\neg x \vee z) \wedge (y \vee z) \wedge (\neg x \vee \neg w \vee \neg z) \wedge (w \vee \neg z)$$

Let $X = \{w, x, y, z\}$ denote the set of variables in φ. To project φ onto $Y_1 = \{w, x, y\}$, i.e. remove all information pertaining to the variables $Y_2 = X \setminus Y_1 = \{z\}$, we introduce fresh sets of variables $Y_1^+ = \{v^+ \mid v \in Y_1\}$ and $Y_1^- = \{v^- \mid v \in Y_1\}$. Each occurrence of the literal v in φ is replaced with v^+ if $v \in Y_1$ and each occurrence of $\neg v$ is replaced with v^- if $v \in Y_1$. The transformed formula is augmented with a constraint $\neg v^+ \vee \neg v^-$ for each $v \in Y_1$ so as to prevent v^+ and v^- holding simultaneously. Let t_{Y_1} denote this transformation, hence:

$$t_{Y_1}(\varphi) = \begin{cases} (x^- \vee z) \wedge (y^+ \vee z) \wedge (x^- \vee w^- \vee \neg z) \wedge (w^+ \vee \neg z) \wedge \\ (\neg w^+ \vee \neg w^-) \wedge (\neg x^+ \vee \neg x^-) \wedge (\neg y^+ \vee \neg y^-) \end{cases}$$

Then the formula $t_{Y_1}(\varphi)$ is defined over the set of variables $X' = Y_1^+ \cup Y_1^- \cup Y_2$, and a model of $t_{Y_1}(\varphi)$ is a map $\mathcal{M} : X' \to \mathbb{B}$ such as:

$$\mathcal{M} = \{w^+ \mapsto 1, w^- \mapsto 0, x^+ \mapsto 0, x^- \mapsto 1, y^+ \mapsto 0, y^- \mapsto 0, z \mapsto 1\}$$

The model $\mathcal{M}$ can be equivalently represented by the set $\{v \in X' \mid \mathcal{M}(v) = 1\}$, and henceforth we shall use the map and set representation interchangeably. The variables of $\mathcal{M} \cap (Y_1^+ \cup Y_1^-)$ define a cube (a conjunction of literals) that is given by $\nu = (\bigwedge_{v^+ \in \mathcal{M} \cap Y_1^+} v) \wedge (\bigwedge_{v^- \in \mathcal{M} \cap Y_1^-} \neg v)$. Therefore $\nu = (\neg x \wedge w)$. Observe that $\nu \models \exists Y_2 : \varphi$ hence ν is a so-called implicant of $\exists Y_2 : \varphi$ which constitutes an under-approximation of $\exists Y_2 : \varphi$. This can be seen since ν is free from any variables of Y_2 and the conjunction $\neg\varphi \wedge \nu$ is unsatisfiable. To converge onto $\exists Y_2 : \varphi$ from below, we augment $t_{Y_1}(\varphi)$ with the blocking clause $(\neg x^- \vee \neg w^+)$ which suppresses the previously derived solution. The blocking clause ensures that any cube that is subsequently found does not entail ν. Then $t_{Y_1}(\varphi) \wedge (\neg x^- \vee \neg w^+)$ is checked for satisfiability, yielding a model:

$$\mathcal{M}' = \{w^+ \mapsto 0, w^- \mapsto 0, x^+ \mapsto 0, x^- \mapsto 1, y^+ \mapsto 1, y^- \mapsto 0, z \mapsto 0\}$$

which defines another implicant $(\neg x \wedge y)$ of $\exists Y_2 : \varphi$, hence the refined under-approximation $(\neg x \wedge y) \vee (\neg x \wedge w)$. Adding another blocking clause and passing $t_{Y_1}(\varphi) \wedge (\neg x^- \vee \neg w^+) \wedge (\neg x^- \vee \neg y^+)$ to a SAT solver reveals the formula to be unsatisfiable. Convergence onto $\exists Y_2 : \varphi$ has thus been achieved and $\exists Y_2 : \varphi = (\neg x \wedge y) \vee (\neg x \wedge w)$. This can be checked by applying Schröder-expansion [22, Sect. 9.2.3] to compute $\exists Y_2 : \varphi = \varphi[z \mapsto 0] \vee \varphi[z \mapsto 1] = ((\neg x) \wedge (y)) \vee ((\neg x \vee \neg w) \wedge (w)) = (\neg x \wedge y) \vee (\neg x \wedge w)$.

2.2 Over-Approximation Using Implicants

To derive an over-approximation of $\exists Y_2 : \varphi$, a formula κ is constructed which is equisatisfiable to $\neg \varphi$:

$$\kappa = \begin{cases} (x \vee t_1) & \wedge (\neg z \vee t_1) \wedge \\ (\neg y \vee t_2) & \wedge (\neg z \vee t_2) \wedge \\ (x \vee t_3) & \wedge (w \vee t_3) \wedge (z \vee t_3) & \wedge \\ (\neg w \vee t_4) & \wedge (z \vee t_4) \wedge (\neg t_1 \vee \neg t_2 \vee \neg t_3 \vee \neg t_4) \end{cases}$$

The formula κ is obtained by a standard CNF translation [28] which introduces fresh variables $T = \{t_1, \ldots, t_4\}$ such that $\neg \varphi \equiv \exists T : \kappa$. The variable t_i indicates whether a truth assignment violates the ith clause of φ. Applying the transformation introduced previously then gives:

$$t_{Y_1}(\kappa) = \begin{cases} (x^+ \vee t_1) & \wedge (\neg z \vee t_1) & \wedge \\ (y^- \vee t_2) & \wedge (\neg z \vee t_2) & \wedge \\ (x^+ \vee t_3) & \wedge (w^+ \vee t_3) & \wedge (z \vee t_3) & \wedge \\ (w^- \vee t_4) & \wedge (z \vee t_4) & \wedge (\neg t_1 \vee \neg t_2 \vee \neg t_3 \vee \neg t_4) \wedge \\ (\neg w^+ \vee \neg w^-) & \wedge (\neg x^+ \vee \neg x^-) \wedge (\neg y^+ \vee \neg y^-) \end{cases}$$

To see how $t_{Y_1}(\kappa)$ can be applied to find an over-approximation $\neg \nu$ of $\exists Y_2 : \varphi$ observe that $\nu \models \forall Y_2 : \exists T : \kappa$ iff $\neg \forall Y_2 : \exists T : \kappa \models \neg \nu$ iff $\exists Y_2 : \neg \exists T : \kappa \models \neg \nu$ iff $\exists Y_2 : \varphi \models \neg \nu$. Hence to find an over-approximation of $\exists Y_2 : \varphi$ it suffices to find an implicant of $\forall Y_2 : \exists T : \kappa$. To find such an implicant observe that $\forall Y_2 : \exists T : \kappa \models \exists Y_2 : \exists T : \kappa$ hence every implicant of $\forall Y_2 : \exists T : \kappa$ is also an implicant of $\exists Y_2 : \exists T : \kappa$. This suggests a strategy in which the implicants of $\exists Y_2 : \exists T : \kappa$ are filtered to find the implicants of $\forall Y_2 : \exists T : \kappa$, that is, the implicants $\nu \models \exists Y_2 : \exists T : \kappa$ are filtered by checking $\exists Y_2 : \varphi \models \neg \nu$. Moreover, the check $\exists Y_2 : \varphi \models \neg \nu$ amounts to deciding whether the conjoined formula $\varphi \wedge \nu$ is unsatisfiable. Thus an unsatisfiability check can be used for filtering. To illustrate, suppose that a SAT solver produces the following solution to the formula $t_{Y_1}(\kappa)$:

$$\mathcal{M} = \begin{cases} w^+ \mapsto 0, w^- \mapsto 1, x^+ \mapsto 0, x^- \mapsto 0, y^+ \mapsto 0, y^- \mapsto 0 \\ z \mapsto 1, t_1 \mapsto 1, t_2 \mapsto 1, t_3 \mapsto 1, t_4 \mapsto 0 \end{cases}$$

The cube $\nu = (\neg w)$ is an implicant of $\exists Y_2 : \exists T : \kappa$ and therefore it remains to check whether $\exists Y_2 : \varphi \models \neg \nu$. Since $\varphi \wedge \nu$ is satisfiable, the cube is discarded. However, before doing so, the formula $t_{Y_1}(\kappa)$ is augmented with $\neg w^- \vee x^+ \vee x^- \vee y^- \vee y^+$ to avoid the cube being found again. This blocking clause can be interpreted as an implication $w^- \rightarrow (x^+ \vee x^- \vee y^- \vee y^+)$ which ensures that any cube subsequently found that entails ν also has more literals than ν. Applying a SAT solver then yields a model:

$$\mathcal{M}' = \left\{ \begin{array}{l} w^+ \mapsto 0,\, w^- \mapsto 0,\, x^+ \mapsto 1,\, x^- \mapsto 0,\, y^+ \mapsto 0,\, y^- \mapsto 0 \\ z \quad \mapsto 0,\, t_1 \quad \mapsto 0,\, t_2 \quad \mapsto 1,\, t_3 \quad \mapsto 1,\, t_4 \quad \mapsto 0 \end{array} \right\}$$

and hence $\nu' = (x)$. Since $\varphi \wedge \nu'$ is unsatisfiable, we conclude that $\exists Y_2 : \varphi \models \neg \nu'$, hence $\neg \nu'$ constitutes an over-approximation of $\exists Y_2 : \varphi$. The blocking clause $\neg x^+$ is then added to $t_{Y_1}(\kappa)$ to prevent any cube which entails ν' being found. Note too that this blocking clause differs in structure from the one imposed previously, and indeed the number of literals in the clause is merely n where n is the number of literals in the cube. In the previous case, the number of literals in the blocking clause is $2|Y_1| - n$. Reapplying a SAT solver yields a further model:

$$\mathcal{M}'' = \left\{ \begin{array}{l} w^+ \mapsto 0,\, w^- \mapsto 1,\, x^+ \mapsto 0,\, x^- \mapsto 0,\, y^+ \mapsto 0,\, y^- \mapsto 1 \\ z \quad \mapsto 1,\, t_1 \quad \mapsto 1,\, t_2 \quad \mapsto 1,\, t_3 \quad \mapsto 1,\, t_4 \quad \mapsto 0 \end{array} \right\}$$

which defines the cube $\nu'' = \neg w \wedge \neg y$. Since $\varphi \wedge \nu''$ is unsatisfiable, it again follows that $\exists Y_2 : \varphi \models \neg \nu''$, which refines the over-approximation of $\exists Y_2 : \varphi$ to the conjunction $(\neg \nu') \wedge (\neg \nu'')$. The blocking clause $\neg w^- \vee \neg y^-$ is then added to the augmented formula at which point one final application of the solver indicates that the conjoined formula is unsatisfiable. Hence convergence onto $\exists Y_2 : \varphi$ has been obtained from above where $\exists Y_2 : \varphi = \neg \nu' \wedge \neg \nu'' = (\neg x) \wedge (w \vee y)$. Terminating the procedure early, before ν'' is computed, would yield the over-approximation $\neg \nu' = \neg x$ which, though safe, has strictly more models than $(\neg x) \wedge (w \vee y)$. Thus the method is diametrically opposed to resolution: In the resolution based scheme, the projection is found in the last step only when all variables have been eliminated one after the other. In the above SAT based scheme, a clause in the projection space is obtained in the first step, as in a parallel form of elimination, which is subsequently refined by adding further clauses.

2.3   Approximation Using Prime Implicants

Thus far we have seen how upper- and lower-approximation can be reduced to finding an implicant c of a formula f where c is a cube, namely a conjunction of literals. Suppose $c_1 \models f$ and $c_2 \models f$ where the cubes c_1 and c_2 are related by $c_1 \models c_2$. Then $\neg f \models \neg c_2 \models \neg c_1$ where $\neg c_2$ and $\neg c_1$ are clauses. Furthermore, if c_2 is shorter than c_1, that is, if c_2 is constructed from fewer literals than c_1, then $\neg c_2$ constitutes a stronger (more descriptive) approximation than $\neg c_1$. Rather than using any implicant to approximate $\neg f$, it is better to use a shorter one, and better still to use one that is said to be prime. The implicant c_2 of f is prime (or irreducible) if there is no shorter implicant c_3 of f such that $c_2 \models c_3 \models f$. The best approximations are thus constructed from the shortest prime implicants.

To derive shortest prime implicants, we turn to sorting networks [14,21]. Examples of sorting networks for 3 and 4 bits are given in Fig. 1. The 3-bit sorter

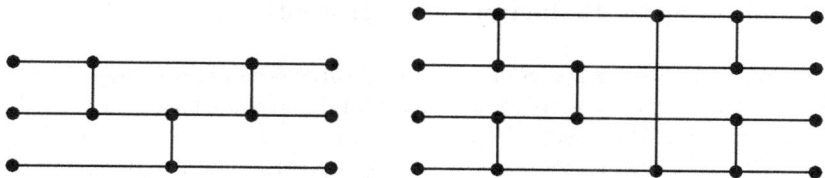

Fig. 1. Sorting networks for 3 and 4 bits

has 3 input bits on the left and 3 output bits on the right. It also has 3 comparison operations, indicated with vertical bars, which compare and if necessary swap bits. A comparator assigns its outgoing upper bit to the maximum of its two incoming bits and its outgoing lower bit to the minimum. A comparator with incoming bits i_1 and i_2 with outgoing bits u and ℓ can be encoded propositionally as the formula $(u \leftrightarrow i_1 \vee i_2) \wedge (\ell \leftrightarrow i_1 \wedge i_2)$. The value of a sorting network is that it can be applied to compute the sum of a series of 0/1 values [14] where the sum is represented in a unary fashion. Moreover, by instantiating the output bits to fixed unary value, a cardinality constraint can be obtained. For example, by constraining the output bits of the 4-bit sorter to 1100, the cardinality constraint is derived which ensures that exactly two of the input bits to the sorter are set. Constraining the output bits to 1110 would ensure that exactly three input bits are set. Such cardinality constraints can be imposed in conjunction with the formula $t_{Y_1}(\kappa)$ to rule out the discovery of implicants that are not prime.

Let us return to the formula $t_{Y_1}(\kappa)$ from Sect. 2.2 where $Y_1 = \{w, x, y\}$. The construction proceeds by introducing variables, denoted $v^{\pm}$ for each $v \in Y_1$, which serve as input to the sorting network. Each $v^{\pm}$ indicates whether v or $\neg v$ appear in the implicant, hence the relationship $v^{\pm} \leftrightarrow (v^+ \vee v^-)$. A 3-bit network is then used to constrain the output bits o_1, o_2, o_3 (top-to-bottom) to the unary sum of the inputs $w^{\pm}, x^{\pm}, y^{\pm}$ (again oriented top-to-bottom). Overall, this construction yields the following propositional encoding, where h_1, h_2, h_3 are intermediate variables computed by the comparators:

$$\mu = t_{Y_1}(\kappa) \wedge (w^{\pm} \leftrightarrow w^+ \vee w^-) \wedge (x^{\pm} \leftrightarrow x^+ \vee x^-) \wedge (y^{\pm} \leftrightarrow y^+ \vee y^-) \wedge$$
$$(h_1 \leftrightarrow w^{\pm} \vee x^{\pm}) \wedge (h_2 \leftrightarrow w^{\pm} \wedge x^{\pm}) \wedge (h_3 \leftrightarrow h_2 \vee y^{\pm}) \wedge$$
$$(o_1 \leftrightarrow h_1 \vee h_3) \wedge (o_2 \leftrightarrow h_1 \wedge h_3) \wedge (o_3 \leftrightarrow h_2 \wedge y^{\pm})$$

To enforce the cardinality constraint, we set $\mu_{k=1} = \mu \wedge o_1 \wedge \neg o_2 \wedge \neg o_3$. Invoking a SAT solver on $\mu_{k=1}$ yields candidates $\neg w$, x and $\neg y$, but only x is implied by $\exists Y_2 : \varphi$. Then $\mu_{k=1}$ is unsatisfiable, and we derive implicants for $\mu_{k=2} = \mu \wedge o_1 \wedge o_2 \wedge \neg o_3$, which yields the clause $w \vee y$ that is implied by $\exists Y_2 : \varphi$. Enumerating implicants by their size may require more SAT instances, but it ensures that the upper-approximation is always conjoined with a clause that is as short as possible. Short clauses are likely to remove more models from the approximation than long ones, thereby encouraging rapid convergence.

2.4   Solution-Space Reduction Using Instantiation

In the example in Sect. 2.2, a SAT solver generates several false candidates ν for implicants, which are then refuted by checking $\varphi \models \neg\nu$. This scheme is based on the observation that every implicant of $\forall Y_2 : \exists T : \kappa$ is also an implicant of $\exists Y_2 : \exists T : \kappa$, where in the case of the example $Y_2 = \{z\}$. However, observe that $\forall Y_2 : \exists Y : \kappa \models \exists Y : \kappa_{z\leftarrow 0}$ where $\kappa_{z\leftarrow 0}$ denotes the formula obtained by replacing each occurrence of z in κ with the truth value 0 (instantiation). Therefore every implicant of $\forall Y_2 : \exists T : \kappa$ is also an implicant of $\exists T : \kappa_{z\leftarrow 0}$. The formula $\exists T : \kappa_{z\leftarrow 0}$ is not only a simplification of $\exists T : \kappa$ but $\exists T : \kappa_{z\leftarrow 0}$ will possess fewer models and hence fewer implicants than $\exists Y_2 : \exists T : \kappa$ provided $\kappa \not\models \neg z$.

Consider again the formula $t_{Y_1}(\kappa)$ given in Sect. 2.2 and consider $t_{Y_1}(\kappa_{z\leftarrow 0}) = t_{Y_1}(\kappa)_{z\leftarrow 0}$. Recall that originally the candidate implicant $\nu = (\neg w)$ was derived which was then refuted because $\varphi \not\models \neg\nu$. This candidate is suppressed by the instantiation and is not a solution of $t_{Y_1}(\kappa)_{z\leftarrow 0}$. It turns out that 13 SAT instances are required to converge onto $\exists Y_2 : \varphi$ whereas operating on $t_{Y_1}(\kappa)_{z\leftarrow 0}$ and $t_{Y_1}(\kappa)_{z\leftarrow 1}$ only requires 9 and 10 SAT instances, respectively. Interestingly, the formulae derived for these cases are equivalent but different. For $t_{Y_1}(\kappa)_{z\leftarrow 0}$ we obtain the limit $(\neg x) \wedge (w \vee y)$ as expected, but operating on $t_{Y_1}(\kappa)_{z\leftarrow 1}$ yields $(w \vee y) \wedge (w \vee \neg x) \wedge (\neg w \vee \neg x)$ which is equivalent to $(\neg x) \wedge (w \vee y)$.

2.5   Solution-Space Reduction Using Multiple Instantiations

Instantiating the variables of Y_2 with truth values can decrease the number of spurious implications that are generated. This suggests instantiating κ in several different ways and then combining the instantiations so as to limit the search space a priori. Thus the basic idea is to derive multiple instantiations, say, $t_{Y_1}(\kappa)_{z\leftarrow 0}$ and $t_{Y_1}(\kappa)_{z\leftarrow 1}$ and solve the conjunction $\mu = t_{Y_1}(\kappa)_{z\leftarrow 0} \wedge t_{Y_1}(\kappa)_{z\leftarrow 1}$. In actuality, care is needed to avoid accident coupling between the T variables in the different instantiations. This can be avoided by introducing fresh, disjoint sets of variables $T_1 = \{t_{i,1} \mid t_i \in T\}$ and $T_2 = \{t_{i,2} \mid t_i \in T\}$ by applying renamings $\rho_1(t_i) = t_{1,i}$ and $\rho_2(t_i) = t_{2,i}$ to $\kappa_{z\leftarrow 0}$ and $\kappa_{z\leftarrow 1}$, respectively. By applying these renamings, combining and then applying simplification we obtain:

$$\mu = \begin{cases} (x^+ \vee t_{1,1}) & \wedge\ (y^- \vee t_{1,2}) & \wedge & (\neg t_{1,1} \vee \neg t_{1,2})\ \wedge \\ (x^+ \vee t_{2,3}) & \wedge\ (w^+ \vee t_{2,3}) & \wedge\ (w^- \vee t_{2,4}) & \wedge\ (\neg t_{2,3} \vee \neg t_{2,4})\ \wedge \\ (\neg w^+ \vee \neg w^-) & \wedge\ (\neg x^+ \vee \neg x^-) \wedge (\neg y^+ \vee \neg y^-) \end{cases}$$

When solving for μ, the sequence of upper-approximations converges onto the limit $(w \vee y) \wedge (w \vee \neg x) \wedge (\neg w \vee \neg x)$ without encountering any spurious implicants. Observe too that μ consists of 10 clauses whereas the $t_{Y_1}(\kappa)$ formula given in Sect. 2.2 has 13 clauses. This is because instantiating the variables of Y_2 often confers significant opportunities for simplification, offering scope for applying multiple instantiation without generating a formula that is unwieldy.

3   Correctness of the Transformation

The techniques presented thus far for computing under- and over-approximations of existentially quantified formula all rest on finding an implicant of a formula of the form $\exists Y_2 : \varphi$ (Sect. 2.1) or $\exists Y_2 : \exists T : \kappa$ (Sect. 2.2 onwards). The transformation t_{Y_1} reduces this problem SAT. This section is concerned with correctness of this transformation. The style of presentation is necessarily formal and a reader who is concerned with the application of the technique (rather than establishing its correctness) can proceed onto the following section.

3.1   Transforming Clauses

Let $Bool_X$ denotes the class of propositional formulae over the set of variables X and suppose X is partitioned into two disjoint subsets Y_1 and Y_2. We shall consider the problem of computing an implicant of $\exists Y_2 : f$ where the formula $f \in Bool_X$ is presented in CNF. The transformation is formalised as a map t_{Y_1} on the set of literals $Lit_X = \{x, \neg x \mid x \in X\}$. This map is, in turn, defined in terms of sets of propositional variables $Y_1^+ = \{x^+ \mid x \in Y_1\}$ and $Y_1^- = \{x^- \mid x \in Y_1\}$ for which we assume that $Y_1^+ \cap Y_1^- = \emptyset$ and $(Y_1^+ \cup Y_1^-) \cap X = \emptyset$.

Definition 1. The literal transformation map $t_{Y_1} : Lit_X \to Lit_{Y_1^+ \cup Y_1^- \cup Y_2}$ (and its inverse $t_{Y_1}^{-1}$) are defined as follows:

$$
t_{Y_1}(l) = \begin{cases} x^+ & \text{if } l = x \wedge x \in Y_1 \\ x^- & \text{if } l = \neg x \wedge x \in Y_1 \\ l & \text{otherwise} \end{cases} \qquad t_{Y_1}^{-1}(l) = \begin{cases} x & \text{if } l = x^+ \wedge x \in Y_1 \\ \neg x & \text{if } l = x^- \wedge x \in Y_1 \\ l & \text{otherwise} \end{cases}
$$

A clause is considered to be a set of literals to simplify the lifting of the literal transformation map from single literals to clauses. Thus if a clause is merely a set $C \subseteq Lit_X$ then $t_{Y_1}(C) = \{t_{Y_1}(l) \mid l \in C\}$.

3.2   Transforming Cubes

The literal transformation map is lifted to cubes and implicants (an implicant is a merely a particular type of cube) by likewise considering these to be sets of (implicitly conjoined) literals. The transformation relates cubes with literals drawn from Lit_X to cubes with literals drawn from $Y_1^+ \cup Y_1^- \cup Lit_{Y_2}$. Our interest is in cubes that are non-trivial, that is, they do not contain opposing literals. These classes of non-trivial cubes are defined below:

Definition 2.

$$
Cube_X = \{C \subseteq Lit_X \mid \forall x \in X : \{x, \neg x\} \not\subseteq C \}
$$
$$
Cube_{Y_1, Y_2} = \left\{ C \cup C' \,\middle|\, \begin{array}{l} C \in Cube_{Y_2} \qquad\qquad\qquad \wedge\ C' \subseteq Y_1^+ \cup Y_1^- \wedge \\ \forall x \in Y_1 : \{x^+, x^-\} \cap C' \neq \emptyset \wedge \{x^+, x^-\} \not\subseteq C' \end{array} \right\}
$$

We transform between these two types of cubes with the following map:

Definition 3. The mapping $c_{Y_1} : Cube_X \to Cube_{Y_1,Y_2}$ is defined:

$$c_{Y_1}(C) = t_{Y_1}(C) \cup \{\neg x^+, \neg x^- \mid x \in Y_1 \wedge \{x, \neg x\} \cap C = \emptyset\}$$

Observe that c_{Y_1} is both injective and surjective, hence it possesses an inverse $c_{Y_1}^{-1} : Cube_{Y_1,Y_2} \to Cube_X$.

3.3   Equivalence

With the t_{Y_1} and c_{Y_1} maps defined on clauses and cubes, we can now state an equivalence result which details how implicants are preserved by transformation. Note that a formula f represented in CNF can be considered to be a set of implicitly conjoined clauses F.

Proposition 1 (equivalence). Let $f = \bigwedge \{\bigvee C \mid C \in F\}$ where $F \subseteq \wp(Lit_X)$ and put $f' = \bigwedge \{\bigvee t_{Y_1}(C) \mid C \in F\}$. Then

- If $D \in Cube_X$ and $(\bigwedge D) \models f$ then $(\bigwedge c_{Y_1}(D)) \models f'$
- If $D' \in Cube_{Y_1,Y_2}$ and $(\bigwedge D') \models f'$ then $(\bigwedge c_{Y_1}^{-1}(D')) \models f$

Proof.

- Let $C \in F$. Since $(\bigwedge D) \models f$ it follows $(\bigwedge D) \models (\bigvee C)$.
  - Suppose $x \in D \cap C$ and $x \in Y_1$. Then $x^+ \in t_{Y_1}(C) \cap c_{Y_1}(D)$.
  - Suppose $\neg x \in D \cap C$ and $x \in Y_1$. Then $x^- \in t_{Y_1}(C) \cap c_{Y_1}(D)$.
  - Suppose $x \in D \cap C$ and $x \in Y_2$. Then $x \in t_{Y_1}(C) \cap c_{Y_1}(D)$.
  - Suppose $\neg x \in D \cap C$ and $x \in Y_2$. Then $\neg x \in t_{Y_1}(C) \cap c_{Y_1}(D)$.

  Hence $(\bigwedge c_{Y_1}(D)) \models (\bigvee t_{Y_1}(C))$ whence $(\bigwedge c_{Y_1}(D)) \models f'$ as required.
- Let $C \in F$. Since $(\bigwedge D') \models f'$ it follows $(\bigwedge D') \models (\bigvee t_{Y_1}(C))$.
  - Suppose $x^+ \in D' \cap t_{Y_1}(C)$ and $x \in Y_1$. Then $x \in C \cap c_{Y_1}^{-1}(D')$.
  - Suppose $x^- \in D' \cap t_{Y_1}(C)$ and $x \in Y_1$. Then $\neg x \in C \cap c_{Y_1}^{-1}(D')$.
  - Suppose $x \in D' \cap t_{Y_1}(C)$ and $x \in Y_2$. Then $x \in C \cap c_{Y_1}^{-1}(D')$.
  - Suppose $\neg x \in D' \cap t_{Y_1}(C)$ and $x \in Y_2$. Then $\neg x \in C \cap c_{Y_1}^{-1}(D')$.

  Hence $(\bigwedge c_{Y_1}^{-1}(D')) \models (\bigvee C)$ whence $(\bigwedge c_{Y_1}^{-1}(D')) \models f$ as required.

The following corollary of the above relates implicants with literals drawn from Lit_{Y_1} to the satisfiability of the transformed clause set:

Corollary 1. Suppose f and f' are defined as above. Then

- If $D \in Cube_{Y_1}$ and $\wedge D \models f$ then $(\bigwedge c_{Y_1}(D)) \wedge f'$ is satisfiable
- If $D' \in Cube_{Y_1,\emptyset}$ and $(\bigwedge D') \wedge f'$ is satisfiable then $(\bigwedge c_{Y_1}^{-1}(D')) \models f$

To present the final result, let $[\![f]\!] \subseteq \wp(X)$ denote the set of models of the Boolean function f. (Recall the set-based representation of a model given in Sect 2.1, for example, if $X = \{x, y\}$ then $[\![x \vee y]\!] = \{\{x\}, \{y\}, \{x, y\}\}$.) We can now that state how a prime implicant of the existentially quantifier formula (whose literals are drawn from Lit_{Y_1}) fulfills two satisfiability conditions:

Corollary 2. Suppose f, f' and $F \subseteq \wp(Lit_X)$ are defined as above. Put $g' = f' \wedge \{\neg x^+ \vee \neg x^- \mid x \in Y_1\}$. Then $D \in Cube_{Y_1}$ is a prime implicant of $\exists Y_2 : f$ iff $D = c_{Y_1}^{-1}(M^* \cap (Y_1^+ \cup Y_1^-))$ where

– $M^* \in [\![g']\!]$
– $|M^* \cap (Y_1^+ \cup Y_1^-)| \leq |M \cap (Y_1^+ \cup Y_1^-)|$ for all $M \in [\![g']\!]$

Note that g' does not include any cardinality constraint on the set $M^* \cap (Y_1^+ \cup Y_1^-)$, hence the need to define a prime implicant in terms of an implicant no longer than any other. The above result can straightforwardly adapted to specify how an implicant of a given size can be defined as a SAT instance.

4   Experimental Results

We have implemented the techniques described in this paper in JAVA using the SAT4J solver [23] so as to integrate with our analysis framework for machine code, [MC]SQUARE [32], which is also coded in JAVA. To encode sorting propositionally, we implemented optimal networks for 9 or fewer variables and resorted to bitonic sorting for larger networks [21]. All experiments were performed on a MacBook Pro equipped with a 2.6 GHz dual-core processor and 4 GB of RAM, but only a single core was used in our experiments. The results obtained for deriving upper-approximations using the combination of methods described in Sect. 2.2 and Sect. 2.3 (without applying instantiation) are summarised in Tab. 1.

The formulae originated from the ISCAS benchmark set [17]. For some of these benchmarks, quantifier elimination by model enumeration is intractable due to the large numbers of models presented in column *#models*, and so is resolution. This is highlighted by the benchmark 74L85b, which describes a 4-bit magnitude comparator. Whereas model enumeration required more than 6 minutes for 74182b and 74283b, it ran out of memory for 74L85b after approximately 10 minutes. Column *#vars/clauses* shows the number of propositional variables and clauses in the original formula, whereas column *trans* gives these numbers

Table 1. Experimental results without instantiation

Formula	models	#vars/clauses	trans.	length	#primes	#SAT	runtime
74182b	262,144	227/526	780/1281	2/5	4	52	0.81
				5/5	4	170	1.80s
74283b	262,144	266/646	966/1,633	4/8	13	1590	5.63s
				6/8	20	4053	14.49s
				8/8	20	4881	16.71s
74L85b	>390,752	412/1084	1582/2747	4/10	6	4496	18.91s
				5/10	14	12349	57.22s
				6/10	30	24960	125.99s
				8/10	30	47536	292.59s
				10/10	30	51522	352.95s

Table 2. Experimental results with a single instantiation

Formula	length	runtime	speedup		Formula	length	runtime	speedup
74182b	2/5	0.50s	38%			4/10	12.61s	23%
	5/5	0.85s	52%			5/10	38.85s	32%
74283b	4/8	4.26s	24%		74L85b	6/10	84.68s	33%
	6/8	10.54s	27%			8/10	203.45s	30%
	8/8	12.34s	26%			10/10	84.68s	33%

after applying the transformation t_{Y_1}. The column *length* first contains the maximum length of prime implicants that were enumerated, followed by the size of Y_1. Thus in the 8/8 case the algorithm was run to completion, whereas the 2/8 case was terminated prematurely. Then *#primes* gives the number of implicants found and *#SAT* the total number of calls to a SAT solver. The overall runtime is given in the last column.

It is important to appreciate that the projection of the 74185b formula does not contain any implicants with size between 7 and 10. Likewise 74283b does not contain any implicants of size 7 and 8. This size distribution has been observed elsewhere [19], though not in the context of projection, which suggests that enumerating implicants up to a size threshold can achieve a good approximation of the projection. The ratio of number of calls to the solver to the number of primes is largely due to spurious candidates (in our experiments, it roughly doubled by increasing the prime length by one or two), which motivates investigating the impact of instantiating variables. Circuits can be simplified after applying instantiation, which involves removing false literals from clauses and removing all clauses that were already satisfied. The effects of single instantiation based on a model of the original formula are highlighted in Tab. 2. The results shown in column *speedup* suggest that instantiation can significantly increase performance.

Finally, we study applying multiple instantiation, accompanied with simplification, for different instances of the 74L85b circuit. Note that simplification reduces the size of the SAT instance which compensates somewhat for multiple instantiation. The instantiations themselves were generated from various models of the formula that were themselves found by applying blocking clauses. By choosing 6 instantiations that constrain the solution space in the 6/10 case a priori, the number of SAT instances reduced from 24960 to 16954, and the runtime decreased to 61.59s. This is a reduction of 32% in terms of the number of calls to a SAT solver and an overall speedup of 51%. Using 10 instantiations, reduced the number of calls to the solver was still further to 14273 and took the runtime down to 52.45s yielding a speedup of 58%. The key point is that a reduction occurs in the ratio of the number of calls to the SAT solver and the number of primes. This is a measure of the effectiveness of the technique, that is, how much effort is needed, on average, to find another implicant and thereby refine the approximation. However, we conjecture, that it is not prudent to apply too many instantiations simultaneously, because at some point the size of the combined SAT instance will become unmanageable (this would correspond to a flattening of quantified bit-vector logic, which can be prohibitively expensive).

5   Related Work

The consensus method has been independently proposed by a number of researchers [3,29,31] as a way of enumerating all the prime implicants of a propositional function in disjunctive normal form (DNF). If f is in CNF, then it is straightforward to derive a DNF representation of $\neg f$, to which the consensus procedure can be applied to find its prime implicants. Then $\exists Y : f$ can be found by conjoining all clauses $\neg c$ where c is a prime implicant of $\neg f$ which has no variables in common with Y. One might think that this provides a way to compute projection, but the key step of the consensus method combines two elementary conjunctions of $\neg f$, say, $x \wedge C$ and $(\neg x) \wedge D$, to form the conjunction $C \wedge D$, which is isomorphic to resolution. Hence the consensus method shares the inefficiency problems associated with applying resolution to a formula in CNF. The complexity of the shortest implicant problem for DNF formulae has been studied by Umans [34] who showed that it is $GC(\log^2(n), coNP)$-complete. Even though this result is not directly transferrable to CNF, it substantiates our application of SAT solvers to the derivation of shortest implicants. Integer linear programming techniques have also be used to find shortest implications, as have SAT engines which have been modified to support inequalities [24]. In this work a transformation is described which is similar to t_{Y_1}. However, the work is not concerned with quantifier elimination, hence pairs of 0-1 variables are introduced for each variable in the formula rather than merely those in Y_1.

Operating on negated formulae has applications in bounded model checking [8], in particular when using Craig interpolants [25]. Given two inconsistent formulae φ and ψ, that is, $\varphi \wedge \psi$ is unsatisfiable, a smaller upper-approximation ξ of φ can be derived from the proof of unsatisfiability of $\varphi \wedge \psi$ in linear time. This approach is sound in the sense that ξ over-approximates φ, and at the same time serves tractability, and thus can be regarded as a form of widening. Prime implicants have been directly applied to widening Boolean functions represented as ROBDDs [19]. By appling a recursive meta-product construction [12] collections of short primes can be used to derive an ROBDD that is an upper-approximation of the input. Our work on applying SAT to projection was motivated by the emperical finding that collections of short primes, for instance those up to length 5, often yield good approximations of Boolean formulae [19]. Note that SAT-based enlargement of cubes also appears in the work of McMillan [26], who uses SAT-based enumeration for existential quantification. The idea of instantiating (multiple) instances of Boolean formulae with models can be seen as a form of circuit co-factoring as described by Ganai et al. [15]. A recent contribution to reasoning about quantified bit-vector formulae was made by Wintersteiger et al. [36], who most notably used word-level simplifications and template instantiations.

Another approach to quantifier elimination (of linear systems) was recently proposed by Monniaux [27]. In his approach, satisfiability tests of quantified formulae are used to derive witnesses (models). Rather than computing quantifier-free formulae directly, his algorithm uses substitution of witnesses to extend the original system towards a quantifier-free formula. Comparing this technique to our method, a similarity is in the use of witnesses to guide the elimination process. His method, however, is not anytime, and thus, cannot be stopped prematurely.

6 Conclusions

Synopsis. This paper advocates using SAT to derive upper-approximations of existentially quantified propositional formulae The approach is designed to be anytime so that it can be stopped early without compromising correctness. This can be considered to be a pragmatic response to the complexity of projection [11]. Further, the technique avoids the blow-up in the number of clauses in an intermediate representation that is associated with eliminating variables with resolution.

Future Work. This work calls for further investigations of ways to reduce the number of spurious candidates that appear when implicants of negations are enumerated, possibly based on the recent work described in [36].

Acknowledgment. We thank Olivier Coudert for discussions on the complexity of finding the smallest prime implicant. This work was funded, in part, by a Royal Society travel grant, reference TG092357, and a Royal Society Industrial Fellowship, reference IF081178. Furthermore, we thank Professor Stefan Kowalewski for his generous financial support that was necessary to initiate our collaboration.

References

1. Armstrong, T., Marriott, K., Schachte, P., Søndergaard, H.: Two Classes of Boolean Functions for Dependency Analysis. Science of Computer Programming 31(1), 3–45 (1998)
2. Bender, E.A.: Mathematical Methods in Artificial Intelligence. IEEE Computer Society Press, Los Alamitos (1996)
3. Blake, A.: Canonical expressions in Boolean algebra. University of Chicago, Chicago (1938)
4. Brauer, J., King, A.: Automatic Abstraction for Intervals using Boolean Formulae. In: Cousot, R., Martel, M. (eds.) SAS 2010. LNCS, vol. 6337, pp. 167–183. Springer, Heidelberg (2010)
5. Bryant, R.E.: Symbolic Boolean Manipulation with Ordered Binary-Decision Diagrams. ACM Computing Surveys 24(3), 293–318 (1992)
6. Bryant, R.E.: A View from the Engine Room: Computational Support for Symbolic Model Checking. In: Grumberg, O., Veith, H. (eds.) 25 Years of Model Checking. LNCS, vol. 5000, pp. 145–149. Springer, Heidelberg (2008)
7. Burch, J.R., Clarke, E.M., McMillan, K.L.: Symbolic model checking: 10^{20} states and beyond. Information and Computation 98, 142–170 (1992)

8. Clarke, E., Kröning, D., Lerda, F.: A tool for checking ANSI-C programs. In: Jensen, K., Podelski, A. (eds.) TACAS 2004. LNCS, vol. 2988, pp. 168–176. Springer, Heidelberg (2004)
9. Clarke, E.M., Biere, A., Raimi, R., Zhu, Y.: Bounded model checking using satisfiability solving. Formal Methods in System Design 19(1), 7–34 (2001)
10. Cook, B., Kroening, D., Rümmer, P., Wintersteiger, C.: Ranking Function Synthesis for Bit-Vector Relations. In: Esparza, J., Majumdar, R. (eds.) TACAS 2010. LNCS, vol. 6015, pp. 236–250. Springer, Heidelberg (2010)
11. Coste-Marquis, S., Le Berre, D., Letombe, F., Marquis, P.: Complexity Results for Quantified Boolean Formulae Based on Complete Propositional Languages. JSAT (1), 61–88 (2006)
12. Coudert, O., Madre, J.C.: Implicit and Incremental Computation of Primes and Essential Primes of Boolean Functions. In: DAC, pp. 36–39. IEEE, Los Alamitos (1992)
13. Duesterwald, E., Gupta, R., Soffa, M.L.: A Practical Framework for Demand-Driven Interprocedural Data Flow Analysis. ACM TOPLAS 19(6), 992–1030 (1997)
14. Eén, N., Sörensson, N.: Translating Pseudo-Boolean Constraints into SAT. JSAT 2(1-4), 1–26 (2006)
15. Ganai, M.K., Gupta, A., Ashar, P.: Efficient SAT-based unbounded symbolic model checking using circuit cofactoring. In: ICCAD, pp. 510–517. IEEE, Los Alamitos (2004)
16. Genaim, S., Giacobazzi, R., Mastroeni, I.: Modeling Secure Information Flow with Boolean Functions. In: IFIP WG 1.7, ACM Workshop on Issues in the Theory of Security, Barcelona, Spain, pp. 55–66 (2004)
17. Hansen, M.C., Yalcin, H., Hayes, J.P.: Unveiling the iscas-85 benchmarks: A case study in reverse engineering. IEEE Design & Test of Computers 16(3), 72–80 (1999)
18. Howe, J.M., King, A.: Positive Boolean Functions as Multiheaded Clauses. In: Codognet, P. (ed.) ICLP 2001. LNCS, vol. 2237, pp. 120–134. Springer, Heidelberg (2001)
19. Kettle, N., King, A., Strzemecki, T.: Widening ROBDDs with Prime Implicants. In: Hermanns, H. (ed.) TACAS 2006. LNCS, vol. 3920, pp. 105–119. Springer, Heidelberg (2006)
20. King, A., Søndergaard, H.: Automatic Abstraction for Congruences. In: Barthe, G., Hermenegildo, M. (eds.) VMCAI 2010. LNCS, vol. 5944, pp. 197–213. Springer, Heidelberg (2010)
21. Knuth, D.E.: Sorting and Searching. In: The Art of Computer Programming, vol. 3, Addison-Wesley, Reading (1997)
22. Kroening, D., Strichman, O.: Decision Procedures. Springer, Heidelberg (2008)
23. Le Berre, D.: SAT4J: Bringing the power of SAT technology to the Java platform (2010), http://www.sat4j.org/
24. Manquinho, V.M., Flores, P.F., Silva, J.P.M., Oliveira, A.L.: Prime implicant computation using satisfiability algorithms. In: International Conference on Tools with Artificial Intelligence, pp. 232–239. IEEE Press, Los Alamitos (1997)
25. McMillan, K.: Interpolation and SAT-based model checking. In: Hunt Jr., W.A., Somenzi, F. (eds.) CAV 2003. LNCS, vol. 2725, pp. 1–13. Springer, Heidelberg (2003)
26. McMillan, K.L.: Applying SAT methods in unbounded symbolic model checking. In: Brinksma, E., Larsen, K.G. (eds.) CAV 2002. LNCS, vol. 2404, pp. 250–264. Springer, Heidelberg (2002)

27. Monniaux, D.: Quantifier Elimination by Lazy Model Enumeration. In: Touili, T., Cook, B., Jackson, P. (eds.) CAV 2010. LNCS, vol. 6174, pp. 585–599. Springer, Heidelberg (2010)
28. Plaisted, D.A., Greenbaum, S.: A structure-preserving clause form translation. Journal of Symbolic Computation 2(3), 293–304 (1986)
29. Quine, W.V.: A Way to Simplify Truth Functions. American Mathematical Monthly 62(9), 627–631 (1995)
30. Reps, T., Sagiv, M., Yorsh, G.: Symbolic Implementation of the Best Transformer. In: Steffen, B., Levi, G. (eds.) VMCAI 2004. LNCS, vol. 2937, pp. 252–266. Springer, Heidelberg (2004)
31. Samson, E.W., Mills, B.E.: Circuit minimization: Algebra and Algorithms for new Boolean canonical expressions. Technical Report TR 54-21, United States Air Force, Cambridge Research Lab (1954)
32. Schlich, B.: Model checking of software for microcontrollers. ACM Trans. Embedded Comput. Syst. 9(4) (2010); Article Number 36
33. Tseitin, G.S.: On the complexity of derivation in the propositional calculus. In: Slisenko, A.O. (ed.) Studies in Constructive Mathematics and Mathematical Logic, vol. Part II, pp. 115–125 (1968)
34. Umans, C.: The Minimum Equivalent DNF Problem and Shortest Implicants. In: FOCS, pp. 556–563. IEEE Press, Los Alamitos (1998)
35. Whittemore, J., Kim, J., Sakallah, K.: SATIRE: a new incremental satisfiability engine. In: Design Automation Conference, pp. 542–545. ACM, New York (2001)
36. Wintersteiger, C.M., Hamadi, Y., de Moura, L.: Efficiently solving quantified bitvector formulas. In: FMCAD (2010) (to appear)

Towards Flight Control Verification Using Automated Theorem Proving

William Denman, Mohamed H. Zaki, Sofiène Tahar, and Luis Rodrigues

Department of Electrical & Computer Engineering Concordia University,
Montreal, Quebec, Canada
{w_denm,mzaki,tahar,luisrod}@encs.concordia.ca

Abstract. To ensure that an aircraft is safe to fly, a complex, lengthy and costly process must be undertaken. Current aircraft control systems verification methodologies are based on conducting extensive simulations in an attempt to cover all worst-case scenarios. A Nichols plot is a technique that can be used to conclusively determine if a control system is stable. However, to guarantee stability within a certain margin of uncertainty requires an informal visual inspection of many plots. To leverage the safety verification problem, we present in this paper a method for performing a formal Nichols Plot analysis using the MetiTarski automated theorem prover. First the transfer function for the flight control system is extracted from a Matlab/Simulink design. Next, using the conditions for a stable dynamical system, an exclusion region of the Nichols Plot is defined. MetiTarski is then used to prove that the exclusion region is never entered. We present a case study of the proposed approach applied to the lateral autopilot of a Model 24 Learjet.

1 Introduction

Modern commercial passenger aircraft are extremely complex systems and their designs must meet strict design and safety requirements. The Federal Aviation Administration (FAA) specifies that the catastrophic failure rate of a passenger aircraft digital flight-control system must be *extremely improbable* (less than 10^{-9} faults per hour) [1]. However, the system must be built using embedded computers, sensors, actuators and control components each with individual failure rates several orders of magnitude higher than that of the level set by the FAA. A combination of redundancy and fault tolerance must therefore be used to achieve this strict reliability requirement.

In general, aircraft are verified using simulation methods. A mathematical model based on the physical equations of flight is constructed and then simulated. An extensive analysis of the experimental results is necessary to ensure a robust result. There are several graphical aids such as Nyquist diagrams and Nichols plots [9] that are commonly used to simplify this task. These techniques provide easily identifiable zones for which the plot should not pass near or enter, clearly indicating the control system's margin of stability [11]. However these graphical methods still require visual analysis to process the information.

M. Bobaru et al. (Eds.): NFM 2011, LNCS 6617, pp. 89–100, 2011.

Even though there are over 78,000 flights without incident per day [10], we cannot assume that the current verification methods are perfectly sound. The first issue with this conclusion is that with simulation alone it is not possible to give 100% safety assurance due to the great number of variations of the model components and parameters. There will always be the possibility of a catastrophic failure due to design errors. Second, to achieve the FAA's failure rate a complex multi-domain, labour intensive and costly process must be undertaken. It is therefore quite important to investigate methods that will reduce the effort and cost of the verification process while ensuring the reliability of the results.

Formal verification is a method where logical reasoning can be used to prove that the implementation of a system correctly matches its design specification. Unlike simulation, a formal proof is valid regardless of the input test cases. There have been several breakthroughs in formal analysis of discrete systems. Systems of large orders of magnitude can now be verified. The tools and methods available for the formal verification of continuous and hybrid-systems cannot handle systems at the same level of complexity. This is one major hurdle that has limited the application of formal methods to the physical portion of aeronautical models.

MetiTarski [2] is an automatic theorem prover for real-valued analytical functions, including trigonometric and exponential functions. It works by a combination of resolution inference and algebraic simplification, invoking a decision procedure (QEPCAD) [5] to prove polynomial inequalities over the real closed filed (RCF). The output of MetiTarski is a complete proof that contains algebraic simplification and decision procedure calls that can be verified using other tools.

This paper illustrates a methodology for ensuring the stability of a flight control system by performing a formal analysis of a Nichols plot using the MetiTarski automated theorem prover. A Nichols plot is a transfer function's gain plotted versus its phase. Information about the stability of a system can be deduced from a visual inspection of the plot. The formal analysis we present removes the need for drawing and checking the Nichols plot visually. We present our investigations on verifying the lateral autopilot of a Model 24 Learjet subsonic business jet (SBJ) [4]. The control system model was implemented in Simulink and the goal of our proposed verification methodology is to supplement design work-flows that depend on the Matlab/Simulink Control Systems Toolbox [15].

The rest of the paper is organized as follows, we first discuss related work in Sect. 2. A description of MetiTarski and its syntax is presented in Sect. 3. Details of the proposed methodology are given in Sect. 4. This is followed by the case study in Sect. 5, before concluding the paper with Sect. 6.

2   Related Work

The bulk of the work on formal verification for aeronautical systems has been on the software components of flight control. Nevertheless, there have been several interesting advancements on the verification of hybrid systems [16]. From those latest results and experiments, it is obvious that they will ultimately play a strong role in the complete formal verification of aircraft autopilots.

Hardy [7] developed and implemented a decision procedure to reason about functions that have a finite number of inflection points. This decision procedure was implemented in the Nichols plot Requirements Verifier (NRV) to perform an automated formal Nichols plot analysis. The tool was developed using the computer algebra system Maple, the formal theorem prover PVS and the quantifier elimination system QEPCAD [5]. NRV was successfully applied to two classic control system examples: an inverted pendulum and a disk drive reader. Our work is closely related to that of Akbarpour and Paulson [3] who successfully formally verified these two examples using MetiTarski. Our main contribution is to remove the required inflection point analysis. We prove over all frequency values that the exclusion region is not entered, not just at single points. This is particularly important when dealing with exclusion regions that are not bounded by linear constraints. In particular, in the analysis of ellipsoid exclusion regions Hardy's [7] inflection point analysis does not hold.

SOSTOOLS [14] is a Matlab toolbox that can convert difficult optimization problems into a sum of squares formulation that can then be analyzed by a convex optimization technique known as semi-definite programming. It has widespread use in the nonlinear control field. In particular, it can be used to search for a Lyapunov function that can be used to verify the stability of dynamical systems. For a particular equilibrium to be stable, it is required that the candidate Lyapunov function V be positive definite and its derivative with respect to time be negative semi-definite [8]. SOSTOOLS can be used to prove the un-satisfiability of systems of non-linear polynomial equations and inequalities over the real numbers [12]. For many problems, SOSTOOLS could replace QEPCAD as the polynomial reasoning engine under MetiTarski. This would not be trivial to implement effectively. Nevertheless, improvements to the theory behind SOSTOOLS would have the potential to enhance MetiTarski.

3   MetiTarski : An Automated Theorem Prover

There exist few methods to automatically prove statements involving inequalities of elementary functions such as *arctan*, *ln* and *sqrt* that commonly appear in flight control verification problems. MetiTarski replaces the functions with upper and lower bounds in an attempt to reduce the problem to one that is decidable over the real closed fields. It consists of a resolution theorem prover (Metis) combined with a decision procedure (QEPCAD). The theorem prover is supplied with axioms approximating the functions with continued fraction expansions which in many cases are extremely accurate.

3.1   MetiTarski Input Syntax

MetiTarski operates on the first-order formula in the Thousands of Problems for Theorem Provers (TPTP) format that includes the corresponding axioms. Take for instance the code in Fig. 1. The *"fof"* keyword indicates to MetiTarski that the logic language used is a first-order formula. It is then followed by a label of the proof as well as the keyword *"conjecture"* indicating that the following formula is to be proved with the included axioms. The conjecture is read as follows: For all (!) X between 0 and 2.39×10^{-9} the formula is always less than 0.03. For a syntax guide see Table 1.

```
fof(
    example1,conjecture, ! [X] :
    (
      (0 <= X & X <= 2.39*10^(-9)) =>
      -0.0059 - 0.000016*exp(-2.55*10^8*X) + 0.031*exp(-5.49*10^7*X)
        < 0.03
    )
  ).

include('Axioms/general.ax').
include('Axioms/exp-upper.ax').
include('Axioms/exp-lower.ax').
```

Fig. 1. MetiTarski Syntax

Table 1. TPTP Syntax Guide for Figure 1

fof	First-Order Logic Formula
!	Universal Quantifier ($\forall$)
X	Quantified Variable
&	Logical AND
exp	e (Exponential Function)
<	Less Than
<=	Less Than Or Equal
=>	Logical Implication

3.2   Axioms

In addition to the problem definition, the required axioms must be chosen using the *'include'* keyword. It is critical that only axioms files for functions in the problem definition are included. Each additional set of axioms can greatly increase the time taken by MetiTarski to complete the proof. For example, there are two sets of axiom declarations for the exponential function. One for *regular bounds* and one for *extended bounds*. The extended bounds are used in cases where a higher level of precision is needed.

There have been cases where including the extended bounds will make the inequality test run until manually stopped. In that specific example, removing the extended axioms allowed MetiTarski to complete the proof in seconds. The inverse can also happen, if for instance the TPTP description contains trigonometric functions and those axioms are not included, then MetiTarski will never terminate. To mitigate this situation when running MetiTarski on a set of problems, as is done in the case study investigated in this paper, a CPU time limit can be set. Deeper analysis is then required to choose the correct axioms for those problems that were not proved.

There are automated scripts included in the MetiTarski distribution that can insert the axioms directly into the TPTP file description. This enables a low level analysis of the problem where specific axioms can be isolated and removed. This axiom weeding out procedure is currently manual, but by doing so has led to proofs for functions with extremely large arguments, such as $arctan(10^{25} \times X^{16})$.

4 Proposed Methodology

An important verification property is to ensure that a system under design is stable. Negative feedback is commonly used to achieve this. In this configuration, the difference between the system's current output and what is required is used to steer the output to the correct value. Time delays around the feedback loop can still cause the system to remain unstable. An in-depth stability analysis of the feedback system is thus quite essential in the design process.

Classic control theory provides several graphical methods to assess the stability of feedback systems: the Bode diagram, the Nyquist plot and the Nichols plot. The idea behind these graphical methods is to show visually how much margin the system has against instability [9]. Note that it is the analysis of the open-loop response that reveals information on the stability of the closed-loop system. The feedback loop must be "broken" to analyze how the signal is processed along the signal loop path.

In this paper, we are concerned with the analysis of a Nichols plot. This type of plot is commonly used in the analysis of flight control laws [6] and requires repeated visual inspection. Our goal is to automate this analysis and provide a formal proof guaranteeing the results.

A Nichols plot is constructed by plotting the gain (in decibels) on the x-axis and the phase shift (in radians) on the y-axis of a Cartesian plane. If the system is described using the transfer function $G(jw)$ then the following equations are used to construct the Nichols plot.

$$x = \arctan \frac{Im(G(jw))}{Re(G(jw))} \tag{1}$$

$$y = 20 \log_{10} |G(jw)| \tag{2}$$

where Re and Im represent, respectively, the real and imaginary parts of the complex value and $|G(jw)|$ represents the magnitude. When calculating the values of the phase shift, the *arctan* function will only return values between $-\frac{\pi}{2}$ and $\frac{\pi}{2}$. It is therefore required to adjust the value by $\pm n\pi$ to get the correct phase-shift. When $Re(G(jw)) = 0$ the phase shift is defined as being equal to $\frac{\pi}{2} \pm n\pi$.

In the Nichols plot, the required gain and phase margins can be described as exclusion regions. If the Nichols plot does not pass through this region, then the system is considered stable. For aeronautical systems, tighter and more descriptive exclusion regions can be chosen to define such properties as a slow or uncomfortable flight response [6].

The most basic exclusion region for aeronautical systems is a hexagon centered at the point $(-\pi, 0)$, see Fig. 2.

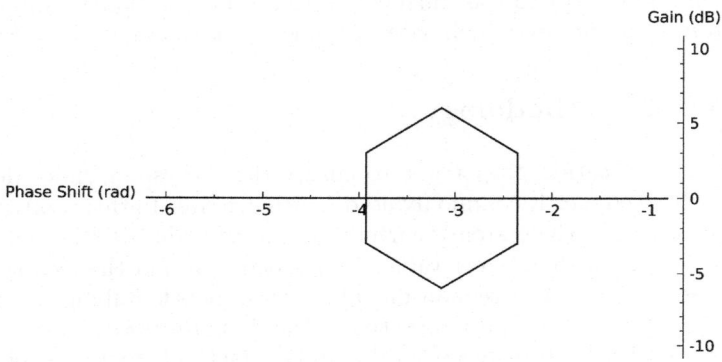

Fig. 2. Nichols Exclusion Region for a Stable System

The conditions to remain outside of the edges of the exclusion region are defined as

$$y > \frac{12}{\pi} + 18 \text{ from } (-\tfrac{5}{4}\pi, 3) \text{ to } (-\pi, 6)$$
$$y < -\frac{12}{\pi} - 18 \text{ from } (-\tfrac{5}{4}\pi, -3) \text{ to } (-\pi, -6)$$
$$y > -\frac{12}{\pi} - 6 \text{ from } (-\tfrac{3}{4}\pi, 3) \text{ to } (-\pi, 6)$$
$$y < \frac{12}{\pi} + 6 \text{ from } (-\tfrac{3}{4}\pi, -3) \text{ to } (-\pi, -6)$$
$$x < -\tfrac{5}{4}\pi$$
$$x > -\tfrac{3}{4}\pi$$

To perform the verification of a flight control system, we propose the methodology described in Fig. 3. First, the flight control system is modeled in Simulink. This will require that the complete dynamics of the aircraft also be modeled. Then using MATLAB's *linmod* [15] function, the open-loop transfer function of the system can be automatically extracted.

An exclusion region of the Nichols plot is then chosen. In general, the exclusion region is chosen from previous experience; depending on the response required from the aircraft, different exclusion region bounds can be chosen. The basic exclusion region is one that assures that the system is stable. In addition,

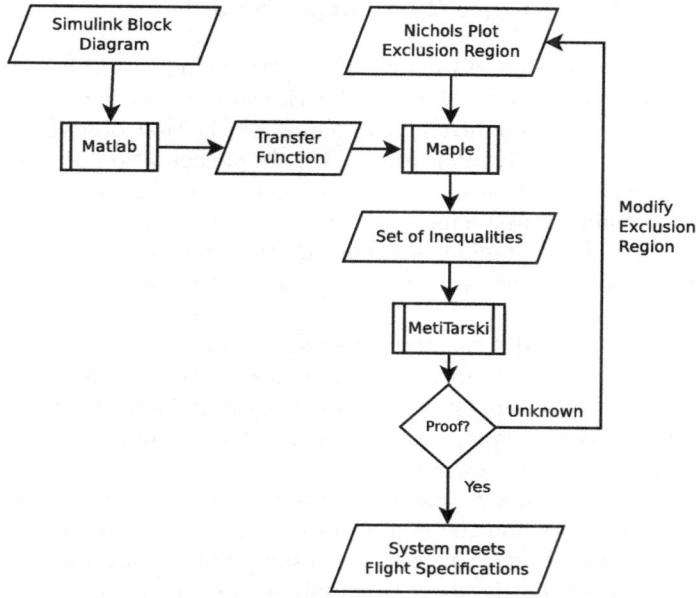

Fig. 3. Verification Methodology

the bounds can be even more tightly chosen to determine the quality of the flight control in terms of handling and response to pilot commands. This will be discussed in more detail below.

The following step is the conversion of the bounds of the exclusion region (in terms of decibels and radians) into inequalities described in terms of the transfer function (frequency domain) using Maple. MetiTarski is first used to verify the results that Maple produces. The resulting expressions for each boundary of the exclusion region are then processed by MetiTarski which automatically generates a proof if it can determine that the inequality holds. This resulting proof indicates that the Nichols plot curve never enters the defined exclusion region.

If MetiTarski is successful, it delivers a proof and we are done. If unsuccessful, it will run until terminated by the user. In the most recent version of MetiTarski (v1.8) it is possible for the user to specify a CPU time limit on the proof. In the event of the CPU limit being reached, we must consider modifying the exclusion region. This has the effect of reducing the required stability margins. A relaxation of the exclusion region can be performed automatically when the CPU limit is reached.

The benefit of this method compared to other aeronautical verification methods is two-fold. First, there is no need to visually inspect any of the plots. If MetiTarski returns that the proof is true then we can be sure that the specification is met. MetiTarski also operates automatically on the continuous range of variables.

5   Case Study : Model 24 Learjet SBJ

To illustrate the application of the proposed methodology, we consider a part of a lateral autopilot design for a Model 24 Learjet subsonic business jet (SBJ) [4]. The SBJ is modeled in Simulink by combining blocks that describe rigid body dynamics and lateral aerodynamic forces. This implementation uses 3 degrees of motion (DOM) equations that have been decoupled from the longitudinal motion terms. This is possible by assuming that derivatives of lateral forces dependent on longitudinal forces are negligible and that all other force and torque derivatives are at trim. At trim, there is no rotation about the center of gravity of the aircraft.

For a pilot, it is often difficult to control an aircraft at high altitude because of high frequency yaw oscillations. Yaw is defined as the side to side motion of an aircraft's nose. In this case study, we are analyzing the SBJ model described above that uses a yaw damper, also commonly known as a washout filter, to augment the stability of the system.

Figure 4 shows a simplified view of the system. The block SBJ4 encapsulates the rigid body dynamics and the lateral aerodynamic forces and moments of an aircraft. We are specifically analyzing the response of the heading angle phi to a deflection of the aileron da, dr is the input to the rudder deflection, r is the yaw rate, p is the roll rate and psi is the heading angle. The washout filter was then place around this block in a feedback configuration.

The first step in the analysis is to extract the transfer function from the Simulink model. In this case study we focus only on the analysis of the response between the aileron displacement da and the roll angle phi. Using Matlab's $linmod$ function, the following transfer function $G(s)$ is extracted from the model,

$$G(s) = \frac{1.065 \times 10^{-14} s^6 + 3.776 s^5 + 19.0633 s^4 + 24.543 s^3 + 21.7634 s^2 - 7.263 \times 10^{-15} s}{s^7 + 7.695 s^6 + 20.3724 s^5 + 26.492 s^4 + 22.0224 s^3 + 0.0442 s^2}$$

The input to $linmod$ is the Simulink design where an input port and an output port have been explicitly defined. The output is a state space model, $\dot{x} = Ax +$

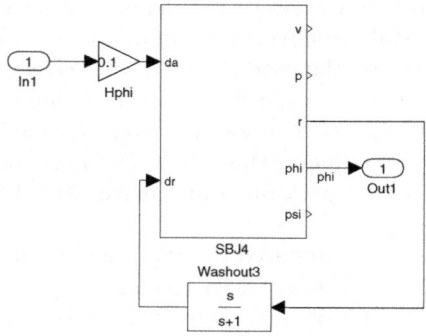

Fig. 4. Flight Control Simulink Model

$Bu, y = Cx + Du$. Then using the Matlab command $ss2tf$, the state space model is converted into a transfer function $G(s)$ and $G(jw)$ is obtained by replacing instances of the variable s with jw.

The gain and phase of the system with the transfer function $G(jw)$ are computed as described in (1) and (2), see Sect. 4.

The next step is to select the exclusion region of the Nichols plot as described before. At the most basic level, we can choose a hexagonal region that is centered around the point $(-\pi, 0)$ which is shown in Fig. 5.

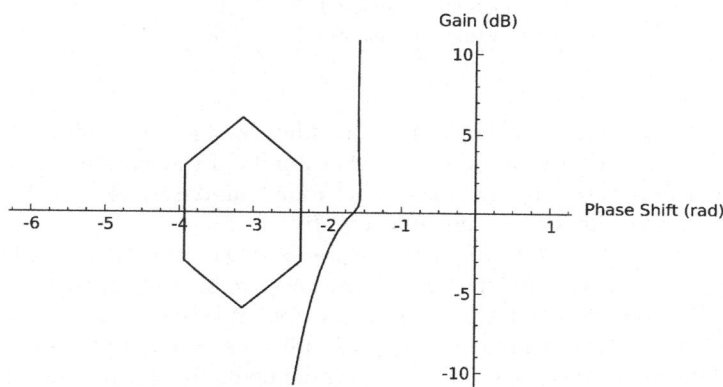

Fig. 5. Nichols Plot of the System G(s)

Now that the Nichols exclusion region has been defined, Maple is used to solve for the frequencies where the Nichols plot passes through the endpoints of the exclusion region. We use MetiTarski to ensure that Maple's computations are in fact correct.

The interval $[-3, 3]$ of the gain (y-axis of Nichols plot), corresponds to the interval $w \in [23080/32333, 75843/46168]$ in the frequency domain. MetiTarski is used to show that outside this frequency interval, we have $(y \geq 3) \vee (y \leq -3)$. Then to show that the exclusion region is never entered from the right middle segment, MetiTarski proves that

$$\forall w. \quad w > 23080/32333 \wedge w < 75843/46168 \Rightarrow x > -3\pi/4$$

The interval $[-\pi, -3\pi/4]$ of the phase (x-axis of the Nichols plot) corresponds to the interval $w \in [42049/14953, 978208/3695]$ in the frequency domain. MetiTarski is used to show that outside this frequency interval, we have $(x \geq -3\pi/4) \vee (x \leq -\pi)$. Then to show that the exclusion region is never entered from the bottom right segment, MetiTarski proves that

$$\forall w. \quad w < 978208/3695 \wedge w > 42049/14953 \Rightarrow y < -\frac{12}{\pi}x + 6$$

Table 2. Case Study Proof Times

Experiment	Time (s)
right-middle-gain-check-U	1.546
right-middle-gain-check-L	0.259
right-middle-exclusion-1	0.221
right-middle-exclusion-2	0.996
right-middle-exclusion-3	0.221
right-middle-exclusion-4	2.322
right-bottom-phase-check	0.221
right-bottom-exclusion-1	3.56
right-bottom-exclusion-2	9.064

From the results obtained from Maple and MetiTarski, we can infer that the Nichols plot does not pass through any other points of the exclusion region and thus does not pass through any of the other four boundaries. A snapshot of the code used to prove this fact is shown in Fig. 6.

The experimental results are shown in Table 2. For the "right-middle" experiments, U and L indicate the upper and lower points at which the transfer function could possibly enter the exclusion region. When an experiment is split into multiple sub-experiments (1,2,3,4), this indicates that the phase function is taking on different values due to $arctan$ being defined only over $(-\pi/2, \pi/2)$. The "check" experiments are verifying Maple's output. The "exclusion" experiments are verifying that the transfer function does not enter the exclusion region. The runtimes were measured on a 2.8 GHz Dual Quad-Core Mac Pro, with 4GB of RAM. The middle boundary proofs completed faster because they are defined using only the ln function, about which MetiTarski can reason very efficiently. The right bottom boundary is defined using a combination of both the $arctan$ and ln functions, which is more difficult to reason about primarily because of the extremely large values that their arguments take. The difference between proof times is not problematic because the final positive result is eventually obtained. Further improvements to the axioms used by MetiTarski, will ultimately improve the proof times.

```
fof(Nichols-Exclusion,! [X] :
((X > 0.9582 & X < 2.86) =>
   10/ln(10)*ln(0.25*10^(-24)*
   (3862622500*X^20+0.3566432250*10^41*X^18+
   ... + 0.8478030764*10^17*X^8)))
   < -6+(12/pi)*arctan(0.2*10^(-3)*(-6100459+
   ...+0.246*10^25*X^16)))
)).
```

Fig. 6. MetiTarski Input for Proving Lower Right Edge of the Exclusion Region

6 Conclusion and Future Directions

In this paper, we have shown that it is possible to use an automated theorem prover, MetiTarski, to verify properties of Nichols plots directly. The inequalities analyzed contain instances of *ln*, *sqrt* and *arctan* functions that take on very large values. This indicates that we will be able to further apply the methodology to similar sized aeronautical systems where the verification of stability is dependent on phase and gain margins.

Building on the ideas demonstrated in this paper, there are directions we are planning to investigate. In advanced flight control verification methods such as μ-analysis, ν-gap analysis and Quantitative Feedback Theory (QFT), the exclusion regions are defined as circles, ellipses, and complex polygons of varying sizes. Since no assumption is made on the number of inflection points of the transfer function, MetiTarski would be able to handle these types of problems. On the other hand, previous methods would have difficulty. This is because MetiTarski can handle inequalities containing transcendental and other special functions over a real valued domain.

One way to guarantee safety of a dynamical system is to find a function called a "barrier certificate" [13]. If a barrier certificate can be found for a specified system, then it is possible to say that starting in some initial state, some unsafe state will never be reached. By using barrier certificates, it is not necessary to calculate the flows of the system directly. Such is the case with several reachability analysis methods. Finding a barrier certificate is not easy, but this problem can be reformulated as a sum-of-squares search problem [14], and we believe MetiTarski will be quite useful for refuting incorrect sum-of-squares formulas during this search.

We would like to have a more realistic model of the aircraft dynamics. Analyzing the non-linear system using qualitative methods is one possible solution. It will also be necessary to consider parameter variations and perturbation effects. We also need to extend the methodology to other potential methods for stability verification such as Lyapunov based methods.

Acknowledgments

We would like to thank the following people: Kyungjae Baik from Concordia University for providing us with the Simulink control system design we used for the case study. Dr. Lawrence Paulson from the University of Cambridge for his help with MetiTarski.

References

1. Advisory Circular: System design and analysis. Tech. rep., Federal Aviation Administration (1988)
2. Akbarpour, B., Paulson, L.C.: MetiTarski: An automatic prover for the elementary functions. In: Autexier, S., Campbell, J., Rubio, J., Sorge, V., Suzuki, M., Wiedijk, F. (eds.) AISC 2008, Calculemus 2008, and MKM 2008. LNCS (LNAI), vol. 5144, pp. 217–231. Springer, Heidelberg (2008)

3. Akbarpour, B., Paulson, L.C.: Applications of MetiTarski in the verification of control and hybrid systems. In: Majumdar, R., Tabuada, P. (eds.) HSCC 2009. LNCS, vol. 5469, pp. 1–15. Springer, Heidelberg (2009)
4. Baik, K.: Flight control systems - final project. Tech. rep., Concordia University (2008)
5. Brown, C.W.: QEPCAD B: A program for computing with semi-algebraic sets using CADs. SIGSAM Bulletin 37(4), 97–108 (2003)
6. Fielding, C., Varga, A., Bennani, S., Selier, M. (eds.): Advanced techniques for clearance of flight control laws. LNCIS, vol. 283. Springer, Heidelberg (2002)
7. Hardy, R.: Formal methods for control engineering: A validated decision procedure for Nichols Plot analysis. Ph.D. thesis, School of Computer Science - University of St. Andrews (February 2006)
8. Khalil, H.: Nonlinear Systems. Prentice Hall, Englewood Cliffs (1996)
9. Langton, R.: Stability and Control of Aircraft Systems. Wiley, Chichester (2006)
10. National Air Traffic Controllers Association: Air trafic control: By the numbers (2009), http://www.natca.org/mediacenter/bythenumbers.msp
11. Padfield, G.D.: The birth of flight control: An engineering analysis of the Wright brothers 1902 glider. University of Liverpool, The Aeronautial Journal (2003)
12. Parrilo, P.A.: Structured Semidefinite Programs and Semialgebraic Geometry Methods in Robustness and Optimization. Ph.D. thesis, California Institute of Technology (May 2000)
13. Prajna, S., Jadbabaie, A.: Safety verification of hybrid systems using barrier certificates. In: Hybrid Systems: Computation and Control, pp. 477–492. Springer, Heidelberg (2004)
14. Prajna, S., Papachristodoulou, A., Seiler, P., Parrilo, P.A.: SOSTOOLS and Its Control Applications. LNCIS, vol. 312, ch. 3, pp. 273–292. Springer, Heidelberg (2005)
15. The MathWorks: Simulink 7 reference (March 2010), http://www.mathworks.com/access/helpdesk/help/pdf_doc/simulink/slref.pdf
16. Tomlin, C., Mitchell, I., Bayen, A.M., Oishi, M.: Computational techniques for the verification of hybrid systems. Proceedings of the IEEE 91(7) (July 2003)

Generalized Rabin(1) Synthesis with Applications to Robust System Synthesis*

Rüdiger Ehlers

Reactive Systems Group
Saarland University

Abstract. Synthesis of finite-state machines from linear-time temporal logic (LTL) formulas is an important formal specification debugging technique for reactive systems and can quickly generate prototype implementations for realizable specifications.

It has been observed, however, that automatically generated implementations typically do not share the robustness of manually constructed solutions with respect to assumption violations, i.e., they typically do not degenerate nicely when the assumptions in the specification are violated. As a remedy, robust synthesis methods have been proposed. Unfortunately, previous such techniques induced obstacles to their efficient implementation in practice and typically do not scale well.

In this paper, we introduce generalized Rabin(1) synthesis as a solution to this problem. Our approach inherits the good algorithmic properties of generalized reactivity(1) synthesis but extends it to also allow co-Büchi-type assumptions and guarantees, which makes it usable for the synthesis of robust systems.

1 Introduction

The problem of synthesizing finite-state systems from specifications written in linear-time temporal logic has recently received an increase in interest. Algorithmic advances in the solution of the synthesis problem have strengthened the practical applicability of synthesis algorithms and consequently, solution quality considerations that are common in the manual engineering process of reactive systems start to appear in the scope of synthesis as well.

In practice, many specifications consist of a set of assumptions the system to be synthesized can assume about the behavior of its environment, and a set of guarantees that it in turn has to fulfill. Such a situation is typical for cases in which a part of a larger system is to be synthesized. In this context, one particularly well-known solution quality criterion is the *robustness* of a system, i.e., how well it behaves under violations of the assumptions. As an example, a bus arbiter system could be designed to work in an environment in which not all clients request access to the bus as the same time. This assumption might however be violated if a part of the system breaks at runtime or errors were made in the engineering process. To counter these problems, manually constructed safety-critical systems are typically built in a way such that at least some

---

* An earlier version of this paper appeared as arXiv/CoRR document no. 1003.1684.

M. Bobaru et al. (Eds.): NFM 2011, LNCS 6617, pp. 101–115, 2011.

guarantees are still fulfilled in such a case. Automatically synthesized systems however typically do not exhibit robust behavior in such situations. As an example, the bus arbiter synthesized under this assumption could stop giving grants to clients completely once too many requests have occurred in a computation cycle.

To remedy these problems, a few techniques especially geared towards the synthesis of robust systems have been proposed. In [7], a robustness criterion and a synthesis algorithm based on cost automata have been defined, with the specification being restricted to consist of only safety properties. On the other hand, in [4], a robustness criterion based on the number of guarantees that still hold if assumptions are violated is defined, which connects robust synthesis to solving generalized Streett games. In both cases, the scalability of these techniques appears to be limited.

In this paper, we propose *generalized Rabin(1) synthesis*, an extension of the *generalized reactivity(1)* synthesis principle, originally proposed by Piterman, Pnueli and Sa'ar [20]. Our approach extends the expressivity of the latter approach while retaining its good algorithmic properties.

In particular, while generalized reactivity(1) synthesis is applicable to all specifications whose assumptions and guarantees are representable as deterministic Büchi automata, we extend its expressivity by allowing also one-pair Rabin-type and, as a special case, co-Büchi-type assumptions and guarantees, which are useful for representing *persistence* requirements [25]. Equally important, these extensions make the class of specifications that can be handled closed under applying a fairly straight-forward robustness criterion based on the number of computation cycles witnessing temporary violations of the assumptions and guarantees. At the same time, our approach inherits the good algorithmic properties of generalized reactivity(1) synthesis. Additionally, we show that any further non-trivial expressivity extension would result in losing these.

In the following, we describe two algorithms solving the robust synthesis problem. We start by showing how the generalized Rabin(1) synthesis problem can be reduced to solving a parity game with 5 colors and describe its use for robust system synthesis. Then, we discuss the fact that it is sometimes desirable to restrict the system to be synthesized to having some upper time bound between a temporary violation of a safety assumption and the final following violation of a safety guarantee afterwards, i.e., to return to normal operation after some upper time bound. For such cases, we present an adapted algorithm that has the additional advantage of extracting implementations having an extra output signal that reports whether the system is currently in the recovery mode after a violation of the assumption.

1.1 Related Work

Automatically synthesizing implementations from specifications given in linear-time temporal logic (LTL) is a well-studied problem in the literature. Its solutions can be classified into two sorts: (1) approaches that aim at handling the full expressivity of LTL, and (2) techniques that trade the full expressivity of LTL against algorithmic advantages. One particularly well-known approach of the latter kind, which we also build upon in this paper, is *generalized reactivity(1) synthesis* [20]. Its applicability in practice is witnessed by the existence of several successful case studies [5,6,17,25].

Synthesis of robust controllers is a relatively recent topic. In [3], Arora and Gouda describe the *closure* and *convergence* robustness criteria. Bloem et al. [4] solve the synthesis problem for the former criterion, where the number of guarantees that still hold in case of assumption violations is to be maximized. The corresponding synthesis problem is reduced to solving generalized Streett games. In [7], a quantitative approach for safety assumptions and guarantees is proposed, where the robustness definition is based on the number of computation cycles in which violations of the assumptions and guarantees are witnessed. This approach thus uses the convergence robustness criterion. Our work follows this line of research, but extends the idea to also allow liveness parts in the specification. The restriction to qualitative robustness improves the scalability of the proposed approach.

In the area of hybrid systems and control theory, work has been performed on robust synthesis where only the continuous part of the controller to be synthesized is to be made robust [25], while for the discrete part, generalized reactivity(1) synthesis is used. This paper can be seen as being orthogonal to that work as it solves the problem of introducing robustness in the discrete part of the controller. Thus, combining the two approaches leads to robustness of the overall solution synthesized. Also, in [25], so-called *stability* or *persistence* properties, which correspond to co-Büchi-type properties in the framework here, are used but applied to the generalized reactivity(1) synthesis approach in a way that leads to incompleteness of the overall procedure. Thus, our methods also increase the scope of properties expressible in that approach.

Closely related to robust synthesis is also the field of *fault-tolerant* synthesis. Here, fault models and fall-back specifications that need to hold in case of fault occurrences are explicitly given as input to the synthesis process. Most works in this area are concerned with adding robustness to completely specified systems, with a few exceptions (e.g., [11]). Our work follows the line of research in which the system to be synthesized should work in a reasonable way in case of assumption violations even if no explicit such fall-back specifications are given [7,4]. Thus, the techniques presented in this paper are even applicable if the fault model is unknown or it is not desired to invest time in writing the additional fall-back specifications.

2 Preliminaries

Words, Languages and natural numbers: Let Σ be a finite set. By Σ^*/Σ^ω we denote the set of all of its finite/infinite sequences, respectively. Such sequences are also called *words* over Σ. Sets of words are also called *languages*. For some sequence $w = w_0 w_1 \ldots$, we denote by w^j the suffix of w starting with the jth symbol, i.e., $w^j = w_j w_{j+1} \ldots$ for all $j \in \mathbb{N}$.

Mealy machines: *Reactive systems* are usually described using a *finite state machine* description. Formally, we define *Mealy machines* as five-tuples $\mathcal{M} = (S, \Sigma_I, \Sigma_O, \delta, s_0)$ where S is some finite set of states, Σ_I and Σ_O are input/output alphabets, respectively, $s_0 \in S$ is the initial state and $\delta : S \times \Sigma_I \to S \times \Sigma_O$ is the transition function of $\mathcal{M}$. The computation steps of a Mealy machine are called *cycles*.

For the scope of this paper, we set $\Sigma_I = 2^{\mathsf{AP}_I}$ and $\Sigma_O = 2^{\mathsf{AP}_O}$ for some sets of input/output atomic propositions AP_I and AP_O.

The languages induced by Mealy machines: Given a Mealy machine $\mathcal{M} = (S, \Sigma_I,$ $\Sigma_O, \delta, s_0)$ and some input word $i = i_0 i_1 \ldots \in \Sigma_I^\omega$, $\mathcal{M}$ induces a *run* $\pi = \pi_0 \pi_1 \ldots$ and some *output word* $o = o_0 o_1 \ldots$ over i such that $\pi_0 = s_0$ and for all $j \in \mathbb{N}$: $\delta(\pi_j, i_j) = (\pi_{j+1}, o_j)$. Formally, we define the language of $\mathcal{M}$, written as $\mathcal{L}(\mathcal{M})$, to be the set of words $w = w_0 w_1 \ldots \in \Sigma^\omega$ with $\Sigma = 2^{\mathsf{AP}_I \uplus \mathsf{AP}_O}$ such that $\mathcal{M}$ induces a run π over the input word $i = w|_{\Sigma_I} = (w_0 \cap \Sigma_I)(w_1 \cap \Sigma_I) \ldots$ such that $w|_{\Sigma_O} = (w_0 \cap \Sigma_O)(w_1 \cap \Sigma_O) \ldots$ is the output word corresponding to π.

Linear-time temporal logic: For the description of the specification of a system, *linear-time temporal logic* (LTL) is a commonly used logic. Syntactically, LTL formulas are defined inductively as follows (over some set of atomic propositions AP):

- For all atomic propositions $x \in \mathsf{AP}$, x is an LTL formula.
- Let ϕ_1 and ϕ_2 be LTL formulas. Then $\neg\phi_1$, $(\phi_1 \vee \phi_2)$, $(\phi_1 \wedge \phi_2)$, $\mathsf{X}\phi_1$, $\mathsf{F}\phi_1$, $\mathsf{G}\phi_1$, and $(\phi_1 \mathsf{U} \phi_2)$ are also valid LTL formula.

The validity of an LTL formula ϕ over AP is defined inductively with respect to an infinite trace $w = w_0 w_1 \ldots \in (2^{\mathsf{AP}})^\omega$. Let ϕ_1 and ϕ_2 be LTL formulas. We set:

- $w \models p$ if and only if (iff) $p \in w_0$ for $p \in \mathsf{AP}$
- $w \models \neg\psi$ iff not $w \models \psi$
- $w \models (\phi_1 \vee \phi_2)$ iff $w \models \phi_1$ or $w \models \phi_2$
- $w \models (\phi_1 \wedge \phi_2)$ iff $w \models \phi_1$ and $w \models \phi_2$
- $w \models \mathsf{X}\phi_1$ iff $w^1 \models \phi_1$
- $w \models \mathsf{G}\phi_1$ iff for all $i \in \mathbb{N}$, $w^i \models \phi_1$
- $w \models \mathsf{F}\phi_1$ iff there exists some $i \in \mathbb{N}$ such that $w^i \models \phi_1$
- $w \models (\phi_1 \mathsf{U} \phi_2)$ iff there exists some $i \in \mathbb{N}$ such that for all $0 \le j < i$, $w^j \models \phi_1$ and $w^i \models \phi_2$

We use the usual precedence rules for LTL formulas in order to be able to omit unnecessary braces and also allow the abbreviations typically used for Boolean logic, e.g., that $a \rightarrow b$ is equivalent to $\neg a \vee b$ for all formulas a, b.

Labeled parity games: A labeled parity game is a tuple $\mathcal{G} = (V_0, V_1, \Sigma_0, \Sigma_1, E_0, E_1, v_0, \mathcal{F})$ with V_0 and V_1 being the sets of vertices of the two players 0 and 1, Σ_0 and Σ_1 being their sets of actions, and $E_0 : V_0 \times \Sigma_0 \rightarrow V_1$ and $E_1 : V_1 \times \Sigma_1 \rightarrow V_0$ being their edge functions, respectively. We abbreviate $V = V_0 \uplus V_1$ and only consider finite games here, for which V_0, V_1, Σ_0 and Σ_1 are finite. The initial vertex v_0 is always a member of V_0. The coloring function $\mathcal{F} : V_0 \rightarrow \mathbb{N}$ assigns to each vertex in V_0 a color. For the scope of this paper, we only assign colors to vertices of player 0. We introduce the notation $\mathcal{F}^{-1}$ to denote the set of vertices of V_0 having a given color, i.e., for $c \in \mathbb{N}$, $\mathcal{F}^{-1}(c) = \{v \in V_0 : \mathcal{F}(v) = c\}$.

A decision sequence in $\mathcal{G}$ is a sequence $\rho = \rho_0^0 \rho_0^1 \rho_1^0 \rho_1^1 \ldots$ such that for all $i \in \mathbb{N}$, $\rho_i^0 \in \Sigma_0$ and $\rho_i^1 \in \Sigma_1$. A decision sequence ρ induces an infinite play $\pi = \pi_0^0 \pi_0^1 \pi_1^0 \pi_1^1 \ldots$ if $\pi_0^0 = v_0$ and for all $i \in \mathbb{N}$ and $p \in \{0, 1\}$, $E_p(\pi_i^p, \rho_i^p) = \pi_{i+p}^{1-p}$.

Given a play $\pi = \pi_0^0 \pi_0^1 \pi_1^0 \pi_1^1 \ldots$, we say that π is winning for player 0 if $\max\{\mathcal{F}(v) \mid v \in V_0, v \in \inf(\pi_0^0 \pi_1^0 \ldots)\}$ is even for the function inf mapping a sequence onto the set of elements that appear infinitely often in the sequence. If a play is not winning for player 0, it is winning for player 1.

Given some parity game $\mathcal{G} = (V_0, V_1, \Sigma_0, \Sigma_1, E_0, E_1, v_0, \mathcal{F})$, a strategy for player 0 is a function $f_0 : (\Sigma_0 \times \Sigma_1)^* \to \Sigma_0$. Likewise, a strategy for player 1 is a function $f_1 : (\Sigma_0 \times \Sigma_1)^* \times \Sigma_0 \to \Sigma_1$. In both cases, a strategy maps prefix decision sequences to an action to be chosen next. A decision sequence $\rho = \rho_0^0 \rho_0^1 \rho_1^0 \rho_1^1 \ldots$ is said to be in correspondence with f_p for some $p \in \{0, 1\}$ if for every $i \in \mathbb{N}$, we have $\rho_i^p = f_p(\rho_0^0 \rho_0^1 \ldots \rho_{i+p-1}^{1-p})$. A strategy is winning for player p if all plays in the game that are induced by some decision sequence that is in correspondence to f_p are winning for player p. It is a well-known fact that for parity games, there exists a winning strategy for precisely one of the players (see, e.g., [15]). We call a state $v \in V_0$ winning for player p if changing the initial state to v makes or leaves the game winning for player p. Likewise, a state $v' \in V_1$ is called winning for player p if a modified version of the game, that results from introducing a new initial state with only one transition to v' is (still) winning for player p.

If a strategy f_p for player p is a *positional strategy*, then $f_p(\rho_0^0 \rho_0^1 \ldots \rho_{n+p-1}^{1-p}) = f_p'(E_{1-p}(\ldots E_1(E_0(v_0, \rho_0^0), \rho_0^1), \ldots, \rho_{n+p-1}^{1-p}))$ for some function $f_p' : V_p \to \Sigma_p$. By abuse of notation, we call both f_p' and f_p positional strategies. Note that such a function f_p' is finitely representable as both domain and co-domain are finite. For parity games, it is known that there exists a winning positional strategy for a player if and only if there exists some winning strategy for the same player.

Note that a translation between this model and an alternative model where the coloring function is defined for both players is easily possible with only a slight alteration of the game structure.

ω-**automata:** An ω-automaton $\mathcal{A} = (Q, \Sigma, q_0, \delta, \mathcal{F})$ is a five-tuple consisting of some finite state set Q, some finite alphabet Σ, some initial state $q_0 \in Q$, some transition function $\delta : Q \times \Sigma \to 2^Q$ and some *acceptance component* $\mathcal{F}$ (to be defined later). We say that an automaton is deterministic if for every $q \in Q$ and $x \in \Sigma$, $|\delta(q, x)| \leq 1$. Given an ω-automaton $\mathcal{A} = (Q, \Sigma, q_0, \delta, \mathcal{F})$, we also call (Q, Σ, q_0, δ) the *transition structure* of $\mathcal{A}$.

Given an infinite word $w = w_0 w_1 \ldots \in \Sigma^\omega$ and an ω-automaton $\mathcal{A} = (Q, \Sigma, q_0, \delta, \mathcal{F})$, we say that some sequence $\pi = \pi_0 \pi_1 \ldots$ is a run for w if $\pi_0 = q_0$ and for all $i \in \mathbb{N}$, $\pi_{i+1} \in \delta(\pi_i, w_i)$. The language of $\mathcal{A}$, written as $\mathcal{L}(\mathcal{A})$, is the set of all words for which an accepting run through $\mathcal{A}$ exists. The acceptance of π by $\mathcal{F}$ is defined with respect to the type of $\mathcal{F}$, for which many have been proposed in the literature [15].

- For a *safety winning condition*, all infinite runs are accepting. In this case, the $\mathcal{F}$-symbol can also be omitted from the automaton definition.
- For a *Büchi acceptance condition* $\mathcal{F} \subseteq Q$, π is accepting if $\inf(\pi) \cap \mathcal{F} \neq \emptyset$. Here, $\mathcal{F}$ is also called the set of *accepting states*.
- For a *co-Büchi acceptance condition* $\mathcal{F} \subseteq Q$, π is accepting if $\inf(\pi) \cap \mathcal{F} = \emptyset$. Here, $\mathcal{F}$ is also called the set of *rejecting states*.
- For a *parity acceptance condition*, $\mathcal{F} : Q \to \mathbb{N}$ and π is accepting in the case that $\max\{\mathcal{F}(v) \mid v \in \inf(\pi)\}$ is even.
- For a *Rabin acceptance condition* $\mathcal{F} \subseteq 2^Q \times 2^Q$, π is accepting if for $\mathcal{F} = \{(E_1, F_1), \ldots, (E_n, F_n)\}$, there exists some $1 \leq i \leq n$ such that $\inf(\pi) \cap E_i = \emptyset$ and $\inf(\pi) \cap F_i \neq \emptyset$.

- For a *Streett acceptance condition* $\mathcal{F} \subseteq 2^Q \times 2^Q$, π is accepting if for $\mathcal{F} = \{(E_1, G_1), \ldots, (E_n, F_n)\}$ and for all $1 \leq i \leq n$, we have $\inf(\pi) \cap E_i \neq \emptyset$ or $\inf(\pi) \cap F_i = \emptyset$.

For a one-pair Rabin automaton $\mathcal{A} = (Q, \Sigma, q_0, \delta, \{(E, F)\})$, we call F the *Büchi acceptance component* of the automaton while E is denoted as being the *co-Büchi acceptance component*. This terminology is justified by the fact that a one-pair Rabin automaton $\mathcal{A} = (Q, \Sigma, q_0, \delta, \{(E, F)\})$ accepts some word if and only if it is accepted by the co-Büchi automaton $\mathcal{A}_C = (Q, \Sigma, q_0, \delta, E)$ and the Büchi automaton $\mathcal{A}_B = (Q, \Sigma, q_0, \delta, F)$. Whenever a deterministic Rabin automaton $\mathcal{A} = (Q, \Sigma, q_0, \delta, (E, F))$ does not accept a word, we say that its *Büchi part is violated* if the states in F are visited only finitely often along the unique run, and say that its *co-Büchi part is violated* if some state in E is visited infinitely often along this run. Henceforth, we assume that all Büchi, co-Büchi, parity, Rabin and Streett automata are deterministic and without loss of generality, for all of their states $q \in Q$ and input symbols $x \in \Sigma$, we have $|\delta(q, x)| = 1$. We say that a parity automaton is weak if all states that are in the same strongly connected component have the same color.

Parity automata and parity games: Given a deterministic parity automaton $\mathcal{A} = (Q, \Sigma, q_0, \delta, \mathcal{F})$ with $\Sigma = 2^{(\mathsf{AP}_I \uplus \mathsf{AP}_O)}$, it is well-known that $\mathcal{A}$ can be converted to a parity game $\mathcal{G}$ such that $\mathcal{G}$ admits a winning strategy for player 1 (the so-called *system player*) if and only if there exists a Mealy machine $\mathcal{M}$ reading $\Sigma_I = 2^{\mathsf{AP}_I}$ and outputting $\Sigma_O = 2^{\mathsf{AP}_O}$ such that the language induced by $\mathcal{M}$ is a subset of the language of $\mathcal{A}$ (see, e.g., [23]). Furthermore, from a winning positional strategy in $\mathcal{G}$, such a Mealy machine $\mathcal{M}$ can easily be extracted.

Game solving and symbolic techniques: Many algorithms have been proposed for solving parity games, of which some are implementable symbolically, i.e., the sets of vertices and the edge functions can be represented implicitly by using, for example, binary decision diagrams (BDDs) [8,14,10]. In practice, BDDs have been shown to be useful when representing and computing properties of systems that are composed of many components that run in parallel [20,12,2]. One particularly important operation that needs to be performed in game solving is the computation of *attractor sets*. Given two sets of vertices A and B, we define $attr_p(A, B)$ to be the set of game vertices from which player p can enforce that eventually some vertex in B is visited while along the way, the set of vertices A is not left. For the scope of this paper, we let $attr_p$ deal only with vertices of player 0, i.e., A and B may only contain vertices of player 0 and we do not restrict visits to vertices of player 1. The attractor set and a corresponding strategy for player p can be computed symbolically [2,1].

Specifications: In this paper, we consider specifications of the form $\psi = (a_1 \wedge a_2 \wedge \ldots \wedge a_{n_a}) \rightarrow (g_1 \wedge g_2 \wedge \ldots \wedge g_{n_g})$. By abuse of notation, we allow both LTL formulas and deterministic automata as *assumptions* $\{a_1, \ldots, a_{n_a}\}$ and *guarantees* $\{g_1, \ldots, g_{n_g}\}$. A word $w \in (2^{\mathsf{AP}})^\omega$ satisfies ψ if either for some LTL assumption a_i, $w \not\models \psi$, for some assumption automaton a_i, $w \notin \mathcal{L}(a_i)$, or for all LTL and automata guarantees g_i, $w \models g_i$ and $w \in \mathcal{L}(g_i)$, respectively. We assume that all LTL formulas in ψ range over the same set of atomic propositions AP and all automata use 2^{AP} as alphabet. Converting an LTL formula to an equivalent deterministic automaton is a classical topic in the literature, is explained in [15] and nowadays, suitable tools are available [16]. We say that an LTL formula has a Rabin index of one if it can be converted to Rabin automata with one acceptance pair.

In this paper, we are especially interested in specifications in which the assumption and guarantee conjuncts are of the forms ψ, $G\psi$, $GF\psi$, $F\psi$ and $FG\psi$ with the only LTL temporal operator occurring in ψ being X. These are called *initialization, basic safety, basic liveness, eventuality* and *persistence* properties.

3   Generalized Rabin(1) Synthesis

In this section, we present the core construction of the generalized Rabin(1) synthesis approach (abbreviated by GRabin(1) in the following) and then prove that its scope cannot be extended without losing its good properties. Afterwards, we discuss its application to the synthesis of robust systems. We start with a specification of the form

$$\psi = (a_1 \wedge a_2 \wedge \ldots \wedge a_{n_a}) \rightarrow (g_1 \wedge g_2 \wedge \ldots \wedge g_{n_g})$$

for some set of assumptions $\{a_1, \ldots, a_{n_a}\}$ and some set of guarantees $\{g_1, \ldots, g_{n_g}\}$. We assume that these are given in form of deterministic one-pair Rabin automata. Most specifications found in practice can be converted to such a form using commonly known techniques [21,18,16].

Our construction transforms such a specification to a deterministic parity automaton with at most 5 colors that accepts precisely the words that satisfy ψ. The number of states of the generated automaton is polynomial in the product of the state numbers of the individual Rabin automata $a_1, \ldots, a_{n_a}, g_1, \ldots, g_{n_g}$. The generated parity automaton can then be syntactically transformed into a parity game (taking into account the partitioning of the atomic propositions into input and output bits) that is winning for player 1 if and only if there exists a Mealy machine over the given sets of inputs and outputs such that all of its runs satisfy the specification. By using for example the parity game solving algorithm by McNaughton/Zielonka [19], the realizability problem is then solvable symbolically. This algorithm is constructive, i.e., it is able to produce a winning strategy that can be used as a prototype implementation.

Let A be the set of assumption one-pair Rabin automata and G be the set of such guarantee automata. For improved readability of the following description of the algorithm, by abuse of notation, we introduce δ, Q, q_0, Σ, and $\mathcal{F}$ as functions mapping automata onto their components. For example, given some automaton $\mathcal{A} = (\tilde{Q}, \tilde{\Sigma}, \tilde{q}_0, \tilde{\delta}, (\tilde{E}, \tilde{F}))$, we have $\delta(\mathcal{A}) = \tilde{\delta}$.

For $A = \{a_1, \ldots, a_{n_a}\}$ and $G = \{g_1, \ldots, g_{n_g}\}$, we construct the deterministic parity automaton $\mathcal{A}' = (Q', \Sigma', \delta', q'_0, \mathcal{F}')$ that accepts precisely the words on which ψ is satisfied as follows:

- Σ' is chosen such that for all $a \in A \uplus G$: $\Sigma' = \Sigma(a)$
- $Q' = Q(a_1) \times \ldots \times Q(g_{n_g}) \times \{0, 1, \ldots, n_a\} \times \{0, 1, \ldots, n_g\} \times \mathbb{B}$
- For all $q = (q_1^a, \ldots, q_{n_g}^g, q^W, q^R, q^V) \in Q'$ and $x \in \Sigma'$, we define $\delta'(q, x) = (q_1'^a, \ldots, q_{n_g}'^g, q'^W, q'^R, q'^V)$ such that:
  - For all $1 \leq i \leq n_a$: $\delta(a_i)(q_i^a, x) = q_i'^a$

- For all $1 \leq i \leq n_g$: $\delta(g_i)(q_i^g, x) = q_i'^g$
- $q'^W = (q^W + 1) \mod (n_a + 1)$ if $q_{q^W}'^a \in F(a_{q^W})$ or $q^W = 0$, otherwise $q'^W = q^W$.
- $q'^R = (q^R + 1) \mod (n_g + 1)$ if $q_{q^R}'^g \in F(g_{q^R})$ or $q^R = 0$, otherwise $q'^R = q^R$.
- $q'^V = \textbf{true}$ if and only if (at least) one the following two conditions hold:
  * $q^W = 0$
  * for all $1 \leq i \leq n_g$, $q_i'^g \notin E(g_i)$ and $q^V = \textbf{true}$
- For all $q = (q_1^a, \ldots, q_{n_g}^g, q^W, q^R, q^V) \in Q'$, we have that $\mathcal{F}'$ maps q to the least value in $c \in \{0, 1, 2, 3, 4\}$ such that:
  - $c = 4$ if for some $1 \leq i \leq n_a$: $q_i^a \in E(a_i)$
  - $c \geq 3$ if $q^V = \textbf{true}$ and for some $1 \leq i \leq n_g$, $q_i^g \in E(g_i)$
  - $c \geq 2$ if $q^R = 0$.
  - $c \geq 1$ if $q^W = 0$
- $q_0' = (q_0(a_1), \ldots, q_0(g_{n_g}), 0, 0, \textbf{false})$

The components $q_1^a, \ldots, q_{n_g}^g$ in a state tuple $q = (q_1^a, \ldots, q_{n_g}^g, q^W, q^R, q^V) \in Q'$ represent the automata of $A \uplus G$ running in parallel. The remaining part of the state tuples corresponds to some additional *control structure* for checking if the overall specification is satisfied. Note that adding the control structure only results in a polynomial blow-up. The parts of the control structure have the following purposes:

- The counter q^W keeps track of the assumption automaton number for which an accepting state in its Büchi component is to be visited next. The construction is essentially the same as for de-generalizing generalized Büchi automata (see, e.g., [22]).
- The counter q^R does the same for the guarantees.
- The bit q^V tracks if accepting states for the Büchi components of all automata in A have been visited since the last visit to a rejecting state for the co-Büchi component of some guarantee.

A full proof of the correctness of the construction can be found in [13].

3.1 Extending Generalized Rabin(1) Synthesis

The generalized Rabin(1) synthesis approach presented above is capable of handling all assumptions and guarantees that have a Rabin index of one. A natural question to ask at this point is whether the approach can be extended in order to be also able to handle specifications with conjuncts of a higher Rabin index without losing its good properties. These are:

- the fact that the state space of the generated parity automaton is the product of the state spaces of the individual automata and some polynomially sized control structure, which makes the automaton state space amenable to an encoding using symbolic techniques, and
- the constant number of colors, which allows the application of efficient symbolic parity game solving algorithms.

Unfortunately, the approach cannot be extended while retaining these advantages. To see this, consider Streett game solving, which is known to be co-NP-complete [15]. If we were able to accommodate one-pair Streett automata (which are a special case of two-pair Rabin automata) as guarantees in the synthesis approach presented, we could decompose a Streett automaton with n acceptance pairs (for some $n \in \mathbb{N}$) into n one-pair Streett automata, each having the transition structure of the original Streett automaton, take these as guarantees and use no assumptions. The specification is then realizable if and only if it is realizable for the original Streett automaton. Since however, the individual one-pair Streett automata have the same transition structure and thus transition in a synchronized manner, if we were able to build a parity game having the properties stated above, the parity game would only have a number of vertices polynomial in the size of the original Streett automaton and thus, due to the constant number of colors, the realizability problem for the original Streett automaton would be solvable in polynomial time. So the existence of a similar algorithm for *generalized Streett(1) synthesis* or *generalized Rabin(2) synthesis* would imply P=NP.

3.2   Application to Synthesize Robust Systems

Assume that we have a specification of the form $(a_1 \wedge a_2 \wedge \ldots \wedge a_{n_a}) \rightarrow (g_1 \wedge g_2 \wedge \ldots \wedge g_{n_g})$, where all assumptions and guarantees are either initialization, basic safety, basic liveness or persistence properties. During the run of a system satisfying ψ, the assumptions may be violated temporarily. A common criterion for the robustness of a system is that in such a case, it must at some point return to *normal operation mode* after such a temporary assumption violation [7,3]. In the scope of synthesis, implementing such a convergence [3] criterion requires fixing a definition of temporary assumption violations. Taking a specification of the form stated above, only a violation of the initialization or basic safety assumptions can be detected during the run of the system. Moreover, only the basic safety properties can be violated temporarily as an initialization property is only evaluated at the start of a system run. Thus we define:

Definition 1. *Given a word $w = w_0 w_1 \ldots \in (2^{AP})^\omega$ and an LTL formula $\psi = G\phi$, we say that position $i \in \mathbb{N}$ in the word witnesses the non-satisfaction of ψ if there exists some $j \leq i$ such that for no $w' \in (2^{AP})^\omega$, $w_j \ldots w_i w' \models \phi$. Furthermore, given a specification of the form $(a_1 \wedge a_2 \wedge \ldots \wedge a_{n_a}) \rightarrow (g_1 \wedge g_2 \wedge \ldots \wedge g_{n_g})$, where all assumptions are initialization, basic safety, basic liveness or persistence properties, we say that the assumptions/guarantees are temporarily violated on a word w at position i if position i witnesses the non-satisfaction of some basic safety assumption/guarantee in the specification, respectively.*

In [3], convergence has been defined for the safety case. In this paper, we extend the definition to the liveness and persistence cases:

Definition 2. *Given a specification of the form $(a_1 \wedge a_2 \wedge \ldots \wedge a_{n_a}) \rightarrow (g_1 \wedge g_2 \wedge \ldots \wedge g_{n_g})$, where all assumptions and guarantees are initialization, basic safety, basic liveness or persistence properties, we say that a system converges if the following conditions hold for all words in the language of the system:*

- *there exists a bound on the number of temporary basic safety guarantee violations in between any two temporary basic safety assumption violations and after the last temporary basic safety assumption violation, and*
- *If w is a word in the language of the system satisfying the initialization assumptions and for some $j \in \mathbb{N}$, w^j satisfies all non-initialization assumptions, then w satisfies the initialization guarantees and for some $j' \geq j$, $w^{j'}$ satisfies all non-initialization guarantees.*

In this definition, there is no requirement that a converging system also performs some progress on its liveness (and persistence) properties in between two temporary assumption violations even if they are sufficiently sparse, which in practice a robust system should surely do. Nevertheless, we argue that for the scope of synthesis, this definition is still useful. The reason is that all synthesis procedures used nowadays produce finite-state solutions. Thus, if temporary assumption violations stop occurring for a couple of computation cycles and at the same time, some progress is made with respect to liveness assumptions (i.e., accepting states are visited for their automata), the system has, after a finite period of time, also continue to work towards fulfilling the liveness guarantees. If this was not the case, we could find a loop in the finite-state machine description of the system that witnesses non-convergence, which would be a contradiction. Given that our synthesis procedure only produces finite-state solutions (as parity game solving algorithms typically do), it is thus enough to require that after the *last* temporary assumption violation the system converges in order to ensure that the system converges in general. We can easily express convergence after the last temporary assumption violation in LTL by prefixing the guarantees and the basic safety assumptions in the specification using the F (finally) operator of LTL.

Definition 3. *Given a specification of the form $\psi = (a_1 \wedge a_2 \wedge \ldots \wedge a_{n_a}) \rightarrow (g_1 \wedge g_2 \wedge \ldots \wedge g_{n_g})$, where all assumptions and guarantees are initialization, basic safety, basic liveness or persistence properties, and $a_s, \ldots, a_{n_a}$ are precisely the basic safety assumptions, we define the ruggedized version of ψ to be:*

$$\psi' = (a_1 \wedge \ldots \wedge a_{s-1} \wedge \mathsf{F}(a_s) \wedge \ldots \mathsf{F}(a_{n_a})) \rightarrow (\mathsf{F}(g_1) \wedge \mathsf{F}(g_2) \wedge \ldots \wedge \mathsf{F}(g_{n_g}))$$

Note that when taking a generalized Rabin(1) specification consisting only of initialization, basic safety, basic liveness and persistence properties, ruggedizing it does not change its membership in this class, as basic safety properties are converted to persistence properties, basic liveness properties stay untouched (as $\mathsf{F}\mathsf{G}\mathsf{F}(\phi)$ is equivalent to $\mathsf{G}\mathsf{F}(\phi)$ for all LTL formulas ϕ) and likewise, persistence properties are not altered. This property does not hold for generalized reactivity(1) specifications, as the ruggedization process converts pure safety properties to persistence properties.

Thus, we can solve the robust synthesis problem, where converging systems are to be found that satisfy a given specification, by ruggedizing the specification and using the generalized Rabin(1) synthesis approach. The convergence criterion above however does not state that under no safety assumption violation, also no guarantee violation should be performed by the system to be synthesized, which is also required in the vast majority of practical cases where synthesis can be applied. In order to incorporate this requirement to the synthesis process, we can build a deterministic weak automaton from the original specification but using only safety assumptions and guarantees. By taking the conjunction of this automaton with the parity automaton obtained from the ruggedized specification using the construction stated above, we obtain a five-color parity game for which all implementations realizing the specification also have this additional property.

So far, we have required all assumption and guarantee conjuncts to be initialization, basic safety, basic liveness and persistence properties. We leave the question how to ruggedize specifications consisting of arbitrary one-pair Rabin automata as assumptions and guarantees open, as there is no extension to the ruggedization concept that is suitable in general. As an example, in general we could have the assumption in a specification that at precisely every second computation cycle, some input bit should be set to **true**. If during the execution of the system, the environment flips the phase of the signal (e.g., from setting the bit to **true** every even cycle to setting it to **true** every odd cycle), whether this should count as only a temporary violation or a permanent one depends on the application. Thus, a generic ruggedization construction for specifications that do not only have initialization, basic safety or liveness and persistence conjuncts cannot be given.

4   Bounded-Transition-Phase Generalized Rabin(1) Synthesis

In the preceding section, we defined generalized Rabin(1) synthesis and its application to synthesize robust systems. These systems have the property that after a temporary violation of some assumption, the system returns to *normal operation mode* after a finite period of time. The system might however have the drawback that the length of the period is not under its control and might grow arbitrarily. Consider the following example, which is a realizable specification over $\mathsf{AP}_I = \{i\}$ and $\mathsf{AP}_O = \{o\}$:

$$(i \wedge \mathsf{GF}i \wedge \mathsf{G}(i \leftrightarrow \mathsf{X}i)) \rightarrow (\mathsf{G}(o \leftrightarrow i) \wedge \mathsf{G}o)$$

The environment can temporarily violate its specification by switching from continuously choosing $i = $ **true** to $i = $ **false** and vice versa. While the environment plays a stream of $i = $ **false**, the system has to violate some of its guarantees. However, the environment has to switch back to setting $i = $ **true** at some point in order not to violate its liveness property. Thus, also the ruggedized version of the specification is realizable. However, we cannot give a time bound on the duration of a phase in which i is set to **false** continuously and thus, there is no implementation of the specification that has a time bound on the length of the *transition phase* in which the system switches back to normal operation mode after the last temporary assumption violation.

Definition 4. *Given a specification $\psi = (a_1 \wedge a_2 \wedge \ldots \wedge a_{n_a}) \rightarrow (g_1 \wedge g_2 \wedge \ldots \wedge g_{n_g})$, where all assumptions and guarantees are initialization, basic safety, basic liveness or persistence properties, we say that a robust system is in normal operation mode with respect to ψ after the input/output prefix word $w \in (2^{\mathsf{AP}_I \uplus \mathsf{AP}_O})^*$ if the system can enforce that the postfix word $w' \in (2^{\mathsf{AP}_I \uplus \mathsf{AP}_O})^\omega$ representing the following input/output either has the property that the initialization assumptions are not satisfied in ww', or no safety guarantee temporary violation is witnessed in ww' from position $|w| + 1$ onwards before the next safety assumption violation.*

We say that a system is in *recovery mode* whenever it is not in normal operation mode. Undoubtedly, systems that guarantee an upper time bound on the number of computation cycles being in recovery mode after a temporary assumption violation has occurred (provided that no further such violation occurs during the recovery process) are more desirable [9]. In this section, we show how to obtain these with the generalized Rabin(1) synthesis approach, whenever possible. As a side-result, the systems synthesized also have an additional output bit that always indicates whether normal operation mode has already been restored.

We solve the *bounded-transition-phase* generalized Rabin(1) synthesis problem by borrowing ideas from [9], where finitary winning conditions for parity games are introduced, but only apply these to the co-Büchi part of the specification that has been introduced when ruggedizing the original specification. In contrast to [9], we thus avoid that specifications in which the system to be synthesized is required to wait for some external events for fulfilling its obligations become unrealizable, which applies to the majority of the industrial case studies available for generalized reactivity(1) synthesis at the moment (see, e.g., [5,6,26]).

Let ψ be a generalized reactivity(1) specification. We start by ruggedizing ψ and convert the resulting LTL formula to a deterministic parity automaton $\mathcal{A}'$ as described in the previous section, but this time modify the construction slightly to always set the q^V flag to **true**. With this modification, if some rejecting state for the co-Büchi-part of some guarantee Rabin automaton is visited along a run, this results in an occurrence of color 3, except if at the same time some state in the co-Büchi-part of an assumption is visited (which leads to color 4). When computing the automata from the specification conjuncts, for the persistence properties, we make sure that their automata are *co-Büchi-tight*. We say that a deterministic co-Büchi word automaton is co-Büchi-tight for some LTL formula $\psi = \mathsf{FG}\phi$ if a run of the automaton visits a rejecting state precisely at the positions in the corresponding word that witness the non-satisfaction of $\mathsf{G}\phi$. We define:

Definition 5. *Given a parity game $\mathcal{A}$ with colors $\{0, 1, 2, 3, 4\}$, we say that some strategy f for player 1 is a winning **color-3 bounded-transition-phase strategy** if there exists some constant $c \in \mathbb{N}$ such that on every run of $\mathcal{A}$ that is in correspondence to f, a visit to color 3 along the run can only occur within c steps after an occurrence of color 4.*

Whenever we have a winning color-3 bounded-transition-phase strategy for the game induced by $\mathcal{A}'$, the strategy represents an implementation realizing ψ using only a bounded transition-phase before returning to normal operation mode after a temporary assumption violation, as in $\mathcal{A}'$, all visits to color 3 signal visits to co-Büchi states in some Rabin guarantee automaton, which in turn witness temporary guarantee violations. Thus, such a strategy represents a system implementation for which the number of computation steps in which it can be in recovery mode between any two temporary assumption violations is limited.

Let $\mathcal{G} = (V_0, V_1, \Sigma_0, \Sigma_1, E_0, E_1, v_0, c)$ be the game built from $\mathcal{A}'$. We use a function $parity_p(B, C)$ for computing the set of winning vertices in a parity game, i.e., for $p \in \{0, 1\}$, it maps the sets of V_0-vertices B and C onto the set of vertices from which player p can win the game, assuming all states in B to be winning for this player and all states in C to be winning for the other player. We assume that the $parity_p$ function also computes a winning strategy. Note that the commonly used parity game solving algorithms can easily be modified to handle such additional parameter sets B and C and to compute such a strategy [14,10].

At any point during a run in which player 1 plays a bounded-transition strategy, she is either in the transition phase or in a vertex from which she can win without being in the transition phase, i.e., from which she has a winning strategy that does not visit a vertex with color 3 before a vertex with color 4 is visited and at the same time, if a color-4-vertex is never visited again, is winning using only vertices with colors 0, 1 and 2. Using this observation, we can compute the set of V_0-vertices from which player 1 can win while being in the transition phase using a fixed point characterization of the set:

$$W = \nu X. X \wedge attr_1(\mathcal{F}^{-1}(4) \vee parity_1(\mathcal{F}^{-1}(4) \wedge X, \mathcal{F}^{-1}(3))) \qquad (1)$$

The subformula $parity_1(\mathcal{F}^{-1}(4) \wedge X, \mathcal{F}^{-1}(3))$ computes from which vertices the game can be won when not being in the transition phase, assuming that the states in X are winning during the transition phase, as a visit to color 4 allows switching to the transition phase, and a vertex with color 3 must not be visited beforehand. Taking the attractor set of $parity_1(\mathcal{F}^{-1}(4) \wedge X, \mathcal{F}^{-1}(3))$ allows finding the vertices from which player 1 can enforce that either a vertex with color 4 is visited after a finite period of time or that in a finite number of steps, vertices are reached which are winning even when not being in the transition phase. Using this set W, the set of vertices from which player 1 can win when not being in the transition phase can then be obtained by computing:

$$Y := parity_1(\mathcal{F}^{-1}(4) \wedge W, \mathcal{F}^{-1}(3) \wedge W) \qquad (2)$$

Theorem 1. *Given a parity game $\mathcal{G}$, the set of the vertices of player 0 in a game from which a winning color-3 bounded-transition-phase strategy exists for player 1 is equal to the set of vertices computed by Equation 2.*

Note that in this setting, the *parity* function only needs to deal with three-color parity games, as the vertices with color 4 and 3 are assumed to be winning and losing for the system player, respectively, so they can be remapped to color 0 for the scope of the *parity* function. This allows the usage of specialized three-color parity game solving algorithms such as the one described in [10], which has been shown to work well for symbolic game solving in practice. Also, as the computations of attractors, conjunctions, disjunctions and fixed points can be done symbolically [14], we obtain a fully symbolic algorithm for computing the states from which a winning color-3 bounded-transition-phase strategy exists.

Extracting such a strategy is also simple. While being in Y, the system player plays the strategy computed by the *parity* function. Whenever this is not the case, it moves along the attractor towards Y. Since an implementation can track its current vertex in the game, it can easily output a signal stating whether it is in Y or not. This signal then serves as an indicator whether the system is in normal operation or recovery mode.

The description of the algorithm established in this section does not immediately generalize to the case that the original specification also contains co-Büchi objectives, as co-Büchi rejecting states may be visited before the first occurrence of a safety assumption violation in practice, but after the ruggedization of the specification and the conversion to a parity game, such a visit is not allowed by a color-3 bounded-transition-phase strategy. It is however easy to adapt our algorithm to fix this problem: we additionally introduce the colors 5 and 6 and map violations of the co-Büchi parts of Rabin guarantee and assumption automata that correspond to safety conjuncts in the original specification to these colors. The colors 3 and 4 are then still used for the persistence properties of the original specification, using the q^V bit of the construction from Sect. 3. Then, we only need to search for color-5 bounded-transition-phase strategies instead of color-3 ones and alter the equations 1 and 2 to $W = \nu X.X \wedge attr_1(\mathcal{F}^{-1}(5) \vee parity_1(\mathcal{F}^{-1}(6) \wedge X, \mathcal{F}^{-1}(5)))$ and $Y := parity_1(\mathcal{F}^{-1}(6) \wedge W, \mathcal{F}^{-1}(5))$.

5   Conclusion

In this paper, we presented generalized Rabin(1) synthesis. We have shown that our approach is the maximally possible extension to generalized reactivity(1) synthesis that has the same good algorithmic properties. As an application, we have defined a robustness criterion suitable for specifications consisting of *initialization*, *basic safety*, *basic liveness*, and *persistence* conjuncts and shown that the set of generalized Rabin(1) specifications consisting only of these conjuncts is closed under the process of ruggedization, which automatically transforms a specification into one for a system that needs to be robust against environment assumption violations. By applying a special algorithm for bounded-transition-phase generalized Rabin(1) synthesis, we can furthermore search for implementations that are even more robust in the sense that the transition phase between the normal operation mode and the recovery mode after a temporary assumption violation has to be bounded in length by some constant.

The practical applicability of the techniques described in this paper is witnessed by the fact that in robotics, where generalized reactivity(1) synthesis starts to be applied, the inability of generalized reactivity(1) synthesis to handle co-Büchi-type specifications is discussed in some publications in that field (see, e.g., [17,26]). Our techniques allow specifying such properties and are implementable in a fully symbolic manner. When applied to only Büchi-type assumptions and guarantees, the basic GRabin(1) synthesis approach is equivalent to the one for generalized reactivity(1) synthesis described in [4] and as a lightweight modification to the former algorithm, it inherits its good algorithmic properties.

Acknowledgements. This work was supported by the German Research Foundation (DFG) within the program "Performance Guarantees for Computer Systems" and the Transregional Collaborative Research Center "Automatic Verification and Analysis of Complex Systems" (SFB/TR 14 AVACS).

The author wants to thank Roderick Bloem and Krishnendu Chatterjee for interesting discussions and ideas on [13].

References

1. Alur, R., Henzinger, T.A., Mang, F.Y.C., Qadeer, S., Rajamani, S.K., Tasiran, S.: Mocha: Modularity in model checking. In: Hu, A.J., Vardi, M.Y. (eds.) CAV 1998. LNCS, vol. 1427, pp. 521–525. Springer, Heidelberg (1998)

2. Alur, R., Madhusudan, P., Nam, W.: Symbolic computational techniques for solving games. STTT 7(2), 118–128 (2005)
3. Arora, A., Gouda, M.G.: Closure and convergence: A foundation of fault-tolerant computing. IEEE Trans. Software Eng. 19(11), 1015–1027 (1993)
4. Bloem, R., Chatterjee, K., Greimel, K., Henzinger, T.A., Jobstmann, B.: Robustness in the presence of liveness. In: [24], pp. 410–424
5. Bloem, R., Galler, S., Jobstmann, B., Piterman, N., Pnueli, A., Weiglhofer, M.: Interactive presentation: Automatic hardware synthesis from specifications: a case study. In: Lauwereins, R., Madsen, J. (eds.) DATE, pp. 1188–1193. ACM, New York (2007)
6. Bloem, R., Galler, S., Jobstmann, B., Piterman, N., Pnueli, A., Weiglhofer, M.: Specify, compile, run: Hardware from PSL. Electr. Notes Theor. Comput. Sci. 190(4), 3–16 (2007)
7. Bloem, R., Greimel, K., Henzinger, T.A., Jobstmann, B.: Synthesizing robust systems. In: FMCAD, pp. 85–92. IEEE, Los Alamitos (2009)
8. Bryant, R.E.: Graph-based algorithms for boolean function manipulation. IEEE Trans. Computers 35(8), 677–691 (1986)
9. Chatterjee, K., Henzinger, T.A., Horn, F.: Finitary winning in omega-regular games. ACM Trans. Comput. Log. 11(1) (2009)
10. de Alfaro, L., Faella, M.: An accelerated algorithm for 3-color parity games with an application to timed games. In: Damm, W., Hermanns, H. (eds.) CAV 2007. LNCS, vol. 4590, pp. 108–120. Springer, Heidelberg (2007)
11. Dimitrova, R., Finkbeiner, B.: Synthesis of Fault-Tolerant Distributed Systems. In: Liu, Z., Ravn, A.P. (eds.) ATVA 2009. LNCS, vol. 5799, pp. 321–336. Springer, Heidelberg (2009)
12. Ehlers, R.: Symbolic bounded synthesis. In: [24], pp. 365–379
13. Ehlers, R.: Generalised Rabin(1) synthesis. arXiv/CoRR abs/1003.1684 (2010)
14. Emerson, E.A., Jutla, C.S.: Tree automata, mu-calculus and determinacy (extended abstract). In: FOCS, pp. 368–377. IEEE, Los Alamitos (1991)
15. Grädel, E., Thomas, W., Wilke, T. (eds.): Automata, Logics, and Infinite Games: A Guide to Current Research. LNCS, vol. 2500. Springer, Heidelberg (2002)
16. Klein, J., Baier, C.: Experiments with deterministic ω-automata for formulas of linear temporal logic. Theor. Comput. Sci. 363(2), 182–195 (2006)
17. Kress-Gazit, H., Fainekos, G.E., Pappas, G.J.: Temporal-logic-based reactive mission and motion planning. IEEE Transactions on Robotics 25(6), 1370–1381 (2009)
18. Krishnan, S.C., Puri, A., Brayton, R.K., Varaiya, P.: The Rabin index and chain automata, with applications to automatas and games. In: Wolper, P. (ed.) CAV 1995. LNCS, vol. 939, pp. 253–266. Springer, Heidelberg (1995)
19. McNaughton, R.: Infinite games played on finite graphs. Ann. Pure Appl. Logic 65(2), 149–184 (1993)
20. Piterman, N., Pnueli, A., Sa'ar, Y.: Synthesis of reactive(1) designs. In: Emerson, E.A., Namjoshi, K.S. (eds.) VMCAI 2006. LNCS, vol. 3855, pp. 364–380. Springer, Heidelberg (2005)
21. Safra, S.: Complexity of Automata on Infinite Objects. PhD thesis, Weizmann Institute of Science, Rehovot, Israel (March 1989)
22. Thomas, W.: Automata on Infinite Objects. In: Handbook of Theoretical Computer Science. Formal Models and Semantics, vol. B, pp. 133–191. MIT Press, Cambridge (1994)
23. Thomas, W.: Church's problem and a tour through automata theory. In: Avron, A., Dershowitz, N., Rabinovich, A. (eds.) Pillars of Computer Science. LNCS, vol. 4800, pp. 635–655. Springer, Heidelberg (2008)
24. Touili, T., Cook, B., Jackson, P. (eds.): CAV 2010. LNCS, vol. 6174. Springer, Heidelberg (2010)
25. Wongpiromsarn, T., Topcu, U., Murray, R.M.: Automatic synthesis of robust embedded control software. In: AAAI Spring Symposium on Embedded Reasoning (2010)
26. Wongpiromsarn, T., Topcu, U., Murray, R.M.: Receding horizon control for temporal logic specifications. In: Johansson, K.H., Yi, W. (eds.) HSCC, pp. 101–110. ACM, New York (2010)

Integrating an Automated Theorem Prover
into Agda

Simon Foster and Georg Struth

Department of Computer Science, University of Sheffield, UK
{s.foster,g.struth}@dcs.shef.ac.uk

Abstract. Agda is a dependently typed functional programming language *and* a proof assistant in which developing programs and proving their correctness is one activity. We show how this process can be enhanced by integrating external automated theorem provers, provide a prototypical integration of the equational theorem prover Waldmeister, and give examples of how this proof automation works in practice.

1 Introduction

The ideal that programs and their correctness proofs should be developed hand-in-hand has influenced decades of research on formal methods. Specification languages and formalisms such as Hoare logics, dynamics logics and temporal logics have been developed for analysing programs, protocols, and other computing systems. They have been integrated into tools such as theorem provers, SMT/SAT solvers and model checkers and successfully applied in the industry. Most of these formalisms do not analyse programs directly on the code, but use external tools and techniques with their own notations and semantics. This usually leaves a formalisation gap and the question remains whether the underlying program semantics has been faithfully captured.

But there are, in fact, programming languages in which the development of a program and its correctness proof can truly be carried out as one and the same activity within the language itself. An example are functional programming languages such as Agda [7] or Epigram [15], which are based on dependent constructive type theory. Here, programs are obtained directly from type-level specifications and proofs via the Curry-Howard isomorphism. These languages are therefore, in ingenious ways, programming languages *and* interactive theorem provers. Program development can be based on the standard methods for functional languages, but the need of formal proof adds an additional layer of complexity. It requires substantial mathematical skill and user interaction even for trivial tasks. Increasing proof automation is therefore of crucial importance.

Interactive theorem provers such as Isabelle [17] are showing a way forward. Isabelle is currently being transformed into a versatile proof environment by integrating external automated theorem proving (ATP) systems, SMT solvers, decision procedures and counterexample generators [5,6,4]. Proof tasks can be delegated to these tools, and the proofs they provide are internally reconstructed

M. Bobaru et al. (Eds.): NFM 2011, LNCS 6617, pp. 116–130, 2011.

to increase trustworthiness. But all this proof technology is based on classical logic. This has two main consequences. First, on the programming side the proofs-as-programs approach is not available in Isabelle, hence programs cannot be extracted from Isabelle proofs. Second, because of the absence of the law of excluded middle in constructive logic, proofs from ATP systems and SMT solvers are not generally valid in dependently typed languages. An additional complication is that proof reconstruction in dependently typed languages must be part of type-checking. This makes an integration certainly not straightforward, but at least not impossible.

Inspired by Isabelle we provide the first ATP integration into Agda. To keep it simple we restrict ourselves to pure equational reasoning, where the rule of excluded middle plays no role and the distinction between classical and constructive proofs vanishes. We integrate Waldmeister [10], the fastest equational ATP system in the world[1]. Waldmeister also provides detailed proofs and supports simple sorts/types. Our main contributions are as follows.

- We implement the basic data-types for representing equational reasoning within Agda. Since Agda needs to manipulate these objects during the type checking process, a reflection layer is needed for the implementation.
- Since Agda provides no means for executing external programs before compile time, the reflection-layer theory data-types are complemented by a Haskell module which interfaces with Waldmeister.
- We implement equational logic at Agda's reflection layer together with functions that parse Waldmeister proofs into reflection layer proof terms. We verify this logic within Agda and link it with the level of Agda proofs. This allows us to reconstruct Waldmeister proofs step-by-step within Agda.
- Mapping Agda types into Waldmeister's simple sort system requires abstraction. Invalid proofs are nevertheless caught during proof reconstruction.
- We provide a series of small examples from algebra and functional programming that show the integration at work.

While part of the integration is specific to Waldmeister, most of the concepts implemented are generic enough to serve as templates for integrating other, more expressive ATP systems. Our integration can also be used as a prototype for further optimisation, for instance, by providing more efficient data structures for terms, equations and proofs, and by improving the running time of proof reconstruction. Such issues are further discussed in the final section of this paper.

Formal program development can certainly be split into creative and routine tasks. Our integration aims at empowering programmers to perform proofs at the level of detail they desire, thus making program development cleaner, faster and less error-prone.

This paper aims to explain the main ideas and features of our approach to a formal methods audience. Its more idiosyncratic aspects, which are mainly of interest for Agda developers, are contained in a technical report [8]; the complete code for our implementation can be found at our website[2].

---

[1] http://www.cs.miami.edu/~tptp/CASC/, 15/02/2011

[2] http://simon-foster.staff.shef.ac.uk/agdaatp

2   Agda

Agda [7] is a dependently typed programming language and proof-assistant. It is strongly inspired by Haskell and offers a similar syntax. In this section we briefly introduce Agda as a programming language, whereas the next section focusses on theorem proving aspects. Additional information about Agda, including libraries and tutorials, can be found at the Agda Wiki[3].

The data-types featured in this section come from Agda's standard library. The following inductive data-type declaration introduces vectors.

```
data Vec (A : Set) : ℕ → Set where
    []    : Vec A zero
    _::_  : ∀ {n} (x : A) (xs : Vec A n) → Vec A (suc n)
```

In contrast to most other functional programming languages, Agda supports dependent data-types. The data-type of vectors is defined depending on their length n. In Agda syntax the parameters before the colon are *constants*, whose values cannot be changed by the constructors. Parameters after the colon are *indices*; their definition depends on the particular constructor. In this example, the element type A of a vector is fixed, whereas the size varies. Vectors have two constructors: The empty vector [] has type Vec A zero and zero length. The operation :: (cons) takes, for each n, an element x : A and a vector xs : Vec A n of length n, and yields a vector Vec A (suc n) of length n + 1. Instances of this data-type need not explicitly supply the parameter n, such *hidden* parameters are indicated by braces. One can now define functions as usual.

```
head : ∀ {n} {A : Set} → Vec A (1 + n) → A
head (x :: xs) = x
```

Agda only accepts *total functions*, but head should only be defined when $n \neq 0$. The dependent type declaration captures this constraint. It thus allows a fine control of data validity in specifications. Predicates can also be data-types:

```
data _≤_ : ℕ → ℕ → Set where
    z≤n : ∀ {n} → zero ≤ n
    s≤s : ∀ {m n} (m≤n : m ≤ n) → suc m ≤ suc n
```

The expressions z≤n and s≤s are *names*. Agda is white-space sensitive, so they are parsed as one token, whereas zero ≤ n is parsed as three tokens. The elements of this data-type are *inductive proofs* of ≤. For instance, s≤s (s≤s z≤n) is a proof of $2 \leq 3$. Hence, Agda data-types capture proofs as well as objects such as numbers or vectors. Similarly, one can define n < m as suc n ≤ m.

Agda provides two definitions of equality. *Propositional equality*, ≡, holds when two values and their types have the same normal forms. *Heterogeneous equality*, ≅, only requires equality of values. Two vectors xs : Vec A (m + n) and ys : Vec A (n + m) have different types in Agda, hence xs ≡ ys is not well typed. But xs ≅ ys would hold if xs and ys have same normal form.

As a constructively typed language, Agda uses the Curry-Howard Isomorphism to extract programs from proofs. The above data-types provide examples

---

[3] http://wiki.portal.chalmers.se/agda/pmwiki.php

of how proofs yield programs for their inhabitants. A central tool for program development by proof is the *meta-variable*; a "hole" in a program which can be instantiated to an executable program by step-wise refinement.

```
greater : ∀ (n : ℕ) → ∃ (λ (m : ℕ) → n < m)
greater n = ?
```

The type of greater specifies that for every natural number n there exists a natural number m greater than n. In the function body, ? indicates a meta-variable for which a program must be constructed through proof. More precisely, Agda requires a natural number m constructed in terms of n and a proof that n < m. Agda provides a variety of tools for proof support. If the user invokes the *case-split* command, two proof obligations are generated from the inductive definition of natural numbers:

```
greater zero    = { } 0
greater (suc n) = { } 1
```

Each contains a meta-variable indicated by the braces and number. The first one requires a value of type $\exists(\lambda\ m \to 0 < m)$. The second one requires a value of type $\exists(\lambda\ m \to suc\ n < m)$ for the parameter suc n, assuming $\exists(\lambda\ m \to n < m)$ for the parameter n. In the first case, *meta-variable refinement* further splits the goal into two meta-variables.

```
greater zero    = { } 0, { } 1
```

This is now a pair consisting of a natural number m and a proof that this witness satisfies zero < m. The following code displays a value and proof:

```
greater zero    = 1, s≤s z≤n
```

In this case, m = 1 and s≤s z≤n are the names of the inference rules needed for the proof. By the first rule, zero ≤ zero, by the second rule, therefore, suc zero ≤ suc zero, whence zero < suc zero by the definition of <.

This proof style lends itself naturally to incremental program construction, where writing a program and proving its correctness are one activity. To further automate this, Agda provides the proof-search tool *Agsy* [14], which can sometimes automatically construct programs and proofs. The remaining proof goal in the example above can be solved automatically by calling Agsy.

```
greater (suc n) = (suc (proj₁ (greater n)), s≤s (proj₂ (greater n))
```

The functions $proj_1$ and $proj_2$ project on the value and the proof of the proof goal. However, Agsy struggles with non-trivial proof goals. Increasing the degree of automation is therefore highly desirable to free programmers from trivial proof and construction tasks.

3   Integration of Automated Theorem Proving

ATP systems have already significantly increased proof automation in interactive theorem provers. Isabelle [17], for instance, can use a tactic called Sledgehammer to call external ATP systems. In contrast to interactive provers, which consist of a relatively small inference engine, ATP systems are complex tools that depend on a large number of heuristics. They are less trustworthy than interactive provers. Consequently, Isabelle internally *reconstructs* all ATP proofs with the internally verified ATP system Metis [11]. Since Metis is less efficient than the external ATP systems, a relevance filter minimises the number of hypotheses given to it. Metis then performs proof search to derive the goal from the hypotheses. In practice, however, this *macro-step* proof reconstruction sometimes fails.

Evidently, Agda could benefit from a similar approach, but all state of the art ATP systems are designed for classical predicate logic. The resolution principle, which underlies most of these systems, is directly based on the law of excluded middle. Since constructive proofs are needed for Agda, we have based a first integration on Waldmeister [10], an ATP system for pure equational logic, where the difference between classical and constructive proof disappears. Equational logic needs only rules for reflexivity, symmetry, transitivity, congruence, substitution and a number of structural rules that are all present in constructive logic.

Our integration can serve as a basis for integrating full first-order ATP systems, which could still be used on subclasses of constructive formulae, such as Harrop formulae [9] where classical and constructive proofs coincide.

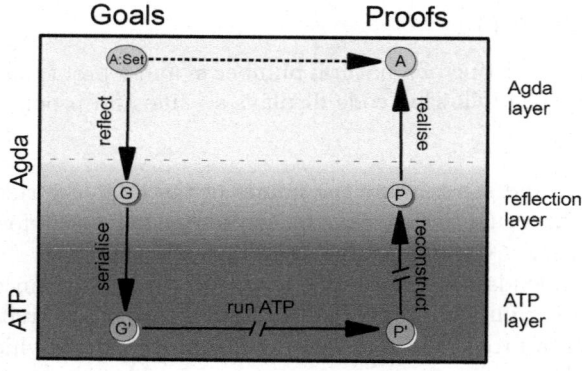

Fig. 1. Overview of automated theorem prover integration in Agda

A Waldmeister integration into Agda is still not straightforward, for two main reasons. Firstly, the built-in Agda normaliser changes hypotheses and proof goals, but these are hidden within the proof state and cannot be accessed from within Agda. Secondly, for both goal extraction and proof reconstruction, Agda syntax must explicitly be manipulated as part of the type-checking process. We see two main approaches to integrating an ATP system into Agda. The *internal approach* performs most of the goal extraction and proof reconstruction within Agda itself. Reflection, which we explain below, provides a way of mediating Agda proofs with ATP proofs. In the *external approach*, proof proceeds by accessing the Agda proof state with an external tool, for instance Haskell, passing this state to the ATP system, and writing an ATP proof back into Agda.

Each approach offers advantages and disadvantages. In this paper, we use the former because it is conceptually cleaner and all the steps of proof reconstruction are internally verified in Agda itself. An additional design decision is whether to use Metis style macro-step proof reconstruction, or micro-step reconstruction of individual proof steps. Again, we take the latter approach. The former would require an internally verified equational prover, and we expect Agda's internal proof tools to be efficient enough to handle single proof steps. Fortunately, Waldmeister output is sufficiently detailed for this purpose.

Our internal approach uses the mechanism of *reflection* which is similar to a quoting mechanism in programming languages, lifting syntax to an internal meta-level, protecting it from evaluation and allowing manipulation within Agda. Therefore all Agda data-types needed in proofs are represented at Agda's reflection level before and after passing them to Waldmeister. Our approach therefore uses the three layers illustrated in Figure 1: The *Agda layer* contains the initial proof goal and realises the final proof. The *reflection layer* represents the Agda goal and reconstructed ATP proof output. The *ATP layer* runs the serialised proof goal and outputs an ATP proof.

Agda's quoting mechanism, however is still experimental. Currently we can reflect and realise a large class of equational problem specifications and proofs. The serialisation of the reflected proof input into an ATP input is obtained by a Haskell module. It requires abstraction because Agda's type system is much more powerful than the simple sorts supported by Waldmeister. In general, types can often be encoded as predicates in ATP systems. State of the art ATP systems can prove quite complex mathematical theorems but they often fail. The integration must be able to cope with this situation. The same holds for proof reconstruction. Ultimately, if Agda succeeds in realising a proof of an initial proof goal, it is guaranteed that this proof is correct in Agda.

4   Proof Cycle Example

This section provides an overview of our Waldmeister integration. It shows how a simple inductive proof is passed from the Agda layer through the reflection layer to Waldmeister, and how the proof obtained by Waldmeister is passed back and reconstructed within Agda.

Consider the following Agda proof goal:

assoc : $\forall\,(x\,y\,z\,:\,\mathbb{N}) \to (x + y) + z \equiv x + (y + z)$

We first perform a case-split on the first argument, yielding two meta-variables.

```
assoc zero y z  =  { } 0
assoc (suc n) y z  =  { } 1
```

Waldmeister can solve each individual goal. Within Agda, they must first be lifted to a reflected signature Σ-Nat for natural numbers and their operations.

Nat : HypVec   -- Proof environment for natural numbers
Nat = HyVec Σ-Nat axioms   -- Construct it from signature and axioms
  where

  +-zero = Γ1, '0 '+ $\alpha \approx \alpha$ -- Quotes indicate reflection layer
  +-suc = Γ2, 'suc α '+ $\beta \approx$ 'suc (α '+ β)

  axioms = (+-zero :: +-suc :: [ ])

assoc-zero : Nat, [ ], Γ2 ⊢ [ ] $\Rightarrow$ ('0 '+ α) '+ $\beta \approx$ '0 '+ (α '+ β)
assoc-zero = ?

assoc-suc : Nat, [ ], Γ3 ⊢ (α '+ β) '+ $\gamma \approx \alpha$ '+ (β '+ γ) :: [ ]
        $\Rightarrow$ ('suc α '+ β) '+ $\gamma \approx$ 'suc α '+ (β '+ γ)
assoc-suc = ?

The reflection layer types of assoc-zero and assoc-suc represent the proof goals including environments for axioms, lemmas and variables used. The question marks indicate meta-variables which need to be instantiated with a proof term. This reflection layer data-type provides sufficient information for generating a Waldmeister input file for the first proof obligation:

```
NAME          agdaProof
MODE          PROOF
SORTS         Nat
SIGNATURE     suc: Nat -> Nat
              plus: Nat Nat -> Nat
              zero,a,b:  -> Nat
ORDERING      LPO a > b > zero > suc > plus
VARIABLES     x,y: Nat
EQUATIONS     plus(zero,x) = x
              plus(suc(x),y) = suc(plus(x,y))
CONCLUSION    plus(plus(zero,a),b) = plus(zero,plus(a,b))
```

Waldmeister instantly returns with the following proof:

```
Axiom 1: plus(zero,x1) = x1
Theorem 1: plus(plus(zero,a),b) = plus(zero,plus(a,b))
Proof:
  Theorem 1: plus(plus(zero,a),b) = plus(zero,plus(a,b))
    plus(plus(zero,a),b)
 =    by Axiom 1 LR at 1 with {x1 <- a}
    plus(a,b)
 =    by Axiom 1 RL at e with {x1 <- plus(a,b)}
    plus(zero,plus(a,b))
```

Non-trivial proofs can easily have hundreds of steps. Waldmeister's output contains sufficient information to reconstruct this proof step-by-step at the reflection layer and instantiate the first meta-variable:

assoc-zero =
  fromJust (reconstruct ((inj_1 (# 0), true, eq-step (0 ::^{l} [ ] ^{l}) (con (# 3) ([ ] ^{x})
        ::^{s} [ ] ^{s})) ::^{l} (inj_1 (# 0), false, eq-step ([ ] ^{l}) (con (# 2) (con (# 3)
        ([ ] ^{x}) ::^{x} con (# 4) ([ ] ^{x}) ::^{x} [ ] ^{x}) ::^{s} [ ] ^{s})) ::^{l} [ ] ^{l}))

The syntax in this proof need not concern us in this paper; it essentially expresses the equational steps above. The function reconstruct uses the reflection layer inference rules for equational logic, which we have internally proved to be sound. To realise this proof in Agda it needs to be translated back to ℕ.

ℕ-⟦Σ⟧ : ⟦Signature⟧ Σ-Nat -- code omitted
ℕ-Nat : ⟦HypVec⟧ Nat ℕ-⟦Σ⟧   -- code omitted

These functions link the reflection layer signature to concrete Agda functions and ground the axioms. The first function instantiates the reflection layer signature, the second one instantiates the hypotheses. The reflection layer proof term is instantiated to a valid Agda proof corresponding to the first meta-variable {}0:

assoc zero y z = ≅-to-≡ (⊢≈-to-≅ _ [] ⊢ {⟦Σ⟧ = add-∃-vars-⟦Σ⟧ {Γ = Γ2}
ℕ-TermModel (y, z, tt)} ℕ-Nat assoc-zero [] f (λ x → z))

The proof cycle for the second meta-variable, {}1, is similar.

5   The Reflection Layer

Agda data-types and proofs must be lifted to the reflection layer to enable their manipulation within Agda. This section shows how data-types, theories and proofs that enable ATP proofs can be implemented at the reflection layer. First we describe the data-types and theories.

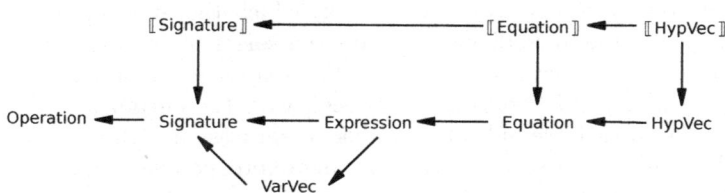

Fig. 2. Dependency graph of reflective components

Figure 2 shows the data-types required. They provide the reflection layer syntax for the terms and equations used in the proofs and the interpretations of these objects within Agda. We will now describe each of them. The basis of our reflection layer syntax are *operations* and *signatures*. Operations can either be Agda functions or data-type constructors; we need not distinguish them.

First we define operations, signatures and variable sets at the reflection layer.

```
record Operation (sorts : FinSet) : Set where
  field
    arity  : ℕ
    args   : Vec (El sorts) arity
    output : El sorts
record Signature : Set where
```

```
    field
        sorts, ops  : FinSet
        operations  : Vec (Operation sorts) ops

record VarVec (Σ : Signature) : Set where
    open Signature Σ
    field
        vars  : FinSet
        vvec  : Vec (El sorts) vars
```

FinSet is a finite set and El indicates an element of a finite set. An n-ary operation is represented as a record parametrised over the set of sorts in its signature. It consists of its arity, its input sorts args and its output sort. A signature is a finite set of sorts together with a vector of operations. We also provide a data-type for variables. Their sorts are determined by the signature under which they appear. Therefore, variable vectors are parametrised by signatures. Next we define terms.

```
mutual
    data Expr Σ (Γ : VarVec Σ) : VarSet Γ → Sort Σ → Set where
        con : (i : El (ops Σ)) {ν : VarSet Γ} (es : ExprVec Γ ν (opArgs Σ i))
              → Expr Σ Γ ν (opOutput Σ i)
        var : (x : Var Γ) → Expr Σ Γ { x } (varSort Γ x)

    data ExprVec   -- code omitted
```

The expression data-type has four parameters: (i) the signature Σ, (ii) the variable context Γ which lists all variables available for building a term, (iii) the subset of variables ν drawn from the context which the expressions contains, represented by shorthand VarSet Γ and (iv) the sort s of the expression. Constructor con takes the operation index i, and an expression vector parametrised over its argument sorts; it yields an expression with the output sort of i. Constructor var takes a variable index and yields an expression with a singleton free variable {x} of the correct sort. Expressions and expression vectors are mutually inductive. It is now possible to define equations and hypotheses sets.

```
record Equation (Σ : Signature) : Set where
    constructor _≈_
    field
        {Γ}      : VarVec Σ
        {sort}   : Sort Σ
        {ν₁ ν₂}  : VarSet Γ
        lhs      : Expr Σ Γ ν₁ sort
        rhs      : Expr Σ Γ ν₂ sort

record HypVec : Set where
    constructor HyVec
    field
        Σ            : Signature
        {hyps}       : FinSet
        hypotheses   : Vec (Equation Σ) hyps
```

We now move to the upper level of Figure 2, where corresponding data-types are implemented. The complete code can, again, be found at our website.

```
record [[Signature]] (Σ : Signature) : Set₁ where    -- code omitted
sem : ∀ {Σ} ([[Σ]] : [[Signature]] Σ) {Γ : VarVec Σ} {s} {ν} ([[ρ]] : [[Subst]] Γ [[Σ]])
      (e : Expr Σ Γ ν s) → [[Signature]].types [[Σ]] s    -- code omitted
record [[Equation]]    -- code omitted
record [[HypVec]]    -- code omitted
```

[[Signature]], [[Equation]] and [[HypVec]] map the reflection layer objects indicated into the Agda layer, and provide soundness proofs. Function sem realises reflection layer terms, using an Agda layer signature [[Σ]] and substitution [[ρ]].

This implementation enables us to represent all elements of Waldmeister input files in a well-typed way at the reflection layer, and to realise these elements within Agda. Since Agda cannot run external program before compile-time, we have written a Haskell module which interfaces with Waldmeister. It provides a function which serialises a Waldmeister input file, as shown in Section 4, executes the prover and parses the resulting Waldmeister proof output back into Agda.

To reconstruct Waldmeister proofs within Agda, we must provide data-types for equational proofs at the reflection layer. First, the parsed proof output provided by Waldmeister must be translated into an inhabitant of a proof data-type. Second, it must be proved that all inhabitants of this data-type are correct with respect to heterogeneous equality.

At the core of proof reconstruction is an implementation of equational reasoning, as performed by Waldmeister. We need some notation and concepts from term rewriting (cf. [19]). A *substitution* is a map ρ from variables to terms, which extends to a function on terms. A term t *matches* a term s (or s *subsumes* t) if $s\rho = t$ for some substitution ρ. We write $t \sqsubseteq s$ if s subsumes t. More specifically, to denote the ternary relation $s\rho = t$ between s, ρ and t, we write $t \sqsubseteq_\rho s$.

With subsumption we can model one-step rewrites of equational logic. Let E be a set of equations $l_i \approx r_i$. We write $E \vdash s = t$ if there is a substitution ρ, a context C and an equation $l \approx r \in E$ such that $s = C[l\rho]$ and $t = C[r\rho]$. Hence

$$E \vdash s =^1 t \iff s \sqsubseteq_\rho C[l] \wedge t \sqsubseteq_\rho C[r] \tag{1}$$

for some substitution ρ, context C and $l \approx r \in E$. We extend this one-step rewrite relation by inductively defining $=$ as the transitive closure of $=^1$. To implement these concepts, we first provide a data-type for substitutions. The expression Subst Γ_1 Γ_2 ν represents a substitution map from the variables in Γ_2 to expressions with variables in Γ_1. The finite subset ν of Γ_2 indicates all those variables that are changed by the substitution. This now allows us to implement the relation $\sqsubseteq_\rho$.

```
mutual
  data _[_⊑_] {Σ} {Γ₁ Γ₂} {ν} (ρ : Subst Γ₁ Γ₂ ν) : ∀ {ν₁} {ν₂} {s₁ s₂}
             → Expr Σ Γ₁ ν₁ s₁ → Expr Σ Γ₂ ν₂ s₂ → Set where
      -- Code omitted

  data _[_⊑*_]    -- Code omitted
```

The inhabitants of this data-type are proofs that two terms match under a given substitution. If a user provides two terms and a substitution, this data-type yields the proof obligations that the user must fulfill to establish the matching relation. These obligations correspond to the inductive definition of terms. The case of a term $f(t_1 \cdots t_n)$ requires mutual induction, as defined by $\sqsubseteq^*$, over the set of subterms. The complete code can be found at our website.

Using the subsumption data structure we can prove the following fact.

Lemma 1. $s \sqsubseteq_\rho t \implies \forall \sigma \,.\, [\![s]\!]\sigma \cong [\![t]\!](\sigma \circ [\![\rho]\!])$.

This lemma states that subsumption implies heterogeneous equality, where the additional substitution σ can be used for further instantiating the resulting expression in equational proofs. The proof of this lemma has been formalised in Agda. As an example, we show the Agda function type corresponding to the lemma.

$$\sqsubseteq\text{-to-}\cong \;:\; \forall\, \{\Sigma\} \, \{\Gamma_1\, \Gamma_2\} \, \{[\![\Sigma]\!]\} \, \{\nu_1\, \nu_2\} \, \{\mathsf{s}_1\, \mathsf{s}_2\} \, (\mathsf{f} \,:\, \mathsf{Expr}\, \Sigma\, \Gamma_2\, \nu_2\, \mathsf{s}_2)$$
$$(\mathsf{e} \,:\, \mathsf{Expr}\, \Sigma\, \Gamma_1\, \nu_1\, \mathsf{s}_1) \, (\rho \,:\, \mathsf{Subst}\, \Gamma_1\, \Gamma_2\, \mathsf{full}) \to \rho\, [\mathsf{e} \sqsubseteq \mathsf{f}]$$
$$\to (\forall\, [\![\sigma]\!] \to \mathsf{sem}\, [\![\Sigma]\!]\, [\![\sigma]\!]\, \mathsf{e} \cong \mathsf{sem}\, [\![\Sigma]\!]\, (\mathsf{sem\text{-}subst}\, \{[\![\Sigma]\!] \;=\; [\![\Sigma]\!]\}\, \rho\, [\![\sigma]\!])\, \mathsf{f})$$

In the next steps we have implemented one-step equational reasoning with and without contexts and n-step equational reasoning, using the subsumption datatype. We have proved soundness of the resulting procedure within Agda.

Theorem 1. *Rewriting implies hetereogeneous equality.*

1. $u \approx v \vdash s = t \implies \forall \rho.[\![s]\!]\sigma \cong [\![t]\!]\sigma$.
2. $E, L \vdash s = t \implies [\![E]\!], [\![L]\!] \models [\![s]\!] \cong [\![t]\!]$.

The Agda proofs can be found at our website. The antecedent of the first statement expresses that s can be rewritten to t using the equation $u \approx v$ at the reflection layer. Its left-hand side essentially states that the interpretation of the reflection layer terms within Agda yield a valid equation. The interpretation of the equation $u \approx v$ is part of the proof state and therefore not visible on the right-hand side. The second statement lifts the first one to sets E of equational axioms, sets L of additional equational hypotheses and n-step proofs. This theorem provides the formal underpinning for proof reconstruction.

Although the data-types presented have been designed predominantly for equational reasoning, they can nevertheless easily be extended to full first-order logic by adding quantifiers, the usual boolean connectives and predicate symbols on top of our existing Equation data-type. Our implementation shows how Agda's reflection layer can be used to achieve such extensions.

6   Proof Reconstruction

We now show how Waldmeister proofs can be reconstructed within Agda as part of the type-checking process. Proof reconstruction can fail since Waldmeister can fail, types can be overabstracted, or Waldmeister introduces constants which have not been accounted for in Agda.

As shown in Section 4, a Waldmeister proof consists of a list of equational steps, each augmented by an axiom number, a term position at which the axiom is applied, its orientation (left-right or right-left), and the substitution used. All these features have been implemented at the reflection layer in Section 5. Execution of an individual rewrite from term e to term f proceeds in two stages which correspond to the definition in Equation (1). Assume the equational proof step $e =^1 f$ is obtained by applying the equation $u \approx v$ under the context g of e, and using substitution ρ.

1. The function build-split uses the term position to split $e = g[e']$ and verify this equality.
2. The function build-$\vdash \approx^1$ takes an equation, substitution and split term, and rewrites e to f. A matching algorithm for computing subsumptions is used by this function.

These one-step reconstruction functions are then used together in reconstruct.

```
reconstruct : ∀ {E : HypVec} {Γ} {ν₁ ν₂} {s} {n} {L : Vec (Equation (Σ E)) n}
            → {e : Expr (Σ E) Γ ν₁ s} {f : Expr (Σ E) Γ ν₂ s}
            → EqProof E Γ L → Maybe (E, L, Γ ⊢ e ≈ f)
```

The type EqProof represents the raw input from Waldmeister which has been reformatted by the Haskell module. A Waldmeister proof output may have been subdivided into a number of lemmas and these are currently flattened out to produced a single sequence of rewrites. reconstruct applies the one-step reconstruction functions iteratively to yield the complete reflection layer proof.

Most first-order theorem provers support the TPTP format[4] as a standard input syntax. Some of them also support TSTP[5] as a proof output standard. For future implementations, Haskell modules supporting these standards should be used instead of the current proprietary Waldmeister ones.

7   Examples

This section shows the Waldmeister integration at work. We have tested it on simple examples about natural numbers (cf. Section 4), groups, Boolean algebras and lists. The following code implements group theory.

```
Group : HypVec
Group = HyVec Σ-Group axioms
  where
    assoc = Γ3, (α · β) · γ ≈ α · (β · γ)
    ident = Γ1, e · α ≈ α
    inv = Γ1, α⁻¹ · α ≈ e
    axioms = (assoc :: ident :: inv :: [])
```

As an example, we show one of the most basic facts.

```
ident-var : Group, [], Γ2 ⊢ α⁻¹ · (α · β) ≈ β
ident-var = fromJust (reconstruct ((inj₁ (# 0), false, eq-step ([] ')
```

---

[4] http://www.cs.miami.edu/~tptp/
[5] http://www.cs.miami.edu/~tptp/TSTP/

$$(\text{con} \ (\# \ 2) \ (\text{var} \ (\# \ 0) ::^x \ [] \ ^x) ::^s \ \text{var} \ (\# \ 0) ::^s \ \text{var} \ (\# \ 1)$$
$$::^s \ [] \ ^s)) ::^l \ (\text{inj}_1 \ (\# \ 2), \text{true}, \text{eq-step} \ (0 ::^l \ [] \ ^l) \ (\text{var}$$
$$(\# \ 0) ::^s \ [] \ ^s)) ::^l \ (\text{inj}_1 \ (\# \ 1), \text{true}, \text{eq-step} \ ([] \ ^l) \ (\text{var}$$
$$(\# \ 1) ::^s \ [] \ ^s)) ::^l \ [] \ ^l))$$

The first line states that a certain equation follows from the group axioms, with
no additional hypotheses and a two variable context. The second line shows how
the Waldmeister proof output, parsed into Agda, is reconstructed. The function
fromJust lifts a Maybe A type to an A type in the case that the proof is successfully
reconstructed, otherwise the proof does not type-check. Additional lemmas can
be found at our website. On such very simple lemmas, Waldmeister returned
almost instantaneously. Proof reconstruction required several seconds.

Proofs in Boolean algebra are more complex, and proof-search is more involved. In our experiments, Waldmeister returned within seconds. But reflection
layer proofs tended to become very long, and their reconstruction sometimes
took several minutes. There are some theorems that Waldmeister could easily
verify, but where proof reconstruction failed, e.g., when Waldmeister chose to
introduce new undeclared constants for non-obvious reasons. Further discussion
can be found in our extended version [8].

Finally we show some simple proofs about lists. This is especially interesting for two reasons. First, lists are two-sorted structures and it is shown that
Waldmeister can handle this situation. Second, proofs require induction, which
is beyond first-order logic.

```
'List : HypVec
'List = HyVec Σ-List axioms
  where
    ++-nil = Γ1, '[] '++ α ≈ α
    ++-cons = Γ3, (α ':: β) '++ γ ≈ α ':: (β '++ γ)
    rev-nil = Γ0, 'rev '[] ≈ '[]
    rev-cons = Γ2, 'rev (α ':: β) ≈ 'rev β '++ (α ':: '[])
    axioms = (++-nil :: ++-cons :: rev-nil :: rev-cons :: [])
```

Lists are essentially monoids with respect to append and nil and we first show
that the empty list is indeed a right identity.

```
rident-nil : 'List, [], Γ0 ⊢ [] ⇒ '[] '++ '[] ≈ '[]
rident-nil = fromJust (reconstruct ((inj₁ (# 0), true, eq-step ([]ˡ)
              (con (# 0) ([] ˣ) ::ˢ [] ˢ)) ::ˡ []ˡ))
rident-cons : 'List, [], Γ2 ⊢ β '++ '[] ≈ β :: []
              ⇒ (α ':: β) '++ '[] ≈ (α ':: β)
rident-cons = fromJust (reconstruct ((inj₁ (# 1), true, eq-step
              ([]ˡ) (con (# 4) ([]ˣ) ::ˢ con (# 5) ([]ˣ) ::ˢ con (# 0)
              ([]ˣ) ::ˢ []ˢ)) ::ˡ (inj₁ (# 4), true, eq-step (1 ::ˡ
              []ˡ)([]ˢ)) ::ˡ []ˡ))
```

The base case and the induction step can be tied together by a case split at the
Agda layer. The induction step goes beyond pure equational reasoning, but can

still be handled by Waldmeister. The implication in the proof goal is skolemised, which yields constants, and the antecedent of the resulting ground formula is then added to the list of axioms. This is captured in our implementation by the derived type $E, L, \Gamma \vdash H \Rightarrow s = t$, where H contains the ground equations resulting from the inductive hypothesis.

Additional lemmas are proved in a similar way. Previously proved lemmas can be added as hypotheses to prove goals. Again, this is managed automatically by Agda. Finally, we can automatically prove a classic.

```
rev-rev-nil  : ‘List, [], Γ0 ⊢ [] ⇒ ‘rev (‘rev ‘[]) ≈ ‘[]
rev-rev-nil  =    -- proof omitted
rev-rev-cons : ‘List, ((Γ2, ‘rev (α ‘⧺ β) ≈ ‘rev β ‘⧺ ‘rev α) :: []),
                    Γ2 ⊢ ((‘rev (‘rev β) ≈ β) :: [])) ⇒ (‘rev (‘rev (α ‘:: β))) ≈ (α ‘:: β)
rev-rev-cons =    -- proof omitted
```

As previously, Waldmeister was very efficient with these proofs. Proof reconstruction succeeded within seconds on these examples, too.

8   Conclusions and Future Work

We have presented a framework for integrating external ATP systems into Agda. Some parts of it are generic while others are specific to the Waldmeister implementation. The main purpose of this work is to explore how such integrations could be achieved by providing a prototype for one particular ATP system. Initial experiments show that our integration works, but should further be optimised to make proof reconstruction faster and more powerful.

First, reflection is experimental in Agda. It has already been used for integrating domain specific solvers and decision procedures [7], but does not suffice for automatically constructing ATP input from a metavariable and a proof state.

Second, a simple command, integrated with Agsy [14], and like Isabelle's Sledgehammer could greatly simplify ATP invocation and proof representation.

Third, full first-order theorem provers should be integrated and syntax checks (for Harrop formulae) could be used for applying them on safe fragments of constructive logic. A theoretical framework for this has already been provided [1].

Fourth, proof reconstruction requires further optimisation. As an alternative to the micro-step approach the unfailing completion procedure [2] underlying Waldmeister could be implemented. The external approach mentioned in Section 3 should also be explored. A similar integration of SAT solvers into Agda is currently undertaken [13]. Its main difference is that proof reconstruction is sacrificed for the sake of efficiency in the tradition of provers such as PVS [18].

Fifth, functional program development methodology [3] has already been integrated into Agda [16,12]. Automating it could lift program development in dependently typed languages to a new level.

Acknowledgements. We would like to thank Thorsten Altenkirch, Nils Anders Danielsson, John Derrick, Peter Dybjer, Ulf Norrell and Makoto Takeyama for helpful discussions. This work has been funded by EPSRC grant EP/G031711/1.

References

1. Abel, A., Coquand, T., Norell, U.: Connecting a logical framework to a first-order logic prover. In: Gramlich, B. (ed.) FroCos 2005. LNCS (LNAI), vol. 3717, pp. 285–301. Springer, Heidelberg (2005)
2. Bachmair, L., Dershowitz, N., Plaisted, D.A.: Completion Without Failure. In: Resolution of Equations in Algebraic Structures, pp. 1–30. Academic Press, London (1989)
3. Bird, R., de Moor, O.: The Algebra of Programming. Prentice-Hall, Englewood Cliffs (1997)
4. Blanchette, J.C., Nipkow, T.: Nitpick: A counterexample generator for higher-order logic based on a relational model finder. In: Kaufmann, M., Paulson, L.C. (eds.) ITP 2010. LNCS, vol. 6172, pp. 131–146. Springer, Heidelberg (2010)
5. Böhme, S., Nipkow, T.: Sledgehammer: Judgement day. In: Giesl, J., Hähnle, R. (eds.) IJCAR 2010. LNCS, vol. 6173, pp. 107–121. Springer, Heidelberg (2010)
6. Böhme, S., Weber, T.: Fast LCF-style proof reconstruction for Z3. In: Kaufmann, M., Paulson, L.C. (eds.) ITP 2010. LNCS, vol. 6172, pp. 179–194. Springer, Heidelberg (2010)
7. Bove, A., Dybjer, P., Norell, U.: A brief overview of agda – A functional language with dependent types. In: Berghofer, S., Nipkow, T., Urban, C., Wenzel, M. (eds.) TPHOLs 2009. LNCS, vol. 5674, pp. 73–78. Springer, Heidelberg (2009)
8. Foster, S., Struth, G.: Integrating an automated theorem prover into Agda. Tech. Rep. CS-10-06, Department of Computer Science, University of Sheffield (2010)
9. Harrop, R.: On disjunctions and existential statements in intuitionistic systems of logic. Mathematische Annalen 132, 347–361 (1956)
10. Hillenbrand, T., Buch, A., Vogt, R., Löchner, B.: Waldmeister: High performance equational deduction. Journal of Automated Reasoning 18(2), 265–270 (1997)
11. Hurd, J.: System description: The Metis proof tactic. In: Benzmüller, C., Harrison, J., Schürmann, C. (eds.) ESHOL 2005, pp. 103–104, arXiv.org (2005)
12. Kahl, W.: Dependently-typed formalisation of relation-algebraic abstractions. In: de Swart, H.C.M. (ed.) RaMiCS 2011. LNCS, Springer, Heidelberg (to appear 2011)
13. Kanso, K., Setzer, A.: Integrating automated and interactive theorem proving in type theory. In: Bendisposto, J., Leuschel, M., Roggenbach, M. (eds.) AVOCS 2010 (2010)
14. Lindblad, F., Benke, M.: A tool for automated theorem proving in Agda. In: Filliâtre, J.-C., Paulin-Mohring, C., Werner, B. (eds.) TYPES 2004. LNCS, vol. 3839, pp. 154–169. Springer, Heidelberg (2006)
15. McBride, C., McKinna, J.: The view from the left. Journal of Functional Programming 14(1), 69–111 (2004)
16. Mu, S.C., Ko, H.S., Jansson, P.: Algebra of programming using dependent types. In: Audebaud, P., Paulin-Mohring, C. (eds.) MPC 2008. LNCS, vol. 5133, pp. 268–283. Springer, Heidelberg (2008)
17. Nipkow, T., Paulson, L.C., Wenzel, M.T.: Isabelle/HOL – A Proof Assistant for Higher-Order Logic. LNCS, vol. 2283. Springer, Heidelberg (2002)
18. Owre, S., Rushby, J.M., Shankar, N.: PVS: A prototype verification system. In: Kapur, D. (ed.) CADE 1992. LNCS (LNAI), vol. 607, pp. 748–752. Springer, Heidelberg (1992)
19. Terese: Term Rewriting Systems. Cambridge University Press, Cambridge (2003)

Efficient Predicate Abstraction of Program Summaries

Arie Gurfinkel, Sagar Chaki, and Samir Sapra

Carnegie Mellon University

Abstract. Predicate abstraction is an effective technique for scaling Software Model Checking to real programs. Traditionally, predicate abstraction abstracts each basic block of a program $\mathcal{P}$ to construct a small finite abstract model – a Boolean program BP, whose state-transition relation is over some chosen (finite) set of predicates. This is called Small-Block Encoding (SBE). A recent advancement is Large-Block Encoding (LBE) where abstraction is applied to a "summarized" program so that the abstract transitions of BP correspond to loop-free fragments of $\mathcal{P}$. In this paper, we expand on the original notion of LBE to promote flexibility. We explore and describe efficient ways to perform CEGAR bottleneck operations: generating and solving predicate abstraction queries (PAQs). We make the following contributions. First, we define a general notion of program summarization based on loop cutsets. Second, we give a linear time algorithm to construct PAQs for a loop-free fragment of a program. Third, we compare two approaches to solving PAQs: a classical AllSAT-based one, and a new one based on Linear Decision Diagrams (LDDs). The approaches are evaluated on a large benchmark from open-source software. Our results show that the new LDD-based approach significantly outperforms (and complements) the AllSAT one.

1 Introduction

Predicate abstraction is a well-established technique for scaling Software Model Checking to real systems [1]. Through predicate abstraction, model checking has been successfully applied to the verification of device drivers, hardware designs, and communication protocols. A core operation in predicate abstraction is the *predicate abstraction query* (PAQ): given a set of quantifier-free predicates P, and a quantifier-free formula e in some first-order theory, compute the strongest formula $\mathcal{G}_P(e)$ over P that is implied by e. It is used to over-approximate sets of states (when e and P are over program variables V), and transition relations (when e and P are over V and V').

Traditionally [1], PAQs are used to abstract transition relations of each individual basic block of an input program – this is called a Small-Block Encoding (SBE) [2]. Since transition relations of a basic blocks are simple (a few conjunctions of equalities) the corresponding PAQs are computationally simple as well [13]. Furthermore, SBE works well under a very coarse over-approximation of PAQs (e.g., via Cartesian abstraction [1] combined with an aggressive refinement [9]) simplifying PAQs even further. On the downside, SBE leads to a very

M. Bobaru et al. (Eds.): NFM 2011, LNCS 6617, pp. 131–145, 2011.

large number of PAQs, a large number of predicates (often a different set for each basic block), and does not take advantage of the state-of-the-art in decision procedures. For example, a safety of a loop-free program can be proved with a single call to an SMT-solver, but with SBE often requires a large number of predicates and many iterations of the CounterExample Guided Abstraction Refinement (CEGAR) loop.

Beyer et al. [2] have proposed an alternative to SBE called the *Large-Block Encoding (LBE)*. LBE lifts predicate abstraction to program summaries (i.e., loop-free program fragments). This leads to fewer PAQs, but the PAQs are more complex, harder to solve, and should not be over-approximated [2]. Overall, [2] shows that LBE is more efficient than SBE, and, even provably exponentially more efficient in some cases. While it is not clear whether LBE is preferable to SBE in all cases, LBE by itself presents three new problems for predicate abstraction. In this paper, we propose an expanded notion of LBE and present solutions to these problems:

(1) What types of program summaries are compatible with LBE? We show that LBE is compatible with a broad notion of a program summary. We argue that a *loop cutset summary* where where all loop-free fragments are summarized is a reasonable (but not the only) choice.

(2) How to efficiently generate PAQs? This problem is unique to LBE. We present a novel algorithm for generating queries for a summary that avoids constructing the summary itself. The algorithm takes a program in SSA form [8] and generates PAQs directly from the program's syntax. The size and the complexity of generating each query are linear in the size of the SSA.

(3) How to efficiently solve PAQs? With LBE, PAQs have a rich propositional structure. We present experiments with two algorithms: an AllSAT-based algorithm due to Lahiri et al. [16] (as implemented in MATHSAT4), and a novel algorithm based on Linear Decision Diagrams (LDDs) [5]. The two algorithms are evaluated on a benchmark derived from open-source programs. Surprisingly, we find that, on the whole benchmark, the LDD-based approach is superior to the AllSAT-based one. Interestingly, the approaches are complementary: we found that the "MIN" combination of the approaches (i.e., run both in parallel, stop as soon as one completes) is much more effective than either one in isolation.

To evaluate end-to-end performance of our approach in the CEGAR framework, we have built a safety checker for C and checked several classical examples from the literature. Our experiments indicate that LBE is more effective than SBE, and that the "MIN" combination of the AllSAT- and LDD-based approaches is most effective. We leave further comparison between LBE and SBE and the effect of LBE on the overall verification process to future work.

We envision that the algorithms proposed here will form a part of a complete CEGAR-based software analysis infrastructure. In particular, we do not argue for an exclusive use of any particular LBE or SBE. Instead, this work provides the flexibility necessary for an analyzer to (heuristically) choose a good block encoding and contributes efficient techniques to solve complex PAQs.

Related work. LBE for predicate abstraction was proposed by Beyer et al. [2]. They show that LBE significantly reduces the size of the abstract state space, the number of required predicates, and the verification time. They observe that success of LBE depends on precise predicate abstraction (as opposed to approximations such as Cartesian abstraction [1]). They use the AllSAT-based predicate abstraction [16] as implemented in MATHSAT4 [3]. We build on this work with a formal and general definition of LBE, new algorithms for efficiently constructing PAQs directly from an SSA program and for solving PAQs, and an extensive empirical evaluation on a large and challenging benchmark.

A naïve predicate abstraction algorithm – enumerating all satisfiable minterms – is exponential. Many heuristics have been proposed to improve its best-case complexity (e.g., [9,10]), and worst-case complexity at expense of completeness (e.g., [1,16]). For example, symbolic predicate abstraction [14,13] avoids exponentially many calls to an SMT solver by generating a symbolic proof from which the result is extracted by Boolean quantification, an exponential step.

Predicate abstraction is reducible to quantifier elimination. This leads to several solutions. In [16], the quantification is delegated to an AllSAT SMT solver. In [4], solutions are enumerated by a BDD and are discharged by an incremental SMT solver. Clarke et al. [6] use a SAT-solver for Boolean quantification for predicate abstraction over propositional logic. Lahiri et al. [15] give an algorithm for first-order logic via a reduction to propositional logic and Boolean quantification with either SAT- or BDD-based method.

In this paper, we propose another alternative: predicate abstraction is reduced to quantifier elimination over first-order logic, and the quantifiers are eliminated using LDDs [5]. On our benchmark this is much more efficient than the corresponding AllSAT-based solution.

The rest of the paper is structured as follows. Sec. 2 provides the necessary background. Sec. 3 describes program summarization. Sec. 4 presents algorithms to generate and solve PAQs. Sec. 5 presents experimental results. Sec. 6 concludes the paper.

2   Background

For a set of variables V, we write V' for $\{v' \mid v \in V\}$. For a binary relation ρ, we write $(s_1, s_2) \models \rho$ for $(s_1, s_2) \in \rho$. We write ρ^* for reflexive transitive closure, and $\rho \circ \rho$ for relational composition. We often represent sets and binary relations in the standard way by Boolean expressions over primed and unprimed variables. For an expression e, we write $e[V/V']$, or e', to mean the expression obtained by replacing each variable v in e with v'.

A *program* $\mathcal{P}$ is a tuple $(V, \mathcal{L}, \ell_0, \mathcal{T}, \mathcal{L}_{\mathcal{E}})$, where V is a set of variables, $\mathcal{L}$ a set of control locations, $\ell_0 \in (\mathcal{L} \setminus \mathcal{L}_{\mathcal{E}})$ a designated entry point, $\mathcal{T}$ a set of transitions, and $\mathcal{L}_{\mathcal{E}} \subset \mathcal{L}$ a set of exit locations. A *program state* is a valuation of all of the variables in V. The set of all states is denoted by Σ. Each *transition* $\tau \in \mathcal{T}$ is a triple (ℓ_1, ρ, ℓ_2), where $\ell_1, \ell_2 \in \mathcal{L}$ and $\rho \subseteq \Sigma \times \Sigma$ is a non-empty relation on program states. By convention, the entry location ℓ_0 and all exit locations in $\mathcal{L}_{\mathcal{E}}$ have no incoming and outgoing transitions, respectively. The *control flow graph* (CFG) of $\mathcal{P}$, $CFG(\mathcal{P})$, is the graph $(\mathcal{L}, E)$, where $E = \{(\ell_1, \ell_2) \mid \exists \rho \cdot (\ell_1, \rho, \ell_2) \in \mathcal{T}\}$.

A *trace* in $\mathcal{P}$ is a finite sequence $\langle \ell_1, s_1 \rangle, \dots, \langle \ell_n, s_n \rangle$ of location-state pairs such that $\forall 1 \leq i \leq (n-1) \cdot \exists \rho \cdot (\ell_i, \rho, \ell_{i+1}) \in \mathcal{T} \wedge (s_i, s_{i+1}) \models \rho$. A *computation*[1] is trace such that $\ell_1 = \ell_0$. A state s is *reachable* at location ℓ iff there exists a computation such that $\ell_n = \ell \wedge s_n = s$; a location ℓ is reachable iff there exists a state s reachable at ℓ; an *invariant* of $\mathcal{P}$ at ℓ is any superset of the states reachable at ℓ.

A predicate is any ground formula. A *cube* over a set of predicates P is a formula of the form $p_1 \wedge \cdots \wedge p_n \wedge \neg q_1 \wedge \cdots \wedge \neg q_m$, where $p_i, q_j \in P$ and every predicate appears at most once. A *minterm* is a cube of size $|P|$.

Let ψ be a quantifier-free first-order expression. A fundamental operation of predicate abstraction is to compute $\mathcal{G}_P(\psi)$ – a strongest Boolean combination of P that is implied by ψ. $\mathcal{G}_P(\psi)$ can be characterized as the set of all minterms that do not contradict ψ:

$$\mathcal{G}_P(\psi) = \bigvee \{c \mid c \text{ is a minterm over } P \text{ and } c \wedge \psi \text{ is satisfiable}\}.$$

$\mathcal{G}_P(\psi)$ can be computed by enumerating all minterms and using a decision procedure to decide satisfiability. Alternatively, the computation can be reduced to quantifier elimination as follows. With each $p \in P$ associate a unique Boolean variable b_p; let V be the set of all free variables in ψ and P; and let F_P be the formula $\psi \wedge (\bigwedge_{p \in P} b_p \Leftrightarrow p)$. Then, $\mathcal{G}_P(\psi)$ is given by the result of eliminating all existential quantifiers in $\exists V \cdot F_P$, and then replacing every b_p with the predicate p.

Let $\mathcal{P} = (V, \mathcal{L}, \ell_0, \mathcal{T}, \mathcal{L}_{\mathcal{E}})$ be a program, and μ a *predicate map* that assigns to each location ℓ a set of predicates denoted $\mu.\ell$. The *(most precise) predicate abstraction* of $\mathcal{P}$ with respect to μ is a program $\mathcal{P}_\mu = (V, \mathcal{L}, \ell_0, \mathcal{T}_\mu, \mathcal{L}_{\mathcal{E}})$, where

$$\mathcal{T}_\mu = \{(\ell_1, \mathcal{G}_P(\rho), \ell_2) \mid (\ell_1, \rho, \ell_2) \in \mathcal{T} \text{ and } P = \mu.\ell_1 \cup \mu.\ell_2'\}.$$

Note that if μ is finite for every program location, then $\mathcal{P}_\mu$ is finite as well.

3   Program Summary

Large-Block Encoding applies predicate abstraction to a *summary* of a program. The original definition of LBE [2] uses a specific notion of summary, which we call *rule summary*. In this section, we present a more general concept of summaries. In particular, we define a *loop cutset summary* as the most general summary that summarizes all loop-free program fragments. Cutset summaries subsume useful classes of summaries, including (as we show later) the rule summary.

Let $\mathcal{P} = (V, \mathcal{L}, \ell_0, \mathcal{T}, \mathcal{L}_{\mathcal{E}})$ be a program; let $\mathcal{L}' \subseteq \mathcal{L}$ such that $\ell_1, \ell_n \in \mathcal{L}'$. A trace $\langle \ell_1, s_1 \rangle, \dots, \langle \ell_n, s_n \rangle$ of $\mathcal{P}$ is $\mathcal{L}'$-*free* iff $\mathcal{L}' \cap \{\ell_2, \dots, \ell_{n-1}\} = \emptyset$. The $\mathcal{L}'$-free (ℓ_1, ℓ_n) fragment of $\mathcal{P}$ comprises locations appearing on $\mathcal{L}'$-free (ℓ_1, ℓ_n) traces of $\mathcal{P}$, with ℓ_1 and ℓ_n as entry and exit locations, respectively.

---

[1] In this paper, we only consider finite computations.

Definition 1 (Summary). *A program* $\mathcal{P}' = (V, \mathcal{L}', \ell_0, \mathcal{T}', \mathcal{L}_{\mathcal{E}})$ *is a summary of a program* $\mathcal{P} = (V, \mathcal{L}, \ell_0, \mathcal{T}, \mathcal{L}_{\mathcal{E}})$ *iff: (i)* $\mathcal{L}' \subseteq \mathcal{L}$, *and (ii)* $\forall \ell_1, \ell_n \in \mathcal{L}'$ *there exists a* $\mathcal{L}'$-free (ℓ_1, ℓ_n) *trace* $\langle \ell_1, s_1 \rangle, \ldots, \langle \ell_n, s_n \rangle$ *of* $\mathcal{P}$ *iff* $\exists \rho. (\ell_1, \rho, \ell_n) \in \mathcal{T}' \wedge (s_1, s_n) \models \rho$.

A program and its summary share the same variables, entry and exit locations, and state space Σ. A summary also preserves reachability of locations, as stated by Theorem 1.

Theorem 1. *Let* $\mathcal{P}' = (V, \mathcal{L}', \ell_0, \mathcal{T}', \mathcal{L}_{\mathcal{E}})$ *be a summary of* $\mathcal{P} = (V, \mathcal{L}, \ell_0, \mathcal{T}, \mathcal{L}_{\mathcal{E}})$. *Then,* $\forall \ell \in \mathcal{L}'$, $s \in \Sigma$ *is reachable at* ℓ *in* $\mathcal{P}$ *iff* s *is reachable at* ℓ *in* $\mathcal{P}'$.

As a corollary, since an invariant is a set of states, a summary also preserves invariants: I is an invariant of $\mathcal{P}'$ at $\ell \in \mathcal{L}'$ iff it is an invariant of $\mathcal{P}$ at ℓ. Thus, any program summary can be used for LBE. Ideally, we want the smallest summary possible since it leads to smaller abstract models. In particular, we'd like a unique minimal summary – when $\mathcal{L}' = \{\ell_0\} \cup \mathcal{L}_{\mathcal{E}}$. Unfortunately, it is not computable since its computation requires summarizing program loops. Instead, we want the smallest summary that summarizes only loop-free program fragments.

Let $G = (V, E)$ be a graph. A set $S \subseteq V$ is a *cycle cutset* (or simply a cutset) of G iff S contains a vertex from every cycle in G, i.e., the graph $(V \setminus S, E \setminus ((S \times V) \cup (V \times S)))$ is acyclic. We call an element $s \in S$ a *cutpoint*.

Definition 2 (Loop Cutset Summary). *A program* $\mathcal{P}' = (V, \mathcal{L}', \ell_0, \mathcal{T}', \mathcal{L}_{\mathcal{E}})$ *is a cutset summary of* $\mathcal{P}$ *iff* $\mathcal{P}'$ *is a summary of* $\mathcal{P}$ *and* $\mathcal{L}'$ *is a cutset of* $CFG(\mathcal{P})$.

The cutset summary of a program is not unique. Finding a minimal one is hard since it requires solving the *minimal feedback vertex set*, which is known to be NP-complete [12]. However, in practice, a good approximation is obtained in polynomial time by letting $\mathcal{L}'$ be the set of destinations of all back-edges discovered by a DFS of $CFG(\mathcal{P})$, together with ℓ_0 and $\mathcal{L}_{\mathcal{E}}$. Given a cutset of $CFG(\mathcal{P})$, the corresponding cutset summary of $\mathcal{P}$ is effectively computable since, by definition, each edge in it corresponds to a loop-free fragment of $\mathcal{P}$.

In the rest of this section, we compare cutset summaries with rule summaries [2]. A rule summary is based on two program transformations, SEQ and CHOICE. Let $\mathcal{P} = (V, \mathcal{L}, \ell_0, \mathcal{T}, \mathcal{L}_{\mathcal{E}})$, and $\ell_1, \ell_2 \in \mathcal{L}$ be two locations. The preconditions of SEQ$(\mathcal{P}, \ell_1, \ell_2)$ are (a) $\ell_1 \neq \ell_2$, (b) there is an edge from ℓ_1 to ℓ_2, (c) ℓ_2 has no other incoming edges, and (d) ℓ_2 has at least one successor. The output is the program $\mathcal{P}' = (V, \mathcal{L}', \ell_0, \mathcal{T}', \mathcal{L}_{\mathcal{E}})$, where $\mathcal{L}' = \mathcal{L} \setminus \{\ell_2\}$ and

$$\mathcal{T}' = (\mathcal{T} \cup \{(\ell_1, \rho \circ \rho_i, \ell_i) \mid (\ell_2, \rho_i, \ell_i) \in \text{out}(\ell_2)\}) \setminus (\text{out}(\ell_2) \cup \text{in}(\ell_2)) ,$$

where out(ℓ) and in(ℓ) are the sets of all outgoing and incoming transitions of ℓ, respectively. The precondition of CHOICE$(\mathcal{P}, \ell_1, \ell_2)$ is that there are two distinct edges (ℓ_1, ρ_1, ℓ_2) and (ℓ_1, ρ_2, ℓ_2) in $\mathcal{T}$. The output is $\mathcal{P}' = (V, \mathcal{L}, \ell_0, \mathcal{T}', \mathcal{L}_{\mathcal{E}})$, where

$$\mathcal{T}' = (\mathcal{T} \setminus \{(\ell_1, \rho_1, \ell_2), (\ell_1, \rho_2, \ell_2)\}) \cup \{(\ell_1, \rho_1 \cup \rho_2, \ell_2)\} .$$

Intuitively, SEQ removes a location with a single incoming edge, and CHOICE replaces multiple edges between the same locations with a single one.

Definition 3 (Rule Summary). *A* rule summary *of a program* $\mathcal{P} = (V, \mathcal{L}, \ell_0, \mathcal{T}, \mathcal{L}_{\mathcal{E}})$, *is a limit of the sequence* $\mathcal{P}_0, \mathcal{P}_1, \ldots$, *where* $\mathcal{P}_0 = \mathcal{P}$, *and* $\mathcal{P}_{i+1}$ *is* SEQ$(\mathcal{P}_i, \ell_1, \ell_2)$ *if* SEQ *is applicable,* CHOICE$(\mathcal{P}_i, \ell_1, \ell_2)$ *if* CHOICE *is applicable, and* $\mathcal{P}_i$ *otherwise.*

The advantage of a cutset summary is that it is not restricted to a particular cutset computation procedure. For example, suppose we construct a cutset of $\mathcal{P}$ by taking the destinations of all back-edges in a topological ordering of $CFG(\mathcal{P})$. Let us call this *back-edge* summary and compare it to rule summary. In both cases, the complexity of constructing a summary is polynomial in the size of $\mathcal{P}$. However, the locations of a rule summary of $\mathcal{P}$ subsumes those of a back-edge summary of $\mathcal{P}$. This is because: (i) the destination of a back-edge always has at least two incoming edges, and hence can never be removed by SEQ, and (ii) CHOICE never eliminates locations. Thus, back-edge summary is never larger than a rule summary. Thus, a rule summary is a custset summary as well.

4   Predicate Abstraction of Program Fragments

A cutset C of a program $\mathcal{P}$ has a BACK-EDGE-AT-END property if for every $\ell_1, \ell_2 \in C$, ℓ_2 is the sole destination of all back-edges in the C-free fragment $\mathcal{P}_{\ell_1,\ell_2}$. Note that a cutset C of any back-edge (or rule) summary satisfies this since in any C-free fragment $\mathcal{P}_{\ell_1,\ell_2}$, $\{\ell_1, \ell_2\}$ are the only possible destinations of back-edges, but ℓ_1 has no incoming edges at all. Let $\mathcal{P}$ be a program and C its BACK-EDGE-AT-END cutset. In this section, we show how to compute a predicate abstraction of a cutset summary of $\mathcal{P}$ (w.r.t. C), without explicitly constructing the summary.

   Our algorithm is called SUMMARYPA, and is shown in Fig. 1. We assume that $\mathcal{P}$ is given in Static Single Assignment (SSA), and work directly on its syntax. Function EDGEQUERY takes $\mathcal{P}$ and two locations $\ell_1, \ell_2 \in C$ and generates a PAQ for the C-free (ℓ_1, ℓ_2) fragment, $\mathcal{P}_{\ell_1,\ell_2}$, of $\mathcal{P}$. The size of the PAQ is linear in the size of $\mathcal{P}_{\ell_1,\ell_2}$. Function SOLVE takes this query and returns a predicate abstraction of $\mathcal{P}_{\ell_1,\ell_2}$. These two functions are applied to every pair of connected cutpoints from C. The output of SUMMARYPA is the predicate abstraction of the summary of $\mathcal{P}$ w.r.t. C. In the rest of this section, we give a brief overview of SSA, describe the formula constructed by EDGEQUERY, and give two strategies for SOLVE: one based on an AllSAT SMT-solver, and one based on LDDs.

4.1   Single Static Assignment

We give here a brief overview of SSA. More details can be found elsewhere [8]. A program is in SSA form if an assignment to each variable appears at most once in its syntax. Any program can be put efficiently into SSA. As an example, an SSA program corresponding to the C program in Fig. 2(a) is shown in Fig. 2(b).

   In addition to normal assignments, SSA uses special ϕ-*assignments*. Their syntax is x := PHI$(v_1 : \ell_1, \ldots, v_n : \ell_n)$, where x is a variable, $\ell_1, \ldots, \ell_n$ are locations, and $v_1, \ldots, v_n$ are values. The PHI-function evaluates to value v_i if it

```
1: Input: SSA program P = (V, L, ℓ₀, T, Lε); a cutset C of P; a predicate map μ
2: Output: Pμ the most precise predicate abstraction of P w.r.t. μ
3: function SUMMARYPA (P, C, μ)
4:     Tμ = ∅
5:     for all ℓ₁, ℓ₂ ∈ C s.t. ∃ a C-free (ℓ₁, ℓ₂)-path in CFG of P do
6:         Q = EDGEQUERY(P, C, ℓ₁, ℓ₂, μ.ℓ₁, μ.ℓ₂)
7:         ρ = SOLVE(Q);    Tμ = Tμ ∪ {(ℓ₁, ρ, ℓ₂)}
8:     Pμ = (V, C, ℓ₀, Tμ, Lε)
```

Fig. 1. Algorithm SUMMARYPA

is reached via location ℓ_i. In our example, PHI(0:0, x_0:4) on line 2 evaluates to 0 when reached from location 0 and to x_0 when reached from location 4.

We model an SSA program as a tuple $(V, \mathcal{L}, E, \phi, G, Act, \ell_0, \mathcal{L}_\mathcal{E})$ where: V, $\mathcal{L}$, ℓ_0, and $\mathcal{L}_\mathcal{E}$ are same as in programs, $E \subseteq \mathcal{L} \times \mathcal{L}$ is the set of control flow edges, Act maps locations to assignments, and ϕ and G map edges to ϕ-assignments and guards, respectively. Intuitively, each $\ell \in \mathcal{L}$ corresponds to a basic block; each basic block is a sequence of assignments terminated by a branch; the branch condition is stored on the edge, and each ϕ-assignment is replaced by the corresponding assignments on the edges. This is a variant of the traditional compiler SSA format, where ϕ-assignments and guards are pushed into the source and destination blocks of their edges, respectively. Fig. 2(c) graphically shows the SSA program from Fig. 2(b).

Operationally, an edge (ℓ_1, ℓ_2) in an SSA program is executed by: (a) executing the assignments $Act(\ell_1)$, (b) validating the guard $G(\ell_1, \ell_2)$, and (c) executing ϕ-assignments $\phi(\ell_1, \ell_2)$. Formally, for a set of assignments A, let $\alpha(A)$ be $\bigwedge_{v:=e \in A} v = e$, and $\alpha'(A)$ be $\bigwedge_{v:=e \in A} v' = e$. The semantics of an SSA program $\mathcal{P} = (V, \mathcal{L}, E, \phi, G, Act, \ell_0, \mathcal{L}_\mathcal{E})$ is a program $\mathcal{P}' = (V, \mathcal{L}, \ell_0, T', \mathcal{L}_\mathcal{E})$, s.t. $(\ell_1, \rho, \ell_2) \in T'$ iff $\rho = \alpha'(Act(\ell_1)) \wedge Skip(K) \wedge G(\ell_1, \ell_2)' \wedge \alpha(\phi(\ell_1, \ell_2))'$, where $K = \{v \in V \mid \neg\exists e \bullet (v := e) \in A \vee (v := e) \in \phi(\ell_1, \ell_2)\}$ and $Skip(U)$ is $\bigwedge_{u \in U} u' = u$. For example, the semantics of SSA program in Fig. 2(c) is shown in Fig. 2(d). The semantics of the edge (3, 2) is $y_0' = y+1 \wedge y' = y_0' \wedge x' = x \wedge x_0' = x_0$. Note that this definition depends on several properties of the SSA: assignments in a block have no circular dependencies, guards do not depend on following ϕ-assignments, etc.

4.2   Generating Predicate Abstraction Queries

EDGEQUERY$(\mathcal{P}, C, \ell_1, \ell_2, P_1, P_2)$ takes an SSA program $\mathcal{P}$, a cutset C, locations ℓ_1, and ℓ_2 in C, and two sets of predicates P_1 and P_2, and generates a PAQ for C-free (ℓ_1, ℓ_2)-fragment, $\mathcal{P}_{\ell_1, \ell_2}$, of $\mathcal{P}$.

The result of EDGEQUERY is similar to a typical "reachability query" , e.g., as in CBMC [7]. It is linear in the size of $\mathcal{P}_{\ell_1, \ell_2}$ and computable in linear time. However, EDGEQUERY works directly on SSA (as opposed to a more expensive Gated SSA used in [7]). The resulting PAQ separates control- and data-flows,

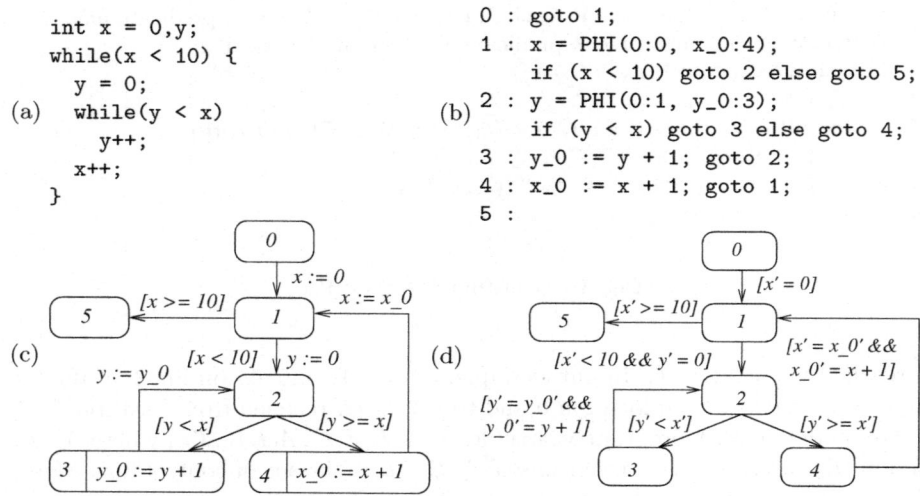

```
int x = 0,y;
while(x < 10) {
    y = 0;
(a)  while(y < x)
        y++;
    x++;
}
```

```
0 : goto 1;
1 : x = PHI(0:0, x_0:4);
    if (x < 10) goto 2 else goto 5;
(b) 2 : y = PHI(0:1, y_0:3);
    if (y < x) goto 3 else goto 4;
3 : y_0 := y + 1; goto 2;
4 : x_0 := x + 1; goto 1;
5 :
```

Fig. 2. Representation of a C program: (a) traditional, (b) SSA, (c) graphical SSA, (d) semantic. In (d), all expressions of the form $v' = v$ have been omitted.

and preservers control-flow structure in the query. These features are crucial for our approach to discharging PAQs (see *LDD-based approach* in Section 4.3).

For ease of understanding, we present the query in parts. Let $\mathcal{L}_f$ denote the set of all locations of $\mathcal{P}_{\ell_1,\ell_2}$. Let A be a formula for all of the simple assignments in the fragment, $A = \bigwedge_{\ell \in \mathcal{L}_f \setminus \{\ell_2\}} \alpha(Act(\ell))$, where α is as defined in Sec. 4.1. Intuitively, a complete satisfying assignment to A corresponds to executing, in parallel, all assignments of all of the locations in $\mathcal{L}_f$. A is always satisfiable because, by assumption, $\mathcal{P}_{\ell_1,\ell_2}$ has no back-edges (except possibly to ℓ_2), and hence no circularly dependent assignments.

For each $\ell \in \mathcal{L}$, let B_ℓ be a Boolean variable corresponding to ℓ, and V_ℓ the set of all such variables. Let R_ℓ be a formula defined for a location ℓ as follows:

$$R_\ell = \left(B_\ell \Rightarrow \bigvee_{\ell' \in Preds(\ell) \cap \mathcal{L}_f} B_{\ell'} \wedge G(\ell', \ell) \wedge \alpha(\phi(\ell', \ell)) \right),$$

where $Preds(\ell)$ is the set of all CFG-predecessors of ℓ. Intuitively, B_ℓ represents whether ℓ is visited in an execution, i.e., is reachable. R_ℓ states that if ℓ is reachable then at least one (but possibly more) of its predecessors ℓ' must be reachable, and the guards and the ϕ-assignments on the (ℓ', ℓ)-edge must be true.

For the final location ℓ_2, we need a variant of R_ℓ, denoted $\hat{R}_\ell$ and defined as:

$$\hat{R}_\ell = \left(B_\ell \Rightarrow \bigvee_{\ell' \in Preds(\ell) \cap \mathcal{L}_f} B_{\ell'} \wedge G(\ell', \ell) \wedge \alpha'(\phi(\ell', \ell)) \right),$$

where α' is as defined in Sec. 4.1. Since ℓ_2 can be the destination of a back-edge, the ϕ-assignment on that edge might be circularly dependent on another assignment in $\mathcal{P}_{\ell_1,\ell_2}$. Such dependencies are eliminated by using α' instead of α.

Next, we define a formula CFG as follows:

$$CFG = \left(B_{\ell_2} \wedge \hat{R}_{\ell_2} \wedge \bigwedge_{\ell \in \mathcal{L}_f \setminus \{\ell_1,\ell_2\}} R_\ell \right).$$

Every satisfying assignment to CFG corresponds to one (or several) paths of $\mathcal{P}_{\ell_1,\ell_2}$. B_{ℓ_2} guarantees that ℓ_2 is visited, the implications in R_ℓ create the path, and the guard and ϕ-assignment constraints ensure that the path is feasible (i.e., can always be elaborated into a concrete computation).

Consider the formula $A \wedge CFG$. Each satisfying assignment to it corresponds to at least one concrete execution from ℓ_1 to ℓ_2. Furthermore, note that any assignment that corresponds to multiple non-contradicting executions can be transformed into a satisfying assignment for a single execution. This is done by picking one of the corresponding executions, setting B_ℓ to true for every location ℓ on that execution, and setting all other B_ℓ variables to false.

Next, we need formulas for predicates. With each predicate $p \in P_1$ we associate a Boolean variable b_p, and with each predicate $p \in P_2$ a Boolean variable b'_p. Let Src and Dst be formulas defined as:

$$Src = \left(\bigwedge_{p \in P_1} b_p \Leftrightarrow p \right) \qquad Dst = \left(\bigwedge_{p \in P_2} b'_p \Leftrightarrow \Phi(p) \right),$$

where $\Phi(p) = p[v/v' \mid \exists \ell \in (Preds(\ell_2) \cap \mathcal{L}_f) . v \in LHS(\phi(\ell, \ell_2))]$. Note that this renaming in Dst corresponds to the renaming in $\hat{R}_\ell$.

Finally, the PAQ produced by EDGEQUERY is

$$\exists V, V', V_\ell . \ A \wedge CFG \wedge Src \wedge Dst .$$

This formula is linear in $|\mathcal{L}_f|$ and can be computed in linear time. Theorem 2 asserts the correctness of EDGEQUERY.

Theorem 2. Let $\rho \subseteq \Sigma \times \Sigma$ be the summary of $\mathcal{P}_{\ell_1,\ell_2}$. Then, EDGEQUERY$(\mathcal{P}, C, \ell_1, \ell_2, P_1, P_2)$ is equivalent to:

$$\exists V, V' . \rho \wedge \left(\bigwedge_{p \in P_1} b_p \Leftrightarrow p \right) \wedge \left(\bigwedge_{p \in P_2} b'_p \Leftrightarrow p' \right)$$

Example 1. Let $\mathcal{P}$ be the SSA program from Fig. 2(c) and $C = \{0,1,2,5\}$ its loop cutset. Consider the C-free $(2,2)$ fragment of $\mathcal{P}$ and predicates $y < 0$, $x < 0$.

$$\mathcal{L}_f = \{2,3\} \qquad\qquad A = (y_0 = y + 1)$$
$$Src = (b_y \Leftrightarrow y < 0) \wedge (b_x \Leftrightarrow x < 0) \qquad \hat{R}_2 = (B_2 \Rightarrow B_3 \wedge y' = y_0)$$
$$Dst = (b'_y \Leftrightarrow y' < 0) \wedge (b'_x \Leftrightarrow x < 0) \qquad R_3 = (B_3 \Rightarrow B_2 \wedge y < x)$$

The overall predicate abstraction query is:

$$\exists y_0, y, y', x, B_2, B_3 \cdot (y_0 = y + 1) \wedge (B_3 \Rightarrow B_2 \wedge y < x) \wedge (B_2 \Rightarrow B_3 \wedge y' = y_0) \wedge$$
$$B_2 \wedge (b_y \Leftrightarrow y < 0) \wedge (b_y' \Leftrightarrow y' < 0) \wedge (b_x \Leftrightarrow x < 0) \wedge (b_x' \Leftrightarrow x < 0) \,.$$

4.3   Solving Predicate Abstraction Queries

SOLVE takes a PAQ of the form $\exists V, V', V_\ell \cdot \Psi$, eliminates the quantifiers, and then replaces the Boolean variables introduced by EDGEQUERY by the corresponding predicates. In this section, we describe two strategies for the quantifier elimination step: *AllSAT-based* – based on the approach of Lahiri et al. [16], and *LDD-based* – based on a recently developed decision diagrams LDDs [5].

AllSAT-based approach. An SMT solver decides satisfiability of a quantifier-free first-order formula F (over a theory T). An AllSAT SMT solver takes a formula F and a subset V_{Imp} of *important* Boolean terms of F and returns the set M of all minterms over V_{Imp} that can be extended to a satisfying assignment to F.

A PAQ of the form $\exists V, V', V_\ell \cdot \Psi$ is solved by giving an AllSAT solver a formula Ψ and setting V_{Imp} to the set of all Boolean variables b_p and b_p' in Ψ. The output is a set M of minterms such that $\bigvee M$ is equivalent to $\exists V, V', V_\ell \cdot \Psi$.

The key advantage of this approach is that all of the reasoning is delegated to an AllSAT solver. Thus, it applies to queries in any SMT-supported theory and leverages advancements in SMT-solvers. The main limitation – it enumerates all minterms of the solution, which can be exponentially larger than the smallest DNF representation. We illustrate this limitation further in Sec. 5.

LDD-based approach. An LDD is a Binary Decision Diagram (BDD) with nodes labeled with atomic terms from Linear Arithmetic (LA). An LDD represents a LA formula in the same way a BDD represents a Boolean formula. LDDs support the usual Boolean operations (conjunction, disjunction, negation, etc.), Boolean quantification, and variable reordering. Additionally, they provide quantification over numeric variables via direct Fourier-Motzkin elimination on the diagram.

Let ℓ_1 and ℓ_k be two cutpoints. Recall that the query Q computed by EDGEQUERY for (ℓ_1, ℓ_k) is of the form $\exists V, V', V_\ell \cdot \Psi$, where

$$\Psi = A \wedge R_{\ell_2} \wedge \cdots \wedge \hat{R}_{\ell_k} \wedge B_{\ell_k} \wedge Src \wedge Dst \,,$$

and each R_{ℓ_i} is of the form $B_{\ell_i} \Rightarrow \theta_i$. A naïve way to solve Q is to first compute and LDD for Ψ, and then use numeric and Boolean quantification to eliminate the variables in V, V', and V_ℓ. Note that the result is a BDD (since all of the remaining variables are Boolean).

Unfortunately, the naïve approach does not scale. The bottleneck is constructing a diagram for Ψ. In large part, this is due to the variables B_ℓ used in encoding control-flow constraints. The solution is to re-arrange the query to eliminate these variables as early as possible.

Let $\mathcal{L}_f = \langle \ell_1, \ldots, \ell_k \rangle$ be the set of locations in the cutpoint-free – and hence, loop-free – (ℓ_1, ℓ_k) fragment of $\mathcal{P}$. Let *Topo* be a topological order on $\mathcal{L}_f$. Without

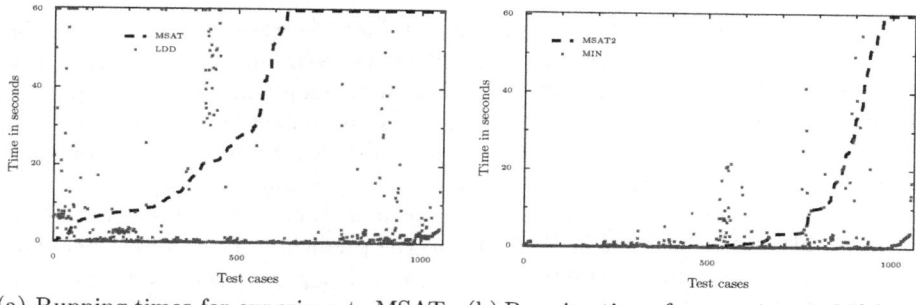

(a) Running times for experiments MSAT   (b) Running times for experiments MSAT2
and LDD   and MIN

Fig. 3. Running times

loss of generality, assume that the locations are numbered such that $i \leq j$ iff ℓ_i precedes ℓ_j in some fixed linearization of *Topo*. Then, a variable B_{ℓ_i} appears in a constraint R_{ℓ_j} iff $i \leq j$. Therefore, Q is equivalent to $\exists V, V' . \Psi'$, where

$$\Psi' = A \wedge Src \wedge Dst \wedge \left(\exists B_{\ell_1} . \exists B_{\ell_2} . R_{\ell_2} \wedge \cdots \wedge \exists B_{\ell_k} . \hat{R}_{\ell_k} \wedge B_{\ell_k} \right) .$$

In summary, our overall solution is to compute an LDD for Ψ', and then use numeric quantification to eliminate V and V' variables. Note that it is possible to apply early quantification to the numeric variables as well. However, we did not explore this direction.

The main advantage of our approach is that the solution is computed directly as an LDD. Thus, its running time is independent of the number of minterms in the solution. Unlike the AllSAT-based approach, it is limited to Linear Arithmetic and does not directly benefit from advances in SMT-solving. However, in our experiments, it significantly outperformed the AllSAT-based approach.

We are not the first to use decision diagrams for predicate abstraction. However, previous approaches use BDDs by reducing numeric reasoning to propositional reasoning. This reduction introduces a large number of Boolean variables, which makes the problem hard for BDDs. For example, Lahiri et al. [15] find a SAT-based approach superior to a BDD-based one. In contrast, we use decision diagrams that are aware of Linear Arithmetic. This avoids the need for additional constraints, and makes the solution very competitive.

5   Experimental Results

We evaluated our approach on a large benchmark of PAQs and as a part of a software model checker. We used the MATHSAT4 SMT-solver [3] for the AllSAT-based solution, and our implementation of LDDs [5] for the LDD-based solution. All PAQs were restricted to two-variables-per-inequality logic (TVPI), i.e., linear constraints with at most two variables. The benchmark and our tools are available at `lindd.sf.net`.

The benchmark. To evaluate our approach on large queries, we constructed the benchmark from C programs using the following technique: (1) convert a program into LLVM bitcode [17] and optimize with loop unrolling and inlining; (2) for each function, use all loop headers as the cutset summary; (3) over-approximate the semantics of statements by TVPI constraints (e.g., loads from memory and function calls are replaced by non-determinism); (4) for each location ℓ, take the atomic formulas that appear in the weakest precondition of some conditional branch reachable from ℓ as the predicates at ℓ; (5) for each pair of locations ℓ and ℓ' in the summary, generate a PAQ, as described in Sec. 4.2, using the predicates at ℓ and ℓ'.

In our view, the benchmark is quite realistic: steps 1-3 are a common pre-processing techniques in program analysis; the choice of predicates is guided by our experience with predicate abstraction.

The benchmark consists of over 20K PAQs. We report the results on the top 1061 cases (exactly the ones that required $\geq$ 5s to solve with at least one approach). These PAQs are from bash, bison, ffmpeg, gdb, gmp, httpd, imagemagick, mplayer, and tar. As formulae in SMT-LIB format, they range in size from 280B to 57KB (avg. 11KB, med. 8KB). The number of predicates per query ranges from 10 to 56 (avg. 22, med. 19). Each experiment was limited to 60s CPU and 512MB of RAM, and was done on a 3.4GHz Pentium D with 2GB of RAM.

The experiments. The results of the experiments are summarized in the first three rows of Table 4a. The first column indicates the experiment as follows – *MSAT*: queries are solved using MATHSAT4; *LDD*: queries are solved using LDDs with dynamic variable order (DVO); and *LDD2*: queries are solved using LDDs with static variable order (SVO). For LDD, diagram nodes were reordered by the diagram manager based on memory utilization. For LDD2, a static order was selected such that terms that appeared earlier in the *query AST* would appear earlier in the diagram order. A query AST is $((A \wedge CFG) \wedge Src \wedge Dst)$.

For each experiment, we report the total time to solve all 1061 queries (*Total*), number of unsolved cases (*Failed*), average time per a solved instance (*Avg. per Solved*), total time for all solved instances (*Total Solved*), total time for all instances solved by MATHSAT4 (*Total MSAT Solved*), and total time for all instances solved by MATHSAT4 with predicates in each query restricted to those that appear in the support of the solution computed by LDD (*Total MSAT2 Solved*). All "Total" times include 60s for each failure.

Surprisingly, the AllSAT-based approach is the worst. It was only able to solve 60% of queries and was 7 times slower compared to the LDD-based solutions. Even restricted to queries that it could solve, it is almost 4 times slower than LDD, and 9 times slower than LDD2. Fig. 3a shows a detailed comparison between the MSAT and LDD experiments. In the chart, test case indices are on the x-axis (sorted first by MSAT time, and then by LDD time), time per test case is on the y-axis. There are several exceptional cases where MATHSAT4 significantly outperforms LDD. However, overall, most test-cases appear to be easy for LDD (solved in under 5s), but are much more evenly distributed for MATHSAT4.

Name	Total (min)	Failed	Avg. per Solved (sec)	Total Solved (min)	Total MSAT Solved (min)	Total MSAT2 Solved (min)
MSAT	610.00	429	17.12	180.29	180.29	523.86
LDD	83.54	35	2.84	48.48	60.64	72.15
LDD2	83.98	64	1.19	19.81	44.34	72.79
MIN	**28.40**	6	1.27	22.39	10.64	**19.98**
LDD3	85.66	74	**0.70**	**11.51**	57.64	77.08
MSAT2	188.14	91	6.87	102.00	**9.04**	102.00

(a) PAQ benchmark

Name	LBE			It	Pr	CP	SBE			
	T						T	It	Pr	BB
	LDD	MSAT	MIN							
floppy.ok	0.18	0.16	0.16	1	0	3	0.44	4	6	83
tst_lck_50	0.5	0.48	0.5	1	0	3	++	++	++	255
diamond-4	2.0	++	1.7s	4	42	4	++	++	++	24
ssl-srv-D	98.96	6.26	5.65	5	60	4	++	++	++	155

(b) End-to-end. T = times in sec; It = # of CE-GAR iterations; CP = # of cutpoints; BB = # of blocks; Pr = total # of preds.

Fig. 4. Summary of experimental results

The two LDD-based experiments clearly highlight the virtues and vices of DVO: DVO makes an LDD-approach more robust (35 failures for LDD v.s. 64 for LDD2) but less efficient (about twice as slow on average). Out of 64 failures for LDD2, 39 where due to memory running out. Coincidentally, with our choice of using 60s for each failure, faster running times balance out more failures for LDD2, and its overall time is very similar to that of LDD.

In our benchmark, LDD-based solution significantly outperforms the AllSAT-based one. We conjecture that the two are complementary: AllSAT-based solution performs well when number of models to enumerate is small, and LDD-based solution performs well when the intermediate (and final) diagrams are small. To validate this conjecture, we computed the best-of time needed to solve a test-case by either of the three techniques. This is equivalent to running the three approaches in parallel and stopping as soon as one was successful. The results are summarized in the fourth row (MIN) of Table 4a. The combination is extremely effective: taking only 28 minutes (3 times better than previous best) for the benchmark and solving all but 6 instances. The improvement is even more significant when restricted to instances that MATHSAT4 could solve.

Oracle experiments. To put our results into perspective, we conducted two experiments against "oracle" solvers. Finding good variable ordering is the bottleneck for LDD-based solution. With DVO most time is spend reordering, but without it many cases run out of memory. We experimented with using the last ordering found by DVO during LDD experiment as a static ordering. The results are shown in the fifth row (LDD3) of Table 4a. We classify this experiment as "oracle" since we don't know how to achieve this variable order other than by repeating the LDD experiment.

Interestingly, the order did not help as much as we expected. The average time per solved test-case did drop to 0.7s (2× and 4× better than LDD2 and LDD, respectively). However, fewer instances could be solved, with 55 out of the 74 failures being memory outs. We believe this indicates that an order that is good for the final result is not necessarily good for the intermediate steps (and hence, overall) of the computation.

The bottleneck for the AllSAT-based solution is in enumerating all the minterms (as opposed to cubes or prime implicants). We found that in many cases that were hard for MATHSAT4, many of the predicates did not appear in the support of the LDD-based solution. That is, many predicates were not part of any prime implicant. To evaluate the effect of this on MATHSAT4, we repeated the MSAT experiment, but restricted the predicates in each query to those that appeared in the support of the solution (as computed by LDD). The results are shown in the last row (MSAT2) of Table 4a. Note that determining variables in the support of a Boolean formula is NP-complete. We do not know how to compute the support other than by solving the problem with LDDs first.

Overall, the running time has improved dramatically. There is a significant improvement on the cases solved by MSAT, even compared to LDD-based solutions. However, overall it is much slower than any of the LDD-based solutions, even when restricted to cases it could solve. Overall, there are 91 failures (all timeouts). Fig. 3b shows the details from the MSAT2 and MIN experiments. There are two interesting points. First, the best-of LDD and MSAT is significantly better than the idealized AllSAT-based solution. Second, there are cases where the idealized AllSAT-based solution is an order-of-magnitude better.

End-to-end experiments. To evaluate the end-to-end performance of our approach, we implemented a CEGAR-based safety checker for C programs following Jhala et al. [11]. Fig. 4b is a sample of our results: `floppy.ok` is derived from a device driver, `test_locks_50` is based on the example from Beyer et al. [2], `diamond-4` is a program with a "diamond-shaped" CFG, and `ssl-srv-D` is derived from OpenSSL. We observe that LBE scales much better than SBE. The performances of LDD and AllSAT are more evenly balanced. LDD scales better for diamond-4. For ssl-srvr-D, LDD by itself is much worse than AllSAT. This is due to a single PAQ that is very hard for LDD. However, LDD outperforms AllSAT elsewhere, as seen by the MIN column.

Summary. Overall, our results show that the AllSAT-based solution is not competitive for solving PAQs of a large program fragment, while the LDD-based solution performs surprisingly well. Moreover, the MIN of the LDD- and the AllSAT-based approaches is the clear winner, even compared to an oracle-based solution.

6   Conclusion

Large-Block Encoding (LBE) [2] is a flavor of predicate abstraction applied to a summarized program. In this paper, we present solutions to three problems for predicate abstraction in the context of LBE. First, we define a general notion of program summarization, called *a loop cutset summary*, that is compatible with LBE and is efficiently computable. We show that it generalizes the rule-based summary of Beyer et al. [2]. Second, we present a linear time algorithm to construct PAQs for a loop-free program fragment. Our algorithm works directly on the SSA representation of the program, and constructs a query that separates control- and data-flow, while preserving both in its structure. Third, we study two approaches to solving PAQs: a classical AllSAT-based, and a new based on LDDs. The approaches are evaluated on a benchmark from open-source software.

Our approach builds on many existing components: SSA, loop-free program fragments, and early quantification – all well known; LDDs are used in [5] for image computation. However, the combination of the techniques is novel, the benchmarks are realistic and challenging, and the results show that our new LDD-based solution outperforms (and complements) the AllSAT-based one.

References

1. Ball, T., Podelski, A., Rajamani, S.K.: Boolean and Cartesian Abstraction for Model Checking C Programs. In: Margaria, T., Yi, W. (eds.) TACAS 2001. LNCS, vol. 2031, pp. 268–283. Springer, Heidelberg (2001)
2. Beyer, D., Cimatti, A., Griggio, A., Keremoglu, M.E., Sebastiani, R.: Software Model Checking via Large-Block Encoding. In: FMCAD 2009 (2009)
3. Bruttomesso, R., Cimatti, A., Franzén, A., Griggio, A., Sebastiani, R.: The Math-SAT4 SMT Solver. In: Gupta, A., Malik, S. (eds.) CAV 2008. LNCS, vol. 5123. Springer, Heidelberg (2008)
4. Cavada, R., Cimatti, A., Franzén, A., Kalyanasundaram, K., Roveri, M., Shyama-sundar, R.K.: Computing Predicate Abstractions by Integrating BDDs and SMT Solvers. In: FMCAD 2007 (2007)
5. Chaki, S., Gurfinkel, A., Strichman, O.: Decision Diagrams for Linear Arithmetic. In: FMCAD 2009 (2009)
6. Clarke, E., Kroening, D., Sharygina, N., Yorav, K.: Predicate Abstraction of ANSI-C Programs using SAT. FMSD 25(2-3) (2004)
7. Clarke, E., Kroening, D., Lerda, F.: A Tool for Checking ANSI-C Programs. In: Jensen, K., Podelski, A. (eds.) TACAS 2004. LNCS, vol. 2988, pp. 168–176. Springer, Heidelberg (2004)
8. Cytron, R., Ferrante, J., Rosen, B.K., Wegman, M.N., Zadeck, F.K.: Efficiently Computing Static Single Assignment Form and the Control Dependence Graph. TOPLAS 13(4) (1991)
9. Das, S., Dill, D.: Successive Approximation of Abstract Transition Relations. In: LICS 2001, pp. 51–60 (2001)
10. Flanagan, C., Qadeer, S.: Predicate Abstraction for Software Verification. In: POPL 2002, pp. 58–70 (2002)
11. Henzinger, T.A., Jhala, R., Majumdar, R., McMillan, K.L.: Abstractions From Proofs. In: POPL 2004 (2004)
12. Karp, R.M.: Reducibility Among Combinatorial Problems. In: Complexity of Computer Computations, pp. 85–103 (1972)
13. Kroening, D., Sharygina, N.: Approximating Predicate Images for Bit-Vector Logic. In: Hermanns, H. (ed.) TACAS 2006. LNCS, vol. 3920, pp. 242–256. Springer, Heidelberg (2006)
14. Lahiri, S.K., Ball, T., Cook, B.: Predicate Abstraction via Symbolic Decision Procedures. In: Etessami, K., Rajamani, S.K. (eds.) CAV 2005. LNCS, vol. 3576, pp. 24–38. Springer, Heidelberg (2005)
15. Lahiri, S.K., Bryant, R.E., Cook, B.: A Symbolic Approach to Predicate Abstraction. In: Hunt Jr., W.A., Somenzi, F. (eds.) CAV 2003. LNCS, vol. 2725, pp. 141–153. Springer, Heidelberg (2003)
16. Lahiri, S.K., Nieuwenhuis, R., Oliveras, A.: SMT Techniques for Fast Predicate Abstraction. In: Ball, T., Jones, R.B. (eds.) CAV 2006. LNCS, vol. 4144, pp. 424–437. Springer, Heidelberg (2006)
17. Lattner, C., Adve, V.: LLVM: A Compilation Framework for Lifelong Program Analysis & Transformation. In: CGO 2004 (2004)

Synthesis for PCTL in Parametric Markov Decision Processes

Ernst Moritz Hahn[1], Tingting Han[2], and Lijun Zhang[3]

[1] Saarland University, Saarbrücken, Germany
[2] Oxford University Computing Laboratory, United Kingdom
[3] DTU Informatics, Technical University of Denmark, Denmark

Abstract. In parametric Markov decision processes (PMDPs), transition probabilities are not fixed, but are given as functions over a set of parameters. A PMDP denotes a family of concrete MDPs. This paper studies the synthesis problem for PCTL in PMDPs: Given a specification Φ in PCTL, we synthesise the parameter valuations under which Φ is true. First, we divide the possible parameter space into hyper-rectangles. We use existing decision procedures to check whether Φ holds on each of the Markov processes represented by the hyper-rectangle. As it is normally impossible to cover the whole parameter space by hyper-rectangles, we allow a limited area to remain undecided. We also consider an extension of PCTL with reachability rewards. To demonstrate the applicability of the approach, we apply our technique on a case study, using a preliminary implementation.

1 Introduction

Markov processes [6, 26] have been applied successfully to reason about quantitative properties in networked, distributed, and recently biological systems. This paper considers *parametric* Markov processes [24], in which transition probabilities are not fixed, but depend on a set of parameters. As an example, consider a communication network with a lossy channel, where whenever a package is sent, it is received with probability x but lost with probability $1 - x$. In this context, we are interested in, for instance, determining the parametric reachability probability with respect to a given set of states. This probability is a function in x. By inserting an appropriate value for x in the function, we will obtain a concrete model without parameters. The synthesis problem asks, for example, what are the possible parameter valuations such that the reachability probability is below the a priori specified threshold.

Daws has devised a language-theoretic approach to solve the reachability problem in parametric Markov chains [11]. In this approach, the transition probabilities are considered as letters of an alphabet. Thus, the model is viewed as a finite automaton. Based on the *state elimination approach* [21], the regular expression describing the language of such an automaton is computed. In a post-processing step, this regular expression is transformed into a rational function over the parameters of the model. In previous works [17], we have improved this

M. Bobaru et al. (Eds.): NFM 2011, LNCS 6617, pp. 146–161, 2011.
© Springer-Verlag Berlin Heidelberg 2011

method by intertwining the state elimination and the computation of the rational function. Briefly, in a state elimination step, we label the edges directly with the appropriate rational function representing the flow of probabilities. Once all states—except the initial one and the goal states—have been eliminated, we can obtain the probabilities directly from the remaining edges. This improved algorithm is implemented in our tool PARAM [16]. The tool also supports bounded reachability, relying on matrix-vector multiplication with rational function entries, and *reachability rewards* [7,13]. For the latter, we extended the model with parametric rewards assigned to both states and transitions, and considered the expected accumulated reward until a given set of states is reached.

In this paper, we extend our approach to solve the PCTL synthesis problem for parametric Markov decision processes (PMDPs). PCTL (Probabilistic CTL) [6,19] is a probabilistic extension of the logic CTL for reasoning about properties over Markov models. In this paper, we extend the PCTL formulae with the reachability reward properties [20, 23] and can express properties like:

> *"The probability is larger than 0.99, that in the next step we move to a state where the accumulated reward until we are able to reach a state in which the property 'a' holds is less than 5."*

as $\mathcal{P}_{>0.99}(\mathcal{X}\,\mathcal{R}_{<5}(\Diamond a))$ in PCTL. We are interested in synthesising the concrete models fulfilling a given specification. Markov decision processes contain both probabilistic choices and nondeterministic choices. The notion of schedulers is used to resolve nondeterminism, leading to a parametric Markov chain. Previously [17], we considered a method for PMDPs by encoding nondeterminism in additional parameters. This method turned out to be limited by the number of nondeterministic choices, and can not be extended to treat nested properties. To handle the PCTL synthesis problem on PMDPs, we propose to divide the parameter valuations into *regions*, which are hyper-rectangles in the dimension of the number of variables. A region represents a family of concrete models. We aim at computing regions that subsume models with the same truth value of the specification. In general, it is not possible to cover the whole space completely. Thus, we stop as soon as the size of regions is below a pre-specified threshold, where it is unknown whether the specification is satisfied or not. To be on the safe side, the unknown regions are usually assumed not to fulfil the specification. To decide properties of a region of parameter valuations, we can use an approximate but fast method [18] which might derive false positive or false negative results. It can thus be used to get a quick overview for which areas the formula may hold, but should not be used if this information is critical. We can also use slower decision procedures with correctness guarantees [14,25,27].

In Fig. 1, we give an example for illustration. Zeroconf [9] is a protocol allowing the dynamic configuration of a network. When a new host enters a network, it randomly chooses an ID and asks the existing members of the network whether the ID is already in use. The request is conducted maximally n times, to minimise the probability of not getting an answer in an unreliable network even though the

ID is used. If the host does not get an answer within n tries, it assumes the ID to be unused. Here, we assume $n = 10$. The parameter p denotes the probability that the host gets no answer in case of a collision, and q denotes the probability that a chosen ID is already in use. We ask whether the expected number of requests till the protocol terminates (with an either unique or duplicate ID) is below 11. In PCTL (with reward extensions), this property can be expressed as $\mathcal{R}_{<11}(\lozenge \, IDConfirmed)$. In Fig. 1, regions for which this holds (resp. does not hold) are given as white (resp.

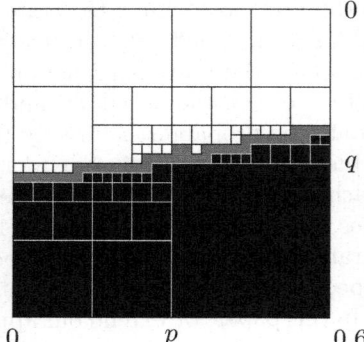

Fig. 1. Dividing the parameter space into regions in the Zeroconf example

black) boxes, while the gray boxes are unknown regions. As we can see, an increase of p or q leads to an increase of the expected number of trials.

To the best of our knowledge, parameter synthesis for PCTL properties in PMDPs has not been handled before. The most closely related work is due to Fribourg and André [15]: For a given PMDP and an instantiation of the parameters, they compute a scheduler for this instantiation which is optimal for a certain (non-nested) property. Afterwards, they compute the set of parameter evaluations for which the scheduler is still optimal. Compared to their work, we can deal with nested formulae and do not have a fixed scheduler a priori, but use different optimising schedulers for different regions if necessary.

Organisation of the paper. In Section 2 we give some preliminaries and define the parametric models and the variant of PCTL used in this paper. Then, in Section 3, we describe our parameter synthesis algorithm. We provide experimental results in Section 4. Finally, Section 5 concludes the paper.

2   Preliminaries

In this section, we first introduce the definitions of non-parametric Markov models and the logic PCTL. Afterwards, we introduce our parametric extensions and hyper-rectangles needed later for the synthesis problem.

2.1   Non-parametric Models

Definition 1. *A* Markov chain (MC) *is a tuple* $\mathcal{D} = (S, s_0, \mathbf{P}, L)$ *where* S *is a finite set of states,* s_0 *is the initial state,* $\mathbf{P} : S \times S \rightarrow [0, 1]$ *denotes the probability matrix, where for all* $s \in S$ *we require that* $\sum_{s' \in S} \mathbf{P}(s, s') = 1$. *Finally,* $L : S \rightarrow 2^{AP}$ *is a state labelling, mapping states to a subset of a given set of* atomic propositions AP.

Markov chains are the most basic model class. Next, we consider Markov decision processes which extend MCs by nondeterministic decisions.

Definition 2. *A* Markov decision process (MDP) *is defined as a tuple* $\mathcal{M} = (S, s_0, Act, \mathbf{P}, L)$ *where* S, s_0 *and* L *are as for MCs, and* Act *is a finite set of actions. The transition probability matrix* $\mathbf{P}$ *is a function* $\mathbf{P} : S \times Act \times S \to [0, 1]$. *For all states* $s \in S$ *and actions* $\alpha \in Act$, *we require that* $\sum_{s' \in S} \mathbf{P}(s, \alpha, s') \in \{0, 1\}$. *We also require that for each* $s \in S$ *there is at least one* $\alpha \in Act$ *with* $\sum_{s' \in S} \mathbf{P}(s, \alpha, s') = 1$.

With $Act(s) = \{\alpha \mid \sum_{s' \in S} \mathbf{P}(s, \alpha, s') = 1\}$ we specify the set of *enabled* actions of a state. The nondeterministic choices are resolved by the notion of *schedulers*. A *simple scheduler* is a function $\delta : S \to Act$ assigning one enabled action to each state. A *counting scheduler* is a function $\delta : S \times [1, n] \to Act$, for some $n \in \mathbb{N}$. Notice that for each $i \in \{1, \dots, n\}$ we have that $\delta(\cdot, i)$ is a simple scheduler. For our purposes, simple and counting schedulers suffice. A simple scheduler induces an MC from an MDP as follows.

Definition 3. *Given an MDP* $\mathcal{M} = (S, s_0, Act, \mathbf{P}, V)$ *and a simple scheduler* δ, *the MC induced by* δ *is defined as* $\mathcal{M}^{\delta} := (S, s_0, \mathbf{P}^{\delta}, V)$ *where the transition matrix* $\mathbf{P}^{\delta} : S \times S \to [0, 1]$ *is defined by* $\mathbf{P}^{\delta}(s, s') := \mathbf{P}(s, \delta(s), s')$.

For MDPs with exactly one enabled action for each state, there is a one-to-one correspondence to MCs, so we can consider MCs as a special case of MDPs.

When model checking PCTL formulae, we will have to consider modified versions of our models, in which certain states are made absorbing.

Definition 4. *Let* sink $: S \to \{false, true\}$ *be a function mapping states to boolean values. For the transition matrix* $\mathbf{P}$ *of an MDP, we define a transition matrix* $\mathbf{P}[\text{sink}]$ *where states* s *with* $\text{sink}(s) = true$ *are made absorbing by setting*

$$\mathbf{P}[\text{sink}](s, \alpha, s') := \begin{cases} \mathbf{P}(s, \alpha, s') & \text{if } \text{sink}(s) = false, \\ 1 & \text{if } \text{sink}(s) = true \wedge s = s', \\ 0 & \text{else.} \end{cases}$$

By skipping the action α *above we get the definition for MCs.*

We now extend our models by *rewards*, which can be interpreted as either costs or bonuses, depending on the model under consideration.

Definition 5. *A reward structure for an MDP with state space* S *and action set* Act *is a partial function* $\mathbf{r} : S \times Act \rightharpoonup \mathbb{R}_{\geq 0}$ *assigning a reward to each state and enabled action. For an MC, a reward structure is a function* $\mathbf{r} : S \to \mathbb{R}_{\geq 0}$ *assigning a reward to each state.*

Similar to the probability matrices, if $\mathbf{r}$ is a reward structure for an MDP and δ is a simple scheduler, we define $\mathbf{r}^{\delta}$ such that $\mathbf{r}^{\delta}(s) = \mathbf{r}(s, \delta(s))$. Given a function sink $: S \to \{true, false\}$ and a reward structure $\mathbf{r}$ for an MDP, we let $\mathbf{r}[\text{sink}](s, \alpha) = 0$ if $\text{sink}(s) = true$ and $\mathbf{r}[\text{sink}](s, \alpha) = \mathbf{r}(s, \alpha)$ otherwise. For an MC, we define $\mathbf{r}[\text{sink}](s) = 0$ if $\text{sink}(s) = true$ and $\mathbf{r}[\text{sink}](s) = \mathbf{r}(s)$ otherwise.

2.2 Probabilistic CTL

To specify properties, we consider the logic Probabilistic CTL (PCTL) [6, 19]. The syntax is given by:

$$\Phi = true \mid a \mid \neg\Phi \mid \Phi \wedge \Phi \mid \mathcal{P}_{\bowtie p}(\varphi) \mid \mathcal{R}_{\bowtie m}(\Diamond\,\Phi), \qquad \varphi = \mathcal{X}\,\Phi \mid \Phi\,\mathcal{U}\,\Phi \mid \Phi\,\mathcal{U}^{\leq n}\,\Phi,$$

where $\bowtie \in \{<, \leq, \geq, >\}$, $n \in \mathbb{N}$, $p \in [0, 1]$, $m \in \mathbb{R}$ and $a \in AP$. Here, Φ is a formula which has a boolean value in a state, whereas φ is interpreted on paths. PCTL can be interpreted on MDPs [6].

The truth values of $true$, a and $\wedge$ in a state are straightforward. For state s, the formula $\mathcal{P}_{\bowtie p}(\varphi)$ is fulfilled if for all schedulers the probability of paths which start in s and fulfil φ meets the bound $\bowtie p$. For $\bowtie \in \{<, \leq\}$, this is equivalent to asking whether the *maximal* probability fulfils $\bowtie p$, whereas for $\bowtie \in \{\geq, >\}$ we only need to consider the *minimal* probability.

Given a path, the *next* formula $\mathcal{X}\,\Psi$ asks whether on the second state of this path Ψ holds. The *unbounded until* formula $\Psi_1\,\mathcal{U}\,\Psi_2$ requires that a state on the path fulfils Ψ_2, and for all states on the path before that point, Ψ_1 must hold. The *bounded until* formula $\Psi_1\,\mathcal{U}^{\leq n}\,\Psi_2$ is similar, but additionally requires that Ψ_2 occurs at latest n steps after the first state of the path. The formal semantics of PCTL on MDPs has been introduced by Bianco and De Alfaro [6]. We write $\mathcal{M} \models \Phi$ if the initial state of an MDP fulfils the PCTL state formula Φ. The reachability reward formula [20, 23] $\mathcal{R}_{\bowtie m}(\Diamond\,\Psi)$ states that the expected accumulated reward until a state satisfying Ψ is reached should meet the bound $\bowtie m$. The formula holds, if under all schedulers this expectation fulfils $\bowtie m$.

2.3 Parametric Models

We fix $V = \{x_1, \ldots, x_n\}$ as the set of variables with domain $\mathbb{R}$. With each variable x, we associate a closed interval $\mathrm{range}(x) = [L_x, U_x]$ specifying which values of x are valid. An *evaluation* v is a function $v : V \to \mathbb{R}$ respecting the variable ranges. A *polynomial* g over V is a sum of monomials

$$g(x_1, \ldots, x_n) = \sum_{i_1, \ldots, i_n} a_{i_1, \ldots, i_n} x_1^{i_1} \cdots x_n^{i_n},$$

where each $i_j \in \mathbb{N}_0$ and each $a_{i_1, \ldots, i_n} \in \mathbb{R}$. A *rational function* f over a set of variables V is a fraction $f(x_1, \ldots, x_n) = \frac{g_1(x_1, \ldots, x_n)}{g_2(x_1, \ldots, x_n)}$ of two polynomials g_1, g_2 over V. Let $\mathcal{F}_V$ denote the set of rational functions from V to $\mathbb{R}$. Given $f \in \mathcal{F}_V$ and an evaluation v, we let $f\langle v \rangle := f(v(x_1), \ldots, v(x_n))$ denote the rational number obtained by substituting each occurrence of x_i with $v(x_i)$.

We now extend MCs to parametric models [11, 24]. The difference to the original model lies in the extension by parameters and the definition of the probability matrix.

Definition 6. *A parametric Markov chain (PMC) is defined as a tuple $\mathcal{D} = (S, s_0, \mathbf{P}, L, V)$ where S, s_0 and L are as in Definition 1, $V = \{x_1, \ldots, x_n\}$ is a finite set of parameters and $\mathbf{P}$ is the probability matrix $\mathbf{P} : S \times S \to \mathcal{F}_V$.*

A parameter evaluation induces a non-parametric MC from a PMC.

Definition 7. *Let* $\mathcal{D} = (S, s_0, \mathbf{P}, L, V)$ *be a PMC. The MC* $\mathcal{D}_v$ *induced by an evaluation* v *is defined as* $\mathcal{D}_v := (S, s_0, \mathbf{P}_v, L)$ *where the transition matrix* $\mathbf{P}_v : S \times S \to [0, 1]$ *is given by* $\mathbf{P}_v(s, s') := \mathbf{P}(s, s')\langle v \rangle$ *if this matrix fulfils the requirements of Definition 1.*

We already considered [16, 17] how to compute a rational function which represents the unbounded reachability probability from the initial state to a set of target states in a PMC. Evaluating this rational function with a certain parameter evaluation leads to the same result as first computing the induced PMCs and then computing the probability in this model. With a simple extension of our previous techniques, we can compute reachability values for all states of the model at the same time, which is necessary when checking nested formulae. Below we define parametric MDPs.

Definition 8. *A* parametric Markov decision process (PMDP) *is a tuple* $\mathcal{M} = (S, s_0, Act, \mathbf{P}, L, V)$ *where* S, s_0, L *and* V *are as for PMCs, and* Act *is a finite set of actions. The transition matrix* $\mathbf{P}$ *is of the form* $\mathbf{P} : S \times Act \times S \to \mathcal{F}_V$.

As for PMCs, we introduce the MDP induced by a valuation function.

Definition 9. *Given a PMDP* $\mathcal{M} = (S, s_0, Act, \mathbf{P}, L, V)$ *and an evaluation* v, *the MDP induced by* v *is defined by* $\mathcal{M}_v := (S, s_0, Act, \mathbf{P}_v, L)$ *where* $\mathbf{P}_v : S \times Act \times S \to [0, 1]$ *is defined by* $\mathbf{P}_v(s, \alpha, s') := \mathbf{P}(s, \alpha, s')\langle v \rangle$. *For* $\mathbf{P}_v$, *the requirements of Definition 2 must be fulfilled.*

The notions of making a state absorbing as well as models and rewards induced by a scheduler are defined as in the non-parametric models. We allow reward structures to take rational functions as values. In turn, for an evaluation v, we define $\mathbf{r}_v$ as $\mathbf{r}_v(s, \alpha) := \mathbf{r}(s, \alpha)\langle v \rangle$ or $\mathbf{r}_v(s) := \mathbf{r}(s)\langle v \rangle$ respectively. As required by the optimality equation used in the later model checking algorithm, we assume nonnegative rewards, i.e., $\mathbf{r}_v \geq 0$, for all evaluations under consideration.

We assume that our evaluation functions fulfil the following assumption.

Assumption 1. *Let* v *be an evaluation function and let* $\mathbf{P}$ *be the probability matrix of a PMC or PMDP. Then no transition probability of* $\mathbf{P}_v$ *is zero or one, except this entry is zero or one for any evaluation.*

Our assumption guarantees that the structure of the underlying graph of $\mathbf{P}$ remains unchanged from v. In other words, a transition with a parameter should not disappear (due to the null probability) no matter what value it takes. It excludes extreme cases such

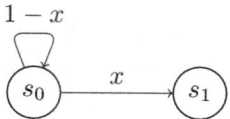

Fig. 2. Example PMC of Assumption 1

as when $x = 0$ or $x = 1$ (see Fig. 2). This is not a severe restriction, as such cases are seldom interesting in practice: They correspond to cases where an error happens either not at all or with certainty. These corner cases can be treated separately, with exponential blow-up in the number of variables, by fixing each such possible evaluation combinations before applying our approach.

2.4   Hyper-rectangles

A *region* is a high-dimensional rectangle $r = \times_{x \in V}[l_x, u_x]$ such that for all $x \in V$ it is $[l_x, u_x] \subseteq \text{range}(x)$. A region represents those evaluations v with $v(x) \in [l_x, u_x]$ for all $x \in V$: In this case, we write $v \in r$. We define the *centre* of a region $r = \times_{x \in V}[l_x, u_x]$ by $\text{centre}(r)(x) := \frac{l_x + u_x}{2}$ for $x \in V$. Later on, we might have to split a region r into several smaller parts, provided r is too coarse with respect to a property, i.e., we are not sure whether the property holds for all evaluations it represents. For this, we introduce the splitting function. For $A \subseteq V$, we let

$$\text{INT}_2(r, x, A) := \begin{cases} \{[l_x, u_x]\} & \text{if } x \notin A, \\ \{[l_x, \text{centre}(r)(x)], [\text{centre}(r)(x), u_x]\} & \text{if } x \in A \end{cases}$$

be the function dividing the interval $[l_x, u_x]$ (of region r) on dimension $x \in A$ into two halves. Define

$$\text{split}(r, 2, A) := \left\{ \underset{x \in V}{\times} Int_x \;\middle|\; \forall x \in V.\ Int_x \in \text{INT}_2(r, x, A) \right\}$$

as the set of split (small) regions, or sub-regions. Moreover, $\text{INT}_m(r, x, A)$ and $\text{split}(r, m, A)$ for $m > 2$ can be defined in a very similar way, where they equally divide the interval in each dimension for each $x \in A$ into m sub-intervals and compute the set of m-divided regions, respectively. The set A will be skipped in case of $A = V$, and we also write $\text{split}(r)$ for $\text{split}(r, 2)$.

We define the *volume* μ of a region $r = \times_{x \in V}[l_x, u_x]$ in a straight-forward way by setting $\mu(r) := \prod_{x \in V} \frac{u_x - l_x}{U_x - L_x}$. This way, the volume is the product of the relative lengths of sides of the hyper-rectangle. For a set $K = \{r_1, \ldots, r_n\}$ of regions, we define $\mu(K) := \sum_{i=1}^{n} \mu(r_i)$.

A *decision procedure* is a tool deciding the validity of formulae for a given region. There exist both approximate decision procedures [18] as well as precise ones [14, 25, 27]. Consider a predicate $constraint := f \bowtie q$ where f is a rational function over the variables in V, $\bowtie \in \{<, \leq, \geq, >\}$ and $q \in \mathbb{R}$. Let r be a given region. For an evaluation v, with $constraint\langle v \rangle$ we denote $f\langle v \rangle \bowtie q$, i.e., the constraint obtained under the valuation v. We assume that we are given a decision procedure CHECK($constraint, r$) which

– returns *true* only if for all $v \in r$ we have that $constraint\langle v \rangle$ is true, and
– returns *false* in case this does not hold or the result can not be decided.

3   Synthesis for PCTL

In this section we present the algorithm for synthesising PCTL formulae against PMDPs. The main routine of the algorithm is given in Algorithm 1. It maintains a set *unprocessed* of regions for which the result is still unknown, initially containing only $\times_{x \in V} \text{range}(x)$. Then, it takes a largest region out of this set and tries to decide its value using the procedure CHECKSTATE. If CHECKSTATE

Algorithm 1. MAIN($\mathcal{M} = (S, s_0, Act, \mathbf{P}, L, V), \Phi, \varepsilon$)

$unprocessed := \{\bigtimes_{x \in V} \text{range}(x)\}$
$result := \emptyset$
while $\mu(unprocessed) \geq \varepsilon$ **do**
    choose one largest $r \in unprocessed$
    $unprocessed := unprocessed \setminus \{r\}$
    $b := \text{CHECKSTATE}(r, \Phi)$
    if $b =$? **then**
        | $unprocessed := unprocessed \cup \text{split}(r)$
    else
        | $result := result \cup \{(r, b)\}$
return $result$

---

returns a definite answer, the pair (r, b) is added to *result*. In this case, the truth value for a state s is the same for all non-parametric MDPs represented by r. Then, $b(s)$ maps each state s to this truth value, which is constant within the region. If ? is obtained, we split the region and add the newly generated regions to *unprocessed*. The procedure is repeated until the volume of *unprocessed* is smaller than ε.

Algorithm 2 describes the procedure CHECKSTATE discussed above. If successful, it returns a function mapping each state to either *true* or *false*. It may also return ? if either the truth value is different for certain parts of the region, or the truth values can not be decided for the whole region at once. Notice that for two functions $b, b' : S \to \{true, false, ?\}$, the boolean connectors $b \wedge b'$, etc. are to be understood state-wise, that is $(b \wedge b')(s) = b(s) \wedge b'(s)$, etc. For $\wedge$ and $\neg$ operations, the result is always ? in case one of the operands is ?. For $\bowtie \in \{<, \leq, \geq, >\}$, we define the negation as $\overline{<} := \geq$, $\overline{\leq} := >$, $\overline{\geq} := <$, $\overline{>} := \leq$. Boolean formulae are trivial. Below we discuss the probabilistic and reward formulae.

Since the maximal and minimal probabilities are dual, in the rest of the paper we will only consider the minimal properties and set $\bowtie \in \{>, \geq\}$ for simplicity.

3.1 Reward Formula $\mathcal{R}_{\bowtie m}(\Diamond \Psi)$

Recursively, we first compute $reach := \text{CHECKSTATE}(r, \Psi)$. Then, we instantiate the PMDP with reward structure at centre(r) where r is the region under consideration. We obtain a non-parametric MDP, from which we compute the minimising scheduler. It is well-known that simple schedulers are sufficient to minimise (or maximise) reachability rewards for MDPs [8,10,29,30]. The procedure MINREACHREWSCHED returns this simple scheduler δ such that the reachability reward is minimised for each state in the induced MDP with respect to the evaluation centre(r). A PMC $\mathcal{M}^\delta$ is further induced under this simple scheduler δ, with the corresponding matrix $\mathbf{P}^\delta$. Using REACHREW, we compute the parametric reachability rewards function $optRew : S \to \mathcal{F}_V$ in this induced PMC (as in a previous publication [17]).

Algorithm 2. CHECKSTATE(r, Φ)

---

switch Φ **do**

    case a **return** b such that $b(s) = ($**if** $a \in AP(s)$ **then** *true* **else** *false*$)$

    case $\neg\Psi$ **return** $\neg$CHECKSTATE(r, Ψ)

    case $\Psi_1 \wedge \Psi_2$ **return** CHECKSTATE(r, Ψ_1) $\wedge$ CHECKSTATE(r, Ψ_2)

    case $\mathcal{P}_{\bowtie p}(\varphi)$

        $val := $ COMPUTEPROB(r, φ)

        if $val = ?$ **then return** ?

        for $s \in S$ **do** $b(s) := \begin{cases} true & \text{if } \text{CHECK}(val(s) \bowtie p, r), \\ false & \text{if } \text{CHECK}(val(s) \overline{\bowtie} p, r), \\ ? & \text{else} \end{cases}$

        if $\exists s.b(s) = ?$ **then return** ? **else return** b

    case $\mathcal{R}_{\bowtie m}(\Diamond \Psi)$

        $reach := $ CHECKSTATE(r, Ψ)

        if $reach = ?$ **then return** ?

        $c := $ centre(r)

        $\delta := $ MINREACHREWSCHED($\mathbf{P}_c[reach], \mathbf{r}_c[reach], reach$)

        $optRew := $ REACHREW($\mathbf{P}^\delta[reach], \mathbf{r}^\delta[reach], reach$)

        for $s \in S, \alpha \in Act(s)$ **do**

            $checkRew(s) := \mathbf{r}[reach](s, \alpha) + \sum_{s' \in S} \mathbf{P}[reach](s, \alpha, s') \cdot optRew(s')$

            $valid := valid \wedge $ CHECK($optRew(s) \le checkRew(s), r$)

        if $\neg valid$ **then return** ?

        for $s \in S$ **do** $b(s) := $ CHECK($optRew(s) \bowtie m, r$)

        if $\exists s.b(s) = ?$ **then return** ? **else return** b

---

Recall that the scheduler δ is minimising with respect to the evaluation centre(r). Our next **for** loop checks whether this is also the case for all evaluations in the region r. It works in a similar way as the *optimality equation* [26,30]. For all states s and enabled actions α, we check whether δ is indeed minimising, but this time for all concrete models represented by the region, through the decision procedure CHECK. In more detail, if the obtained reward *checkRew(s)* in the **for** loop satisfies the constraint *optRew(s)* $\le$ *checkRew(s)* for each concrete model in the region, then indeed $\delta(s)$ is minimising. In this case we have proven that δ is locally optimal for each state, which induces global optimality of the current scheduler.

3.2 Probabilistic Formula $\mathcal{P}_{\bowtie p}(\varphi)$

The function COMPUTEPROB, in Algorithm 3, returns a function mapping each state s to the minimal probability of all paths which fulfil φ when starting in s. Again, if this value can not be decided, the result is ?. The functions work recursively: The cases for atomic propositions, negation and conjunction are as for usual model checking procedures. For $\mathcal{P}_{\bowtie p}(\varphi)$, we use the procedure CHECK discussed in Section 2.4 to decide the truth value for each state, if this is possible.

Algorithm 3. COMPUTEPROB(r, φ)

---

switch φ **do**

    case $\Psi_1 \, \mathcal{U} \, \Psi_2$

        $left := $ CHECKSTATE(r, Ψ_1), $right := $ CHECKSTATE(r, Ψ_2)

        if *(left = ? or right = ?)* **then return** ?

        $c := $ centre(r)

        $\delta := $ MINUREACHSCHED$(\mathbf{P}_c[\neg left \vee right], right)$

        $optProb := $ UREACHPROB$(\mathbf{P}^\delta[\neg left \vee right], right)$

        $valid := true$

        for $s \in S, \alpha \in Act(s)$ **do**

            $checkProb(s) := \sum_{s' \in S} \mathbf{P}[\neg left \vee right](s, \alpha, s') \cdot optProb(s')$

            $valid := valid \wedge $ CHECK$(optProb(s) \leq checkProb(s), r)$

        if *valid* **then return** *optProb* **else return** ?

    case $\Psi_1 \, \mathcal{U}^{\leq n} \, \Psi_2$

        $left := $ CHECKSTATE(r, Ψ_1), $right := $ CHECKSTATE(r, Ψ_2)

        if *(left = ? or right = ?)* **then return** ?

        $c := $ centre(r)

        $\delta := $ MINBREACHSCHED$(\mathbf{P}_c[\neg left \vee right], right)$

        forall the s **do** $optProb(s) := $ **if** $right(s)$ **then** 1 **else** 0

        $valid := true$

        for $step = n, \ldots, 1$ **do**

            $optProb' := \mathbf{P}^{\delta(\cdot, step)}[\neg left \vee right] \cdot optProb$

            for $s \in S, \alpha \in Act(s)$ **do**

                $checkProb(s) := \sum_{s' \in S} \mathbf{P}[\neg left \vee right](s, \alpha, s') \cdot optProb(s')$

                $valid := valid \wedge $ CHECK$(optProb'(s) \leq checkProb(s), r)$

            $optProb := optProb'$

        if *valid* **then return** *optProb* **else return** ?

In COMPUTEPROB, for $\Psi_1 \, \mathcal{U} \, \Psi_2$ we compute a minimising scheduler to fulfil the unbounded until formula for the centred parameter evaluation, using standard means, by calling MINUREACHSCHED. Notice that the minimising scheduler is a simple scheduler, which suffices for minimal reachability probabilities [6]. By UREACHPROB, we compute the reachability probability of the PMC induced by this scheduler (as in a previous work [17]). Note that the probability obtained this way is only valid for parameter evaluations which fulfil Assumption 1. Afterwards, we use another optimality equation [4] to check whether the decision is minimal for all parameter evaluations of the region.

For the bounded until $\Psi_1 \, \mathcal{U}^{\leq n} \, \Psi_2$, we need to consider the minimum over all *counting* schedulers. We compute the minimising scheduler for one instantiation. Afterwards, we use a recursive (backward) characterisation [2] to prove that for each step the choices the scheduler takes are indeed optimal for the whole parameter region. We leave out the case $\mathcal{X} \, \Psi$, as it can be handled by a simpler variant of the algorithm for the bounded until.

3.3   Termination and Correctness

To guarantee termination of our algorithm, we need the following assumption.

Assumption 2. *Let* $r_0 := \bigtimes_{x \in V} \text{range}(x)$ *denote the initial region, and* ε *the given precision. We assume that there exists* $m \in \mathbb{N}$ *with the following property. There exists a set* $K \subseteq \text{split}(r_0, m)$ *of regions such that 1.) for all regions* $r \in K$, *either for all evaluations* $v \in r$ *it is* $\mathcal{M}_v \models \Phi$, *or for all evaluations* $v \in r$ *it is* $\mathcal{M}_v \not\models \Phi$, *2.)* $\mu(K) > 1 - \varepsilon$ *and 3.) the decision procedure is able to decide all constraints occurring during the parameter synthesis of all regions* $r \in K$.

The assumption requires that by repeated splitting we arrive at a sufficiently large set of regions (with volume larger than $1 - \varepsilon$) in which each state has a constant truth value, decidable by the (possibly incomplete) decision procedure. It is similar to an assumption used to reason about the quasi-decidability of hybrid systems [28]. In case the assumption is valid, the following lemma guarantees termination.

Lemma 1. *Let* $\mathcal{M}$ *be a PMDP,* Φ *be a PCTL formula and* $\varepsilon > 0$ *the analysis precision. Then Algorithm 1 terminates in finite time with this input, given that Assumption 2 holds.*

Lemma 1 follows by a simple structural induction on the formula, provided Assumption 2 holds. We now state the correctness of the algorithm.

Lemma 2. *Let* $\mathcal{M} = (S, s_0, Act, \mathbf{P}, V)$ *be a PMDP,* Φ *be a PCTL state formula and* $\varepsilon > 0$ *the analysis precision. Further, assume we are using a precise decision procedure. Then Algorithm 1 is correct in the following sense. For each tuple* (r, b) *of the result, and for each* $v \in r$ *for which* $\mathcal{M}_v$ *is a valid MDP and for which Assumption 1 is valid, we have* $\mathcal{M}_v \models \Phi$ **iff** $b(s_0)\langle v \rangle = true$.

Notice that its correctness does not depend on Assumption 2, thus the result is correct also in case termination is not guaranteed. The proof of the correctness of Lemma 2 also follows by structural induction. For atomic propositions and boolean connectors, the induction step is trivial. For until and reachability rewards, we use the correctness of the corresponding optimality equations.

4   Experiments

We implemented the model checking procedure of Algorithm 1 in a prototypical way in our tool PARAM $2.0\,\alpha$. For the analysis to be feasible, it was necessary to implement a number of optimisations. We minimise induced PMCs using weak [3] or strong [12] bisimulation. We use a caching technique to avoid computing reachability probabilities in PMCs twice, in case the same PMCs are induced from several calls to CHECKSTATE or CHECKREGION. We also reuse known truth values of constraints. Because we usually have to split large regions into smaller ones anyway, we do some pre-checks whether the truth value may be constant. To minimise the number of regions to be considered, and thus the overall time, we split regions along one widest side, i.e., $\text{split}(r, 2, \{x\})$ with variable

x representing the (or a) widest side. For the case study under consideration, we extended an approximate decision method [18], which does not guarantee correctness. Initial experiments with exact solvers have not been successful, as verifying a single region did not terminate within several minutes. In the method used, for a constraint $f \bowtie q$ we evaluate f in the corners of the region as well as some randomly chosen points inside. The more points we evaluate, the more unlikely is a wrong result, but still correctness cannot be guaranteed formally.

We applied the implementation on a randomised consensus shared coin protocol by Aspnes and Herlihy [1], based on an existing PRISM model [22]. In this case study, there are N processes sharing a counter c, which initially has the value 0. In addition, a value K is fixed for the protocol. Each process i decides to either decrement the counter with probability p_i or to increment it with probability $1 - p_i$. In contrast to the original PRISM model, we do not fix the p_i to $\frac{1}{2}$, but use them as parameters of the model. After writing the counter, the process reads the value again and checks whether $c \leq -KN$ or $c \geq KN$. In the first case, the process votes 1, in the second it votes 2. In both cases, the process stops afterwards. If neither of the two cases hold, the process continues its execution. As all processes which have not yet voted try to access the counter at the same time, there is a nondeterministic choice on the access order.

A probabilistic formula. As the first property, we ask whether for each execution of the protocol the probability that all processes finally terminate with a vote of 2 is at least $\frac{K-1}{2K}$. With appropriate atomic propositions *finished* and *allCoinsEqualTwo*, this property can be expressed as $\mathcal{P}_{\geq \frac{K-1}{2K}}(true \, \mathcal{U} \, (finished \, \wedge \, allCoinsEqualTwo))$. For the case $N = 2$ and $K = 2$, we give results in Fig. 3.

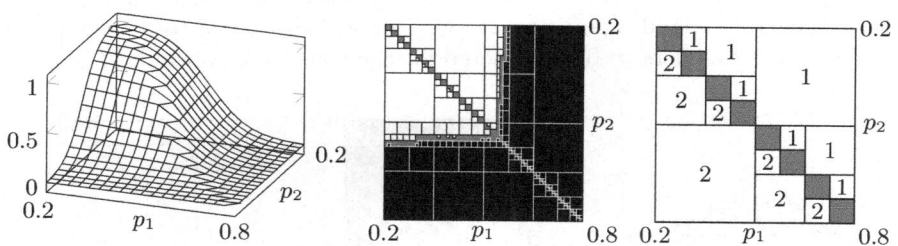

Fig. 3. Randomised consensus: $\mathcal{P}_{\geq \frac{K-1}{2K}}(true \, \mathcal{U} \, (finished \, \wedge \, allCoinsEqualTwo))$

The leftmost part of the figure provides the minimal probabilities among all schedulers that all processes terminate with a vote of 2, depending on the parameters p_i. With decreasing p_i, the probability that all processes vote 2 increases, since it becomes more likely that a process increases the counter and thus also the chance that finally $c \geq KN$ holds. The plot is symmetric, because both processes are independent and have an identical structure.

On the right part of the figure, we give an overview which schedulers are optimal for which parameter values. Here, boxes labelled with the same number share the same minimising scheduler. In case $p_1 < p_2$, to obtain the minimal probability the nondeterminism must be resolved such that the first process is activated if it has not yet voted. Doing so maximises the probability that we have $c \leq -KN$ before $c \geq KN$, and in turn minimises the probability that both processes vote 2. For $p_1 > p_2$, the second process must be preferred.

In the middle part of the figure, we give the truth values of the formula. White boxes correspond to regions where the property holds, whereas in black boxes it does not hold. In gray areas, the truth value is undecided. To keep the gray areas viewable, we chose a rather high tolerance of 0.15. The truth value decided is as expected by inspecting the plot on the left part of the figure, except for the gray boxes along the diagonal of the figure. In the gray boxes enclosed by the white area, the property indeed holds, while in the gray areas surrounded by the black area, it does not hold. The reason that these areas remain undecided is that the minimising scheduler changes at the diagonals, as discussed in the previous paragraph. If the optimal scheduler in a box is not constant for the region considered, we have to split it. Because the optimal scheduler always changes at the diagonals, there are always some gray boxes remaining.

A reward formula. As a second property, we ask whether the expected number of steps until all processes have voted is above 25, expressed as $\mathcal{R}_{>25}(\lozenge\, \textit{finished})$. Results are given in Fig. 4. On the left part, we give the expected number of steps. This highest value is at $p_i = \frac{1}{2}$. Intuitively, in this case the counter does not have a tendency of drifting to either side, and is likely to stay near 0 for a longer time. Again, gray boxes surrounded by boxes of the same colour are those regions in which the minimising scheduler is not constant. We see from the right part of the figure that this happens along four axes. For some values of the parameters, the minimising scheduler is not always the one which always prioritises one of the processes. Instead, it may be necessary to schedule the first process, then the second again, etc. As we can see, this leads to a number of eight different schedulers to be considered for the considered variable ranges.

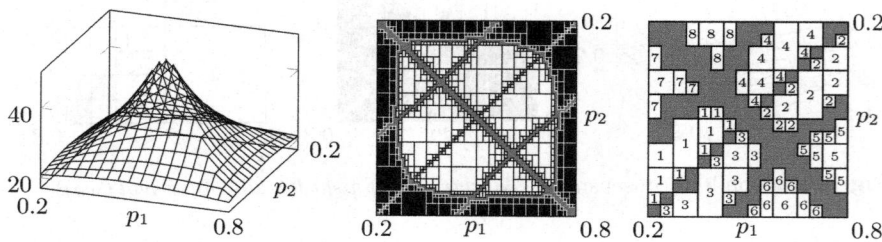

Fig. 4. Randomised consensus: $\mathcal{R}_{>25}(\lozenge\, \textit{finished})$

Runtime. In Table 1 we give the runtime of our tool (on an Intel Core 2 Duo P9600 with 2.66 GHz running on Linux) for two processes and different constants K. Column "States" contains the number of states. The columns labelled with "Until" contain results of the first property while those labelled with "Reward" contain those of the second. Columns labelled with "min" contain just the time to compute the minimal values whereas those labelled with "truth" also include the time to compare this value against the bound of the formula. For all analyses, we chose a tolerance of $\varepsilon = 0.05$. The time is given in seconds, and "–" indicates that the analyses did not terminate within 90 minutes.

As we see, the performance drops quickly with a growing number of states. For reward-based properties, the performance is worse than for unbounded until. These analyses are more complex, as rewards have to be taken into account, and weak bisimulation can not be applied for minimisation of the induced models. In addition, a larger number of different schedulers has to be considered to obtain minimal values, which also increases the analysis time. We are however optimistic that we will be able to improve these figures, using a more advanced implementation.

Table 1. Randomised consensus: performance statistics

K	States	Until		Reward	
		min	truth	min	truth
2	272	4.7	22.8	287.8	944.7
3	400	13.7	56.7	4610.1	–
4	528	31.7	116.1	–	–
5	656	65.5	215.2	–	–
6	784	123.4	374.6	–	–
7	912	272.6	657.4	–	–

5 Conclusion

In this paper, we have studied the parameter synthesis problem of PCTL formulae for PMDPs. We have demonstrated the principal applicability of the method, using a prototypical implementation. As future work we aim to make the method applicable to models with larger state space. It will be necessary to improve the technique, from both the theory and implementation perspective. To guarantee correctness of the results, we intend to try out different solver tools, and to bring the rational functions into a form which is easier to be handled by the respective solver. Another possible future work is to extend the recent interesting work about model repair systems for PMCs [5] to PMDPs.

Acknowledgements. This work was supported by the SFB/TR 14 AVACS, FP7-ICT Quasimodo, NWO-DFG ROCKS, DAAD-MinCyT QTDDS, ERC Advanced Grant VERIWARE, MT-LAB—a VKR Centre of Excellence.

We thank Alexandru Mereacre for many comments and insightful discussions.

References

1. Aspnes, J., Herlihy, M.: Fast randomized consensus using shared memory. Journal of Algorithms 11(3), 441–461 (1990)
2. Baier, C.: On algorithmic verification methods for probabilistic systems. Mannheim University, Habilitationsschrift (1998)
3. Baier, C., Hermanns, H.: Weak bisimulation for fully probabilistic processes. In: Grumberg, O. (ed.) CAV 1997. LNCS, vol. 1254, pp. 119–130. Springer, Heidelberg (1997)
4. Baier, C., Katoen, J.P.: Principles of Model Checking (Representation and Mind Series). The MIT Press, Cambridge (2008)

5. Bartocci, E., Grosu, R., Katsaros, P., Ramakrishnan, C.R., Smolka, S.A.: Model repair for probabilistic systems. In: Abdulla, P.A., Leino, K.R.M. (eds.) TACAS 2011. LNCS, vol. 6605, pp. 326–340. Springer, Heidelberg (2011)
6. Bianco, A., Alfaro, L.D.: Model checking of probabilistic and nondeterministic systems. In: Thiagarajan, P.S. (ed.) FSTTCS 1995. LNCS, vol. 1026, pp. 499–513. Springer, Heidelberg (1995)
7. Blackwell, D.: On the functional equation of dynamic programming. Journal of Mathematical Analysis and Applications 2(2), 273–276 (1961)
8. Blackwell, D.: Positive dynamic programming. In: Proceedings of the 5th Berkeley Symposium on Mathematical Statistics and Probability, pp. 415–418 (1967)
9. Bohnenkamp, H.C., van der Stok, P., Hermanns, H., Vaandrager, F.W.: Cost-optimization of the IPv4 Zeroconf protocol. In: DSN, pp. 531–540. IEEE Computer Society, Los Alamitos (2003)
10. van Dawen, R.: Finite state dynamic programming with the total reward criterion. Mathematical Methods of Operations Research 30, A1–A14 (1986)
11. Daws, C.: Symbolic and parametric model checking of discrete-time Markov chains. In: Liu, Z., Araki, K. (eds.) ICTAC 2004. LNCS, vol. 3407, pp. 280–294. Springer, Heidelberg (2005)
12. Derisavi, S., Hermanns, H., Sanders, W.H.: Optimal state-space lumping in Markov chains. IPL 87(6), 309–315 (2003)
13. Dubins, L.E., Savage, L.: How to Gamble If You Must. McGraw-Hill, New York (1965)
14. Fränzle, M., Herde, C., Teige, T., Ratschan, S., Schubert, T.: Efficient solving of large non-linear arithmetic constraint systems with complex boolean structure. JSAT 1(3-4), 209–236 (2007)
15. Fribourg, L., André, É.: An inverse method for policy iteration based algorithms. In: INFINITY, pp. 44–61. Open Publishing Association, EPTCS (2009)
16. Hahn, E.M., Hermanns, H., Wachter, B., Zhang, L.: PARAM: A model checker for parametric Markov models. In: Touili, T., Cook, B., Jackson, P. (eds.) CAV 2010. LNCS, vol. 6174, pp. 660–664. Springer, Heidelberg (2010)
17. Hahn, E.M., Hermanns, H., Zhang, L.: Probabilistic reachability for parametric Markov models. STTT 13, 3–19 (2010)
18. Han, T.: Diagnosis, synthesis and analysis of probabilistic models. Ph.D. thesis, RWTH Aachen University/University of Twente (2009)
19. Hansson, H., Jonsson, B.: A logic for reasoning about time and reliability. FAC 6, 102–111 (1994)
20. Haverkort, B.R., Cloth, L., Hermanns, H., Katoen, J.P., Baier, C.: Model checking performability properties. In: DSN, pp. 103–112 (2003)
21. Hopcroft, J.E., Motwani, R., Ullman, J.D.: Introduction to automata theory, languages, and computation. SIGACT News, 2nd edn. 32(1), 60–65 (2001)
22. Kwiatkowska, M., Norman, G., Segala, R.: Automated verification of a randomized distributed consensus protocol using Cadence SMV and PRISM. In: Berry, G., Comon, H., Finkel, A. (eds.) CAV 2001. LNCS, vol. 2102, pp. 194–206. Springer, Heidelberg (2001)
23. Kwiatkowska, M.Z., Norman, G., Parker, D.: Stochastic model checking. In: Bernardo, M., Hillston, J. (eds.) SFM 2007. LNCS, vol. 4486, pp. 220–270. Springer, Heidelberg (2007)
24. Lanotte, R., Maggiolo-Schettini, A., Troina, A.: Parametric probabilistic transition systems for system design and analysis. FAC 19(1), 93–109 (2007)

25. Passmore, G.O., Jackson, P.B.: Combined decision techniques for the existential theory of the reals. In: Carette, J., Dixon, L., Coen, C.S., Watt, S.M. (eds.) MKM 2009, Held as Part of CICM 2009. LNCS, vol. 5625, pp. 122–137. Springer, Heidelberg (2009)
26. Puterman, M.L.: Markov decision processes: Discrete stochastic dynamic programming. John Wiley and Sons, Chichester (1994)
27. Ratschan, S.: Efficient solving of quantified inequality constraints over the real numbers. CoRR cs.LO/0211016 (2002)
28. Ratschan, S.: Safety verification of non-linear hybrid systems is quasi-semidecidable. In: Kratochvíl, J., Li, A., Fiala, J., Kolman, P. (eds.) TAMC 2010. LNCS, vol. 6108, pp. 397–408. Springer, Heidelberg (2010)
29. Strauch, R.E.: Negative dynamic programming. Annals of Mathematical Statistics 37(4), 871–890 (1966)
30. van der Wal, J.: Stochastic dynamic programming. The Mathematical Centre, Amsterdam (1981)

Formalizing Probabilistic Safety Claims

Heber Herencia-Zapana[1,*], George Hagen[2], and Anthony Narkawicz[2]

[1] National Institute of Aerospace, Hampton, VA
[2] NASA Langley Research Center, Hampton, VA

Abstract. A safety claim for a system is a statement that the system, which is subject to hazardous conditions, satisfies a given set of properties. Following work by John Rushby and Bev Littlewood, this paper presents a mathematical framework that can be used to state and formally prove probabilistic safety claims. It also enables hazardous conditions, their uncertainties, and their interactions to be integrated into the safety claim. This framework provides a formal description of the probabilistic composition of an arbitrary number of hazardous conditions and their effects on system behavior. An example is given of a probabilistic safety claim for a conflict detection algorithm for aircraft in a 2D airspace. The motivation for developing this mathematical framework is that it can be used in an automated theorem prover to formally verify safety claims.

1 Introduction

In [9,5], Rushby and Littlewood present a framework for formalizing safety claims for systems, which is illustrated with probabilistic safety claims in an automated theorem prover. In this paper, the mathematics behind their ideas is formalized. The mathematical framework presented will equip the reader to formalize a probabilistic safety claim about a system with an *arbitrary number* of hazardous conditions in a precise mathematical formula that can be proved in a theorem prover. One advantage that this adds to Rushby's approach is that it provides a formal way for new hazardous conditions to be considered without changing the overall structure of the safety argument.

A *safety claim* is a statement that a system will behave in a desired manner with an acceptable probability. A *hazard* is a state or set of conditions that, together with other conditions in the environment, will cause a system to enter an undesirable state. For more on terminology related to safety analyses and system hazards, see [4]. In this paper, a *potentially hazardous condition*, referred to hereafter simply as a *hazardous condition*, is anything that may cause a system to behave in an unexpected or undesired manner. Examples of hazardous conditions may include such things as signal noise, timing delays, or interruptions of service. The number of hazardous conditions in a safety argument typically depends on the available expertise in analyzing the system, and it is important to

---

* This work was supported in part by the National Aeronautics and Space Administration under NASA Cooperative Agreement NCC-1-02043.

M. Bobaru et al. (Eds.): NFM 2011, LNCS 6617, pp. 162–176, 2011.

allow the safety claim to evolve as new factors are uncovered. Hazardous Conditions typically have uncertainties associated with them, and they can therefore be modeled as random variables. This paper proposes a formal mathematical framework for modeling hazardous conditions as random variables in a way that makes it possible to also model interactions between different hazardous conditions. The underlying concepts are due to Rushby [9], but this paper gives precise mathematical definitions of probabilistic safety claims and provides a concrete example of such a claim. The example presented is for a state based conflict detection system.

In general, a probabilistic safety claim can be expressed as a mathematical formula stating that the probability of a certain event occurring is bounded in a specific range. Since new factors affecting system behavior may become known in the future, is desirable for the safety argument to be easily updated without reconstructing the entire argument. The mathematical formalism presented in this paper allows hazardous conditions to be modeled in a way that is modular and can handle the addition of new hazardous conditions.

The interdependency between random variables, e.g., hazardous conditions, is modeled by *probabilistic kernels*, which uses the fact that the set of all hazardous conditions can be modeled via a concatenation of σ-algebras, as seen in [10]. A σ-algebra is a set of sets where it is possible to assign probabilities to elements in a consistent way, and is often used to model events. See Section 2.5 for more compete discussion of probabilistic kernels.

The composition of hazardous conditions is formalized through the concatenations of Lebesgue integrals. This allows hazardous conditions and assumptions to be incorporated into the formula in a modular fashion. The majority of the complexity is encapsulated in sub-formulas specific to the assumption or hazardous condition in question, while the main safety claim formula need only be modified in a limited and systematic fashion. The mathematics behind this formalization is presented in following sections.

2 Systems

Systems of interest are those that can modeled as well-defined functions with inputs and outputs. In this formalization, a system is a function S with n *parameters* and m *variables*:

$$S : (K_1 \times \ldots \times K_n; L_1 \times L_2 \times \ldots \times L_m) \to \mathcal{T}_0,$$

where $K_1, \ldots, K_n$ and $L_1, \ldots, L_m$ are the *types* of the n parameters and m variables of S, respectively. The type $\mathcal{T}_0$ consists of the possible outputs of S, and if $k_i \in K_i$ and $l_j \in L_j$, then $S(k_1, \ldots, k_n; l_1, \ldots, l_m)$ is an element of $\mathcal{T}_0$. It will sometimes be useful to view the system S as only a function on its m variables $l_1, \ldots, l_m$, where the n parameters $k_1, \ldots, k_n$ are fixed, the notation $S_{k_1, \ldots, k_n}(l_1, \ldots, l_m)$ is used in place of $S(k_1, \ldots, k_n; l_1, \ldots, l_m)$. Because the system S will be modeled as a random variable in order to reason about it probabilistically, it is assumed that $\mathcal{T}_0$ is a measure space with σ-algebra $\sigma(\mathcal{T}_0)$.

The values $k_1, \ldots, k_n$ of the parameters of the system are predetermined and their values, without any errors, are known to the system. In a real system, the values of the input variables $l_1, \ldots, l_m$ are measured by the system, and the measurements can have errors. These errors may be due to either expected accuracy problems with instruments or faulty components in other systems from which the instruments receive data. In either case, events that can cause such measurement errors in the system are referred to as *hazardous conditions*, which are formally modeled in this context in Section 2.2.

For a system described in this way, a probabilistic safety claim is a statement that, given some set of possible hazardous conditions, the probability that the value of the system S lies in a predetermined subtype Z_0 of T_0 is contained in particular range $[p_0, p_1]$.

2.1   Modeling Uncertainty in System Variables

As noted above, the values of the n parameters $k_1, \ldots, k_n$ of the system S are known to the system without errors. The errors in the measurements of the input variables $l_1, \ldots, l_m$ can be modeled as random variables

$$l_i : \; \Omega \to L_i$$

where $(\Omega, \sigma(\Omega))$ is a probability space ($\sigma(\Omega)$ is a σ-algebra on the set Ω). Thus, given a fixed value $\kappa = \{k_1, \ldots, k_m\}$ for the set of parameters, the system S becomes a random variable as well:

$$S_\kappa : \; (\Omega, \sigma(\Omega)) \to (T_0, \sigma(T_0))$$
$$\chi \mapsto S(\kappa, l_1(\chi), l_2(\chi), \ldots, l_m(\chi)) \in T_0.$$

Thus, if Z_0 is any measurable subset of T_0 (i.e. an element of $\sigma(T_0)$), and if the distributions of the random variables l_i are known, then the probability that the output of S_κ lies in Z_0 can be computed.

2.2   Modeling Hazardous Conditions

As noted in Section 2, the errors in the variables $l_1, \ldots, l_m$ of the system S may be due to either expected accuracy problems with instruments or faulty components in other systems from which the instruments receive data. Conditions in the environment of a system that can cause such measurement errors in the system are referred to as *hazardous conditions*.

In a model of the environment of the system S, which includes the output of possible hazardous conditions, these conditions can be modeled as random variables

$$H_i : (\Omega, \sigma(\Omega)) \to (T_i, \sigma(T_i)),$$

where $i \geq 1$, T_i is an arbitrary type, and $\sigma(T_i)$ is a σ-algebra on T_i. This modeling framework allows for the computation of the probability that a hazardous condition H_i takes values in a particular subtype of T_i.

2.3   Modeling All Possible Hazardous Conditions

It is possible that the environment of a system S has an arbitrary number of hazardous conditions. Further, it may be the case that when developing a model of system behavior, only a few of these possible hazardous conditions are understood. Even in this case, the environment of the system can be modeled as an infinite product

$$\mathcal{T} = \prod_{i=0}^{\infty} \mathcal{T}_i,$$

where $\mathcal{T}_0$ is the type of the output values of S, and for $i \geq 1$, $\mathcal{T}_i$ is the type of the output of the i-th hazardous condition H_i. This is a measure space with σ−algebra $\sigma(\mathcal{T}) = \prod_{i=0}^{\infty} \sigma(\mathcal{T}_i)$. This type of model is possible even though there are only finitely many hazardous conditions, because for i large enough, $\mathcal{T}_i$ can be defined to be a singleton set, and $H_i \colon \Omega \to \mathcal{T}_i$ as the trivial function.

In general, for any choice $\kappa = \{k_1, \ldots, k_m\}$ of system parameters, there is a random variable

$$S_\kappa \times H_1 \times H_2 \times \ldots \colon \Omega \to \mathcal{T} \tag{1}$$

given by $\chi \mapsto (S \circ (\kappa \times l_1 \times l_2 \times \ldots \times l_m))(\chi), \times H_1(\chi), H_2(\chi), \ldots)$. Thus, the type $\mathcal{T}$ inherits the structure of a probability space from $(\Omega, \sigma(\Omega))$ and from the random variable (1).

Definition 1. *Since the random variable* (1) *depends on the choice κ of parameters for the system S, the probability distribution of $\mathcal{T}$ depends on κ as well. Thus, the probability function on $\mathcal{T}$ induced by S and κ will be denoted P_κ.*

If β is a subtype of $\mathcal{T}$, then the probability $P_\kappa[\beta]$ can be defined and possibly computed.

2.4   Probabilistic Safety Claims

Suppose that the r hazardous conditions $H_1, \ldots, H_r$, the corresponding types $\mathcal{T}_1, \ldots, \mathcal{T}_r$, and the probability distributions of the random variables H_i are all known. Let $\beta_i \in \sigma(\mathcal{T}_i)$ be events in $\mathcal{T}_i$. That is, each β_i is a subtype of $\mathcal{T}_i$, and the probability that the value of H_i is an element of β_i can be computed.

In general, the probability that the value of every H_i (for $i = 1, \ldots, n$) is in β_i and that the system S takes a value in β is given by

$$P_\kappa[\sigma(\beta_0, \beta_1, \ldots, \beta_r)],$$

where $\sigma(\beta_0, \beta_1, \ldots, \beta_r)$ is the concatenation of σ-algebras given by

$$\sigma(\beta_0, \beta_1, \ldots, \beta_2) = \{\omega \in T | \omega_0 \in \beta_0, \ \omega_1 \in \beta_1, \ \ldots, \ \text{and} \ \omega_n \in \beta_r\}.$$

An introduction to concatenations of σ-algebras can be found in [10]. As more sigma algebras are concatenated, the concatenation becomes smaller:

$$\sigma(\beta_0) \supseteq \sigma(\beta_0, \beta_1) \supseteq \sigma(\beta_0, \beta_1, \beta_2) \supseteq \sigma(\beta_0, \beta_1, \beta_2, \beta_3) \supseteq \ldots,$$

and the sequence of associated probabilities is decreasing:

$$P_\kappa[\sigma(\beta_0)] \geq P_\kappa[\sigma(\beta_0, \beta_1)] \geq P_\kappa[\sigma(\beta_0, \beta_1, \beta_2)] \geq P_\kappa[\sigma(\beta_0, \beta_1, \beta_2, \beta_3)] \geq \ldots$$

With this formalism, it is possible to formally state a safety claim in a way that can be specified in an automated theorem prover. Let p_0 and p_1 be any two probabilities, and let β_0 and α_0 be two subtypes of T_0.

Definition 2. *A probabilistic safety claim on the system S is a statement of the following form: If $l_1^{meas}, \ldots, l_m^{meas}$ are measured values for the variables of the system S such that the system output value $S(\kappa'; l_1^{meas}, \ldots, l_m^{meas})$ is an element of α_0, then the probability that the system S, with parameter set κ, takes values in β_0 is between p_0 and p_1. i.e.*

$$P_\kappa[\sigma(\beta_0)] \in [p_0, p_1]. \tag{2}$$

It should be noted that the hypothesis that $S(\kappa; l_1^{meas}, \ldots, l_m^{meas})$ is an element of α_0 is not needed to formally state a safety claim in a theorem prover. However, such a hypothesis will often be required to prove such a safety claim, because the expected values of the random variables $l_1, \ldots, l_m$ are often equal to $l_1^{meas}, \ldots, l_m^{meas}$, respectively. Thus, the computation of the probability (2) often depends on these measured values.

Another important property of this definition is that the set of system parameters κ' is different than the set κ. In practice, the parameter set κ' may be chosen so that if $S(\kappa; l_1^{meas}, \ldots, l_m^{meas})$ is an element of T_0, then the probability $P_\kappa[\sigma(\beta_0)]$ is more likely to be between p_0 and p_1. An example of this is given below in Section 3.2, where the radius of the protected zone around an aircraft and the lookahead time for conflict detection are artificially increased to ensure that if a conflict detection probe returns `False`, then the probability that the two aircraft are actually in conflict (using the correct radius and lookahead time) is reduced.

It is also important to note that neither the infinite product T nor concatenations of sigma algebras are required to make a safety claim on a system. However, as illustrated in Section 2.4, both of these concepts are necessary when developing a formal proof of such a safety claim.

An example of such a safety claim, for a conflict detection probe, is presented in Section 3.

2.5   Dependence of System Variables on Hazardous Conditions

In general, the hazardous conditions H_i for the system S may have an impact on the accuracy of the variables of S, which are modeled as random variables $l_1, \ldots, l_m$, as in Section 2.1. It is possible to model the dependence of the random variables l_i on the random variables H_i using probabilistic kernels. This section provides a brief introduction to probabilistic kernels, and the construction follows that in [10].

Probabilistic Kernels. Suppose that the distribution of the random variable $S_\kappa\colon \Omega \to T_0$ (the output of the system S) depends on the value of $H_1\colon \Omega \to T_1$. That is, if $\omega_1 \in T_1$, then there is an associated random variable $\Omega \to T_0 \times T_1$ given by

$$\chi \mapsto (S_\kappa(\chi), \omega_1), \tag{3}$$

for $\chi \in \Omega$ and the distribution of this random variable depends on the choice of ω_1. If this is the case, then there is an induced probability function

$$p\colon T_1 \times \sigma(T_0) \to [0,1].$$

Since this function depends on the parameter κ of the system S, it will be written as p_κ. Given $\omega_1 \in T_1$ and $\beta_0 \in T_0$, the corresponding output of p_κ is written $p_\kappa(\omega_1; \beta_0)$, which is the probability that the random variable (3) takes a value in $\beta_0 \times \{\omega_1\}$. If β_0 and β_1 are elements of $\sigma(T_0)$ and $\sigma(T_1)$, respectively, then the probability $P_\kappa[\sigma(\beta_0, \beta_1)]$, defined in Section 2.4, is given by the Lebesgue integral

$$P_\kappa[\sigma(\beta_0, \beta_1)] = \int_{\omega_1 \in \beta_1} \int_{\omega_0 \in \beta_0} p_\kappa(\omega_1; d\omega_0) p(d\omega_1).$$

It is important to note that there is no assumption of independence required for this equation. In order to compute this integral, it is necessary to know how the random variable S_κ depends on the random variable H_1.

Probabilistic Kernels with Several Variables. The construction of this probabilistic kernel can be generalized to handle multiple hazardous conditions as follows. Suppose as above that the random variable $S_\kappa\colon \Omega \to T_0$ depends on the random variables $H_1, \ldots, H_r$. Suppose further that for all $i = 1, \ldots, r$, the random variable $H_i\colon \Omega \to T_i$ depends on the values of the random variables $H_{i+1}, \ldots, H_r$. That is, S_κ depends on $H_1, \ldots, H_r$; H_1 depends on $H_2, \ldots, H_r$; H_2 depends on $H_3, \ldots, H_r$; etc. As above, this means that if $i \geq 0$, then for $\omega_{i+1} \in T_{i+1}, \ldots, \omega_r \in T_r$, the distribution of the random variable $\Omega \to T_i \times \ldots T_r$, given by

$$\chi \mapsto (H_i(\chi), \omega_{i+1}, \ldots, \omega_r), \tag{4}$$

depends on the values of $\omega_{i+1}, \ldots, \omega_r$ (by abuse of notation, $H_0 = S_\kappa$ in this equation). Further, there is an induced probability function

$$p\colon T_r \times \cdots \times T_{i+1} \times \sigma(T_i) \to [0,1]$$

given by $(\omega_r, \ldots, \omega_{i+1}; \beta_i) \mapsto p(\omega_r, \ldots, \omega_{i+1}; \beta_i)$, which is the probability that the random variable (4) takes a value in $\beta_i \times \{\omega_1\} \times \cdots \times \{\omega_r\}$. This probability is written with a subscript of κ if $i = 0$ to indicate the dependence on the system parameter κ. If β_i is an element of the $\sigma-$algebra $\sigma(T_i)$ for $i = 0, \ldots, r$, then the probability $P_\kappa[\sigma(\beta_0, \ldots, \beta_r)]$ (cf. Section 2.4) is given by the Lebesgue integral

$$\int_{\omega_r \in \beta_r} \cdots \int_{\omega_0 \in \beta_0} p_\kappa(\omega_r, \ldots, \omega_1; d\omega_0) p(\omega_r, \ldots, \omega_1; d\omega_1) \cdots p(\omega_r; d\omega_{r-1}) p(d\omega_r).$$

An example of such an integral is given in Section 3.2, where this integral is explicitly computed to prove a safety claim for a conflict detection system.

3   A Proved Safety Claim for Conflict Detection

This section illustrates the framework presented in the previous sections with an example of a safety claim for a conflict detection probe in a 2D airspace. This is an algorithm that detects conflicts between two aircraft, referred to here as the *ownship* and the *intruder*. Its variables include the *state* information of the aircraft, which consists of their current positions and velocities, which are represented by points and vectors in $\mathbb{R}^2$, respectively.

Aircraft trajectories are represented by a point moving at constant linear speed, i.e., if the current state of an aircraft is given by the position $\mathbf{s}$ and the velocity vector $\mathbf{v}$, then its predicted position at time t is $\mathbf{s} + t\,\mathbf{v}$. In this paper, the vectors $\mathbf{s}_o, \mathbf{v}_o, \mathbf{s}_i$, and $\mathbf{v}_i$ represent the ownship's position and velocity and the intruder's position and velocity, respectively. The formalization presented here usually considers a relative view where the intruder is fixed at the origin of the coordinate system. The vectors $\mathbf{s}$ and $\mathbf{v}$ will denote the relative position $\mathbf{s}_o - \mathbf{s}_i$ and the relative velocity $\mathbf{v}_o - \mathbf{v}_i$, respectively.

In the airspace, it is required that aircraft maintain a certain horizontal separation, specified by a minimum horizontal distance D. Typically, D is 5 nautical miles. A conflict detection probe detects conflicts between the aircraft over some given lookahead time T, usually less than five minutes. A *conflict* between the ownship and the intruder aircraft occurs when there is a time $t \in [0, T]$ at which the horizontal distance between the aircraft is projected to be less than D, i.e.,

$$\|(\mathbf{s}_o + t\,\mathbf{v}_o) - (\mathbf{s}_i + t\,\mathbf{v}_i)\| < D.$$

Since $(\mathbf{s}_o + t\,\mathbf{v}_o) - (\mathbf{s}_i + t\,\mathbf{v}_i) = (\mathbf{s}_o - \mathbf{s}_i) + t\,(\mathbf{v}_o - \mathbf{v}_i)$, the predicate that characterizes conflicts can be defined in terms of the relative vectors $\mathbf{s} = \mathbf{s}_o - \mathbf{s}_i$ and $\mathbf{v} = \mathbf{v}_o - \mathbf{v}_i$, i.e., the relative position and velocity vectors, respectively, of the ownship with respect to the intruder. The predicate *horizontal_conflict?*, parametric on the lookahead time T and the horizontal distance D, is formally defined as follows.

$$horizontal\_conflict?(D, T, \mathbf{s}, \mathbf{v}) \equiv \exists t \in [0, T] : \|\mathbf{s} + t\,\mathbf{v}\| < D.$$

A conflict detection probe is an algorithm that computes whether the predicate *horizontal_conflict?* holds for the current states of two aircraft. One example of such an algorithm is cd2d, developed at NASA Langley [6]. Formally, a conflict detection probe is defined as a function

$$\texttt{cd}: \mathbb{R}^+ \times \mathbb{R}^+; \mathbb{R}^2 \times \mathbb{R}^2 \longrightarrow \{\texttt{True}, \texttt{False}\}.$$

It is designed so that $\texttt{cd}(D, T; \mathbf{s}, \mathbf{v}) \Longleftrightarrow horizontal\_conflict?(D, T, \mathbf{s}, \mathbf{v})$, for all $D, T \in \mathbb{R}^+$ and $\mathbf{s}, \mathbf{v} \in \mathbb{R}^2$. Such a conflict detection probe is a system, as described above. The distance D and time T are parameters of cd because their values are typically known to the aircraft without error. For instance, the airspace may have a 5 nautical mile minimum horizontal separation, and a standards document may define the lookahead time T to be 3 minutes.

3.1   GPS and ADS-B Hazardous Conditions

If the ownship is using the conflict probe cd to detect conflicts, it must depend on broadcast signals from the intruder to determine the intruder's position and velocity vectors. In this example, the aircraft use Automatic Dependent Surveillance Broadcast (ADS-B)[8] messages to communicate their positions and velocities, and it is assumed that ADS-B messages with state information are sent by each aircraft once per second. When the ownship uses the algorithm cd, it is possible that several consecutive position and velocity updates from the intruder have been dropped due to signal attenuation, which results in greater uncertainty in the values of s_i and v_i. Thus, ADS-B message loss due to signal attenuation can be modeled as a hazardous condition:

$$H_{2,adsb} \colon \Omega \to T_{2,adsb} \qquad T_{2,adsb} = \{0, 1, 2, 3, \dots\}.$$

The random variable $H_{2,adsb}$ returns the number of consecutive ADS-B messages from the intruder that were not received by the ownship, since the last received message from the intruder. At a given instant of time when a conflict detection probe is used, τ will be used to represent this number of consecutive dropped messages. The number τ is easy for the ownship to compute, since it just has to know when the last ADS-B update from the intruder was received. The number τ is an integer, and τ_s will be used to represent the time period τ *seconds*.

In addition, if the conflict detection probe cd is being used by the ownship, then the position and velocity vectors s_o, s_i, v_o, and v_i will be estimated using instruments such as GPS. These instruments can be faulty or have expected errors. For instance, there may be some error in the position predicted by a GPS device. The effects of uncertainty in positions and velocities of aircraft on conflict detection have been studied before [3].

Error in GPS is modeled as a hazardous condition as follows. The vectors s_i^m and v_i^m represent the intruder's reported position and velocity vectors, respectively, from the last ADS-B signal that was received by the ownship, and the vectors s_o^m and v_o^m represent the ownship's measured position and velocity at that time. The relative vectors s^m and v^m are defined by $s^m = s_o^m - s_i^m$ and $v^m = v_o^m - v_i^m$. The true positions of the ownship and the intruder at the time when the vectors s^m and v^m were measured (τ seconds ago) are given by $s_o - \tau_s v_o$ and $s_i - \tau_s v_i$, respectively. It is clear that if the measured vectors s_o^m, v_o^m, s_i^m, and v_i^m have no error, then $s^m = s - \tau_s v$ and $v^m = v$. In this case, if $cd(D, T + \tau_s; s^m, v^m) = $ False, then $cd(D, T; s, v) = $ False as well. Thus, the symbol e (called *GPS error*) denotes the fact that one of the following inequalities is satisfied.

$$(i)\ \ ||(s_o - \tau_s v_o) - s_o^m|| \geq a_o \qquad ||(s_i - \tau_s v_i) - s_i^m|| \geq a_i\ \ (iii)$$
$$(ii)\ \ ||v_o - v_o^m|| \geq b_o \qquad\qquad\qquad ||v_i - v_i^m|| \geq b_i\ \ (iv)$$

Here, the distances a_o and a_i and the speeds b_o and b_i are predetermined parameters. For instance, one set of these parameters that is used in the proof of a safety claim in Section 3.3 is $a_o, a_i = 30$ m and $b_o, b_i = 0.3$ m/s, which correspond to certain navigation accuracy categories (NAC_P 9 and NAC_V 4, respectively), as specified by RTCA, Inc. in DO-242A for precision in ADS-B messages [8]. This specification is for 95 percent confidence intervals on the position and velocity vectors of aircraft, within the given ranges. Other choices for $a_o, a_i, b_o,$ and b_i may be considered, and thus in the next few sections they are simply treated as variables.

With this construction, GPS error is modeled as a hazardous condition

$$H_{1,gps} \colon \Omega \to \mathcal{T}_{1,gps} \qquad \text{(where} \quad \mathcal{T}_{1,gps} \ = \{e, \neg e\}).$$

The return type $\mathcal{T}_{2,adsb}$ of the second hazardous condition $H_{2,adsb}$ represents the number of seconds since the last ADS-B update from the intruder aircraft. If d is any non-negative integer, it is possible to formally define the probability that the most recent ADS-B message that was sent by the intruder and detected/decoded by the ownship occurred within the last d seconds.

As noted above, inaccuracies in the measurements of the positions s_o and s_i and the velocities v_o and v_i imply that the conflict detection probe cd can be modeled as a random variable:

$$cd_{D,T} \colon \ \Omega \to \mathcal{T}_0 = \{\texttt{True}, \texttt{False}\}$$
$$\chi \mapsto cd(D, T; s(\chi), v(\chi))$$

This random variable depends on the hazardous conditions $H_{1,gps}$ and $H_{2,adsb}$.

3.2   Probabilistic Kernels in Conflict Detection

It is clear that the random variable $S_{D,T}$, which takes values in $\{\texttt{False}, \texttt{True}\}$, depends on the hazardous conditions $H_{1,gps}$ and $H_{2,adsb}$. Thus, as in Section 2.5, if $\beta_2 \subset \mathcal{T}_{2,adsb}, \beta_1 \subset \mathcal{T}_{1,gps}$, and $\beta_0 \subset \mathcal{T}_0 = \{\texttt{False}, \texttt{True}\}$, then the probability that $H_{2,adsb}$ and $H_{1,gps}$ take values in β_2 and β_1, respectively, and that $cd_{D,T}$ takes a value in β_0, is given by

$$P_{D,T}[\sigma(\beta_0, \beta_1, \beta_2)] = \int_{\omega_2 \in \beta_2} \int_{\omega_1 \in \beta_1} \int_{\omega_0 \in \beta_0} p_{D,T}(\omega_1, \omega_2; d\omega_0) p(\omega_2; d\omega_1) p(d\omega_2).$$

As a simple example of this, if $i \in \mathcal{T}_{2,adsb}$, then the probability that the random variable (conflict probe) $cd_{D,T}$ returns $\texttt{True}$, that there is no error in GPS, and that the last ADS-B signal from the intruder aircraft was exactly i seconds ago is given by

$$P_{D,T}[\sigma(\{\texttt{True}\}, \{\neg e\}, \{i\})]$$
$$= \int_{\omega_2 \in \{i\}} \int_{\omega_1 \in \{\neg e\}} \int_{\omega_0 \in \{\texttt{True}\}} p_{D,T}(\omega_1, \omega_2; d\omega_0) p(\omega_2; d\omega_1) p(d\omega_2)$$
$$= \int_{\omega_1 \in \{\neg e\}} \int_{\omega_0 \in \{\texttt{True}\}} p_{D,T}(\omega_1, i; d\omega_0) p(i; d\omega_1) p(\{i\})$$
$$= \int_{\omega_0 \in \{\texttt{True}\}} p_{D,T}(\neg e, i; d\omega_0) p(i; \{\neg e\}) p(\{i\})$$
$$= p_{D,T}(\neg e, i; \{\texttt{True}\}) p(i; \{\neg e\}) p(\{i\})$$

The random variables $cd_{D,T}$, $H_{1,gps}$, and $H_{2,adsb}$ are all discrete, so the probability that $cd_{D,T}$ returns True, which is given by $P_{D,T}[\sigma(\{\text{True}\})]$, can be computed as an infinite sum as follows.

$$P_{D,T}[\sigma(\{\text{True}\})]$$

$$= \int_{\omega_2 \in \{0,1,2,\dots\}} \int_{\omega_1 \in \{e, \neg e\}} \int_{\omega_0 \in \{\text{True}\}} p_{D,T}(\omega_1, \omega_2; d\omega_0) p(\omega_2; d\omega_1) p(d\omega_2)$$

$$= \sum_{i=0}^{\infty} \int_{\omega_1 \in \{e, \neg e\}} \int_{\omega_0 \in \{\text{True}\}} p_{D,T}(\omega_1, i; d\omega_0) p(i; d\omega_1) p(i)$$

$$= \sum_{i=0}^{\infty} \left(\int_{\omega_0 \in \{\text{True}\}} p_{D,T}(e, i; d\omega_0) p(i; \{e\}) p(\{i\}) \right. \tag{5}$$

$$+ \left. \int_{\omega_0 \in \{\text{True}\}} p(\neg e, i; d\omega_0) p(i; \{\neg e\}) p(\{i\}) \right)$$

$$= \sum_{i=0}^{\infty} (p_{D,T}(e, i; \{\text{True}\}) p(i; \{e\}) p(\{i\})$$

$$+ \ p_{D,T}(\neg e, i; \{\text{True}\}) p(i; \{\neg e\}) p(\{i\}))$$

Distribution of the ADS-B Hazardous Condition. Under the assumption that there is no ADS-B signal interference due to multiple intruder aircraft, the distribution of the hazardous condition $H_{2,adsb}$ follows a Poisson distribution, as discussed in [2]. In that paper, the probability that a given ADS-B message from the intruder aircraft will not be detected and decoded by the ownship, which is equal to $p(\{0\})$, is (approximately) given by $p(\{0\}) = 1 - \left(\frac{r}{r_0}\right)^k$ with $r \leq r_0$, where $k = 6.4314$ and $r_0 = 96.6$ nmi [2]. The number r is the current distance between the two aircraft. Thus, if it is known that the ownship and the intruder are no greater than 60 nmi apart, a reasonable distance for most commercial aircraft given short lookahead times such as 3 minutes, then $p(\{0\}) \geq \eta$, where

$$\eta = 0.953.$$

The key assumption that can be used to deduce that $H_{2,adsb}$ follows a Poisson distribution is that whether any particular ADS-B message from the intruder aircraft is received by the ownship is independent from whether any other, different, ADS-B message from the intruder is received. Under this assumption,

$$p(\{i\}) = \eta(1 - \eta)^i \qquad \text{for } i \geq 0.$$

This is because the last i messages (sent $0, 1, \dots$ and $i-1$ seconds ago) have been dropped, which has a probability of $(1 - \eta)^i$ of occurring, and the message sent exactly i-seconds ago was not dropped, which has a probability of η of occurring. The equation above can be used to replace $p(\{i\})$ in Equation (5).

Probability of GPS Error. A key assumption in this example is that probabilities p_{so}, p_{si}, p_{vo} and p_{vi} are known that satisfy the following properties.

- At any given time, the probability, that the distance between the ownship's predicted position (by GPS) and its actual position is at least a_o, is bounded above by p_{so}.
- At any given time, the probability, that the difference (speed) between the ownship's predicted velocity (by GPS) and its actual velocity is at least b_o, is bounded above by p_{vo}.
- At any given time, the probability, that the distance between the intruder's predicted position (by GPS) and its actual position is at least a_i, is bounded above by p_{si}.
- At any given time, the probability, that the difference (speed) between the intruder's predicted velocity (by GPS) and its actual velocity is at least b_i, is bounded above by p_{vi}.

Specific examples of such numbers can be found in the RTCA, Inc. document DO-242A [8], which provides examples for the analyses in Section 3.3.

At a given instant of time, the actual positions of the ownship and the intruder τ seconds ago were given by $\mathbf{s}_o - \tau_s \mathbf{v}_o$ and $\mathbf{s}_i - \tau_s \mathbf{v}_i$, respectively. The positions at that time, as predicted by GPS, are by definition given by $\mathbf{s}_o^m$ and $\mathbf{s}_i^m$, respectively. Thus, the following four equations hold.

$$P[||(\mathbf{s}_o - \tau^m \mathbf{v}_o) - \mathbf{s}_o^m|| \geq a_o] \leq p_{so} \qquad P[||(\mathbf{s}_i - \tau^m \mathbf{v}_i) - \mathbf{s}_i^m|| \geq a_i] \leq p_{si}$$
$$P[||\mathbf{v}_o - \mathbf{v}_o^m|| \geq b_o] \leq p_{vo} \qquad P[||\mathbf{v}_i - \mathbf{v}_i^m|| \geq b_i] \leq p_{vi}$$

By the definition of the error $\mathbf{e}$ in Section 3.1, $p(i; \{\mathbf{e}\}) \leq p_{so} + p_{vo} + p_{si} + p_{vi}$. Set $p_{error} = p_{so} + p_{vo} + p_{si} + p_{vi}$. Equation (5) implies that if d is any integer (a specific number of seconds), then

$$\begin{aligned}
&P_{D,T}[\sigma(\{\texttt{True}\})] \\
&= \sum_{i=0}^{\infty} (p_{D,T}(\mathbf{e}, i; \{\texttt{True}\})p(i; \{\mathbf{e}\})p(\{i\}) \\
&\qquad\qquad\qquad + p_{D,T}(\neg\mathbf{e}, i; \{\texttt{True}\})p(i; \{\neg\mathbf{e}\})p(\{i\})) \\
&\leq \sum_{i=0}^{\infty} (P_{error}\, \eta(1-\eta)^i + p_{D,T}(\neg\mathbf{e}, i; \{\texttt{True}\})p(i; \{\neg\mathbf{e}\})p(\{i\})) \\
&= P_{error} + \sum_{i=0}^{\infty} p_{D,T}(\neg\mathbf{e}, i; \{\texttt{True}\})p(i; \{\neg\mathbf{e}\})p(\{i\}) \\
&\leq P_{error} + \sum_{i=0}^{\infty} p_{D,T}(\neg\mathbf{e}, i; \{\texttt{True}\})\eta(1-\eta)^i \\
&\leq P_{error} + \sum_{i=d+1}^{\infty} \eta(1-\eta)^i + \sum_{i=0}^{d} p_{D,T}(\neg\mathbf{e}, i; \{\texttt{True}\})\eta(1-\eta)^i \\
&= P_{error} + (1-\eta)^{d+1} + \sum_{i=0}^{d} p_{D,T}(\neg\mathbf{e}, i; \{\texttt{True}\})\eta(1-\eta)^i
\end{aligned}$$

(6)

The number d, which is an element of $\mathcal{T}_{2,adsb}$ can chosen so that the finite sum is a good approximation to the infinite sum (since $(1 - \eta)^{d+1}$ is quite small). This equation is true for any choice of d.

An Upper Bound on the Probability of Failure. Equation (6) implies that if $p_{D,T}(\neg e, i; \{\texttt{True}\}) = 0$ for $i \in \{0, \ldots, d\}$, then the probability that $\texttt{cd}(D, T; \mathbf{s}, \mathbf{v}) = \texttt{True}$, which is given by $P_{D,T}[\sigma(\{\texttt{True}\})]$, is bounded above by $P_{error} + (1 - \eta)^{d+1}$. As noted in Section 2.4, to mitigate the effect of measurement errors on the conflict detection probe $\texttt{cd}$, a positive distance ψ and a positive time λ can be artificially added to the distance D and the time T when they are used as parameters in $\texttt{cd}$. The important question here is how large do ψ and λ need to be so that if $\texttt{cd}(D + \psi, T + \lambda; \mathbf{s}^m, \mathbf{v}^m) = \texttt{False}$, then $p_{D,T}(\neg e, i; \{\texttt{True}\}) = 0$ for $i \in \{0, \ldots, d\}$. This question is answered by the following lemma. It refers to the distances a_o and a_i and the speeds b_o and b_i that define the probabilities $p_{so}, p_{vo}, p_{si}, p_{vi}$ (cf. Section 3.1).

Lemma 1. *If $\lambda = d$ seconds, $\psi = a_o + a_i + (T + \lambda)(b_o + b_i)$, and $\texttt{cd}(D + \psi, T + \lambda; \mathbf{s}^m, \mathbf{v}^m) = \texttt{False}$, then $p_{D,T}(\neg e, i; \{\texttt{True}\}) = 0$ for $i \in \{0, \ldots, d\}$.*

Proof. Suppose that $\neg e$ holds, and recall from Section 3.1 that τ denotes the number of seconds since the ownship successfully received position and velocity updates from the intruder aircraft's ADS-B device. Suppose that $\tau = i$, where $i \leq d$. Then in order to show that $p_{D,T}(\neg e, i; \{\texttt{True}\}) = 0$, it suffices to prove that $\texttt{cd}(D, T; \mathbf{s}, \mathbf{v}) = \texttt{False}$. Since $\tau \leq d$, it follows from the hypotheses of the lemma that $\texttt{cd}(D + \psi, T + \tau_s; \mathbf{s}^m, \mathbf{v}^m) = \texttt{False}$. Further, since $\neg e$ holds, the equations $||(\mathbf{s}_o - (i \text{ sec})\mathbf{v}_o) - \mathbf{s}_o^m|| < a_o$ and $||(\mathbf{s}_i - (i \text{ sec})\mathbf{v}_i) - \mathbf{s}_i^m|| < a_i$ and $||\mathbf{v}_o - \mathbf{v}_o^m|| < b_o$ and $||\mathbf{v}_i - \mathbf{v}_i^m|| < b_i$ are all satisfied.

By contradiction, suppose that $\texttt{cd}(D, T; \mathbf{s}, \mathbf{v}) = \texttt{True}$, and choose $t^* \in [0, T]$ such that $||\mathbf{s} + t^*\mathbf{v}|| < D$. Then $t^* + \tau_s \in [0, T + \lambda]$ and since $\mathbf{s} = \mathbf{s}_o - \mathbf{s}_i$ and $\mathbf{v} = \mathbf{v}_o - \mathbf{v}_i$, it follows that

$$
\begin{aligned}
&||\mathbf{s}^m + (t^* + \tau_s)\mathbf{v}^m|| \\
&= ||(\mathbf{s}_o^m - \mathbf{s}_i^m) + (t^* + (i \text{ sec}))(\mathbf{v}_o^m - \mathbf{v}_i^m)|| \\
&= ||(\mathbf{s}_o^m - \mathbf{s}_i^m) + (t^* + (i \text{ sec}))(\mathbf{v}_o^m - \mathbf{v}_i^m) - (\mathbf{s} + t^*\mathbf{v}) + (\mathbf{s} + t^*\mathbf{v})|| \\
&= ||(\mathbf{s}_o^m - (\mathbf{s}_o - (i \text{ sec})\mathbf{v}_o)) - (\mathbf{s}_i^m - (\mathbf{s}_i - (i \text{ sec})\mathbf{v}_i)) + (t^* + (i \text{ sec}))(\mathbf{v}_o^m - \mathbf{v}_o) \\
&\quad - (t^* + (i \text{ sec}))(\mathbf{v}_i^m - \mathbf{v}_i) + (\mathbf{s} + t^*\mathbf{v})|| \\
&\leq ||\mathbf{s}_o^m - (\mathbf{s}_o - (i \text{ sec})\mathbf{v}_o)|| + ||\mathbf{s}_i^m - (\mathbf{s}_i - (i \text{ sec})\mathbf{v}_i)|| + (t^* + (i \text{ sec}))||\mathbf{v}_o^m - \mathbf{v}_o|| \\
&\quad + (t^* + (i \text{ sec}))||\mathbf{v}_i^m - \mathbf{v}_i|| + ||\mathbf{s} + t^*\mathbf{v})|| \\
&< a_o + a_i + (t^* + \lambda)b_o + (t^* + \lambda)b_i + D \\
&\leq a + (t^* + (i \text{ sec}))b + D \\
&\leq \psi + D.
\end{aligned}
$$

This is a contradiction, since $\texttt{cd}(D + \psi, T + \lambda; \mathbf{s}^m, \mathbf{v}^m) = \texttt{False}$ and $\lambda = d$ seconds. This completes the proof. ◻

3.3 The Safety Claim for Conflict Detection

The safety claim that can be proved by using Lemma 1 is stated below. It has not been formally proved in a theorem prover, but the formal mathematics has been developed in this paper that enables a standard mathematical proof. It follows trivially from that Lemma and from Equation 6 in Section 3.2.

Proved Safety Claim for the Conflict Probe cd. *Let* $\lambda = d$ *seconds,* $\psi = a_o + a_i + (T+\lambda)(b_o + b_i)$. *Suppose that* $cd(D+\psi, T+\lambda; \mathbf{s}^m, \mathbf{v}^m) = False$ *and that the ownship and the intruder aircraft are no greater than 60 nmi apart. Then the probability that the aircraft are in conflict, i.e. that* $cd(D, T; \mathbf{s}, \mathbf{v}) = True$, *is no greater than* $p_{so} + p_{vo} + p_{si} + p_{vi} + (1 - \eta)^{d+1}$.

A *missed alert* is a conflict that is not detected. Artificially increasing the distance D and the lookahead time T in the conflict probe cd will make missed alerts less likely. The proved safety claim above gives a formula that returns the amount that D and T must be increased, as well as an upper bound on the probability of a missed alert if D is increased in this way, assuming that the ownship and the intruder aircraft are within 60 nmi of each other. The inputs to these formulas are the distances a_o and a_i, the speeds b_o and b_i, the probabilities p_{so}, p_{vo}, p_{si} and p_{vi}, and the number of seconds d that T is to be increased in the conflict probe cd. Equation (6) expresses the relationships between $a_o, a_i, b_o, b_i, p_{so}, p_{vo}, p_{si}$ and p_{vi}. Given these inputs, the associated upper bound for the probability of a missed alert is

$$p_{missed-alert} = p_{so} + p_{vo} + p_{si} + p_{vi} + (1 - \eta)^{d+1}, \tag{7}$$

where, as in Section 3.2, η is a lower bound for the probability that a given ADS-B message from the intruder aircraft will not be detected and decoded by the ownship, and in this example $\eta = 0.953$.

In the equation above, the amount ψ that D should be artificially increased to ensure that the probability of a missed alert is less than $p_{missed-alert}$ is given by

$$\psi = a_o + a_i + (T + \lambda)(b_o + b_i), \tag{8}$$

where $\lambda = d$ second. It should be noted that Equations (8) and (7) imply that if the velocity b dominates the calculation of ψ, then as ψ increases, d increases as well, and so the probability of a missed alert decreases.

Computing Actual Probabilities. DO-242A [8] specifies several system performance confidence-levels that are to be included in ADS-B messages detailing how precise and trusted the contained state information is. The relevant ones here are the navigation accuracy categories for position and velocity (NAC$_P$ and NAC$_V$). NAC$_P$ is a maximum distance for errors in position; similarly NAC$_V$ is a maximum velocity error. That is, these numbers specify the parameters a_0, a_i and b_o, b_i, respectively. Both NAC$_P$ and NAC$_V$ specify that the stated values will fall within a 95% confidence interval, which is equivalent to saying that p_{so}, p_{vo}, p_{si} and p_{vi} are all equal to 0.05. Table 1 uses these numbers along with

Table 1. Horizontal uncertainty, lookahead, and buffer sizes. The < 30 m position error corresponds to the NAC$_P$ 9 error category (NAC$_P$ 11 is the most accurate) and the < 0.3 m/s velocity error corresponds to the NAC$_V$ 4 (most accurate) error category. The velocity error dominates in calculating ψ these cases. When the position error is < 185.2 m (NAC$_P$ 7) and the velocity error is < 1.0 m/s (NAC$_V$ 3) the position error dominates the calculation of ψ for lookahead times less than 186 seconds.

Position Error	Velocity Error	Time $+\lambda$	Buffer ψ	$p_{missed-alert}$
< 30 m	< 0.3 m/s	180+0 sec	+0.09 nmi (168 m)	0.24700
< 30 m	< 0.3 m/s	180+1 sec	+0.09 nmi (169 m)	0.20221
< 30 m	< 0.3 m/s	180+2 sec	+0.09 nmi (169 m)	0.20010
< 30 m	< 0.3 m/s	180+3 sec	+0.09 nmi (170 m)	0.20000
< 185.2 m	< 1.0 m/s	180+0 sec	+0.39 nmi (730 m)	0.24700
< 185.2 m	< 1.0 m/s	180+3 sec	+0.40 nmi (736 m)	0.20000

Equations (7) and (8) to compute the amount the distance that D needs to be increased, as well the associated upper bounds on the probabilities of missed alerts for different choices of the number of seconds d.

It should be noted that the upper bounds on the probabilities of missed alerts in this table are quite high, but that this is not due to imprecision in the presented methods. This is mostly due to the fact that the confidence intervals specified in DO-242A are for 95% confidence and provide little knowledge of what is happening the other 5% of the time. It is quite possible that these formulas could calculate the probability of missed alerts to be less than 4×10^{-9}, if $1 - (10^{-9})$-confidence intervals were available for the positions and velocities of the aircraft.

4   Conclusion and Future Work

This paper has built on Rushby and Littlewood's framework [9,5] for formalizing safety claims, specifically providing a mathematical basis for dealing with certain probabilistic safety claims. The mathematics behind this is based on the notion of probabilistic kernels, which were illustrated in a safety claim for a conflict detection system for aircraft. The framework presented allows for an arbitrary number of potentially hazardous conditions. Future work in this area will include formalizing the mathematics presented here in a theorem prover such as PVS [7]. Many of the tools needed for this task already exist, including PVS libraries for Riemann integration [1] and Riemann-Stieltjes integration, as well as a Lebesgue measure and integration library developed by David Lester. Some additions are needed to these libraries to facilitate manipulations of multiple integrals.

An additional area for future work would be to incorporate a degree of assumption checking into the framework. This may include formally capturing the assumptions of independence between hazardous conditions, which could be formed into a verification condition that can be automatically checked for inconsistencies by a satisfiability checker (a SAT-solver).

References

1. Butler, R.: Formalization of the integral calculus in the PVS theorem prover. Journal of Formalized Reasoning 2(1) (2009)
2. Chung, W.W., Staab, R.: A 1090 extended squitter automatic dependent surveillance broadcast (ADS-B) reception model for air-traffic-management simulations. In: AIAA Modeling and Simulation Technologies Conference and Exhibit (2006)
3. Herencia-Zapana, H., Jeannin, J.B., Muñoz, C.: Formal verification of safety buffers for state-based conflict detection and resolution. In: Proceedings of 27th International Congress of the Aeronautical Sciences, ICAS 2010, Nice, France (2010)
4. Holloway, C.M.: Safety case notations: alternatives for the non-graphically inclined? In: 3rd IET International Conference on System Safety (2008)
5. Littlewood, B., Rushby, J.: Reasoning about the realiability of diverse two channel systems in which one channel is possibly perfect. In: Tech report SRI-CSL-09-02 (2010)
6. NASA Langley Formal Methods Team: Airborne coordinated conflict resolution and detection (2010), http://shemesh.larc.nasa.gov/people/cam/ACCoRD/
7. Owre, S., Rushby, J.M., Shankar, N.: PVS: A prototype verification system. In: Kapur, D. (ed.) CADE 1992. LNCS, vol. 607, pp. 748–752. Springer, Heidelberg (1992), http://www.csl.sri.com/papers/cade92-pvs/
8. Minimum aviation system performance standards for automatic dependent surveillance broadcast (ADS-B). DO-242A, RTCA (June 2002), section 2.1.2.12–2.1.2.15
9. Rushby, J.: Formalism in safety cases. In: Proceedings of the Eighteenth Safety-critical Systems Symposium (2010)
10. Shiryaev, A.N.: Probability. Springer, Heidelberg (1995)

The OpenTheory Standard Theory Library

Joe Hurd

Galois, Inc.
joe@gilith.com
http://www.gilith.com

Abstract. Interactive theorem proving is tackling ever larger formalization and verification projects, and there is a critical need for theory engineering techniques to support these efforts. One such technique is cross-prover package management, which has the potential to simplify the development of logical theories and effectively share theories between different theorem prover implementations. The OpenTheory project has developed standards for packaging theories of the higher order logic implemented by the HOL family of theorem provers. What is currently missing is a standard theory library that can serve as a published contract of interoperability and contain proofs of basic properties that would otherwise appear in many theory packages. The core contribution of this paper is the presentation of a standard theory library for higher order logic represented as an OpenTheory package. We identify the core theory set of the HOL family of theorem provers, and describe the process of instrumenting the HOL Light theorem prover to extract a standardized version of its core theory development. We profile the axioms and theorems of our standard theory library and investigate the performance cost of separating the standard theory library into coherent hierarchical theory packages.

1 Introduction

Interactive theorem proving has grown from toy examples to major formalization and verification projects in mathematics and computer science. Recent examples include: the 20 man-year verification of the seL4 operating system kernel [24]; the CompCert project, which verified an optimizing compiler from a large subset of C to PowerPC assembly code [25]; and the Flyspeck project, which aims to mechanize a proof of the Kepler sphere-packing conjecture [14].

Just as the term software engineering was coined in 1968 [26] to give a name to techniques for developing increasingly large programs, there is now a need for *theory engineering* techniques to develop increasingly large proofs (*"proving in the large"*). One software engineering technique that can be applied to proof development is effective package management. Modern operating systems [8] and programming languages [6] bundle software into packages that carry their dependencies, supporting easy distribution and automatic checking at installation time to ensure that the system can properly support the package. The goal of

M. Bobaru et al. (Eds.): NFM 2011, LNCS 6617, pp. 177–191, 2011.

the OpenTheory project is to transfer the benefits of package management to aid the development of logical theories[1].

The initial case study of the OpenTheory project is to develop the infrastructure necessary to port theories between three related interactive theorem provers: HOL Light [15], HOL4 [28] and ProofPower [23]. These three theorem provers implement the same higher order logic, namely Church's simple theory of types extended with Hindley-Milner style type variables [10]. They also have a similar design of an interactive interface where the user invokes proof tools to prove subgoals, built on top of a small logical kernel that enforces soundness. The logical kernel design is inherited from Milner's pioneering work on the LCF theorem prover [11], which Gordon reused to implement higher order logic in the HOL theorem prover [12], and from which the three chosen theorem provers are all descended [13].

Even though HOL Light, HOL4 and ProofPower implement the same logic using the same conceptual design, they each contain significant theory formalizations that are not accessible to each other. For example, HOL Light has a formalization of complex analysis [16], HOL4 has a formalization of probability theory [18], and ProofPower has a formalization of the Z specification language [2]. The reason that these useful theories are not available in all of the theorem provers is that it requires significant human effort to port a theory to a new environment, due to differences in the native theories and proof tools[2].

To overcome the differences between the name and behavior of proof tools between the theorem provers, the OpenTheory project has developed a standard *article* file format for serializing proofs of higher order logic [21]. Proofs are reduced to a standard set of primitive inferences that are precisely specified and can be simulated by any theorem prover in the HOL family. This bypasses the differences in the proof tools, at the cost of archiving proofs in a format that is hard to modify.

Once the differences between the proof tools have been removed as an obstacle, the challenge that remains is to reconcile the differences between the native theories available in each theorem prover. To illustrate the need for this, suppose we desire to port the theory of complex numbers from HOL Light to HOL4. One way to do this is to export every theory that the HOL Light complex numbers depend on as proof articles, and then import these into HOL4. However, now we have two copies of the theory of real numbers inside HOL4: the original real number theory of HOL4 and the real number theory imported from HOL Light that the complex numbers depend on. Because of this, we cannot easily combine the new theory of complex numbers with other HOL4 theories that depend on the original real number theory, such as the theory of probability.

To avoid this *duplicate theory* problem, when we speak of porting theories between theorem provers we usually have in mind the following procedure:

---

[1] The OpenTheory project homepage is `http://gilith.com/research/opentheory`
[2] The author has first-hand experience of this: his introduction to theorem proving was porting a theory of real numbers from HOL Light to HOL4.

1. Export the theory of complex numbers from HOL Light, leaving the references to native theories as uninterpreted type operators, constants, and assumptions.
2. Import the theory of complex numbers into HOL4, binding the references to native theories to native type operators, constants and theorems.

Note that the success of this porting procedure depends on there being a degree of alignment between the native theories of HOL Light and HOL4. The native theories do not have to be identical: type operators and constants may have different names in the two theorem provers; and HOL4 may contain additional theorems beyond the set required for the import of the HOL Light theory to succeed. But beyond these superficial differences, to port a theory from the theorem prover context A to B there must be a semantic embedding $A \rightarrow B$ mapping type operators and constants from A to ones in B with properties that are at least as logically strong. This notion of semantic embeddings between theorem prover contexts has been formalized in category theory as *theory morphisms* [32], and this provides a theoretical foundation for the OpenTheory project.

To support the use case of porting theories between the HOL Light, HOL4 and ProofPower, we will need semantic embeddings from the core theories of each theorem prover to the core theories of the others. As an alternative to explicitly maintaining these semantic embeddings, we instead take the set of core theories that the theorem provers share and release a *standard theory library* of them in OpenTheory format. The advantage of a standard theory library are as follows:

- Each theorem prover is responsible for maintaining mappings between its core theories and the OpenTheory standard library, reducing the number of semantic embeddings that must be maintained from $O(n^2)$ to $O(n)$ (where n is the number of theorem provers that wish to share theories).
- The standard theory library is a published contract of interoperability: *"If your theory uses only the standard theory library, we promise it will work on all of the supported theorem provers."*
- If a property such as associativity of addition is in the standard theory library, it does not need to be proved in every theory that relies on it. This is analogous to dynamic linking of programs to standard libraries.
- Constructions in the standard library can serve as standard specifications. A formal proof of Fermat's Last Theorem that uses the version of the natural numbers in the standard theory library is much easier to check than one that uses a custom version.

This paper presents the OpenTheory standard theory library and describes the process of identifying and extracting the core theory set of the HOL family of theorem provers. The proof articles that result from this process are combined to form higher level theories such as *natural numbers* or *lists*, and the final step is to combine these to form the standard theory library.

The remainder of the paper is structured as follows: Section 2 reviews the OpenTheory formats and infrastructure that we used and extended to support this work; Section 3 identifies the core theories that are included in the standard theory library; Section 4 describes the process of extracting standard proof articles by instrumenting an existing theorem prover; Section 5 profiles the result of combining proof articles into the standard theory library; and finally Sections 6–8 examine related work, summarize and consider future directions.

2   The OpenTheory Proof Archive

In this section we review the OpenTheory proof article [19] and theory package [20] formats, which are used to represent the standard theory library. These formats are now stable, and tools for processing theory packages are included with the OpenTheory toolset[3]. Tools exist for displaying meta-information, querying dependencies, pretty-printing assumptions and theorems, and compiling theory packages to proof articles.

2.1   Articles of Proof

The unit of composition in OpenTheory is a higher order logic theory $\Gamma \triangleright \Delta$, which consists of:

1. A set Γ of assumption sequents.
2. A set Δ of theorem sequents.
3. A proof that the theorems in Δ logically derive from the assumptions in Γ.

An article is a compact representation of a higher order logic theory, encoded as instructions for a stack-based virtual machine. The format was designed to simplify the process of importing theories into theorem prover implementations: all that is required is to execute the article instructions in the desired context.

The initial version of the proof article format [19] contained instructions for constructing types and terms, but the inference rules were system dependent. However, after receiving comments from the interactive theorem proving community, this system dependence was replaced with a set of 10 article instructions for executing precisely specified primitive inferences. These new instructions are shown in Figure 1.

2.2   Theory Packages

The proof article format supports a theorem prover independent representation of theories. The theory package format is a domain-specific language for combining theories, supporting the following operations:

1. Renaming type operators and constants in theories, either to avoid namespace clashes or to bind the arguments of a *parametric theory*.
2. Forming compound theories by satisfying the assumptions of one theory with the theorems of others.

---

[3] The OpenTheory toolset is available for download at
http://gilith.com/software/opentheory

$$\frac{}{\vdash t = t} \text{ refl } t \qquad \frac{}{\{\phi\} \vdash \phi} \text{ assume } \phi \qquad \frac{\Gamma \vdash \phi = \psi \quad \Delta \vdash \phi}{\Gamma \cup \Delta \vdash \psi} \text{ eqMp}$$

$$\frac{\Gamma \vdash t = u}{\Gamma \vdash (\lambda v.\ t) = (\lambda v.\ u)} \text{ absThm } v \qquad \frac{\Gamma \vdash f = g \quad \Delta \vdash x = y}{\Gamma \cup \Delta \vdash f\ x = g\ y} \text{ appThm}$$

$$\frac{\Gamma \vdash \phi \quad \Delta \vdash \psi}{(\Gamma - \{\psi\}) \cup (\Delta - \{\phi\}) \vdash \phi = \psi} \text{ deductAntisym} \qquad \frac{\Gamma \vdash \phi}{\Gamma[\sigma] \vdash \phi[\sigma]} \text{ subst } \sigma$$

$$\frac{}{\vdash (\lambda v.\ t)\ u = t[u/v]} \text{ betaConv } ((\lambda v.\ t)\ u) \qquad \frac{}{\vdash c = t} \text{ defineConst } c\ t$$

$$\frac{\vdash \phi\ t}{\vdash abs\ (rep\ a) = a \quad \vdash \phi\ r = (abs\ (rep\ r) = r)} \text{ defineTypeOp } n\ abs\ rep\ vs$$

Fig. 1. The OpenTheory logical kernel

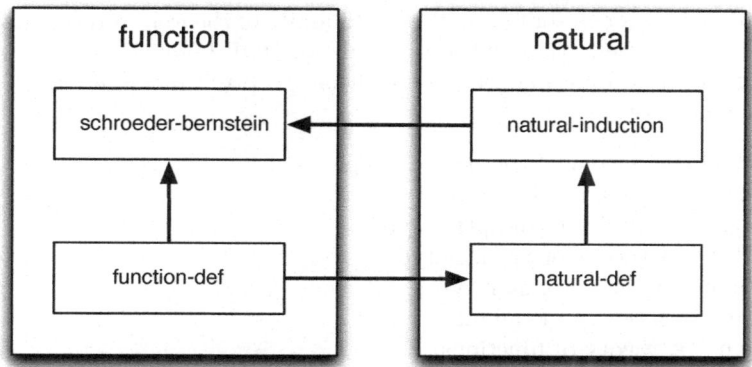

Fig. 2. Example theory dependency graph

Theory packages are hierarchical, using the above operations to build up from basic theory packages containing proof articles to more complex theories. An important concept for a standard theory library is the *compilation theory package*, which is designed to help construct coherent theory packages in the face of the complex dependency structures that often arise in theory development.

An example of compilation theories is shown in Figure 2, where four theory packages are contained in two compilation theory packages, and the arrows indicate package dependencies. The statement of the Schroeder-Bernstein theorem depends only on the function theory definitions, but the proof also depends on natural number induction. Natural numbers in turn are constructed using function theory definitions. The most coherent function theory package would contain both the function theory definitions and the Schroeder-Bernstein theorem, but this package would then have a cyclic dependency with the natural number theory package. Defining the function theory package as a compilation of two theory packages allows finer grained theory package dependencies, which removes the offending cycle.

Early experimentation with the theory package language revealed some desirable properties of a reusable theory package:

1. a clear topic (e.g., trigonometric functions);
2. assumptions that are satisfied by the theorems of other reusable theory packages;
3. a carefully chosen set of theorems, presenting an abstract interface to the theory (hiding construction details).

We will refer to these guiding principles when describing the construction details of the OpenTheory standard theory library.

3   Identifying Core Theories

The first step in the construction of the OpenTheory standard theory library is to identify the core theories shared by the HOL family of theorem provers. Looking at the system documentation and source code for HOL Light, HOL4 and Proof-Power turns up the following set of core theories, sorted into the OpenTheory standard namespace:

- Data.Bool – A theory of the boolean type
- Data.List – A theory of list types
- Data.Option – A theory of option types
- Data.Pair – A theory of product types
- Data.Sum – A theory of sum types
- Data.Unit – A theory of the unit type
- Function – A theory of functions
- Number.Natural – A theory of natural numbers
- Number.Numeral – A theory of natural number numerals
- Relation – A theory of relations

This is not intended to be a complete list, but sufficient to demonstrate the practicality of building an OpenTheory standard theory library, and full-featured enough upon which to build some non-trivial theories. The above theories are all present in version 1.0 of the standard theory library, and future versions can standardize other shared theories such as integers, reals, sets, characters and strings.

4   Extracting Standard Articles

The next step in the construction of the standard theory library is to represent the core theories as a set of proof articles that can be turned into basic theory packages. One approach to this would be to create an OpenTheory version of the standard theory library from scratch, proving everything using the standard inference rules. However, since the standard theory library is (by definition) shared by each member of the HOL family of theorem provers, an alternative to this is to instrument one of the theorem provers to emit its version of the standard theory library in proof article format. We take this latter approach and choose the HOL Light theorem prover as having the simplest logical kernel to instrument. The remainder of this section describes the experience of extracting standard proof articles from HOL Light theories.

4.1 Granularity

With the primitive inferences of HOL Light instrumented to emit proof articles, the next choice to be made is the granularity of the proof articles. At the coarsest extreme of the granularity spectrum, the whole standard theory library could be emitted as one big proof article. However, this would violate Guideline 1 of constructing reusable theories from Section 2.2, because the resulting theory would not have a clear topic. At the finest extreme, we could put every exported theorem into its own proof article file, with the caveat that proof articles that make definitions need to export enough theorems to form a minimal abstract interface. This would result in a set of theories that score well according to the reusability guidelines (except possibly for Guideline 3 that asks for a carefully chosen set of theorems), but introduces myriad theory packages to be stored and processed.

We choose an intermediate point on the granularity spectrum where theory packages that make definitions export enough theorems to form a minimal abstract interface, and theory packages that make no definitions can export any number of theorems so long as they form a coherent topic. This design choice is made to maximize the reusability of the resulting theory packages while respecting performance goals.

Another issue is that there are two kinds of proved theorems in HOL Light: visually appealing theorems designed for the user to apply as lemmas in future proofs; and auxiliary pro-forma theorems designed to be used internally by proof tools. The reusable theory guidelines dictate that only the visually appealing theorems should appear in the standard theory library. This is achieved by collecting together the auxiliary theorems as they are proved and storing them in a separate proof article, which is 'statically linked' to standard proof articles as they are generated. When the whole standard theory library has been harvested, the auxiliary proof article is packaged as a special theory to support theory development building on the standard theory library using the HOL Light proof tools.

4.2 Standardization

In addition to statically linking auxiliary theorems, we used other methods to standardize the proof articles generated from HOL Light. As a simple example, the names of HOL Light type operators and constants are mapped into the OpenTheory standard namespace (as described in Section 3).

Another source of system dependence is the presence of 'tags' in terms. For example, in HOL Light every natural number numeral is a term of the form NUMERAL t, where the constant NUMERAL is defined as a synonym for the identity function. The presence of NUMERAL has no logical significance, but is a tag to help proof tools and other theorem prover infrastructure. Different theorem prover implementations may have different tagging schemes, so we remove tags from theorems that we add to the standard theory library. The scheme we use to do this is to rename the tag constant NUMERAL to be called Unwanted.id, and then rewrite all generated proof articles to remove all type operators and constants in the Unwanted namespace.

Finally, during the process of extracting proof articles from HOL Light we discovered many improvements to HOL Light that would simplify the extraction process, including: removing duplicate theorems; simplifying the definition of numerals; universally quantifying theorems with free variables. We submitted these as patches to the HOL Light developer, and several have already been incorporated into the upstream version.

4.3 Partial Functions

Partial functions require special handling in a classical two-valued logic such as higher order logic. For example, the natural number div and mod functions are not mathematically defined when the denominator is zero, but since every function in higher order logic is total the term 1 div 0 must be some natural number. In this case the solution we adopt is for the theory defining div and mod to export the single theorem

$$\vdash \ \forall m, n. \ n \neq 0 \implies m = (m \ \mathsf{div} \ n) * n + m \ \mathsf{mod} \ n \ \wedge \ m \ \mathsf{mod} \ n < n \ ,$$

preventing client theories from deducing anything about the value 1 div 0 (that could not be deduced about every natural number).

There are a few situations when this information-hiding approach cannot be used. Both HOL Light and HOL4 (but not ProofPower) define a predecessor function pre as an inverse to the successor function, and set pre $0 = 0$ even though the inverse of successor is not mathematically defined for zero. The value of pre 0 is subsequently relied on, among other things to define cut-off subtraction, and so we choose to 'grandfather' the value of pre 0 into the standard library. However, in the theory that defines the predecessor function we separate the definition into the two theorems

$$\vdash \ \mathsf{pre} \ 0 = 0 \quad \text{and} \quad \vdash \ \forall n. \ \mathsf{pre} \ (\mathsf{suc} \ n) = n \ ,$$

to encourage client theories to rely only on the standard domain.

5   The Standard Theory Library

After identifying the core theories of the HOL family of theorem provers and extracting them as proof articles, the final step is to use them to construct the standard theory library.

5.1   Construction

This is the procedure for converting the proof articles extracted from HOL Light into the standard theory library:

1. Create a basic theory package for each proof article.
2. Create theory packages for higher-level topics, such as bool or list, which are compilations of lower-level theory packages.
3. Create a theory package called base, which is a compilation of the highest-level theory packages.

Although the standard theory library consists of the whole collection of these theory packages, the base theory package exports all the theorems needed to build client theories on top of the standard theory library. The other theory packages can be regarded as scaffolding by most users of the standard theory library, and safely ignored.

As we expected, it was straightforward to carry out Steps 1 and 2 of the above procedure. We expected the difficulty to appear in Step 3, when compiling highest level theories with potentially complex dependencies between them. Surprisingly, it turned out that there was a natural order to arrange the highest-level theories where each one only depended on the previous ones: bool; unit; function; pair; natural; relation; sum; option; and list. Because of this, there was no need to unpack the compilation theories to eliminate cyclic dependencies as described in Section 2.2. The real-life example shown in Figure 2 demonstrates that this functionality will be required for some theories, but the current version of the standard theory library is naturally acyclic.

5.2   Axioms

Before looking at the theorems and proofs of the standard theory library, it is worth examining what it depends on. The OpenTheory primitive inference rules, shown in Figure 1, were taken from HOL Light and refer only to:

- the type operator bool;
- the function space type operator $\cdot \rightarrow \cdot$; and
- the equality constant $= : \alpha \rightarrow \alpha \rightarrow$ bool.

These two type operators and one constant can be considered to be implicitly axiomatized by the primitive inferences. In addition, the standard theory library explicitly asserts the following three axioms, each of which is contained in its own theory package:

$\vdash \forall t. (\lambda x. t\, x) = t$ (axiom-extensionality)

$\vdash \forall P, x.\ P\, x \implies P\, (\text{select } P)$ (axiom-choice)

$\vdash \exists f : \text{ind} \rightarrow \text{ind. injective } f \wedge \neg\text{surjective } f$ (axiom-infinity)

The formulation of these three axioms is taken from HOL Light, but can be proved as theorems in HOL4 and ProofPower.

The standard theory library offers a simple way to check that a theory package developed on a HOL family theorem prover will easily port to other theorem provers. Just statically link the new theory package to the **base** theory package, and any system dependent behavior will appear as extra axioms (beyond the standard three). This static linking procedure could also be used to develop theories that avoid the axiom of choice, while still making use of theorems from the standard theory library that do not use choice in their proof.

5.3   Theorems

The current version of the standard theory library exports 450 theorems, containing 64 defined constants and 6 defined type operators. For reasons of space the theorems cannot all be shown here, but an HTML version of the **base** package can be viewed at the following URL:

http://opentheory.gilith.com/?pkg=base-1.0

The standard theory library comprises 139 packages: 102 of which are basic theory packages wrapping proof articles; 36 of which are higher-level theory packages; and one is the **base** package. The left side of Table 1 shows the primitive inference count of replaying all the proofs in the standard theory library, for a total of 211,058 inferences. The cost of separating the standard theory library into 102 basic theory packages is highlighted by the axiom count of 1,672, which is increased whenever one basic theory uses a theorem proved by an imported theory.

It is possible to compile the whole of the standard theory library into a single proof article (with 965,433 commands), and then repeat the experiment of replaying all of the proofs and counting the primitive inferences: the results are shown on the right side of Table 1. The total number of primitive inferences drops by 40%, and the axiom count is only three: one for each of the standard axioms. It is also remarkable that the number of auxiliary defined type operators and constants (used in proofs but that do not appear in any theorems)

Table 1. The primitive inference count of replaying all the proofs in the standard theory library, when split into theories (left) and compiled into a single article (right)

Primitive Inference	Count		Primitive Inference	Count
eqMp	55,209		eqMp	32,386
subst	45,651		subst	27,949
appThm	44,130		appThm	27,796
deductAntisym	28,625		deductAntisym	17,300
refl	17,388		refl	9,332
betaConv	8,035		absThm	6,313
absThm	7,765		betaConv	3,646
assume	2,455		assume	1,169
axiom	1,672		defineConst	85
defineConst	119		defineTypeOp	7
defineTypeOp	9		axiom	3
Total	**211,058**		**Total**	**125,986**

drops from 55 and 3 to 21 and 1 (respectively). This provides some concrete data quantifying the performance cost of splitting the standard theory library into coherent hierarchical theory packages.

6 Related Work

Extracting proofs from LCF theorem provers is not new: Wong's pioneering *Recording and checking HOL proofs* in 1995 appears to be the first [35]. More recently, Obua and Skalberg [29] instrumented HOL4 and HOL Light to export theories in XML format that could be imported into the Isabelle/HOL theorem prover. The present work differs from this line of proof recording work by its focus on the theory as the central concept, independent of any particular theorem prover implementation.

From this point of view, the most related work is the AWE project [5], which builds on the explicit proof terms in Isabelle [4]. Though tied to one theorem prover, it nevertheless focuses on the theory as the central concept, and has developed sophisticated mechanisms for theory interpretation based on rewriting proof terms. The present work differs from AWE by being theorem prover independent, and also by its technique of processing proofs one step at a time rather than requiring the whole proof to be in memory, which may allow it to scale up more effectively.

The HOL Zero project [1] has aims similar to OpenTheory of making proofs portable between different implementations of the HOL family of theorem provers, by creating a minimal theorem prover "for checking and/or consolidating proofs created on other theorem provers", "designed with trustworthiness as its top priority". OpenTheory differs from HOL Zero by its focus on proofs that have been reduced to the object format of primitive inferences, and the theory packaging mechanisms that can be built on top of this starting point. HOL Zero complements OpenTheory by encouraging portability at the earlier stage of proof source files.

Many theorem provers implement a theory infrastructure that offers functionality similar to the theory operations in the OpenTheory package format. ProofPower has a sophisticated system for building and navigating a hierarchy of theories which contain both logical data and information for tools such as parsers, pretty printers and proof tools [23]. Theory interpretations are implemented in the EVES [7], IMPS [9], PVS [31] and Specware [34] theorem provers. Called locales in the Isabelle theorem prover [22], they are integrated with its declarative proof language [3]. The present work differs from these efforts by pursuing a theorem prover independent approach to theory combination and interpretation.

Another approach to higher order logic theory operations is to extend the logic so that theories can be directly represented with theorems [33,17]. The goal of the present work is to implement a theory infrastructure on top of the existing logic, but extending the logic has the significant advantage of supporting theory operations without replaying proofs.

7 Summary

In this paper we motivated the need for a standard library of higher order logic theories to support large-scale logical theory development and increase portability of theories between the HOL family of theorem provers.

The core contribution of this paper is the presentation of a theorem prover independent standard theory library, represented as an OpenTheory package. We identified the core theory set of the HOL family of theorem provers, and described the process of instrumenting the HOL Light theorem prover to extract a standardized version of its core theory development.

The OpenTheory package language is suitable to package proof articles extracted from HOL Light, and we showed how to combine these first into higher level theory packages, and then into a single package representing the user interface to the whole standard theory library. Finally, we profiled the axioms and theorems of the standard theory library, and investigated the performance cost of separating the standard theory library into coherent hierarchical theory packages.

8 Future Work

The current version of the standard theory library is not fixed, and in fact is expected to evolve as more theories are standardized between the HOL family of theorem provers. One desirable goal would be keep later versions of the standard theory library backwards compatible with earlier versions, which implies that we should exercise caution when adding theorems, because they might be hard to remove later.

The current version of the standard theory library does not make any use of parametric theories containing assumptions about uninterpreted type operators and constants, which users are expected to interpret to defined type operators and constants in their proof context. The advantage of using parametric theories is that proving a theorem once in a parametric theory makes it available 'for free' in every context in which it is used. The use of parametric theories has the potential to reduce the effort required to extend the standard theory library, while giving users more tools to use in their theory developments.

The standard theory library is based on the simple version of higher order logic implemented by the HOL family of theorem provers. There is a straightforward semantic embedding from this logic to the more complex versions of higher order logic implemented by the Isabelle/HOL [27] and PVS [30] theorem provers, making it technically possible to import OpenTheory packages into these systems. However, there is an interesting line of research in designing importers that result in 'natural-looking' theories in the target system. Such an importer could modify the theories as they were processed (similar to the de-tagging rewriting described in Section 4.2) to use logical features of the target system such as Isabelle/HOL type classes or PVS subtypes.

Acknowledgements

The OpenTheory project was initiated in 2004 as a result of discussions between Rob Arthan and the author, and the work since then has been guided by feedback from many other people, including John Harrison, Rebekah Leslie, John Matthews, Michael Norrish and Konrad Slind. This paper was greatly improved by comments from Rob Arthan, Ramana Kumar, Lee Pike and the anonymous referees.

References

1. Adams, M.: Introducing HOL zero. In: Fukuda, K., van der Hoeven, J., Joswig, M., Takayama, N. (eds.) ICMS 2010. LNCS, vol. 6327, pp. 142–143. Springer, Heidelberg (2010)
2. Arthan, R.D., Jones, R.B.: Z in HOL in ProofPower. In: BCS FACS FACTS (January 2005)
3. Ballarin, C.: Locales and locale expressions in isabelle/Isar. In: Berardi, S., Coppo, M., Damiani, F. (eds.) TYPES 2003. LNCS, vol. 3085, pp. 34–50. Springer, Heidelberg (2004)
4. Berghofer, S., Nipkow, T.: Proof terms for simply typed higher order logic. In: Aagaard, M.D., Harrison, J. (eds.) TPHOLs 2000. LNCS, vol. 1869, pp. 38–52. Springer, Heidelberg (2000)
5. Bortin, M., Johnsen, E.B., Lüth, C.: Structured formal development in Isabelle. Nordic Journal of Computing 13, 1–20 (2006)
6. Coutts, D., Potoczny-Jones, I., Stewart, D.: Haskell: Batteries included. In: Gill, A. (ed.) Haskell 2008: Proceedings of the first ACM SIGPLAN symposium on Haskell, pp. 125–126. ACM, New York (2008)
7. Craigen, D., Kromodimoeljo, S., Meisels, I., Pase, B., Saaltink, M.: EVES: An overview. Technical Report CP-91-5402-43, ORA Corporation (1991)
8. Dolstra, E., Löh, A.: Nixos: a purely functional linux distribution. In: Hook, J., Thiemann, P. (eds.) Proceedings of the 13th ACM SIGPLAN International Conference on Functional Programming (ICFP 2008), pp. 367–378. ACM, New York (2008)
9. Farmer, W.M.: Theory interpretation in simple type theory. In: Heering, J., Meinke, K., Möller, B., Nipkow, T. (eds.) HOA 1993. LNCS, vol. 816, pp. 96–123. Springer, Heidelberg (1994)
10. Farmer, W.M.: The seven virtues of simple type theory. Journal of Applied Logic 6, 267–286 (2008)
11. Gordon, M., Milner, R., Wadsworth, C.: Edinburgh LCF. LNCS, vol. 78. Springer, Heidelberg (1979)
12. Gordon, M.J.C., Melham, T.F. (eds.): Introduction to HOL (A theorem-proving environment for higher order logic). Cambridge University Press, Cambridge (1993)
13. Gordon, M.J.C.: Proof, Language, and Interaction: Essays in Honour of Robin Milner. From LCF to HOL: A Short History, ch. 6. MIT Press, Cambridge (2000)
14. Hales, T.C.: Introduction to the Flyspeck project. In: Coquand, T., Lombardi, H., Roy, M.-F. (eds.) Mathematics, Algorithms, Proofs. Dagstuhl Seminar Proceedings, vol. 05021, Internationales Begegnungs- und Forschungszentrum fuer Informatik (IBFI), Schloss Dagstuhl (2006)

15. Harrison, J.: HOL light: A tutorial introduction. In: Srivas, M., Camilleri, A. (eds.) FMCAD 1996. LNCS, vol. 1166, pp. 265–269. Springer, Heidelberg (1996)
16. Harrison, J.: Formalizing basic complex analysis. In: Matuszewski, R., Zalewska, A. (eds.) From Insight to Proof: Festschrift in Honour of Andrzej Trybulec, Studies in Logic, Grammar and Rhetoric, vol. 10(23), pp. 151–165. University of Białystok (2007)
17. Homeier, P.V.: The HOL-Omega Logic. In: Berghofer, S., Nipkow, T., Urban, C., Wenzel, M. (eds.) TPHOLs 2009. LNCS, vol. 5674, pp. 244–259. Springer, Heidelberg (2009)
18. Hurd, J.: A formal approach to probabilistic termination. In: Carreño, V.A., Muñoz, C.A., Tahar, S. (eds.) TPHOLs 2002. LNCS, vol. 2410, p. 230-245. Springer, Heidelberg (2002)
19. Hurd, J.: OpenTheory: Package management for higher order logic theories. In: Reis, G.D., Théry, L. (eds.) PLMMS 2009: Proceedings of the ACM SIGSAM 2009 International Workshop on Programming Languages for Mechanized Mathematics Systems, pp. 31–37. ACM, New York (2009)
20. Hurd, J.: Composable packages for higher order logic theories. In: Aderhold, M., Autexier, S., Mantel, H. (eds.) Proceedings of the 6th International Verification Workshop (VERIFY 2010) (July 2010)
21. Hurd, J.: OpenTheory Article Format (August 2010), Available for download at http://gilith.com/research/opentheory/article.html
22. Kammüller, F.: Modular reasoning in Isabelle. In: McAllester, D.A. (ed.) CADE 2000. LNCS, vol. 1831. Springer, Heidelberg (2000)
23. King, D.J., Arthan, R.D.: Development of practical verification tools. ICL Systems Journal 11(1) (May 1996)
24. Klein, G., Elphinstone, K., Heiser, G., Andronick, J., Cock, D., Derrin, P., Elkaduwe, D., Engelhardt, K., Kolanski, R., Norrish, M., Sewell, T., Tuch, H., Winwood, S.: seL4: Formal verification of an OS kernel. In: Matthews, J.N., Anderson, T.E. (eds.) Proceedings of the 22nd ACM Symposium on Operating Systems Principles, pp. 207–220. ACM, New York (2009)
25. Leroy, X.: Formal certification of a compiler back-end or: programming a compiler with a proof assistant. In: Morrisett, J.G., Jones, S.L.P. (eds.) Proceedings of the 33rd ACM SIGPLAN-SIGACT Symposium on Principles of Programming Languages (POPL 2006), pp. 42–54. ACM, New York (2006)
26. Naur, P., Randell, B. (eds.): Software Engineering. Scientific Affairs Division, NATO (October 1968)
27. Nipkow, T., Paulson, L.C., Wenzel, M.: Isabelle/HOL. LNCS, vol. 2283. Springer, Heidelberg (2002)
28. Norrish, M., Slind, K.: A thread of HOL development. The Computer Journal 41(1), 37–45 (2002)
29. Obua, S., Skalberg, S.: Importing HOL into Isabelle/HOL. In: Furbach, U., Shankar, N. (eds.) IJCAR 2006. LNCS (LNAI), vol. 4130, pp. 298–302. Springer, Heidelberg (2006)
30. Owre, S., Shankar, N., Rushby, J.M., Stringer-Calvert, D.W.J.: PVS System Guide. Computer Science Laboratory, SRI International, Menlo Park, CA (September 1999)
31. Owre, S., Shankar, N.: Theory interpretations in PVS. Technical Report SRI-CSL-01-01, SRI International (April 2001)
32. Rabe, F.: Representing Logics and Logic Translations. PhD thesis, Jacobs University Bremen (May 2008)

33. Völker, N.: HOL2P - A System of Classical Higher Order Logic with Second Order Polymorphism. In: Schneider, K., Brandt, J. (eds.) TPHOLs 2007. LNCS, vol. 4732, pp. 334–351. Springer, Heidelberg (2007)
34. Westfold, S.: Integrating Isabelle/HOL with Specware. In: Schneider, K., Brandt, J. (eds.) Theorem Proving in Higher Order Logics: Emerging Trends Proceedings. Department of Computer Science, University of Kaiserslautern Technical Reports, vol. 364/07 (August 2007)
35. Wong, W.: Recording and checking HOL proofs. In: Schubert, E.T., Windley, P.J., Alves-Foss, J. (eds.) HUG 1995. LNCS, vol. 971, pp. 353–368. Springer, Heidelberg (1995)

Instantiation-Based Invariant Discovery*

Temesghen Kahsai, Yeting Ge, and Cesare Tinelli

The University of Iowa

Abstract. We present a general scheme for automated instantiation-based invariant discovery. Given a transition system, the scheme produces k-inductive invariants from templates representing decidable predicates over the system's data types. The proposed scheme relies on efficient reasoning engines such as SAT and SMT solvers, and capitalizes on their ability to quickly generate counter-models of non-invariant conjectures. We discuss in detail two practical specializations of the general scheme in which templates represent partial orders. Our experimental results show that both specializations are able to quickly produce invariants from a variety of synchronous systems which prove quite useful in proving safety properties for these systems.

1   Introduction

The automated verification of hardware or software systems benefits greatly from the specification of invariants, state properties that hold over all iterations of a program loop or over all reachable states of a transition system. Since invariants are notoriously difficult or time-consuming to specify manually, a lot of research in verification over the years has been dedicated to their automatic generation.

In much of previous work, invariants are synthesized from a system's description (formal specification or source code), using sophisticated algorithms guided by the semantics of the description language. In this paper, we propose a complementary approach based on a somewhat brute-force invariant *discovery* scheme which has proven quite effective in our experimental evaluation. The approach looks for possible invariants by sifting through a large set of automatically generated formulas. These formulas are all instances of the same template, the parameter of the scheme, representing a decidable relation over one of the system's data types.

Our approach relies on efficient reasoning engines such as SAT and SMT solvers, and capitalizes on their ability to quickly generate counter-models. For the invariant discovery scheme to be practical, they key point is to encode large sets of candidate invariants compactly and process them efficiently. One case when this is possible is when the chosen template represents a partial order, that is, a reflexive, transitive and antisymmetric relation. This paper investigate two specializations of the scheme, one for general partial order sets (posets) and one for binary posets.

Our primary intended use for the discovered invariants is to assist the automatic verification of safety properties. To illustrate the effectiveness of our approach, we developed a new tool based on it. While our invariant discovery scheme can be applied to any

---

* This work was partially supported by AFOSR grant #AF9550-09-1-0517.

M. Bobaru et al. (Eds.): NFM 2011, LNCS 6617, pp. 192–206, 2011.

transition system, our tool applies to programs in the synchronous data flow language Lustre [8] and generates invariants over their Boolean and integer variables. We have carried extensive experiments with a large set of Lustre programs and annotated with safety properties. Our experimental results indicate that our techniques are quite effective in practice. As we discuss later, the generated invariants considerably increase the number of provable safety properties; moreover, they do not slow down the processing of safety properties already provable without those invariants.

Related work. Automatic invariant generation has been intensively investigated since the 1970s, producing a large body of literature. Manna and Pnueli [10] provide a compendium of this research and an extensive set of references. They present a number of methods for generating invariants to prove safety properties, which have been later extended by others (e.g., [13,2]). These methods could be classified as either *top-down* or *bottom-up*. Top-down invariant generation begins with a property to be verified for a particular system. When attempts to prove the property fail, various heuristics are applied to strengthen it. Bottom-up methods look at the system and use it to deduce properties of it. Until recently, invariants generated with these methods tended to be simple properties and not very useful. The invariant discovery scheme described in this paper could be classified as a bottom-up method. Its major distinction with respect to previous approaches is its ability to produce more complex invariants efficiently.

Counterexample guided refinement is a popular technique in model checking that has also been used for invariant generation [15,3,11]. Thalmaier *et al.* propose an induction-guided refinement process to approximate reachability analysis by providing inductive invariants to a SAT-based property checker [14]. Such analysis is based on BDD techniques. Another line of research on invariant generation builds on predicate abstraction techniques [6,11]. De Moura *et al.* describe invariant strengthening techniques based on quantifier elimination. That work is one of the first to use modern SMT solvers as reasoning engines for the verification of safety properties. There is recent interest in using SMT-solvers for generating inductive loop invariants. Srivastava and Gulwani describe a technique combining templates and predicate abstraction [12].

The work by Hunt *et al.* [9] is more closely related to ours, and was in fact its main inspiration. They propose a SAT-based method to prove safety properties of circuits that uses induction to identify equivalent sub-circuits inexpensively before attempting to prove the given property. This equivalence information either implies the property directly or can be used to decrease the amount of state space traversal by the main model checking procedure. Compared with that work, our approach is more general, with respect to both the transition systems it applies to and the relations it discovers between sub-circuits.

Synopsis. In the next sub-section we give a brief description of the notions and notations that will be used throughout the paper. Section 2 presents a general scheme for invariant discovery using k-induction. Section 3 describes two specializations of the general scheme. Experimental results are reported in Section 4. Section 5 concludes with a discussion of further research.

Formal Preliminaries. We denote finite tuples (or vectors) by letters in bold font. If t is an n-tuple, $t(i)$ is the i-th element of t for $i = 1, \ldots, n$.

For generality, we consider here an arbitrary logic $\mathcal{L}$ (with classical semantics) extending propositional logic. We employ $\mathcal{L}$'s notion of variable, term, formula, free variable, model, and formula satisfiability in a model. Relevant examples of $\mathcal{L}$ are propositional logic or any of the logics used in SMT: linear arithmetic, linear arithmetic with uninterpreted function symbols, and so on. If Γ is a set of formulas in $\mathcal{L}$, a model $\mathcal{M}$ *satisfies* Γ if it satisfies every formula in it; Γ is *$\mathcal{L}$-(un)satisfiable* in $\mathcal{L}$ if some (no) model of $\mathcal{L}$ satisfies it. We define an entailment relation $\models_{\mathcal{L}}$ in $\mathcal{L}$ as usual: for any set $\Gamma \cup \{F\}$ of formulas in $\mathcal{L}$, we have that $\Gamma \models_{\mathcal{L}} F$ iff every model of $\mathcal{L}$ that satisfies Γ satisfies F as well. Two formulas F and G are *$\mathcal{L}$-equivalent* if $F \models_{\mathcal{L}} G$ and $G \models_{\mathcal{L}} F$.

If F is a formula with free variables $x_1, \ldots, x_m$, and $t_1, \ldots, t_m$ are any terms in the logic, we use $F[t_1, \ldots, t_m]$ to denote the formula obtained from F by simultaneously replacing each occurrence of x_i in F by t_i, for all $i = 1, \ldots, m$. Abusing the notation, we will write $F[x_1, \ldots, x_m]$ also to denote that F has free variables $x_1, \ldots, x_m$, and sometimes just $F[\_, \ldots, \_]$ when the name of the free variables is unimportant.

Let Q be a set of *states*, a *state space*. A *transition system* $\mathcal{S}$ *over* Q is a pair $(\mathcal{S}_I, \mathcal{S}_T)$ where $\mathcal{S}_I \subseteq Q$ is the set of $\mathcal{S}$'s *initial states*, and $S_T \subseteq Q \times Q$ is $\mathcal{S}$'s *transition relation*. A state $q \in Q$ is *0-reachable* if $q \in \mathcal{S}_I$; it is *k-reachable* with $k > 0$ if it is $(k-1)$-reachable or $(s, q) \in S_T$ for some $(k-1)$-reachable state s. A state is *($\mathcal{S}$-)reachable* if it is k-reachable for some $k \geq 0$. We assume some encoding of the state space Q in terms of n-tuples of ground terms in $\mathcal{L}$, for some fixed n[1]. Then, we say that (the encoding of) a state q *satisfies* a formula $F[x]$, where x is an n-tuple of distinct variables, if $F[x]$ is satisfied by every model of $\mathcal{L}$ interpreting x as q. This terminology extends to formulas over several n-tuples of free variables in the obvious way.

Let $\mathcal{S} = (\mathcal{S}_I, \mathcal{S}_T)$ be a transitions system. A *(state) property* is any formula $P[x]$ over an n-tuple x of variables. It is *invariant (for $\mathcal{S}$)* if it is satisfied by all $\mathcal{S}$-reachable states. An *$\mathcal{L}$-encoding* of $\mathcal{S}$ is a pair $(I[x], T[x, y])$ of formulas of $\mathcal{L}$ respectively over the n-tuples of variables x and x, y, where

- $I[x]$ is a formula satisfied exactly by the initial states of $\mathcal{S}$;
- $T[x, y]$ is a formula satisfied by two reachable states q, q' iff $(q, q') \in \mathcal{S}_T$.

For any formula F over a single state and formula G over two states, we will write F_i and G_{i+1} as an abbreviation of $G[x_i]$ and $G[x_i, x_{i+1}]$, respectively, where x_i and x_{i+1} are n-tuples of distinct variables.

Definition 1. *A state property $P[x]$ is k-inductive (wrt T) for some $k \geq 0$ if*

$$I_0 \wedge T_1 \wedge \cdots \wedge T_k \models_{\mathcal{L}} P_0 \wedge \cdots \wedge P_k \tag{1}$$

$$T_1 \wedge \cdots \wedge T_{k+1} \wedge P_0 \wedge \cdots \wedge P_k \models_{\mathcal{L}} P_{k+1} \tag{2}$$

A property is inductive in the usual sense if it is 0-inductive. Every property that is k-inductive for some k is invariant (but not vice versa). An invariant $P[x]$ is *trivial* if $T_1 \models_{\mathcal{L}} P_1$. Note that this includes all properties $P[x]$ that are *valid* in $\mathcal{L}$.

---

[1] Depending on $\mathcal{L}$, states may be encoded for instance as n-tuples of Boolean constants or as n-tuples of integer constants, and so on.

2   A General Scheme for Invariant Discovery

Given an $\mathcal{L}$-encoding $S = (I[\boldsymbol{x}], T[\boldsymbol{x}, \boldsymbol{y}])$ of a system $\mathcal{S} = (\mathcal{S}_I, \mathcal{S}_T)$, we are interested in discovering invariants for $\mathcal{S}$ automatically. We describe here a general scheme for doing so. The scheme is parameterized by a *template formula* $R[\_, \_]$ and produces invariants for $\mathcal{S}$ that are conjunction of instances $R[s, t]$ of R where s, t are in principle arbitrary terms over a single state[2]. The scheme relies on the existence of an $\mathcal{L}$-*solver*, a decision procedure for $\mathcal{L}$-satisfiability, which for each $\mathcal{L}$-satisfiable formula $F[\boldsymbol{x}_1, \ldots, \boldsymbol{x}_m]$ is also able to return a state list $\boldsymbol{q}_1, \ldots, \boldsymbol{q}_m$ that satisfies $F[\boldsymbol{x}_1, \ldots, \boldsymbol{x}_m]$[3]. The scheme also relies on a procedure that can generate from S a non-empty *instantiation set* U of terms over $\boldsymbol{x}$ to be used to generate the instance of R. In this setting, a naive approach would be to check every possible instance $R[s, t]$ individually for invariance. This would be highly impractical since the number of instances of R is quadratic in the size of the instantiation set U. In our approach, we check the satisfiability of all instances at the same time and rely on the model generation ability of the $\mathcal{L}$-solver to weed out several non-invariant instances at once.

The general scheme consists of a simple two-phase procedure, with an optional third phase. Given the formula $R[\_, \_]$ and the term set U, the first phase starts with the optimistic conjecture that the property

$$C[\boldsymbol{x}] = \bigwedge_{s,t \in U} R[s, t]$$

is invariant. Then, it uses the $\mathcal{L}$-solver to weaken that conjecture by eliminating from it as many conjuncts $R[s, t]$ as possible—specifically, all conjuncts falsified by a k-reachable state, for some heuristically determined k. The resulting formula C is passed to the second phase, which attempts to prove C k-inductive by establishing the entailment (2) in Definition 1. Counterexamples to (2), i.e., models that falsify the entailment, are used to weaken C further by eliminating additional conjuncts until (2) holds. The final formula—the empty conjunction in the worst case—is guaranteed to be invariant. That formula can be further processed in the optional third phase by removing from it any conjunct that is a trivial invariant. The rationale for the last phase is that trivial invariants are never needed, for being directly implied by the formula encoding the transition relation, and including them could put extra burden on the $\mathcal{L}$-solver.

The pseudo-code for the procedure sketched above is provided in Figure 1. The termination condition for Phase 1 is a heuristic one: the search for the value k stops when C is falsified by no k-reachable states. Furthermore, every conjunct of C that does not pass the test in Phase 2 is conservatively assumed not to be invariant (even if it may be k'-inductive for some $k' > k$) and removed. It is not difficult to show that both phases are terminating. The final C is invariant because, by construction, it is k-inductive for the final k.

The practical feasibility of this invariant discovery scheme depends on the possibility of representing the conjecture C compactly, i.e., by an equivalent formula using less

---

[2] The restriction to binary templates is used here only to simplify the exposition.

[3] Modern SAT or SMT solvers are of course examples of $\mathcal{L}$-solvers for specific $\mathcal{L}$'s.

Require: a template formula $R[\_, \_]$ and a term set U
Ensure: P is invariant
$\quad i := 0$
$\quad C := \bigwedge \{R[s, t] \mid s, t \in U\}$
$\quad$ -------------- Phase 1 ---------------
$\quad$ **repeat**
$\quad\quad i := i + 1; \ \ \textit{refined} := \textit{FALSE}$
$\quad\quad$ **repeat**
$\quad\quad\quad a := \mathsf{SAT}(I_0 \wedge T_1 \wedge \cdots \wedge T_{i-1} \wedge \neg C_{i-1})$
$\quad\quad\quad$ **if** $a = (q_0, \ldots, q_{i-1})$ **then**
$\quad\quad\quad\quad C := \mathsf{filter}(C, q_{i-1}); \ \ \textit{refined} := \textit{TRUE}$
$\quad\quad$ **until** $a = \mathsf{unsat}$
$\quad$ **until** $\neg \textit{refined}$
$\quad$ -------------- Phase 2 ---------------
$\quad k := i - 1$
$\quad$ **repeat**
$\quad\quad a := \mathsf{SAT}(T_1 \wedge \cdots \wedge T_k \wedge C_0 \wedge \cdots \wedge C_{k-1} \wedge \neg C_k)$
$\quad\quad$ **if** $a = (q_0, \ldots, q_k)$ **then**
$\quad\quad\quad C := \mathsf{filter}(C, q_k)$
$\quad$ **until** $a = \mathsf{unsat}$
$\quad P := C$
$\quad$ -------------- Phase 3 ---------------
$\quad$ **repeat**
$\quad\quad a := \mathsf{SAT}(T_1 \wedge \neg C_1)$
$\quad\quad$ **if** $a = (q_0, q_1)$ **then**
$\quad\quad\quad C := \mathsf{filter}(C, q_1)$
$\quad$ **until** $a = \mathsf{unsat}$
$\quad P := P \setminus C$

Fig. 1. Pseudo-code for the general invariant discovery scheme. The function SAT implements the $\mathcal{L}$-solver. It takes a formula F over n states and returns either unsat or a sequence of n states that satisfies F. The function filter takes a conjunctive property P and a state q and returns the property obtained from P by removing all conjuncts that are falsified by q. In the last statement, $P \setminus C$ denotes the conjunction of the conjuncts of P that do not occur in C.

than $O(n^2)$ space with n being the size of the instantiation set U, and refining it efficiently, i.e., in less than $O(n^2)$ time . This may not be the case in general for arbitrary template formulas $R[\_, \_]$. Hence, we focus on a class of templates for which in practice, if not in theory, these space and time costs are sub-quadratic in n: *$\mathcal{L}$-formulas denoting a partial order*. Common useful examples of partial orders include implication over the Booleans, the usual orderings over numeric domains, set inclusion over finite sets, as well as equality over any domain.

3   Partial Order Templates

In this section, we describe two specializations of the general invariant discovering scheme provided in Figure 1. Both specializations rely on the properties of partial orders in order to represent the conjunctive conjecture C compactly and process it efficiently. We start with one that works for any domain $\mathbb{D}$ and partial order $\preceq \subseteq \mathbb{D} \times \mathbb{D}$

provided that both the identity relation $\approx$ (i.e., equality) over $\mathbb{D}$ and the partial order $\preceq$ are expressible in a logic $\mathcal{L}$ with a decidable satisfiability problem. For example, this is the case when $\mathcal{L}$ is rational (resp., linear integer) arithmetic and $\preceq$ is $\leq$ or $\geq$ over the rational numbers (resp., the integers). Then, we discuss a further specialization for binary domains. For simplicity, in both cases we assume that $\approx$ and $\preceq$ are built-in symbols of $\mathcal{L}$. As a consequence, the template $R[\_, \_]$ will be just $\_ \preceq \_$.

Let U be again the given instantiation set, and let M be a sequence $(q_1, \ldots, q_m)$ of $m \geq 0$ states from Q. To each $t \in U$ we associate an m-vector $\boldsymbol{v}_t$ where, for $i = 1, \ldots, m$, $\boldsymbol{v}_t(i)$ is the *value* of t in state q_i, i.e., the element of $\mathbb{D}$ that t evaluates to in q_i. The state sequence M induces an equivalence relation $\equiv_M$ over the terms in U where $s \equiv_M t$ iff $\boldsymbol{v}_s = \boldsymbol{v}_t$.

Definition 2. *Let M be a state sequence. Suppose $\equiv_M$ has m equivalence classes and let $r_1, \ldots, r_m$ be their respective representatives. Let the point-wise extension of $\preceq$ to m-vectors over $\mathbb{D}$ be denoted by $\preceq$ as well[4]. The strongest conjecture C_M consistent with M is the smallest conjunction of $\approx$- and $\preceq$-atoms that satisfies the following.*

1. *For each $i = 1, \ldots, m$ and $t \in U \setminus \{r_i\}$, C_M contains $t \approx r_i$ if $t \equiv_M r_i$.*
2. *For each distinct $i, j = 1, \ldots, m$, C_M contains the atom $r_i \preceq r_j$ if $\boldsymbol{v}_{r_i} \preceq \boldsymbol{v}_{r_j}$.*

We can specialize the procedure described in Figure 1 by using the formula C_M above instead of C where M is a sequence of states produced by the $\mathcal{L}$-solver. We describe this specialization in the following. We consider just Phase 1 since the other phases are analogous.

Specializing the general scheme (Phase 1). For each iteration of the repeat loop in Phase 1 let M be the sequence of all the states generated until then (those passed to filter in Figure 1). Initially, M is the empty sequence, which means that $\equiv_M$ is $U \times U$ and so C_M has the form $t_2 \approx t_1 \wedge \cdots \wedge t_m \approx t_1$ with $\{t_1, \ldots, t_m\} = U$. Calls to filter now amount to computing the formula C_M for the most recent M. This specialization maintains the following (meta-)invariants on M: for all $s, t \in U$, (i) $s \equiv_M t$ iff none the models generated by the $\mathcal{L}$-solver so far falsifies the formula $s \approx t$, i.e., contradicts the conjecture that $s \approx t$ is invariant; (ii) $\boldsymbol{v}_s \preceq \boldsymbol{v}_t$ iff at least one of the models so far falsifies the formula $t \preceq s$ but none falsify $s \preceq t$; in other words, the evidence so far disproves the conjecture that $s \approx t$ is invariant but not that $s \preceq t$ is.

Relying on the two properties above it possible to show that, at each step of Phase 1, the formula C_M is $\mathcal{L}$-equivalent to the formula C in Figure 1. The formula C_M is more compact than C because it replaces the quadratically many $\preceq$-atoms between distinct $\equiv_M$-equivalent terms by linearly-many equality atoms between these terms and their equivalence class representative (e.g., $\{t_2 \approx t_1, t_3 \approx t_1\}$ in place of $\{t_1 \preceq t_2, t_1 \preceq t_3, t_2 \preceq t_3, t_2 \preceq t_1, t_3 \preceq t_1, t_3 \preceq t_2\}$).

An even more compact version of C_M is possible by exploiting the transitivity of $\preceq$. In concrete, this can be done by computing a minimal, or close to minimal, base for the poset $(\mathcal{V}_M, \preceq)$ where $\mathcal{V}_M = \{\boldsymbol{v}_{r_1}, \ldots, \boldsymbol{v}_{r_m}\}$ ($\boldsymbol{v}_{r_1}, \ldots, \boldsymbol{v}_{r_m}$ are again the representatives of $\equiv_M$'s classes). A *base* for the poset is a binary relation B on $\mathcal{V}_M$ whose

---

[4] So $(u_1, \ldots, u_n) \preceq (v_1, \ldots, v_n)$ iff $u_i \preceq v_i$ for $i = 1, \ldots, n$.

transitive closure B^+ coincides with $\preceq$ over $\mathcal{V}_M$[5]. A base is *minimal* if no strict subset of B is a base for the poset. Then, given a base B, Requirement 2 in the definition of C_M (Definition 2) can be relaxed to having C_M contain $r_i \preceq r_j$ only if $(\boldsymbol{v}_{r_i}, \boldsymbol{v}_{r_j}) \in B$.

Partial order sorting. One way to compute a base B for the poset $(\mathcal{V}_M, \preceq)$ is to use a procedure for partial order sorting. We describe here a procedure that, while probably not as efficient in general as those in the most recent literature (see, e.g., [4]), is much simpler to describe and implement, and is explicitly geared towards posets with many incomparable elements such as those generated by our invariant discovery scheme.

A *chain* over $\mathcal{V}_M$ is a list $[\boldsymbol{v}_1, \boldsymbol{v}_2, \ldots, \boldsymbol{v}_p]$ of members of $\mathcal{V}_M$ such as $\boldsymbol{v}_1 \preceq \boldsymbol{v}_2 \preceq \ldots \preceq \boldsymbol{v}_p$. Our sorting procedure takes the set $\mathcal{V}_M$ as input, and computes a set $\mathcal{C}$ of chains over $\mathcal{V}_M$ as well as a mapping σ from $\mathcal{V}_M$ to $2^{\mathcal{V}_M}$ such that $\boldsymbol{v} \preceq \boldsymbol{v}'$ for all $\boldsymbol{v}' \in \sigma(\boldsymbol{v})$. In essence, $\mathcal{C}$ is a selection of chains in the partial order, and for each element $\boldsymbol{v}$ in a chain, $\sigma(\boldsymbol{v})$ collects all the immediate successors of $\boldsymbol{v}$ in chains of $\mathcal{C}$ that do not contain $\boldsymbol{v}$. The base B for the poset $(\mathcal{V}_M, \preceq)$ is obtained by collecting all pairs $(\boldsymbol{v}, \boldsymbol{v}')$ such that $\boldsymbol{v}' \in \sigma(\boldsymbol{v})$ or $\boldsymbol{v}$ and $\boldsymbol{v}'$ occur consecutively in a chain of $\mathcal{C}$.

The procedure, shown in Figure 2, works as follows. For each $\boldsymbol{v} \in \mathcal{V}_M$ and for each existing chain $c \in \mathcal{C}$, it inserts $\boldsymbol{v}$ into c if possible. That is the case if, with respect to $\preceq$, $\boldsymbol{v}$ is smaller than the first value of c, greater than the last, or in between two consecutive elements of c. Otherwise, if c contains elements smaller than $\boldsymbol{v}$, it adds $\boldsymbol{v}$ to the set $\sigma(\boldsymbol{v}_i)$ where $\boldsymbol{v}_i$ is the greatest of these elements; also, if c contains elements greater than $\boldsymbol{v}$, it adds $\boldsymbol{v}_j$ to the set $\sigma(\boldsymbol{v})$ where $\boldsymbol{v}_j$ is the least of these elements. If the procedure is unable to add $\boldsymbol{v}$ to any existing chain, it puts $\boldsymbol{v}$ in its own chain and adds that to $\mathcal{C}$.

Example 1. We briefly illustrate the partial order sorting procedure where $\mathbb{D}$ is the domain of the integers and $\preceq$ is the usual $\leq$ relation. Consider a sequence M with two states. Let $s, t, q, r, p \in U$ be terms, and let the associated poset $(\mathcal{V}_M, \preceq)$ be $(\{\boldsymbol{v}_s, \boldsymbol{v}_t, \boldsymbol{v}_q, \boldsymbol{v}_r, \boldsymbol{v}_p\}, \leq)$ where

$$\boldsymbol{v}_s = (6,5), \quad \boldsymbol{v}_t = (5,2), \quad \boldsymbol{v}_q = (5,3), \quad \boldsymbol{v}_r = (10,2), \quad \boldsymbol{v}_p = (2,4) \,.$$

Initially, the chain $\mathcal{C}$ and the mapping σ are empty. The following table shows the value of $\mathcal{C}$ and σ after each main iteration of the sorting procedure.

	$\mathcal{C}$	σ
1	$[\boldsymbol{v}_s]$	$\boldsymbol{v}_s \mapsto \emptyset$
2	$[\boldsymbol{v}_t, \boldsymbol{v}_s]$	$\boldsymbol{v}_s \mapsto \emptyset, \boldsymbol{v}_t \mapsto \emptyset$
3	$[\boldsymbol{v}_t, \boldsymbol{v}_q, \boldsymbol{v}_s]$	$\boldsymbol{v}_s \mapsto \emptyset, \boldsymbol{v}_t \mapsto \emptyset, \quad \boldsymbol{v}_q \mapsto \emptyset, \boldsymbol{v}_r \mapsto \emptyset$
4	$[\boldsymbol{v}_t, \boldsymbol{v}_q, \boldsymbol{v}_s], [\boldsymbol{v}_r]$	$\boldsymbol{v}_s \mapsto \emptyset, \boldsymbol{v}_t \mapsto \{\boldsymbol{v}_r\}, \boldsymbol{v}_q \mapsto \emptyset, \boldsymbol{v}_r \mapsto \emptyset$
5	$[\boldsymbol{v}_t, \boldsymbol{v}_q, \boldsymbol{v}_s], [\boldsymbol{v}_r], [\boldsymbol{v}_p]$	$\boldsymbol{v}_s \mapsto \emptyset, \boldsymbol{v}_t \mapsto \{\boldsymbol{v}_r\}, \boldsymbol{v}_q \mapsto \emptyset, \boldsymbol{v}_r \mapsto \emptyset, \boldsymbol{v}_p \mapsto \{\boldsymbol{v}_s\}$

□

Analysis of the sorting procedure. Our sorting procedure is trivially terminating because the input $\mathcal{V}_M$ is finite and the set $\mathcal{C}$ and map σ are initially empty. It is correct

---

[5] That is, for all distinct $\boldsymbol{v}, \boldsymbol{v}' \in \mathcal{V}_M$, $\boldsymbol{v} \preceq \boldsymbol{v}'$ iff $(\boldsymbol{v}, \boldsymbol{v}') \in B^+$.

Require: $(\mathcal{V}_M, \preceq)$ is a poset,
Ensure: $\mathcal{C}$ is set of chains over $\mathcal{V}_M$, $\sigma : \mathcal{V}_M \to 2^{\mathcal{V}_M}$, and $\boldsymbol{v} \preceq \boldsymbol{v}'$ for all $\boldsymbol{v}' \in \sigma(\boldsymbol{v})$
  $\mathcal{C} := \emptyset$; $\sigma := \emptyset$
  for $\boldsymbol{v} \in \mathcal{V}_M$ **do**
    $\sigma := \sigma \cup \{\boldsymbol{v} \mapsto \emptyset\}$
    for $c \in \mathcal{C}$ **do**
      $i :=$ greatestBelow$(\boldsymbol{v}, c)$
      $j :=$ leastAbove$(\boldsymbol{v}, c)$
      if $j = 1$ **then**
        insert $\boldsymbol{v}$ at the beginning of c
      else
        if $i = j - 1$ **then**
          insert $\boldsymbol{v}$ at position j in c
        else
          if $i =$ the length of c **then**
            append $\boldsymbol{v}$ at the end of c
          else
            if $0 < i$ **then**
              add $\boldsymbol{v}$ to $\sigma(c(i))$
            if $0 < j$ **then**
              add $c(j)$ to $\sigma(\boldsymbol{v})$
    if $\boldsymbol{v}$ was not inserted into any chain **then**
      $\mathcal{C} := \mathcal{C} \cup \{[\boldsymbol{v}]\}$

Fig. 2. A partial order sorting procedure. The call greatestBelow$(\boldsymbol{v}, c)$ returns the position in the chain c of its greatest element smaller than $\boldsymbol{v}$, if any; otherwise, it returns 0. The call leastAbove$(\boldsymbol{v}, c)$ returns the position of the least element of c larger than $\boldsymbol{v}$, if any; otherwise, it returns 0. The notation $c(i)$ stands for the i-th element of c.

in the sense that the set B determined by $\mathcal{C}$ and σ is a base of $(\mathcal{V}_M, \preceq)$. It is not optimal because it may produce non-disjoint chains, giving rise to non-minimal bases; but it seemed to work fairly well during the experimental evaluation we describe in Section 4.

A coarse-grained worst-case complexity analysis shows that the procedure has time complexity $O(nwh)$, where w is the *width* of the poset $(\mathcal{V}_M, \preceq)$, the cardinality of the largest anti-chain in it, h is the *height* of the poset, the length of its longest chain, and n is the cardinality of $\mathcal{V}_M$[6]. This analysis assumes that comparing two elements of $\mathcal{V}_M$ for $\preceq$ takes constant time and that we store chains into arrays, which allows the functions greatestBelow and leastAbove in Figure 2 to be implemented by binary search. The former assumption does not generally hold because $\preceq$ is a point-wise ordering over vectors. One can make it only with a careful implementation based on the fact that the elements of $\mathcal{V}_M$ are built incrementally at each round of the invariant discovery procedure: vectors of length $k + 1$ are obtained by adding a new component to vectors of length k. Since $(u_1, \ldots, u_{k+1}) \preceq (v_1, \ldots, v_{k+1})$ iff $(u_1, \ldots, u_k) \preceq (v_1, \ldots, v_k)$ and $u_{k+1} \preceq v_{k+1}$, by caching in a hash table the results of vector comparisons at round k, vector comparisons at round $k + 1$ can be reduced to two constant time operations[7].

---

[6] Note that $h \leq n - w + 1$, $h = n$ when $w = 1$, and $h = 1$ when $w = n$.
[7] The hash table will have quadratic size only in the worst case when a linear number of vectors are all pairwise comparable.

A recent and efficient partial sorting algorithm by Daskalakis *et al.* based on merge sort [4] has complexity $O(w^2 n \log \frac{n}{w})$, where again n is the cardinality of the poset and w its width. This complexity and that of our procedure do not easily compare in general. But we note that the posets we work with tend to have a small height, because most value vectors are incomparable. Now, with an upper bound on a poset's height, the poset's width grows proportionally with its cardinality. This makes our procedure quadratic in n and the one by Daskalakis *et al.* more than cubic.

3.1 Binary Domains

When the domain $\mathbb{D}$ has cardinality 2, for example in the Boolean case, there is a better way to compute a base B for the poset $(\mathcal{V}_M, \preceq)$. Instead of a partial order sorting procedure, we can use one that represents B more directly as a directed acyclic graph (dag) G_M whose nodes are the equivalence classes of $\equiv_M$, and whose edges represent (selected) pairs in $\preceq$. More precisely, the set of edges is such that for all distinct equivalence classes S and T of $\equiv_M$ with respective representatives s and t, S and T are connected in G_M iff $v_s \preceq v_t$. The graph for the initial, empty state sequence is simply the graph with no edges and a single node, the whole instantiation set U.

Graph generation. We developed a procedure to compute the graph G_M for state sequences M, relying on the fact that each M is built incrementally, by appending a new state q to a previous sequence L. Given a sequence L and its graph G_L, and a new state q, the procedure computes the graph G_M for the sequence M obtained by appending q to L. We do not describe the procedure in detail here for space constraints. Instead, we give a general intuition on how it works.

Assume for concreteness that $\mathbb{D} = \{0, 1\}$ and $0 \preceq 1$, and let X be an arbitrary node of the old graph G_L. For $i = 0, 1$, let X_i be the set consisting of all the terms in X that evaluate to i in the new state q. The set X_i becomes a node of the new graph G_M iff $X_i \neq \emptyset$. In other words, G_M gets a node identical to X if all the terms of X have the same value in q, and gets two new nodes, partitioning X, otherwise. Whenever both X_0 and X_1 are added to G_M, the edge $X_0 \longrightarrow X_1$ is also added. Edges between old nodes in G_L are inherited by the corresponding new nodes consistently with the ordering induced by M. In general, every edge $X_i \longrightarrow Y_j$ of G_M (where X and Y are nodes of G_L) comes from a path of length at most 2 from X to Y in G_L; moreover, $i \leq j$. The effect of the procedure is best illustrated with an example.

Example 2. Let $M = (L, q)$ and suppose G_L is the following dag.

$$ \boxed{A} \longrightarrow \boxed{B} \longrightarrow \boxed{C} $$

Suppose B_1 is empty, and A_0, A_1, B_0, C_0, C_1 are all non-empty. The procedure starts by creating G_M with the following nodes and edges.

$$ \boxed{A_0} \longrightarrow \boxed{A_1} \qquad \boxed{B_0} \qquad \boxed{C_0} \longrightarrow \boxed{C_1} $$

Then it adds edges derived from G_L, returning the following dag as G_M.

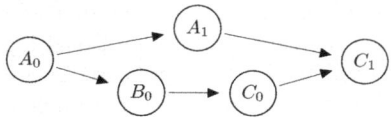

The edge $A_0 \longrightarrow B_0$ comes from $A \longrightarrow B$. Similarly, $B_0 \longrightarrow C_0$ comes from $B \longrightarrow C$. In contrast, $A_1 \longrightarrow C_1$ comes from the path $A \longrightarrow B \longrightarrow C$, because of the absence of B_1.   □

The procedure works in three phases. In the first phase, it scans G_L's node set to generate the nodes of G_M and the edges between nodes X_0 and X_1. It also builds a map from each node of G_L to its corresponding node(s) in G_M. In the second phase, it traverses the dag G_L bottom up (from leaves to roots), to determine for each of its nodes which nodes of G_M should inherit the node's incoming edges, and how. A marking mechanism is used to visit each node of G_L only once. In the third phase, it scans G_L's edge set to generate the corresponding edges in G_M.

Analysis of the graph generation procedure. From the above high-level description of the procedure it is perhaps already clear that its time complexity is linear in the number of nodes and edges of G_L. The linearity, however, comes at the cost of suboptimality. Since in the second phase each node of G_L is visited only once, it is possible for G_M to end up containing *redundant edges*, edges connecting directly two nodes also connected by a longer path. Redundant edges lead to non-minimal bases for the associated poset because the inequations they generate are implied by other inequations in the base.

For example, the edge $A \longrightarrow C$ is redundant if $A \longrightarrow B$ and $B \longrightarrow C$ are also in G_M. By the transitivity of $\preceq$, the inequation $r_A \preceq r_B$ between the A's and C's representatives is then superfluous. In our implementation, discussed next, redundant edges are removed in an optional post-processing step on the final dag.

4   Experimental Evaluation

To evaluate experimentally the specialized invariant discovery procedures described in the previous section we implemented two instances of the general invariant discovery scheme: one for the domain of linear integer arithmetic, with $\leq$ as the partial order,[8] and one for the Boolean domain, with implication as the partial order. The instances are implemented in a new tool, called KIND-INV[9], built with components of the KIND model checker [7]. Kind was developed to check safety properties of Lustre programs. Lustre [8] is a synchronous data-flow language with infinite streams of values of three basic types: bool, int, and real. It is typically used to model circuits at a high level or control software in embedded devices.

KIND is a k-induction-based model checker for programs in an idealized version of Lustre that uses (mathematical) integers in place of machine integer values, and rational

---

[8] This instance works with rational numbers as well, but we ignore that here for simplicity.

[9] System and experimental data can be found at http://clc.cs.uiowa.edu/Kind/.

numbers in place of floating values. The underlying logic of KIND, and of KIND-INV, is a quantifier-free logic that includes both propositional logic and linear arithmetic. We'll refer to it as $\mathcal{IL}$ (for Idealized Lustre logic) here. Lustre programs can be readily encoded in $\mathcal{IL}$ as transition systems of the sort we use here (see [7] for more details). The SMT solvers CVC3 [1] and Yices [5] are used, in alternative, as satisfiability solvers for this logic. A Lustre program can be structured as a set of modules called *nodes* which can be understood as macros. KIND-INV currently takes a single-node Lustre program as input. A multi-node program can be treated by expanding it in advance to a behaviorally equivalent single-node one. The invariants discovered by KIND-INV are then added to the Lustre input program as "assertions." Contrary to other languages, such as C, assertions in Lustre are expressions of type bool that are *assumed* to be true at each execution step of the program.

KIND-INV accepts two options for generating invariants: **bool** and **int**. The first option produces invariants of the form $s \rightarrow t$ or $s = t$ where s and t are Lustre Boolean terms. The second produces invariants of the form $s \leq t$ or $s = t$ where s and t are integer terms. The instantiation set U currently consists of heuristically selected terms from the input Lustre program plus some distinguished constant terms such as true and false. Note that bool terms may contain int terms, as in $(x + y > 0)$ or done, and vice versa, as in $x + ($if $y > 0$ then y else $1)$.

KIND-INV provides three binary options affecting invariant generation. The first two work only with the **bool** invariant option, the last one with both options:

No_Ands : When this flag is turned on, KIND-INV will not consider candidate terms of the form $s \wedge t$. The rationale behind this flag is that, conjunctive terms lead to many trivial invariants, for instance, those of the form $(s \wedge t) \rightarrow s$. Having too many of these unnecessary invariants can be burdensome for the SMT-solver, limiting the effectiveness of the non-trivial invariants in the generated assertion.

No_Redundant_Edges : When this flag is on, KIND-INV will remove redundant edges from the final dag storing the computed poset (see Section 3.1).

No_Trivial_Invariants : This flags governs whether the third phase of the invariant discovery procedure is performed or not. Its rationale is that the third phase is expensive and may not be worthwhile.

Evaluation setup. To evaluate KIND-INV, we used a benchmark set derived from the one used in [7], which consists of a variety of benchmarks from several sources. Each benchmark in the original set is a Lustre program together with a single property to check, expressed as a Lustre bool term. Our derived set discards some duplicate benchmarks—included in the original set by mistake—and converts each program to a single-node one using the *pollux* tool from the Lustre 4 distribution.

Let us call a benchmark *valid* if its safety property holds for the associated program, and *invalid* otherwise. KIND is able to prove 438 of the 941 benchmarks in our set invalid by returning a (independently verified) counter-example trace for the program. KIND reports 309 of the remaining benchmarks as valid, and diverges on the remaining 194 benchmarks, even with very large timeout values. We conjecture that those 194 *unsolved* benchmarks are all valid but contain a property that is either k-inductive for an extremely large k or, more plausibly, not k-inductive for any k.

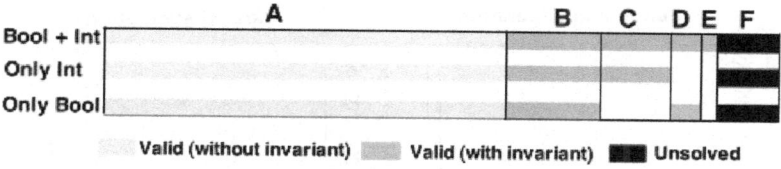

	A	B	C	D	E	F
Bool + Int						
Only Int						
Only Bool						

Valid (without invariant)   Valid (with invariant)   Unsolved

Fig. 3. Distribution of solved and unsolved benchmarks for three classes of invariants. Green bars indicate the percentage of benchmarks solvable only with invariants. All bars are drawn to scale.

For the experiments described here the benchmark set consists of the valid and the unsolved benchmarks, 503 in total. Our main goal was to evaluate how effective the invariants generated by KIND-INV are at improving KIND's *precision*, measured as the percentage of solved benchmarks. The experiments were run on a small dedicated cluster of identical machines with a 3.0 GHz Intel Pentium 4 cpu, 1GB of memory and Redhat Enterprise Linux 4.0. Version 1.0.9 of the Yices solver was used both for KIND-INV and KIND.

In a first step, we ran KIND-INV on the benchmark set twice, once for the bool and once with the `int` invariant generation option. For each of the benchmarks where KIND-INV did not time out, we obtained a set of invariants, and added them to the benchmark as a single conjunctive assertion. The added assertion was the constant true when KIND-INV timed out or ended up discarding all conjectures from the initial set. To make sure that the added assertions were indeed invariants, we verified each of them independently by formulating it as a safety property and asking KIND to prove it[10]. In a second step, we ran KIND in *inductive mode* on each benchmark, with and without the assertion that collects the discovered invariants. In that mode, KIND attempts to prove the benchmark's property by k-induction, using any assertion in the program to strengthen the k-induction hypothesis with the invariant in the assertion. The timeout for KIND-INV was set to 300 seconds and that for KIND to 120 seconds.

We did an extensive evaluation over our benchmarks with various configurations. By and large, all configurations are comparable in terms of the precision achieved by KIND when using their generated invariants. The only significant differences are with respect to invariant generation speed. A statistical analysis of the results obtained with the various configurations, not reported here, indicated that the following configuration is superior to the others: **No_Ands = on, No_Redundant_Edges = on**, and **No_Trivial_Invariants = off**. Hence, we report our results just for that configuration.

Precision results. The size of the generated invariants, measured as their number of conjuncts, varies from 0 to 1150, with a median value of 133. With the **bool** option, KIND-INV times out in 19 cases (out of 503), and terminates normally but with an empty invariant in 2 cases. With the **int** option, it times out in 62 cases and terminates with an empty invariant in 12 cases.

Using only bool *invariants*, i.e., invariants generated by KIND-INV with the **bool** option, KIND is able to prove 40% of the 194 previously unsolved benchmarks; using

---

[10] Since KIND-INV and KIND used the same SMT solver it is possible that we missed incorrect assertions because of a bug in the solver, but we believe this to be unlikely.

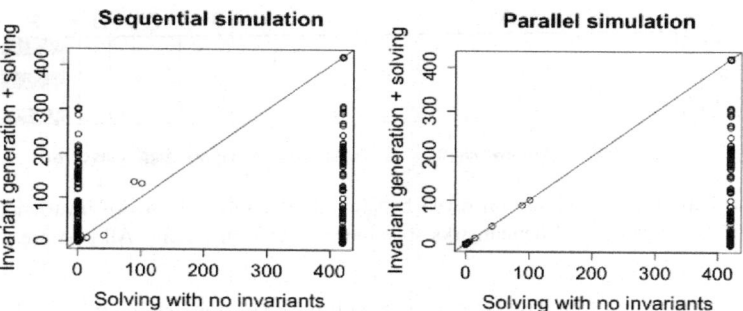

Fig. 4. Solving times without invariants versus int invariant generation times plus solving times with int invariants. In the parallel simulation, solving without invariants is attempted during invariant generation. Invariants are then used once available, and only if still needed.

int *invariants*, invariants generated with the `int` option, it proves 53% of the unsolved benchmarks; using both bool and int invariants, it proves 63% of the unsolved benchmarks. In the three cases above, Kind's precision over all 513 benchmarks grows from 61% (without invariants) to 77%, 82%, and 85%, respectively. For all the newly solvable benchmarks the properties goes from (most likely) not k-inductive for any k to k-inductive with some $k \leq 16$. The set of new benchmarks solved with bool invariants and that solved with int invariants have a large overlap, which we find somewhat surprising. Less surprising is that using bool and int invariants together allows KIND to solve all the benchmarks solvable with either type alone, and more.

The addition of invariants preserves the set of benchmarks proved valid by KIND without them. Furthermore, it often shortcuts the k-induction process. In fact, without invariants, 14.5% of the previously valid benchmark have a safety property that is k-inductive for some $k > 1$; that percentage goes down respectively to 6.7%, 8.7% and 3.8%, with only bool, only int and both bool and int invariants.

Figure 4 summarizes graphically the various effects achieved with bool and int invariants, alone and in combination. For each of these three cases, column A represents benchmarks solvable by KIND without invariants; columns B to E represent benchmarks solvable with the generated invariants; column F represents benchmarks that remain unsolved, either because KIND-INV was not able to generate an invariant for them or because the generated invariant is not helpful. Columns C and D represent the benchmarks solved only with int and only with bool invariants, respectively. Column E represents the benchmarks solved only with both bool and int invariants together.

Runtime results. Adding invariants to previously solvable benchmarks systematically makes them slightly faster to solve. The total time to solve them decrease from 305.7 to 246.5 seconds. Individual solving times in the presence of invariants are very small; on average just 0.95s for all solvable benchmarks. In addition to the substantial increase in precision, this provides further evidence that our invariant discovery procedure produces high quality invariants. Invariant generation has of course its own, non-insignificant cost. Over the whole benchmark set, KIND-INV runtimes vary from less than a second to hundreds of seconds, to timing out at 300s. However, their median value is fairly small: 22.4s for int invariants and just 6.3s for bool ones. For the great majority of benchmarks (84%) bool invariant generation takes less than a minute per benchmark.

Evaluating invariant generation costs against the increase in precision is a difficult task because it also depends on the relative importance of precision versus prompt response. A supporting argument is that invariant generation and k-induction model checking can be done in parallel—with invariants fed to the k-induction loop as soon as they are generated—mitigating this way the cost of invariant generation. Developing a parallel model checker integrating KIND and KIND-INV was beyond the scope of this work. An approximate analysis, however, can be provided with a rough conceptual simulation of such a concurrent system.

Since the synchronization overhead in the parallel model checker would be arguably very small, we can ignore it here for simplicity. Then we can imagine the parallel checker's runtimes to be, for each benchmark, the minimum between the following two values: (i) the time KIND takes to prove the property without invariants and (ii) the sum of the times KIND-INV takes to output an invariant and KIND takes to prove the property using that invariant. The scatter plots in Figure 4 illustrate this comparison with int invariants—the results are similar for bool invariants. The first plot compares for each benchmark the runtime of KIND with no invariants and a 420s timeout[11] against the runtime of a hypothetical sequential checker that uses KIND-INV with a timeout of 120s, to add an invariant to the program, and then calls KIND with a timeout of 300s. The considerable invariant generation time penalty paid by the sequential checker (illustrated by all the points above the diagonal lines in the first plot) essentially disappears with the parallel checker, as shown in the second plot.

5 Conclusion and Future Work

We presented a novel scheme for discovering invariants in transition systems. The scheme is parametrized by a formula template representing a decidable relation over the system's datatypes, and by a set of terms used to instantiate the template. Its main features are that it checks all template instances for invariance at the same time and makes heavy use of a satisfiability solver for the logic in which the system and the instances are encoded. We described two specializations of the scheme to templates representing partial orders where we can exploit the properties of posets to achieve space and time efficiencies. Initial experimental results are very encouraging in terms of the speed of invariant generation and the effectiveness of the generated invariants in automating the verification of safety properties.

In the implementation discussed in the previous section, invariant generation is done off-line. We are developing a parallel model checking architecture and implementation in which k-induction and invariant generation are done concurrently, with invariants fed to the k-induction loop as soon as they are produced.

Our invariant discovery scheme lumps together, in a single invariant produced at the end, instances of the template that may be k-inductive for different values of k. We believe that the effectiveness of the parallel model checking architecture would increase

---

[11] Increasing the timeout from 300s to 420s does not change the set of solved benchmarks.

if invariant instances were identified and output progressively—with k-inductive instances produced before $(k + 1)$-inductive ones. We are working on a new version of the scheme based on this idea.

We are also investigating techniques for compositional reasoning with synchronous systems based on the invariant discovery method presented in this paper. The main idea is to generate invariants separately for each module of a multi-module system, and then use them to aid the verification of properties of the entire system.

References

1. Barrett, C.W., Tinelli, C.: CVC3. In: Damm, W., Hermanns, H. (eds.) CAV 2007. LNCS, vol. 4590, pp. 298–302. Springer, Heidelberg (2007)
2. Bensalem, S., Lakhnech, Y.: Automatic generation of invariants. Form. Methods Syst. Des. 15(1), 75–92 (1999)
3. Das, S., Dill, D.L.: Counter-example based predicate discovery in predicate abstraction. In: Aagaard, M.D., O'Leary, J.W. (eds.) FMCAD 2002. LNCS, vol. 2517, pp. 19–32. Springer, Heidelberg (2002)
4. Daskalakis, C., Karp, R.M., Mossel, E., Riesenfeld, S., Verbin, E.: Sorting and selection in posets. In: ACM-SIAM Symposium on Discrete Algorithms, pp. 392–401 (2009)
5. Dutertre, B., de Moura, L.: The YICES SMT solver. Technical report, SRI International (2006)
6. Gulwani, S., Srivastava, S., Venkatesan, R.: Constraint-based invariant inference over predicate abstraction. In: Jones, N.D., Müller-Olm, M. (eds.) VMCAI 2009. LNCS, vol. 5403, pp. 120–135. Springer, Heidelberg (2009)
7. Hagen, G., Tinelli, C.: Scaling up the formal verification of lustre programs with SMT-based techniques. In: FMCAD 2008, Piscataway, NJ, USA, 2008, pp. 1–9. IEEE Press, Los Alamitos (2008)
8. Halbwachs, N., Caspi, P., Raymond, P., Pilaud, D.: The synchronous data-flow programming language LUSTRE. Proceedings of the IEEE 79(9), 1305–1320 (1991)
9. Hunt, W., Johnson, S., Bjesse, P., Claessen, K.: SAT-based verification without state space traversal. In: Johnson, S.D., Hunt Jr., W.A. (eds.) FMCAD 2000. LNCS, vol. 1954, pp. 409–426. Springer, Heidelberg (2000)
10. Manna, Z., Pnueli, A.: Temporal Verification of Reactive Systems: Safety. Springer, Heidelberg (1995)
11. Pandav, S., Slind, K., Gopalakrishnan, G.: Counterexample guided invariant discovery for parameterized cache coherence verification. In: Borrione, D., Paul, W. (eds.) CHARME 2005. LNCS, vol. 3725, pp. 317–331. Springer, Heidelberg (2005)
12. Srivastava, S., Gulwani, S.: Program verification using templates over predicate abstraction. SIGPLAN Not. 44, 223–234 (2009)
13. Su, J.X., Dill, D.L., Barrett, C.W.: Automatic generation of invariants in processor verification. In: Srivas, M., Camilleri, A. (eds.) FMCAD 1996. LNCS, vol. 1166, pp. 377–388. Springer, Heidelberg (1996)
14. Thalmaier, M., Nguyen, M.D., Wedler, M., Stoffel, D., Bormann, J., Kunz, W.: Analyzing k-step induction to compute invariants for SAT-based property checking. In: DAC 2010, pp. 176–181. ACM, New York (2010)
15. Tiwari, A., Rueß, H., Saïdi, H., Shankar, N.: A technique for invariant generation. In: Margaria, T., Yi, W. (eds.) TACAS 2001. LNCS, vol. 2031, pp. 113–127. Springer, Heidelberg (2001)

Stuttering Mostly Speeds Up Solving Parity Games

Sjoerd Cranen, Jeroen J.A. Keiren, and Tim A.C. Willemse

Department of Mathematics and Computer Science,
Technische Universiteit Eindhoven,
P.O. Box 513, 5600 MB Eindhoven, The Netherlands

Abstract. We study the process theoretic notion of stuttering equivalence in the setting of parity games. We demonstrate that stuttering equivalent vertices have the same winner in the parity game. This means that solving a parity game can be accelerated by minimising the game graph with respect to stuttering equivalence. While, at the outset, it might not be clear that this strategy should pay off, our experiments using typical verification problems illustrate that stuttering equivalence speeds up solving parity games in many cases.

1 Introduction

Parity games [6,13,22] are played by two players (called *even* and *odd*) on a directed graph in which vertices have been assigned *priorities*. Every vertex in the graph belongs to exactly one of these two players. The game is played by moving a token along the edges in the graph indefinitely; the edge that is moved along is chosen by the player owning the vertex on which the token currently resides. Priorities that appear infinitely often along such infinite plays then determine the winner of the play.

Solving a parity game essentially boils down to computing the set of vertices that, if the token is initially placed on a vertex in this set, allows player *even* (resp. *odd*) to win. This problem is known to be in NP ∩ co-NP; it is still an open problem whether a polynomial time algorithm exists for the problem, but even in case such an algorithm is found, it may not be the most efficient algorithm in practice.

Parity games play a crucial role in verification; the model checking problem for the modal μ-calculus can be reduced to the problem of solving a given parity game. It is therefore worthwhile to investigate methods by which these games can be solved efficiently in practice. In [7], Friedman and Lange describe a meta-algorithm that, combined with a set of heuristics, appears to have a positive impact on the time required to solve parity games. Fritz and Wilke consider more-or-less tried and tested techniques for *minimising* parity games using novel refinement and equivalence relations, see [9]. The delayed simulation they introduce, and its induced equivalence relation, however, are problematic for quotienting, which is why they go on to define two variations of delayed simulations that do not suffer from this problem. As stated in [8], however, "Experiments

M. Bobaru et al. (Eds.): NFM 2011, LNCS 6617, pp. 207–221, 2011.

indicate that simplifying parity games using our approach before solving them is not faster than solving them outright in practice".

Despite the somewhat unsatisfactory performance of the delayed simulation in practice, we follow a methodology similar to the one pursued by Fritz and Wilke. As a basis for our investigations, we consider *stuttering equivalence* [3], which originated in the setting of Kripke Structures. Stuttering equivalence has two qualities that make it an interesting candidate for minimising parity games. Firstly, vertices with the same player and priority are only distinguished on the basis of their future branching behaviour, allowing for a considerable compression. Secondly, stuttering equivalence has a very attractive worst-case time complexity of $\mathcal{O}(n \cdot m)$, for n vertices and m edges, which is in stark contrast to the far less favourable time complexity required for delayed simulation, which is $\mathcal{O}(n^3 \cdot m \cdot d^2)$, where d is the number of different priorities in the game. In addition to these, stuttering equivalence has several other traits that make it appealing: quotienting is straightforward, distributed algorithms for computing stuttering equivalence have been developed (see *e.g.* [2]), and it admits efficient, scalable implementations using BDD technology [21].

On the basis of the above qualities, stuttering equivalence is likely to significantly compress parity games that stem from typical model checking problems. Such games often have a rather limited number of priorities (typically at most three), and appear to have regular structures. We note that, as far as we have been able to trace, quotienting parity games using stuttering equivalence has never been shown to be sound. Thus, our contributions in this paper are twofold.

First, we show that stuttering equivalent vertices are won by the same player in the parity game. As a side result, given a winning strategy for a player for a particular vertex, we obtain winning strategies for all stuttering equivalent vertices. This is of particular interest in case one is seeking an explanation for the solution of the game, for instance as a means for diagnosing a failed verification.

Second, we experimentally show that computing and subsequently solving the stuttering quotient of a parity game is in many cases *faster* than solving the original game. In our comparison, we included several competitive implementations of algorithms for solving parity games, including several implementations of *Small Progress Measures* [11] and McNaughton's *recursive* algorithm [13]. Moreover, we also compare it to quotienting using *strong bisimulation* [15]. For an up-to-date overview of experiments we refer to [5], which we plan to keep updated with new results. While we do not claim that stuttering equivalence minimisation should always be performed prior to solving a parity game, we are optimistic about its effects in practical verification tasks.

Structure. The remainder of this paper is organised as follows. Section 2 briefly introduces the necessary background for parity games. In Section 3 we define both strong bisimilarity and stuttering equivalence in the setting of parity games; we show that both can be used for minimising parity games. Section 4 is devoted to describing our experiments, demonstrating the efficacy of stuttering equivalence minimisation on a large set of verification problems. In Section 5, we briefly discuss future work and open issues.

2   Preliminaries

We assume the reader has some familiarity with parity games; therefore, the main purpose of this section is to fix terminology and notation. For an in-depth treatment of these games, we refer to [13,22].

2.1   Parity Games

A parity game is a game played by players *even* (represented by the symbol 0) and *odd* (represented by the symbol 1). It is played on a total finite directed graph, the vertices of which can be won by either 0 or 1. The objective of the game is to find the partitioning that separates the vertices won by 0 from those won by 1. In the following text we formalise this definition, and we introduce some concepts that will make it easier to reason about parity games.

Definition 1. *A parity game $\mathcal{G}$ is a directed graph $(V, \rightarrow, \Omega, \mathcal{P})$, where*

- V *is a finite set of vertices,*
- $\rightarrow \subseteq V \times V$ *is a total edge relation (i.e., for each $v \in V$ there is at least one $w \in V$ such that $(v, w) \in \rightarrow$),*
- $\Omega : V \rightarrow \mathbb{N}$ *is a priority function that assigns priorities to vertices,*
- $\mathcal{P} : V \rightarrow \{0, 1\}$ *is a function assigning vertices to players.*

Instead of $(v, w) \in \rightarrow$ we will usually write $v \rightarrow w$. Note that, for the purpose of readability later in this text, our definition deviates from the conventional one: instead of requiring a partitioning of V into vertices owned by player even and vertices owned by player odd, we achieve the same through the function $\mathcal{P}$.

Paths. A sequence of vertices $v_1, \ldots, v_n$ for which $v_i \rightarrow v_{i+1}$ for all $1 \leq i < n$ is called a *path*, and may be denoted using angular brackets: $\langle v_1, \ldots, v_n \rangle$. The concatenation $p \cdot q$ of paths p and q is again a path. We use p_n to denote the nth vertex in a path p. The set of paths of length n, for $n \geq 1$ starting in a vertex v is defined inductively as follows.

$$\Pi^1(v) = \{\langle v \rangle\}$$
$$\Pi^{n+1}(v) = \{\langle v_1, \ldots, v_n, v_{n+1} \rangle \mid \langle v_1, \ldots, v_n \rangle \in \Pi^n(v) \land v_n \rightarrow v_{n+1}\}$$

We use $\Pi^\omega(v)$ to denote the set of infinite paths starting in v. The set of all paths starting in v, both finite and infinite is defined as follows:

$$\Pi(v) = \Pi^\omega(v) \cup \bigcup_{n \in \mathbb{N}} \Pi^n(v)$$

Winner. A game starting in a vertex $v \in V$ is played by placing a token on v, and then moving the token along the edges in the graph. Moves are taken indefinitely according to the following simple rule: if the token is on some vertex v, player $\mathcal{P}(v)$ moves the token to some vertex w such that $v \rightarrow w$. The result is an infinite path p in the game graph. The *parity* of the lowest priority that occurs infinitely often on p defines the *winner* of the path. If this priority is even, then player 0 wins, otherwise player 1 wins.

Strategies. A *strategy* for player i is a partial function $\phi : V^* \to V$, that for each path ending in a vertex owned by player i determines the next vertex to be played onto. A path p of length n is *consistent* with a strategy ϕ for player i, denoted $\phi \Vdash p$, if and only if for all $1 \leq j < n$ it is the case that $\langle p_1, \ldots, p_j \rangle \in \mathsf{dom}(\phi)$ and $\mathcal{P}(p_j) = i$ imply $p_{j+1} = \phi(\langle p_1, \ldots, p_j \rangle)$. The definition of consistency is extended to infinite paths in the obvious manner. We denote the set of paths that are consistent with a given strategy ϕ, starting in a vertex v by $\Pi_\phi(v)$; formally, we define:

$$\Pi_\phi(v) = \{p \in \Pi(v) \mid \phi \Vdash p\}$$

A strategy ϕ for player i is said to be a *winning strategy* from a vertex v if and only if i is the winner of every path that starts in v and that is consistent with ϕ. It is known from the literature that each vertex in the game is won by exactly one player; effectively, this induces a partitioning on the set of vertices V in those vertices won by player 0 and those vertices won by player 1.

Orderings. We assume that V is ordered by an arbitrary, total ordering $\sqsubseteq$. The minimal element of a non-empty set $U \subseteq V$ with respect to this ordering is denoted $\sqcap(U)$. Let $|v, u|$ denote the least number of edges required to move from vertex v to vertex u in the graph. We define $|v, u| = \infty$ if u is unreachable from v. For each vertex $u \in V$, we define an ordering $\prec_u \subseteq V \times V$ on vertices, that intuitively orders vertices based on their proximity to u, with a subjugate role for the vertex ordering $\sqsubseteq$:

$$v \prec_u v' \text{ iff } |v, u| < |v', u| \text{ or } (|v, u| = |v', u| \text{ and } v \sqsubset v')$$

Observe that $u \prec_u v$ for all $v \neq u$. The minimal element of $U \subseteq V$ with respect to $\prec_u$ is written $\lambda_u(U)$.

3   Strong Bisimilarity and Stuttering Equivalence

Process theory studies refinement and equivalence relations, characterising the differences between models of systems that are observable to entities with different observational powers. Most equivalence relations have been studied for their computational complexity, giving rise to effective procedures for deciding these equivalences. Prominent equivalences are *strong bisimilarity*, due to Park [15] and *stuttering equivalence* [3], proposed by Browne, Clarke and Grumberg.

Game graphs share many of the traits of the system models studied in process theory. As such, it is natural to study refinement and equivalence relations for such graphs, see *e.g.*, *delayed simulation* [9]. In the remainder of this section, we recast the bisimilarity and stuttering equivalence to the setting of parity games, and show that these are finer than *winner equivalence*, which we define as follows.

Definition 2. *Let* $\mathcal{G} = (V, \to, \Omega, \mathcal{P})$ *be a parity game. Two vertices* $v, v' \in V$ *are said to be* winner equivalent, *denoted* $v \sim_w v'$ *iff* v *and* v' *are won by the same player.*

Because every vertex is won by exactly one player (see, *e.g.*, [22]), winner equivalence partitions V into a subset won by player 0 and a subset won by player 1. Clearly, winner equivalence is therefore an equivalence relation on the set of vertices of a given parity game. The problem of deciding winner equivalence, is in NP ∩ co-NP; all currently known algorithms require time exponential in the number of priorities in the game.

We next define strong bisimilarity for parity games; basically, we interpret the priorities and players of vertices as state labellings.

Definition 3. *Let* $\mathcal{G} = (V, \rightarrow, \Omega, \mathcal{P})$ *be a parity game. A symmetric relation* $R \subseteq V \times V$ *is a* strong bisimulation *relation if* $v \, R \, v'$ *implies*

- $\Omega(v) = \Omega(v')$ *and* $\mathcal{P}(v) = \mathcal{P}(v')$;
- *for all* $w \in V$ *such that* $v \rightarrow w$, *there should be a* $w' \in V$ *such that* $v' \rightarrow w'$ *and* $w \, R \, w'$.

Vertices v *and* v' *are said to be* strongly bisimilar, *denoted* $v \sim v'$, *iff a strong bisimulation relation* R *exists such that* $v \, R \, v'$.

Strong bisimilarity is an equivalence relation on the vertices of a parity game; quotienting with respect to strong bisimilarity is straightforward. It is not hard to show that strong bisimilarity is strictly finer than winner equivalence. Moreover, quotienting can be done effectively with a worst-case time complexity of $\mathcal{O}(|V| \log |V|)$.

Strong bisimilarity quotienting prior to solving a parity game can in some cases be quite competitive. One of the drawbacks of strong bisimilarity, however, is its sensitivity to counting (in the sense that it will not identify vertices that require a different number of steps to reach a next equivalence class), preventing it from compressing the game graph any further.

Stuttering equivalence shares many of the characteristics of strong bisimilarity, and deciding it has only a slightly worse worst-case time complexity. However, it is insensitive to counting, and is therefore likely to lead to greater reductions. Given these observations, we hypothesise (and validate this hypothesis in Section 4) that stuttering equivalence outperforms strong bisimilarity and, in most instances, reduces the time required for deciding winner equivalence in parity games stemming from verification problems.

We first introduce stuttering bisimilarity [14], a coinductive alternative to the stuttering equivalence of Browne, Clarke and Grumberg; we shall use the terms stuttering bisimilarity and stuttering equivalence interchangeably. The remainder of this section is then devoted to showing that stuttering bisimilarity is coarser than strong bisimilarity, but still finer than winner equivalence. The latter result allows one to pre-process a parity game by quotienting it using stuttering equivalence.

Definition 4. *Let* $\mathcal{G} = (V, \rightarrow, \Omega, \mathcal{P})$ *be a parity game. Let* $R \subseteq V \times V$. *An infinite path* p *is* R-divergent, *denoted* $\mathsf{div}_R(p)$ *iff* $p_1 \, R \, p_i$ *for all* i. *Vertex* $v \in V$ *allows for divergence, denoted* $\mathsf{div}_R(v)$ *iff there is a path* p *such that* $p_1 = v$ *and* $\mathsf{div}_R(p)$.

We generalise the transition relation $\to$ to its reflexive-transitive closure, denoted $\Rightarrow$, taking a given relation R on vertices into account. The generalised transition relation is used to define stuttering bisimilarity. Let $\mathcal{G} = (V, \to, \Omega, \mathcal{P})$ be a parity game and let $R \subseteq V \times V$ be a relation on its vertices. Formally, we define the relations $\to_R \subseteq V \times V$ and $\Rightarrow_R \subseteq V \times V$ through the following set of deduction rules.

$$\frac{v \to w \qquad v \, R \, w}{v \to_R w} \qquad \frac{}{v \Rightarrow_R v} \qquad \frac{v \to_R w \qquad w \Rightarrow_R v'}{v \Rightarrow_R v'}$$

We extend this notation to paths: we sometimes write $\langle v_1, \dots, v_n \rangle \to u$ if $v_n \to u$; similarly, we write $\langle v_1, \dots, v_n \rangle \to_R u$ and $\langle v_1, \dots, v_n \rangle \Rightarrow_R u$.

Definition 5. *Let $\mathcal{G} = (V, \to, \Omega, \mathcal{P})$ be a parity game. Let $R \subseteq V \times V$ be a symmetric relation on vertices; R is a* stuttering bisimulation *if $v \, R \, v'$ implies*

- $\Omega(v) = \Omega(v')$ *and* $\mathcal{P}(v) = \mathcal{P}(v')$;
- $\mathsf{div}_R(v)$ *iff* $\mathsf{div}_R(v')$;
- *If $v \to u$, then either $(v \, R \, u \wedge u \, R \, v')$, or there are u', w, such that $v' \Rightarrow_R w \to u'$ and $v \, R \, w$ and $u \, R \, u'$;*

Two states v and v' are said to be stuttering bisimilar, *denoted $v \approx v'$ iff there is a stuttering bisimulation relation R, such that $v \, R \, v'$.*

Note that stuttering bisimilarity is the largest stuttering bisimulation. Moreover, stuttering bisimilarity is an equivalence relation, see *e.g.* [14,3]. In addition, quotienting with respect to stuttering bisimilarity is straightforward.

Stuttering bisimilarity between vertices extends naturally to finite paths. Paths of length 1 are equivalent if the vertices they consist of are equivalent. If paths p and q are equivalent, then $p \cdot \langle v \rangle \approx q$ iff v is equivalent to the last vertex in q (and analogously for extensions of q), and $p \cdot \langle v \rangle \approx q \cdot \langle w \rangle$ iff $v \approx w$. An infinite path p is equivalent to a (possibly infinite) path q if for all finite prefixes of p there is an equivalent prefix of q and *vice versa*.

We next set out to prove that stuttering bisimilarity is finer than winner equivalence. Our proof strategy is as follows: given that there is a strategy ϕ for player i from a vertex v, we define a strategy for player i that from vertices equivalent to v schedules only paths that are stuttering bisimilar to a path starting in v that is consistent with ϕ.

If after a number of moves a path p has been played, and our strategy has to choose the next move, then it needs to know which successors for p will yield a path for which again there is a stuttering bisimilar path that is consistent with ϕ. To this end we introduce the set $\mathsf{reach}_{\phi,v}(p)$.

Let ϕ be an arbitrary strategy, v an arbitrary vertex owned by the player for which ϕ defines the strategy, and let p be an arbitrary path. We define $\mathsf{reach}_{\phi,v}(p)$ as the set of vertices in new classes, reachable by traversing ϕ-consistent paths that start in v and that are stuttering bisimilar to p.

$$\mathsf{reach}_{\phi,v}(p) = \{u \in V \mid \exists q \in \Pi_\phi(v) : p \approx q \wedge \phi \Vdash q \cdot \langle u \rangle \wedge q \cdot \langle u \rangle \not\approx q\}$$

Observe that not all vertices in $\mathsf{reach}_{\phi,v}(p)$ have to be in the same equivalence class, because it is not guaranteed that all paths $q \in \Pi_\phi(v)$, stuttering bisimilar to p, are extended by ϕ towards the same equivalence class.

Suppose the set $\mathsf{reach}_{\phi,v}(p)$ is non-empty; in this case, our strategy should select a *target class* to which p should be extended. Because stuttering bisimilar vertices can reach the same classes, it does not matter which class present in $\mathsf{reach}_{\phi,v}(p)$ is selected as the target class. We do however need to make a unique choice; to this end we use the total ordering $\sqsubset$ on vertices.

$$\mathsf{targetclass}_{\phi,v}(p) = \{u \in V \mid u \approx \sqcap(\mathsf{reach}_{\phi,v}(p))\}$$

Not all vertices in the target class need be reachable from p, but there must exist at least one vertex that is. We next determine a *target vertex*, by selecting a unique, reachable vertex from the target class. This target of p, given a strategy ϕ and a vertex v is denoted $\tau_{\phi,v}(p)$; note that the ordering $\sqsubset$ is again used to uniquely determine a vertex from the set of reachable vertices.

$$\tau_{\phi,v}(p) = \sqcap\{u \in \mathsf{targetclass}_{\phi,v}(p) \mid \exists w \in V : p \Longrightarrow_\approx w \to u\}$$

Definition 6. *We define a strategy* $\mathsf{mimick}_{\phi,v}$ *for player* i *that, given some strategy* ϕ *for player* i *and a vertex* v, *allows only paths to be scheduled that have a stuttering bisimilar path starting in* v *that is scheduled by* ϕ. *It is defined as follows.*

$$\mathsf{mimick}_{\phi,v}(p) = \begin{cases} \curlywedge_t\{u \in V \mid p \to_\approx u\}, & \begin{array}{l} t = \tau_{\phi,v}(p) \\ p \not\to \tau_{\phi,v}(p) \\ \mathsf{reach}_{\phi,v}(p) \neq \emptyset \end{array} \\[2em] \tau_{\phi,v}(p) & \begin{array}{l} p \to \tau_{\phi,v}(p) \\ \mathsf{reach}_{\phi,v}(p) \neq \emptyset \end{array} \\[1.5em] \sqcap\{u \in V \mid p \to_\approx u\}, & \mathsf{reach}_{\phi,v}(p) = \emptyset \end{cases}$$

Lemma 1. *Let* ϕ *be a strategy for player* i *in an arbitrary parity game. Assume that* $v, w \in V$ *and* $v \approx w$, *and let* $\psi = \mathsf{mimick}_{\phi,v}$. *Then*

$$\forall l \in \mathbb{N} : \forall p \in \Pi_\psi^{l+1}(w) : \ \exists k \in \mathbb{N} : \exists q \in \Pi_\phi^k(v) : \ p \approx q$$

Proof. We proceed by induction on l. For $l = 0$, the desired implication follows immediately. For $l = n + 1$, assume that we have a path $p \in \Pi_\psi^{n+1}(w)$. Clearly, $\langle p_1, \ldots, p_n \rangle$ is also consistent with ψ. The induction hypothesis yields us a $q \in \Pi_\phi^k(v)$ for some $k \in \mathbb{N}$ such that $\langle p_1, \ldots, p_n \rangle \approx q$. Let q be such. We distinguish the following cases:

1. $p_n \approx p_{n+1}$. In this case, clearly $p \approx \langle p_1, \ldots, p_n \rangle \approx q$, which finishes this case.
2. $p_n \not\approx p_{n+1}$. We again distinguish two cases:
   (a) Case $\mathcal{P}(p_n) \neq i$. Since $p_n \approx q_k$, we find that there must be states $u, w \in V$ such that $q_k \Longrightarrow_\approx w \to u$ and $p_{n+1} \approx u$. So there must be a path r and vertex u such that $p \approx q \cdot r \cdot \langle u \rangle$, for which we know that $r \approx q_k$. Therefore,

all vertices in r are owned by $\mathcal{P}(q_k) = \mathcal{P}(p_n)$, so ϕ is not defined for the extensions of q along p. We can therefore conclude that $\phi \Vdash q \cdot r \cdot \langle u \rangle$.

(b) Case $\mathcal{P}(p_n) = i$. Then it must be the case that $p_{n+1} = \tau_{\phi,v}(\langle p_1, \ldots, p_n \rangle)$. By definition, that means that there is a ϕ-consistent path $r \in \Pi_\phi(v)$, such that $r \approx p$. □

In the following lemma we extend the above obtained result to infinite paths.

Lemma 2. *Let ϕ be a strategy for player i in an arbitrary parity game. Assume that $v, w \in V$ and $v \approx w$, and let $\psi = \mathsf{mimick}_{\phi,v}$. Then*

$$\forall p \in \Pi_\psi^\omega(w) : \exists q \in \Pi_\phi^\omega(v) : p \approx q.$$

Proof. Suppose we have an infinite path $p \in \Pi_\psi^\omega(w)$. Using Lemma 1 we can obtain a path q starting in v that is stuttering bisimilar, and that is consistent with ϕ. The lemma does not guarantee, however, that q is of infinite length. We show that if q is finite, it can always be extended to an infinite path that is still consistent with ϕ.

Notice that paths can be partitioned into subsequences of vertices from the same equivalence class, and that two stuttering bisimilar paths must have the same number of partitions. This also follows from the original definition of stuttering equivalence given in [3].

Suppose now that q is of finite length, say $k + 1$. Then p must contain such a partition that has infinite size. In particular, there must be some $n \in \mathbb{N}$ such that $p_{n+j} \approx p_{n+j+1}$ for all $0 \le j \le |V|$. We distinguish two cases.

1. $\mathcal{P}(p_n) = i$. We show that then also $\mathsf{reach}_{\phi,v}(\langle p_0, p_1, \ldots p_n \rangle) = \emptyset$. Suppose this is not the case. Then we find that for some $u \in V$, $u = \tau_{\phi,v}(p)$ exists, and therefore $p_{n+j} \prec_u p_{n+j+1}$ for all $j \le |V|$. Since $\prec_u$ is total, this means that the longest chain is of length $|V|$, which contradicts our assumptions. So, necessarily $\mathsf{reach}_{\phi,v}(\langle p_0, p_1, \ldots p_n \rangle) = \emptyset$, meaning that no path that is consistent with ϕ leaves the class of p_n. But this means that the infinite path that stays in the class of p_n is also consistent with ϕ.

2. $\mathcal{P}(p_n) \ne i$. Since $p_n \approx q_k$, also $\mathcal{P}(q_k) \ne i$. Since $p_n \approx p_{n+j}$ for all $j \le |V|+1$, this means that there is a state u, such that $u = p_{n+l} = p_{n+l'}$. But this means that u is divergent. Since $\mathcal{P}(u) \ne i$, and $u \approx q_k$, we find that also q_k is divergent. Therefore, there is an infinite path with prefix q that is consistent with ϕ and that is stuttering bisimilar to p. □

Theorem 1. *Stuttering bisimilarity is strictly finer than winner equivalence, i.e., $\approx \subseteq \sim_w$.*

Proof. The claim follows immediately from Lemma 2 and the fact that two stuttering bisimilar infinite paths have the same infinitely occurring priorities. Strictness is immediate. □

Note that strong bisimilarity is strictly finer than stuttering bisimilarity; as a result, it immediately follows that strong bisimilarity is finer than winner equivalence, too.

As an aside, we point out that our proof of the above theorem relies on the construction of the strategy mimick$_{\phi,v}$; its purpose, however, exceeds that of the proof. If, by solving the stuttering bisimilar quotient of a given parity game $\mathcal{G}$, one obtains a winning strategy ϕ for a given player, mimick$_{\phi,v}$ defines the winning strategies for that player in $\mathcal{G}$. This is of particular importance in case an explanation of the solution of the game is required, for instance when the game encodes a verification problem for which a strategy helps explain the outcome of the verification (see *e.g.* [18]). It is not immediately obvious how a similar feature could be obtained in the setting of, say, the delayed simulations of Fritz and Wilke [9], because vertices that belong to different players and that have different priorities can be identified through such simulations.

4   Experiments

We next study the effect that stuttering equivalence minimisation has in a practical setting. We do this by solving parity games that originate from three different sources (we will explain more later) using three different methods: direct solving, solving after bisimulation reduction and solving after stuttering equivalence reduction. Parity games are solved using a number of different algorithms, *viz.* a naive C++ implementation of the *small progress measures* algorithm [11] due to Jurdziński, and the optimised and unoptimised variants that are implemented in the PGSolver tool [7] of the small progress measures algorithm, the recursive algorithm due to McNaughton [13], the bigstep algorithm due to Schewe [16] and a strategy improvement algorithm due to Vöge [20]. We compare the time needed by these methods to solve the parity games, and we compare the sizes of the parity games that are sent to the solving algorithms.

To efficiently compute bisimulation and stuttering equivalence for parity games we adapted a single-threaded implementation of the corresponding reduction algorithms by Blom and Orzan [2] for labelled transition systems.

All experiments were conducted on a machine consisting of 28 Intel® Xeon® E5520 Processors running at 2.27GHz, with 1TB of shared main memory, running a 64-bit Linux distribution using kernel version 2.6.27. None of our experiments employ multi-core features.

4.1   Test Sets

The parity games that were used for our experiments are partitioned into three test sets, of which we give a brief description below.

Test set 1. Our main interest is in the practical implications of stuttering equivalence reduction on solving model checking problems, so a number of typical model checking problems have been selected and encoded into parity games.

Five properties of the Firewire Link-Layer protocol (1394) [12] were considered, as they are described in [17]. They are numbered I–V in the order in which they can be found in that document.

Four properties are checked on the specification of a lift in [10]; a liveness property (I), a property that expresses the absence of deadlock (II) and two safety properties (III and IV). These typical model checking properties are expressed as alternation-free μ-calculus formulae.

On a model of the sliding window protocol [1], a fairness property (I) and a safety property (II) are verified, as well as 7 other fairness, liveness and safety properties.

Note that some of the properties are described by alternation free μ-calculus formulae, whereas others use alternation. The parity games induced by the alternation free μ-calculus formulae have different numbers of priorities, but the priorities along the paths in the parity games are ascending. In contrast, the paths in the parity games induced by alternating properties have no such property and are therefore computationally more challenging. Note that the parity games generated for these problems only have limited alternations between vertices owned by player 0 and 1 in the paths of the parity games.

Test set 2. The second test set was taken from [7] and consists of several instances of the elevator problem and the Hanoi towers problem described in that paper. For the latter, a different encoding was devised and added to the test set.

Test set 3. This test set consists of a number of equivalence checking problems encoded into parity games as described in [4].

The problems taken from [7], as well as some of the equivalence checking problems, give rise to parity games with alternations between both players and priorities.

4.2   Results

To analyse the performance of stuttering equivalence reduction, we measured the number of vertices and the number of edges in the original parity games, the bisimulation-reduced parity games and the stuttering-reduced parity games. Some of the results for test set 1 are shown in Table 1. For the Elevator model from [7], the results are shown in Table 2.

Figure 1.a compares these sizes (and those not shown in the tables) graphically; each plot point represents a parity game, of which the position along the y-axis is determined by its stuttering-reduced size, and the position along the x-axis by its original size and its bisimulation-reduced size, respectively. The plotted sizes are the sum of the number of vertices and the number of edges.

In addition to these results, we measured the time needed to reduce and to solve the parity games. The time needed to solve a parity game using stuttering equivalence or bisimulation reduction is computed as the time needed to reduce the parity game, plus the time needed to solve the reduced game. Also, the time needed to solve these games directly was measured. The solving time for a game is the time that the *fastest* of the solving algorithms needs to solve it. The results are plotted in Figure 1.b. Again, every data point is a parity game, of which the solving times determine the position in the scatter plot.

(a) Parity game sizes

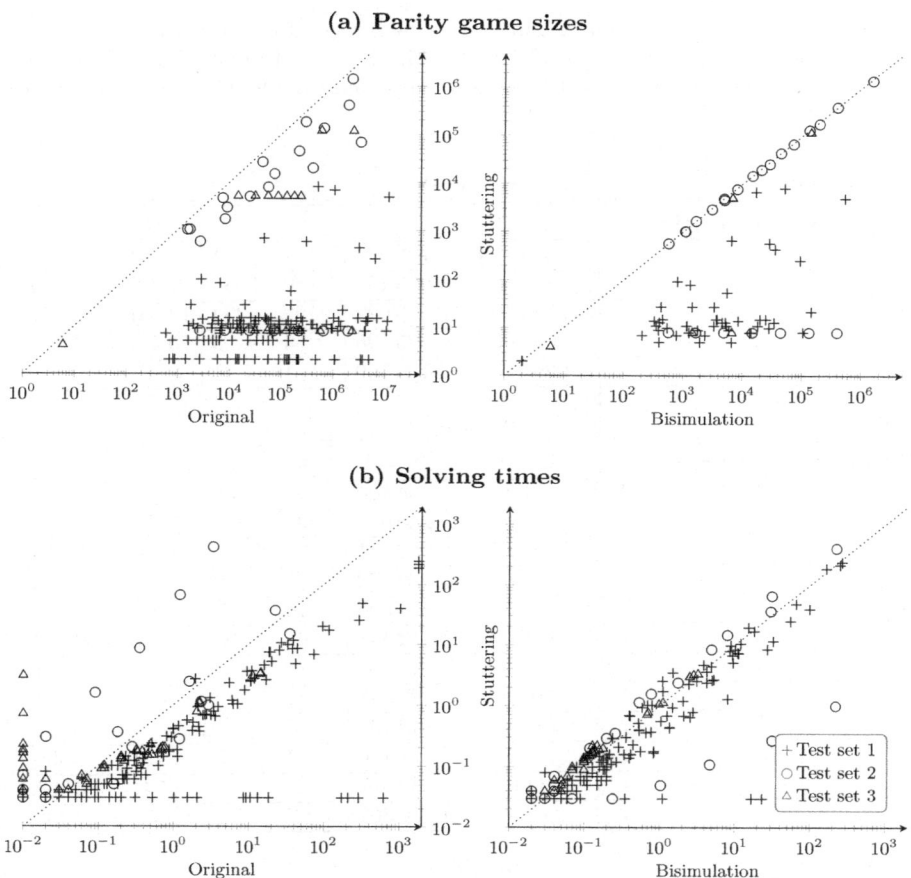

(b) Solving times

Fig. 1. Sizes and solving times (in seconds) of the stuttering-reduced parity games set out against sizes and solving times of the original games and of the bisimulation-reduced games. The vertical axis is shared between the plots in each subfigure. The dotted line is defined as $x = y$ and serves as a reference. Note that axes are in log scale.

4.3   Discussion

At a glance, stuttering reduction seems a big improvement on bisimulation reduction in terms of size reduction. Figure 1.a shows clearly that stuttering equivalence gives a better size reduction than bisimulation equivalence in the majority of cases. The difference is often somewhere between a factor ten and a factor thousand. Looking at solving times, the results also seem promising. In Figure 1.b we see that in most cases reducing the game and then solving it costs significantly less time. We will discuss the results in more detail for each test set separately.

Table 1. Statistics for the parity games for experiments from test set 1. In the Lift case, N denotes the number of distributed lifts; in the case of SWP, N denotes the size of the window. The number of priorities in the original (and minimised) parity games is listed under Priorities.

IEEE 1394		original		$\approx$		$\sim$	
Property	Priorities	$|V|$	$|\rightarrow|$	$|V|$	$|\rightarrow|$	$|V|$	$|\rightarrow|$
I	1	346 173	722 422	1	1	1	1
II	1	377 028	679 157	3 730	3 086	5 990	11 180
III	4	1 190 395	2 025 022	102	334	13 551	22 166
IV	2	524 968	875 296	4	6	10 814	17 590
V	1	1 295 249	2 150 590	1	1	1	1

Lift			original		$\approx$		$\sim$	
Property	Priorities	N	$|V|$	$|\rightarrow|$	$|V|$	$|\rightarrow|$	$|V|$	$|\rightarrow|$
I	4	2	1 691	4 825	22	58	333	1 021
I	4	3	63 907	240 612	131	450	5 148	23 703
I	4	4	1 997 579	9 752 561	929	4 006	74 059	462 713
II	2	2	846	2 172	5	9	94	240
II	2	3	31 954	121 625	16	39	1 092	4 514
II	2	4	998 790	5 412 890	64	193	14 353	80 043
III	1	2	763	1 903	1	1	1	1
III	1	3	26 996	99 348	1	1	1	1
III	1	4	788 879	4 146 139	1	1	1	1
IV	2	2	486	1 126	4	6	151	396
IV	2	3	11 977	39 577	5	9	1 741	6 951
IV	2	4	267 378	1 257 302	7	15	23 526	122 230

SWP			original		$\approx$		$\sim$	
Property	Priorities	N	$|V|$	$|\rightarrow|$	$|V|$	$|\rightarrow|$	$|V|$	$|\rightarrow|$
I	3	1	1 250	3 391	4	7	314	849
I	3	2	14 882	47 387	4	7	1 322	4 127
I	3	3	84 866	291 879	4	7	4 190	14 153
I	3	4	346 562	1 246 803	4	7	11 414	40 557
II	2	1	1 370	4 714	5	8	90	316
II	2	2	54 322	203 914	5	8	848	3 789
II	2	3	944 090	3 685 946	5	8	5 704	28 606
II	2	4	11 488 274	45 840 722	5	8	34 359	183 895

Test set 1. For these cases, we see that the size reduction is always better than that of bisimulation reduction, unless bisimulation already compressed the parity game to a single state. Solving times using stuttering equivalence are in general better than those of direct solving.

The experiments indicate that minimising parity games using stuttering equivalence before solving the reduced parity games is at least as fast as directly solving the original games.

Table 2. Statistics for the parity games for the FIFO and LIFO Elevator models taken from [7]. **Floors** indicates the number of floors.

Elevator Models			original		$\approx$		$\sim$	
Model	**Floors**	**Priorities**	$|V|$	$|\rightarrow|$	$|V|$	$|\rightarrow|$	$|V|$	$|\rightarrow|$
FIFO	3	3	564	950	351	661	403	713
FIFO	4	3	2 688	4 544	1 588	2 988	1 823	3 223
FIFO	5	3	15 684	26 354	9 077	16 989	10 423	18 335
FIFO	6	3	108 336	180 898	62 280	116 044	71 563	125 327
FIFO	7	3	861 780	1 431 610	495 061	919 985	569 203	994 127
LIFO	3	3	588	1 096	326	695	363	732
LIFO	4	3	2 832	5 924	866	2 054	963	2 151
LIFO	5	3	16 356	38 194	2 162	5 609	2 403	5 850
LIFO	6	3	111 456	287 964	5 186	14 540	5 763	15 117
LIFO	7	3	876 780	2 484 252	16 706	51 637	18 563	53494

The second observation we make is that stuttering equivalence reduces the size quite well for this test set, when compared to the other sets. This may be explained by the way in which the parity games were generated. As they encode a μ-calculus formula together with a state space, repetitive and deterministic parts of the state space are likely to generate fragments within the parity game that can be easily compressed using stuttering reduction.

Lastly, we observe that solving times using bisimulation reduction are not in general much worse than those using stuttering reduction. The explanation is simple: both reductions compress the original parity game to such an extent that the resulting game is small enough for the solvers to solve it in less than a tenth of a second.

Test set 2. Both stuttering equivalence and strong bisimulation reduction perform poorly on a reachability property for the Hanoi towers experiment, with the reduction times vastly exceeding the times required for solving the parity games directly. A closer inspection reveals that this is caused by an unfortunate choice for a new priority for vertices induced by a fixpoint-free subformula. As a result, all paths in the parity game have alternating priorities with very short stretches of the same priorities, because of which hardly any reduction is possible. We included an encoding of the same problem which does not contain the unfortunate choice, and indeed observe that in that case stuttering equivalence does speed up the solving process.

The LIFO Elevator problem shows results similar to those of the other model checking problems. The performance with respect to the FIFO Elevator however is rather poor. This seems to be due to three main factors: the relatively large number of alternating fixed point signs, the alternations between vertices owned by player 0 and vertices owned by player 1, and the low average branching degree in the parity game. This indicates that for alternating μ-calculus formulae with nested conjunctive and disjunctive subformulae, stuttering equivalence reduction generally performs suboptimal. This should not come as a surprise, as stuttering equivalence only allows one to compress sequences of vertices with equal priorities and owned by the same player.

Test set 3. The results for these experiments indicate that reduction using stuttering equivalence sometimes performs poorly. The subset where performance is especially poor is an encoding of branching bisimilarity, which gives rise to parity games with alternations both between different priorities as well as different players. As a result, little reduction is possible.

5   Conclusions

We have adapted the notion of stuttering bisimilarity to the setting of parity games, and proven that this equivalence relation can be safely used to minimise a parity game before solving the reduced game.

Experiments were conducted to investigate the effect of quotienting stuttering bisimilarity on parity games originating from model checking problems. In many practical cases this reduction leads to an improvement in solving time, however in cases where the parity games involved have many alternations between odd and even vertices, stuttering bisimilarity reduction performs only marginally better than strong bisimilarity reduction. Although we did compare our techniques against a number of competitive parity game solvers, using other solving algorithms, or even other implementations of the same algorithms, may give different results.

The fact that stuttering bisimilarity does not deal at all well with alternation leads us to believe that weaker notions of bisimilarity, in which vertices with different players can be related under certain circumstances, may resolve the most severe performance problems that we saw in our experiments. We regard the investigation of such weaker relations as future work.

Stuttering bisimilarity has been previously studied in a distributed setting [2]. It would be interesting to compare its performance to a distributed implementation of the known solving algorithms for parity games. However, we are only aware of a multi-core implementation of the *Small Progress Measures* algorithm [19].

References

1. Badban, B., Fokkink, W., Groote, J.F., Pang, J., van de Pol, J.: Verification of a sliding window protocol in μCRL and PVS. Formal Aspects of Computing 17, 342–388 (2005)
2. Blom, S., Orzan, S.: Distributed branching bisimulation reduction of state spaces. ENTCS 89(1) (2003)
3. Browne, M.C., Clarke, E.M., Grumberg, O.: Characterizing finite Kripke structures in propositional temporal logic. Theor. Comput. Sci. 59, 115–131 (1988)
4. Chen, T., Ploeger, B., van de Pol, J., Willemse, T.A.C.: Equivalence checking for infinite systems using parameterized boolean equation systems. In: Caires, L., Vasconcelos, V.T. (eds.) CONCUR 2007. LNCS, vol. 4703, pp. 120–135. Springer, Heidelberg (2007)

5. Cranen, S., Keiren, J.J.A., Willemse, T.A.C.: Stuttering equivalence for parity games, arXiv:1102.2366 [cs.LO] (2011)
6. Emerson, E.A., Jutla, C.S.: Tree automata, mu-calculus and determinacy. In: SFCS 1991, Washington, DC, USA, pp. 368–377. IEEE Computer Society, Los Alamitos (1991)
7. Friedmann, O., Lange, M.: Solving parity games in practice. In: Liu, Z., Ravn, A.P. (eds.) ATVA 2009. LNCS, vol. 5799, pp. 182–196. Springer, Heidelberg (2009)
8. Fritz, C.: Simulation-Based Simplification of omega-Automata. PhD thesis, Christian-Albrechts-Universität zu Kiel (2005)
9. Fritz, C., Wilke, T.: Simulation relations for alternating parity automata and parity games. In: Ibarra, O.H., Dang, Z. (eds.) DLT 2006. LNCS, vol. 4036, pp. 59–70. Springer, Heidelberg (2006)
10. Groote, J.F., Pang, J., Wouters, A.G.: Analysis of a distributed system for lifting trucks. J. Log. Algebr. Program 55, 21–56 (2003)
11. Jurdziński, M.: Small progress measures for solving parity games. In: Reichel, H., Tison, S. (eds.) STACS 2000. LNCS, vol. 1770, pp. 290–301. Springer, Heidelberg (2000)
12. Luttik, S.P.: Description and formal specification of the link layer of P1394. In: Workshop on Applied Formal Methods in System Design, pp. 43–56 (1997)
13. McNaughton, R.: Infinite games played on finite graphs. Annals of Pure and Applied Logic 65(2), 149–184 (1993)
14. De Nicola, R., Vaandrager, F.W.: Three logics for branching bisimulation. J. ACM 42(2), 458–487 (1995)
15. Park, D.: Concurrency and automata on infinite sequences. Theor. Comput. Sci. 104, 167–183 (1981)
16. Schewe, S.: Solving parity games in big steps. In: Arvind, V., Prasad, S. (eds.) FSTTCS 2007. LNCS, vol. 4855, pp. 449–460. Springer, Heidelberg (2007)
17. Sighireanu, M., Mateescu, R.: Verification of the Link Layer Protocol of the IEEE-1394 Serial Bus (FireWire): An Experiment with E-LOTOS. STTT 2(1), 68–88 (1998)
18. Stevens, P., Stirling, C.: Practical model checking using games. In: Steffen, B. (ed.) TACAS 1998. LNCS, vol. 1384, pp. 85–101. Springer, Heidelberg (1998)
19. van de Pol, J., Weber, M.: A multi-core solver for parity games. ENTCS 220(2), 19–34 (2008)
20. Vöge, J., Jurdziński, M.: A discrete strategy improvement algorithm for solving parity games. In: Emerson, E.A., Sistla, A.P. (eds.) CAV 2000. LNCS, vol. 1855, pp. 202–215. Springer, Heidelberg (2000)
21. Wimmer, R., Herbstritt, M., Hermanns, H., Strampp, K., Becker, B.: Sigref— a symbolic bisimulation tool box. In: Graf, S., Zhang, W. (eds.) ATVA 2006. LNCS, vol. 4218, pp. 477–492. Springer, Heidelberg (2006)
22. Zielonka, W.: Infinite games on finitely coloured graphs with applications to automata on infinite trees. Theor. Comp. Sci. 200(1-2), 135–183 (1998)

Counterexample-Based Error Localization of Behavior Models

Tsutomu Kumazawa and Tetsuo Tamai

Graduate School of Arts and Sciences, The University of Tokyo,
Tokyo, Japan
{kumazawa,tamai}@graco.c.u-tokyo.ac.jp

Abstract. Behavior models are often used to describe behaviors of the system-to-be during requirements analysis or design phases. The correctness of the specified model can be formally verified by model checking techniques. Model checkers provide counterexamples if the model does not satisfy the given property. However, the tasks to analyze counterexamples and identify the model errors require manual labor because counterexamples do not directly indicate where and why the errors exist, and when liveness properties are checked, counterexamples have infinite trace length, which makes it harder to automate the analysis. In this paper, we propose a novel automated approach to find errors in a behavior model using an infinite counterexample. We find similar witnesses to the counterexample then compare them to elicit errors. Our approach reduces the problem to a single-source shortest path search problem on directed graphs and is applicable to liveness properties.

Keywords: Requirements Analysis, Design, Model Checking, Error Localization.

1   Introduction

Model Driven Engineering (MDE) is being accepted as a practical approach to develop reliable software efficiently [24]. Following MDE, a model of the software-to-be is built first, which goes through a series of model transformations to derive final code. It is obvious that the whole scheme crucially depends on correctness and appropriateness of the initial model.

As widely acknowledged, *model checking* [5] is one of the most powerful methods for formally validating correctness and appropriateness of a given model. The mostly used type of model checking technique takes behavior models represented as state machines as its target and checks if a given set of properties hold, employing graph searching algorithms or symbolic logical formula decision algorithms. Comparing the model checking approach to the theorem proving approach, one of the advantages of the former is often attributed to its capability of presenting counterexamples when verification fails. E. Clarke writes "It is impossible to overestimate the importance of the counterexample feature [4]."

But in practice, difficulties arise after counterexamples are obtained. Counterexamples do not directly indicate where in the model the errors that cause

M. Bobaru et al. (Eds.): NFM 2011, LNCS 6617, pp. 222–236, 2011.

them exist. It is up to the developer's effort and intuition to find the part of the model that should be fixed to prevent occurrence of the counterexamples.

In this paper, we propose a new method and a tool that help developers fix errors in their models based on the detected counterexamples. In the research field of software model checking and debugging of source code, there exist a certain number of techniques that explain counterexamples and localize errors [1,13,3,12,11,16]. However, those existing methods for programs [1,13,12,11] only treat the violation of *safety properties* due to the limitation of software model checkers for source programs. Counterexamples for safety properties are by nature composed of event traces with finite length. Then it is relatively easy to automate identification of bad events in the trace. On the other hand, error localization for *liveness properties* poses a greater challenge, because in general it requires analysis of infinite-length counterexamples.

The state space of a program at the source code level is in general quite huge and when there is a loop structure, it is hard to decide when to stop expanding the state model graph. Techniques such as predicate abstraction are used to circumvent the problem. It is all right if a counterexample against some safety property is found in the current abstraction, because it can safely be concluded that the program violates the property. Otherwise, to explore the unsearched space, loops have to be expanded and it is not easy to decide when to stop the search. To deal with liveness properties, it induces much harder problems, because it is essential to identify precise loop structures (strongly connected components) and moreover even if counterexamples are found in an abstracted model, it is not sound to conclude that the original concrete model also violates the liveness property.

There are some pieces of work trying to treat liveness properties [3,16]. However, they have limitations such that it involves highly expensive computational complexity [3] or only specific kinds of liveness properties are supported [16].

In this paper, we propose LLL-S, a novel error localization technique in the given behavior model. We address the problem of analyzing an infinite trace, which takes the form of a finite prefix followed by an infinite cycle. Our idea is to find infinite and lasso-shaped witnesses (traces that satisfy the property) that resemble the given counterexample, and identify events to be modified by comparing each witness with the counterexample. We report all transitions that trigger the differences as candidate errors and the corresponding witnesses as their explanations. We use a Büchi automaton recognizing the target property as a set of witnesses, and adopts the edit distance between strings to measure distances between infinite and lasso-shaped traces. We find appropriate witnesses based on the distance by solving a single-source shortest path search problem on the Büchi automaton. LLL-S can be applied to the safety property class as well, where the length of counterexamples is finite and that of witness traces is infinite.

The main contributions of LLL-S are as follows. LLL-S can be applied to any Linear Temporal Logic formulas [9,21], including both liveness properties and safety properties. LLL-S focuses on errors in the behavior models that are used in MDE. We do not have to prepare a set of witnesses in advance, because it is given as a Büchi automaton. Since LLL-S is based on well-established techniques combined together, it is easily automated.

Section 2 presents a motivating example. Section 3 explains the background. We explain LLL-S in Section 4. In Section 5, we report the results of the tool implementation of LLL-S and some case studies. We discuss some issues concerning our work and introduce related work in Section 6, and conclude in Section 7.

2   Motivating Example

Consider a concurrent system with a semaphore [21], CSys, whose LTS is shown in Fig. 1 (a). A LTS is a finite state machine described in terms of events. CSys consists of three processes: p.1, p.2 and Sema. The initial states of the processes are labeled 0. Two processes p.1 and p.2 repeatedly enter and leave the critical region by $p.\{1,2\}.enter$ and $p.\{1,2\}.exit$, respectively. Their exclusive access to the critical region is controlled by the mutual exclusion mechanism of the semaphore process Sema such that $p.i.mx.down$ $(i = 1, 2)$ lets p.i enter the critical region, and blocks the entrance of the other process until $p.i.mx.up$ occurs. The transitions sharing source and destination states are depicted by a single arrow. For example, the transition $(0, p.\{1,2\}.mx.up, 1)$ of Sema denotes $(0, p.1.mx.up, 1)$ and $(0, p.2.mx.up, 1)$. The behavior of CSys is presented by *parallel composition* [21] of three processes, which is based on interleaving of unshared events and simultaneous executions of shared events.

To verify the correctness of CSys's behavior, consider the *fluent Linear Temporal Logic* property EXIT1= $\mathbf{G}(p.1.enter \Rightarrow \mathbf{F}p.1.exit)$, where $\mathbf{G}$, $\mathbf{F}$ and $\Rightarrow$ respectively denote *always*, *eventually* and *implication*. EXIT1 says that when p.1 enters the critical region, p.1 eventually leaves it. However, CSys does not satisfy EXIT1, because when p.1 stays in the critical region, p.2 can access it infinitely many times. The flaw in CSys is the incorrect mutual exclusion mechanism realized by Sema. A counterexample is $\pi^c = PC^\omega$, where the prefix P and the cycle C are finite event sequences shown in Table 1. The problem is to identify erroneous transitions of Sema.

We will find a witness $\tau = P'C'^\omega$ that are closest to π^c. A set of witnesses is given by a Büchi automaton recognizing EXIT1, $B(\text{EXIT1})$ in Fig. 1 (b). Its initial state is b_0 and its event set is A_1, identical to the event set of CSys. Term (p, A, q) represents the transitions that share the source state p and destination

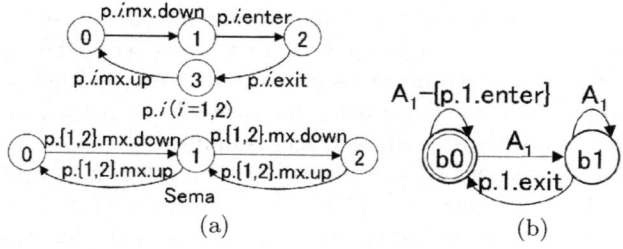

Fig. 1. Concurrent System CSys (a) and $B(\text{EXIT1})$ (b)

Table 1. Counterexample of CSys for EXIT1 (π^c), and Witnesses (from τ^1 to τ^4)

$\pi^c = [p.1.mx.down,\ p.1.enter\ (\ p.2.mx.down,\ p.2.enter,\ p.2.exit,\ p.2.mx.up)^\omega]$
$\tau^1 = [p.1.mx.down,\ p.1.enter,\ p.1.exit\ (p.2.mx.down,\ p.2.enter,\ p.2.exit,\ p.2.mx.up)^\omega]$
$\tau^2 = [p.1.mx.down,\ p.1.enter\ (p.2.mx.down,\ p.1.exit,\ p.2.exit,\ p.2.mx.up)^\omega]$
$\tau^3 = [p.1.mx.down,\ p.1.enter,\ p.1.exit\ (\ p.2.mx.down,\ p.2.enter,\ p.2.exit\ )^\omega]$
$\tau^4 = [(p.1.mx.down,\ p.1.enter,\ p.1.exit,\ p.2.mx.down,\ p.2.enter,\ p.2.exit,\ p.2.mx.up)^\omega]$

state q, and whose events constitute the set A. To explain the advantage of using the closest witnesses to the counterexample for error localization and show the limitation of existing techniques, consider witnesses accepted by $B(\mathsf{EXIT1})$ shown in Table 1. One of the simplest distances between π^c and τ is the number of edit operations required in transforming PC into $P'C'$ ignoring the cycling of C and C' (denoted by $d_e(\pi^c, \tau)$), whose concept is almost the same as those proposed by Chaki et al. [3] and Groce et al. [12].

The witness τ^1 is the closest to π^c because at least one insertion of $p.1.exit$ after the second event $p.1.enter$ of π^c should be applied to make π^c satisfy EXIT1 $(d_e(\pi^c, \tau^1) = 1)$. The witness τ^1 tells us that p.1 should leave the critical region before p.2 enters there. Thus, τ^1 indicates that Sema does not correctly control p.1 and p.2's access to the critical region, and that the events other than $p.1.enter$ have nothing to do with making π^c satisfy EXIT1. By comparing π^c with τ^1, we know the erroneous transitions $(0,\ p.1.mx.down, 1)$ and $(1,\ p.2.mx.down, 2)$ of Sema that are the nearest to the inserted event $p.1.enter$. These transitions show that Sema allows p.2 to enter the critical region by $p.2.mx.down$ after it allows p.1 to enter there by $p.1.mx.down$ but without following $p.1.mx.up$.

Another witness τ^2 is also the closest to π^c, i.e. $d_e(\pi^c, \tau^2) = 1$. Their difference is interpreted that $p.2.enter$ of π^c should be forbidden. Therefore, p.2 should not enter the critical region infinitely many times when p.1 stays there, and the transition $(1,\ p.2.mx.down, 2)$ of Sema enables p.2 to enter the region.

However, the witness τ^3 is unsuitable for showing the errors because τ^3 additionally requires the deletion of $p.2.mx.up$ from τ^1, which is an unnecessary operation to make π^c satisfy EXIT1 (i.e. $d_e(\pi^c, \tau^3) = 2$). This deletion may mislead the developers into believing that $p.2.mx.up$ should not occur. The distance d_e appropriately shows that τ^1 and τ^2 are closer to π^c than τ^3, and that τ^3 must not be used for error localization.

The witness τ^4 is the closest to π^c according to d_e because τ^1 and τ^4 consist of the same finite event sequence. However, τ^4 does not provide useful information to determine whether every event in P should be repeated infinitely many times, or P does not contain errors and $p.1.exit$ is the only significant event to modify the violation of EXIT1 as the case of τ^1. Thus, we wish to judge that τ^4 is *not* as close to π^c as τ^1, but d_e does not work for our purpose. The cause of the problem is that d_e does not separate differences between the prefixes and the cycles of π^c and τ^4. Other existing methods [15,27,1,13,23,6,11] have the similar limitation due to their assumption that traces are of finite-length.

To summarize, it is desirable for error localization to obtain τ^1 and τ^2, but not τ^3 or τ^4 . In Section 4, we present a novel method to automatically find such witnesses based on a specific distance and the errors in CSys.

3   Background

A LTS is a tuple $L = (S, A, \Delta, s_0)$, where S is a finite set of states, A is a set of events, $\Delta \subseteq S \times A \times S$ is a transition relation, and $s_0 \in S$ is the initial state. A *trace* of L is a sequence of events $\pi = [a_0, a_1, \ldots, a_{n-1}]$ ($\forall 0 \leq i < n.(s_i, a_i, s_{i+1}) \in \Delta$). For π, the sequence of states $[s_0, s_1, \ldots, s_n]$ is called a *path* of π. If $n = \infty$, we call π an infinite trace. Otherwise, we call π a finite trace. A set of all traces of L is denoted by $Tr(L)$. The suffix of a trace $\pi \in Tr(L)$ from a_i is denoted by $\pi[i]$. A transition $(s, a, s) \in \Delta$ is called a *self transition*.

Concerning model checking on L, a Büchi automaton-based technique has been proposed for FLTL [9]. A fluent is an atomic proposition whose truth value is determined over occurrence of events appearing in a trace. A fluent is a tuple $\mathsf{fl} = (I_{\mathsf{fl}}, T_{\mathsf{fl}}, b_{\mathsf{fl}})$, where $I_{\mathsf{fl}}, T_{\mathsf{fl}} \in A$ are a set of initiating and terminating events respectively such that $I_{\mathsf{fl}} \cap T_{\mathsf{fl}} = \emptyset$, and $b_{\mathsf{fl}} \in \{\mathbf{t}, \mathbf{f}\}$ is the initial truth value. For $\pi \in Tr(L)$, $\pi[i]$ satisfies fl ($\pi[i] \models \mathsf{fl}$) iff one of the following conditions holds: either $b_{\mathsf{fl}} \wedge (\forall j \in \mathcal{N}.0 \leq j \leq i \Rightarrow a_j \notin T_{\mathsf{fl}})$, or $\exists j \in \mathcal{N}.(j \leq i \wedge a_j \in I_{\mathsf{fl}}) \wedge (\forall k \in \mathcal{N}.j < k \leq i \Rightarrow a_k \notin T_{\mathsf{fl}})$. The set of fluents considered is denoted by FL.

A FLTL formula is defined inductively with the boolean and temporal operators as follows: $\phi, \psi = \mathbf{t} \mid \mathsf{fl} \in \mathsf{FL} \mid \phi \wedge \psi \mid \neg\phi \mid \mathbf{X}\phi \mid \phi\mathbf{U}\psi$. Given a trace $\pi \in Tr(L)$, the satisfaction operator $\models$ is defined inductively as follows:

$$\pi \models \mathbf{t}, \qquad \pi \models \mathsf{fl} \in \mathsf{FL} \text{ iff } \pi[0] \models \mathsf{fl}, \qquad \pi \models \neg\phi \text{ iff } \pi \not\models \phi,$$
$$\pi \models \phi \wedge \psi \text{ iff } (\pi \models \phi) \text{ and } (\pi \models \psi), \qquad \pi \models \mathbf{X}\phi \text{ iff } \pi[1] \models \phi,$$
$$\pi \models \phi\mathbf{U}\psi \text{ iff } \exists j \geq 0.\pi[j] \models \psi \text{ and } \forall 0 \leq i < j.\pi[i] \models \phi.$$

Other operators are derived from the above operators: $\phi \vee \psi = \neg(\neg\phi \wedge \neg\psi)$, $\phi \Rightarrow \psi = \neg\phi \vee \psi$, $\mathbf{F}\phi = \mathbf{t}\mathbf{U}\phi$ and $\mathbf{G}\phi = \neg\mathbf{F}\neg\phi$. We define $L \models \phi$ (L satisfies ϕ) iff $\forall\pi \in Tr(L).\pi \models \phi$. FLTL formulas are classified into safety and liveness properties [21]. A safety property such as $\mathbf{G}\neg p$ asserts that nothing bad ever happens, while a liveness property such as $\mathbf{F}p$ asserts that something good will eventually happen. EXIT1 is an instantiation of liveness properties.

Model checking on LTS L for FLTL formula ϕ is conducted as follows [9]: 1) build a Büchi automaton that accepts all traces satisfying $\neg\phi$, $B(\neg\phi)$, 2) build the parallel composition of L and $B(\neg\phi)$, and 3) search for an accepting trace, which is a *counterexample*. A Büchi automaton $B = (S_b, A_b, \Delta_b, s_0, S_b^a)$ is a LTS augmented with a set of accepting states, where $S_b^a \subseteq S_b$ is an accepting state set and the other constructs are the same as those of a LTS. A trace π is accepted by B if π passes some accepting state infinitely many times.

Parallel composition ($\|$) [21] captures the concurrent and interactive execution of LTSs. Let $L^1 = (S^1, A^1, \Delta^1, s_0^1)$ and $L^2 = (S^2, A^2, \Delta^2, s_0^2)$ be LTSs. $L^1 \parallel L^2 = (S^1 \times S^2, A^1 \cup A^2, \Delta, (s_0^1, s_0^2))$, where $\Delta \subseteq (S^1 \times S^2) \times (A^1 \cup A^2) \times (S^1 \times S^2)$ is computed as follows: $\Delta = \{((s^1, s^2), a, (t^1, t^2)) | (s^1, a, t^1) \in \Delta^1, (s^2, a, t^2) \in$

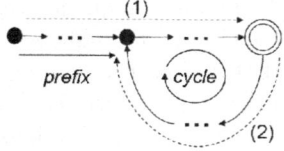

Fig. 2. Shape of Counterexample and Witness

$\Delta^2\} \cup \{((s^1, s^2), a, (t^1, s^2)) \mid (s^1, a, t^1) \in \Delta^1, a \notin A^2\} \cup \{((s^1, s^2), a, (s^1, t^2)) \mid (s^2, a, t^2) \in \Delta^2, a \notin A^1\}$.

At step 3 of the checking procedure for liveness properties, algorithms to search for strongly connected components such as nested depth-first search [14] are used by many existing model checkers (e.g. SPIN [14]). They find a counterexample that forms an infinite and lasso-shaped trace $\pi = PC^\omega$ (see Fig. 2), where the prefix P and the cycle C are finite event sequences whose subsequences contain no cycle. C passes some accepting state of $B(\neg\phi)$ depicted as a double circle in Fig. 2. Hence, we assume $\pi = PC^\omega$. An example is π^c in Table 1. A witness is a trace satisfying ϕ and is assumed to have a form $\tau = P'C'^\omega$.

4   Error Localization Procedure

This section presents an error localization technique LLL-S. The idea is that we find the closest (i.e. the most similar) witnesses to π, and then detect their differences. The inputs to LLL-S are a LTS $L = (S, A, \Delta, s_0)$, a FLTL formula ϕ where $L \not\models \phi$, and a counterexample $\pi = PC^\omega$, where $P = [a_0, a_1, \ldots, a_{m-1}]$ and $C = [b_0, b_1, \ldots, b_{n-1}]$ for $0 \le m$ and $1 \le n$. Let $B(\phi) = (S_\phi, A_\phi, \Delta_\phi, u_0, S_\phi^a)$. We assume that $A_\phi \subseteq A$ and a witness to be searched has a form $\tau = P'C'^\omega$.

If we consider an event as a character, a trace is regarded as an infinite string. We define the distance D between π and τ using the edit distance between finite strings on the edit operations *insertion, deletion* and *replacement* [20]. The edit distance between finite strings s_1 and s_2, denoted by $d(s_1, s_2)$, is the minimum cost to change one string to the other. We assume that the cost of each edit operation is 1. The distance D is defined as follows: $D(\pi, \tau) = d(P, P') + d(C, C')$.

D meets all properties of a metric, i.e. positive definiteness, symmetry and triangle inequality when we define $\pi = \tau$ iff $P = P'$ and $C = C'$. D is appropriate for our goal because it distinguishes the distance of prefixes and cycles. For example, in Table 1, $D(\pi^c, \tau^1) = D(\pi^c, \tau^2) = 1$, $D(\pi^c, \tau^3) = 2$ and $D(\pi^c, \tau^4) = 5$. D judges that both τ^1 and τ^2 are the closest to π^c while τ^3 and τ^4 are not.

4.1   Outline

As a set of witnesses is given by traces accepted by $B(\phi)$, we find every witness τ in $B(\phi)$ such that $D(\pi, \tau)$ is the smallest. In order to make τ meet the Büchi's acceptance condition (see Fig. 2), we divide the procedure to find τ into two steps: 1) finding a sequence that ends in an accepting state $s_\phi^a \in S_\phi^a$ (i.e. the sequence (1) in Fig. 2) and 2) finding a sequence that leaves s_ϕ^a and returns to a state on the path from u_0 to s_ϕ^a (i.e. the sequence (2) in Fig. 2).

First we construct a model W_π^A from the counterexample π, embedding edit operations and their costs. W_π^A is a *Weighted Transition System* (WTS) [19], a LTS augmented with a cost function ζ_w : Transitions $\rightarrow$ Cost. As π has a structure PC^ω, W_π^A consists of a linear path corresponding to P, followed by a cycle corresponding to C. For the $P = [a_0, \ldots, a_{m-1}]$ part, states p_i and transitions (p_i, a_i, p_{i+1}) $(i = 0, \ldots, m-1)$ are generated. For the $C = [b_0, \ldots, b_{n-1}]$ part, states c_i and transitions (c_i, b_i, c_{i+1}) $(i = 0, \ldots, n-1)$ are generated, where c_n is identical to c_0. All the transitions thus generated have cost 0. The transitions are augmented by the following three types of new transitions with cost 1.

1. *Replace*: for a pair (p_i, p_{i+1}), transitions (p_i, a, p_{i+1}) where $a \in (A - \{a_i\})$, meaning replacing the event a_i with the event a. Likewise, for a pair (c_i, c_{i+1}), transitions (c_i, b, c_{i+1}) where $b \in (A - \{b_i\})$.
2. *Delete*: for a pair (p_i, p_{i+1}), transition (p_i, ϵ, p_{i+1}) meaning deleting a_i. ϵ is a null event. Likewise, for a pair (c_i, c_{i+1}), transition (c_i, ϵ, c_{i+1}).
3. *Insert*: for a state p_i, transitions (p_i, a, p_i) where $a \in A$, meaning inserting a at p_i. Likewise, for a state c_i, transitions (c_i, b, c_i) where $b \in A$.

Next, we build a product model $W_\bowtie = B(\phi) \bowtie W_\pi^A$. The problem of finding witnesses of the property that are the most similar to the counterexample is reduced to the problem of finding the shortest paths in the graph of $W_\bowtie$, starting from the initial vertex, visiting a vertex corresponding to an accepting state of $B(\phi)$ and ending in a vertex that closes the path to make a cycle. The vertex of the accepting state should be included in the cycle. We can employ a shortest path algorithm such as Dijkstra's method [7] to solve this problem. In the first step, the shortest paths from the initial vertex to the accepting vertices are obtained. Then, for each accepting vertex that has been reached from the initial vertex, the second shortest path problem is solved starting from the accepting vertex, ending in the vertices on the shortest path from the initial vertex to the accepting vertex, so as to close a cycle. Thus, we need to solve the single-source shortest path problem $v_a + 1$ times, where v_a is the number of accepting states.

The differences between τ and π indicate potential errors. LLL-S detects every difference and extracts every transition that has the erroneous event.

4.2 Constructing WTS Models

We define a WTS as an extension of a LTS [19]. A WTS is a tuple $W = (S_w, A_w, \Delta_w, q_0, \zeta, M_w)$, where S_w is a finite set of states, A_w is a set of event labels, $\Delta_w \subseteq S_w \times A_w \times S_w$ is a transition relation, and $q_0 \in S_w$ is the initial state, the total function $\zeta : \Delta_w \rightarrow \mathcal{R}$ is a *weight* to every transition, and $M_w \subseteq S_w$ is a set of *end states*. We use the terms on a LTS also for a WTS, e.g. traces.

A WTS W_π^A made from π consists of two parts: the part constructed from P and that from C which respectively show edit operations and their costs applied to P and C. Finite traces of W_π^A that pass the P and C part respectively provide P' and C'. A set of end states includes all states of the C part to indicate that a

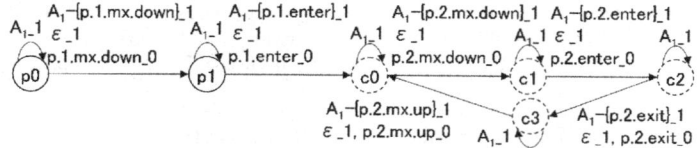

Fig. 3. WTS $W_{\pi^c}^{A_1}$ Constructed from π^c

finite trace in W_π^A ends in any element of the set, and that an accepting state of $B(\phi)$ appearing in the element is the destination of the sequence (1) in Fig. 2.

The WTS $W_\pi^A = (S_w, A \cup \{\epsilon\}, \Delta_w, q_0, \zeta_w, M_w)$ of π is constructed as follows. The state set $S_w = \{p_i | 0 \le i < m\} \cup \{c_i | 0 \le i < n\}$. The initial state $q_0 = p_0$ if $m \ne 0$; otherwise, $q_0 = c_0$. The transition relation $\Delta_w = \Delta_p \cup \Delta_b \cup \Delta_c$, where Δ_p, Δ_b and Δ_c are defined as follows. $\Delta_p = \{(p_i, a, p_{i+1}) | \ 0 \le i < m - 1, a \in A \cup \{\epsilon\}\} \cup \{(p_i, a, p_i) | 0 \le i < m, a \in A\}$. $\Delta_b = \{(p_{m-1}, a, c_0) | a \in A \cup \{\epsilon\}\}$ if $m \ne 0$; otherwise, $\Delta_b = \emptyset$. $\Delta_c = \{(c_i, a, c_{i+1}) | 0 \le i < n - 1, a \in A \cup \{\epsilon\}\} \cup \{(c_{n-1}, a, c_0) | a \in A \cup \{\epsilon\}\} \cup \{(c_i, a, c_i) | 0 \le i < n, a \in A\}$. For each $\delta \in \Delta_w$, $\zeta_w(\delta) = 0$ if either of the following conditions holds: $\delta = (p_i, a_i, p_{i+1})(i = 0, \dots, m - 2)$, $\delta = (p_{m-1}, a_{m-1}, c_0)$, $\delta = (c_i, b_i, c_{i+1})$ $(i = 0, \dots, n - 2)$, or $\delta = (c_{n-1}, b_{n-1}, c_0)$; otherwise, $\zeta_w(\delta) = 1$. The set of end states $M_w = \{c_i | 0 \le i < n\}$.

Fig. 3 shows the WTS model $W_{\pi^c}^{A_1}$ constructed from π^c in Table 1. The initial state is p_0 and end states are states with dashed circles c_i $(i = 0, \dots, 3)$. A weight to each transition is written after the event. The set of transitions (p, A_1, q) with weight w indicates that every transition in (p, A_1, q) has the same weight w. The P part of $W_{\pi^c}^{A_1}$ consists of the states $p_i(i = 0, 1)$ and c_0, and the transitions defined by Δ_p and Δ_b. For example, a transition $(p_0, p.1.mx.down, p_1)$ with weight 0 shows $p.1.mx.down$ in P. Transitions $(p_0, A_1 - \{p.1.mx.down\}, p_1)$ mean that $p.1.mx.down$ is *replaced* by another event. A transition (p_0, ϵ, p_1) represents that $p.1.mx.down$ is *deleted*. Self transitions (p_0, A_1, p_0) show *insertion* operations just before $p.1.mx.down$. Likewise, the C part of $W_{\pi^c}^{A_1}$ consists of the states c_i $(i = 0, \dots, 3)$ and the transitions defined by Δ_c.

Finite traces that pass from p_0 to c_0 present P'. Similarly, finite traces that pass from c_0 to itself via $c_i(i = 1, \dots, 3)$ present C'.

4.3   Finding Witnesses

We next find a witness τ such that $D(\pi, \tau)$ is the smallest by conducting the single-source shortest path search twice.

We first find a sequence that ends in an accepting state $s_\phi^a \in S_\phi^a$. We compute the product of the WTS W_π^A and $B(\phi)$ so that such event sequences can be obtained by the shortest path from the initial state of the product graph.

We extend the parallel composition operation of LTSs to the operation ($\bowtie$) of a LTS and a WTS [19]. Let $B = (S_b, A_b, \Delta_b, s_0, S_b^a)$ and $W = (S_w, A_w, \Delta_w, q_0, \zeta, M_w)$ be a Büchi automaton and a WTS such that $A_b \subseteq A_w$, respectively. Their product is a WTS $B \bowtie W = (S_b \times S_w, A_w, \Delta'_w, (s_0, q_0), \zeta', S_b^a \times M_w)$, where $\Delta'_w \subseteq (S_b \times S_w) \times A_w \times (S_b \times S_w)$ is a transition relation such that

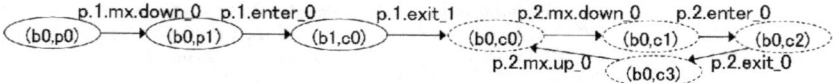

Fig. 4. Fragment of $B(\mathsf{EXIT1}) \bowtie W_{\pi^c}^{A_1}$

$\Delta'_w = \Delta^1_w \cup \Delta^2_w$ where $\Delta^1_w = \{((s_b, s_w), a, (s'_b, s'_w))|(s_b, a, s'_b) \in \Delta_b, (s_w, a, s'_w) \in \Delta_w\}$ and $\Delta^2_w = \{((s_b, s_w), a, (s_b, s'_w))|(s_w, a, s'_w) \in \Delta_w, a \notin A_b\}$. For each $\delta = ((s_b, s_w), a, (s'_b, s'_w)) \in \Delta'_w$, we define $\zeta' : \Delta'_w \to \mathcal{R}$ by $\zeta'(\delta) = \zeta((s_w, a, s'_w))$.

Intuitively, the WTS $B(\phi) \bowtie W_\pi^A$ labels transitions of $B(\phi)$ with costs of edit operations applied to π. Each end state $(s_\phi^a, c_M) \in S_\phi^a \times M_w$ is both an accepting state s_ϕ^a of $B(\phi)$ and an end state c_M of W_π^A. The shortest paths from the initial state (u_0, q_0) to (s_ϕ^a, c_M) present the event sequences that end in s_ϕ^a.

For each end state (s_ϕ^a, c_M), we conduct the second shortest path search to find sequences ending in a state on each shortest path from (u_0, q_0) to (s_ϕ^a, c_M). The witness $\tau = P'C'^\omega$ is finally generated by combining the sequences computed by the two shortest path searches. We remove τ whose C' is an empty sequence from candidate witnesses. We collect every witness such that the sum of the distances obtained by the first and second search is the smallest of all possible witnesses.

Consider $B(\mathsf{EXIT1})$ and $W_{\pi^c}^{A_1}$. A fragment of their product is shown in Fig. 4, where the only relevant information to find τ^1 are written. One of the shortest paths from the initial state (b_0, p_0) to an end state (b_0, c_0) presents the sub-sequence of τ^1 to the accepting state b_0 of $B(\mathsf{EXIT1})$: $H^1 = [$ *p.1.mx.down*, *p.1.enter*, *p.1.exit*]. Next, the sequence $T^1 = [$ *p.2.mx.down*, *p.2.enter*, *p.2.exit*, *p.2.mx.up*] is presented by the shortest path from (b_0, c_0) to itself, which is one of the states on the shortest path from (b_0, p_0) to (b_0, c_0). We find τ^1 by combining H^1 and T^1. Another witness τ^2 is obtained using the same procedure.

4.4   Identifying Errors

To find errors in L, we compute the differences between $\pi = [a_0, a_1, \ldots]$ and τ. We can assume that the different events between π and τ directly or indirectly designate causes of the property violation. If L consists of r processes, each of which is denoted by $L^h = (S^h, A^h, \Delta^h, s_0^h)$ where $0 \le h < r$ and $A = \cup_{0 \le h < r} A^h$, we identify a set of transitions over the processes triggered by the events as error candidates. However, some of the processes might not have transitions corresponding to the events to be modified. For such processes, we take a set of last transitions that occur before the differences due to the assumption that the events of these transitions trigger the events to be modified.

If an event a_d is replaced or deleted, we say that a_d is a *mismatched event*. A candidate error in L^h is its transition with the mismatched event a_d if $a_d \in A^h$; otherwise, the last transition that occurs before a_d. LLL-S returns the transition $(s, a_j, t) \in \Delta^h$ such that $0 \le j \le d$ and $a_j \in A^h \wedge \forall l \in \mathcal{N}.(j < l \le d \Rightarrow a_l \notin A^h)$.

Consider π^c and τ^2. *p.2.enter* is the mismatched event as it is replaced by *p.1.exit*. LLL-S finds error candidates for p.1, p.2 and Sema using the mismatched event. For Sema, we have to examine the preceding events in Sema in exploring the cause of error because $p.2.enter$ does not belong to the event set of Sema. The Sema's last event occurring before $p.2.enter$ is $p.2.mx.down$. LLL-S reports Sema's error candidate $(1, p.2.mx.down, 2)$, which is interpreted that $p.2.mx.down$ triggers $p.2.enter$ of p.2. In addition, LLL-S respectively returns $(1, p.1.enter, 2)$ of p.1 and $(1, p.2.enter, 2)$ of p.2 as the other error candidates.

If an event is inserted between a_{d-1} and a_d, LLL-S reports a pair of transitions of L^h that *enclose* the inserted event as follows. 1) Return the transition of L^h with a_{d-1} if $a_{d-1} \in A^h$; otherwise, its last transition occurring before a_{d-1} using the procedure above by regarding a_{d-1} as a mismatched event. 2) Return the transition of L^h with the event a_d if $a_d \in A^h$; otherwise, its first transition occurring after a_d. LLL-S returns the transition $(s, a_j, t) \in \Delta^h$ such that $j \geq d$ and $a_j \in A^h \wedge \forall l \in \mathcal{N}.(d \leq l < j \Rightarrow a_l \notin A^h)$.

Consider finding the error candidate of Sema using τ^1. The events that enclose the inserted event $p.1.exit$ in π^c are $p.1.mx.down$ and $p.2.mx.down$. LLL-S returns the transitions $(0, p.1.mx.down, 1)$ and $(1, p.2.mx.down, 2)$ as a candidate cause of the violation. The inserted event $p.1.exit$ may be demanded by the preceding event $p.1.mx.down$ or the succeeding event $p.2.mx.down$ or both.

Of all transitions computed by LLL-S, developers decide which transitions appropriately capture the erroneous behavior of L with the help of the witnesses. For example, the erroneous mutual exclusion realized by Sema is captured by the transitions given above, and both τ^1 and τ^2 show how this behavior is avoided. The presentation of witnesses and error candidates enable developers to easily identify the incorrect processes, which is an important character of LLL-S.

5   Implementation and Case Studies

We implemented a prototype tool in Java that automatically executes LLL-S. The inputs to the tool are a LTS model, the Büchi automaton of a property and a counterexample. The tool outputs a list of potential erroneous transitions and the corresponding witnesses. To enhance its performance, we have implemented some heuristics, e.g., the tool does not conduct the second search in Section 4.3 if the edit distance obtained by the first search is larger than the smallest value of distance D computed in previous iterations. The tool also supports error localization for safety property violation, which produces finite counterexamples [9]. Since the cycle of a counterexample, in this case, is regarded as an empty sequence $[\epsilon]$, we revise the way of synthesizing the WTS model in Section 4 so that ϵ can be replaced by another event [19]. The witnesses to be searched are infinite and lasso-shaped because they satisfy the Büchi's acceptance condition.

We conducted seven case studies with the prototype tool: the microwave oven (MOvn) [5], the Andrew File System (AFS-1) [26], CSys, the mine pump (MPmp) [25], and the distributed databases (DDb1, DDb2 and DDb3) [21]. Each case study was conducted as follows: 1) we made a LTS model consisting of one or more processes, 2) we prepared a FLTL property that the model did not satisfy, 3) we obtained a counterexample and a Büchi automaton recognizing the property using the model checker LTSA [21], and 4) we executed our tool and

Table 2. The Number of Generated Witnesses Indicating Errors (A) out of Total Number of Generated Witnesses (B) and Execution Time for Each Case

System		Büchi Automaton		Counterexample	Witnesses		Time
Model	States/Trans.	Property	States/Trans.	Prefix/Cycle	(A)	(B)	[s]
MOvn	7/21	HEAT	7/91	0/4	2	10	0.23
AFS-1	16/21	VALID	4/28	5/-	2	8	0.05
CSys	16/32	MUTEX	4/16	4/-	7	7	0.04
		EXIT2	6/99	5/4	3	12	0.34
MPmp	22/56	EMG	2/30	3/4	2	9	0.16
DDb1	160/402	QUIS	10/897	12/1	2	23	0.25
DDb2	6460/18537	QUIS	10/890	26/33	10	40	5.35
DDb3	-	SAFE	452/33900	18/-	57	467	37.14

manually investigated whether its result contained the transitions that were the causes of violations and the witnesses that appropriately explained the causes or not. Table 2 shows the results of the case studies. When the target property is a safety property, the length of the counterexample cycle is written as "-" in the table. We executed our tool ten times for each case on 3.4GHz Pentium 4 with 2GB RAM (JDK 1.6.0), and its average is shown as execution time. The DDb3 model has more than 2 million states and 60 million transitions, but its size could not be computed due to the heap memory limitation (shown as "-").

MOvn and MPmp are models with a single process. Although in the case of MOvn, we had to change the shape of the counterexample beforehand because LLL-S generates witnesses based on its shape, LLL-S successfully pointed out transitions that include the erroneous ones in both cases. Compared to manual error search, LLL-S made the search space for errors reduced. AFS-1, CSys, DDb1, DDb2 and DDb3 models consist of multiple processes. To find errors in component processes by hand, we have to investigate the behavior of all processes. The composite behavior analysis of all processes requires a complex composite model and makes it hard to manually identify errors whose cause is rooted in concurrency. LLL-S generated error candidates for all the processes we investigated and in all the cases, real errors were located from its subset. The task of examining the error candidates saves us much effort in locating errors compared to the case of analyzing the counterexample without any other clues.

Let us see the DDb1 case. It consists of a ring of three database nodes and a controller that allows a single update of the local data of each node. QUIS requires that every node become inactive, i.e., each node is not engaged in an update [21]. LLL-S found error candidates and the corresponding witnesses showing that an inactive node should not update or the controller should not terminate. LLL-S also reported the appropriate cause of violation that the controller terminates before checking inactivity of all nodes. We selected the error of the controller guided by two witnesses indicating how its incorrect behavior was avoided.

We next investigated how execution time of our tool respectively scaled according to the size of the Büchi automaton, and the prefix and cycle length of the counterexample using the MPmp case (shown in Table 3 and Fig. 5).

Table 3. Execution Time vs. Büchi automaton Size

Büchi Automaton	States	4	8	11	14	29	29	35	35	52	50	64	64
	Trans.	58	176	216	302	534	650	751	855	931	1190	1327	1471
Time [s]		0.16	0.69	0.40	0.66	0.35	1.13	0.97	1.42	0.60	2.59	1.00	1.05

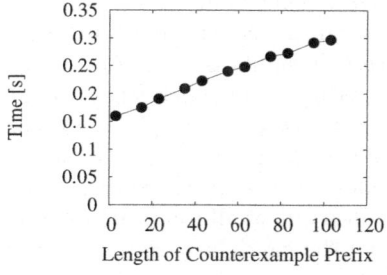 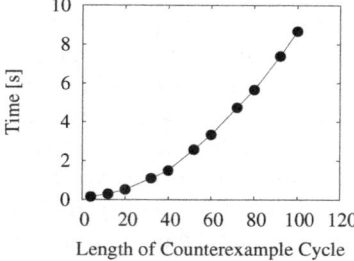

Fig. 5. Execution Time vs. Length of Counterexample Prefix (*left*) and Cycle (*right*)

In Table 3, each Büchi automaton was made by adding safety properties to EMG that the model satisfied. In Fig. 5, we expanded the cycle of the counterexample in Table 2 to make longer counterexamples to be used as samples of different size. Both results indicates that LLL-S practically handles large Büchi automata, and counterexamples with long prefixes or cycles. The execution time for large automata is almost the same as that for medium-sized ones due to the heuristics explained at the beginning of this section. Fig. 5 shows that the cycle length of a counterexample has a larger impact on the execution time of LLL-S than its prefix length. This is because the cycle length influences on the running time of the first search in Section 4.3 as well as the second search, whereas the prefix only influences on the first search.

6   Discussions and Related Work

Computational Complexity. We estimate the time to find witnesses using LLL-S. Let the counterexample $\pi = PC^\omega$ where $|P| = m$ and $|C| = n$, and the Büchi automaton of the property ϕ be $B(\phi) = (S_\phi, A_\phi, \Delta_\phi, u_0, S_\phi^a)$ where $|S_\phi| = v_\phi$ and $|S_\phi^a| = v_\phi^a$. WTS W_π^A has $m+n$ states and n end states, and $B(\phi) \bowtie W_\pi^A$ has $v_\phi(m+n)$ states and $v_\phi^a n$ end states. The first shortest path search in Section 4.3 requires $O(v_\phi(m + n) \log(v_\phi(m + n)))$ time. The second search is conducted $v_\phi^a n$ times because a source of the search is an end state of $B(\phi) \bowtie W_\pi^A$. Each search is conducted on the subgraph of the product consisting of $v_\phi n$ states because only the C part of W_π^A is used for the search. For each end state, a shortest path search requires $O(v_\phi n \log(v_\phi n))$ time. Thus, the total time of the second search is $O(v_\phi^a v_\phi n^2 \log(v_\phi n))$. If $m \approx n$, the running time is dominated by the total time of the second search. Thus, LLL-S requires $O(v_\phi^a v_\phi n^2 \log(v_\phi n))$ time.

On Fairness Constraints. When we verify a liveness property on a LTS, we often assume a kind of fairness constraints, *fair choice* [10]. Fair choice asserts that if a choice over a set of transitions is executed infinitely often, every transition in the set will be executed infinitely often. Model checking with fair choice finds an infinite and lasso-shaped counterexample under the constraint, which is the same assumption of LLL-S. Thus, LLL-S is applicable to the case.

On Property Patterns. It is useful to investigate what kinds of witnesses LLL-S produces for each property pattern [8]. Some of the liveness properties used in our case studies are written in the response pattern formula $\mathbf{G}(p \Rightarrow \mathbf{F}q)$ [8]. In this case LLL-S generated two kinds of witnesses: witnesses in which p never holds, or in which q holds after or at the same time as p holds. For example, the witnesses for π^c are classified into either those in which p.1 never enters the critical region, or those in which p.1 leaves the critical region after entering there. Although developers need to identify the appropriate ones out of all found witnesses, this information may enable them to narrow down the candidate errors.

Related Work. We previously proposed a method to find behavior model errors with infinite counterexamples [19]. Although it finds the witnesses that resemble the counterexample analogous to LLL-S, it is not based on a solid criterion to measure distances between infinite traces and may miss witnesses that appropriately point out errors. LLL-S solves the problem using the distance D.

J. Beer *et al.* [2] proposes a way to explain counterexamples for LTL model checking. While its goal is not error localization, it complements LLL-S.

Our work is related to the debugging techniques for programs as a result of model checking. A way to identify C program errors and their causes was developed by Groce and Visser with multiple counterexamples leading to the same error state [13], and later by Groce *et al.* with a single counterexample [12]. Chaki *et al.* extends the work [12] to abstracted programs [3]. Griesmayer *et al.* proposed an error localization technique for C programs [11]. Ball *et al.* proposed a technique to isolate causes of errors using counterexamples [1].

In software testing, Zeller proposed a way to find the cause-effect chains of errors in C programs [27]. Cleve and Zeller later developed a complementary technique that identifies when failure causes propagate to faults [6]. Spectrum-based fault localization techniques collect faulty runs and correct runs and compare them with certain criteria to locate faults in programs [15,23].

The above approaches resemble ours in that the comparison of a faulty run with a correct run tells us errors where the correct run is the closest to the faulty run based on the specific distance between finite runs. However, even if these distances are adapted to our context, they do not distinguish between prefixes and cycles of infinite runs and cannot overcome the problem discussed in Section 2. Although Chaki *et al.* [3] tackles the error localization problem of liveness properties, their technique reduced the problem to the SAT, which is NP-complete. LLL-S solves the classical graph search problem and performs much more efficiently. Finally, the existing methods [13,12,1,3,27,6] assume the existence of at least one correct run. LLL-S uses Büchi automata to build witnesses and does not require any correct runs supplied by the user.

Killian *et al.* developed a model checker for C++ programs, MaceMC, and its debugger MDB [16]. MaceMC supports verification of liveness properties. MDB helps developers understand errors by returning a comparison of a faulty run obtained by MaceMC with a correct run which shares a common prefix. The idea resembles ours, but only focuses on a certain kind of liveness properties.

Mohri [22] and Konstantinidis and Silva [17] developed graph-based methods that compute the edit distance between finite regular languages. LLL-S focuses on infinite strings, whose similarity cannot be computed by their methods.

7   Conclusions

In this paper, we have presented a novel automated technique to locate errors in behavior models based on the result of fluent model checking. We adopt a counterexample-based and model-based approach, which require only the model composition and classical graph search techniques. In particular, we can generate infinite-length witnesses that fix the given infinite counterexample to satisfy the property, which, we believe, is a major breakthrough.

There is much future work including integration of fluent model checking [9] with LLL-S, further practical case studies and generation of domain-specific witnesses. The last issue extends our work to help developers fix model errors [18]. Since witnesses are searched on Büchi automata, they do not reflect knowledge of the whole range of the problem domain. One of the possible solutions to this problem is to introduce the properties that hold in the target model. The introduced properties are formal descriptions of the domain knowledge.

References

1. Ball, T., Naik, M., Rajamani, S.K.: From symptom to cause: localizing errors in counterexample traces. In: POPL 2003, pp. 97–105 (2003)
2. Beer, I., Ben-David, S., Chockler, H., Orni, A., Trefler, R.: Explaining counterexamples using causality. In: Bouajjani, A., Maler, O. (eds.) CAV 2009. LNCS, vol. 5643, pp. 94–108. Springer, Heidelberg (2009)
3. Chaki, S., Groce, A., Strichman, O.: Explaining abstract counterexamples. In: SIGSOFT 2004/FSE 12, pp. 73–82 (2004)
4. Clarke, E.M.: The birth of model checking. In: Grumberg, O., Veith, H. (eds.) 25 Years of Model Checking. LNCS, vol. 5000, pp. 1–26. Springer, Heidelberg (2008)
5. Clarke, E.M.J., Grumberg, O., Peled, D.A.: Model checking. MIT Press, Cambridge (1999)
6. Cleve, H., Zeller, A.: Locating causes of program failures. In: ICSE 2005, pp. 342–351 (2005)
7. Dijkstra, E.W.: A note on two problems in connection with graphs. Numerische Mathematik, 269–271 (1959)
8. Dwyer, M.B., Avrunin, G.S., Corbett, J.C.: Patterns in property specifications for finite-state verification. In: ICSE 1999, pp. 411–420 (1999)
9. Giannakopoulou, D., Magee, J.: Fluent model checking for event-based systems. In: ESEC/FSE 2003, pp. 257–266 (2003)

10. Giannakopoulou, D., Magee, J., Kramer, J.: Checking progress with action priority: Is it fair? In: ESEC/FSE 1999, pp. 511–527 (1999)
11. Griesmayer, A., Staber, S., Bloem, R.: Automated fault localization for C programs. Elec. Notes in Theor. Comp. Sci. 174, 95–111 (2007)
12. Groce, A., Chaki, S., Kroening, D., Strichman, O.: Error explanation with distance metrics. STTT 8(3), 229–247 (2006)
13. Groce, A., Visser, W.: What went wrong: explaining counterexamples. In: Ball, T., Rajamani, S.K. (eds.) SPIN 2003. LNCS, vol. 2648, pp. 121–135. Springer, Heidelberg (2003)
14. Holzmann, G.J.: The SPIN model checker: primer and reference manual. Addison-Wesley, Reading (2004)
15. Jones, J.A., Harrold, M.J., Stasko, J.: Visualization of test information to assist fault localization. In: ICSE 2002, pp. 467–477 (2002)
16. Killian, C., Anderson, J.W., Jhala, R., Vahdat, A.: Life, death, and the critical transition: finding liveness bugs in systems code. In: NSDI 2007, pp. 243–256 (2007)
17. Konstantinidis, S., Silva, P.V.: Computing maximal error-detecting capabilities and distances of regular languages. Technical report, CMUP 2008-28 (2008)
18. Kumazawa, T., Tamai, T.: Iterative model fixing with counterexamples. In: APSEC 2008, pp. 369–376 (2008)
19. Kumazawa, T., Tamai, T.: Localizing errors and presenting alternatives: a model-based approach. In: SES 2009, pp. 55–62 (2009) (in Japanese)
20. Levenshtein, V.I.: Binary codes capable of correcting deletions, insertions and reversals. Soviet Physics Doklady 10(8), 707–710 (1966)
21. Magee, J., Kramer, J.: Concurrency: state models & Java programming, 2nd edn. John Wiley & Sons, Chichester (2006)
22. Mohri, M.: Edit-distance of weighted automata: general definitions and algorithms. Int. J. of Found. of Comp. Sci. 14(6), 957–982 (2003)
23. Renieris, M., Reiss, S.P.: Fault localization with nearest neighbor queries. In: ASE 2003, pp. 30–39 (2003)
24. Schmidt, D.C.: Model-driven engineering. IEEE Computer 39(2), 25–31 (2006)
25. Uchitel, S., Brunet, G., Chechik, M.: Synthesis of partial behaviour models from properties and scenarios. IEEE TSE 35(3), 384–406 (2009)
26. Wing, J.M., Vaziri-Farahani, M.: A case study in model checking software systems. Sci. of Comp. Prog. 28, 273–299 (1997)
27. Zeller, A.: Isolating cause-effect chains from computer programs. In: SIGSOFT 2002/FSE 10, pp. 1–10 (2002)

Call Invariants

Shuvendu K. Lahiri and Shaz Qadeer

Microsoft Research

Abstract. Program verifiers based on first-order theorem provers model the program heap as a collection of mutable maps. In such verifiers, preserving unmodified facts about the heap across procedure calls is difficult because of scoping and modification of possibly unbounded set of heap locations. Existing approaches to deal with this problem are either too imprecise, require introducing untrusted assumptions in the verifier, or resort to unpredictable reasoning using quantifiers. In this work, we propose a new approach to solve this problem. The centerpiece of our approach is the *call invariant*, a new annotation for procedure calls. A *call invariant* allows the user to specify at a call site an assertion that is inductively preserved across an arbitrary update to a heap location modified in the call. Our approach allows us to leverage existing techniques for reasoning about call-free programs to precisely and predictably reason about programs with procedure calls. We have implemented the approach and applied it to the verification of examples containing dynamic memory allocations, linked lists, and arrays. We observe that most call invariants have a fairly simple shape and discuss ways to reduce the annotation overhead.

1 Introduction

Floyd-Hoare logic is a framework for decomposing the partial correctness checking of a program into smaller proof obligations, where a Floyd-Hoare triple $\{P\}$ s $\{Q\}$ is associated with each statement s in the program [10]. Verification condition (VC) generation based on Dijkstra's *weakest liberal precondition* (wp) predicate transformer allows precise reasoning about Floyd-Hoare triples without requiring intermediate assertions for loop-free and call-free statements [7]. The use of automated theorem provers (including *satisfiability modulo theories* (SMT) solvers [20]) for checking the verification conditions provide a scalable and precise approach to program verification, and forms the basis of several tools (e.g. ESC/Java [8], Spec# [5], HAVOC [12]).

However, this framework is not as effective in the presence of the heap and procedure calls. The main issue is to preserve unmodified facts about the part of the heap in the caller's scope that is not in scope of the callee. More formally, a procedure specification comprises of (a) preconditions, (b) postconditions and (c) set of variables modified by the procedure. Since the heap is modeled as a collection of maps, a procedure that modifies a location in a map has to specify that the entire map is potentially modified. The only facts related to such a modified map, known after a procedure call has to come from the postconditions

M. Bobaru et al. (Eds.): NFM 2011, LNCS 6617, pp. 237–251, 2011.

of the callee procedure. However, the postconditions can only refer to part of the map *in scope*, i.e. the locations reachable from globals and parameters. This means that unmodified facts at caller's scope about a modified map may not be preserved across a procedure call. Matters are further complicated as a procedure call might update an unbounded number of locations in a map. Efficiently decidable SMT-based logics (e.g. linked lists [12], arrays [6]) that deal with a bounded number of heap updates (for a loop-free, call-free program fragment) are rendered ineffective in the presence of the unbounded number of updates.

An existing approach to address this problem has been to introduce *frame axioms* to allow preserving certain unmodified facts [14]. These axioms are not verified in the same spirit as the rest of the user annotations, and may introduce unsoundness in the verifier. Moreover, these frame axioms are encoded using complex quantified facts in the verification condition that severely compromises the predictability of the underlying theorem prover. Verification of such quantified formulas require expert users to be able to guide the theorem provers. These shortcomings make *wp* based Floyd-Hoare reasoning less appealing for reasoning about programs with scoping and the heap.

In this work, we present an alternative approach based on the following insight:

> For any statement *s* in the program, if an assertion *R* is preserved by an abstract (re-)execution of *s* in which the heap locations modified by *s* are updated nondeterministically, then *R* is preserved across the statement *s*.

We use an instance of this general rule for procedure calls to deal with the imprecision due to scoping. We also provide a new annotation called *call invariant* for the user to specify an inductive hypothesis when the abstract execution could be unbounded. Given a program with call invariant annotations, we perform a source-to-source transformation to create another program that can be reasoned with any existing technique for call-free programs. In particular, this allows a user to leverage existing *wp*-based verifiers to analyze programs with procedure calls with unbounded heap updates.

One can also view our approach as a strategy to augment the underlying first-order theorem provers with an induction scheme to verify formulas containing unbounded number of heap updates. The call invariant (provided by the user) plays the role of an inductive hypothesis and the underlying theorem prover is used to discharge the proof obligations for establishing the inductive hypothesis. However, there are several advantages of formalizing the inductive hypothesis at the program level instead of at the level of a formula:

1. The call invariants are specified as program annotations independent of the underlying prover. Therefore, the user does not need to interact with the specific syntax of the underlying theorem provers.

2. We formulate the call invariants as loop invariants. This opens the possibility to leverage existing loop invariant synthesis techniques to infer call invariants in many cases.

We have augmented the `Boogie` [3] verifier with call invariant annotations, and have applied it to verify a set of examples containing dynamic memory allocation, linked lists and arrays. These examples were already annotated with preconditions and postconditions — we discuss the additional annotation burden due to call invariants. We introduce useful syntactic sugars and observe the common shape of most call invariants and additional specification required to prove these examples. We also discuss tradeoffs in reducing the additional burden at the cost of slight complication of the assertion logic, without sacrificing soundness.

2 Motivation

Consider two versions of a program in Figure 1 written in a variant of the `Boogie` language [3]. The example is an abstraction of a real-life device driver `kbdclass` [21] that uses multiple lists of device extensions. The first version (on the left) has single procedure with no procedure calls, and the second version (on the right) has a procedure call.

2.1 Program without Procedure Calls

Let us first look at the example in Figure 1(a). Initially ignore the lines starting with **pre**, **post**, **inv** and **modifies**, which denote annotations. The example contains two map (or array) variables N and D to model two fields in an object. The procedure $Proc1$ takes two pointers p and q to denote the heads of the two disjoint acyclic lists $\{p, N[p], N[N[p]], \ldots, nil\}$ and $\{q, N[q], N[N[q]], \ldots, nil\}$ respectively. The procedure first initializes the D field of all the pointers in the linked list starting at p in a while loop, and then non-deterministically deletes some entries from the list starting at q — this mimics removing elements from a list that satisfy some criteria. We would like to prove that the D field has been correctly initialized for the list from p.

The assertions in **pre** and **post** denote preconditions and postconditions of a procedure. The precondition states that the two lists are disjoint and acyclic: we use the set constructor $Btwn(m, u, v)$, where m is a map value of type int $\rightarrow$ int and u and v are values of type int, to denote the set of values $\{u, m[u], m[m[u]], \ldots, v\}$ when v lies in the set, or $\{\}$ otherwise [17,12]. The postcondition states that the value of D map at all the elements of list from p is 1. The "modifies" clause in **modifies** says that the maps N and D are modified by the procedure, possibly at all locations. Loop invariant assertions are provided using **inv** annotations. The expression **old**(x) denotes the value of a variable x at the entry to a procedure (when used in a postcondition), or at the entry of a loop (when used in a loop invariant). The loop invariants on the first loop states that the variable iter points to the list from p, and all the entries upto iter have been initialized to 1. The first loop establishes the postcondition of the procedure on exit from the loop — the problem is to preserve it across the second loop. The first loop invariant for the second loop states that the set of pointers in the list from p remains unchanged. The second loop invariant says that the iterator variable iter points to elements in the list from q.

var N : int → int;   var N : int → int;
var D : int → int;   var D : int → int;

pre Btwn(N, p, nil) ∩ Btwn(N, q, nil) = {nil}   pre Btwn(N, p, nil) ∩ Btwn(N, q, nil) = {nil}
post ∀u ∈ Btwn(N, p, nil).u = nil ∨ D[u] = 1   post ∀u ∈ Btwn(N, p, nil). u = nil ∨ D[u] = 1
modifies D, N   modifies D, N
proc $Proc1$(p : int, q : int) : void =   proc $Proc1$(p : int, q : int) : void =
  var iter : int;   var iter : int;
  iter := p;   iter := p;

  inv iter ∈ Btwn(N, p, nil)   //Loop invariants omitted
  inv ∀u : int ∈ Btwn(N, p, nil).   while (iter) do D[iter] := 1; iter := N[iter];
      u ∈ Btwn(N, iter, nil) ∨ D[u] = 1
  while (iter) do D[iter] := 1; iter := N[iter];   cinv cframe(Btwn(N, p, nil))
                                                  call $Proc2$(q);
  iter := q;
                                                  updates N @ Btwn(N, t, nil)
  inv Btwn(N, p, nil) = Btwn(old(N), p, nil)   proc $Proc2$(t : int) : void =
  inv iter ∈ Btwn(old(N), q, nil)   var iter : int;
  while (iter ∧ N[iter]) do   iter := t;
    if (∗) N[iter] := N[N[iter]];
    iter := N[iter];   inv iter ∈ Btwn(old(N), t, nil)
                                                    while (iter ∧ N[iter]) do
                                                      if (∗) N[iter] := N[N[iter]];
                                                      iter := N[iter];

Fig. 1. Example with (a) no procedure calls and (b) a procedure call

These annotations are sufficient to prove the postcondition, since the map D does not change in the second loop. The annotated program can be encoded precisely in the assertion logic if the logic is closed under weakest (liberal) pre-condition [7] of statements in the programming language. Such logics with decision procedures have been proposed in [12], thereby providing an algorithm for checking such annotated programs.

2.2 Program with Procedure Calls

Now let us look the second version in Figure 1(b), where the second loop has been moved to a procedure $Proc2$. Let us initially ignore the annotation in **cinv**, and the **updates N @ .** annotation on $Proc2$. Instead, let us pretend that we only have a **modifies N** annotation for $Proc2$. Since $Proc2$ modifies the map N, any fact involving N will be invalidated after the call to $Proc2$. Therefore, the postcondition of $Proc1$ will not be provable. It is not hard to see that we cannot write any specification about the heap in the scope of $Proc2$ (namely the pointers in the list reachable from t) that would allow us to prove the postcondition for $Proc1$.

One approach to address the imprecision has been to use *frame axioms* that allow the user to specify how to preserve certain unmodified facts [14]. However, the use of frame axioms can lead to unsoundness as they are not verified. Besides, these frame axioms have complex quantified structure that may destroy the predictability of the underlying theorem provers. In the rest of this section, we show how our approach helps retain precision in the presence of procedure calls, without requiring the use of frame axioms.

First, let us look at the new annotation on *Proc2*. The annotation updates X @ $\mathcal{S}$ denotes that the map X could have been modified only at locations in $\mathcal{S}$ by the procedure., where the set-expression $\mathcal{S}$ is interpreted at entry to a procedure. This annotation is actually a syntactic sugar for a particular postcondition that we explain in Section 3.2, and does not introduce any new annotation construct. In this example, the annotation is used to specify that the map N is only modified in the locations present in the list from t at the start of the procedure.

Second, we introduce a new annotation construct called *call invariants* (using **cinv**) that allows the user to annotate a call site of a procedure. A user can specify an assertion R inside **cinv** at a call site of a procedure — with the intention that R is preserved across modifications to the maps in the callee. We use the syntactic sugar cframe(e) to denote the assertion that the value of the expression e is preserved across the call. In this example, the assertion in **cinv** states that the set of pointers in the list from p is preserved across the call to *Proc2*.

These annotations suffice to prove the postcondition of *Proc1*. Indeed, we can prove the specifications of this example (including the new proof obligations for showing that assertions in **cinv** are really preserved across a call). Not only that, the proof obligations can be encoded using the same logic that was used to prove the example without a procedure call in Figure 1(a).

3   Call Invariants

3.1   Source and Assertion Language

Figure 2 shows a simple programming language. The language supports scalar and map variables (*Scalars* and *Maps* respectively) and various operations on them. Let *Vars* = *Scalars* ∪ *Maps*. The type of any variable x ∈ *Scalars* is integer (int), and the type of any variable X ∈ *Maps* is a map from integers to integers (int → int). The standard assignment statement for scalars is extended with assignment statements for maps. The statement for variable introduction var x in s endvar introduces a variable x with an arbitrary value in s (the variable introduction rule for a map variable is similar). The statement **assert** ϕ behaves as a skip when the formula ϕ evaluates to true in the current state; else the execution of the program *fails*. The statement **assume** ϕ behaves as a skip when the formula ϕ evaluates to true in the current state; else the execution of the program is *blocked*. Expression terms of the statements and formulas are denoted by *Expr*, and include scalar variables, constants, arithmetic expressions and map lookups. The language also supports sequential composition, procedure calls, conditional statements and while loops. Allocation and deallocation can be modeled by introducing a special map Alloc to track the allocation status of objects, but is not built into the language. We show an example of dymamic allocation in the next section 4.1.

$$
\begin{array}{lll}
\mathsf{x,y} & \in Scalars \\
\mathsf{X,Y} & \in Maps \\
e & \in Expr & ::= \mathsf{x} \mid c \mid e \pm e \mid \mathsf{X}[e] \\
s,t & \in Stmt & ::= \mathbf{skip} \mid \mathbf{assert}\ \phi \mid \mathbf{assume}\ \phi \mid \mathsf{x} := e \mid \mathsf{X} := \mathsf{Y} \mid \mathsf{X}[e] := e \mid \\
& & \quad\ \mathbf{var}\ \mathsf{x}\ \mathbf{in}\ s\ \mathbf{endvar} \mid \mathbf{var}\ \mathsf{X}\ \mathbf{in}\ s\ \mathbf{endvar} \mid s; s \mid \mathsf{x} := \mathbf{call}\ f(e) \mid \\
& & \quad\ \mathbf{if}\ (e)\ \mathbf{then}\ s\ \mathbf{else}\ t \mid \mathbf{while}\ (e)\ \mathbf{do}\ s \\
P,Q,R,\phi,\psi & \in Formula & ::= e \le e \mid \phi \wedge \phi \mid \neg\phi \mid e \in \mathcal{S} \mid \mathcal{S} \subseteq \mathcal{S} \mid \dots \\
\mathcal{S} & \in SetExpr & ::= \mathsf{Btwn}(\mathsf{X},e,e) \mid \mathsf{Inverse}(\mathsf{X},e) \mid [e,e) \mid \dots \mid \mathcal{S} \cup \mathcal{S} \mid \mathcal{S} \setminus \mathcal{S} \mid \dots
\end{array}
$$

Fig. 2. A simple programming language and assertion logic

The formulas in *Formula* constitute the *assertion logic* for specifying contracts for programs in this language. The language of formulas in *Formula* is extensible, and includes relational, Boolean operations and set operations. *SetExpr* represent set-valued expressions and can be constructed from various set constructors such as Btwn and $[e, e)$ and other operations on sets. The weakest (liberal) precondition of an assertion $\phi \in Formula$ with respect to a statement $s \in Stmt$ is denoted as $wp(s, \phi)$. Intuitively, $wp(s, \phi)$ is a formula that represents the set of states for which executing s does not fail any assertions in s and moreover if the execution terminates, it does so in a state satisfying ϕ. The weakest precondition for the simple statements in our language are described in the appendix of the detailed report [13], and are fairly standard [4]. The assertion logic in *Formula* is *closed under wp*, when for any $\phi \in Formula$ and $s \in Stmt$, $wp(s, \phi) \in Formula$. The following proposition relates checking partial correctness of statements using Floyd-Hoare triples [10] and provability in a logic. Let us refer to loop-free and call-free statements as *simple* statements.

Proposition 1. *If the assertion logic in Formula is closed under wp for simple statements in Stmt, then for any simple statement s, (i) the logical formula $(P \implies wp(s, Q))$ is in Formula, and (ii) is valid if and only if the Floyd-Hoare triple $\{P\}\ s\ \{Q\}$ holds.*

In such a case, an automated theorem prover for checking assertions in *Formula* provides a method (an algorithm when the theorem prover is complete and terminating) to check Floyd-Hoare triples expressed in the logic. In the presence of loops that may have unbounded updates, the user decomposes the problem by specifying a loop invariant.

However, the presence of procedure calls and heap makes reasoning in Floyd-Hoare logic imprecise because it introduces the challenge of preserving unmodified facts about the heap in the callers scope (and not in the callee's scope) across (a possibly unbounded) update to the heap in the callee.

3.2   Call Invariant and Program Instrumentation

In this section, we instrument the source program with additional (ghost) variables to track the set of locations modified by a procedure and then introduce a new annotation called *call invariants* at the call site of a procedure. The purpose of the call invariants at a call site is to specify the assertions that are preserved across the procedure call.

First, for each global map $X \in Maps$, we introduce a state variable MS_X, whose interpretation is a set of locations. Intuitively, the value of MS_X at exit from a procedure captures the set of locations where X was modified between the entry and exit to the procedure. The source program is automatically instrumented to update the MS_X variables as follows:

- For any explicit update to the map X, $X[e_1] := e_2;$, we insert $MS_X := MS_X \cup \{e_1\};$ before the update to X.
- Every procedure has a precondition **pre** $MS_X = \{\}$.
- Before any procedure call that has a **modifies** X annotation, we save the current value of MS_X into a caller local variable, set MS_X to $\{\}$. Upon return from the procedure, we union the saved set with the value of the set after the procedure call.

It is not hard to see that the value of MS_X at exit from a procedure captures the set of locations where X was modified between the entry and exit to the procedure. For this instrumented program, the user can specify preconditions, postconditions and loop invariants in terms of MS_X variable just as any other state variable.

Next, we introduce an annotation construct called *call invariant* at the call site of a procedure specified using **cinv** R, where $R \in Formula$. For a call site that may (transitively) modify locations in the map X, we provide the following instrumentation in addition to the updates for MS_X:

$$
\begin{aligned}
&\textsf{var } X_{pre}, X_{post} \textsf{ in} \\
&\quad X_{pre} := X; \\
&\quad \textsf{call } Foo(e); //\text{procedure call} \\
&\quad X_{post} := X; \\
&\quad X := X_{pre}; \\
\\
&\quad \textbf{inv } R[X_{pre}/\textbf{old}(X)] \\
&\quad \textsf{while } (*) \textsf{ do} \\
&\qquad \textsf{var } u, v \textsf{ in } \textbf{assume } u \in MS_X; X[u] := v; \textsf{ endvar} \\
\\
&\quad \textbf{assume } X = X_{post}; \\
&\textsf{endvar}
\end{aligned}
$$

First, it copies the value of X before and after the call into local variables X_{pre} and X_{post} respectively. It restores X to the value before the call, and introduces a non-deterministic loop that updates X at one of the locations in MS_X. Finally, the assume relates the value of X after the loop with the value at the end of the procedure call X_{post}. The purpose of the loop is to model the abstract re-execution of the procedure call that nondeterministically modifies the heap locations modified by the callee, starting from the state before the call. The call invariant R (specified using **cinv** R) is checked as a loop invariant for this loop; any occurrence of **old**(X) is replaced with copy X_{pre}, the value of X just before the procedure call. Although we have described the instrumentation for a single map variable, our implementation allows for multiple maps and is presented in the detailed report [13].

Syntactic sugars. We introduce two syntactic sugars to make the specifications concise and readable:

1. We introduce updates X @ $\mathcal{S}$ for a set-valued expression $\mathcal{S}$, as a syntactic sugar for the following annotations:

$$\textbf{modifies } X, MS_X;$$
$$\textbf{post } MS_X \subseteq \textbf{old}(\mathcal{S})$$

2. The use of **old**(X) in a call invariant R refers to the value of X prior to the procecudure call (same as X_{pre} at the time of the call). We provide a sugar cframe(.) to denote that an expression is preserved by the procedure. For an expression e, cframe(e) expands to $e = \textbf{old}(e)$. This is most useful when specifying that the value of a scalar expression or a set-valued expression is preserved.

Bounded updates. Recall that the scoping problem exists even when a callee modifies a bounded number of heap locations. Our program instrumentation is still essential to preserve facts at the callers scope. However, if callee has a postcondition that bounds MS_X to a bounded number of locations (say n), then one does not require the user to specify a call invariant R. In such cases, it suffices to unroll the loop that modifies locations in MS_X n times, and eliminate the need for the call invariant.

4   Evaluation

We have built a prototype implementation of call invariant annotations over the Boogie program verifier [3]. In addition to specifying procedure preconditions, postconditions and loop invariants, the user can specify call invariants at call sites using the **cinv** annotation. We highlight the annotations required related to call invariants in the program in addition the already provided preconditions and postconditions. We discuss more about the **cinv** . call invariants (the call invariants that are both highlighted and underlined) later in Section 4.4.

We perform the program instrumentation to create the transformed program where the call invariants are desugared as loop invariants as described earlier. The resultant annotated program is verified by Boogie by generating a logical formula (verification condition) and checking the formula with *satisfiability modulo theories* (SMT) solvers. The assertion logics used in these programs use sophisticated set constructors in addition to the usual theories of uninterpreted functions, select-update arrays and arithmetic supported by most SMT solvers. In spite of the complexity, the assertion logics used in these examples are closed

```
var Alloc : int → bool;
var D : int → int;
```

updates D @ Inverse(Alloc, false)
```
proc Proc4() : void =
  var y : int;
  while (*) do
    y := new; D[y] := 0;
    /* Add y to a list */
```

```
proc Proc3() : void =
  var x : int;
  x := new;
  D[x] := 5;
```

cinv cframe(D[x])
```
  call Proc4();

  assert D[x] = 5;
```

Fig. 3. Example with dynamic memory allocation and the Inverse set constructor

under the wp predicate transformer with respect to call-free and loop-free statements in the language. For a few cases, we even have decision procedures (i.e. a sound, complete and terminating procedure) for deciding formulas in the assertion logic [6,12].

4.1   Dynamic Allocation

In the example in Figure 3, we consider a map Alloc whose range contains two Boolean values true and false. The map is used to track the set of *allocated* elements of the domain; Alloc[u] = true if and only if u is an allocated element. The statement x := new is a sugar for the following statements: {var u in x := u endvar; **assume** Alloc[x] = false; Alloc[x] := true; }. In this example, the map D is mutated at an unbounded set of freshly allocated locations in procedure *Proc4* — namely at the locations u for which Alloc[u] = false at entry to *Proc4*. In this case, this modified set excludes x in *Proc3*.

To specify the modified set, we use the set constructor Inverse : (int → int) ∗ int → 2^{int}, which takes a map and returns all elements of the domain that map to a given value; i.e. Inverse(X, v) $\doteq$ {u | X[u] = v}. For this set constructor, wp(X[x] := y, $u \in$ Inverse(X, v)) is given by [12]:

$$(y = v \wedge u \in \mathsf{Inverse}(\mathsf{X}, v) \cup \{\mathsf{x}\}) \vee (y \neq v \wedge u \in \mathsf{Inverse}(\mathsf{X}, v) \setminus \{\mathsf{x}\})$$

4.2   Linked Lists

Reverse.   Consider the recursive implementation of list reversal in Figure 4(a). This implementation performs an in-place reversal of an input list. The precondition requires the argument h to point to a nonempty acyclic list. The procedure may modify the N map only at the pointers in this list. The first postcondition asserts that the set of elements in the output list is the same as the set of elements in the input list. The second postcondition strengthens this assertion to ensure that the ordering in the output list is the reverse of the ordering in the input list.

The recursive call to *reverse* requires a call invariant stating that the value of N[h] remains unchanged by the call. This assertion is crucial for ensuring that subsequent updates to N first, do not trash the list reversal performed by the recursive call itself and second, successfully reverse the link from h to N[h].

```
var N : int → int;                          var N : int → int;
                                            var D : int → int;
pre h ≠ nil
pre nil ∈ Btwn(N, h, nil)
updates N @ Btwn(N, h, nil)                 pre r ∈ Btwn(N, l, r)
post Btwn(old(N), h, nil) = Btwn(N, r, nil) updates N @ ROS(N, l, r);
post ∀u ∈ Btwn(old(N), h, nil).             post Sorted(N, D, hd, r)
        u = nil ∨                           post r ∈ Btwn(N, hd, r)
        Btwn(old(N), h, u) = Btwn(N, u, h)  post ROS(old(N), l, r) = ROS(N, hd, r)
proc reverse(h : int) : (r : int) =         proc quick_sort(l : int, r : int) returns (hd : int) =
  if (N[h] = nil) {                            var ret : int;
    r := h;
  } else {                                     if (l = r ∨ N[l] = r) {
    cinv cframe(N[h])                            hd := l;
    r := reverse(N[h]);                        } else {
    N[N[h]] := h;                                hd := partition(l, r);
    N[h] := nil;                                 cinv cframe(ROS(N, hd, N[l]));
  }                                              cinv cframe(N[l]);
                                                 ret := quick_sort(N[l], r);
                                                 N[l] := ret;
                                                 cinv cframe(ROS(N, N[l], r));
                                                 cinv Sorted(N, D, N[l], r);
                                                 ret := quick_sort(hd, N[l]);
                                                 hd := ret;
                                               }
                                            }
```

Fig. 4. List examples (a) reverse, (b)list-based quick sort

List sort. We have also verified an implementation of quick sort for lists. This example (present in Figure 4(b)), required nontrivial call invariants. We have used the following helper predicates to define the annotations for this example:

$$
\begin{aligned}
ROS(X, u, v) &\doteq \mathsf{Btwn}(X, u, v) \setminus \{v\} \\
UpperBound(X, Y, l, r, d) &\doteq \forall u \in ROS(X, l, r) : Y[u] \leq d \\
LowerBound(X, Y, l, r, d) &\doteq \forall u \in ROS(X, l, r) : Y[u] \geq d \\
Sorted(X, Y, l, r) &\doteq \forall u \in ROS(X, l, r) : \forall v \in ROS(X, u, r) : Y[u] \leq Y[v]
\end{aligned}
$$

The example requires reasoning about shapes of lists, properties of a collection of pointers, and arithmetic relationship on the data elements of the list. The recursive nature of the procedure makes the proof highly non-trivial — which explains the complexity of the call invariants. The proof illustrates the benefits of combination frameworks present in first-order theorem provers for precise reasoning of such examples.

Merge and append. In addition to these programs, we have also successfully verified recursive implementations for appending and merging two lists. These implementations and their specifications are described in the appendix of the detailed report [13]. Interestingly, in spite of the presence of recursive calls

```
var A : int → int;

post idx ∈ [l, r)
post Deref(A, [l, r)) = Deref(old(A), [l, r))
updates A @ [l, r)
proc Partition(l : int, r : int, pivot : int) : (idx : int) =
    /* Partitions A and returns the final index of pivot */

pre 0 < l ≤ r
post Deref(A, [l, r)) = Deref(old(A), [l, r))
updates A @ [l, r)
proc QuickSort(l : int, r : int) : void =
    var pivot : int, idx : int;
    if (l = r) return;
    pivot := A[l];
    idx := Partition(l, r, pivot);
    cinv cframe(Deref(A, [idx, r)));
    QuickSort(l, idx);
    cinv cframe(Deref(A, [l, idx + 1)));
    QuickSort(idx + 1, r);
```

Fig. 5. Example of quicksort over an array A

and unbounded number of updates, no call invariants were required for proving the correctness of these examples.

4.3 Arrays

In this example, we illustrate the use of two new set constructors, (i) the range set constructor $[i, j)$, and (ii) a set constructor Deref to collect the content of a set of locations. Consider the quicksort algorithm in Figure 5 where the array map A is being sorted with recursive invocations to *QuickSort*. The procedure *QuickSort* sorts the indices of the array A in the range $[l, r)$. The procedure *Partition* (we omit the procedure body) takes a value pivot and returns an index idx such that idx $\in [l, r)$.

Let us check a simple property that the algorithm preserves the contents of A, assuming distinct elements in the array. Since the postcondition of *Partition* establishes this constraint, the main challenge is in establishing this fact across the two recursive calls to *QuickSort*. The two call invariants serve to preserve facts at the call-site required to establish the postcondition. For example, the first call invariant states that the contents of A in the range $[idx, r)$ is preserved by the call to *QuickSort*(l, idx) since the procedure does not modify these locations.

The specifications for this example refer to a dependent set constructor Deref : (int → int) $* 2^{int} → 2^{int}$ that takes a map and a set and constructs a set with the union of values of the map at elements in the set. Formally, Deref$(X, S) \doteq$ $\{X[u] \mid u \in S\}$. Interestingly, this assertion logic is still closed under *wp* with respect to statements of our language. For this set constructor, $wp(X[x] := y, u \in$ Deref$(X, S))$ is defined as follows:

$$\bigvee x \notin S \wedge u \in \mathsf{Deref}(\mathsf{X}, S)$$
$$\bigvee x \in S \wedge \mathsf{X}[\mathsf{x}] \in \mathsf{Deref}(\mathsf{X}, S \setminus \{\mathsf{x}\}) \wedge u \in \mathsf{Deref}(\mathsf{X}, S) \cup \{\mathsf{y}\}$$
$$\bigvee x \in S \wedge \mathsf{X}[\mathsf{x}] \notin \mathsf{Deref}(\mathsf{X}, S \setminus \{\mathsf{x}\}) \wedge u \in (\mathsf{Deref}(\mathsf{X}, S) \setminus \{\mathsf{X}[\mathsf{x}]\}) \cup \{\mathsf{y}\}$$

The first disjunct corresponds to the case when $\mathsf{Deref}(\mathsf{X}, S)$ remains unchanged; the second (and the third) disjunct corresponds to the cases when the value $\mathsf{X}[\mathsf{x}]$ is contained (respectively not contained) in X at an index other than x.

4.4   Discussion

We now discuss the call invariant annotations highlighted as   **cinv** . These call invariants are interesting because their specification can be obviated by adding the following postcondition automatically for each procedure:

$$\textbf{post } \forall u : \mathsf{int} :: u \in \mathsf{MS}_\mathsf{X} \vee \mathsf{X}[u] = \mathbf{old}(\mathsf{X})[u]$$

This postcondition ensures that the map X is preserved at a location disjoint from MS_X. However, the postcondition introduces a quantifier, which may compromise the termination of a theorem prover. For example, the decision procedure for reasoning about linked lists with sets [12] may not terminate when reasoning with quantified facts of the above form. The invariants marked with   **cinv** . are the ones that do not have to be specified, when the theorem prover can prove the program with this quantified postcondition. The reader may observe that these are precisely those call invariants that do not contain a set constructor or a predicate that depends on the map being modified by the callee. This explains why most call invariants using Btwn are not removed.

The example of the list implementation of quick sort uses a call invariant which is not specified using the cframe(.) syntactic sugar. We needed to specify a single-state predicate to specify the sortedness of the list, instead of a set or a scalar expression inside cframe(.) — one may view this as a way to preserve a relation (in this case sortedness). Finally, several examples did not need a call invariant (e.g. list append and merge) even in the presence of unbounded updates in callees. This is due to the fact that these recursive procedures are actually tail-recursive, where one does not need to carry facts before the procedure call.

5   Related Work

In this work, we provide a simple approach for leveraging precise verifiers for call-free programs to reason about programs with procedure calls. An important benefit of our approach is that it does not require adding additional unchecked or complex frame axioms [14] to the verification conditions. However, the benefit comes at the cost of additional user-specified annotations. Our contribution is complementary to the large body of work on modular verification in the presence of data hiding [5,11,2]. The implementations of these approaches invariably use complex quantifiers to encode variants of frame axioms that make verification unpredictable [1]; our work could potentially be used to eliminate or simplify these frame axioms.

Separation logic [18] is a specialization of Floyd-Hoare logic that requires a specialized assertion logic. The assertion logic contains formulas to describe heaps, where the formula $\phi * \psi$ denotes a heap with two disjoint subheaps for which ϕ and ψ hold respectively. The following *frame rule* [16] allows preservation of the fact R whose domain does not intersect with the modified locations in the statement s.

$$[\text{FRAME RULE}]$$
$$\frac{\{P\}\ s\ \{Q\}}{\{P * R\}\ s\ \{Q * R\}}$$

However, the specialized extension prevents leveraging existing tools for doing precise verification of call-free programs. Many of the inference rules for Floyd-Hoare logic do not apply in a straightforward manner; for example, the rule of constancy is no longer sound [18] in this extension. In addition, the specifications of s has to precisely describe the locations in the heap that s reads or write from, which is not required in Floyd-Hoare logic. Besides, one cannot use the standard *wp*-based methods to generate verification conditions. Although scalable and automatic shape analysis engines have been recently developed based on separation logic [22], proof obligations for expressive separation logic properties are often checked with higher order theorem provers [15].

The frame problem due to procedure scoping has also been explored in the context of automatic shape analysis [19,9], where the caller's heap is separated from the callee's heap by identifying a set of *cutpoints* which are dominators for any location in the heap of the callee. These cutpoints are treated as ghost parameters, but precision is lost when when the set of cutpoints can be unbounded. We believe that call invariants may alleviate the need to introduce cutpoints to pass a local heap to a callee.

6   Conclusion

This paper makes two important contributions. First, we provide an automatic program instrumentation to address the imprecision due to procedure scoping. The call invariant annotation allows a user to specify an inductive hypothesis while dealing with unbounded updates in a callee. Such an instrumented and annotated program can be verified using any off-the-shelf verifier without any need to interact with the lower-level theorem prover to specify the inductive hypothesis. This has allowed us to leverage existing precise verifiers for call-free programs to verify non-trivial examples with a small annotation burden. Second, we have separated the problem of specifying the frame from inferring the frame automatically for procedure calls. We imagine exploiting various loop invariant inference algorithms to synthesize most call invariants, given their restricted shape in practice. However, it still allows the user to explicitly specify the frame when the inference algorithm fails to discover the necessary frame.

References

1. Banerjee, A., Barnett, M., Naumann, D.A.: Boogie meets regions: A verification experience report. In: Shankar, N., Woodcock, J. (eds.) VSTTE 2008. LNCS, vol. 5295, pp. 177–191. Springer, Heidelberg (2008)
2. Banerjee, A., Naumann, D.A., Rosenberg, S.: Regional logic for local reasoning about global invariants. In: Ryan, M. (ed.) ECOOP 2008. LNCS, vol. 5142, pp. 387–411. Springer, Heidelberg (2008)
3. Barnett, M., Chang, B.E., DeLine, R., Jacobs, B., Leino, K.R.M.: Boogie: A modular reusable verifier for object-oriented programs. In: de Boer, F.S., Bonsangue, M.M., Graf, S., de Roever, W.-P. (eds.) FMCO 2005. LNCS, vol. 4111, pp. 364–387. Springer, Heidelberg (2006)
4. Barnett, M., Leino, K.R.M.: Weakest-precondition of unstructured programs. In: Program Analysis For Software Tools and Engineering (PASTE 2005), pp. 82–87 (2005)
5. Barnett, M., Leino, K.R.M., Schulte, W.: The Spec# programming system: An overview. In: Barthe, G., Burdy, L., Huisman, M., Lanet, J.-L., Muntean, T. (eds.) CASSIS 2004. LNCS, vol. 3362, pp. 49–69. Springer, Heidelberg (2005)
6. Bradley, A.R., Manna, Z., Sipma, H.B.: What's decidable about arrays? In: Emerson, E.A., Namjoshi, K.S. (eds.) VMCAI 2006. LNCS, vol. 3855, pp. 427–442. Springer, Heidelberg (2006)
7. Dijkstra, E.W.: Guarded commands, nondeterminacy and formal derivation of programs. Communications of the ACM 18, 453–457 (1975)
8. Flanagan, C., Leino, K.R.M., Lillibridge, M., Nelson, G., Saxe, J.B., Stata, R.: Extended static checking for Java. In: Programming Language Design and Implementation (PLDI 2002), pp. 234–245 (2002)
9. Gotsman, A., Berdine, J., Cook, B.: Interprocedural shape analysis with separated heap abstractions. In: Yi, K. (ed.) SAS 2006. LNCS, vol. 4134, pp. 240–260. Springer, Heidelberg (2006)
10. Hoare, C.A.R.: An axiomatic basis for computer programming. Commun. ACM 12(10), 576–580 (1969)
11. Kassios, I.T.: Dynamic frames: Support for framing, dependencies and sharing without restrictions. In: Misra, J., Nipkow, T., Karakostas, G. (eds.) FM 2006. LNCS, vol. 4085, pp. 268–283. Springer, Heidelberg (2006)
12. Lahiri, S.K., Qadeer, S.: Back to the Future: Revisiting Precise Program Verification using SMT Solvers. In: Principles of Programming Languages (POPL 2008), pp. 171–182 (2008)
13. Lahiri, S.K., Qadeer, S.: Call invariants. Technical Report MSR-TR-2009-13, Microsoft Research (2009)
14. Leino, K.R.M., Nelson, G.: Data abstraction and information hiding. ACM Trans. Program. Lang. Syst. 24(5), 491–553 (2002)
15. McCreight, A., Shao, Z., Lin, C., Li, L.: A general framework for certifying garbage collectors and their mutators. In: Programming Language Design and Implementation (PLDI 2007), pp. 468–479 (2007)
16. O'Hearn, P.W., Reynolds, J.C., Yang, H.: Local reasoning about programs that alter data structures. In: Fribourg, L. (ed.) CSL 2001 and EACSL 2001. LNCS, vol. 2142, pp. 1–19. Springer, Heidelberg (2001)
17. Rakamarić, Z., Bingham, J., Hu, A.J.: An inference-rule-based decision procedure for verification of heap-manipulating programs with mutable data and cyclic data structures. In: Cook, B., Podelski, A. (eds.) VMCAI 2007. LNCS, vol. 4349, pp. 106–121. Springer, Heidelberg (2007)

18. Reynolds, J.C.: Separation logic: A logic for shared mutable data structures. In: Logic in Computer Science (LICS 2002), pp. 55–74 (2002)
19. Rinetzky, N., Bauer, J., Reps, T.W., Sagiv, S., Wilhelm, R.: A semantics for procedure local heaps and its abstractions. In: Symposium on Principles of Programming Languages (POPL 2005), pp. 296–309. ACM, New York (2005)
20. Satisfiability Modulo Theories Library (SMT-LIB),
    http://goedel.cs.uiowa.edu/smtlib/
21. Windows Driver Kit,
    http://www.microsoft.com/whdc/devtools/wdk/default.mspx
22. Yang, H., Lee, O., Berdine, J., Calcagno, C., Cook, B., Distefano, D., O'Hearn, P.W.: Scalable shape analysis for systems code. In: Gupta, A., Malik, S. (eds.) CAV 2008. LNCS, vol. 5123, pp. 385–398. Springer, Heidelberg (2008)

Symmetry for the Analysis of Dynamic Systems

Zarrin Langari and Richard Trefler*

David R. Cheriton School of Computer Science
University of Waterloo, ON, Canada
{zlangari,trefler}@cs.uwaterloo.ca

Abstract. Graph Transformation Systems (GTSs) provide visual and explicit semantics for dynamically evolving multi-process systems such as network programs and communication protocols. Existing symmetry reduction techniques that generate a reduced, bisimilar model for alleviating state explosion in model checking are not applicable to dynamic models such as those given by GTSs. We develop symmetry reduction techniques applicable to evolving GTS models and the programs that generate them. We also provide an on-the-fly algorithm for generating a symmetry-reduced quotient model directly from a set of graph transformation rules. The generated quotient model is GTS-bisimilar to the model under verification and may be exponentially smaller than that model. Thus, analysis of the system model can be performed by checking the smaller GTS-bisimilar model.

1 Introduction

Model checking is used to analyze finite state program models. Many of these models are composed of similar components. In practice, the number of components in these models may be dynamically changing within a given upper bound. For instance, for many communication protocols, the given bound arises naturally due to inherent limitations on system size. Examples of dynamic systems composed of similar components include communication protocols such as IP-telephony protocols where telephony features are dynamically assembled in a call over the Internet [15], network programs with a variable number of clients, and object-oriented systems such as dynamic heap allocation programs [12].

Due to the use of similar components, symmetry is often a feature of the above system models that can be exploited to reduce the state space of a model under verification. Unfortunately, existing symmetry-reduction methods [13,7,9] are not applicable to dynamic systems. In addition, they may offer only limited reduction to system models that are not fully symmetric. *Full symmetry* causes the system model to be invariant under arbitrary rearranging of the components, resulting in an exponential reduction by defining an equivalence relation on symmetric states of the system model. An example of a fully symmetric system model

---

* The authors' research is supported in part by the NSERC of Canada. Zarrin Langari is currently in McMaster University, Hamilton, ON, Canada; and is supported in part by Mathematics of Information Technology and Complex Systems (MITACS).

M. Bobaru et al. (Eds.): NFM 2011, LNCS 6617, pp. 252–266, 2011.

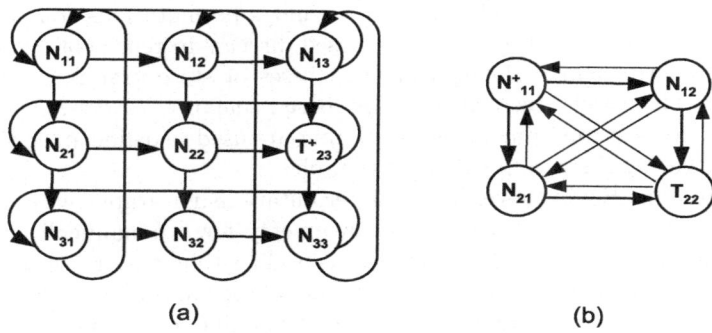

(a)   **(b)**

Fig. 1. a) A non-fully symmetric 3×3 and b) a fully symmetric 2×2 toroidal mesh

with four components is illustrated in Figure 1-b. We propose a symmetry reduction method for analyzing visual models of dynamically evolving systems. Our symmetry reduction approach is applicable to non-fully symmetric system architectures such as hypercube, ring, and torus (used in metropolitan area networks that need high scalability) used for modelling next-generation communication and hardware protocols.

Motivation: Graphs provide visual and explicit operational semantics for presenting states and demonstrating structural symmetries of a system. GTSs, which use this graph-based semantics, are straight forward formalisms that offer several key advantages over naive methods in modelling the dynamic evolution of multi-process systems [15,16]. Recently, the GTS formalism has been used to perform reasoning, including verification and error detection, on multi-component, reactive systems [16,11,2,5]. Our motivation is to exploit the advantages that graph-based models provide for the modelling and analysis of dynamically evolving systems.

When systems are composed of several similar components, it is often convenient to identify the various components by their process indices. In a Kripke model of these systems, a state consists of the values of all global variables and the local states of each process. For example, consider a 3×3 toroidal mesh network of processes, as in Figure 1-a. A toroidal mesh is a grid network with wrap-around links, where each process can communicate to two other processes. A shared token is used to show the access of processes to some resource. In this example, the local state T_{23}^+ describes that the process in row 2, column 3 possesses a token (denoted by a plus sign) and is trying to access a shared resource (denoted by T), and the other processes are in their non-trying modes (denoted by N). Symmetries in these models are then represented as permutations of the process indices. Symmetry-reduction methods [13,7,9,21] use the index permutation to build a symmetry-reduced quotient model that is equivalent, up to permutation, to the behaviour of the original model.

In Kripke models, the labelling of each state does not explicitly show the architecture of the system. On the contrary, in a GTS model of the system, each global state is represented by a graph that explicitly provides the architecture in which processes are connected together. Since index permutations do not respect the architecture of states, they cannot be used directly to represent symmetries of graph semantics and build equivalence classes of state graphs in non-fully symmetric GTS models. Instead, in graph-based semantic models, symmetries are represented as graph isomorphisms [19] that are used to define an equivalence relation on the set of states presented as graphs.

Contribution: Having several sets of permutations for graphs with different number of nodes, we define a notion of symmetry for a dynamically evolving symmetric multi-process system modelled as a GTS that may grow to a given maximum size. The explicit GTS semantic modelling can directly be exploited for reducing symmetric systems. Our symmetry reduction technique is based on generating a reduced state space directly from the set of graph transformation rules that define the model under verification. For this purpose, we define the notions of *GTS symmetry*, and *GTS bisimulation* based on graph isomorphism. With GTS bisimulation, we describe an on-the-fly algorithm that builds a symmetry-reduced model using the set of graph transformation rules that describe the full dynamic behaviour of the system.

To improve the reduction for symmetric GTS models, we define *vertex bisimulation*. Vertex bisimulation describes an equivalence relation on state graphs based on their set of vertices and can be used in our algorithm for symmetry reduction resulting in an exponential state space saving (*cf* [21]). We also show that two vertex-bisimilar GTS models can prove the same reachability properties given by a subset of CTL. In our method, we use *proposition graphs* to indicate Boolean expressions of atomic propositions. We use proposition graphs, which provide an abstraction of the process indices, to encode symmetric Boolean expressions describing local system states.

Related Work: Ip and Dill [13], Emerson and Sistla [9], and Clarke et al. [7] have been the first who explored symmetry reduction for systems with a fixed number of similar processes. These methods offers only polynomial reductions for most non-fully symmetric systems; thus, in [10,21] the authors have addressed those systems, however, those methods do not apply to graph-based models and, furthermore, are restricted to models with a fixed number of components.

Our approach is also different than approaches such as regular model checking [6] or parameterized verification (*cf* [1]). These methods provide abstractions that generally are not an equivalent representation of the original model. Our method provides an abstraction with an equivalence between the models.

In the area of GTS models, it is only Rensink's [18,19] work that has directly addressed symmetry in GTS models. In [18], a generalized definition of bisimulation is used. This bisimulation is defined for graphs and for developing efficient algorithms to check if two graphs are isomorphic, and not for the GTSs.

The rest of the paper is organized as follows: an overview of GTS modelling is given in Section 2 and is followed by definitions of GTS symmetry and GTS bisimulation in sections 3 and 4. We present vertex bisimulation for symmetric dynamic GTSs in Section 5 and conclude in Section 6.

2   Graph Transformation System Modelling

GTS is a powerful formalism for modelling the semantics of distributed reactive systems [20,8]. In this formalism, graphs are used as the most natural representation of a system [11], where each node represents a process in the system and edges show the direct communication between processes. In previous work [15], GTS modelling was used to represent the dynamic behaviour of a telecommunication system that included the creation and deletion of processes. In a GTS model, each state of the system is specified as a graph. Transformation rules are then used to describe how one state may change to another. The GTS formalism that we use to describe multi-process systems is defined below.

Definition 1 (Graph). *A graph $G = (V, E, Src, Trg, Lab)$ consists of a set V of nodes, a set E of edges, and functions $Src, Trg : E \rightarrow V$, that define the source and the target of a graph edge, and the labelling function $Lab : E, V \rightarrow l$, where l belongs to a set of labels.*

Definition 2 (Graph Morphism). *Let $G = (V_G, E_G, Src_G, Trg_G, Lab_G)$ and $H = (V_H, E_H, Src_H, Trg_H, Lab_H)$. A graph morphism $f : G \rightarrow H$ maps nodes (V) and edges (E) of graph G to nodes and edges of graph H where $f = (f_v, f_e)$, $f_v : V_G \rightarrow V_H$, and $f_e : E_G \rightarrow E_H$ are structure-preserving functions. That is, we have for all edges $e \in E_G$, $f_v(Src_G(e)) = Src_H(f_e(e))$, $f_v(Trg_G(e)) = Trg_H(f_e(e))$, and $Lab_H(f_e(e)) = Lab_G(e)$, $Lab_H(f_v(v)) = Lab_G(v)$. If f_v, f_e are total functions, then we have a total morphism, and if these are partial functions, and f_e is defined on e, i.e. there is an $e' \in E_H$, such that $f_e(e) = e'$, we have a partial morphism.*

Definition 3 (Graph Isomorphism). *In the above definition, if f, respectively f_v and f_e, are bijective functions, then we have a graph isomorphism. We write $G \cong H$ if there exists an isomorphism between graphs G and H.*

If f_v and and f_e map the set of all nodes and edges of graph G respectively, then the morphism is called a *total morphism*. On the other hand, f_v and f_e are *partial morphisms* iff the mapping is not from the whole source graph nodes and edges. Note that in a structure-preserving mapping, the shape and the edge labelling of the original graph are preserved.

Definition 4 (Graph Transformation Rule). *A transformation rule r is defined as $r : L \rightarrow R$, where L and R are graphs, called the left side graph and the right side graph of the rule, and there is a partial morphism between them.*

To transform a graph, a rule is applied to the graph. The application of a rule r to a graph G, is based on a total morphism between L and G. We write $G_0 \xrightarrow{r} G_1$ to show that the graph G_0 is transformed to G_1 by the application of rule r. In general, the result of applying a rule to a graph is as follows: everything in the left side graph (L) but not in the right side graph (R) will be *deleted*, everything in R which is not in L will be *created*, and everything that is in both sides will be *preserved* [20]. A total match between the left side subgraph of a rule and a subgraph in the source graph is made, and then the source subgraph is deleted and replaced by the right side subgraph R.

To describe how the states of a system defined as graphs transform as the transformation rules are applied to them repeatedly starting from the initial state, we give the definition of a graph transition system $\mathcal{G} = \langle S, T, I \rangle$.

Definition 5 (Graph Transition System). *A graph transition system is defined as, $\mathcal{G} = \langle S, T, I \rangle$, such that:*

1. *S is a set of states, where each state $s \in S$ has a graph structure denoted as G_s .*
2. *T is a set of transitions : $T \subseteq S \times P \times S$ where P is a set of transformation rules and for all $t \in T$, t is given by $s_1 \xrightarrow{r} s_2$, there is a graph transformation rule $r \in P$ that transforms G_{s_1} to G_{s_2}.*
3. *I is a set of initial state graphs.*

The transformation sequence $s_0 \xrightarrow{r_1} s_1 \xrightarrow{r_2} \dots \xrightarrow{r_n} s_n$ is called a GTS derivation. We write $s_0 \xrightarrow{r^*} s_n$ to denote that such a derivation from s_0 to s_n exists. Since in our modelling all the transitions are made by the application of rules, we sometimes omit the r superscript and show a transition as $s_1 \to s_2$ and a sequence of transitions as a *path*, denoted by $\rightsquigarrow$, e.g. $s_0 \rightsquigarrow s_n$ shows that there is a path between the state s_0 and s_n in the graph transition system.

Later, in sections 4 and 5 we need to prove that a GTS and its bisimilar quotient satisfy the same set of properties. Thus, at first, it is required that we describe how these properties and their propositional formulas are expressed in terms of graphs, and how the property satisfaction is defined for graphs. In our previous work [16], we have defined the notion of *graph satisfaction* and extended the definitions of graph and graph morphism to *regular expression graph (REG)* and *regular expression graph morphism*. Here, we briefly present these definitions again. REGs are used for expressing Boolean expressions of propositions as graphs (called *proposition graphs*) with edges labelled as regular expressions (e.g. Kleene-star labels). Using regular expression graphs in the proposition graphs and the transformation rule graphs makes these graphs more expressive. REGs are used to compactly express component connectivity patterns, for instance, to show that between two components of interest there may be an arbitrary length sequence of intervening components.

Definition 6 (Regular Expression Graph (REG) [16]). *An REG is a graph G where for a set of labels, L, the labelling function Lab is defined as $Lab : E_G \to \{l^+ \mid l \in L\} \cup \{l^* \mid l \in L\} \cup L$ where l^* and l^+ represent Kleene closure and the positive Kleene closure of l.*

For REG morphism we need to define the notion of a *graph path*. On a graph, a path is defined as a sequence of nodes connected by edges. Hence, the sequence of edge labels in a graph path specifies a string (language).

Definition 7 (REG Morphism [16]). *An REG morphism between G and H, when either G or H is an REG or a graph is defined as, for a path $p = \{v_1, ..., v_n\}$ in G there is a path $q = \{u_1, ..., u_n\}$ in H such that:*

- *There is a graph morphism $m : V_G \rightarrow V_H$ between the beginning and the end nodes of these two paths.*
- *For $2 \leq i \leq n - 1$, these cases may occur:*
  1. *If both G and H are REGs, then the language specified by the sequence of corresponding labels over the edges connecting nodes v_i in p is a subset of the language specified by the sequence of labels over the edges connecting nodes u_i in q.*
  2. *If H is an REG, and G is a graph without Kleene-star-labelled elements, then the string specified by the sequence of corresponding labels over the edges connecting nodes v_i in p is a member of the language specified by the sequence of labels over the edges connecting nodes u_i in q.*

We have total or partial REG morphisms, if the mappings are respectively total or partial.

Definition 8 (Graph Satisfaction [16]). *An REG or a graph G satisfies an REG or a graph ϕ, written as $G \models \phi$, iff there exists a total graph or REG morphism m between ϕ and G written as $m : \phi \rightarrow G$.*

We adopt a GTS model with attributed graphs and node identification [20,3], in which nodes are uniquely identified by their attributes.

3   Symmetry in Dynamic GTS Models

We define symmetry for dynamic GTS models of systems which may not be fully symmetric, but that show some symmetry in their structure. Traditionally, for a fixed size system, symmetries are represented by a group of index permutations [7,9]. For GTS systems, we consider states to be symmetric if their associated graphs are isomorphic.

Definition 9 (Graph Permutation). *A permutation $\pi : G \rightarrow H$ is an isomorphism between a graph G and a graph H, $G \cong H$.*

Since π is an isomorphism, it associates vertices and edges of H (V_H, E_H) to vertices and edges of G (V_G, E_G), such that $V_H = V_G$. For example, for a ring graph with three labelled nodes 1, 2, 3, with edges: $1 \rightarrow 2$, $2 \rightarrow 3$, $3 \rightarrow 1$, a permutation π that maps nodes 1 to 2, 2 to 3, and 3 to 1 permutes the graph to the one that has the same set of vertices. Also π associates the edges $1 \rightarrow 2$, $2 \rightarrow 3$, and $3 \rightarrow 1$, respectively, to the edges $2 \rightarrow 3$, $3 \rightarrow 1$, and $1 \rightarrow 2$ in the permuted graph.

In dynamic systems, where the number of components may change, we consider sets of such permutations to define symmetries for different state sizes. In fact, there are different groups of permutations for graphs with different sizes. The state graph permutation implicitly considers the number of nodes in a graph because graph isomorphism is used to define these permutations and isomorphism is based on a bijection on the sets of nodes and edges of the graph. For specific graphs of n nodes, we use the notion π_n to show a permutation on those graphs. For a ring of size n this permutation is a rotation on an n-node ring. For a $k \times k$ toroidal mesh (where $n = k \times k$), a permutation is either the rotation of k horizontal rings, the rotation of k vertical rings, or a mix of these rotations. $k - 1$ horizontal rotations followed by a vertical one or $k - 1$ vertical rotations followed by a horizontal one is actually a flip for the $k \times k$ toroidal mesh, where the flip α is defined as $\alpha(i, j) = (j, i)$. These permutations are automorphisms of a toroidal mesh network.

For a specific topology, consider a set composed of a disjoint union of graphs with different sizes. We use the notation $\mathcal{A}_i$ to show a group of graph symmetries, where i denotes size of the graph. The number of groups is finite as we work with GTS models with an upper-bound $maxsize$ on the number of graph nodes. Γ is defined as a new generalized group of symmetries built from the product of groups of permutations of graphs with different sizes. For details on this product and for the reason on why this product forms a group we refer the reader to [14]. Each element of Γ is a tuple $(\pi_1, \pi_2, ..., \pi_n)$ where $\pi_i \in \mathcal{A}_i$. Each π_i can be an identity permutation indicated as e_i, which is a morphism that maps each graph of size i to itself, where $1 \le i \le maxsize(\mathcal{G})$. Note that the group $\mathcal{A}_k$ is isomorphic to the subgroup of elements $(e_1, e_2, ..., \pi_k, ..., e_n)$; therefore, for simplicity we indicate $(e_1, e_2, ..., \pi_k, ..., e_n)$ as π_k from now on.

Definition 10 (GTS Symmetry). *A GTS $\mathcal{G} = \langle S, T, I \rangle$ is symmetric with respect to the set of graph permutations Γ if:*

1. *For all $s_1, s_2 \in S$, where s_1 has an associated n-node graph and s_2 has an associated m-node graph, if t is a transition in T such that $t : s_1 \to s_2$, then for π_n, an n-node symmetry in Γ, there is a path $p \in T^+$, $p : \pi_n(s_1) \rightsquigarrow \pi_m(s_2) \in T^+$ where π_m is an m-node symmetry in Γ.*
2. *For all $s_0 \in I$ where G_{s_0} is an n-node graph associated with state s_0 and for all $\pi_n \in \Gamma$, $\pi_n(G_{s_0}) \in I$.*

A GTS model is fully symmetric if for all transitions and for all arbitrary index permutations on state graphs (not just isomorphisms), the GTS model is invariant. GTS symmetry differs from architectural symmetry defined for fixed-size systems in [21], because in the case that G_{s_1} (an n-node graph associated with state s_1) and G_{s_2} (an m-node graph associated with state s_2) are of the same size, then in the above definition, $m = n$ and $\pi_n = \pi_m$, which means that we have the same permutation for graphs with the same number of nodes. The reason is that both π_n and π_m are isomorphisms on graphs of the same size and architecture. In this case, the path p would be of length one, because for each transition between two state graphs, there is one symmetric transition between their isomorphic state graphs. In addition, the set of symmetries in architectural symmetry differs from those in GTS symmetry, which are based on graph isomorphisms.

In dynamic GTSs, it is important to describe the way the system evolves within a maximum bound. Our methods are applicable when evolution of the system does not change the architecture describing the model structure. For example, the basic building block of a toroidal mesh is a ring, and the toroidal mesh evolves by the addition of these building blocks. Therefore, the dynamic evolution is done by adding a certain number of k nodes to form a new vertical or horizontal ring to keep a balanced toroidal mesh network. Therefore, in toroidal mesh, $m = n$ (when the toroidal mesh is not dynamic), or $m = n + k$ (when k nodes are added), or $m = n - k$ (when k nodes are deleted).

We use graph isomorphism to build a bisimilar quotient of a GTS model. It is notable that graph isomorphism requires that graphs be of the same size and structure. We can use graph isomorphism as an equivalence relation on a GTS model with state graphs of different sizes. Thus, in state-space reduction, we are looking to cut down the number of isomorphic state graphs belonging to the same equivalence class that are represented during verification.

4   GTS Bisimulation

Using graph isomorphism (Definition 3), we now define GTS bisimulation, and then give an algorithm to generate a reduced bisimilar quotient of a GTS model. Isomorphism provides a strong equivalence relation for generating the quotient, because the same set of transformation rules are applicable to a state in the quotient and the isomorphic state in the original model.

Definition 11 (GTS Bisimulation). *Given two GTSs $\mathcal{G}_1 = \langle S_1, T_1, I_1 \rangle$ and $\mathcal{G}_2 = \langle S_2, T_2, I_2 \rangle$, a relation $\sim \subseteq S_1 \times S_2$ is a GTS bisimulation if $s_1 \sim s_2$ implies:*

1. *$G_{s_1} \cong G_{s_2}$.*
2. *For every $t_1 \in T_1$, $t_1 : s_1 \to s_1'$, there is a path $p_2 \in T_2$ of length at least one, such that $p_2 : s_2 \leadsto s_2'$ and $s_1' \sim s_2'$.*
3. *For every $t_2 \in T_2$, $t_2 : s_2 \to s_2'$, there is a path $t_1 \in T_1$ of length at least one such that $p_1 : s_1 \leadsto s_1'$ and $s_2' \sim s_1'$.*

Quotient of a GTS: Let $\mathcal{G} = \langle S, T, I \rangle$ be a GTS with a set P of transformation rules and $\equiv$ is an equivalence relation on S such that $s_1 \equiv s_2$ implies $G_{s_1} \cong G_{s_2}$. If each equivalence class of state graphs is shown as $[s]$, then the quotient structure of a GTS is represented by $\bar{\mathcal{G}} = \langle \bar{S}, \bar{T}, \bar{I} \rangle$ such that

$$\bar{S} = \{[s] : s \in S\}, G_{[s]} \cong G_s,$$
$$\bar{T} = \{[s] \xrightarrow{r} [t] \in \bar{S} \times P \times \bar{S} : \exists s_0 \in [s], t_0 \in [t] : s_0 \xrightarrow{r} t_0 \in T\}, \text{ and}$$
$$\bar{I} = \{[s] : s \in I\}.$$

4.1   Generating a Bisimilar Quotient

In this section, we present an algorithm for generating a symmetry-reduced GTS model of a dynamically evolving multi-process system. This algorithm (*cf* [9]) provides an on-the-fly generation of the symmetry-reduced model of a GTS-based labelled-transition system. The algorithm may provide an exponential savings in the cost of system analysis for fully symmetric GTS models, but for GTS models with some symmetry we get a polynomial-size reduction.

GENERATEQUOTIENT(**state** s_0, T, **int** n)
Input: s_0: initial state, T: set of GTS rules, n: initial number of processes
Output: E: equivalence classes of states, R: quotient transition relation

1 $E[1].st \leftarrow s_0$
2 $CurrentState \leftarrow 1$, $LastState \leftarrow 1$
   // loops over E to apply the transformation rules
3 **while** $CurrentState \leq LastState$ **do**
4   **forall** $r \in T$ *applicable* **to** $E[CurrentState].st$ **do**
       // applies rule r to a representative state st in table E
5     $temp \leftarrow Apply(r, E[CurrentState].st)$
6     $\bar{s} \leftarrow temp.st$
7     $\bar{n} \leftarrow temp.n$
8     $stateFound \leftarrow false$
       // checks if the transformed state is a permutation of the
         existing representative states
9   **for** $i \leftarrow 1$ **to** $LastState$ **do**
         // finds the equivalence class based on the graph size
10   **if** $(E[i].n = \bar{n})$ **and** $(\bar{s} = E[i].st$ **or** ISAPERMUTATION$(E[i].st, \bar{s}, E[i].n))$
         then
11         $stateFound \leftarrow true$
12         $AddTransition(R, CurrentState, i)$
13   **exit for loop**
14   **endfor**
       // the newly found equivalence class is inserted in E
15   **if** $stateFound = false$ **then**
16       $LastState \leftarrow LastState + 1$
17       $E[LastState].st \leftarrow \bar{s}$
18       $E[LastState].n = \bar{n}$
19       $AddTransition(R, CurrentState, LastState)$
20   **endforall**
21   $CurrentState \leftarrow CurrentState + 1$
22 **endwhile**
23 **return** E, R

Fig. 2. Quotient Generation Algorithm

The algorithm GENERATEQUOTIENT in Figure 2 accepts a set of graph transformation rules, an initial-state graph labelling, and the initial number of processes as input. As output, it generates a table E: the representatives of the equivalence classes of state graphs, and a table R: the quotient transition relation. Each element in E consists of a single representative state graph (st), and the number of processes in that state graph (n). Table R is a two-dimensional table consisting of pointers to table E. There is a transition between each state in $E[i]$ and the state in $E[R[i,j]]$, where j is an index iterating over all transitions of the state in $E[i]$. By keeping track of the number of processes in each representative state: $E[i].n$, our algorithm works correctly for dynamic architectures in which processes can be added or deleted in the execution path.

In line 10, the algorithm checks that two state graphs with the same size are a permutation of each other. For more clarity, here we consider the node labelling of a state graph instead of the graph itself. The function IsAPermutation iterates over permutations to find the right permutation, and it can be specialized for different topologies. For example, for ring networks, the permutations are circular ones. For a toroidal mesh, they are appropriate horizontal or vertical rotations, or flips. As an example, we have implemented the GTS modelling of mutual exclusion for both a dynamic toroidal mesh and a dynamic token ring in [14].

Theorem 1. *Let $\mathcal{G} = \langle S, T, I \rangle$ be a GTS and symmetric with respect to the set of graph permutations Γ, and $\bar{\mathcal{G}} = \langle \bar{S}, \bar{T}, \bar{I} \rangle$ be the quotient of $\mathcal{G}$, then $\mathcal{G}$ and $\bar{\mathcal{G}}$ are GTS-bisimilar: $\mathcal{G} \sim \bar{\mathcal{G}}$.*

Proof. Consider $\pi_n, \pi_m \in \Gamma$ as graph permutations for a set of state graphs with different number of nodes. The proof considers two claims: 1) for every graph transformation $\bar{s}_0 \to \bar{s}_1 \in \bar{T}$, there is a corresponding path $p = s_0 \rightsquigarrow s_1$ in $\mathcal{G}$, and 2) for every $s_0 \to s_1 \in T$, there is a corresponding path $\bar{p} = \bar{s}_0 \rightsquigarrow \bar{s}_1$ in $\bar{\mathcal{G}}$. We prove the first claim, and the other follows similarly. The proof for each claim is broken into two cases: one for transformations $\bar{s}_0 \to \bar{s}_1$ that do not add or delete components (nodes) to or from the start graph $\bar{s}_0$ of size n. The second case considers a transformation that changes the number of components in the source state graph. For the second case, we only consider the addition of components, as proof for the deletion is similar.

Case 1: Choose an arbitrary reachable state $s_0 \in S$ such that $G_{s_0} \cong G_{\bar{s}_0}$. Using on-the-fly generation of the quotient, we know that there exists a transition $\bar{s}_0 \to \bar{s}_1 \in \bar{T}$ such that $\bar{s}_0$ and $\bar{s}_1$ are equivalence classes of state graphs. Thus, there is a graph $u \in S$ that belongs to the equivalence class of $\bar{s}_0$ and there is a graph $v \in S$ that belongs to the equivalence class of $\bar{s}_1$. Therefore, $u \to v \in T$. Thus, $G_u \cong G_{\bar{s}_0}$ and $G_v \cong G_{\bar{s}_1}$. Since $G_{s_0} \cong G_{\bar{s}_0}$ and $G_u \cong G_{\bar{s}_0}$, by transitivity $G_{s_0} \cong G_u$. Now let G_{s_1} be isomorphic to a permutation of graph v, i.e. $G_{s_1} \cong \pi_n(G_v)$ which implies $G_{s_1} \cong G_v$ and because we had $G_v \cong G_{\bar{s}_1}$, thus $G_{s_1} \cong G_{\bar{s}_1}$. From $u \to v \in T$, $G_{s_0} \cong G_u$, and $G_{s_1} \cong G_v$ we deduce $\pi_n(u) \to \pi_n(v) = s_0 \xrightarrow{r} s_1 \in T$. Inductively, we can prove for each $\bar{s}_i \to s_{i+1} \in \bar{T}$, there is a transformation $s_i \xrightarrow{r} s_{i+1} \in T$.

Case 2: In $\bar{s}_0 \to \bar{s}_1 \in \bar{T}$, we know that $G_{\bar{s}_0}$ is of size n and $G_{\bar{s}_1}$ is of size m, where $m > n$. Choose an arbitrary state $s_0 \in S$ such that $G_{s_0} \cong G_{\bar{s}_0}$. Since $\bar{s}_0 \to \bar{s}_1 \in \bar{T}$, then based on the quotient generation algorithm, there is a transformation rule $u \xrightarrow{r} v \in T$ in GTS $\mathcal{G}$ such that $G_{\bar{s}_0} \cong \pi_n(G_u)$, $G_{\bar{s}_1} \cong \pi_m(G_v)$. Since $\mathcal{G}$ is GTS-symmetric, then for each transition $u \to v \in T$ there exist permutations π'_n and π'_m such that $\pi'_n(G_u) \rightsquigarrow \pi'_m(G_v) \in T$, and since permutation is based on isomorphism then every permutation of a graph is isomorphic to it, so $\pi'_n(G_u) \cong \pi_n(G_u)$. From $G_{s_0} \cong G_{\bar{s}_0}$, $G_{\bar{s}_0} \cong \pi_n(G_u)$, and $\pi'_n(G_u) \cong \pi_n(G_u)$ we have $G_{s_0} \cong \pi_n(G_u)$. Therefore, $s_0 \rightsquigarrow \pi'_m(v)$. Let s_1 be the permutation of graph v; hence, $\pi'_m(G_v) \cong G_{s_1}$. We had $G_{\bar{s}_1} \cong \pi_m(G_v)$, also we know all permutations of a graph are isomorphic with each other, thus $\pi'_m(G_v) \cong \pi_m(G_v)$; hence, we have $G_{\bar{s}_1} \cong G_{s_1}$, and conclude $s_0 \rightsquigarrow s_1$. Inductively, we can prove for each $\bar{s}_i \to s_{i+1} \in \bar{T}$ that there is a path in T. $\square$

As a result, we have a theorem about satisfaction of reachability properties, EFf (eventually a long a path) and ¬EFf, where f is a propositional formula. In GTS-bisimilar models, a transition matches with a path; therefore, neither X (next-time) nor U (until) operators can be expressed in properties.

Theorem 2. *Let ϕ be an EF formula over a set of atomic graph propositions defined as graphs or REGs. For the graph transition system $\mathcal{G}$, and its quotient $\bar{\mathcal{G}}$ and the property ϕ and state graphs $s_1 \in \mathcal{G}$ and $\bar{s}_1 \in \bar{\mathcal{G}}$, where $s_1 \sim \bar{s}_1$ we have $\mathcal{G}, s_1 \models \phi$ iff $\bar{\mathcal{G}}, \bar{s}_1 \models \phi$.*

Proof idea. This theorem is a direct consequence of exploiting symmetry and GTS symmetry, and the proof is done using the bisimulation between the GTS $\mathcal{G}$ and its quotient $\bar{\mathcal{G}}$, and it is similar to the proof given in [21].

5   Vertex Bisimulation

In this section, we introduce vertex bisimulation to be used for GTSs that are not fully symmetric. Vertex bisimulation enables an exponential reduction with respect to full symmetry group for GTS models. We require that the transformations of these GTSs preserve the architecture, which is the case that usually occurs in practice, i.e., if the initial state graph architecture is a toroidal mesh, then this architecture is preserved in all state graphs of the model and in the state graphs of the symmetry-reduced model. Even if the structure dynamically evolves, the evolution of components preserve the overall system structure.

Definition 12 (Vertex Bisimulation). *For GTSs $\mathcal{G}_1 = \langle S_1, T_1, I_1 \rangle$ and $\mathcal{G}_2 = \langle S_2, T_2, I_2 \rangle$ a relation $\sim^v \subseteq S_1 \times S_2$ is a vertex bisimulation if $s_1 \sim^v s_2$ implies:*

1. *G_{s_1} and G_{s_2} have the same set of vertices and the same architecture.*
2. *for every $t_1 \in T_1$, $t_1 : s_1 \to s'_1$, there is a path $p_2 : s_2 \rightsquigarrow s'_2 \in T_2$ and $s'_1 \sim^v s'_2$.*
3. *for every $t_2 \in T_2$, $t_2 : s_2 \to s'_2$, there is a path $p_1 \in T_1$ such that $p_1 : s_1 \rightsquigarrow s'_1$ and $s'_2 \sim^v s'_1$.*

From a GTS-symmetric model with respect to full symmetry group, we derive a vertex-bisimilar quotient. Thus, we can apply all permutations and obtain full symmetry reduction resulting in an exponential reduction. To be able to gain this reduction without the application of the large set of all permutations, there are techniques that allow the representation of full symmetry-reduced state spaces by a program translation into a symmetry-reduced program text [10,4]. Vertex bisimulation for GTS models is comparable to safety-bisimulation for Kripke models [21], but unlike safety-bisimulation it can be used for dynamic graph models. In the following theorem, we show the vertex-bisimilarity of a model and its quotient. The proof of this theorem is similar to the proof of Theorem 1.

Theorem 3. *Let $\bar{\mathcal{G}} = \langle \bar{S}, \bar{T}, \bar{I} \rangle$ be the quotient model of a GTS-symmetric system $\mathcal{G} = \langle S, T, I \rangle$, then $\mathcal{G}$ is vertex-bisimilar to $\bar{\mathcal{G}}$.*

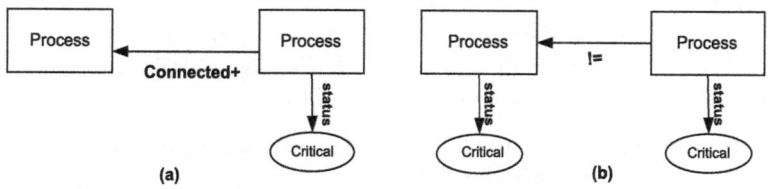

Fig. 3. Two atomic proposition graphs

5.1   Property Preservation

If we prove that the quotient of a GTS-symmetric system is vertex-bisimilar to the original model, then we can use the quotient to prove interesting properties of the system. As stated in [21], one of the problems of verifying properties on the quotient models is that the property should have *symmetric atomic propositions*, that is, permutations of process indices in the property formula leaves the formula and its significant sub-formulas invariant. Expressing Boolean expressions of atomic propositions as graphs and using graph satisfaction (Definition 8) [16] provides an abstraction on the process indices that solves this problem. The reason is that when we use an REG in an atomic proposition and a generic node that represents, for example, any of the processes appearing in the proposition, then we do not need to specify each symmetric part of the atomic proposition explicitly. For example, in a model with three processes, an atomic proposition for expressing that at least one of the processes is in the *Critical* state is: $Critical_1 \lor Critical_2 \lor Critical_3$. Figure 3-a illustrates such an expression in which one process (any of 1, or 2, or 3) is in the *Critical* state and connected to at least one other process. The condition "at least one" has been modelled as an edge labelled with $Connected^+$ between two processes.

As presented in Definition 6, we have used the regular-expression graph in which edges may be labelled with a Kleene-star operator over the set of labels. Therefore, all formulas with the existential process quantifier form, $\lor_i$, can be abstractly modelled as a proposition graph with nodes being an abstraction of process indices. Also, the universal process quantifier form, $\land_i$, in a graphical notation, is implicitly presented as all the process nodes that participate in the $\land_i$ formula connected together. For instance, in the property $\neg EF(\exists i \neq j : Critical_i \land Critical_j)$ in a toroidal mesh or ring, the Boolean expression of propositions can be expressed as a graph illustrated in Figure 3-b. In this figure, two different processes are presented to be in the *Critical* state.

Thus reachability properties and all the properties that can be expressed in terms of EF, such as AG ϕ which is equal to $\neg$ EF $\neg\phi$, are verifiable on the symmetry-reduced GTS model. For these properties, we prove that for a GTS and its quotient that are vertex-bisimilar, they both satisfy the same properties. Based on this theorem, we can use the vertex-bisimilar reduced GTS model of a system to prove interesting properties of it.

Theorem 4. *Let $\mathcal{G} = \langle S, T, I \rangle$ be a symmetric GTS and $\bar{\mathcal{G}} = \langle \bar{S}, \bar{T}, \bar{I} \rangle$ be the quotient of $\mathcal{G}$ and vertex-bisimilar to it, $\mathcal{G} \sim^v \bar{\mathcal{G}}$. For $s_1 \in S$ and $\bar{s}_1 \in \bar{S}$, where $s_1 \sim^v \bar{s}_1$ we have $\mathcal{G}, s_1 \models \phi$ iff $\bar{\mathcal{G}}, \bar{s}_1 \models \phi$ where ϕ is an EF formula over a set of atomic propositions defined as graphs or REGs.*

To prove this theorem, first in the lemma below, we show that there is a matching path between two vertex-bisimilar GTSs for GTS symmetric models.

Lemma 1. *Let $\mathcal{G} = \langle S, T, I \rangle$ be a symmetric GTS and $\bar{\mathcal{G}} = \langle \bar{S}, \bar{T}, \bar{I} \rangle$ be its vertex-bisimilar GTS, $\mathcal{G} \sim^v \bar{\mathcal{G}}$. For $s_1 \in S$, $\bar{s}_1 \in \bar{S}$, if $s_1 \sim^v \bar{s}_1$ then for any GTS derivation in $\mathcal{G}$, $s_1 \xrightarrow{r^*} s_m$, there is a derivation in $\bar{\mathcal{G}}$, $\bar{s}_1 \xrightarrow{r^*} \bar{s}_n$, and vice versa.*

Proof. It is notable that there may not be a one-to-one correspondence between transformations of these two derivations, which means that the lengths of the two derivations may not be the same. We show the proof for ($\Leftarrow$), and the other direction will follow because $\mathcal{G}$ and $\bar{\mathcal{G}}$ are vertex-bisimilar.

For $\bar{p} : \bar{s}_1 \xrightarrow{r^*} \bar{s}_n$ in $\bar{\mathcal{G}}$, we prove that there is $p : s_1 \xrightarrow{r^*} s_m$ in $\mathcal{G}$ such that $s_m \sim^v \bar{s}_n$. The proof is shown by breaking the derivation $\bar{p}$ into individual transformations and matching each graph transformation in the derivation $\bar{p}$ to a sequence of transformations in p. Later we match the concatenated transformations in $\bar{p}$ to the concatenated sequence of transformations in p.

For the first transition in $\bar{p}$, if the length of the GTS derivation p is zero, then $s_1 = s_m$, and we have a mapping to a path of length zero. If the length of the GTS derivation p is greater than or equal to one, then based on Definition 12, we have $s_1 \sim^v \bar{s}_1$ and for one transition $\bar{s}_1 \to \bar{s}_2$ in $\bar{p}$, there is a derivation in p of length at least one, thus $\bar{s}_2 \sim^v s_i$. We proved that for the first transformation in $\bar{p}$, there is a sequence of transformations in $p : s_1 \xrightarrow{r^i} s_i$ where $\bar{s}_2 \sim^v s_i$. The same reasoning can be used for the second and subsequent transformations, e.g. $\bar{s}_2 \to \bar{s}_3$ is matched to a path from s_i to s_j in p.

We now use induction. As to the hypothesis, consider for a sequence of k transformations in $\bar{p} : \bar{s}_1 \xrightarrow{r^k} \bar{s}_k$, there is a sequence of l transformations in $p : s_1 \xrightarrow{r^l} s_l$, such that $s_1 \sim^v \bar{s}_1$ and $s_l \sim^v \bar{s}_k$. Based on the vertex bisimulation definition, for the transformation $\bar{s}_k \to \bar{s}_{k+1}$ in $\bar{\mathcal{G}}$, there is a path $s_l \xrightarrow{r^*} v$ in $\mathcal{G}$, where $\bar{s}_k \sim^v s_l$, and $\bar{s}_{l+1} \sim^v v$, let v be s_{l+1}. Therefore, for $\bar{p} : \bar{s}_k \xrightarrow{r} \bar{s}_{k+1}$ in $\bar{\mathcal{G}}$ there is a derivation $p : s_l \xrightarrow{r^*} s_{l+1}$ in $\mathcal{G}$. We consider the application of the first k transformations and the $k + 1$th transformation in $\bar{\mathcal{G}}$ as one GTS derivation: $\bar{s}_1 \xrightarrow{r^*} \bar{s_{k+1}}$, and also the first l sequences of transformations and the $l + 1$th transformation in $\mathcal{G}$ as the derivation $p : s_1 \xrightarrow{r^*} s_{l+1}$. Let $k+1 = n$ and $l+1 = m$. Thus, we have matched the two derivations. $\square$

Proof (Theorem 4). To prove this theorem, we use the fact that $\mathcal{G}$ is GTS-symmetric, to ensure the preservation of architecture in states of $\mathcal{G}$ and its quotient, even though s_1 and $\bar{s}_1$ only have the same set of vertices. The proof is given for different cases of ϕ. It is sufficient to show the proof for one direction ($\Rightarrow$). The other direction is similar.

Atomic Propositions. The propositional formula is built as a graph with an abstraction on process indices. Therefore, without considering specific indices, if the formula is true for s_1, it is symmetrically true for any other communication graph of processes with the same set of local states. Since s_1 and $\bar{s}_1$ are vertex-bisimilar, they have the same node labelling or the same set of possible local states, and both satisfy the same formula.

EF formula. From $\mathcal{G}, s_1 \models \text{EF } \varphi$, we deduce that there is a derivation $p : s_1 \xrightarrow{r^*} u$ in $\mathcal{G}$, where u is a state graph that satisfies the proposition graph φ. Since $s_1 \sim^v \bar{s}_1$ and based on Lemma 1, we know that for each derivation p in $\mathcal{G}$, there is a matching derivation $\bar{p} : \bar{s}_1 \xrightarrow{r^*} v$ in $\bar{\mathcal{G}}$ such that $u \sim^v v$. Therefore, each property that is satisfied in u is satisfied in v as well, $\bar{\mathcal{G}}, v \models \varphi$, and v is a state along the path starting at $\bar{s}_1$. Hence, $\bar{\mathcal{G}}, \bar{s}_1 \models \text{EF } \varphi$. $\qquad \square$

6 Conclusion

We have described symmetry-reduction techniques for models that provide explicit visual semantics for dynamic multi-process systems. To generalize notions of symmetry for dynamic GTS models, we defined GTS symmetry and GTS bisimulation. Using these notions, we provided an on-the-fly algorithm for generating a symmetry-reduced GTS model based on graph isomorphism.

Determining if two graphs are permutations of each other needs graph isomorphism checking, which is a hard problem for unlabelled graphs, but it can be shown to have a polynomial complexity for deterministic labelled graphs [18]. Also, McKay [17] has developed an algorithm for graph isomorphism that works quite well in practice, handling graphs with up to millions of nodes.

We note that our work requires an upper bound on the number of nodes (components) that can be added to a state, because verification of systems for an arbitrary number of processes is generally undecidable [1]. We also have proved that the generated quotient is GTS-bisimilar to the original GTS model, and thus they both satisfy the same set of properties. To achieve better state-space savings for dynamic GTS models that are not fully symmetric, we have defined vertex bisimulation. The vertex-bisimilar GTS model provides exponential savings over the original model. Vertex bisimulation defines an equivalence relation on state graphs based on their vertices.

We showed that the vertex-bisimilar reduced model can prove an interesting subset of CTL properties satisfied by the original model. This subset includes all the properties expressed with the EF and ¬EF operators. This includes the important class of safety properties that are typically checked in an industrial verification setting. The propositional formula of these properties has been illustrated as a graph. Proposition graphs provide an abstraction on the process indices that take care of the symmetry of propositions. Currently, we are investigating the satisfaction of EF-CTL properties as well. These properties consist of all Boolean connectives and CTL's EF operator, including arbitrary nesting.

References

1. Apt, K., Kozen, D.: Limits for automatic verification of finite-state concurrent systems. Information Processing Letters 22, 307–309 (1986)
2. Baldan, P., Corradini, A., König, B.: Verifying finite-state graph grammars: an unfolding-based approach. In: Gardner, P., Yoshida, N. (eds.) CONCUR 2004. LNCS, vol. 3170, pp. 83–98. Springer, Heidelberg (2004)
3. Baresi, L., Heckel, R.: Tutorial introduction to graph transformation: A software engineering perspective. In: Corradini, A., Ehrig, H., Kreowski, H.-J., Rozenberg, G. (eds.) ICGT 2002. LNCS, vol. 2505, pp. 402–429. Springer, Heidelberg (2002)
4. Basler, G., Mazzucchi, M., Wahl, T., Kroening, D.: Symbolic counter abstraction for concurrent software. In: Bouajjani, A., Maler, O. (eds.) CAV 2009. LNCS, vol. 5643, pp. 64–78. Springer, Heidelberg (2009)
5. Becker, B., Beyer, D., Giese, H., Klein, F., Schilling, D.: Symbolic invariant verification for systems with dynamic structural adaptation. In: ICSE 2006, pp. 72–81 (2006)
6. Bouajjani, A., Jonsson, B., Nilsson, M., Touili, T.: Regular model checking. In: Emerson, E.A., Sistla, A.P. (eds.) CAV 2000. LNCS, vol. 1855, pp. 403–418. Springer, Heidelberg (2000)
7. Clarke, E.M., Enders, R., Filkorn, T., Jha, S.: Exploiting symmetry in temporal logic model checking. Form. Methods in Sys. Des. 9(1-2), 77–104 (1996)
8. Degano, P., Montanari, U.: A model for distributed systems based on graph rewriting. J. ACM 34(2), 411–449 (1987)
9. Emerson, E.A., Sistla, A.P.: Symmetry and model checking. Form. Methods Syst. Des. 9(1/2), 105–131 (1996)
10. Emerson, E.A., Trefler, R.J.: From asymmetry to full symmetry: New techniques for symmetry reduction in model checking. In: Pierre, L., Kropf, T. (eds.) CHARME 1999. LNCS, vol. 1703, pp. 142–157. Springer, Heidelberg (1999)
11. Heckel, R.: Compositional verification of reactive systems specified by graph transformation. In: Astesiano, E. (ed.) ETAPS 1998 and FASE 1998. LNCS, vol. 1382, p. 138. Springer, Heidelberg (1998)
12. Iosif, R.: Symmetry reduction criteria for software model checking. In: Bošnački, D., Leue, S. (eds.) SPIN 2002. LNCS, vol. 2318, pp. 22–41. Springer, Heidelberg (2002)
13. Ip, C.N., Dill, D.L.: Better verification through symmetry. Form. Methods Syst. Des. 9(1-2), 41–75 (1996)
14. Langari, Z.: Modelling and Analysis using Graph Transformation Systems. Ph.D. thesis, University of Waterloo, Waterloo, Canada (2010)
15. Langari, Z., Trefler, R.: Formal modeling of communication protocols by graph transformation. In: Misra, J., Nipkow, T., Karakostas, G. (eds.) FM 2006. LNCS, vol. 4085, pp. 348–363. Springer, Heidelberg (2006)
16. Langari, Z., Trefler, R.: Application of graph transformation in verification of dynamic systems. In: Leuschel, M., Wehrheim, H. (eds.) IFM 2009. LNCS, vol. 5423, pp. 261–276. Springer, Heidelberg (2009)
17. McKay, B.: Practical graph isomorphism. Congressus Numerantium 30, 45–87 (1981)
18. Rensink, A.: Isomorphism checking in groove. ECEASST 1 (2006)
19. Rensink, A.: Explicit state model checking for graph grammars. In: Degano, P., De Nicola, R., Bevilacqua, V. (eds.) Concurrency, Graphs and Models. LNCS, vol. 5065, pp. 114–132. Springer, Heidelberg (2008)
20. Rozenberg, G. (ed.): Handbook of Graph Grammars and Computing by Graph Transformations. Foundations, vol. 1. World Scientific, Singapore (1997)
21. Trefler, R.J., Wahl, T.: Extending symmetry reduction by exploiting system architecture. In: Jones, N.D., Müller-Olm, M. (eds.) VMCAI 2009. LNCS, vol. 5403, pp. 320–334. Springer, Heidelberg (2009)

Implementing Cryptographic Primitives in the Symbolic Model*

Peeter Laud

Cybernetica AS and Tartu University
peeter@cyber.ee

Abstract. When discussing protocol properties in the symbolic (Dolev-Yao; term-based) model of cryptography, the set of cryptographic primitives is defined by the constructors of the term algebra and by the equational theory on top of it. The set of considered primitives is not easily modifiable during the discussion. In particular, it is unclear what it means to define a new primitive from the existing ones, or why a primitive in the considered set may be unnecessary because it can be modeled using other primitives. This is in stark contrast to the computational model of cryptography where the constructions and relationships between primitives are at the very foundation of the theory. In this paper, we explore how a primitive may be constructed from other primitives in the symbolic model, such that no protocol breaks if an atomic primitive is replaced by the construction. As an example, we show the construction of (symbolic) "randomized" symmetric encryption from (symbolic) one-way functions and exclusive or.

1 Introduction

One of the main tasks of cryptographic research is the building of secure and efficient protocols needed in various systems, and the construction of primitives that these protocols need. In the computational model [12,21] of cryptography, where messages are modeled as bit-strings and the adversary as a probabilistic polynomial-time adversary, these primitives are constructed from simpler primitives, all the way down to certain base primitives (one-way functions or trapdoor one-way functions). The security properties of constructed primitives are derived from the properties of the base primitives. In the further development, only derived properties are important, making the whole approach modular (in theory).

The research in the symbolic model of cryptography (also known as the *Dolev-Yao model*, or *perfect cryptography assumption*) [10] has so far almost fully concentrated on the construction and analysis of cryptographic protocols. The messages are modeled as elements of some term algebra where the constructors of that algebra are seen as abstractions of cryptographic algorithms.

* Supported by Estonian Science Foundation, grant #8124, by European Regional Development Fund through the Estonian Center of Excellence in Computer Science, EXCS, and by EU FP7-ICT Project HATS.

M. Bobaru et al. (Eds.): NFM 2011, LNCS 6617, pp. 267–281, 2011.

In these treatments, the set of constructors has been fixed, causing the set of primitives (in our treatment, a primitive is a set of cryptographic algorithms) also to be fixed. There is no notion of implementing a primitive using some already existing primitives. This can hinder the generalization of certain kinds of results. For example, as stated, the impossibility result of [13] only applies to hash functions and XOR operations.

Another obvious application of our result is the modularization of security proofs of protocols in the symbolic model. While the symbolic model is generally more amenable to automatic analysis, certain commonly-used operations (exclusive or, and to lesser extent, Diffie-Hellman computation) are only handled with difficulty. Certain other operations (addition and multiplication, equipped with the theory for rings) are not handled at all. If these primitives are only used in a certain manner (e.g. to define the session key) then the construction of messages containing the uses of those primitives can be seen as a primitive itself, which may have properties that are simpler to handle by the automatic analysis.

The main difficulty in defining that a primitive has been securely implemented by a set of messages with variables is the difference in signatures. In general, the implementation satisfies different (more) equalities than the primitive. Hence the set of meaningful operations is richer and a simple observational equivalence is not a useful definition of security. In this paper, we give a suitable definition that in our opinion precisely captures the intuition of the equivalence of processes using the primitive operation and the processes using the implementation. We propose a technique for proving the security of the implementation. The technique does not require the prover to universally quantify over processes and contexts, but just provide an observationally equivalent process to a specific, the "most powerful" attacker against processes using the implementation of the primitive. We apply the technique by providing a secure implementation for the randomized symmetric encryption primitive in terms of one-way hash functions and exclusive or, thereby generalizing our impossibility result [13].

The paper is structured as follows. Sec. 2 gives the necessary background on process calculi, introducing the applied pi-calculus that we will be working with. Sec. 3 provides the main definitions and proof techniques, while Sec. 4 applies them to the randomized symmetric encryption primitive. Finally, Sec. 5 reviews related work and Sec. 6 concludes.

2   Applied Pi-calculus

Let us recall the syntax and semantics of the applied pi-calculus [2], in which our results will be stated. We have a countable set *Vars* of *variables*, ranged over by $x, y, \ldots$, and a countable set *Names* of *names*, ranged over by $m, n, \ldots$. We let $u, v, \ldots$ range over both names and variables. Additionally, we have a set Σ of *function symbols*, ranged over by $f, g, \ldots$, each with a fixed arity. Function symbols abstract cryptographic and other operations used by protocols. In the current paper, more often occurring function symbols besides tupling and projections are the ternary *randomized encryption* $\mathsf{Enc}(R, K, X)$ and binary *decryption* $\mathsf{Dec}(K, Y)$, as well as the unary one-way (or hash) function $\mathsf{H}(X)$ and the binary XOR (written using infix notation) $X \oplus Y$ together with its nullary neutral element 0. A *term* in signature Σ, ranged over by $M, N, \ldots$ is either a name, a variable, or a function application $f(M_1, \ldots, M_k)$, where the arity of f is k. Let $\mathbf{T}_\Sigma(\textit{Vars} \cup \textit{Names})$ denote all terms over the signature Σ, where the atomic terms belong to the set $\textit{Vars} \cup \textit{Names}$.

An *equational theory* E is a set of pairs of terms in signature Σ. It defines a relation $=_E$ on terms which is the smallest congruence containing E and is closed under the substitution of terms for variables and bijective renaming of names [20]. Equational theories capture the relationships between primitives defined in Σ. The properties of tupling are captured by the equations $\pi_i^n((x_1, \ldots, x_n)) =_E x_i$ for all i and n (we let π_i^n denote the i-th projection from an n-tuple). Encryption and decryption are related by $\mathsf{Dec}(k, \mathsf{Enc}(r, k, x)) =_E x$. The XOR-operation has its own set of equations capturing commutativity $(x \oplus y =_E y \oplus x)$, associativity $((x \oplus y) \oplus z =_E x \oplus (y \oplus z))$, unit $(x \oplus 0 =_E x)$ and cancellation $(x \oplus x = 0)$. No equations are necessary to capture the properties of H. If E is clear from the context, we abbreviate $M =_E N$ as $M = N$.

Processes, ranged over by $P, Q, \ldots$, *extended processes*, ranged over by $A, B, \ldots$, and their structural equivalence are defined in Fig. 1. We use $\overrightarrow{x}$ and $\overrightarrow{M}$ to denote a sequence of variables or terms. In the if-statement, the symbol $=$ denotes equality modulo the theory E, not syntactic equality. The extended process $\{^M/_x\}$ represents a process that has previously output M which is now available to the environment through the handle x. Variable x is free in $\{^M/_x\}$. As indicated by the structural equivalence, $\{^M/_x\}$ can replace the variable x with M in any process it comes to contact with under the scope of νx. Here $A[x \leftarrow M]$ denotes the process A where all free occurrences of the variable x have been replaced with the term M, without capturing any free variables in M. If $\{^M/_x\}$ is outside the scope of νx then we say that the variable x is *exported*. The *domain* of an extended process is the set of variables it exports. An extended process is *closed* if it exports all its free variables. The internal reduction relation describes a single step in the evolution of a process. As usual, we consider only *well-sorted* processes: all variables and names have *sorts* and all operations (conditional checking, communication, substitution) must obey them. In our sort system, there is a sort Data; the inputs and output of any function symbol have this sort. For any sequence of sorts $T_1, \ldots, T_l$ there is also a sort $\mathcal{C}\langle T_1, \ldots, T_l \rangle$ for channels communicating values of that sort. Let $\mathbf{Proc}(\Sigma)$ and $\mathbf{Ctxt}(\Sigma)$ denote the sets of all extended processes and evaluation contexts with function symbols from the set Σ. We refer to [2,20] for details and justifications.

In this paper we want to state that two processes, where the second has been obtained from the first by replacing in it certain term constructors with their "implementations", are somehow indistinguishable. *Observational equivalence*, denoted $\approx$ is the standard notion capturing indistinguishability by all environments. For defining it, we denote with $A \Downarrow c$ the existence of an evaluation context C not binding c, a term M and a process P, such that $A \rightarrow^* C[\overline{c}\langle M \rangle . P]$.

Definition 1. *The* observational equivalence *is the largest symmetric relation* $\mathcal{R}$ *over closed extended processes with the same domain such that* $A \, \mathcal{R} \, B$ *implies*

$$P ::= \mathbf{0} \mid P \mid Q \mid !P \mid \nu n.P \mid \text{if } M = N \text{ then } P \text{ else } Q \mid u(\overrightarrow{x}).P \mid \overline{u}\langle\overrightarrow{M}\rangle.P$$

$$A ::= P \mid A \mid B \mid \nu n.A \mid \nu x.A \mid \{^M/_x\} \qquad C ::= [] \mid A \mid C \mid C \mid A \mid \nu n.C \mid \nu x.C$$

$$A \equiv A \mid \mathbf{0} \qquad A \mid (B \mid C) \equiv (A \mid B) \mid C \qquad A \mid B \equiv B \mid A \qquad !P \equiv P \mid !P$$

$$\nu n.\mathbf{0} \equiv \mathbf{0} \qquad \nu u \nu v.A \equiv \nu v \nu u.A \qquad A \mid \nu u.B \equiv \nu u.(A \mid B) \text{ if } u \text{ not free in } A$$

$$\nu x.\{^M/_x\} \equiv \mathbf{0} \qquad \{^M/_x\} \mid A \equiv \{^M/_x\} \mid A[x \leftarrow M] \qquad \{^M/_x\} \equiv \{^N/_x\} \text{ if } M =_E N$$

$$\overline{c}\langle\overrightarrow{x}\rangle.P \mid c(\overrightarrow{x}).Q \rightarrow P \mid Q \qquad \text{if } N = N \text{ then } P \text{ else } Q \rightarrow P$$

$$\text{if } M = N \text{ then } P \text{ else } Q \rightarrow Q \text{ for ground terms } M \text{ and } N, \text{ where } M \neq_E N$$

$$\frac{P \equiv P' \quad P' \rightarrow Q' \quad Q' \equiv Q}{P \rightarrow Q} \qquad \frac{A \rightarrow B}{C[A] \rightarrow C[B]}$$

Fig. 1. Applied pi calculus processes and extended processes, evaluation contexts, structural equivalence $\equiv$, and internal reduction $\rightarrow$ [20]

(a) if $A \Downarrow c$ for some c, then $B \Downarrow c$; (b) if $A \rightarrow^* A'$, then $B \rightarrow^* B'$ and $A' \mathcal{R} B'$ for some B'; (c) $C[A] \mathcal{R} C[B]$ for all closing evaluation contexts C. [20]

3   Secure Implementation of Primitives

We start by defining the notions of *cryptographic primitive* and *implementing* a cryptographic primitive. A cryptographic primitive Prim is a subset of Σ, e.g. the randomized symmetric encryption primitive is R-ENC = {Enc, Dec, e_r, is$_{enc}$}. Beside the encryption and decryption function we also have a unary *randomness extraction* function with the equality e_r(Enc(r, k, x)) $=_E r$ and the unary *type verifier* with the equality is$_{enc}$(Enc(r, k, x)) $=_E$ true, where true is a nullary operation. Many implementations of symmetric randomized encryption (in the computational model) allow the randomness used in encryption to be recovered from the ciphertext, and our intended implementation has the same property. The possibilities it gives to the adversary must be reflected at the primitive level. Another reason for including e_r and is$_{enc}$ in R-ENC is, that the security proof of our implementation in Sec. 4 makes significant used of these symbols.

An implementation assigns to each function symbol $f \in$ Prim a term f^i over Σ with no free names and with free variables $x_1, \ldots, x_{\text{arity}(f)}$. The implementation defines a mapping (a second-order substitution) tr from terms to terms, replacing each occurrence of each $f \in$ Prim in the term with f^i. Formally,

$$tr(u) = u$$
$$tr(f(M_1, \ldots, M_n)) = f(tr(M_1), \ldots, tr(M_n)) \qquad \text{if } f \notin \text{Prim}$$
$$tr(f(M_1, \ldots, M_n)) = f^i[x_1 \leftarrow tr(M_1), \ldots, x_n \leftarrow tr(M_n)] \qquad \text{if } f \in \text{Prim}.$$

To make the translation well-behaved with respect to the equational theory E we require $tr(M) =_E tr(N)$ for each $(M, N) \in E$. The mapping tr can be straightforwardly extended to processes, extended processes and evaluation contexts. When defining secure implementations, we want to state that A and $tr(A)$ are somehow indistinguishable for all extended processes A.

Example 1. We can give the following implementation to the randomized symmetric encryption primitive R-ENC. Let eq? be a ternary function symbol. Let eq?$(x, x, y) =_E y$ be a pair of terms in the equational theory E. Let the implementation of R-ENC be

$$\mathsf{Enc}^i = (x_1, H(x_2, H(x_2, x_1, x_3)) \oplus x_3, H(x_2, x_1, x_3))$$

$$\mathsf{Dec}^i = \mathsf{eq?}(H(x_1, \pi_1^3(x_2), H(x_1, \pi_3^3(x_2)) \oplus \pi_2^3(x_2)), \pi_3^3(x_2), H(x_1, \pi_3^3(x_2)) \oplus \pi_2^3(x_2))$$

$$\mathsf{e\_r}^i = \pi_1^3(x_1)$$

$$\mathsf{is}_{enc}^i = \mathsf{eq?}((\pi_1^3(x_1), \pi_2^3(x_1), \pi_3^3(x_1)), x_1, \mathsf{true})$$

(recall that the arguments of Enc were the formal randomness x_1, the key x_2 and the plaintext x_3, while the arguments of Dec were the key x_1 and the ciphertext x_2). The application of H to several arguments denotes the application of H to the tuple of these arguments.

Our implementation of $\mathsf{Enc}(r, k, x)$ is similar to the OFB- or CTR-modes of operation of block ciphers [11]. The randomness r (the initialization vector IV) is included in the ciphertext and used, together with the key k, to generate a random-looking sequence which is then reversibly combined with the plaintext x. Hence the result of the OFB- or CTR-mode can be modelled as $(r, H(k, r) \oplus x)$. Formal encryption also provides integrity of the plaintext. Thus we add the third component $H(k, r, x)$ as a formal *message authentication code*. Finally, we note that if the randomness r is reused (this case is ruled out in computational definitions, but in this paper we are considering the most general way of using the primitives), then the adversary is able to find $x \oplus x'$ from the implementations of $\mathsf{Enc}(r, k, x)$ and $\mathsf{Enc}(r, k, x')$ by XOR-ing their second components. Such recovery of $x \oplus x'$ is not possible with primitive Enc. We rule out the reuse of randomness by making it depend also on the key and the plaintext — we replace r in the second component $H(k, r) \oplus x$ by $H(k, r, x)$. In this construction, similarities to the *resettable encryption* [22] can be seen. But the use of $H(k, r, x)$ in two roles seems to be novel to our construction.

The decryption $\mathsf{Dec}(k, y)$ recovers the plaintext by extracting $H(k, r, x)$ from y, XOR-ing the second component of y with $H(k, H(k, r, x))$, and checking the authentication tag.

The only pair in E relating Enc and Dec to other primitives (or each other) is $\mathsf{Dec}(k, \mathsf{Enc}(r, k, x)) =_E x$. It is simple to verify that $\mathsf{Dec}^i(k, \mathsf{Enc}^i(r, k, x) =_E x$. Also, obviously $\mathsf{e\_r}^i(\mathsf{Enc}^i(r, k, x)) =_E r$ and $\mathsf{is}_{enc}^i(k, \mathsf{Enc}^i(r, k, x)) =_E \mathsf{true}$. The implementation of is_{enc} is unsatisfactory because it declares all triples to be ciphertexts, but in the following we'll see how to improve it.

Defining the secureness of the implementation as $A \approx tr(A)$ for all extended processes A immediately leads to problems. Consider e.g. the following process

$$\nu r \nu k \nu x.(\overline{c}\langle \mathsf{Enc}(r, k, x)\rangle | c(y).\mathbf{if} \ y = (\pi_1^3(y), \pi_2^3(y), \pi_3^3(y)) \ \mathbf{then} \ P_{Impl} \ \mathbf{else} \ P_{Prim}) \ . \tag{1}$$

It generates a ciphertext and then proceeds to check whether it is a triple or not. A triple means that we are dealing with the implementation, while a non-triple is a sign of using primitive encryption. Clearly, we have to restrict the processes A if we want to have a meaningful definition. There should be a set of function symbols $\Sigma_{fb} \subseteq \Sigma$ that it is forbidden to apply. These function symbols express the "internal details" of the implementation of the primitive. A process just using the primitive should have no need to use them. Denote $\Sigma_{ok} = \Sigma \backslash \Sigma_{fb}$.

Example 2. It is unreasonable to restrict A from constructing and decomposing triples. We thus introduce $\overline{(\cdot, \cdot, \cdot)}$ as the *tagged* [9,6] version of tripling and $\bar{H}$ as the tagged version of hashing (these can be thought of as normal operations with one extra argument that is fixed as a constant that is used nowhere else). We use these operations to implement R-ENC. The use of these operations, as well as the projections $\bar{\pi}_1$, $\bar{\pi}_2$ and $\bar{\pi}_3$ from tagged triples, should be unnecessary for any process A whose security we care about (and for which we desire $A \approx tr(A)$).

Continuing our running example, we redefine

$$\mathsf{Enc}^i = \overline{(x_1, \bar{H}(x_2, \bar{H}(x_2, x_1, x_3)) \oplus x_3, \bar{H}(x_2, x_1, x_3))}$$

$$\mathsf{Dec}^i = \mathsf{eq}?(\bar{H}(x_1, \bar{\pi}_1(x_2), \bar{H}(x_1, \bar{\pi}_3(x_2)) \oplus \bar{\pi}_2(x_2)), \bar{\pi}_3(x_2), \bar{H}(x_1, \bar{\pi}_3(x_2)) \oplus \bar{\pi}_2(x_2))$$

$$\mathsf{e\_r}^i = \bar{\pi}_1(x_1)$$

$$\mathsf{is}^i_{enc} = \mathsf{eq}?(\overline{(\bar{\pi}_1(x_1), \bar{\pi}_2(x_1), \bar{\pi}_3(x_1))}, x_1, \mathsf{true})$$

The forbidden set of function symbols is $\Sigma_{fb} = \{\bar{H}, \bar{\pi}_1, \bar{\pi}_2, \bar{\pi}_3, \overline{(\cdot)}\}$. The newly introduced symbols are related to each other by $\bar{\pi}_i(\overline{(x_1, x_2, x_3)}) =_E x_i$. No other $(M, N) \in E$ contains those function symbols. We see that the tagging of triples induces the tagging also for encryptions.

In the definition of observational equivalence, the usage of function symbols in Σ_{fb} has to be restricted in evaluation contexts, too, or the test (1) can still be performed in cooperation with the process A (generating the ciphertext) and the context (testing whether it is a tagged triple). We see that the contexts must also be translated — if A is enveloped by $C[]$, then $tr(A)$ should be enveloped by $tr(C)[]$. This models the fact that both A and C implement the primitive Prim in the same manner (otherwise they would be different primitives). Thus we modify the notion of observational equivalence as follows.

Definition 2. *The* observational equivalence modulo implementation, *denoted* $\approx_{tr}$, *is the largest relation $\mathcal{R}$ over closed extended processes with the same domain such that $A \, \mathcal{R} \, B$ implies*
1. *$A \Downarrow c$ if and only if $B \Downarrow c$, for all channel names c;*
2. *if $A \rightarrow^* A'$, then there exists a process B', such that $B \rightarrow^* B'$ and $A' \, \mathcal{R} \, B'$;*
3. *if $B \rightarrow^* B'$, then there exists a process A', such that $A \rightarrow^* A'$ and $A' \, \mathcal{R} \, B'$;*
4. *$C[A] \, \mathcal{R} \, tr(C)[B]$ for all closing evaluation contexts $C \in \mathbf{Ctxt}(\Sigma_{ok})$*

While we can show that $A \approx_{tr} tr(A)$ for all extended processes in the randomized symmetric encryption example, the relation $\approx_{tr}$ also does not satisfactorily capture the meaning of secure implementation. Namely, the context is restricted in the operations it can perform; as it cannot use the function symbols in Σ_{fb}, it cannot attack the implementation of the cryptographic primitive. We would like to have a simulation-based definition — for any attacker D attacking $tr(A)$ there is an attacker S attacking A, such that $A \mid S$ and $tr(A) \mid D$ are indistinguishable [7,17]. This motivates our definition of secure implementation.

Definition 3. *Let Σ be a signature and E an equational theory over it. An im-plementation of a cryptographic primitive* Prim $\subseteq \Sigma = \Sigma_{ok} \ \dot{\cup}\ \Sigma_{fb}$ *with forbidden symbols Σ_{fb} is secure if for any closed process $A \in \mathbf{Proc}(\Sigma_{ok})$ and any closed pro-cess $D \in \mathbf{Proc}(\Sigma)$ (the adversary) there exists a closed process $S \in \mathbf{Proc}(\Sigma_{ok})$ (the simulator), such that $A \mid S \approx_{tr} tr(A) \mid D$. Here the mapping tr is induced by the implementation.*

The definition captures the notion of $tr(A)$ being *at least as secure as A* — any-thing that the environment $tr(C)$ can experience when interacting with $tr(A)$, it (as C) can also experience when interacting with A. Hence, if nothing bad can happen to C when running together with A, then nothing bad can happen to $tr(C)$ when running together with $tr(A)$. The first can be established by an-alyzing A (and possibly C), without considering the implementation details of the cryptographic primitive.

An immediate consequence of the definition is, that the process A we're trying to protect does not have to be quantified over.

Proposition 1. *Let an implementation of a cryptographic primitive* Prim $\subseteq \Sigma = \Sigma_{ok} \dot{\cup} \Sigma_{fb}$ *with forbidden symbols Σ_{fb} be given; let the mapping tr be induced by the implementation. If for any closed process $D \in \mathbf{Proc}(\Sigma)$ there exists a closed process $S \in \mathbf{Proc}(\Sigma_{ok})$, such that $S \approx_{tr} D$, then the implementation is secure.*

Proof. Let $A \in \mathbf{Proc}(\Sigma_{ok})$ and $D \in \mathbf{Proc}(\Sigma)$ be closed processes. By the premise of the proposition, there exists a closed process $S \in \mathbf{Proc}(\Sigma_{ok})$, such that $S \approx_{tr} D$. Consider the context $C[\,] = A \mid [\,]$. It does not use symbols in Σ_{fb}. Item 4 of the definition of $\approx_{tr}$ implies that $A \mid S = C[S] \approx_{tr} tr(C)[D] = tr(A) \mid D$. $\qquad\square$

We propose the following method for showing the security of a certain imple-mentation. We rewrite any process D as $\nu c_q.(D_{ctrl} \mid VM)$ where VM does not depend on D (it only depends on Σ) and D_{ctrl} does not contain any function symbols. Intuitively, the process D_{ctrl} sends computation requests to the process VM (the "virtual machine") which performs those computations and stores their results in its database, responding with *handles* (new names) that the process D_{ctrl} can later use to refer to them. The channel c_q is used for communication between the two processes. We then construct a process VM_{sim} and show that $VM_{sim} \approx_{tr} VM$ (this construction is primitive-specific). As $tr(D_{ctrl}) = D_{ctrl}$, we deduce that $\nu c_q.(D_{ctrl} \mid VM_{sim}) \approx_{tr} \nu c_q.(D_{ctrl} \mid VM)$.

By defining a suitable bisimulation [2], it will be straightforward to show that $D \approx \nu c_q.(D_{ctrl} \mid VM)$. To complete the security proof, we only need refer to the following proposition that is given here in a somewhat more general form.

$$VM = \nu c_{\mathsf{int}}.(!\big(c_{\mathsf{q}}(=\mathrm{put}, x, c_b).\nu n.(\overline{c_b}\langle n\rangle \mid !c_{\mathsf{int}}(=n, c_o).\overline{c_o}\langle x\rangle)\big) \mid$$
$$!\big(c_{\mathsf{q}}(=\mathrm{get}, x, c_b).\overline{c_{\mathsf{int}}}\langle x, c_b\rangle\big) \mid$$
$$!\big(c_{\mathsf{q}}(=\mathrm{comp}_f, (x_1, \ldots, x_k), c_b).\nu c_o.\overline{c_{\mathsf{int}}}\langle x_1, c_o\rangle.c_o(v_1)\ldots\overline{c_{\mathsf{int}}}\langle x_k, c_o\rangle.c_o(v_k).$$
$$\nu n.(\overline{c_b}\langle n\rangle \mid !c_{\mathsf{int}}(=n, c_o).\overline{c_o}\langle f(v_1, \ldots, v_k)\rangle)))))$$

Fig. 2. The process VM

$$\llbracket u\rrbracket_c^{\mathcal{N}} = \overline{c}\langle u\rangle \quad \text{if } u \notin \mathcal{N}$$
$$\llbracket n\rrbracket_c^{\mathcal{N}} = \overline{c_{\mathsf{q}}}\langle\mathrm{put}, n, c\rangle \quad \text{if } n \in \mathcal{N}$$
$$\llbracket f(M_1, \ldots, M_k)\rrbracket_c^{\mathcal{N}} = \nu c_1\ldots\nu c_k.(\llbracket M_1\rrbracket_{c_1}^{\mathcal{N}} \mid\cdots\mid \llbracket M_k\rrbracket_{c_k}^{\mathcal{N}} \mid c_1(x_1)\ldots c_k(x_k).\overline{c_{\mathsf{q}}}\langle\mathrm{comp}_f, (x_1, \ldots, x_k), c\rangle)$$

$$\llbracket 0\rrbracket^{\mathcal{N}} = 0$$
$$\llbracket P \mid Q\rrbracket^{\mathcal{N}} = \llbracket P\rrbracket^{\mathcal{N}} \mid \llbracket Q\rrbracket^{\mathcal{N}}$$
$$\llbracket !P\rrbracket^{\mathcal{N}} = !\llbracket P\rrbracket^{\mathcal{N}}$$
$$\llbracket \nu u.P\rrbracket^{\mathcal{N}} = \nu n\nu c_b.\overline{c_{\mathsf{q}}}\langle\mathrm{put}, n, c_b\rangle.c_b(u).\llbracket P\rrbracket^{\mathcal{N}\setminus\{u\}} \quad \text{if } u \text{ is data}$$
$$\llbracket \nu c.P\rrbracket^{\mathcal{N}} = \nu c.\llbracket P\rrbracket^{\mathcal{N}} \quad \text{if } c \text{ is channel}$$
$$\llbracket \text{if } M = N \text{ then } P \text{ else } Q\rrbracket^{\mathcal{N}} = \nu c_M\nu c_N.(\llbracket M\rrbracket_{c_M}^{\mathcal{N}} \mid \llbracket N\rrbracket_{c_N}^{\mathcal{N}} \mid c_M(x_M).c_N(x_N).$$
$$\nu c_b.(\overline{c_{\mathsf{q}}}\langle\mathrm{get}, x_M, c_b\rangle.c_b(y_M).\overline{c_{\mathsf{q}}}\langle\mathrm{get}, x_N, c_b\rangle.c_b(y_N).\text{if } y_M = y_N \text{ then } \llbracket P\rrbracket^{\mathcal{N}} \text{ else } \llbracket Q\rrbracket^{\mathcal{N}}))$$
$$\llbracket c(u_1, \ldots, u_k, \overrightarrow{c}).P\rrbracket^{\mathcal{N}} = c(x_1, \ldots, x_k, \overrightarrow{c}).$$
$$\nu c_b.\overline{c_{\mathsf{q}}}\langle\mathrm{put}, x_1, c_b\rangle.c_b(u_1)\ldots\overline{c_{\mathsf{q}}}\langle\mathrm{put}, x_k, c_b\rangle.c_b(u_k).\llbracket P\rrbracket^{\mathcal{N}}$$
$$\llbracket \overline{c}\langle M_1, \ldots, M_k, \overrightarrow{c}\rangle.P\rrbracket^{\mathcal{N}} = \nu c_1\cdots\nu c_k.(\llbracket M_1\rrbracket_{c_1}^{\mathcal{N}} \mid\cdots\mid \llbracket M_k\rrbracket_{c_k}^{\mathcal{N}} \mid c_1(x_1)\ldots c_k(x_k).\nu c_b.$$
$$\overline{c_{\mathsf{q}}}\langle\mathrm{get}, x_1, c_b\rangle.c_b(y_1)\ldots\overline{c_{\mathsf{q}}}\langle\mathrm{get}, x_k, c_b\rangle.c_b(y_k).\overline{c}\langle y_1, \ldots, y_k, \overrightarrow{c}\rangle.\llbracket P\rrbracket^{\mathcal{N}})$$

Fig. 3. Transforming out computations

Proposition 2. *Let* A_1, A_2, B_1, B_2 *be four closed extended processes with the same domain. If* $A_1 \approx A_2$, $A_2 \approx_{tr} B_1$ *and* $B_1 \approx B_2$, *then* $A_1 \approx_{tr} B_2$.

Proof. Co-induction over the definition of $\approx_{tr}$. Consider the relation $\approx \circ \approx_{tr} \circ \approx$. It is easy to verify that it satisfies the requirements put on relations $\mathcal{R}$ in Def. 2. Hence $(\approx \circ \approx_{tr} \circ \approx) \subseteq \approx_{tr}$.   □

The process VM is depicted in Fig. 2. We use syntactic sugar $u(=\mathrm{key}, \overrightarrow{x}).P$ for the process $u(z, \overrightarrow{x}).$if $z = \mathrm{key}$ then P else $\overline{u}\langle z, \overrightarrow{x}\rangle$ that reads a tuple of values from the channel u and continues as P, with the restriction that the first component of the tuple must be equal to key. The VM process can "obey" the commands for putting a new value in the database (input: the value; output: a handle to it), getting a value from the database (input: handle; output: corresponding value) and applying a function symbol f to the values in the database (input: handles to arguments; output: handle to result). Here "put", "get", and "comp$_f$" for each function symbol $f \in \Sigma$ are fixed free names. The process VM gives its output on a channel c_b that is given together with the input.

The translation from D to $D_{\text{ctrl}} = [\![D]\!]^{fn(D)}$ is given in Fig. 3. The process $[\![M]\!]_c^{\mathcal{N}}$ causes the handle to the value of M to be sent on the channel c if run in parallel with VM. Here $\mathcal{N}$ is a set of names of sort Data that are supposed to be free in the transformed process. In the transformed process, the values of the names in $\mathcal{N}$ will be the same as in the original process, while the variables and the names not in $\mathcal{N}$ will contain handles to their values in the original process. The notations $c(\overrightarrow{u}, \overrightarrow{c})$ and $\overline{c}\langle\overrightarrow{M}, \overrightarrow{c}\rangle$ indicate the inputs and outputs of sort Data and $\mathcal{C}(\ldots)$, respectively. We see that data is handled by the virtual machine, while the values of sort "channel" are not affected by the translation.

The bisimilarity relates each closed process P to a process $\hat{P} = ([\![P]\!]^{\mathcal{N}} \mid VM \mid$ Store$)$ where $\mathcal{N}$ is a subset of free names in P and Store is a parallel composition of processes of the form $!c_{\text{int}}(=n, c_o).\overline{c_o}\langle M\rangle$ associating the names n to values M. Moreover, the terms occurring in P (except for names in $\mathcal{N}$) must correspond to names in $[\![P]\!]^{\mathcal{N}}$ that are mapped to the same terms by Store. One transition step of P may correspond to several internal steps of $\hat{P}$. The processes at intermediate steps are also related to P.

4   Security of the Implementation of Randomized Symmetric Encryption

We have to present a process VM_{sim}, such that $VM_{\text{sim}} \approx_{tr} VM$. The process VM performs computations on behalf of the processes knowing the channel name c_q. Given handles to values $v_1, \ldots, v_k$, it returns the handle to value $f(v_1, \ldots, v_k)$, where $f \in \Sigma = \Sigma_{\text{ok}} \dot{\cup} \Sigma_{\text{fb}}$. The process VM_{sim} must respond to the same computation (and put/get) queries, but it may not use the operations in Σ_{fb}. These queries must be handled in some other way.

For the R-ENC primitive, the set Σ of function symbols contains at least tupling, projections, Enc, Dec and the operations in Σ_{fb} outlined in Example 2. Any other operations must be handled by VM_{sim}, too. In the following, we are not going to present VM_{sim} as precisely as VM in Fig. 2, but we explain in detail the operations it performs and appeal to the Turing-completeness of π-calculus [16] in order to convince ourselves that such VM_{sim} exists.

The process VM_{sim} responds to the same commands as VM in the same manner (receives a channel for sending its output as part of the input). It keeps a table T_{val} of values it has received or constructed. Each entry (row) R in T_{val} has the fields "handle" (denoted $R.hnd$), "value" (denoted $R.v$) and extra arguments for bookkeeping (denoted $R.args$). For the rows R where $R.v$ has been computed by VM_{sim} after a request to apply a symbol in Σ_{fb}, the extra arguments record that request.

The process VM_{sim} also keeps a second table T_{ct} of ciphertexts it has seen or constructed. Each row R in this table has the fields $R.ct$ (the ciphertext), $R.snd$, $R.thd$ (the second and third component of the ciphertext, considered as a tagged triple), $R.k$ (the correct key) and $R.pt$ (the corresponding plaintext). The field ct is a unique identifier for rows in this table. We let $T_{\text{ct}}[M]$ denote the row of the table T_{ct} where the field ct equals M.

The process $VM_{\sf sim}$ handles the commands as follows.

Storing. To a store a value M, the process $VM_{\sf sim}$ generates a new name n and a new row R in the table $T_{\sf val}$ with $R.{\sf hnd} = n$, $R.{\sf v} = M$ and $R.args = \bot$. If M is a valid ciphertext (checked by comparing ${\sf is}_{enc}(M)$ to ${\sf true}$) and $T_{\sf ct}$ does not yet contain a row $T_{\sf ct}[M]$, then this row is added, the field ct is initialized to M and other fields to $\bot$. The command returns n.

Retrieving. To retrieve a value by handle n, the process $VM_{\sf sim}$ locates the row R of $T_{\sf val}$ with $R.{\sf hnd} = n$, and replies with $R.{\sf v}$. If there is no such R, there will be no answer (This is similar to the behavior of VM).

Computing. To apply a function symbol f to values with handles $n_1, \ldots, n_k$, the process $VM_{\sf sim}$ locates the rows $R_1, \ldots, R_k$ of $T_{\sf val}$ with $R_i.{\sf hnd} = n_i$ for $i \in \{1, \ldots, k\}$. If some R_i cannot be located, or if k is different from the arity of f, there will be no answer. Otherwise, $VM_{\sf sim}$ generates a new name n and a new row R in $T_{\sf val}$ with $R.{\sf hnd} = n$, and replies with n. Before replying, it also defines $R.{\sf v}$ and $R.args$ as follows.

- If $f \in \Sigma_{\sf ok}$, then $R.{\sf v} = f(R_1.{\sf v}, \ldots, R_k.{\sf v})$ and $R.args = \bot$. Additionally,
  - If the operation was ${\sf Enc}$ and the row $T_{\sf ct}[R.{\sf v}]$ was not present, then it is added (and the field ct initialized with $R.{\sf v}$). The field k of this row is set to $R_2.{\sf v}$ and the field pt to $R_3.{\sf v}$.
  - If the operation was ${\sf Dec}$, the row $T_{\sf ct}[R_2.{\sf v}]$ exists (recall that ciphertext was the second argument of ${\sf Dec}$), and $R_1.{\sf v}$ was the correct key (checked by comparing ${\sf Enc}({\sf e\_r}(R_2.{\sf v}), R_1.{\sf v}, R.{\sf v})$ to $R_2.{\sf v}$) then the field k of this row is updated to $R_1.{\sf v}$ and the field pt is updated to $R.{\sf v}$.
- If $f \in \Sigma_{\sf fb}$ and there exists a row R', such that $R'.args$ indicates that the same computation query has been made to $VM_{\sf sim}$ before (i.e. $R'.args$ names the operation f and the arguments $R_1.{\sf v}, \ldots, R_k.{\sf v}$), then let $R.{\sf v} = R'.{\sf v}$ and $R.args = R'.args$. In the following cases, we assume that the same query has not been made before.
- If f is $\bar{H}$ and the argument $R_1.{\sf v}$ is a triple (x, y, z) then check whether $T_{\sf ct}$ contains a row $T_{\sf ct}[M]$, where $M = {\sf Enc}(y, x, z)$. If such row exists, then its fields k and pt are updated to x and z. If such row does not exist, then it is created and its fields k and pt likewise set to x and z. If $T_{\sf ct}[M].thd$ is not $\bot$ then $VM_{\sf sim}$ sets $R.{\sf v}$ to $T_{\sf ct}[M].thd$, otherwise it generates new names $n', \bar{n}$, sets both $R.{\sf v}$ and $T_{\sf ct}[M].thd$ to n', and adds a new row $\bar{R}$ to $T_{\sf val}$ with ${\sf hnd} = \bar{n}$, ${\sf v} = n'$ and $args = (\bar{\pi}_3, M)$. Next, it generates new names $\tilde{n}, \hat{n}$ and adds a two new rows $\tilde{R}, \hat{R}$ to the table $T_{\sf val}$ with $\tilde{R}.{\sf hnd} = \tilde{n}$, $\hat{R}.{\sf hnd} = \hat{n}$, $\tilde{R}.{\sf v} = (x, R.{\sf v})$, $\tilde{R}.args = \bot$, $\hat{R}.args = (\bar{H}, \tilde{R}.{\sf v})$. If $T_{\sf ct}[M].snd$ is not $\bot$ then $VM_{\sf sim}$ sets $\hat{R}.{\sf v}$ to $T_{\sf ct}[M].snd \oplus z$. Otherwise it generates a new name n'', sets $\hat{R}.{\sf v}$ to n'' and $T_{\sf ct}[M].snd$ to $n'' \oplus z$, and adds a new row to $T_{\sf val}$ with ${\sf hnd} = \bot$, ${\sf v} = n'' \oplus z$ and $args = (\bar{\pi}_2, M)$.

We see that if $VM_{\sf sim}$ is requested to create a value that may serve as the third component of a ciphertext then this ciphertext will also appear in the table $T_{\sf ct}$ and the entire row corresponding to it will be initialized. Also, all entries in that row will also appear in $T_{\sf val}$.

- If f is $\bar{H}$ and the argument $R_1.\mathsf{v}$ is a pair (x,y) then check whether there exists a row $T_{\mathsf{ct}}[M]$, such that $T_{\mathsf{ct}}[M].thd = y$ and x is the correct key for M. If there is no such row then y cannot have the form $\bar{H}(a,b,c)$; hence VM_{sim} generates a new name n and sets $R.\mathsf{v} = n$. Otherwise check whether $T_{\mathsf{ct}}[M].snd$ is not $\bot$. If this is the case, then set $R.\mathsf{v} = T_{\mathsf{ct}}[M].snd \oplus \mathsf{Dec}(x,M)$. Otherwise generate a new name n', set $R.\mathsf{v} = n'$ and $T_{\mathsf{ct}}[M].snd = n' \oplus \mathsf{Dec}(x,M)$, and add a new row to T_{val} with $\mathsf{hnd} = \bot$, $\mathsf{v} = n' \oplus \mathsf{Dec}(x,M)$ and $args = (\bar{\pi}_2,M)$.
- If f is $\bar{H}$ and the argument $R_1.\mathsf{v}$ is neither a pair nor a triple then generate a new name n' and set $R.\mathsf{v} = n'$.
- If f is $\bar{(\cdot,\cdot,\cdot)}$ then check whether there exists a row $T_{\mathsf{ct}}[M]$, such that $\mathsf{e\_r}(M) = R_1.\mathsf{v}$, $T_{\mathsf{ct}}[M].snd = R_2.\mathsf{v}$ and $T_{\mathsf{ct}}[M].thd = R_3.\mathsf{v}$. If such row exists then set $R.\mathsf{v} = M$. Otherwise generate new names n_k, n_x and add a new row $T_{\mathsf{ct}}[\mathsf{Enc}(R_1.\mathsf{v}, n_k, n_x)]$ to the table T_{ct}. Initialize the field snd of this row to $R_2.\mathsf{v}$ and field thd to $R_3.\mathsf{v}$. Also set $R.\mathsf{v} = \mathsf{Enc}(R_1.\mathsf{v}, n_k, n_x)$.

  We see that the result of applying the symbol $\bar{(\cdot)}$ is always a ciphertext. If the three components would result in an invalid ciphertext, then we generate this ciphertext using a throw-away key, thereby making its decryption impossible.
- If f is $\bar{\pi}_1$ then set $R.\mathsf{v} = \mathsf{e\_r}(R_1.\mathsf{v})$.
- If f is $\bar{\pi}_2$ and there is a row $T_{\mathsf{ct}}[R_1.\mathsf{v}]$ in T_{ct} and $T_{\mathsf{ct}}[R_1.\mathsf{v}].snd$ is not $\bot$, then let $R.\mathsf{v} = T_{\mathsf{ct}}[R_1.\mathsf{v}].snd$. If $T_{\mathsf{ct}}[R_1.\mathsf{v}].snd$ is $\bot$ then generate a new name n' and let both $R.\mathsf{v}$ and $T_{\mathsf{ct}}[R_1.\mathsf{v}].snd$ equal it. If the row $T_{\mathsf{ct}}[R_1.\mathsf{v}]$ does not exist then $R_1.\mathsf{v}$ is not a ciphertext, because this fact would have been noticed at the time when the row R_1 was added to T_{val}. Generate a new name n' and let $R.\mathsf{v} = n'$.
- If f is $\bar{\pi}_3$ then behave similarly to the case $f = \bar{\pi}_2$.

In all cases of handling a function symbol f from Σ_{fb}, the process VM_{sim} sets the fields $args$ of the newly created row R of T_{val} to $(f, R_1.\mathsf{v}, \ldots, R_k.\mathsf{v})$, where $R_1.\mathsf{hnd}, \ldots, R_k.\mathsf{hnd}$ were the arguments given to f.

Proposition 3. $VM_{\mathsf{sim}} \approx_{tr} VM$.

Proof (Sketch). Both VM and VM_{sim} maintain a database that maps from handles to values, update it according to certain rules, and reveal the values of queried elements. A context $C \in \mathbf{Ctxt}(\Sigma_{\mathsf{ok}})$ trying to distinguish VM and VM_{sim} (i.e. having a channel c, such that $C[VM_{\mathsf{sim}}] \Downarrow c$, but $tr(C)[VM] \not\Downarrow c$) will at some point query for certain elements of the database and then perform a test (check the equality of two terms built from the queried elements), the result of which determines whether the communication on c happens. We will show that at no point in the execution there exists a test that can tell apart the databases of VM and VM_{sim}. Formally, a *test* is a pair of *test messages* $M, N \in \mathbf{T}_{\Sigma_{\mathsf{ok}}}(Names \cup Refs)$, where $Refs$ is the set of "references" to the cells of the database of VM_{sim}. A reference $r \in Refs$ can either be $n.\mathsf{v}$, where n is the handle of a row R in T_{val}, or $T_{\mathsf{ct}}[M_e].field$, where the ciphertext M_e identifies a row in T_{ct} and the value of $field \in \{snd, thd\}$ in this row is different from $\bot$. The context C can access these references by using get-queries, possibly preceded by a $\text{comp}_{\bar{\pi}_2}$ or $\text{comp}_{\bar{\pi}_3}$ query. The names that a test message may contain are free names generated by C.

If the context $C[\,]$ encapsulating VM_{sim} evaluates a test message M, then the value $\langle\!\langle M \rangle\!\rangle^{\text{sim}}$ it learns is obtained by replacing the references to database cells in M with their actual values, as kept by VM_{sim}. If $C[\,]$ encapsulates VM, then it learns the value $\langle\!\langle M \rangle\!\rangle^{\text{real}}$ instead, where the reference $n.\text{v}$ is replaced with the value VM associates with the handle n. In this case, the references $T_{\text{ct}}[M_e].snd$ and $T_{\text{ct}}[M_e].thd$ are replaced with $\bar{\pi}_2(tr(M_e))$ and $\bar{\pi}_3(tr(M_e))$, respectively.

Example 3. Suppose that the context $C[\,]$ issues the following commands: $n_1 \leftarrow \text{put}(r)$; $n_2 \leftarrow \text{put}(k)$; $n_3 \leftarrow \text{put}(x)$; $n_4 \leftarrow \text{comp}_{(,,)}(n_2, n_1, n_3)$; $n_5 \leftarrow \text{comp}_{\bar{H}}(n_4)$; $n_6 \leftarrow \text{comp}_{(,)}(n_2, n_5)$; $n_7 \leftarrow \text{comp}_{\bar{H}}(n_6)$; $n_8 \leftarrow \text{comp}_{\oplus}(n_7, n_3)$; and finally $n_9 \leftarrow \text{comp}_{\bar{(,,)}}(n_1, n_8, n_5)$. Through these queries, the handle n_9 will correspond to the ciphertext $\text{Enc}(r, k, x)$. Let $M = (n_9.\text{v}, n_5.\text{v}, T_{\text{ct}}[\text{Enc}(r, k, x)].thd)$ and n^{hash} be a new name generated by $VM_{\text{sim}}..$ Then $\langle\!\langle M \rangle\!\rangle^{\text{sim}} = (\text{Enc}(r, k, x), n^{\text{hash}}, n^{\text{hash}})$ and $\langle\!\langle M \rangle\!\rangle^{\text{real}} = ((r, \bar{H}(k, \bar{H}(k, r, x)) \oplus x, \bar{H}(k, r, x)), \bar{H}(k, r, x), \bar{H}(k, r, x))$.

We show that the following claim holds.

Claim (*)   At any step of the computation of VM_{sim}, the equivalence $\langle\!\langle M \rangle\!\rangle^{\text{sim}} = \langle\!\langle N \rangle\!\rangle^{\text{sim}} \Leftrightarrow \langle\!\langle M \rangle\!\rangle^{\text{real}} = \langle\!\langle N \rangle\!\rangle^{\text{real}}$ holds for all $M, N \in \mathbf{T}_{\Sigma_{\text{ok}}}(Names \cup Refs)$.

If (*) would not hold for some M, N at some step of VM_{sim}, then we can show that for some $M_{\text{prev}}, N_{\text{prev}} \in \mathbf{T}_{\Sigma_{\text{ok}}}(Names \cup Refs)$ the claim (*) would not hold at the previous step. We will not present the full analysis of cases in this paper, but as an example, consider the computation step where f is $\bar{H}$ and the argument $R_1.\text{v}$ is a triple (x, y, z). If M and N do not refer to any newly created entries of T_{val} or T_{ct}, then we can set $M_{prev} = M$ and $N_{prev} = N$. Otherwise we obtain M_{prev} and N_{prev} by replacing the new entries in M and N as follows. Let $M_e = \text{Enc}(y, x, z)$. We consider four possibilities, depending on whether $T_{\text{ct}}[M_e].thd$ is defined (1&2) or not (3&4), and whether $T_{\text{ct}}[M_e].snd$ is defined (1&3) or not (2&4). Fig. 4 outlines the replacements for references to possibly new entries in T_{ct} and T_{val}. An empty cell indicates that the reference existed before the current computation step, or the row was not created in this computation step. We refer to the description of VM_{sim} for the meaning of new rows and definitions of new names. The table in Fig. 4, describing how M_{prev} is constructed from M, should be understood as follows: if e.g. M contains the reference $\tilde{n}.\text{v}$, and the cells $T_{\text{ct}}[M_e].thd$ and $T_{\text{ct}}[M_e].snd$ are not defined (4th case), then $\tilde{n}.\text{v}$ should be substituted with the pair (x, n'), where n' is a new name.

The replacement gives us M_{prev} and N_{prev}, such that $\langle\!\langle M \rangle\!\rangle^{\text{sim}} = \langle\!\langle M_{prev} \rangle\!\rangle^{\text{sim}}$ and $\langle\!\langle N \rangle\!\rangle^{\text{sim}} = \langle\!\langle N_{prev} \rangle\!\rangle^{\text{sim}}$. The induction assumption states that $\langle\!\langle M_{prev} \rangle\!\rangle^{\text{sim}} = \langle\!\langle N_{prev} \rangle\!\rangle^{\text{sim}}$ if and only if $\langle\!\langle M_{prev} \rangle\!\rangle^{\text{real}} = \langle\!\langle N_{prev} \rangle\!\rangle^{\text{real}}$. The values $\langle\!\langle M \rangle\!\rangle^{\text{real}}$ and $\langle\!\langle N \rangle\!\rangle^{\text{real}}$ are obtained from $\langle\!\langle M_{prev} \rangle\!\rangle^{\text{real}}$ and $\langle\!\langle N_{prev} \rangle\!\rangle^{\text{real}}$ by substituting some names with (possibly more complex) values. If $\langle\!\langle M_{prev} \rangle\!\rangle^{\text{real}} = \langle\!\langle N_{prev} \rangle\!\rangle^{\text{real}}$ then also $\langle\!\langle M \rangle\!\rangle^{\text{real}} = \langle\!\langle N \rangle\!\rangle^{\text{real}}$. If $\langle\!\langle M_{prev} \rangle\!\rangle^{\text{real}} \neq \langle\!\langle N_{prev} \rangle\!\rangle^{\text{real}}$ then we also have $\langle\!\langle M \rangle\!\rangle^{\text{real}} \neq \langle\!\langle N \rangle\!\rangle^{\text{real}}$, because the structure of substituted values cannot be explored using only the function symbols in Σ_{ok}.

ref.	replacement			
$T_{ct}[M_e].snd$		$n'' \oplus z$		$n'' \oplus z$
$T_{ct}[M_e].thd$			n'	n'
$n.v$	$T_{ct}[M_e].thd$	$T_{ct}[M_e].thd$	n'	n'
$\bar{n}.v$			n'	n'
$\tilde{n}.v$	$(x, T_{ct}[M_e].thd)$	$(x, T_{ct}[M_e].thd)$	(x, n')	(x, n')
$\hat{n}.v$	$T_{ct}[M_e].snd \oplus z$	n''	$T_{ct}[M_e].snd \oplus z$	n''

Fig. 4. Replacement of new references in simulating $\bar{H}(x, y, z)$

In such manner we obtain a *bisimilarity modulo implementation* between VM_{sim} and VM even for the case where their entire databases are public. Hence also $VM_{\mathsf{sim}} \approx_{tr} VM$. □

5   Related Work

The implementation of cryptographic primitives is a certain case of *process refinement*. While various aspects of refinement have been explored [3,19,18], mostly concerned with the refinement of possible behaviors of a process, the work reported in this paper has primarily been inspired by the notions of universal composability [7] and (black-box) reactive simulatability [17], both originating in the computational model of cryptography. The notion of indifferentiability by Maurer et al. [15] is similar. These definitions have been recently carried over to the symbolic model by Delaune et al. [9]. In all these definitions, there is a notion of two interfaces — one for the "legitimate" user and one for the adversary — of the process under investigation. Two processes can be equivalent (or in the refinement relation) only if the user's interface stays the same. The adversary's interface can change and a simulator process is used to translate between different interfaces. This is in contrast to our problem, where the user's interface is also naturally considered as changing, and the replacement of the primitive with the implementation is more invasive for the user process. While we think that the definition of secure implementation could be based on the notion of strong simulatability by Delaune et al. [9], the setup would be less natural and possibly the virtual machine process VM has to be included even in the definition.

The question of secure protocol composition is related to the issues of implementability of abstract processes or primitives. In recent work, Ciobâcă and Cortier [8] give sufficient conditions for the security of composition of two protocols using arbitrary primitives to follow from the security of stand-alone protocols. Interestingly, they have a similar restriction on primitives used by different protocols — the sets of primitives have to be disjoint.

Regarding the study of implementability of primitives, it is worth mentioning that in the same paper where they introduced applied pi-calculus, Abadi and Fournet [2, Sec 6.2] also considered an implementation of the MAC primitive, inspired from the HMAC construction [5]. Still, the construction is *ad hoc* and puts restrictions on how the process uses certain values.

6 Conclusions

We have explored the notion of securely implementing a cryptographic primitive in the symbolic model, and presented definitions that are more general and convenient to use than definitions that could be obtained from the application of existing treatment of process refinement and simulation. We have shown the usefulness of the proposed definition by demonstrating a secure implementation for the randomized symmetric encryption primitive. Future work in this topic would involve a systematic treatment of the implementability of common cryptographic primitives from each other. Obtained reductions and simulations may also give new insights to the security proofs in the computational model. The first step in this direction would be the analysis of the Luby-Rackoff construction [14] for constructing pseudorandom permutations ("deterministic" symmetric encryption, where $\mathsf{Enc}(k, \cdot)$ and $\mathsf{Dec}(k, \cdot)$ are inverses of each other) from random functions (modeled in the symbolic model as hashing) and exclusive or. Another line of future work is the application of the results of this paper, as well as [9], to the analysis of complex protocols.

References

1. Proceedings of the 23rd IEEE Computer Security Foundations Symposium, CSF 2010, Edinburgh, United Kingdom, July 17-19. IEEE Computer Society, Los Alamitos (2010)
2. Abadi, M., Fournet, C.: Mobile values, new names, and secure communication. In: POPL, pp. 104–115 (2001)
3. Aceto, L., Hennessy, M.: Towards action-refinement in process algebras. Inf. Comput. 103(2), 204–269 (1993)
4. Backes, M., Pfitzmann, B., Waidner, M.: A Universally Composable Cryptographic Library. In: Proceedings of the 10th ACM Conference on Computer and Communications Security, Washington, DC. ACM Press, New York (2003); Extended version available as Report 2003/015 of Cryptology ePrint Archive
5. Bellare, M., Canetti, R., Krawczyk, H.: Keying Hash Functions for Message Authentication. In: Koblitz, N. (ed.) CRYPTO 1996. LNCS, vol. 1109, pp. 1–15. Springer, Heidelberg (1996)
6. Blanchet, B., Podelski, A.: Verification of Cryptographic Protocols: Tagging Enforces Termination. In: Gordon, A.D. (ed.) FOSSACS 2003. LNCS, vol. 2620, pp. 136–152. Springer, Heidelberg (2003)
7. Canetti, R.: Universally composable security: A new paradigm for cryptographic protocols. In: FOCS, pp. 136–145 (2001)
8. Ciobâca, S., Cortier, V.: Protocol composition for arbitrary primitives. In: CSF [1], pp. 322–336
9. Delaune, S., Kremer, S., Pereira, O.: Simulation based security in the applied pi calculus. In: Kannan, R., Kumar, K.N. (eds.) FSTTCS. LIPIcs, vol. 4, pp. 169–180. Schloss Dagstuhl - Leibniz-Zentrum fuer Informatik (2009)

10. Dolev, D., Yao, A.C.-C.: On the Security of Public Key Protocols. IEEE Transactions on Information Theory 29(2), 198–207 (1983)
11. Dworkin, M.: Recommendation for Block Cipher Modes of Operation. NIST Special Publication 800-38A (2001)
12. Goldwasser, S., Micali, S.: Probabilistic Encryption. Journal of Computer and System Sciences 28(2), 270–299 (1984)
13. Muñiz, M.G., Laud, P.: On the (Im)possibility of Perennial Message Recognition Protocols without Public-Key Cryptography. In: 26th ACM Symposium On Applied Computing, vol. 2, pp. 1515–1520 (March 2011)
14. Luby, M., Rackoff, C.: How to construct pseudorandom permutations from pseudorandom functions. SIAM J. Comput. 17(2), 373–386 (1988)
15. Maurer, U.M., Renner, R., Holenstein, C.: Indifferentiability, Impossibility Results on Reductions, and Applications to the Random Oracle Methodology. In: Naor, M. (ed.) TCC 2004. LNCS, vol. 2951, pp. 21–39. Springer, Heidelberg (2004)
16. Milner, R.: Functions as processes. Mathematical Structures in Computer Science 2(2), 119–141 (1992)
17. Pfitzmann, B., Waidner, M.: A model for asynchronous reactive systems and its application to secure message transmission. In: IEEE Symposium on Security and Privacy, pp. 184–200 (2001)
18. Reeves, S., Streader, D.: Comparison of Data and Process Refinement. In: Dong, J.S., Woodcock, J. (eds.) ICFEM 2003. LNCS, vol. 2885, pp. 266–285. Springer, Heidelberg (2003)
19. Roggenbach, M.: CSP-CASL - a new integration of process algebra and algebraic specification. Theor. Comput. Sci. 354(1), 42–71 (2006)
20. Ryan, M.D., Smyth, B.: Applied pi calculus. In: Cortier, V., Kremer, S. (eds.) Formal Models and Techniques for Analyzing Security Protocols. IOS Press, Amsterdam (2010)
21. Yao, A.C.: Theory and applications of trapdoor functions (extended abstract). In: 23rd Annual Symposium on Foundations of Computer Science, Chicago, Illinois, pp. 80–91. IEEE Computer Society Press, Los Alamitos (1982)
22. Yilek, S.: Resettable Public-Key Encryption: How to Encrypt on a Virtual Machine. In: Pieprzyk, J. (ed.) CT-RSA 2010. LNCS, vol. 5985, pp. 41–56. Springer, Heidelberg (2010)

Model Checking Using SMT and Theory of Lists

Aleksandar Milicevic[1] and Hillel Kugler[2]

[1] Massachusetts Institute of Technology (MIT), Cambridge, MA, USA
aleks@csail.mit.edu
[2] Microsoft Research, Cambridge, UK
hkugler@microsoft.com

Abstract. A main idea underlying bounded model checking is to limit the length of the potential counter-examples, and then prove properties for the bounded version of the problem. In software model checking, that means that only program traces up to a given length are considered. Additionally, the program's input space must be made finite by defining bounds for all input parameters. To ensure the finiteness of the program traces, these techniques typically require that all loops are explicitly unrolled some constant number of times. Here, we show how to avoid explicit loop unrolling by using the SMT Theory of Lists to model feasible, potentially unbounded program traces. We argue that this approach is easier to use, and, more importantly, increases the confidence in verification results over the typical bounded approach. To demonstrate the feasibility of this idea, we implemented a fully automated prototype software model checker and verified several example algorithms. We also applied our technique to a non software model-checking problem from biology – we used it to analyze and synthesize correct executions from scenario-based requirements in the form of Live Sequence Charts.

1 Introduction

We present a finite-state model-checking technique, based on satisfiability solving, that does not require the user to explicitly bound the length of the search traces. We use the SMT *Theory of Lists* [7] to model potentially infinite search traces. A benefit of this approach is that it does not require providing the number of loop unrollings. Similarly, when trying to solve a planning problem, we do not have to specify the maximum number of steps needed to solve the problem. This way, we can achieve most of the benefits of the unbounded case. Unfortunately, in some cases our approach cannot prove that no counter-example exists (e.g., in the presence of infinite loops in the program), so it is not fully unbounded.

We use a list to model an unbounded search path. Every list element represents a single state traversed during the search. In order to find a path to an error state, we impose the following constraints on that list: (1) the first element is a valid initial state, (2) every two consecutive elements represent a valid state transition; and (3) the last element corresponds to one of the states we want to reach (*error states*). Having formulated the problem in this way, we can run an SMT solver, namely Z3 [23], to search for such a list, without constraining its length. If Z3 terminates and reports that the problem is unsatisfiable, we

M. Bobaru et al. (Eds.): NFM 2011, LNCS 6617, pp. 282–297, 2011.
© Springer-Verlag Berlin Heidelberg 2011

have proved that the error states are unreachable; otherwise, we have found a counter-example.

This idea is readily applicable to *software model checking*. In the presence of loops, program traces become infinite. A common resort is to explicitly perform *loop unrolling*, as it is the case with CBMC [10], Forge [16] and [5]. The limitation of this approach is that the number of unrollings must be specified beforehand by the user. Typically, the number of unrollings and the bounds for the input space are specified independently of each other, even though they are almost never independent in practice. For example, in order to verify the "selection sort" algorithm for arrays of length up to N, at least $N-1$ loop unrollings are needed. If the user provides a number less than $N-1$, a tool for bounded verification will typically report that no counter-example can be found within the given bounds, which may trick the user into believing that the algorithm is proven to be correct for all arrays of length up to N. With our approach, to verify the "selection sort" algorithm, the user only specifies the bound for N. Bounds for array elements are not needed in this case, so we can prove the algorithm correct for **all** integer arrays up to the given length N.

The main contributions of this paper are:

- A novel approach to model checking using SMT and the theory of lists: we explain how lists can be used to model unbounded traces;
- Application of this idea to software model checking: we present an optimized encoding of a program, and show that loops need not be explicitly unrolled;
- Execution of Live Sequence Charts case study: we analyzed scenario-based models of biological systems [19], written in the language of Live Sequence Charts (LSC) [15]. We show that declarative scenario-based specifications, written in LSC, can be translated into the logic of SMT, and an off-the-shelf solver can be used to automatically execute them.

2   Background

In order to check whether a safety property holds within some number of states k, one can define k sets of variables, one set for each state, $s_1, s_2, \cdots, s_k$, and then, as with any model-checking problem, assert that the following hold:

1. **Initial State** constraint: $\Theta(s_1)$;
2. **Transition** constraint: $\rho(s_1, s_2) \land \rho(s_2, s_3) \land \cdots \land \rho(s_{k-1}, s_k)$; and
3. **Safety Property** constraint: $\mathcal{P}(s_1) \land \mathcal{P}(s_2) \land \cdots \land \mathcal{P}(s_{k-1}) \land \neg\mathcal{P}(s_k)$.

Θ encodes constraints that must hold in the initial state; $\rho(s_{i-1}, s_i)$ is a transition function which returns **true** if and only if the system is allowed to go from state s_{i-1} to state s_i; finally, $\mathcal{P}(s_i)$ is the safety property that we want to prove. In order to find a counter-example, we assert that the safety property doesn't hold in the last state while holding in all previous states. Additionally, the transition function must hold for every two consecutive states. The conjunction of these three formulas is passed to an off-the-shelf solver, which either returns a model encoding a counter-example, or proves that the formula is unsatisfiable (meaning that the safety property is verified for the given k). This approach is commonly referred to as *bounded model checking using satisfiability solving*.

We focus on how to use the theory of lists to avoid having k copies of the state variables. The theory of lists is currently supported by many state-of-the-art SMT solvers. A description of how other theories can be used to encode programs and why that can be advantageous is presented in [5].

SMT lists are defined recursively: `List<E> = nil | cons (head: E, tail: List)`. For a given list, only two fields, `head` and `tail` are immediately accessible. In addition, predicates `is_cons` and `is_nil` are readily available to check whether a given list variable is `cons` or `nil`. As a consequence, it is not possible to directly access the list element at a given position, or immediately get the length of the list, which is inconvenient when asserting properties about lists.

3   Approach

Our approach is based on the idea of bounded model checking using satisfiability solving, except that instead of explicitly enumerating all state variables (s_1, s_2, $\cdots$, s_k), and thus bounding the length of a potential counter-example, we use only a single variable of type *List of States*. Every list element is of type *State*, which is a tuple of all variables needed to represent the problem state. We still assert the same three constraints, (1) initial state, (2) transition; and (3) safety constraint, but now in terms of a single list variable.

Expressing the initial state constraint is easy, since the first element of the list is immediately accessible. To express the other two constraints, we use an uninterpreted function accompanied with an axiom. More precisely, in order to enforce the transition constraint between every two consecutive elements of the list, we first define an uninterpreted function, named `check_tr`, that takes a list and returns a boolean value. Next we add an axiom (*transition axiom*) to assert that `check_tr` returns `true` when applied to a list if and only if every two consecutive states of that list represent a valid state transition.

A recursive definition of the transition axiom is given in Figure 1. The only case of importance is when the list argument, namely `lst`, is not `nil` and has a non-`nil` next element (`tail`). This is because we only care to assert the transition property between two consecutive elements. We do that by inlining the actual model-checking transition constraint between the current and the next list element. In addition, we have to make sure that all subsequent consecutive elements represent valid state transitions, so we recursively assert that the same `check_tr` function returns `true` for the `tail` of the given list argument.

In order to enforce the safety property on all list elements but the last one, we could similarly define another uninterpreted function and an additional axiom. However, since we already have an axiom that "traverses" the whole list, we decided to include the safety property check in the existing transition axiom. This can simply be done by checking whether the next list element (`tail(lst)`) corresponds to an error state (by inlining the *error condition*, i.e. $\neg\mathcal{P}(s_i)$). If the

DEF check_tr: StateList → bool
ASSERT FORALL lst: StateList
   IF (is_cons(lst) ∧ is_cons(tail(lst))) **THEN**
      transition_condition(head(lst), head(tail(lst))) ∧
      check_tr(tail(lst)) ∧
      IF (*error_condition(tail(lst))*)
      THEN is_nil(tail(tail(lst)))
      ELSE is_cons(tail(tail(lst)))
  :**PAT** {check_tr(lst)}

Fig. 1. Axiom for the check_tr function

DEF states: StateList
ASSERT
  is_cons(states) ∧
  initial_condition(head(states)) ∧
  check_tr(states)
CHECK

Fig. 2. SMT logic context

next element is in fact an error state, we have found a counter-example, so we force the list to end right there (i.e. its next element must be `nil`). Otherwise, we must keep searching, so the next element in the list must be `cons`.

Finally, it is important to stress the purpose of the instantiation pattern (**PAT**: {check_tr (lst)}) in the **FORALL** clause. This axiom states something about *all* lists. However, it would be impossible for the SMT solver to try to prove that the statement indeed holds for all possible lists. Instead, the common approach is to provide an instantiation pattern to basically say in which cases the axiom should be instantiated and therefore enforced by the solver. In our case, we simply say that every time we apply check_tr function to a list, the axiom must be enforced, so that the evaluation of check_tr indeed indicates whether the list satisfies both transition and safety property constraints.

The rest of the SMT logic context is given in Figure 2. It provides a generic template for model-checking problems. For a specific problem, the user only needs to define: (1) the `State` tuple (basically enumerate all state variables), (2) initial condition, (3) transition condition; and (4) error condition.

4   Applicability to Software Model Checking

4.1   The Idea

We observe a program as a traditional *Control Flow Graph* (CFG) [3]. The *state* of the execution of a program consists of the current basic block (at a given moment, the execution is exactly in a single basic block) and the evaluations of relevant program variables. The edges between the basic blocks are called *transitions*. An edge is *guarded* by a logic condition that specifies when the program execution is allowed to go from one basic block to another. The goal of model checking is to find a feasible execution *trace* (a path in the CFG) from the start node to one of the error nodes.

Programs with loops have cyclic control flow graphs, which means that some of their traces are infinite. Using unbounded lists seems like a very natural way to model program traces. Instead of truncating loops up front, we let the satisfiability solver simulate them, by effectively executing loops until the loop condition becomes `false`. Even though some traces may be infinite, the number of basic blocks is always finite, meaning that the transition condition (i.e. the logic expression that defines all valid transitions from a given state) is also finite and can be expressed in a closed form.

4.2   Formal Definitions

Program Graph (PG). We formally introduce Program Graphs, which are a variation of Control Flow Graphs.

A PG is defined over a set of typed variables Var. We will use $Eval(Var)$ to denote the set of possible evaluations of variables, $Expr(Var)$ to denote the set of all expressions over Var (e.g., constants, integer arithmetic, "select" and "store" operations over integer arrays, and boolean expressions), and $Cond(Var)$ to denote the set of all boolean expressions over Var $(Cond(Var) \subset Expr(Var))$. A PG is then defined as a tuple:

$$PG = (\mathsf{L}, \mathsf{Act}, \mathsf{Eff}, \rightarrow, l_0, \mathsf{E})$$

L is a set of program locations (corresponding to basic blocks), l_0 is the start location $(l_0 \in \mathsf{L})$ and E is a set of error locations $(\mathsf{E} \subset \mathsf{L})$. Act is a set of actions (program statements) and function $\mathsf{Eff} : \mathsf{Act} \times Eval(Var) \mapsto Eval(Var)$ defines the effects of actions on variable evaluations. Finally, $\rightarrow: \mathsf{L} \times Cond(Var) \times \mathsf{Act} \times \mathsf{L}$ is the conditional transition relation with side effects (i.e., actions assigned to it). This definition is very similar to the one presented in [6].

The semantics of the $\rightarrow$ relation is defined by the following rule

$$\frac{\eta \models g \quad \eta' = \mathsf{Eff}(\alpha, \eta)}{\langle l, \eta \rangle \xrightarrow{g:\alpha} \langle l', \eta' \rangle}$$

where the notation $l \xrightarrow{g:\alpha} l'$ is a shorthand for $(l, g, \alpha, l') \in \rightarrow$.

4.3   Example

We introduce a simple example that will be used throughout this section to explain optimizations and the actual translation to SMT logic. The code is shown in Figure 6, the algorithm is named `simpleWhile`, the corresponding CFG is shown in Figure 3(a). Blocks with grey background are simply branch conditions, and they do not modify the program state. The red block represents the error state. All steps presented here are fully automated.

4.4   Optimizing Transformations from CFG to PG

We decided to model state changes as transitions between basic blocks, and not between single statements. This is useful because it makes the traces explored by the solver much shorter. While searching for a counter-example, the solver creates a list node for every new state it explores. If every statement caused a state transition (which is what happens in reality), then the solver would have to add a new node to the list after every variable assignment, growing the list rapidly. Instead, we accumulate the effects of all statements of a basic block (by symbolically executing them) and use the resulting effect to define a single state

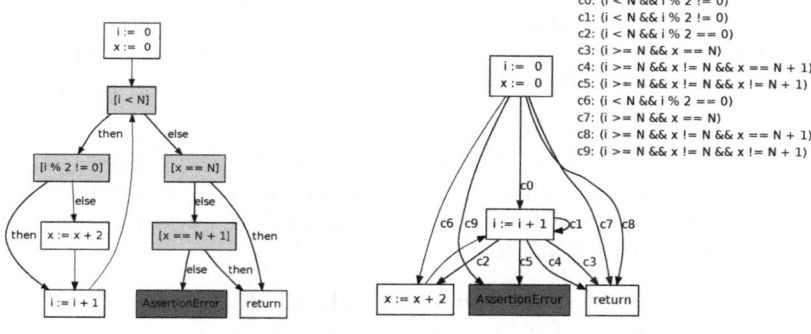

(a) Original CFG   (b) Empty blocks removed

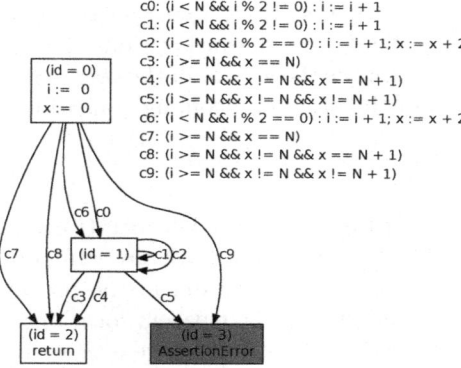

(c) Blocks without self loops eliminated

Fig. 3. Control Flow Graphs for the "SimpleWhile" example

transition. That way we enable the solver to perform more computation in every step (basically execute the entire basic block at once), thus reducing the overall number of states it has to explore, and significantly improving the solving time.

Since the solver can thus execute an entire basic block at once, we can think of the search process as a graph path finding problem: the solver is given a task of finding a path from the start block to one of the error blocks in the CFG. The search traces become sequences of basic blocks. The idea of shortening traces explored by the solver (i.e., reducing the number of basic blocks) is the basic idea behind our optimizations.

Symbolic Execution of Basic Blocks. In order to arrive at the final expression for every variable at the end of a basic block, we must execute the entire basic block symbolically. Since our goal is to formulate how variables are updated when transitioning from one basic block to another, the final expressions must be in terms of symbolic variables in the previous state. For example, the effect of the following code fragment x++; y = 2*x; x--; is x := (x+1)-1; y := 2*(x+1);. Formally, we introduce the *expression update* operator $\diamondsuit$, which takes an expression e and an action α and updates variables in e according to α:

$$e \diamondsuit \alpha = \begin{cases} e, & \text{if } \alpha = \emptyset \\ e[v_1/e_{v_1}, \cdots, v_k/e_{v_k}], & \text{if } \alpha = v_1 := e_{v_1}, \cdots, v_k := e_{v_k} \end{cases}$$

In short, we start with an empty action α, we go through all the basic block instructions of type $v = e$, and for each of them we add $v := e \diamondsuit \alpha$ to α (overwriting the previous assignment, if one existed).

Optimization 1: Empty Location Removal. We do not want the solver to grow the list by exploring basic blocks that do not change the state. Therefore, the first optimization step takes the original CFG and removes all locations that do not have any actions that modify the program state (so-called *empty locations*). For such a location l_x, selected for removal, every incoming transition is split into several new transitions so that, after the transformation, each of the l_x's parents points to all of the l_x's successor locations. The guards of the newly created transitions are the same as the guards of the original outgoing transitions conjoined with the guard of the original incoming transition:

$$(\forall l_p \xrightarrow{g_p} l_x) \ (\forall l_x \xrightarrow{g_s} l_s) \ l_p \xrightarrow{g_p \wedge g_s}{}' l_s$$

Optimization 2: Non-looping Location Elimination. Here, the idea is to completely remove basic blocks that do not have any self-loops. We can split the incoming transitions, similarly to what we did in the previous step. However, we cannot simply move the actions to their parent locations, since they are not to be executed every time the parent locations are executed. The solution is to switch from CFG to PG, since program graphs allow us to associate actions with transitions instead of locations, which is exactly what we need here: we will add the actions of the location to be removed to newly created transitions.

Before this optimization step is performed, the CFG has to be converted to its corresponding PG. This can trivially be done by moving actions associated with states to their incoming transitions. Next, we iteratively keep eliminating locations that do not contain any self-loops (*non-looping locations*) until only locations with self-loops are left in the graph. Elimination of a non-looping location l_x involves three steps: (1) splitting the incoming transitions (similarly as before); (2) merging their actions; and (3) updating their guards:

$$elim((L, Act, \mathsf{Eff}, \rightarrow, l_0, E), l_x) \mapsto (L \setminus \{l_x\}, Act, \mathsf{Eff}, \rightarrow', l_0, E)$$

$$(\forall l_p \xrightarrow{g_1 : \alpha_1} l_x) \ (\forall l_x \xrightarrow{g_2 : \alpha_2} l_s) \ l_p \xrightarrow{g_0 : \alpha_0}{}' l_s \text{ , where } \alpha_0 = \alpha_1 \circ \alpha_2, \ g_0 = g_1 \wedge (g_2 \diamondsuit \alpha_1)$$

We have introduced another operator, the *action merge* operator $\circ$. The idea of merging two actions α_1 and α_2 is to get a new action whose effect is going to be the same as the final effect of α_1 and α_2 when executed in that order on any variable evaluation η: $\mathsf{Eff}(\alpha_1 \circ \alpha_2, \eta) \mapsto \mathsf{Eff}(\alpha_2, \mathsf{Eff}(\alpha_1, \eta))$. In terms of merging actions α_1 and α_2, expressions in α_2 refer to the state after α_1 has been executed,

$initial\_condition \quad \equiv \quad$ head(statesList).$stateId = 0 \wedge$ head(statesList).$x = 0 \wedge$ head(statesList).$i = 0$

$transition\_condition$

$\equiv$ **IF** head(lst).$stateId = 0$ **THEN**
    IF $i < N \wedge i \% 2 \neq 0$ **THEN**
        head(tail(lst)).$stateId = 1 \wedge$ head(tail(lst)).$i =$ head(lst).$i + 1$
    ELSE IF $i < N \wedge i \% 2 = 0$ **THEN**
        head(tail(lst)).$stateId = 1 \wedge$ head(tail(lst)).$x =$ head(lst).$x + 2 \wedge$ head(tail(lst)).$i =$ head(lst).$i + 1$
    ELSE IF $i \geq N \wedge x = N$ **THEN**
        head(tail(lst)).$stateId = 2$
    ELSE IF $i \geq N \wedge x \neq N \wedge x = N + 1$ **THEN**
        head(tail(lst)).$stateId = 2$
    ELSE
        head(tail(lst)).$stateId = 3$
ELSE IF head(lst).$stateId = 1$ **THEN**
    IF $i < N \wedge i \% 2 \neq 0$ **THEN**
        head(tail(lst)).$stateId = 1 \wedge$ head(tail(lst)).$i =$ head(lst).$i + 1$
    IF $i < N \wedge i \% 2 = 0$ **THEN**
        head(tail(lst)).$stateId = 1 \wedge$ head(tail(lst)).$x =$ head(lst).$x + 2 \wedge$ head(tail(lst)).$i =$ head(lst).$i + 1$
    ELSE IF $i \geq N \wedge x = N$ **THEN**
        head(tail(lst)).$stateId = 2$
    ELSE IF $i \geq N \wedge x \neq N \wedge x = N + 1$ **THEN**
        head(tail(lst)).$stateId = 2$
    ELSE
        head(tail(lst)).$stateId = 3$

$error\_condition \quad \equiv \quad$ head(lst).$stateId = 3$

Fig. 4. Translation of the CFG shown in Figure 3(c) to SMT logic

therefore, it would be incorrect to simply append α_2 to α_1. Instead, α_2 has to be updated first ($\star$ operator in the listing below) so that for each variable assignment $v_2 := e_2$ in α_2, expression e_2 is updated with respect to α_1 ($e_2 \diamond \alpha_1$). Once α_2 has been updated, the result of the merge operation is the updated α_2 appended with variable assignments in α_1 that do not already appear in it. A similar intuition holds for updating transition conditions, it is not correct to simply conjoin g_1 and g_2, instead, g_2 has to be updated first.

$$\alpha \star \beta = \begin{cases} \emptyset, & \text{if } \alpha = \emptyset \\ \{v := e_v \diamond \beta\} \cup (\alpha \setminus \{v := e_v\}) \star \beta, & \text{if } \exists (v := e_v) \in \alpha \end{cases}$$
$$\alpha_1 \circ \alpha_2 = (\alpha_1 \setminus \alpha_2) \cup (\alpha_2 \star \alpha_1)$$

Figure 3(c) shows the PG for the "Simple While" example, after all non-looping locations have been eliminated. First, the action `i:=i+1` from the state with `id=1` is moved to its incoming transitions `c0`, `c1`, `c2`, and `c6`. Next, the location with `x:=x+2` action is eliminated, and as a result, edges `c2` and `c6` are redirected and updated to include the `x:=x+2` action.

4.5 Translation of PG to SMT

Figure 4 shows the actual translation of the PG in Figure 3(c) to *initial*, *transition*, and *error* conditions, needed for the template SMT context given in Figures 1 and 2. The translation is pretty straightforward. An extra field, *stateId*, is first added to the state tuple to identify the current location. In this case, the state consists of 3 variables: *stateId*, x, i (the variable N is constant so it is kept outside of the state tuple). The *initial_condition* is a direct representation of the state in the entry block. The *error_condition* is also easy to formulate, since all error states are explicitly known upon the CFG creation. The *transition_condition* contains two big nested *if-then-else* statements. The outer *if-then-else* has a case for every non-leaf location. Inside each such case, there is an inner *if-then-else* that has a case for each of the location's outgoing transition, where it specifies how the state is updated when that transition is taken.

Finally, we need to define the set of possible values for the input variable N (e.g., $N > 0 \wedge N \leq 10$). This additional constraint is necessary because integers are unbounded in SMT theories. Recall that this technique effectively simulates program loops inside SMT. Since the value of N influences the number of loop iterations, if a bound is not provided for N, the solver will try to simulate the loop for all possible values of N, and thus never terminate.

5   Execution of Live Sequence Charts

In this section we show how this model-checking technique can be applied to a non-trivial biological model-checking problem. We use the theory of lists to encode Live Sequence Charts and then run Z3 to analyze and execute them.

5.1   Example

We will use an example to briefly introduce LSCs and their semantics. Figure 5(a) shows the specification of the interaction between a cell phone and the user. A single LSC consists of a number of *Instances* passing messages between them. Instances either belong to the *System* or the *Environment*. Every Instance has an associated *timeline* (represented as vertical bars) which is used to impose the ordering between messages. The upper portion of the chart (bordered with a dotted line) is called the *Pre-Chart*, whereas the rest of the chart is called the *chart body*. Every chart is initially *inactive*. It becomes *active* when its Pre-Chart is satisfied, i.e., when messages that appear in the Pre-Chart occur in the specified order. The semantics of LSCs require that once a chart becomes active, it must finish its execution according to the specification in its body, when it becomes *closed*. The chart specifies only partial ordering of the message occurrences: only messages that have a common timeline as either source or target must happen in the given ordering; messages that do not have a timeline in common may appear in an arbitrary order.

In terms of the example in Figure 5(a), once the chart becomes active, as a result of `open` occuring, there are 3 possible *valid* executions: (1) `SetColor(Grey)`, `SetColor(Green)`, `activate`, (2) `SetColor(Grey)`, `activate`, `SetColor(Green)`, and (3) `activate`, `SetColor(Grey)`, `SetColor(Green)`. Note that it is not allowed that message `SetColor(Green)` appears before `SetColor(Grey)`, that would be considered as an immediate violation of the specification. Also note that it is allowed that some other messages, not shown in this chart are sent at any point during the execution of this chart. The chart's body specifies only messages that must happen, and partial ordering between them, it does not forbid other messages. This way, a formal contract is established saying that every time the user opens the cover (message `open` is sent from `User` to `Cover`), the cell phone must respond as specified in the chart's body.

5.2   Motivation

Single step is defined as a single message sent by the System that does not cause an immediate violation. *Super step* is a sequence of messages that drives all active charts to their completion, without causing any violations. It is allowed for a super step to activate some new charts along the way, but at the end of it, no charts must be active. For example, consider another scenario given in Figure 5(b). The message `activate` activates the `antenna_act` chart. Its body contains a single conditional element that states that the color must be `Grey` after the chart is activated. Adding this additional scenario rules out the first of the three valid executions of the "open cover" scenario given above.

We describe our solution for encoding of LSCs into the logic of SMT with the theory of lists, which allows for using the Z3 SMT solver to automatically find all valid super steps from a given point in the execution of the system. Here we illustrate the applicability and usefulness of our technique to this problem; a more detailed discussion and formal translation is not presented due to space limitations and will be reported in a future paper.

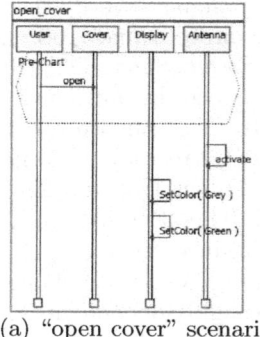   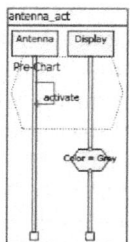

(a) "open cover" scenario   (b) "antenna activated" scenario

Fig. 5. The cell phone LSC example

5.3   Solution

We formulate the problem of finding a super step as a model-checking problem. For every Instance, we keep an integer variable to keep track of its *location* (a point on its timeline) in the current state of the execution. We also maintain variables for object properties (e.g., `Color` as in the example) and a single variable for the message sent by the system in the current step. The initial state is explicitly given and consists of current locations of all instances and evaluation of all properties. In the transition constraint, we let the solver non-deterministically pick a message to be sent by the system and based on that decision we specify how the rest of the state should be updated. We assert that the chosen message must be *enabled* at the current step (i.e., that at least one Instance is at a location where this message can be sent from) and that it must not cause any violations in other charts. The safety property that we want the solver to prove is that the state where all charts are closed can never be reached from the initial state. If the solver proves this property, that means that no valid super step exists. Otherwise, the solver will come back with a counter-example that contains a list of state changes, which lets us decode which message is sent at each step.

Formulating this problem using the theory of lists seems very convenient, since the number of steps needed to find a counter-example is not known in advance. We analyzed several models of biological systems [2] and were able to find valid super steps for systems with more than ten charts within seconds.

6   Evaluation and Results

We implemented a fully automated prototype model checker for Java programs to evaluate the idea of using the SMT theory of lists to model program traces. Currently, we support only a subset of Java programs. We used this tool to verify the correctness of several algorithms. We also applied this technique to solve the *Rush Hour* puzzle [1]. All experiments were conducted on a 64-bit Intel Core Duo CPU @2.4GHz box, with 4GB of RAM, running 32-bit Windows Vista.

Verifying Simple Algorithms. We used this technique to verify the "Simple-While" algorithm, two sorting algorithms, and the integer square root algorithm from Carroll Morgan's book *Programming with Specifications* [22] (Figure 6). We present the comparison of verification times between the optimized and non-optimized translation for several different bounds. We compare our tool to a representative tool from the bounded model-checking category – JForge [16, 26], and a finite model checker that doesn't require explicit loop unrolling – Java PathFinder [25]. The results are shown in Figure 7. The "Related Work" section describes these tools in detail and discusses the obtained results.

Non-monotonicity of some of the graphs in Figure 7 can be explained by the nature of satisfiability solvers. The solving time is highly dependent on internal heuristics (e.g., [24, 20]), so it can happen that a larger problem is solved faster simply because the heuristics worked better (for example, it happened that a large portion of the search space was pruned early on).

Finally, this approach performs quite efficiently when a counter-example exists. For all of the presented benchmarks, our tool was able to find different (manually introduced) bugs within seconds.

Solving the Rush Hour Puzzle. *RushHour* is a well known puzzle where the goal is to get the designated car (the red car in Figure 9) out of the traffic jam. This puzzle is easily expressible as a model-checking problem: the initial state is the given configuration of cars at the starting point, the transition function constrains the allowed movements of the cars so that they do not crash or go over each other, and the safety property is that the red car can never reach the far right side of the stage. If we find a counter-example to this model-checking problem, we have found the way to get the red car out of the jam.

We took several puzzles from [1] and compared the execution times of the two approaches: bounded (the case when we know the optimal number of steps) and unbounded with lists (Figure 8). SMT solvers are optimized to deal with large flat formulas, so the fact that the bounded encoding currently performs better does not come as a surprise. However, we were able to solve the most difficult puzzles (e.g., Jam 38-40 require more than thirty steps) within a minute.

```
void simpleWhile(int N) {
  int x = 0, i = 0;
  while (i < N) {
    if (i % 2 == 0)
      x += 2;
    i++;
  }
  assert x == N || x == N + 1;
}
```

```
void selectSort(int [] a, int N) {
  for (int j=0; j<N-1; j++) {
    int min = j;
    for (int i=j+1; i < N; i++)
      if (a[min] > a[i]) min = i;
    int t = a[j]; a[j] = a[min]; a[min] = t;
  }
  for (int j=0; j<N-1; j++)
    assert a[j] <= a[j+1];
}
```

```
void bubbleSort(int [] a, int N) {
  for (int j=0; j<N-1; j++)
    for (int i=0; i<N-j-1; i++)
      if (a[i] > a[i+1]) {
        int t = a[i];
        a[i] = a[i+1];
        a[i+1] = t;
      }
  for (int j=0; j<N-1; j++)
    assert a[j] <= a[j+1];
}
```

```
int intSqRoot(int N) {
  int r = 1, q = N;
  while (r+1 < q) {
    int p = (r+q) / 2;
    if (N < p*p) q = p;
    else r = p;
  }
  assert r*r <= N && (r+1)*(r+1)>N;
  return r;
}
```

Fig. 6. Benchmark Algorithms

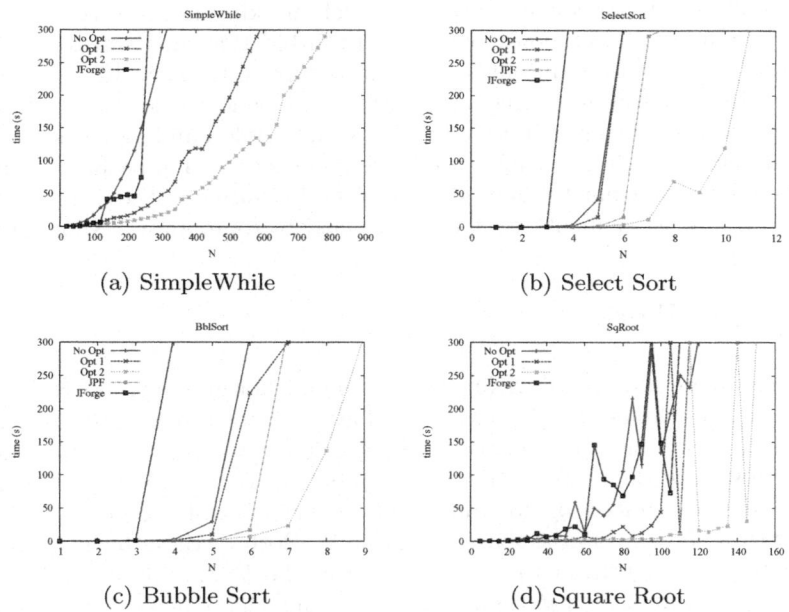

(a) SimpleWhile

(b) Select Sort

(c) Bubble Sort

(d) Square Root

Fig. 7. Benchmark Results

This problem is quite different from the software model-checking problems, because at every step, there are typically several available valid moves, so at every step, the solver has to non-deterministically decide which move to take in order to finally reach an error state (this never happens in software model checking if programs are deterministic). This puzzle is a typical example of how

	B	U
Jam 25	1.20s	1.88s
Jam 30	1.21s	2.17s
Jam 38	4.47s	36.6s
Jam 39	1.90s	14.66s
Jam 40	6.31s	17.89s

Fig. 8. RushHour benchmark (B – Bounded, U – unbounded)   **Fig. 9.** RushHour instance

this technique can be used to solve planning problems without bounding the number of steps in advance.

One limitation of our current implementation is that it is not able to prove it if the solution does not exist. The solver gets stuck exploring the same states over and over again (e.g., moving the red car back and forth between the neighboring cells). However, if a solution exists, this problem is not manifested. Also note that this does not happen in software model checking if the target program always terminates. An obvious solution is to forbid the same states to appear in the states list. This additional constraint is expressible in SMT logic, but in practice it does not perform that well. Instead, we believe that the SMT solver could be tweaked so that it internally knows that while building the states list it should never include the same state twice in a single search path. It would be very efficient to implement this inside the solver, because the state is represented explicitly inside list elements, so it would be easy to compare states for equality.

7   Related Work

Model checking was originally defined as a technique for proving properties about Finite State Machines (FSM) [12]. The pioneering tools had used an explicit representation of the entire state graph, which led to what is known as the *state explosion problem.* To mitigate that problem, Binary Decision Diagrams (BDD) were introduced by McMillan [14] to symbolically represent a set of states with a single propositional logic formula. Both of these techniques used a custom search algorithm to explore paths in the FSM. Infinite traces were supported by computing a *fixpoint,* i.e., not visiting the same state twice on the same search path. The growing popularity and efficiency of satisfiability solvers had influenced another branch of model checking, called *Bounded Model Checking* [8,9,11], which significantly improved the scalability of model checking. The idea was to bound the traces by unrolling the FSM for some number of times k. As a result, the whole problem could be formulated as a single propositional formula, solvable by off-the-shelf SAT solvers. On the other side, Counter-Example Guided Abstraction Refinement [13] was developed to deal with infinite state machines. In comparison, our approach lies somewhere between bounded and unbounded finite state model checking: in many cases, we achieve benefits of the unbounded method, but in some, our tool cannot prove the absence of counter-examples.

JForge is a bounded software model checker that uses SAT. It requires the user to bound the program input space by specifying the bit-width for integers, in addition to providing the number of loop unrollings. In all benchmarks, we used the minimal bit-width needed to represent the bound N, and an appropriate number of loop unrollings needed to verify the code for the given input size. JForge enumerates all integers within the given bit-width so that it has the explicit representation of the whole universe. That turns out to be the reason why JForge does not perform as well as our tool in these benchmarks.

Alloy [17] is a bounded model finder that can be used to search for traces (sequences of events) that satisfy certain logic property, but it also requires that the number of events is specified in advance.

JPF [25] is an extensible plaform for running model checkers for Java programs. The explicit-state version of JPF directly executes the program on all possible inputs, whereas we translate the program into logic and formulate a satisfiability problem. We present results for JPF only for the two sorting algorithms. In the other two examples, JPF is a clear winner. However, the sorting examples show the case where the ability of our tool to symbolically represent array elements brings a significant advantage. To verify the sorting algorithms using JPF on arrays of size exactly n, we ran the algorithm on all possible arrays of size n whose elements are between 1 and n, which turned out to be very expensive in terms of both memory and time. Symbolic JPF [4] can treat the variables symbolically, but it currently does not support arrays.

Armando et al. [5] present a bounded software model-checking technique (requires explicit loop unrolling) based on SMT, and report significant improvement over the traditional SAT-based technique. Other techniques for unbounded model checking with satisfiability solving (e.g., [18, 21]) iteratively invoke the solver until they reach a fixpoint, whereas our approach translates the whole problem into a single formula.

8   Conclusion

We have presented a novel technique for finite-state unbounded model checking using the theory of lists and satisfiability solving. Our technique is a finite-state technique, in the sense that it requires explicit bounds on certain parts of the input state (e.g., those that influence the length of the state machine traces). On the other hand, it can prove properties for infinite-state systems, as shown for the "sorting" examples. We have shown the generic pattern for solving model-checking problems, and also provided detailed explanation of how it can be applied to software model checking in particular. The results of the comparison with some of the existing tools for software model checking seem promising. The applicability of this method to analyzing and executing scenario-based models in the form of Live Sequence Charts seems to have a stong potential and will enable efficiently supporting a larger subset of the LSC language including arithmetic operations that are more natural to handle using SMT solvers.

References

1. Rush Hour Puzzle, http://www.puzzles.com/products/rushhour.htm
2. Microsoft Research Cambridge, Synthesizing Biological Theories (2011), http://research.microsoft.com/SBT/
3. Aho, A.V., Lam, M.S., Sethi, R., Ullman, J.D.: Compilers: Principles, Techniques, and Tools, 2nd edn. (August 2006)
4. Anand, S., Pasareanu, C.S., Visser, W.: JPF–SE: A Symbolic Execution Extension to Java PathFinder. In: Grumberg, O., Huth, M. (eds.) TACAS 2007. LNCS, vol. 4424, pp. 134–138. Springer, Heidelberg (2007)
5. Armando, A., Mantovani, J., Platania, L.: Bounded model checking of software using SMT solvers instead of SAT solvers. STTT 11(1) (2009)
6. Baier, C., Katoen, J.-P.: Principles of Model Checking (Representation and Mind Series) (2008)
7. Barrett, C., Ranise, S., Stump, A., Tinelli, C.: The Satisfiability Modulo Theories Library, SMT-LIB (2008), www.SMT-LIB.org
8. Biere, A., Cimatti, A., Clarke, E.M., Zhu, Y.: Symbolic Model Checking without BDDs. In: TACAS (1999)
9. Clarke, E., Biere, A., Raimi, R., Zhu, Y.: Bounded Model Checking Using Satisfiability Solving. In: Formal Methods in System Design (2001)
10. Clarke, E., Kroening, D., Lerda, F.: A tool for checking ANSI-C programs. In: Jensen, K., Podelski, A. (eds.) TACAS 2004. LNCS, vol. 2988, pp. 168–176. Springer, Heidelberg (2004)
11. Clarke, E., Kroening, D., Yorav, K.: Behavioral Consistency of C and Verilog Programs Using Bounded Model Checking. In: DAC (2003)
12. Clarke, E.M., Emerson, E.A., Sistla, A.P.: Automatic verification of finite-state concurrent systems using temporal logic specifications. ACM Transactions on Programming Languages and Systems (1986)
13. Clarke, E.M., Grumberg, O., Jha, S., Lu, Y., Veith, H.: Counterexample-Guided Abstraction Refinement. In: Emerson, E.A., Sistla, A.P. (eds.) CAV 2000. LNCS, vol. 1855, pp. 154–169. Springer, Heidelberg (2000)
14. Clarke, E.M., McMillan, K.L., Zhao, X., Fujita, M., Yang, J.: Spectral Transforms for Large Boolean Functions with Applications to Technology Mapping. In: DAC (1993)
15. Damm, W., Harel, D.: LSCs: Breathing Life into Message Sequence Charts. In: Formal Methods in System Design (1998)
16. Dennis, G.: A Relational Framework for Bounded Program Verification. PhD thesis, Massachusetts Institute of Technology, Advised by Daniel Jackson (2009)
17. Jackson, D.: Software Abstractions: Logic, language, and analysis. MIT Press, Cambridge (2006)
18. Kang, H.-J., Park, I.-C.: SAT-based unbounded symbolic model checking. In: DAC (2003)
19. Kugler, H., Segall, I.: Compositional Synthesis of Reactive Systems from Live Sequence Chart Specifications. In: Kowalewski, S., Philippou, A. (eds.) TACAS 2009. LNCS, vol. 5505, pp. 77–91. Springer, Heidelberg (2009)
20. Marques-Silva, J.: The impact of branching heuristics in propositional satisfiability algorithms. In: Barahona, P., Alferes, J.J. (eds.) EPIA 1999. LNCS (LNAI), vol. 1695, pp. 62–74. Springer, Heidelberg (1999)
21. McMillan, K.L.: Applying SAT Methods in Unbounded Symbolic Model Checking. In: Brinksma, E., Larsen, K.G. (eds.) CAV 2002. LNCS, vol. 2404, p. 250. Springer, Heidelberg (2002)

22. Morgan, C.: Programming from specifications (1990)
23. Moura, L.D., Bjrner, N.: Z3: An Efficient SMT Solver. In: Ramakrishnan, C.R., Rehof, J. (eds.) TACAS 2008. LNCS, vol. 4963, pp. 337–340. Springer, Heidelberg (2008)
24. Piskac, R., Moura, L., Bjørner, N.: Deciding Effectively Propositional Logic Using DPLL and Substitution Sets. J. Autom. Reason. 44, 401–424 (2010)
25. Visser, W., Havelund, K., Brat, G.: Model Checking Programs. In: ASE (2000)
26. Yessenov, K.: A light-weight specification language for bounded program verification. Master's thesis, Advised by Daniel Jackson (May 2009)

Automated Test Case Generation with SMT-Solving and Abstract Interpretation

Jan Peleska, Elena Vorobev, and Florian Lapschies

Department of Mathematics and Computer Science,
University of Bremen, Germany
{jp,elenav,florian}@informatik.uni-bremen.de

Abstract. In this paper we describe an approach for automated model-based test case and test data generation based on constraint types well known from bounded model checking. Our main contribution consists of a demonstration showing how this process can be considerably accelerated by using abstract interpretation techniques for preliminary explorations of the model state space. The techniques described support models for concurrent synchronous reactive systems under test with clocks and dense-time.

1   Introduction

Motivation and Overview. In this paper we present results for model-based test case and test data generation for concurrent real-time systems. The expected behavior of the system under test is specified by a model whose abstract syntax representation is used to derive suitable *symbolic test cases* which are represented as logical constraints G over model computations. The term "symbolic" is used in the sense that at this stage no concrete test data exists yet in order to stimulate a model computation satisfying G. The concrete test data is gained by handling constraint satisfaction problems (CSPs) of the type

$$tc(c, G) \equiv_{\text{def}} \bigwedge_{i=0}^{c-1} \Phi(\sigma_i, \sigma_{i+1}) \wedge G(\sigma_c) \tag{1}$$

These CSPs are well-known from the field of bounded model checking: σ_0 is a pre-state from where a model exploration should start. $\Phi(\sigma_i, \sigma_{i+1})$ denotes the transition relation, represented as a first order predicate relating pre-states σ_i to possible post-states σ_{i+1}. $G(\sigma_c)$ is a predicate representing the symbolic test case, so solving $tc(c, G)$ yields test data to satisfy G by performing c transitions from the pre-state σ_0.

In the general case G will not only refer to the target state σ_c but to the complete computation $\sigma_0, \ldots, \sigma_c$. By introducing additional *observer components*, however, this more general situation can be reduced to the one captured in (1): the observer runs concurrently with the model and checks whether $G(\sigma_0, \ldots, \sigma_c)$ is satisfied. If this is the case the observer performs an auxiliary transition to a

M. Bobaru et al. (Eds.): NFM 2011, LNCS 6617, pp. 298–312, 2011.

target location ℓ indicating "$G(\sigma_0, \ldots, \sigma_c)$ is satisfied". Then the test case may be re-formulated to "ℓ shall be reached after $c + 1$ transitions". In practice, however, this introduction of observers is only infrequently required, because most test cases can be identified by means of predicates on a model state σ_c alone.

For HW/SW integration and system integration testing it is desirable to find the shortest path from σ_0 to a state satisfying G. Therefore it is tried to consecutively solve $tc(1, G), tc(2, G), \ldots$, and stop as soon as a c has been found for which solution of $tc(c, G)$ exists. Given a collection of test cases $G_1, \ldots, G_k$ it is desirable to find a model computation $\sigma_0, \ldots, \sigma_n$ where all of these G_i are covered (not necessarily in a given order). The existence of such a computation has the advantage that the SUT will be driven into a larger number of internal states, as when testing only one G_i at a time and resetting the SUT in between, since this increases the confidence into the SUT reliability. Moreover, SUT resets are often time consuming when testing integrated HW/SW systems. Therefore G is usually specified as the disjunction of the remaining goals to be covered, and every G_i that is reached is removed from this disjunction. If, however, a test case G_i cannot be covered from a given pre-state σ_p within an acceptable number of steps, it is advisable to perform *backtracking* to a suitable state σ_{p-q} from where it is less time consuming for the SMT solver to reach this goal (recall that in general, the running time of the SMT solver depends exponentially on the number c in formula (1), specifying how many times the transition relation is unrolled). Finding this state represents another challenge, because trying to solve the CSP from some σ_{p-q} where G_i cannot be reached within the given limit of transitions wastes time to an extent where backtracking no longer offers any advantage. For tackling CSPs of the type (1) we use an SMT solver which is sketched in Section 2.

Main Contribution. We present an *abstract interpretation algorithm* for concurrent synchronous real-time models which, given an initial state σ_0 and a test case goal G returns a natural number c_0 such that it is guaranteed that no solution of $tc(i, G)$ exists for $0 < i < c_0$. Additionally the abstract interpretation yields boundary conditions to be fulfilled by every solution of $tc(j, G), j \geq c_0$. These conditions can be exploited by the SMT solver to speed up the solution process. To our best knowledge no abstract interpretation algorithms for the concurrent synchronous real-time system paradigm have been suggested before, in particular not for the objective of speeding up automated test data generation (see paragraph on related work below).

The experiments described in Section 5 show that use of the abstract interpreter accelerates a solution process of $tc(1, G), tc(2, G), \ldots$ by an average factor of 1.44 just by being able to avoid infeasible tries to solve $tc(i, G)$ for $i < c_0$. If backtracking is applied the results are even more significant, since the abstract interpreter is very fast in detecting states from where no solution of $tc(i, G)$ exists within admissible range of i: here the average acceleration is 3.09. Observe that the experiments have not been performed on case studies, but on models developed for real-world testing campaigns in the automotive domain.

While our test automation framework is independent on the concrete modeling language[1], we sketch an UML2-based modeling formalism in Section 3 which is suitable to specify the expected behavior of synchronous concurrent real-time systems, in order to illustrate the main contribution of the paper.

Related Work. Modeling formalisms for synchronous systems are of considerable practical value in the field of safety-critical control systems. The formalism presented here is based on UML2.0. A more powerful formalism is SCADE [8] which is widely used in the avionic domain. Our main contribution would work equally well for the SCADE modeling language, because it does not depend on the concrete syntax "front-end", but only on the synchronous paradigm and the availability of the transition relation.

Our abstract interpretation approach is inspired by Cousot's work [5,4] and uses facts from interval analysis [12]. The Astrée abstract interpreter [6] is specialized on the analysis of embedded C-code and can also handle the effect of concurrent access to global program variables. Our abstract interpretation algorithm does not compete with, but is somewhat complementary to Astrée and its underlying methods: our abstract interpreter aims at the analysis of models on a more abstract level than C code. Similar to Timed Automata, it takes into account the valuations of dense-time clocks ("timers") which is not needed in the domain where Astrée is applied. Moreover, the modeling formalism used in this paper follows closely Harel's Statecharts in the semantics presented in [10] with synchronous execution of enabled transitions in parallel components, while Astrée operates on the semantics of a restricted class of C programs, where concurrency is expressed by interleaving of actions.

The problem of deciding the satisfiability of logical (first order) formulas where propositions may be constraints of certain background theories is commonly referred to as the *Satisfiability Modulo Theories (SMT)* problem. SMT solvers have been developed for numerous theories and combinations thereof. In recent years SMT solvers have become important tools for software verification [14]. Like most other state-of-the-art SMT solvers [2,13] solving these kind of formulas our SMT solver, SONOLAR, is based on the bit-blasting approach that translates an SMT formula to a purely propositional formula and lets a SAT solver decide the satisfiability. Various extensions to pure bit blasting have been proposed [3,1,16] which have inspired the SONOLAR implementation, and our solver was ranked second in the division for solving closed quantifier-free formulas over fixed-size bit vectors (QF_BV) at the Satisfiability Modulo Theories Competition (SMT COMP 2010).

2   SMT Solver

Our SMT solver SONOLAR follows the bit blasting approach, so Boolean, integral and floating-point variables are encoded as fixed-width bit vectors, where

---

[1] An algorithm to generate the transition relation Φ from a given abstract syntax representation of the model suffices in order to support the formalism.

the bit widths are given by the associated data types. Arithmetic and logical operations on these variables are transformed to Boolean constraints that encode the exact relationship of input and output bits. This allows us to have bit-precise results in the presence of modular arithmetic.

To this end the SMT formula is first transformed into a directed acyclic formula graph, where each single arithmetic and logical operation is represented as a single node. Structural hashing ensures that structurally identical terms are shared among expressions. On this formula graph a series of word-level simplifications like the evaluation of constant expressions, normalizations and term rewriting is performed. This word-level formula graph is then transformed to a bit-level, purely propositional *And-Inverter Graph (AIG)*. AIGs are commonly used among recent bit vector SMT solvers for synthesising propositional formulas [2,13]. AIGs represent propositional formulas as directed acyclic graphs (DAGs), where nodes are propositional variables or two-input AND-gates and edges may be optionally inverted. These AIG nodes are structurally hashed, too, and allow us to perform simplifications on bit level.

Although a number of competitive SAT solvers accept AIGs as input [15,11], most SAT solvers require the input to be in CNF. To generate the CNF, for each node of the AIG a boolean variable is introduced. Each node with possibly inverted inputs $n \Leftrightarrow in_1 \wedge in_2$ is then translated to $(\neg n \vee in_1) \wedge (\neg n \vee in_2) \wedge (n \vee \neg in_1 \vee \neg in_2)$. For each root of the AIG an additional unit clause containing the associated variable asserts the corresponding boolean formula to be either true or false, respectively.

SONOLAR has the capability to be called incrementally. This technique allows us to add constraints between solver runs and to add constraints that are only valid for one run (so-called *assumptions*). The SAT solver can then re-use conflict clauses learned in previous runs to speed up the following ones.

3   Modeling Formalism

In this section we sketch a modeling formalism for illustration purposes. It is based on an UML2 profile, and Fig. 1 — 3 present a sample model specifying the operation of an automotive controller handling turn indication and emergency flashing. Each model is structured into hierarchic components operating concurrently. Fig. 1 shows the SUT interacting with the testing environment TE via SUT input interfaces (TurnIndLeft,TurnIndRight) (positions (0,0), (1,0), (0,1) of the turn indicator lever), EmerFlash (=1 if emergency flash button is pressed), Voltage (percentage of the nominal voltage) and outputs (FlashLeft,FlashRight) (state of turn indication lamps left and right). The legal ranges of variables are specified by a model invariant (for example, TurnIndLeft/Right may not both be 1 in a normal behavior test), and optionally the admissible TE behaviors can be further restricted by associating nondeterministic timed state machines with the TE model component.

In our example the SUT is further structured into sub-components FLASH_-and and OUTPUT_CTRL. The former controls the decision whether or not to activate the turn indication lamps on the left-hand, right-hand or both sides.

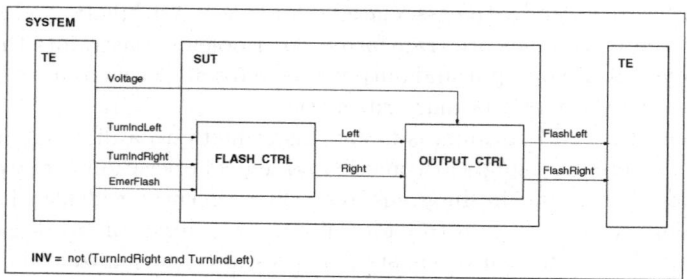

Fig. 1. Complete system consisting of TE and SUT

The latter controls the flashing cycles and automatically switches the lamps off if the actual voltage is less or equal 80% of the nominal voltage. This behavior is encoded by means of state machines S1, S2 as shown in Fig. 2 and 3.

While the EmerFlash button is not pressed, state machine S1 resides in control state EMER_OFF, where the state of the turn indicator lever is simply passed on to OUTPUT_CTRL via internal variables Left and Right, which is expressed by the do-action and its associated assignments. As soon as the EmerFlash button is pressed a state machine transition to basic control state EMER_ACTIVE is performed, where both Left and Right are switched to 1. The state machine transitions inside higher-level control state EMER_ON cope with the situation where the turn indicator lever state changes while emergency flashing is active: turn indication overrides emergency flashing (state TURN_IND_OVERRIDE). When resetting the turn indication lever, emergency flashing is resumed.

State machine S2 reacts on the status of Left, Right and Voltage. As long as Voltage > 80, non-zero states of Left and Right lead to flash cycles with periods of 560 time units. This is controlled by a clock variable t which is reset in basic control states ON and OFF and leads to state machine transitions as soon as the guards $t \geq 340$ or $t \geq 220$ become true. Semantically the clock is encoded as an ordinary real-valued variable, and each clock reset corresponds to storing the current model execution time $\hat{t}$ in t. The guard conditions are then internally evaluated as conditions $\hat{t} \geq t + 340$ and $\hat{t} \geq t + 220$, respectively.

The behavioral semantics of concurrent components is synchronous: both state machines evaluate the same pre-state. If the guard conditions of some transitions between control states evaluate to true a *discrete model transition* is performed by deterministically and simultaneously firing the enabled transitions with the highest priority in each component. The effect of each state machine transition may consist in a change of control states accompanied by a write to internal variables and outputs, while inputs remain unchanged. For calculating these write effects all expressions on the right-hand sides of assignments are evaluated in the pre-state, so that no evaluation order has to be considered. On the other hand, synchronous assignments performed by concurrent components to the same variables have to be consistent, otherwise a racing condition occurs which has to be fixed in order to gain a valid model. Only if discrete state machine transitions are disabled, a *delay model transition* is performed: the model execution time $\hat{t}$

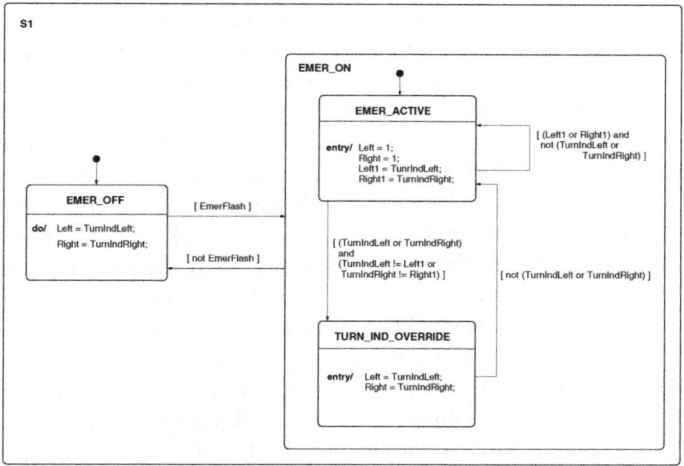

Fig. 2. Statechart S1 associated with component FLASH_CTRL, controlling decisions "flash left" and "flash right"

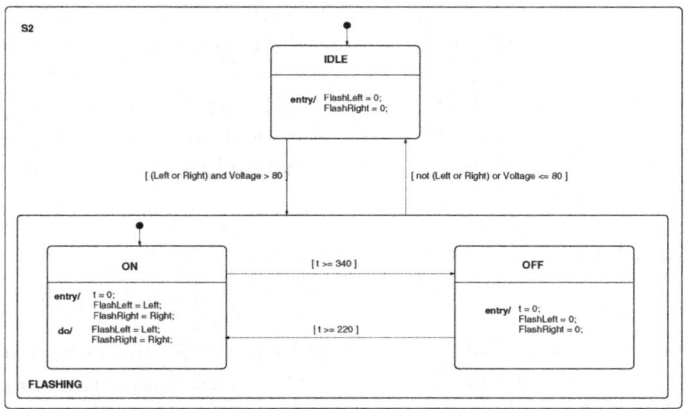

Fig. 3. Statechart S2 associated with component OUTPUT_CTRL managing indication lights and associated flash cycles

is advanced by a positive amount, but at most as up to a value where the next timer condition might become true. New values may be placed on the input interfaces, otherwise the model state remains unchanged.

4   Abstract Interpretation

In this section the detailed specification of the abstract interpretation algorithm is presented. The exposition requires some basic knowledge about lattices and Galois connections, for details readers are referred to [7].

Abstract Domains. Abstract interpretation performs over-approximation on possible model computations. For this approximation we map the concrete data types of state space components to so-called *abstract domains* which are lattices suitable for approximating concrete value sets for each state component. (1) The basic control states $\ell \in Loc(s)$ of each state machine s in the model have concrete data type Boolean, $\sigma(\ell) = 1$ signifying that the state machine resides in ℓ when the system is in state σ. We use the power set lattice $2^{Loc(s)}$ as the associated abstract domain: an element $\{\ell_1, \ldots, \ell_k\} \in 2^{Loc(s)}$ represents the knowledge that the state machine currently resides in one of the basic control state $\ell_1, \ldots, \ell_k$. We use symbol ℓ_A^s to denote this set-valued control state abstraction for state machine s. (2) Model variables of type Boolean are mapped to the lattice $L(\mathbb{B}) = \{\bot, 0, 1, \top\}$ with $\bot \sqsubseteq 0, 1 \sqsubseteq \top$ and $0, 1$ incomparable. Floating point and integer types are mapped to their associated interval lattices. Recall that the lattice join operation is defined by $[x_0, x_1] \sqcup [y_0, y_1] =_{\text{def}} [\min(x_0, y_0), \max(x_1, y_1)]$ for interval lattices, and that the meet operation is just set intersection, $[x_0, x_1] \sqcap [y_0, y_1] =_{\text{def}} [x_0, x_1] \cap [y_0, y_1]$. Model execution time $\hat{t}$ and timer variables are abstracted to intervals over non-negative reals.

Galois Connection. A set $U =_{\text{def}} \{\sigma_1, \ldots, \sigma_n\}$ of concrete model states is mapped to its abstraction $\sigma_A =_{\text{def}} U^{\triangleright}$ by setting $\sigma_A(x) = [\min(\{\sigma(x) \mid \sigma \in U\}), \max(\{\sigma(x) \mid \sigma \in U\})$ for integer and float variable symbols x. For Booleans b we define $\sigma_A(b) = \top$ if $\{\sigma(b) \mid \sigma \in U\} = \{0, 1\}$, $\sigma_A(b) = 0$ if $\{\sigma(b) \mid \sigma \in U\} = \{0\}$ and $\sigma_A(b) = 1$ if $\{\sigma(b) \mid \sigma \in U\} = \{1\}$. Furthermore, $\sigma_A(\ell_A^s) = \{\ell \in Loc(s) \mid \exists \sigma \in U : \sigma(\ell) = 1\}$ for the abstracted locations ℓ_A^s of state machines s. Conversely, each abstract state σ_A may be mapped to a set of concrete states by means of the mapping

$$\sigma_A{}^{\triangleleft} =_{\text{def}} \{\sigma \mid \forall b : \sigma(b) \sqsubseteq \sigma_A(b) \wedge \forall x : \sigma(x) \in \sigma_A(x) \wedge$$
$$\forall s : \forall \ell \in Loc(s) : \sigma(\ell) = 1 \Leftrightarrow \ell \in \sigma_A(\ell_A^s)\}$$

where b denotes Booleans, s state machines and x floating point and integer model variables. The pair of mappings $\triangleright, \triangleleft$ represents a Galois connection and its characteristic property $a^{\triangleright} \sqsubseteq_2 b \Leftrightarrow a \sqsubseteq_1 b^{\triangleleft}$ ensures that the algorithm introduced below really computes an over-approximation of all possible computation states.

Goal of the Abstract Interpretation Algorithm. The abstract interpretation algorithm starts from the abstraction $\sigma_A^0 = \{\sigma_0\}^{\triangleright}$ of a concrete pre-state σ_0 and calculates a single bounded abstract computation sequence $\langle \sigma_A^0, \ldots, \sigma_A^c \rangle$ such that each concrete computation $\langle \sigma_0, \ldots, \sigma_c \rangle$ starting in σ_0 is approximated by the abstract sequence in the sense that

$$\forall i \in \{0, \ldots, c\} : \sigma_i \in \sigma_A^i{}^{\triangleleft}$$

Now suppose that the test case goal G is fulfilled in state σ_c of the concrete computation. Interpreted as a Boolean function on the state space, predicate G may be lifted to the abstract domain by defining

$$[G](\sigma_A) = \begin{cases} 1 \text{ if } \forall \sigma \in \sigma_A{}^\lhd : G(\sigma) = 1 \\ 0 \text{ if } \forall \sigma \in \sigma_A{}^\lhd : G(\sigma) = 0 \\ \top \text{ otherwise} \end{cases}$$

Since $\sigma_c \in \sigma_A^c{}^\lhd$ and $G(\sigma_c) = 1$, evaluation of $[G](\sigma_A^c)$ will result in 1 or $\top$, that is, $1 \sqsubseteq [G](\sigma_A^c)$. Conversely, G will not hold in any σ_i as long as $[G](\sigma_A^i) = 0$. Therefore the objective of the abstract interpretation algorithm is to return the smallest $c_0 \geq 0$ such that $1 \sqsubseteq [G](\sigma_A^{c_0})$ holds. Given this c_0 the SMT solver can try to solve the test case constraint satisfaction problems $tc(c, G)$ specified in (1) with $c = c_0, c_0 + 1, \ldots$, and without having to investigate the feasibility of $tc(m, G)$ for $m < c_0$. Since the abstract interpreter operates significantly faster than the SMT solver, a considerable speed-up can be expected from the fact that the solver skips these $tc(m, G)$.

Abstract Interpretation Algorithm – Introductory Example. To give an intuitive idea of the abstract interpretation algorithm specified formally further below, we assume that our sample system is initialized in a state σ with $\sigma(\hat{t}) = 0$ and $\sigma(\text{TurnIndLeft/Right}) = 0$, $\sigma(\text{EmerFlash}) = 1$, $\sigma(\text{Voltage}) = 85$ and all internal variable and output valuations equal to zero. Suppose further that our test objective is to cover the condition $G \equiv$ S1.ACTIVE.OVERRIDE $\wedge$ S2.FLASHING.OFF starting from this given initial system state. If the abstract interpretation function exploreGoal() is called with $c = 6$ then the algorithm explores abstract interpretation states as shown in the table below, where the columns have the following meaning: **TT** = transition type (DIScrete or DELay or both (DD)); **Si** = sets of possible control states state machines S1, S2 reside in; **TIL, TIR, E, V** = input valuations for TurnIndLeft,...,Voltage; **L, R, L1, R1** = valuations of model variables Left,...,Right1; **t,$\hat{t}$** = valuations of timer variable t and current execution time $\hat{t}$; **FL, FR** valuation of outputs FlashLeft, FlashRight.

The abstract interpretation algorithm starts by mapping the concrete initial state into its abstract counterpart; the result is displayed in row 0 of the table below: control states are mapped to singleton sets because there is no uncertainty which locations are active. Boolean values are represented in $L(\mathbb{B})$ in the same way, and numeric values are mapped to their single-point interval counterparts. As a result of the initial state valuation only discrete transitions are possible until abstract state 2 is reached, from where only a delay transition may occur. After the delay the inputs may assume arbitrary values, so they are marked by $\top$. Moreover, the model time $\hat{t}$ may have been increased by some positive amount less or equal 340, where the next timer is bound to elapse. The next transition leading to abstract state 4 may be discrete or a delay, and – due to the full-range input valuations – all guards depending on inputs evaluate to $\top$. As a consequence abstract state 4 admits arbitrary control states, and $[G]$ evaluates to $\top$, so this is the first state where a solution for G may be found. The abstract interpretation algorithm returns with $c_0 = 4$ and also provides a constraint

$$\beta \equiv \text{ACT}_1 \wedge \text{IDLE}_1 \wedge t_1 = 0 \wedge \hat{t}_1 = 0 \wedge$$
$$\text{ACT}_2 \wedge \text{ON}_2 \wedge t_2 = 0 \wedge \hat{t}_2 = 0 \wedge$$
$$\text{ACT}_3 \wedge \text{ON}_3 \wedge t_3 = 0 \wedge \hat{t}_3 \in (0, 340] \wedge$$
$$t_4 \in [0, 340] \wedge \hat{t}_4 \in (0, 679]$$

indicating the restrictions valid at each concrete computation step. This may be used by the SMT solver to reduce the search space.

#	TT	S1	S2	TIL	TIR	E	V	L	R	L1	R1	t	$\hat{t}$	FL	FR
0.		{OFF}	{IDLE}	0	0	1	[85,85]	0	0	0	0	[0,0]	[0,0]	0	0
1.	DIS	{ACT}	{IDLE}	0	0	1	[85,85]	1	1	0	0	[0,0]	[0,0]	0	0
2.	DIS	{ACT}	{ON}	0	0	1	[85,85]	1	1	0	0	[0,0]	[0,0]	1	1
3.	DEL	{ACT}	{ON}	$\top$	$\top$	$\top$	[0,100]	1	1	0	0	[0,0]	(0,340]	1	1
4.	DD	{OFF, ACT, OVR}	{IDLE, ON, OFF}	$\top$	$\top$	$\top$	[0,100]	$\top$	$\top$	0	0	[0,340]	(0,679]	$\top$	$\top$

Main Function. The top-level function of the abstract interpretation algorithm operates as specified in Fig. 4. Function exploreGoal() is invoked on the current concrete system state σ, and inputs the test case goal G according to Formula (1). Integer $c > 0$ denotes the limit of interpretation steps to be performed. Output β represents a constraint to be constructed by the function. On function return, β contains restrictions about the possible computations states leading to a solution. This auxiliary information may be used by the SMT solver to restrict the search space. The assignment $\sigma_A := \{\sigma\}^{\triangleright}$ creates the abstract start state associated with input σ. In each loop cycle i an abstract interpretation step is performed by means of procedure call absInt(σ_A, σ'_A), creating a new abstract state σ'_A. The knowledge that each concrete computation state σ_i is contained in $\sigma'_A{}^{\triangleleft}$ is exploited by adding conjuncts to constraint β, restricting the possible valuations of σ_i: for each state machine s the disjunction of all possible basic control states ℓ the machine may reside in are added as a conjunct to β. Observe that index i adds version information to the basic location identifier ℓ, since this applies to the ith computation state reachable from start state σ. Further restrictions added to β are the bounds for the model execution time $\hat{t}$ in step i and intervals for admissible variable values in this step.

Condition $(1 \sqsubseteq [G](\sigma'_A))$ is evaluated to check whether there is a chance of solving the test case goal in step i. If this is the case the function returns with value i as the first possible computation step number where G may become true, and β contains the restrictions accumulated up to step i. If $[G](\sigma'_A)$ evaluates to 0, the next interpretation cycle is prepared. If limit c is reached without encountering an abstract state satisfying $(1 \sqsubseteq [G](\sigma'_A))$ the function returns with code -1.

Abstract Interpretation Step Procedure. Fig. 5 shows the procedure absInt() for performing one abstract interpretation step: if the trigger condition for discrete transitions evaluates to 1 in the current abstract state σ_A then only an abstract interpretation of possible discrete transitions takes place. If the condition for a discrete model transition to be enabled, $[\text{trigger}_D](\sigma_A)$, is guaranteed to be false,

```
function exploreGoal(σ : S, G : BExpr, c : ℕ, out β : BExpr) : ℤ
begin
    i := 1; σ_A := {σ}^▷; β := 1; r := -1;
    while i ≤ c do
        absInt(σ_A, σ'_A);
        foreach s ∈ SM do β := β ∧ (⋁_{ℓ∈σ'_A(ℓ^s_A)} ℓ_i); enddo
        β := β ∧ t̂_i ∈ σ'_A(t̂) ∧ (⋀_{x∈I} x_i ∈ σ'_A(x)) ∧ (⋀_{v∈L∪O} v_i ∈ σ'_A(v));
        if (1 ⊑ [G](σ'_A)) then r := i; break; endif
        σ_A := σ'_A; i := i + 1;
    enddo
    exploreGoal := r;
end
```

Fig. 4. Top-level procedure of the state space exploration by means of abstract interpretation. Sets I, L, O denote input, local and output variables, respectively.

only a delay can occur. In that case, function absIntTime() (Fig. 6) calculates the boundaries of the new execution time stamp $\hat{t}$, and the abstractions of all input values x are set to their maximal ranges $D_x{}^▷ \in L(D_x)^2$. If $[\text{trigger}_D](\sigma_A)$ evaluates to $\top$, both discrete and delay transitions have to be taken into account and, consequently, the potential post-state is the maximum $\sigma_A^1 \sqcup \sigma_A^2$ of the post-states resulting from these two transition types.

Abstraction of Delay Transitions. The calculation of the time bounds for a delay transition is subtle, as can be seen in Fig. 6: The maximal delay may be infinite if no active timer is being observed in the current system state abstracted by σ_A. Therefore the variable limit which is used to store intermediate and final upper bounds of the time growth is initialised by ∞^3. If some timers are active, the delay is limited by the shortest value at which some state machine is guaranteed to fire a discrete transition. Therefore a loop over all state machines indexed by $i \in 1, \ldots, p$ is performed, and the maximal delay which may occur in one state machine is stored in smLimit. To determine smLimit, the minimal delay locLimit for each location the state machine may currently reside in, where a timed transition guard is guaranteed to become true is determined. The smallest smLimit-value calculated over all state machines is the global upper bound limit to be returned as the upper bound of the new $\hat{t}$-value[4], because at least one state

---

[2] D_x denotes the concrete data type of x. Operator $\oplus$ used in Fig. 5 denotes functional overriding: function $f \oplus \{x \mapsto y\}$ coincides with $f(z)$ for all arguments $z \neq x$, but maps x to y.

[3] In concrete test equipment implementations some suitable value greater than the longest timeout value defined in the SUT model is used instead of ∞, in order to guarantee new stimuli from test equipment to SUT within a reasonable amount of time.

[4] For variables x interpreted in an interval lattice we use $\underline{\sigma_A(x)}$ and $\overline{\sigma_A(x)}$ to denote the lower and upper bounds of their interval valuation, respectively.

procedure absInt($\sigma_A : L(S)$, **out** $\sigma'_A : L(S)$)
begin
  if $[\text{trigger}_D](\sigma_A) = 1$ **then**
    absIntDisc(σ_A, σ'_A);
  elseif $[\text{trigger}_D](\sigma_A) = 0$ **then**
    $\sigma'_A := \sigma_A \oplus \{\hat{t} \mapsto \text{absIntTime}(\sigma_A)\} \oplus \{x \mapsto D_x{}^{\triangleright} \mid x \in I\}$;
  else
    absIntDisc(σ_A, σ^1_A);
    $\sigma^2_A := \sigma_A \oplus \{\hat{t} \mapsto \text{absIntTime}(\sigma_A)\} \oplus \{x \mapsto D_x{}^{\triangleright} \mid x \in I\}$;
    $\sigma'_A := \sigma^1_A \sqcup \sigma^2_A$;
  endif
end

Fig. 5. Single step abstract interpreter

function absIntTime($\sigma_A : L(S)$) : $\mathbb{R}_+$
begin
  limit := ∞;
  foreach $i \in \{1, \ldots, p\}$ **do**
    smLimit := $\underline{\sigma_A(\hat{t})}$;
    foreach $\ell_0 \in \overline{\sigma_A(\ell^i_A)}$ **do**
      locLimit := ∞;
      foreach $\ell \in \ell_{0..s_i},\ (\ell, g, a, \ell') \in \omega_{s_i}(\ell)$ **do**
        if $(\exists g', t, x : g \equiv \underline{(\hat{t} \geq x + t \wedge g')}) \wedge [g'](\sigma_A) = 1$ **then**
          $m := \underline{\sigma_A(x)} + \overline{\sigma_A(t)}$;
          if $m < $ locLimit **then** locLimit := m; **endif**
        endif
      enddo
      if locLimit $>$ smLimit **then** smLimit := locLimit; **endif**
    enddo
    if smLimit $<$ limit **then** limit := smLimit; **endif**
  enddo
  absIntTime := $(\underline{\sigma_A(\hat{t})}, \text{limit}]$;
end

Fig. 6. Function calculating the maximal time interval associated with a delay transition

machine is guaranteed to fire a discrete transition until limit. Since some time has to pass during delay transitions, the lower bound of the new $\hat{t}$-value has to be greater than the old lower bound $\underline{\sigma_A(\hat{t})}$.

Abstraction of Discrete Transitions. The abstract interpretation of a discrete transitions is specified in Fig. 7. A partial auxiliary function $\zeta : V \nrightarrow \bigcup_{w \in V} L(D_w)$ is used for intermediate recordings of assignments to abstracted variables. For

procedure absIntDisc($\sigma_A : L(S)$, **out** $\sigma'_A : L(S)$)
begin
$\quad \zeta := \varnothing;\ (q_1, \ldots, q_p) := (\varnothing, \ldots, \varnothing);$
$\quad$ **foreach** $i \in \{1, \ldots, p\}$ **do**
$\quad\quad$ **foreach** $\ell_0 \in \sigma_A(\ell_A^i)$ **do**
$\quad\quad\quad$ leave $:= 0;$
$\quad\quad\quad$ **foreach** $\ell \in \ell_0..s_i,\ \tau \in \omega_{s_i}(\ell),\ \tau$ ordered by priority **do**
$\quad\quad\quad\quad$ **if** $1 \sqsubseteq [\text{trigger}_{s_i}(\tau)](\sigma_A)$ **then**
$\quad\quad\quad\quad\quad \sigma_A^1 := \sigma_A;\ C(\text{trigger}_{s_i}(\tau), \sigma_A^1);$
$\quad\quad\quad\quad\quad$ absIntTransEffect$(\sigma_A^1, \tau, \zeta, q_i);$
$\quad\quad\quad\quad\quad$ **if** $1 = [\text{trigger}_{s_i}(\tau)](\sigma_A)$ **then** leave $:= 1;$ **break; endif**
$\quad\quad\quad\quad$ **endif**
$\quad\quad\quad$ **enddo**
$\quad\quad\quad$ **if** $\neg$leave **then**
$\quad\quad\quad\quad \sigma_A^2 := \sigma_A;\ C(\bigwedge_{\ell \in \ell_0..s_i,\ \tau \in \omega_{s_i}(\ell)} \neg\text{trigger}_{s_i}(\tau), \sigma_A^2);$
$\quad\quad\quad\quad$ absIntDoEffect$(\sigma_A^2, \ell_0, \zeta, q_i);$
$\quad\quad\quad$ **endif**
$\quad\quad$ **enddo**
$\quad$ **enddo**
$\quad \sigma'_A := \sigma_A \oplus \{e_i \mapsto q_i \mid i = 1, \ldots, p\} \oplus \{w \mapsto \zeta(w) \mid w \in \text{dom } \zeta\};$
end

Fig. 7. Discrete transition abstract interpreter

each basic control state ℓ_0 a state machine may potentially reside in, all emanating transitions from ℓ_0 and its higher-level locations are investigated. If a transition τ may fire, that is, if its abstracted trigger condition $\text{trigger}_{s_i}(\tau)$ evaluates to 1 or $\top$ in the pre-state σ_A, a copy σ_A^1 of the pre-state is first contracted, using the knowledge that $\text{trigger}_{s_i}(\tau)$ must have evaluated to 1 in order to get the effect of τ^5.

This effect on the abstracted state space is then calculated by procedure absIntTransEffect() which records these results by changing ζ: Suppose the effect of the transition comprises a value assignment $w := \text{expr}$. If w is not yet in the domain of ζ, this means that it is the first potential write to w during this abstracted discrete transition. Therefore ζ's domain is extended by setting $\zeta := \zeta \oplus \{w \mapsto [\text{expr}](\sigma_A^1)\}$, where $[\text{expr}]$ is the lifted version of the assignment's right-hand side expression. The abstract expression evaluation is performed on the contracted abstract state σ_A^1. If w is already in dom ζ, this means that another transition might also write to w. In order to approximate the discrete transition effects in a conservative manner, we build the join of both potential effects, that is, we set $\zeta := \zeta \oplus \{w \mapsto \zeta(w) \sqcup [\text{expr}](\sigma_A^1)\}$. Finally, absIntTransEffect() adds the target basic control state associated with τ to the set q_i of potential target

---

[5] For interval lattices we have natural contractors for arithmetic constraints: for example in $L(\mathbb{Z})$, $C_<(x < y; [\underline{x}, \overline{x}], [\underline{y}, \overline{y}]) =_{\text{def}} ([\underline{x}, \min(\overline{x}, \overline{y} - 1)], [\max(\underline{x} + 1, \underline{y}), \overline{y}])$ defines contractions for x and y under the hypothesis that $x < y$ evaluated to true.

Model/Config	$\#g_t$	$\#s$	d_s	$(\#g_r)$	d_{sa}	$(\#g_r)$	d_{sb}	$(\#g_r)$	d_{sba}	$(\#g_r)$
TURNIND/1	15	35	5.49	(3)	2.95	(3)	18.06	(3)	3.45	(3)
TURNIND/2	27	35	53.26	(7)	20.91	(7)	82.21	(7)	22.08	(7)
TURNIND/3	46	35	11.68	(8)	7.67	(8)	45.15	(8)	9.70	(8)
TURNIND/4	9	35	5.30	(2)	3.17	(2)	21.39	(2)	3.81	(2)
TURNIND/5	17	35	5.19	(3)	2.94	(3)	18.08	(3)	3.56	(3)
TURNIND/6	11	35	5.32	(2)	2.54	(2)	17.43	(2)	3.02	(2)
POWERTRUNK/1	2	50	55.68	(1)	67.90	(2)	109.93	(2)	67.71	(2)
POWERWINDOW/1	58	40	27.99	(9)	18.18	(9)	89.15	(9)	21.58	(9)
STOP-START/1	13	50	269.62	(3)	376.01	(3)	436.06	(13)	546.09	(13)
STOP-START/2	9	50	3.23	(9)	5.83	(9)	3.20	(9)	5.83	(9)
STOP-START/3	19	50	378.67	(15)	434.45	(15)	619.66	(15)	451.08	(15)
STOP-START/4	28	50	10.93	(17)	10.19	(17)	69.99	(17)	14.07	(17)
STOP-START/5	32	50	6.59	(7)	2.96	(7)	18.44	(7)	3.72	(7)
STOP-START/6	36	50	6.60	(7)	2.96	(7)	18.40	(7)	3.71	(7)
STOP-START/7	36	50	217.12	(36)	191.28	(36)	217.58	(36)	191.39	(36)
STOP-START/8	28	50	998.58	(28)	478.49	(28)	995.35	(28)	477.65	(28)
STOP-START/9	4	50	340.88	(4)	365.99	(4)	341.35	(4)	367.15	(4)
STOP-START/10	12	50	331.50	(8)	358.51	(8)	479.75	(8)	356.80	(8)
STOP-START/11	26	50	337.62	(18)	302.26	(18)	508.46	(18)	315.20	(18)
STOP_START_SYS/1	21	50	588.45	(10)	523.12	(10)	833.10	(21)	648.90	(21)

$\#g_t$: number of goals to be covered, $\#s$: maximal number of transition steps, d_s: execution duration [s] with solver, d_{sa}: execution duration [s] with solver and abstract interpretation, d_{sb}: execution duration [s] with solver and backtracking, d_{sba}: execution duration [s] with solver, abstract interpretation and backtracking, $\#g_r$: number of covered goals

Fig. 8. Test generation results

locations. This join of potential write results and target locations ensures that all potential concrete target states σ_i are really contained in $\sigma'_A{}^\triangleleft$.

If no transition emanating from a location in $\ell_0..s_i$ is guaranteed to fire, that is, $\text{trigger}_{s_i}(\tau) \in \{0, \top\}$ for all of these τ and therefore $leave = 0$, the do actions associated with the locations in $\ell_0..s_i$ may be executed. Their effect on the abstract state space is calculated by absIntDoEffect() which works similar to absIntTransEffect(), but adds the source location ℓ_0 to q_i and operates on a copy of the source state contracted with the knowledge that all transition triggers must have evaluated to 0, in order to get the effect of these do-actions. At the end of procedure absIntDisc() the new abstract state σ'_A is constructed by changing the pre-state σ_A with respect to the new sets of potentially active basic control states and the new abstract valuations of variables that have been potentially written to during the abstract interpretation step.

5   Conclusion and Evaluation Results

The evaluation of the combined abstract interpretation, SMT-solving and backtracking approach has been performed using five real-world test models for the system test of automotive control functions which are intellectual property of Daimler[6]: (1) Model TURNIND specifies all automotive functions acting on the turn indication lights, such as turn indication and emergency flashing. (2) Models STOP-START and (3) STOP_START_SYS specify the behavior of the stop-start mechanism controlling automated engine cutoff when stopping at red lights on HW/SW integration and system integration level, respectively. (4) Model POWERWINDOW specifies the functionality of the electronic window regulation, including detection of and reaction on blocking window states, and specialized functions like automated opening of windows for the purpose of ventilation in crash situations and automated closing of windows when entering tunnels. (5) Model POWERTRUNK describes the functionality of the electronic closing mechanism of the trunk lid. Although none of these models involves floating-point arithmetic our system is capable of handling these.

For the evaluation, coverage goals were defined for each model. These goals consisted in specific state machine transitions to be reached, which was equivalent to coverage of certain requirements. Then the test case/test data generation was activated with different techniques, and the execution times have been measured and inserted into the table shown in Fig. 8. This table shows considerable performance improvements for the situations where abstract interpretation is used, with very few outliers where the abstract interpretation leads to a slowdown. Without backtracking the generator was 1.44 times faster on average when using the abstract interpreter. The results were even better with backtracking enabled: with abstract interpretation we observed an average acceleration by a factor of 3.09. This dramatic speed-up when using the abstract interpreter in combination with backtracking can largely be attributed to the fact that the abstract interpreter is very fast at immediately discarding backtracking points from which no new goals can be covered, whereas the solver would spend a lot of time to do so.

While in our current approach the algorithm stops unrolling the transition relation as soon as at least one goal can be satisfied it is generally desirable to satisfy as many goals as possible within a sequence of transitions. Therefore we plan to explore the possibility to extend the present constraint satisfaction problem to an optimization problem that aims to maximize the number of satisfied goals. The necessary means to achieve this are provided by *Partial MAX-SAT* techniques [9].

References

1. Brillout, A., Kroening, D., Wahl, T.: Mixed Abstractions for Floating-Point Arithmetic. In: Proceedings of FMCAD 2009, pp. 69–76. IEEE, Los Alamitos (2009)
2. Brummayer, R.: Efficient SMT Solving for Bit-Vectors and the Extensional Theory of Arrays. Ph.D. thesis, Johannes Kepler University Linz, Austria (November 2009)

---

[6] It is currently discussed with Daimler, whether at least one of these models may be published because this would represent valuable information for tool benchmarking. We hope that this will be the case by the time of the NFM2011 conference.

3. Bryant, R.E., Kroening, D., Ouaknine, J., Seshia, S.A., Strichman, O., Brady, B.: Deciding Bit-Vector Arithmetic with Abstraction. In: Grumberg, O., Huth, M. (eds.) TACAS 2007. LNCS, vol. 4424, pp. 358–372. Springer, Heidelberg (2007)
4. Cousot, P.: Abstract interpretation: Theory and practice (April 11-13, 2000)
5. Cousot, P., Cousot, R.: Abstract interpretation: a unified lattice model for static analysis of programs by construction or approximation of fixpoints. In: Conference Record of the Fourth Annual ACM SIGPLAN-SIGACT Symposium on Principles of Programming Languages, pp. 238–252. ACM Press, New York (1977)
6. Cousot, P., Cousot, R., Feret, J., Mauborgne, L., Miné, A., Monniaux, D., Rival, X.: Combination of abstractions in the ASTRÉE static analyzer. In: Okada, M., Satoh, I. (eds.) ASIAN 2006. LNCS, vol. 4435, pp. 272–300. Springer, Heidelberg (2008)
7. Davey, B.A., Priestley, H.A.: Introduction to Lattices and Order. Cambridge University Press, Cambridge (2002)
8. Esterel Technologies: SCADE Suite Product Description, http://www.estereltechnologies.com
9. Fu, Z., Malik, S.: On solving the partial max-sat problem. In: Biere, A., Gomes, C.P. (eds.) SAT 2006. LNCS, vol. 4121, pp. 252–265. Springer, Heidelberg (2006)
10. Harel, D., Naamad, A.: The statemate semantics of statecharts. ACM Transactions on Software Engineering and Methodology 5(4), 293–333 (1996)
11. Jain, H., Clarke, E.M.: Efficient SAT Solving for Non-Clausal Formulas using DPLL, Graphs, and Watched Cuts. In: 46th Design Automation Conference, DAC (2009)
12. Jaulin, L., Kieffer, M., Didrit, O., Walter, É.: Applied Interval Analysis. Springer, London (2001)
13. Jung, J., Sülflow, A., Wille, R., Drechsler, R.: SWORD v1.0. Tech. rep. (2009), sMTCOMP 2009: System Description
14. Ranise, S., Tinelli, C.: Satisfiability modulo theories. TRENDS and CONTRO VERSIES–IEEE Magazine on Intelligent Systems 21(6), 71–81 (2006)
15. Sörensson, N.: MiniSat 2.2 and MiniSat++ 1.1. Tech. rep. (2010), SAT-Race 2010: Solver Descriptions
16. Wille, R., Fey, G., Große, D., Eggersglüß, S., Drechsler, R.: SWORD: A SAT like Prover Using Word Level Information. In: Proceedings of VLSI-SoC 2007, pp. 88–93 (2007)

Generating Data Race Witnesses by an SMT-Based Analysis*

Mahmoud Said[1], Chao Wang[2], Zijiang Yang[1], and Karem Sakallah[3]

[1] Department of Computer Science, Western Michigan Univerisity,
Kalamazoo, MI 49008
[2] NEC Laboratories America, 4 Independence Way, Suite 200, Princeton, NJ 08540
[3] Department of Electrical Engineering and Computer Science,
University of Michigan, Ann Arbor, Michigan 48109

Abstract. Data race is one of the most dangerous errors in multi-threaded programming, and despite intensive studies, it remains a notorious cause of failures in concurrent systems. Detecting data races is already a hard problem, and yet it is even harder for a programmer to decide *whether* or *how* a reported data race can appear in the actual program execution. In this paper we propose an algorithm for generating debugging aid information called *witnesses*, which are concrete thread schedules that can deterministically trigger the data races. More specifically, given a concrete execution trace, e.g. non-erroneous one which may have triggered a warning in Eraser-style data race detectors, we use a symbolic analysis based on SMT solvers to search for a data race witness among alternative interleavings of events of that trace. Our symbolic analysis precisely encodes the sequential consistency semantics using a scalable predictive model to ensure that the reported witness is always feasible.

Keywords: Data Race, Debug, SMT, Concurrent Programs.

1 Introduction

A data race occurs in a multithreaded program when two threads access the same memory location with no ordering constraints enforced in between, and at least one of the accesses is a write. Programs containing data races are difficult to debug because they may exhibit different behaviors under the same input. In practice, a single synchronization error caused by data race can take weeks for programmers to identify [3,21]. For the Java Memory Model (JMM) and other relaxed memory models, it is absolutely crucial to remove all data races in user applications even if they do not appear to cause logic errors, because these models guarantee sequential consistency only to race-free programs [15].

Stateful model checking is one of the approaches for finding bugs in concurrent programs [10,11,23]. As more scalable exhaustive techniques, stateless model

---

* The work was supported in part by NSF Grants CCF-0811287, CCF-0810865 and ONR Grant N000140910740.

M. Bobaru et al. (Eds.): NFM 2011, LNCS 6617, pp. 313–327, 2011.

chekers [2,16] have been developed. Being exhaustive in nature, model checkers in principle can be used to provide counter-examples. Unfortunately, most existing model checking tools do not scale.

The numerous static and dynamic techniques that have been developed to detect data races [8,1,6,18,17,13,24,5,9], except for exhaustive techniques, can only report data race warnings, often in the form of pairs of program locations. None of these methods provide *witnesses* to help the programmers deterministically reproduce the reported data race during actual program executions. By witness, we mean a concrete thread schedule of the program execution that leads to a program state in which two concurrent events with data conflict are both enabled. It is essential debugging information for programmers to decide whether the race is benign, and subsequently figure out how to fix it.

The problem of generating witnesses is orthogonal to detecting data races. The latter problem, which have been studied extensively, ends with a set of *data race warnings*. The witness generation starts from where the data race detection ends, with the goal of providing a concrete thread schedule to reproduce each data race during execution. The witness generation problem is significantly harder, since it has to concern with the feasibility (or existence) of particular concrete executions. It is also a practically important problem with no satisfying solution yet.

In this paper we present an algorithm to generate data race witnesses in multithreaded Java programs based on analyzing a single execution trace. The key idea is to perform a postmortem analysis on a log of the access events. Here we can use any of the existing data race *detection* algorithms [8,1,6,18,17,13,24,5,9] to compute a set of *potential* data races, which then act as input to our witness generation algorithm. Given a trace and a set of potential data races, we model the access events of that trace using suitable classes of constraints and formulating the witnesses generation problem as constraint solving. What these constraints represent is not just the given trace itself, but a *maximal set* of interleavings of events of that trace, and all these *alternative* traces are guaranteed to be actual program executions. The constraints generated by our algorithm are in a quantifier-free first-order logic. They can be decided by off-the-shelf Satisfiability Modulo Theory (SMT) solvers, and therefore can benefit from the significant performance advances in recent SMT solvers (e.g. [4]).

Our symbolic predictive model improves over the maximal causal model (MCM) proposed by Serbănută, Chen and Rosu [22]. We improve over the MCM based method in the following aspects. First, the MCM considers semaphores as the only synchronization primitives, whereas in this paper, we precisely model a wide range of synchronization primitives in Java, including wait, notify, and notifyall. Second, the search algorithm used in [22] is based on explicitly enumerating the feasible interleavings, which may become a bottleneck for practical uses; in our method, we conduct the search symbolically using an SMT solver.

To further reduce the overhead of the symbolic search, we pre-simplify the SMT formulas by applying a trace-based conservative analysis [14]. Our analysis is based on computing lock acquisition histories and a must-happen-before relation defined by thread creation/join and matching wait/ notify/notifyall. The goal is to reduce the cost of the more precise, but also expensive, symbolic analysis, by quickly weeding out (bogus) data races that do not have concrete witnesses. The constraints derived from this analysis can also be added as hints to speed up the SMT search.

We have implemented the proposed method for multithreaded Java programs. Our trace logging is implemented using an agent interface that captures the Java Virtual Machine Execution events, and our symbolic analysis uses the Yices SMT solver [4]. Our preliminary results on public benchmarks show that the witness generation algorithm is scalable enough as a post-mortem analysis, to help programmers better understand the data races.

2 Multithreaded Trace

2.1 Execution Traces

We consider a multithreaded Java program as a set of concurrently running threads, and use $Tid = \{1, \ldots, n\}$ to denote the set of thread indices. The operations on global or *shared* variables are called visible operations, while those on thread-local variables are called invisible operations. In particular, synchronization primitives such as operations on locks and condition variables are regarded as visible operations. An execution trace π is a sequence of instances of *visible* operations in a concrete execution of the multithreaded program. Each instance is called an *event*. For Java programs, both read/write accesses to shared variables and the synchronization operations are recorded as events, while invisible operations are ignored. An event is represented as a tuple $(tid, type, var, val)$, where tid is the thread index, $type$ is the event type, var is either a shared variable (in read/write) or a synchronization object, val is either a concrete value (in read/write) or the child thread index (in thread creation/join). The event type is one of $\{read, write, fork, join, acquire, release, wait, notify, notifyAll\}$. They can be classified into three categories:

1. *read* and *write* denote the read and write access to a shared variable, where *var* is the variable and *val* is the concrete value;
2. *fork* and *join* denote the creation and termination of a child thread, where $(tid, fork, -, val)$ creates a child thread whose index is *val*, and $(tid, join, -, val)$ joins the child thread back;
3. the rest correspond to synchronization operations over locks and condition variables. The **synchronized** keyword is translated into a pair of *acquire* and *release* events over the lock implicitly associated with an object.

For an event e and its attribute a, we will use $e.a$. In addition, given an execution π and an event e in it, $e.idx$ denote the unique index of event e in π. For example, in event $e_i : (1, fork, -, 2)$, we have $e_i.tid = 1, e_i.type = fork, e_i.val = 2$, and $e_i.idx = i$.

2.2 Partial Order and Linearizations

Let $\pi = e_1 \ldots e_n$ be a concrete execution. The trace can be viewed as a total order of the set $\{e_1, \ldots, e_n\}$ of events. To capture all the alternative and yet feasible interleavings of the events in π, we define a *partially ordered set*, denoted $\mathcal{T}_\pi = (T, \sqsubseteq)$, such that

- $T = \{e \mid e \text{ is an event in } T_\pi\}$.
- $\sqsubseteq$ is a partial order such that
  - if $e_i.tid = e_j.tid$ and e_i appears before e_j in π, then $e_i \sqsubseteq e_j$,
  - if $e_i = (tid_1, fork, -, tid_2)$ and e_j is the first event of thread tid_2 in π, then $e_i \sqsubseteq e_j$,
  - if $e_i = (tid_1, join, -, tid_2)$ and e_j is the last event of thread tid_2 in π, then $e_j \sqsubseteq e_i$.
  - $\sqsubseteq$ is transitively closed.

That is, T_π orders events from the same thread based on their execution order in π, but does not order events from different threads except for fork and join.

In the presence of shared variables and synchronization primitives, not all linearizations (total orders) of T_π correspond to actual program executions. We define a *sequentially consistent linearization* τ_π of T_π as one that satisfies $\sqsubseteq$ as well as the following requirements:

- *Write-Read Consistency*: the value read by an event is always written by the most recent write in τ_π, and
- *Synchronization Consistency*: τ_π does not violate the semantics of the synchronization events.

The set of all linearizations of T_π forms the search space of our witness generation algorithm. That is, we search for a sequentially consistent linearization that leads to a state in which two data-conflict events are both enabled.

Our notion of sequentially consistent linearization is inspired by the maximal causal model in [22]. However, the maximal causal model considers semaphore as the only synchronization primitive, and does not explicitly model thread creation and join (*fork* and *join*), whereas we precisely model a wide range of Java synchronization primitives. Our symbolic method for searching sequentially consistent linearizations is also related to the symbolic predictive analysis [25] based on *concurrent trace programs (CTPs)*. However, in CTPs each event is not a concrete read or write (as in our case) but a symbolic statement derived from the program source code. The concurrent trace program in general captures more feasible interleavings, but it is also more expensive to check.

As an example, consider the Java program in Figure 1. Inside the `main` method, thread $t1$ creates threads $t2$ and $t3$, which execute methods $t1.run()$ and $t2.run()$, respectively. The shared variables are $a.x$ and $b.x$. Note that, according to the Java execution semantics, $a.x$ is aliased to $t2.v1.x$ and $t3.v2.x$, and $b.x$ is aliased to $t2.v2.x$ and $t3.v1.x$.

```
class Value {
1    private int x = 1;
2    public synchronized void add(Value v) {      13   public void run() {
3        x = x+v.get();                           14       v1.add(v2);
4    }                                             15   }}
5    public int get() {                            class Main {
6        return x;                                 16   public static void main (String[] args) {
7    }}                                            17       Value a = new Value();
class Task extends Thread {                        18       Value b = new Value();
8    Value v1; Value v2;                           19       Thread t2 = new Thread (new Task(a, b));
9    public Task(Value v1, Value v2) {             20       Thread t3 = new Thread (new Task(b, a));
10       this.v1 = v1;                             21       t2.start();
11       this.v2 = v2;                             22       t3.start();
12   }                                             23   }}
```

Fig. 1. A Java program with data races

Let $Tid = \{1, 2, 3\}$. Executing the program may result in the following partial trace, i.e. a subsequence of events from threads $t2$ and $t3$ as follows: ... (2,13-14), (2,2-3), (2,5-7), (2,4), (2,15), (3,13-14), (3,2-3), (3,5-7), (3,4), (3,15), where each event is denoted as a pair of the thread index and the line number(s). During this execution, the shared variable $b.x$ is read by thread $t2$ at line 6 (aliased as $t2.v1.x$) and written by thread $t3$ at line 3 (aliased as $t3.v2.x$). However, this trace is not a witness of data race because the two aforementioned accesses to $b.x$ are never simultaneously enabled. There exists an alternative interleaving of the same set of events: ... (2,13-14), (2,2-3), (2,5), (3,13-14), (3,2), **(2,6)**, **(3,3)**, (3,5-7), (3,4),(3,15), (2,7), (2,4), (2,15). It is a data race witness because there exists a state in which the read access by event (2,6) and the write access by event (3,3) are both enabled. It is guaranteed to be an actual program execution because both write-read consistency and synchronization consistency

The goal of our symbolic analysis is to search for witnesses among all sequentially consistent linearizations of $\mathcal{T}_\pi$ derived from the concrete execution π. We formulate the data race witness generation problem as a satisfiability problem. That is, we construct a quantifier-free first-order logic formula ψ_π such that the formula is satisfiable if and only if there exists a sequentially consistent linearization of $\mathcal{T}_\pi$ that leads to a state in which two data-conflict events are both enabled. The formula ψ_π is a conjunction of the following subformulas

$$\psi_\pi := \alpha_\pi \wedge \beta_\pi \wedge \gamma_\pi \wedge \rho_\pi$$

In Section 3 we present algorithms to encode the partial order (α_π), write-read consistency (β_π), and data race property (ρ_π) in first-order logic (FOL) formulas. In Section 4 we discuss the encoding of synchronization consistency (γ_π).

3   Symbolic Encoding of the Write-Read Consistency

3.1   Encoding the Partial Order

Given a multithreaded trace π, let $\pi|_t = \langle e_1^t, \ldots, e_n^t \rangle$ be a sub-sequence that is a projection of π onto the thread t. Let $t.first$ and $t.last$ be the first and last event of thread t in π,i.e., e_1^t and e_n^t, respectively. For each event e, we introduce an event order (EO) variable whose value represents its position in a linearization of $\mathcal{T}_\pi$. To ease our presentation, we assume that an EO variable shares the same unique index with the corresponding event. Therefore $o_{e.idx}$ is the EO variable for e. Let the number of events be $|\pi|$. The domain of o_i, where $1 \leq i \leq |\pi|$, is $[1..|\pi|]$. Furthermore, we have $o_i \neq o_j$ if $i \neq j$.

Equation 1 encodes the partial order requirement of sequentially consistent linearizations of $\mathcal{T}_\pi$. It enforces a total order within each thread-local sequence $\pi|_t (1 \leq t \leq N)$, and enforces the order between the first (or last) event of a thread and the corresponding fork (or join) event, if such event exists. In Equation 1 $FORK$ and $JOIN$ denote the set of $fork$ and $join$ events in $\mathcal{T}_\pi$. For an event $e \in FORK$, $e.val$ gives the child thread index, thus $(t_{e.val}).first.idx$ is the index of the first event in the child thread.

$$\alpha_\pi \equiv \left(\bigwedge_{t=1}^{T} \left(o_{e_1^t.idx} < \cdots < o_{e_n^t.idx} \right) \wedge \bigwedge_{e \in FORK} \left(o_{e.idx} < o_{(t_{e.val}).first.idx} \right) \wedge \bigwedge_{e \in JOIN} \left(o_{(t_{e.val}).last.idx} < o_{e.idx} \right) \right) \tag{1}$$

$$
\begin{array}{ll}
e_0 : (1, fork, -, 2) & \\
e_1 : (1, write, x, 1) & e_6 : (2, read, x, 0) \\
e_2 : (1, acquire, o, -) & e_7 : (2, notifyAll, o, -) \\
e_3 : (1, write, x, 0) & e_8 : (2, release, o, -) \\
e_4 : (1, wait, o, -) & e_9 : (2, read, x, 0) \\
e_5 : (2, acquire, o, -) & e_{10} : (1, release, o, -)
\end{array}
$$

partial order:
$\alpha_1 : o_0 < o_1 < o_2 < o_3 < o_4 < o_{10}$
$\alpha_2 : o_5 < o_6 < o_7 < o_8 < o_9$
$\alpha_3 : o_0 < o_5$
write-read consistency:
$\beta : (o_6 < o_1 \lor o_3 < o_6)$
$\quad \land (o_9 < o_1 \lor o_3 < o_9)$

Fig. 2. An execution with initial value $x = 0$

$$
\beta_\pi \equiv \bigwedge_{e \in \pi \land e.type=read} \left(\begin{array}{l} \left((e.tiwp = null) \land (e.val = e.var.init) \land \bigwedge_{e1 \in e.pws} (o_{e.idx} < o_{e1.idx}) \right) \lor \\ \bigvee_{e1 \in e.pwsv} \left(\begin{array}{l} (o_{e1.idx} < o_{e.idx}) \land \\ \bigwedge_{e2 \in e.pws \land e2 \neq e1} (o_{e.idx} < o_{e2.idx} \lor o_{e2.idx} < o_{e1.idx}) \end{array} \right) \end{array} \right)
\tag{2}
$$

$$
\rho_\pi \equiv \bigvee_{(e1,e2) \in PDR} ((o_{e1'.idx} < o_{e2.idx} < o_{e1''.idx}) \land (o_{e2'.idx} < o_{e1.idx} < o_{e2''.idx}))
\tag{3}
$$

Figure 2 show an execution trace π with 11 events $e_0, \ldots, e_{10}$ generated by two threads. The last column in Figure 2 lists the partial order constraints: α_1 and α_2 enforces a total order on the events from thread 1 and 2, respectively; α_3 ensures that the fork of thread 2 happens before the first event in thread 2.

3.2   Encoding Write-Read Consistency

Given a linearization l, we use $e_1 \prec_l e_2$ to denote that event e_1 happens before e_2 in l. Similarly, we use $e_1 \prec_t e_2$ to denote that e_1 happens before e_2 within the same thread t.

Definition 1. Linearization Immediate Write Predecessor: *Given a read event e in a linearization l, we define its* linearization immediate write predecessor , *denoted as $e.liwp$, to be a write event $e' \prec_l e$ such that $e.var = e'.var$ and there does not exist another write event e'' such that $e' \prec_l e'' \prec_l e$ and $e''.var = e.var$.*

Definition 2. Thread Immediate Write Predecessor: *Let $\pi|_t$ be the projection of execution π onto thread t. The* thread immediate write predecessor *to a read event e, denoted as $e.tiwp$, is a write event $e' \prec_t e$ in $\pi|_t$ such that $e.var = e'.var$ and there does not exist another write event e'' such that $e' \prec_t e'' \prec_t e$ and $e''.var = e.var$.*

Definition 3. Write-Read Consistency: *A linearization l is write-read consistent iff for any read event e (1) if there exists a write event e' such that $e' = e.liwp$, then $e.val = e'.val$; (2) if e' does not exist, then $e.val = e.var.init$. Here $e.var.init$ is the initial value of variable $e.var$.*

Definition 4. Predecessor Write Set: *Given an execution π, the* predecessor write set *of a read event e, denoted as $e.pws$ is a set that includes any write event e' such that $e'.var = e.var$ and (1) $e'.tid \neq e.tid$, or (2) $e'.tid = e.tid$ and $e' = e.tiwp$. The* predecessor write of the same value set *to a read event e, denoted as $e.pwsv$, is a subset of $e.pws$, where for any $e' \in e.pwsv$, we have $e'.val = e.val$.*

Equation 2 considers all the possible linearizations that satisfy the write-read consistency requirement. For each read event e in π, there are two possible cases:

1. e has no thread immediate write predecessor ($e.tiwp = null$), its read value is the same as the variable's initial value ($e.val = e.var.init$), and all the write events in the predecessor write set of e happen after e ($o_e.idx < o_{e1}.idx$). Note that the two equality constraints evaluate to either true or false statically, and therefore will not be added in the SMT formula.
2. e follows a write event $e1$ in its predecessor write of the same value set ($o_e.idx < o_{e1}.idx$), and all other writes to $e.var$ happens either before $e1$ ($o_{e2}.idx < o_{e1}.idx$), or after e ($o_e.idx < o_{e2}.idx$). This constraint guarantees that e reads the value written by $e1$ and no other writes can interfere with this write-read pair.

If all the read events satisfy the above constraints, as specified in Equation 2, the linearizations are write-read consistent. Consider the example in Figure 2. Column 3 shows the write-read constraints, along with some implementation optimizations, described as follows:

1. $o_6 < o_1$ requires that the read event e_6 appears before any write to x. Note that although $o_6 < o_3$ is also required as in Equation 2, it is removed (constant true) because it is implied by ($o_6 < o_1$) together with α_1.
2. $o_3 < o_6$ requires that the read event e_6 happens after e_3. Although the full constraint as in Equation 2 is ($o_3 < o_6$) $\wedge$ ($o_1 < o_3 \vee o_6 < o_1$), we remove the second conjunct because $o_1 < o_3$ is implied by α_1.

3.3   Encoding the Data Race

Definition 5. Data Race Witness: *An execution* $\pi = \pi_1 e_1 e_2 \pi_2$, *where* π_1 *and* π_2 *are the trace prefix and suffix, respectively, has a data race on* e_1 *and* e_2 *if the two events belong to different threads, access the same shared variable and at least one access is a write.*

Let PDR be the set of potential data races in $\mathcal{T}_\pi$, where each data race is represented as a pair ($e1, e2$) of events that belong to different thread ($e1.tid \neq e2.tid$), access the same variable ($e1.var = e2.var$), and at least one access is a write ($e1.type = write \vee e2.type = write$).

Given every event pair ($e1, e2$) $\in PDR$, let $e1'$ and $e1''$ be the events immediately before and after $e1$ in the same thread, and $e2'$ and $e2''$ be the events immediately before and after $e2$ in the same thread. Equation 3 captures the existence of a witness in which $e1$ and $e2$ are simultaneously reachable.

We can further reduce the number of data race constraints (currently 4) into 3 by adding $o_{e1.idx} < o_{e2.idx}$, since it implies the two existing constraints $o_{e1'.idx} < o_{e2.idx}$ and $o_{e1.idx} < o_{e2''.idx}$. A data race exists in an execution π if $e1$ is immediately followed by $e2$ in π. We do not need to consider the dual case that $e1$ immediately follows $e2$ because if such linearization exists, since it is guaranteed that the linearization in which $e2$ follows $e1$ exists as well.

4 Symbolic Encoding of the Synchronization Consistency

4.1 Synchronization Interpretation

The interpretation of the synchronization operations involves replacing object variables with simple-type variables available to SMT solvers, and map the synchronization operations on objects to logic operations on simple-type variables. Although Java allows recursive locks, they happen rarely in executions. An execution π has a recursive lock if there exist two events e_i and e_j in π such that $e_i = e_j = (t, acquire, o, -)$ and there is no event $(t, release, o, -)$ in between; otherwise π is called *recursive-lock-free*. If an execution π is recursive-lock-free, then any sequentially consistent linearization of $\mathcal{T}_\pi$ is also recursive-lock-free (a reorder of events within the same thread is not allowed). In this section we discuss the interpretation for recursive-lock-free executions and defer the discussion for executions with recursive locks until Section 4.3.

We introduce the following simple-type shared variables for each object o.

- An integer variable o_o with domain $[0..N]$, where N is the number of threads. Object o is free if o_o is 0. Otherwise o_o is the thread index that owns object o.
- N Boolean variables $o_{w\_t}(1 \leq t \leq N)$. The value of o_{w_t} is true iff thread t is in object o's wait set.

In the following we list the interpretation of the synchronization operations. For each variable v, we use the normal form v to indicate its current value, and use the primed version v' to indicate its value at the next step.

- Event $(t, acquire, o, -)$ is interpreted as $o_o = 0 \rightarrow o'_o = t$. It requires that the object is free, and then set the owner of object o to thread t.
- Event $(t, release, o, -)$ is interpreted as $o_o = t \rightarrow o'_o = 0$. It requires that the owner of object o is thread t, and then set object o to be free.
- Event $(t, wait, o, -)$ is converted into two consecutive atomic events. The first atomic event is interpreted as $(o_o = t \rightarrow o'_{w\_t} \wedge o_o = 0)$, which requires that the owner of thread o is thread t, and then sets object o to free and the flag o'_{w_t} to true. The second atomic event is interpreted as $(o_o = 0 \wedge \neg o_{w\_t}) \rightarrow o'_o = t$, which requires that object o is free and thread t is no longer waiting. For the *wait* event to complete, a *notify* or *notifyAll* event from another thread needs to interleave in between to reset o_{w_t}.
- Event $(t, notifyAll, o, -)$ is interpreted as $o_o = t \rightarrow \bigwedge_{t1 \in o.wait} \neg o'_{w\_t1}$, where $o.wait$ is the set of threads waiting on object o. It requires that the owner of o is thread t, and then reset o_{w_t1} for any waiting thread $t1$.
- Event $(t, notify, o, -)$ requires that *one and only one* thread waiting on o, if any, is waken up. We introduce N auxiliary variables H_{w_t} with domain $\{0, 1\}$, one for each thread $t \in Tid$, such that (1) H_{w_t} must have value 0 if thread t is not waiting for on o and (2) exactly one H_{w_t} has value 1 if the waiting set for o is not empty. The requirement can be obtained by the following constraints: $\bigwedge_{1 \leq t \leq N}(\neg o_{w\_t} \rightarrow \neg H_{w\_t} = 0), (\bigvee_{1 \leq t \leq N} o_{w\_t}) \rightarrow (\Sigma_{1 \leq t \leq N} H_{w\_t} = 1)$ Finally, the *notify* event is interpreted as $\bigwedge_{t \in Tid}(H_{w\_t} = 1 \rightarrow \neg o'_{w\_t} \wedge H_{w\_t} = 0 \rightarrow o'_{w\_t} = o_{w\_t})$, which states that thread t is no longer waiting on object o if it is chosen; otherwise its waiting status remains the same.

4.2   The Recursive-Lock-Free Encoding

In this section we present the constraints that enforce synchronization consistency for recursive-lock-free multithreaded traces. The first two columns in Table 1 give the interpretation of the synchronization events in Figure 2. The original wait event e_3 is split into two new events: e_3 and and its shadow event e_3'. Correspondingly we introduce an event order variable o_3' and adds partial order constraint $o_3 < o_3' < o_4$.

Definition 6. Initial Value: *The initial value $v.iv$, is defined as follows: (1) the value for a variable o_o that denotes the ownership of an object is 0, i.e. $o_o.iv = 0$, (2) the value for a variable that denotes whether thread t is waiting for an object is false, i.e. $o_{w\_t}.iv = false$ for $1 \leq t \leq N$.*

Assumed Value: *The assumed value of a variable v in a synchronization event e in the format of $assume \rightarrow update$, denoted $v_e.av$, is the value specified in the sub-formula $e.assume$. Here v is called an* assumed variable *in e, and $e.assume$ is the set of assumed variables in e.*

Written Value: *The written value of a variable v in a synchronization event e in the format of $assume \rightarrow update$, denoted as $v_e.wv$, is the value specified in the sub-formula $e.update$. v is called an* updated variable *in e, and $e.updated$ is the set of updated variables in e.*

$$\gamma_e \equiv \bigwedge_{v \in e.assume} \left(\begin{array}{l} \left(v_e.av = v.iv \land v_e.first \land \bigwedge_{e_1 \in v_e.pws} o_{e.idx} < o_{e_1.idx} \right) \lor \\ \bigvee_{e_1 \in v_e.pwsv} \left(\begin{array}{l} (o_{e.idx} < o_{e_1.idx}) \land \\ \bigwedge_{e_2 \in v_e.pws \land e_2 \neq e_1} (o_{e.idx} < o_{e_2.idx} \lor o_{e_2.idx} < o_{e_1.idx}) \end{array} \right) \end{array} \right) \quad (4)$$

Given a synchronization event e, Equation 4 enforces a valid position in any linearization for e with respect to other synchronization events. It considers each assumed variable v in e, and adds constraints on the position of e based on the v's assumed value:

- If v's assumed value in e, $v_e.av$, is the same as v's initial value $v.iv$, then e can be in a position that is before any write to v. That is, $\bigwedge_{e_1 \in v_e.pws} o_{e.idx} < o_{e_1.idx}$.
  Note that if there exist writes to v before e from the same thread, this constraint contradicts the partial order constraint thus becomes false.
- Event e follows an event $e_1 \in v_e.pwsv$. In this case e happens after $e_1(o_{e_1.idx} < o_{e.idx})$ so the assumed value at e can take updated value at e', and other

Table 1. Recursive-lock-free synchronization consistency Interpretation

Synchronization Event	Interpretation	Predecessor Write Set	Predecessor Write Set with Same Value
$e_2 : (1, acquire, o, -)$	$o_o = 0 \rightarrow o_o' = 1$	$o_o : \{e_5, e_8\}$	$o_o : \{e_8\}$
$e_4 : (1, wait, o, -)$	$o_o = 1 \rightarrow o_{w\_1}' \land o_o' = 0$	$o_o : \{e_2, e_5, e_8\}$	$o_o : \{e_2\}$
e_4'	$o_o = 0 \land \neg o_{w\_1} \rightarrow o_o' = 1$	$o_o : \{e_4, e_8, e_5\}$ $o_{w\_1} : \{e_4, e_7\}$	$o_o : \{e_4, e_8\}, o_{w\_1} : \{e_7\}$
$e_5 : (2, acquire, o, -)$	$o_o = 0 \rightarrow o_o' = 2$	$o_o : \{e_2, e_4, e_4', e_{10}\}$	$o_o : \{e_4, e_{10}\}$
$e_7 : (2, notifyAll, o, -)$	$o_o = 2 \rightarrow \neg o_{w\_1}'$	$o_o : \{e_2, e_4, e_4', e_5, e_{10}\}$	$o_o : \{e_5\}$
$e_8 : (2, release, o, -)$	$o_o = 2 \rightarrow o_o' = 0$	$o_o : \{e_2, e_4, e_4', e_5, e_{10}\}$	$o_o : \{e_5\}$
$e_{10} : (1, release, o, -)$	$o_o = 1 \rightarrow o_o' = 0$	$o_o : \{e_4', e_5, e_8\}$	$o_o : \{e_4'\}$

Table 2. Recursive-lock-free synchronization consistency encoding

Event	Encoding	Encoding with Optimization
e_2	$(o_2 < o_5 \wedge o_2 < o_8) \vee ((o_8 < o_2) \wedge (o_5 < o_8 \vee o_2 < o_5))$	$(o_2 < o_5) \vee (o_8 < o_2)$
e_4	$(o_2 < o_4) \wedge (o_5 < o_2 \vee o_4 < o_5) \wedge (o_8 < o_2 \vee o_4 < o_8)$	$(o_5 < o_2 \vee o_4 < o_5) \wedge (o_8 < o_2 \vee o_4 < o_8)$
e_4'	$\left(\begin{array}{l} (o_4 < o_{4'}) \wedge (o_5 < o_4 \vee o_{4'} < o_5) \\ \wedge (o_8 < o_4 \vee o_{4'} < o_8) \\ (o_8 < o_{4'}) \wedge (o_4 < o_8 \vee o_{4'} < o_4) \\ \wedge (o_5 < o_8 \vee o_{4'} < o_5) \end{array} \right) \vee$	$\left(\begin{array}{l} (o_4 < o_{4'}) \wedge (o_5 < o_4 \vee o_{4'} < o_5) \\ \wedge (o_8 < o_4 \vee o_{4'} < o_8) \end{array} \right) \vee \\ ((o_8 < o_{4'}) \wedge (o_4 < o_8))$
	$(o_7 < o_4') \wedge (o_4 < o_7 \vee o_4' < o_4)$	$(o_7 < o_4') \wedge (o_4 < o_7)$
e_5	$\left(\begin{array}{l} o_5 < o_2 \wedge o_5 < o_4 \wedge o_5 < o_4' \wedge o_5 < o_{10}) \vee \\ (o_4 < o_5) \wedge (o_2 < o_4 \vee o_5 < o_2) \wedge \\ (o_4' < o_4 \vee o_5 < o_4') \wedge (o_{10} < o_4 \vee o_5 < o_{10}) \wedge \\ (o_{10} < o_5) \wedge (o_2 < o_{10} \vee o_5 < o_2) \wedge \\ (o_4' < o_{10} \vee o_5 < o_4') \wedge (o_4 < o_{10} \vee o_5 < o_4) \end{array} \right) \vee$	$(o_5 < o_2) \vee (o_{10} < o_5) \vee \\ (o_4 < o_5 \wedge o_5 < o_4' \wedge o_5 < o_{10})$
e_7	$(o_5 < o_7) \wedge (o_2 < o_5 \vee o_7 < o_2) \wedge (o_4 < o_5 \vee o_7 < o_4) \wedge \\ o_4' < o_5 \vee o_7 < o_4') \wedge (o_{10} < o_5 \vee o_7 < o_{10})$	$(o_2 < o_5 \vee o_7 < o_2) \wedge (o_4 < o_5 \vee o_7 < o_4) \wedge \\ (o_4' < o_5 \vee o_7 < o_4') \wedge (o_{10} < o_5 \vee o_7 < o_{10})$
e_8	$(o_5 < o_8) \wedge (o_2 < o_5 \vee o_8 < o_2) \wedge (o_4 < o_5 \vee o_8 < o_4) \wedge \\ o_4' < o_5 \vee o_8 < o_4') \wedge (o_{10} < o_5 \vee o_8 < o_{10})$	$(o_2 < o_5 \vee o_8 < o_2) \wedge (o_4 < o_5 \vee o_8 < o_4) \wedge \\ (o_4' < o_5 \vee o_8 < o_4') \wedge (o_{10} < o_5 \vee o_8 < o_{10})$
e_{10}	$(o_4' < o_{10}) \wedge (o_5 < o_4' \vee o_{10} < o_5) \wedge (o_8 < o_4' \vee o_{10} < o_8)$	$(o_5 < o_4' \vee o_{10} < o_5) \wedge (o_8 < o_4' \vee o_{10} < o_8)$

events that write to v do not interfere by happening either before the write at e_1 or after the read at e.

Column 3 and 4 in Table 1 list the predecessor write set of the shared variables o_o and o_{w_1} and its subset, predecessor write with the same value set, respectively. Table 2 gives the encoding based on Equation 4. Although in Equation 4 there is a constraint $\left(v_e.av = v.iv \wedge \bigwedge_{e_1 \in v_e.pws} o_e.idx < o_{e_1}.idx \right)$, the constraint can be removed if v_e's value is not the same as the initial value, or be reduced to $\bigwedge_{e_1 \in v_e.pws} o_e.idx < o_{e_1}.idx$ if the values are the same. In addition, several other straightforward optimizations can be applied. Column 3 gives more concise encoding than Column 2 due to the following optimizations:

- A sub-formula s that can be implied by partial order constraint. For example, $o_6 < o_9$ in e_1 and $o_1 < o_3$ in e_3. This reduces $s \wedge s'$ to s, and $s \vee s'$ to $true$.
- A sub-formulas s that contradicts partial order constraint. For example, $o_3' < o_3$ in e_4 and $o_5 < o_3$ in e_6. This reduces $s \vee s'$ to s.
- A sub-formula s that is weaker than s' in $s \wedge s'$. For example, in $o_1 < o_6 \wedge o_1 < o_9$ in e_1, $o_1 < o_9$ can be removed because $o_6 < o_9$.

Finally the synchronization consistency constraint is specified by $\gamma_\pi \equiv \bigwedge_e \gamma_e$, where e is a synchronization event in π.

4.3   Encoding with Recursive Locks

If an execution π has recursive locks, we define a variable $depth_o^t$ that denotes the depth of object o that has been locked by thread t. The initial value of $depth_o^t$ is 0. For each sequence $\pi|_t$ that is a projection of π on thread t, we increase the value of $depth_o^t$ by 1 for each $(t, acquire, o, -)$, and decrease the value by 1 for each $(t, release, o, -)$. Depending on the value of $depth_o^t$, acquire and release events are encoded differently as the following:

- An event $e : (t, acquire, o, -)$ is called the *first acquire event* if $e.depth_o^t = 0$. Its corresponding constraint is $o_o = 0 \rightarrow o_o' = t$.

```
(1, acquire, l, -);
(1, write, x, 10);
(1, release , l, -);
                                    (2, acquire, l, -);
                                    (2, read, x, 10);
                                    (2, write, y, 20);
                                    (2, release , l, -);
(1, acquire, l, -);
(1, write, x, 20);
(1, release , l, -);
(1, write, y, 30);
```

Fig. 3. An execution with shared variables x, y

- For event $e : (t, acquire, o, -)$ that is not a first acquire event, its corresponding constraint is $o_o = t \rightarrow o'_o = t$.
- An event $e : (t, release, o, -)$ is called the *last release event* if $e.depth^t_o = 0$. Its corresponding constraint is $o_o = t \rightarrow o'_o = 0$.
- For event $e : (t, release, o, -)$ that is not a last release event, its corresponding constraint is $o_o = t \rightarrow o'_o = t$.

We do not need to explicitly record the depth of recursive locks. It is based on the observation that (1) π is a valid execution, thus the number of acquire and release events must be balanced; and (2) The depths of recursive locks associated with an acquire or release event (a thread-local property) will not be changed by thread interleavings.

4.4 Correctness and Complexity

Theorem 1. *Let π be the given multithreaded trace. There exists a data race witness in a sequentially consistent linearization of $\mathcal{T}_\pi$ iff ψ_π is satisfiable:*

$$\psi_\pi \equiv \alpha_\pi \wedge \beta_\pi \wedge \gamma_\pi \wedge \rho_\pi$$

According to the definitions of partial order constraint α_π, write-read consistency constraint β_π, and synchronization consistency constraint γ_π, a linearization of $\mathcal{T}_\pi$ that satisfies $\alpha_\pi \wedge \beta_\pi \wedge \gamma_\pi$ is sequentially consistent. Since the events are all from a real execution, a sequentially consistent linearization represents events from a valid execution as well. In addition, the definition of data race property enforces that in the linearization there are two adjacent events (at least one is a write event) from different threads accessing the same variable.

Our approach eliminates the bogus warnings reported by typical data race detection algorithms, e.g. those based on lock-set analysis. Consider the execution shown in Figure 3 where x, y are shared variables with initial value 0. A lock-set analysis will reports a data race warning between the two write events to y as one of them is not protected by any lock. Our approach will not produce a data race witness because write-read consistency enforces the read event of x in thread 2 must happen between the two write events to x in thread 1. In addition, each corresponding acquire-release pair is atomic according the synchronization constraints. Therefore the two write events are never enabled at the same time.

For most Java executions the number of synchronization events is very small compared with the number of total events. Since the majority of the constraints are generated from encoding read, write events and data race properties, their complexity determines the scalability of our approach. We note that these constraints are in pure integer difference logic (IDL) – an efficiently decidable subset of FOL where each IDL constraint is of the form $(x - y \leq c)$, where x and y are integer variables and c is 0.

5   Static Optimizations

In the implementation, we use the incremental feature of the Yices SMT solver [4]. We divide the constraints in ψ_π into two parts: $\psi_\pi = (\alpha_\pi \wedge \beta_\pi \wedge \gamma_\pi) \wedge \rho_\pi$, where the first part encodes all the sequentially consistent linearizations, and the second part states that a data race exists. Let ρ_π be a conjunction of subformulas $\rho_\pi(e_i, e_j)$, each of which states the simultaneous reachability of an event pair $(e_i, e_j) \in PDR$. Instead of building and checking ρ_π in one step (same as combining all potential data races in one check), we check each individual event pair in isolation. The incremental SAT procedure is as follows.

1. Within the SMT solver, we first construct the subformula $(\alpha_\pi \wedge \beta_\pi \wedge \gamma_\pi)$.
2. Then for the first data race event pair we construct $\rho_\pi(e_i, e_j)$ and add this subformula as a *retractable* assertion. The retractable assertion can be removed after satisfiability checking, while allowing the SMT solver to retain the lemmas (clauses) learned during the process. If the result is satisfiable, then the SMT solver returns a satisfying assignment (witness); otherwise, such witness does not exists.
3. After retracting the first assertion $\rho_\pi(e_i, e_j)$, we construct $\rho_\pi(e_i', e_j')$ for the second event pair (e_i', e_j') and add it to the SMT solver.

We keep repeating steps 2 and 3 till all the event pairs in PDR are checked. The benefit of using incremental SAT is reducing the overall runtime by sharing the cost of checking different data races. Although it might appear to be costly to call the SMT solver once for each potential data race in PDR, the entire process turns out to be efficient because of incremental SAT[1].

Typical data race detection algorithms (e.g. those based on locksets) have false alarms—sometimes many of them, which means the input to our witness generation algorithm, the set PDR of (potential) data races, may have event pair (e_i, e_j) such that e_i, e_j are not simultaneously reachable. Therefore, it is often advantageous to check, before calling the precise SMT analysis, whether (e_i, e_j) simultaneously reachable by using a conservative analysis. Our analysis is based on statically computing the following information: (1) lock acquisition histories [14]; (2) must-happen-before constraints, where event e_1 must happen before e_2 iff that is the case in every linearization of T_π. This analysis is in general comparable to and sometimes more precise than standard data race detectors (e.g. [8,1,6,18,17,13,24,5,9]).

---

[1] Often the first few SAT calls take a significant portion of the total runtime; after that, the "learned clauses" make the subsequent SAT calls extremely fast.

6   Experiments

We have implemented the proposed method and conducted experiments on some public benchmarks. We collected traces using a Java agent interface that captures the Java Virtual Machine Execution events. Our symbolic analysis is implemented using the Yices SMT solver [4]. All benchmark programs are accompanied by test cases to facilitate the concrete execution. Our experiments were conducted on a workstation with 2.8 GHz processor and 2GB memory.

Table 3 shows the experimental results. Among the benchmarks, `Example` (`run 1`) is the simple example illustrated in Figure 1, `Example` (`run 2`) is the same example except that the `get` method is synchronized. All other benchmarks are publicly available in [12,20,10,19,7]. The first two columns show the statistics of the test program, including the name and the number of threads. The next three columns show the statistics of the given trace, including the length (visible events only), the number of acquire/release events, and the number of wait/notify/notifyAll events. The next three columns show the number of data variables (rw), the number of lock variables (lk) and the number of condition variables (wn) in the trace. The last four columns show the statistics of the symbolic witness generation algorithm, including the number of potential data races after the lock acquisition history analysis (lsa), the number of potential data races after the must-happen-before analysis (mhb), the number of witnesses generated (wtns), and the runtime of our symbolic algorithm in seconds. During symbolic witness generation, we call the SMT solver incrementally, one at a time, only for the potential data races in the column *mhb*. The runtime in seconds is the combined processing time for all these potential data races.

The runtime results show that our witness generation algorithm scale to medium length traces, and is fast enough to be used as a postmortem analysis. In almost all cases, our static pruning based on lock acquisition history and must-happen-before constraints is able to reduce the number of potential data races significantly, therefore reducing the burden on the symbolic algorithm. We also note that, even after pruning, most of the potential data races do not have

Table 3. Performance of the symbolic data race witness generation algorithm

Test Program		Given Trace (events)			Shared Variables			Witness Generation			
name	threads	length	lk-evs	wn-evs	rw	lk	wn	lsa	mhb	wtns	time (s)
Example run1	3	25	4	0	6	2	0	8	2	1	0.01
Example run2	3	29	8	0	6	2	0	6	0	0	0.01
Remote Agent	3	45	12	5	6	3	4	12	4	2	0.01
connectionpool	4	85	16	5	5	1	3	21	0	0	0.01
liveness.BugGen	7	241	44	6	12	9	6	138	10	1	0.36
account #1	6	336	82	10	17	11	5	125	45	4	0.09
account #2	11	651	162	20	32	21	10	250	90	9	0.28
account #3	21	1281	322	40	62	41	20	500	180	19	0.79
SyncBench #1	2	107	22	0	3	2	0	8	2	1	0.01
SyncBench #2	13	722	156	0	16	3	0	805	333	40	18.3
BarrierBench #1	7	407	80	14	10	2	7	229	12	0	0.7
BarrierBench #2	13	653	136	28	16	2	7	361	38	0	2.04
philo	6	1050	126	41	23	6	22	563	0	0	0.0
hedc	10	1457	234	0	85	23	0	508	164	40	57.7
Daisy	3	1998	330	14	34	9	12	328	16	7	5.65
elevator	4	8000	1298	0	121	12	0	12	0	0	0.0
tsp	4	45637	20	5	42	5	3	83	4	3	0.05

concrete witnesses – they are likely to be bogus errors. This result highlights the problem associated with many data race detection algorithms in the literatures. Reporting such data races (warnings) directly to programmers could be counter-productive in practice, since it imposes significant burden (manual effort) on the programmers for deciding whether a reported data race is real.

7   Conclusion

Despite that numerous static and dynamic techniques exist to detect data races, few are capable of providing witnesses to help programmers understand how a data race can happen during program execution. In this paper we propose a SMT-based symbolic method to produce concrete witnesses for data races in concurrent programs. Our tool can be integrated seamlessly with traditional testing procedure because of the following reasons: (1) the inputs to our tool are ordinary program execution traces, (2) our approach amplifies the effectiveness of each testing run by considering all the alternative event interleavings, (3) the witnesses produced by our tool pinpoint data races and thus help programmers better understanding the erroneous behaviors. Our experimental results show that the proposed algorithm is scalable enough for a postmortem analysis.

References

1. Boyapati, C., Rinard, M.C.: A parameterized type system for race-free Java programs. In: OOPSLA 2001. SIGPLAN Notices, vol. 36(11), pp. 56–69. ACM, New York (2001)
2. Wang, C., Mahmoud Said, A.G.: Coverage guided systematic concurrency testing. In: International Conference on Software Engineering, ICSE 2011 (2011)
3. Christey, S. (ed.): Top 25 most dangerous programming errors. CWE/SANS report (2009), http://cwe.mitre.org/top25/
4. Dutertre, B., de Moura, L.: A Fast Linear-Arithmetic Solver for DPLL(T). In: Ball, T., Jones, R.B. (eds.) CAV 2006. LNCS, vol. 4144, pp. 81–94. Springer, Heidelberg (2006)
5. Elmas, T., Qadeer, S., Tasiran, S.: Goldilocks: a race and transaction-aware Java runtime. j-SIGPLAN 42(6), 245–255 (2007)
6. Engler, D., Ashcraft, K.: RacerX: effective, static detection of race conditions and deadlocks. In: ACM Symposium on Operating Systems Principles, pp. 237–252. ACM, New York (2003)
7. Farchi, E., Nir, Y., Ur, S.: Concurrent bug patterns and how to test them. In: Parallel and Distributed Processing, p. 286.2. IEEE Computer Society, Washington, DC (2003)
8. Flanagan, C., Freund, S.: Type-based race detection for Java. In: Programming Language Design and Implementation, pp. 219–232. ACM, New York (2000)
9. Flanagan, C., Freund, S.N.: Fasttrack: efficient and precise dynamic race detection. In: Programming Language Design and Implementation, pp. 121–133. ACM, New York (2009)
10. Havelund, K.: Using runtime analysis to guide model checking of java programs. In: Havelund, K., Penix, J., Visser, W. (eds.) SPIN 2000. LNCS, vol. 1885, pp. 245–264. Springer, Heidelberg (2000)
11. Havelund, K., Pressburger, T.: Model checking JAVA programs using JAVA PathFinder. International Journal on Software Tools for Technology Transfer (STTT) 2(4), 366–381 (2000)

12. Joint cav/issta special even on specification, verification, and testing of concurrent software, http://research.microsoft.com/qadeer/cavissta.htm
13. Kahlon, V., Yang, Y., Sankaranarayanan, S., Gupta, A.: Fast and accurate static data-race detection for concurrent programs. In: Damm, W., Hermanns, H. (eds.) CAV 2007. LNCS, vol. 4590, pp. 226–239. Springer, Heidelberg (2007)
14. Kahlon, V., Ivancic, F., Gupta, A.: Reasoning about threads communicating via locks. In: Etessami, K., Rajamani, S.K. (eds.) CAV 2005. LNCS, vol. 3576, pp. 505–518. Springer, Heidelberg (2005)
15. Manson, J., Pugh, W., Adve, S.V.: The java memory model. In: Principles of Programming Languages (2005)
16. Musuvathi, M., Qadeer, S., Ball, T., Musuvathi, M., Qadeer, S., Ball, T.: Chess: A systematic testing tool for concurrent software. Tech. Rep. MSR-TR-2007-149, Microsoft Research (2007)
17. Naik, M., Aiken, A.: Conditional must not aliasing for static race detection. In: Principles of programming languages. ACM, New York (2007)
18. Pratikakis, P., Foster, J., Hicks, M.: LOCKSMITH: context-sensitive correlation analysis for race detection. In: Programming Language Design and Implementation, pp. 320–331. ACM, New York (2006)
19. von Praun, C., Gross, T.R.: Static detection of atomicity violations in object-oriented programs. Object Technology 3(6) (2004)
20. The java grande forum benchmark suite, http://www2.epcc.ed.ac.uk/computing/research_activities/java_grande/index_1.html
21. Savage, S., Burrows, M., Nelson, G., Sobalvarro, P., Anderson, T.E.: Eraser: A dynamic data race detector for multi-threaded programs. ACM Trans. Comput. Syst. 15(4), 391–411 (1997)
22. Serbănută, T.F., Chen, F., Rosu, G.: Maximal causal models for multithreaded systems. Tech. Rep. UIUCDCS-R-2008-3017, University of Illinois at Urbana-Champaign (2008)
23. Siegel, S.F., Mironova, A., Avrunin, G.S., Clarke, L.A.: Using model checking with symbolic execution to verify parallel numerical programs. In: ISSTA (2006)
24. Voung, J., Jhala, R., Lerner, S.: RELAY: static race detection on millions of lines of code. In: Foundations of Software Engineering, pp. 205–214. ACM, New York (2007)
25. Wang, C., Kundu, S., Ganai, M., Gupta, A.: Symbolic predictive analysis for concurrent programs. In: International Symposium on Formal Methods. ACM, New York (2009)

Applying Atomicity and Model Decomposition to a Space Craft System in Event-B

Asieh Salehi Fathabadi, Abdolbaghi Rezazadeh, and Michael Butler

University of Southampton, UK
{asf08r,ra3,mjb}@ecs.soton.ac.uk

Abstract. Event-B is a formal method for modeling and verifying consistency of systems. In formal methods such as Event-B, refinement is the process of enriching or modifying an abstract model in a step-wise manner in order to manage the development of complex and large systems. To further alleviate the complexity of developing large systems, Event-B refinement can be augmented with two techniques, namely atomicity decomposition and model decomposition. Our main objective in this paper is to investigate and evaluate the application of these techniques when used in a refinement based development. These techniques have been applied to the formal development of a space craft system. The outcomes of this experimental work are presented as assessment results. The experience and assessment can form the basis for some guidelines in applying these techniques in future cases.

1 Introduction

Event-B [2] is a formal method that evolved from the B-Method [8] and Action Systems [10]. Simplicity of notation and structure is one of the primary reasons for choosing Event-B to develop formal models of our case study. Event-B also is proven to be applicable in different domains including distributed systems [2]. Moreover Event-B supports refinement and uses mathematical proofs to verify consistency of models. Furthermore there is good tool support for modeling and proving.

Exploration of the planet Mercury is the main goal of the BepiColombo mission [13], which consist of two orbiters. One of the orbiters is the Mercury Planetary Orbiter (MPO) which performs global remote sensing and radio science investigations. An important part of this orbiter consist of a core and four devices: Solar Intensity X-ray Spectrometer (SIXS-X and SIXS-P) and Mercury Imaging X-ray Spectrometer (MIXS-T and MIXS-C). The whole system is controlled by mission-critical software. The core and the control software are responsible for controlling the power of devices and their operation states and to handle TeleCommand (*TC*) and TeleMessage (*TM*) communications. In the rest of this paper we refer to the core and the devices including control software as the probe system. Our aim is to present a part of the probe system related to the management of TC and TM communications.

M. Bobaru et al. (Eds.): NFM 2011, LNCS 6617, pp. 328–342, 2011.

Modeling a large and complex system such as this probe system, can result in large and complex models with difficult proofs [3]. However Event-B provides some techniques to address this problem. One such technique is refinement that allows us to add details during a sequence of models instead of building a model in a flat manner. The Event-B refinement rules are very general and they do not explicitly represent relationships between abstract events and new events, introduced during refinement. Refinement can be augmented with another technique called atomicity decomposition [4] that provides a structuring mechanism for refinement in Event-B. Atomicity decomposition provides definitions and a diagrammatic notation to explicitly represent relationships between refinement levels. Using atomicity decomposition we can also illustrate the explicit sequencing between events of a model that is not always explicit in Event-B model. Model decomposition [6], [7] is another technique to divide a large model into smaller and more easily manageable sub-models.

Figure 1 presents the development architecture of Event-B model of the probe system. In the abstraction, $M0$, the main goal of the system is modeled. The details of the system are added through three refinement levels, $M1$, $M2$ and $M3$. Then the last model, $M3$, is decomposed to two sub-models, called $Core$ and $Device$. The intention in decomposing $M3$ is to decrease the complexity of the produced Event-B sub-models. Also this model decomposition reflects the structure of target architecture by separating the core from the devices. Finally the core sub-model is refined further in two levels of refinement, $M4$ and $M5$. During the refinement process both before and after model decomposition, the atomicity decomposition technique is employed to explicitly represent the event sequencing and relationships between abstract and refined events.

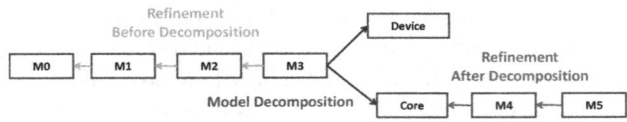

Fig. 1. Development Architecture of Event-B Model

The contribution of this paper is to assess the Event-B atomicity and model decomposition techniques in the development of a complex and large distributed system. In the development process of this case study we will explore how the atomicity decomposition technique will help us to structure refinement steps. After some refinement levels we will see how model decomposition can help us to manage the large model by cutting it into two smaller sub-models. Using the probe system as a carrier, our intention is to identify challenges and provide some solutions by using atomicity and model decomposition techniques. These solutions are presented as assessment results which can lead towards a guideline in using the atomicity decomposition and model decomposition techniques.

This paper is organized into 6 sections. Section 2 outlines the background of this work. Here we overview Event-B method and related techniques, namely atomicity decomposition and model decomposition. In Section 3 we present the Event-B model of the probe system including abstraction and refinement levels. Assessment results are outlined in Section 4 and finally we outline related work in Section 5 and conclude this paper in Section 6.

2   Background

2.1   Event-B and Refinement

The Event-B formal method [2] models the states and events of a system. Variables present the states. Events transform the system from a state to another state by changing the value of variables. The modeling notation is based on set theory and logic. Event-B uses mathematical proof to ensure consistency of a model.

Event-B Structure: An Event-B model [11] is made of several components of these two types, *Context* and *Machine*. Contexts contain the static part(types and constants) of a model while Machines contain the dynamic part(variables and events). A context can be "extended" by other contexts and "referenced" by machines. A Machine can be "refined" by other machines and reference contexts.

Refinement in Event-B: In Event-B development, rather than having a single large model, it is encouraged to construct the system in a series of successive layers, starting with an abstract representation of the system. The abstract model provides a simple view of the system, focusing on main purposes of the system. The details of how the purposes are achieved are ignored in the abstract specification. Details are added gradually to the abstract model in stepwise manner. This process called refinement [3]. In Event-B refinement is used to introduce new functionality or add details of current functionality. One of the important features of Event-B refinement is the ability to introduce new events in a refinement step. From a given machine, *Machine1*, a new machine, *Machine2*, can be built as a refinement of Machine1. In this case, *Machine1* is called an abstraction of *Machine2*, and *Machine2* will said to be a concrete version of *Machine1*.

Event-B Tool: Rodin [9] is an Eclipse-based tool for formal modeling and proving in Event-B. Rodin is an extensible tool that can be extended to include new features.

2.2   Atomicity Decomposition

Although refinement in Event-B provides a flexible approach to modeling, it has the limitation that we cannot explicitly represent the relationship between new events in a refinement and abstract events. To overcome this issue, the atomicity decomposition approach is proposed in [4]. The idea is to augment Event-B refinement with a graphical notation that is capable of representing the relations between abstract and concrete events explicitly. Using the atomicity decomposition approach has another advantage which is that we can represent event

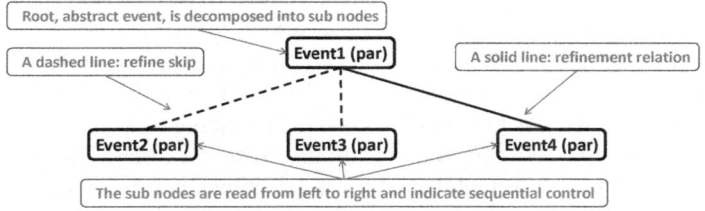

Fig. 2. Atomicity Decomposition Diagram

sequencing explicitly. An example of an atomicity decomposition diagram is presented in Figure 2. This diagram explicitly illustrates that the effect achieved by *Event*1 at the abstract level is realized at the refined level by occurrence of *Event*2 followed by *Event*3 followed by *Event*4. The execution order of the leaf events is always from left to right (this is based on JSD diagrams of Jackson [5]). We say that *Event*1 is a causal event for *Event*2 since it must occur before *Event*2 and so on. The solid line indicates that *Event*4 refines *Event*1 while the dashed lines indicate that *Event*2 and *Event*3 are new events. In standard Event-B refinement, *Event*2 and *Event*3 do not have any explicit connection with *Event*1. Technically, *Event*4 is the only event that refines *Event*1 but the diagram indicates that we break the atomicity of *Event*1 into three (sub-)events in the refinement.

The parameter *par* in the diagram indicates that we are modelling multiple instances of *Event*1 and its refining sub-events. Refined sub-events associated with different values of *par* may be interleaved thus modelling interleaved execution of multiple processes. Further details may be found in [4]. Two more diagrammatic concepts, "*XOR* case splitting" and "*ALL* replicator", are used in development of the case study and they will be explained later. Atomicity decomposition has been applied to a distributed file system in [4] and to a multi media protocol in [12]. The Event-B model for the diagram of Figure 2 is

event Event2	event Event3	event Event4 refines Event1
any *par*	any *par*	any *par*
where	where	where
@grd1 *par* ∈ PARAMETERS \ Event2	@grd1 *par* ∈ Event2 \ Event3	@grd1 *par* ∈ Event3 \ Event4
then	then	then
@act1 Event2 := Event2 ∪ {*par*}	@act1 Event3 := Event3 ∪ {*par*}	@act1 Event4 := Event4 ∪ {*par*}
end	end	end

Fig. 3. Event-B Model

presented in Figure 3. The effect of a refined event with parameter *par* is to add the value of *par* to a set with the same name as the event, i.e., *par* ∈ *Event*1 means that *Event*1 has occured with value *par*. The use of a set means that the same event can occur multiple times with different values for *par*. The guard of an event with value *par* specifies that the event has not already occured for value *par* but has occured for the causal event, e.g., the guard of *Event*3 says that *Event*2 has occurred and *Event*3 has not occurred for value *par*.

2.3   Model Decomposition

The motivation for model decomposition [6], [7] is to decrease the complexity of large models, increase the modularity and reflect the target architecture. After several layers of refinement and as a result of introducing new events, we can end up having to deal with many events and many state variables. The main idea of decomposition is to cut a model into sub-models which can be refined separately and more easily than the initial model. Independent sub-models provides the possibility of team development which seems a very attractive option for the industry.

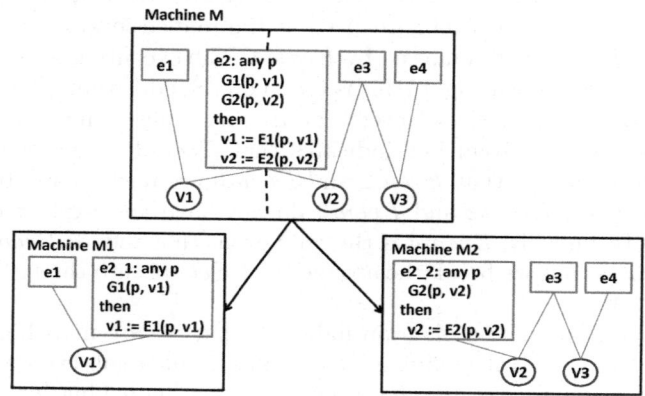

Fig. 4. Model Decomposition, Shared-event Style

In Event-B there are two ways of decomposing a model, shared-variable and shared-event. The shared-event approach is particularly suitable for message-passing in distributed systems, whereas the shared-variable approach is more suitable for concurrent systems. Since the probe system is a distributed system we use the shared-event approach in decomposing its model after three levels of refinement. In the shared-event model decomposition, variables are partitioned among the sub-models, whereas in shared-variable approach, events are partitioned among the sub-models. Shared-event model decomposition is presented graphically in Figure 4. First variables of the initial model M are partitioned among sub-models $M1; ...; Mn$ according to the devised policy. Then events of the initial model, M, are distributed among sub-models $M1; ...; Mn$, according to the variable partitioning. Events that are using variables allocated to different sub-models, called shared events, must be split between these sub-models. For example event $e2$ uses both $v1$ and $v2$ which are going to different sub-models. Therefore as depicted we have split it to $e2\_1$ and $e2\_2$ corresponding to variable $v1$ and $v2$ respectively. In the next stages the sub-models can be refined independently.

3   Event-B Model of the Probe System

3.1   An Overview of System Requirements and Development Process

The core software (CSW) plays a management role over the devices. CSW is responsible for communication with Earth on one hand and with the devices on the other hand. Here is the summary of the system requirements:

- A TeleCommand (*TC*) is received by the Core from Earth.
- The CSW checks the syntax of the received *TC*.
- Further semantic checking has to be carried out on the syntactically validated *TC*. If the *TC* contains a message for one of the devices, it has to be sent to the device for semantic checking, otherwise the semantic checking is carried out in the core.
- For each validate *TC* a control TeleMessage (*TM*) is generated and sent to Earth.
- For some particular types of *TC*, one or more data *TM*s are generated and sent back to Earth.

As mentioned earlier, we only present the part of the probe system that handles TeleCommands and TeleMessages communications. In Figure 1 of Section 1 we diagrammatically presented the development process of the probe system in Event-B. The development process consists of:

- Machine *M0* models the goal of the probe system. Three main events are receiving a *TC*, validating the received *TC*, and generating one or more *TM(s)* if it is needed.
- In machine *M1* the validation phase is refined and further details of validation process are added.
- In machine *M2* we distinguish between validation checking of *TC*s that should be carried out by the core or the devices.
- In machine *M3* we refine the model to introduce the process of sending related *TC*s to the devices for further validation and processing.
- Machine *M4* and *M5* model producing and sending *TM*s carried out in the core.

3.2   Abstract Specification

In the abstract model, the main goal of the system is modeled. Abstract events are illustrated in Figure 5 with a diagram resembling an atomicity decomposition diagram. Note that the top box is the system name rather than an event name (as the case in an atomicity decomposition diagram). In addition to this we only use solid lines to show the events of the abstract specification. After receiving a *TC*, three different scenarios are possible. Scenario(a): the received *TC* is validated and in response to this *TC*, it is necessary to produce some data. This is achieved by the occurrences of the third event. The response is sent back to Earth in the form of some data TMs by the occurrences of the fourth event. Scenario(b): for some TC's type there is no need to generate data TMs in response. Producing a control TM is later done by refining the $TC\_Validation\_Ok$ event. Scenario(c): it shows the case that the validation of a received TC fails. This is modeled by $TC\_Validation\_Fail$ event.

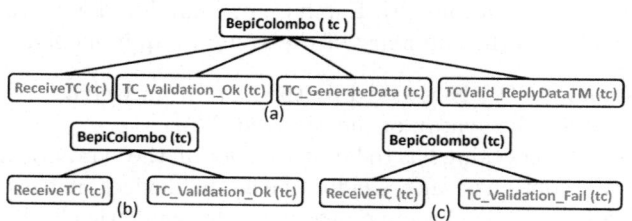

Fig. 5. Abstract Events, Machine $M0$

The sequencing between events is specified by following the rules explained in Section2.2. In abstract machine, $M0$, there are five sets used as control variables. Using sets allows multiple instance of a TC to be processed concurrently in an interleaved fashion. Figure 6 shows variables and invariants of $M0$. For each event there is a variable with the same name as the event, and if one event appears after another one in the sequence, its variable is a subset of the variable associated with the former. For example, as described before $TC\_Validation\_Ok$ event can occur only after occurrence of $ReceiveTC$ event, so invariant $inv2$ describes $TC\_Validation\_Ok$ variable as a subset of $ReceiveTC$ variable.

variables	invariants
ReceiveTC	@inv1 ReceiveTC ⊆ TC
TC_Validation_Ok	@inv2 TC_Validation_Ok ⊆ ReceiveTC
TCValid_GenerateData	@inv3 TCValid_GenerateData ⊆ TC_Validation_Ok
TCValid_ReplyDataTM	@inv4 TCValid_ReplyDataTM ⊆ TCValid_GenerateData
TC_Validation_Fail	@inv5 TC_Validation_Fail ⊆ ReceiveTC
	@inv6 TC_Validation_Ok ∩ TC_Validation_Fail = ∅

Fig. 6. Variables and Invariants of the Abstract Machine $M0$

To enforce the exact ordering of Figure 5.(a), when a TC is received we add it to the variable $ReceiveTC$ of the $ReceiveTC$ event. This event is represented in Figure 7. The guard of $TC\_Validation\_Ok$ event means that only after this stage, it is possible for the $TC\_Validation\_Ok$ event to occur and to add this TC to the list of validated TCs.

3.3   First Level of Refinement: Introducing Validation Steps

In the abstract model, the validation process is carried out in a single stage. The outcome can be either ok or fail which is modeled by $TC\_Validation\_Ok$ and $TC\_Validation\_Fail$ events. However validating a received TC is not an atomic action, accomplished in a single stage. It is done in two steps, checking the

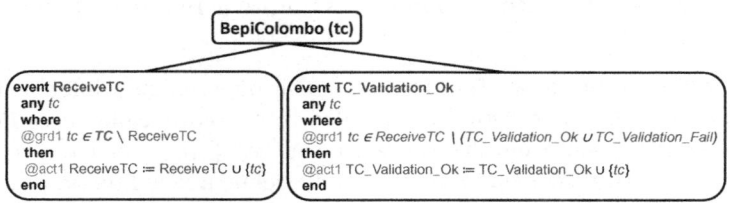

Fig. 7. Event-B Model of Sequencing between Events of the Abstract Machine $M0$

syntax and semantic of a received *TC*. After syntax and semantic checks, in the third step a control *TM* is produced and sent. These details are modeled in the first level of refinement, named machine *M1*. It can be seen in Figure 8 that *TC_Validation_Ok* and *TC_Validation_Fail* are decomposed to sub-events which show further details of the validation process. Checking the syntax of a received *TC* is modeled by *TCCheck_Ok* and *TCCheck_Fail* events. The semantic checking is modeled by *TCExecute_Ok* and *TCExecute_Fail* events. *TCExecOk_ReplyCtrlTM*, *TCExecFail_ReplyCtrlTM* and *TCCheckFail_ReplyCtrlTM* are events for generating control *TM*s. Again the Event-B model can be produced following the rules explained in Section2.2.

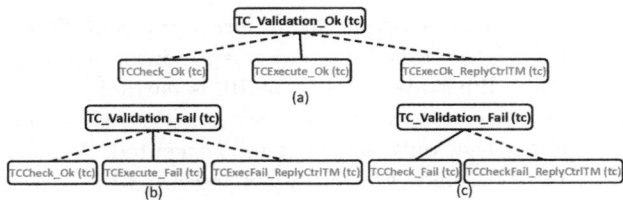

Fig. 8. Atomicity Decomposition of Validation Events, Machine $M1$

For each solid line in atomicity decomposition diagram there is an invariant which shows the relation between the set variable corresponding to abstract event and concrete variable of the refined event. There are three invariants in machine *M1*, shown in Figure 9. For example, *inv9* shows that concrete variable of *TCExecute_Ok* is a subset of abstract variable of *TC_Validation_Ok*, since the *TCExecute_Ok* event refines *TC_Validation_Ok* event.

@inv9 TCExecute_Ok ⊆ TC_Validation_Ok
@inv10 TCExecute_Fail ⊆ TC_Validation_Fail
@inv11 TCCheck_Fail ⊆ TC_Validation_Fail

Fig. 9. Invariants, Machine $M1$

3.4 Second Level of Refinement: Distinguish between Different Types of TCs

In this stage we are in a position to distinguish between two different types of TCs. There are TCs that should be handled by the core, called csw TCs, and TCs that should be sent from the core to the devices, ($mixsc$, $mixst$, $sixsp$, $sixsx$), to be processed. To model this new aspect, we define a new function called PID which maps every TC either to the core or the devices.

So far semantic checking of a received TC is done regardless of considering the type of TC. Now that we have the distinction between the core TCs and the devices TCs. If a received TC belongs to the core, its semantic should be checked in the core, otherwise it should be sent to a one of the devices for validation and processing. It is helpful to emphasis that syntax checking is exclusively carried out in the core.

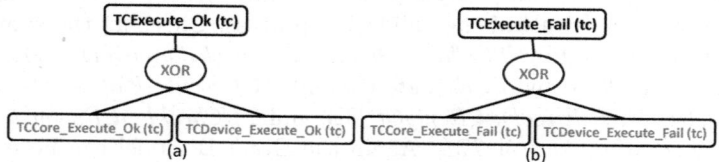

Fig. 10. Case Splitting, Machine $M2$

To model different cases associated with different types of TCs, both $TCExecute\_Ok$ event and $TCExecute\_Fail$ event are split into two sub-events. The splitting of these events, illustrated in Figure 10, is carried out using a special construct, called XOR or case splitting. In case splitting, an event is split into some sub-events in a way that only one of them is executed. As it can be seen in Figure 10, XOR, case splitting is graphically represented by a circle containing an "XOR". We draw the attention of the reader to the fact that XOR refers to mutual exclusion of events' execution, but guards of events do not need to be disjoint.

Figure 11 presents the Event-B model of Figure 10.(a). Note that both sub-events refine the abstract event. In the both sub-events we have added a new guard, $grd2$, which check the type of TCs.

3.5 Third Level of Refinement: Refining TCs Processing by the Devices

In the previous level we introduced the distinction between two types of TCs that are processed by the core and the devices respectively. In this level our aim is to refine the case of processing TCs by the devices. As presented in Figure 12, we applied the atomicity decomposition approach to three events of the previous level. By introducing communication between the core and devices, the abstract event, $TCDevice\_Execute\_ok$, is refined to $SendTC\_Core\_to\_Device$, $CheckTC\_in\_Device\_Ok$ and $SendOkTC\_Device\_to\_Core$ events. These three

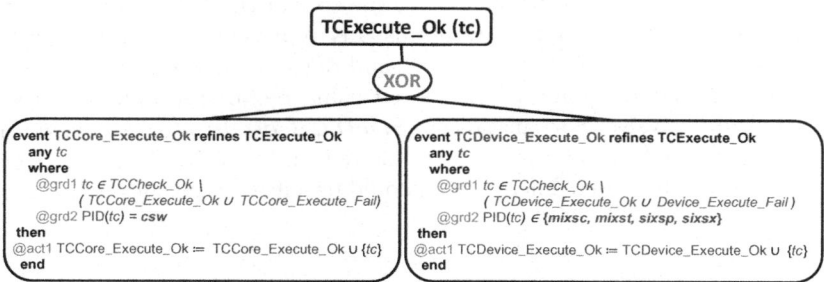

Fig. 11. Event-B Model, Machine $M2$

events model the case where a TC is successfully processed by a device and some response is generated for the core. In Figure 12.(b) a very similar approach is followed for the case when processing of a TC fails in a device and the atomicity of the abstract event, $TCDevice\_Execute\_Fail$, is decomposed to three sub-events based on the atomicity decomposition rules. Note that in Figure 12.(a) and (b) the event with the solid line, which directly refines the abstract event, appears in the middle rather than being the last one. Finally in Figure 12.(c), we show how $TCValid\_GenerateData$ is refined into two events to represent the case where extra data is produced in response to a TC.

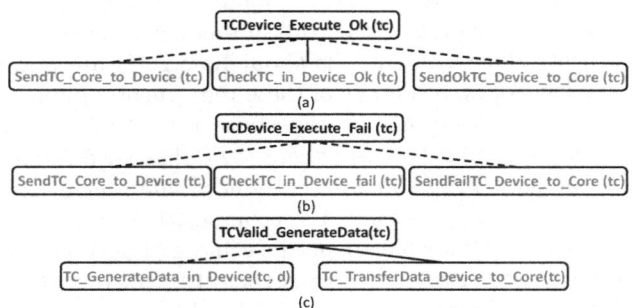

Fig. 12. Atomicity Decomposition Diagrams, Machine $M3$

3.6 Decomposing the Probe Model to the Core and Devices Sub-models

So far by applying atomicity decomposition in a few consecutive steps, we have managed to distinguish between events of the core and devices. Also we have reached the stage that we have a big model consisting of several events and many variables. Therefore it is a good time to take the next step and by applying the model decomposition, divide our current Event-B model to two sub-models, namely core and devices. When it comes to model decomposition we can identify three types of events, events that belong to the core or the devices or events that are shared between them. Shared events usually represent communication links and they should be split between the core and devices sub-models.

In Figure 13 shared events are presented using rectangles and variables are presented using ovals. For instance, *SendTC_Core_to_Device* event uses *TCCheck_ok* variable from the core sub-model and *SendTC_Core_to_Device* from the devices sub-model. Therefore it should be slit between these sub-models.

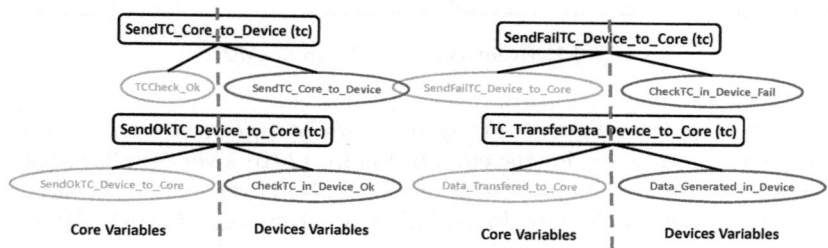

Fig. 13. Shared Events

3.7   Further Refinements of the Core Sub-model

After decomposing our intermediate Event-B model to two sub-models, we have carried out two further refinement of the core sub-model as depicted in Figure 1. These refinements introduce some details about how *TM*s are produced in response to *TC*s. We have omitted details of these refinements. Figure 14 presents the *TCValid_ReplyCtrlTM* event and its two consecutive levels of atomicity decomposition. This is modeling the case where a *TC* has successfully processed and in response some data *TM*s should be produced and sent back to Earth.

Here using Figure 14 an extra atomicity decomposition concept is explained. In response to a *TC*, it is possible to produce more than one data *TM*. To model such a situation we have used a construct [12] called "*ALL* replicator" applied to *TCValid_ProcessDataTM* event. The *ALL*, parameterized by *tm*, means that *TCValid_ProcessDataTM* occurs for multiple values of *tm* and the *TCValid_CompleteDataTM* can only occur when all the values of *tm* associated with a *tc* have occurred. In Event-B we model this by adding a parameter, which is a set containing all possible *TM*s that should be produced in response to a *TC*.

Another interesting aspect in Figure 14 is the sequencing order between leaf events. Based on the atomicity decomposition rules, *Produce_DataTM* event should be completed before *TCValid_CompleteDataTM* event. However there is no sequencing enforced between *Send_DataTM* and *TCValid_CompleteDataTM* events. This means that sending *TM*s to Earth can be carried out before or after occurrence of *TCValid_CompleteDataTM* event. This concept is discussed in more detail in [12].

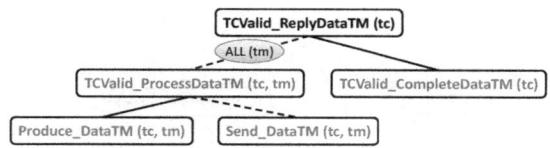

Fig. 14. *"ALL"* Construct, the Core Sub-Model

4   Assessment

In this section we discuss how the atomicity and model decomposition techniques helped us in enhancing the development process of the probe system. We also explain notable effects of these techniques in term of methodological contribution that can form a basis for a set of future guidelines. As a part of our formal modeling, we have developed a substantial set of Event-B models including three levels of refinement before model decomposition and two levels of refinement after it. In total the Rodin tool produced 174 proofs, 158 of them discharged automatically. The remaining proofs are discharged interactively. Atomicity decomposition diagrams enabled us to explore and explain our formal development without getting into technical details of the underlying Event-B models. We consider this as an advantage of the atomicity decomposition technique. The next important advantage of this technique is that we can explicitly represent refinement relations between events of different levels. Another merit of atomicity decomposition technique is the capability of representing sequencing between events of the same model. Further aspects are discussed in the following sections.

4.1   Providing Insight for Appropriate Event Decomposition

During the development process, atomicity decomposition diagrams helped us to spot some flaws in our decomposition approach. For example if the adapted approach did not cover all desired scenarios, we managed to discover this from the diagrams before attempting to produce any Event-B code.

To clarify this further, in Figure 15 we present one possible way of decomposing the atomicity of *TC* validation process. Applying two successive levels of atomicity decomposition to the abstract event *TC_Validation* results in four sub-events. The diagram shows that the possible scenarios are: <TCCheck_OK(tc) and TCExecute_OK(tc)> or <TCCheck_Fail(tc)> or <TCExecute_Fail(tc)>. Clearly this approach does not cover the case where *TCCheck_Ok* and *TCExecute_Fail* events can happen together as described in Section 3.3. This helped us to go back to the abstract level and followed an appropriate of atomicity decomposition which was presented in Section 3.3.

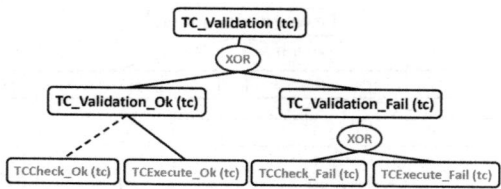

Fig. 15. An Example of Wrong Atomicity Decomposition

4.2   Assessing the Influence of Atomicity Decomposition and Model Decomposition over Each Other

In this case study we have used both atomicity decomposition and model decomposition together. One interesting aspect is to investigate whether by analyzing atomicity decomposition diagrams a decision can be made on a proper point that model decomposition can be applied. Atomicity decomposition diagrams provide an overall visualization of the refinement process. By grouping relevant events together, it is easier to decide about the point at which we can apply model decomposition.

Usually when we develop a system, we have a target architecture in mind. Therefore the outcome of the model decomposition should give us the desired sub-models. To be able to decompose an Event-B model, all events should either belong to one of sub-models or otherwise they should model communication links between its sub-models. In this regard model decomposition can provides us with some hint to which events, atomicity decomposition should be applied as a preparation stage for model decomposition.

To clarify this aspect we use a part of development process presented in Figure 16. As a preparation for model decomposition, we have applied atomicity decomposition to events such as *TCExecute_Ok* to distinguish between functionality of the core and devices. Note that leaf events satisfy the pre-mentioned condition that either should belong to one of the sub-models or represent communication links.

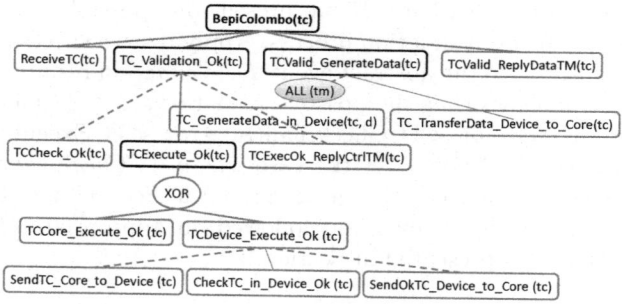

Fig. 16. Overall Refinement Structure before Model Decomposition

5   Related Work

The desire to explicitly model control flow is not restricted to Event-B. To address this issue usually a combination of two or more formal methods are suggested. A good example of such approach is Circus [14] combining CSP [15] and Z [16]. The combination of CSP and Classical B [8] also has been investigated in [17] and [18] with some differences. To explicitly define event sequencing in Event-B the Flows Approach is suggested in [19]. Another method to provide explicit control flow for an Event-B model is presented in [20] which is again based on using CSP alongside Event-B. These methods only deal with event sequencing; they do not support the explicit refinement of atomicity decomposition diagrams. UML-B [21] provides a "UML-like" graphical front-end for Event-B. It adds support for class-oriented and state machine modeling. State machines provide us with a graphical notation to explicitly define event sequencing.

Atomicity decomposition approach provides a graphical front-end to Event-B along other features such as supporting event sequencing and expressing refinement relations between concrete and abstract events. Also it can be combined effectively with other techniques such as model decomposition.

6   Conclusion

In this paper we demonstrated how atomicity decomposition diagrams provide a systematic means of introducing control structure into the Event-B development process. It also provides a means to express refinement relations between events of different refinement levels, through a set of hierarchal diagrams. In addition it can be merged with model decomposition technique to manage the complexity of large models. We have done an assessments of this approach and some merits of it explained in the previous section. In future work we hope that the outcomes of this stage can contribute toward providing some guidelines for atomicity and model decomposition. During the development of this case study, translation from atomicity decomposition diagrams to Event-B was carried out manually. As a continuation of this work, currently we are working on a tool providing support for producing atomicity decomposition diagrams as well as translating them to Event-B. This tool will be developed as a plug-in for the Rodin toolset.

Acknowledgement. This work is partly supported by the EU research project ICT 214158 DEPLOY (Industrial deployment of system engineering methods providing high dependability and productivity) www.deploy-project.eu.

References

1. Abrial, J.-R.: Formal Methods: Theory Becoming Practice. J.UCS 13(5), 619–628 (2007)
2. Abrial, J.-R.: Modeling in Event-B: System and Software Engineering. Cambridge University Press, Cambridge (2010)

3. Abrial, J.-R.: Refinement, Decomposition and Instantiation of Discrete Models. In: Abstract State Machines, pp. 17–40 (2005)
4. Butler, M.: Decomposition structures for event-B. In: Leuschel, M., Wehrheim, H. (eds.) IFM 2009. LNCS, vol. 5423, pp. 20–38. Springer, Heidelberg (2009)
5. Jackson, M.A.: System Development. Prentice-Hall, Englewood Cliffs (1983)
6. Silva, R., Pascal, C., Hoang, T.S., Butler, M.: Decomposition Tool for Event-B. In: ABZ (2010)
7. Pascal, C., Silva, R.: Event-B Model Decomposition. DEPLOY Plenary Technical Workshop (2009)
8. Abrial, J.-R.: The B-book: Assigning Programs to Meanings. Cambridge University Press, Cambridge (1996)
9. Abrial, J.-R., Butler, M., Hallerstede, S., Hoang, T.S., Mehta, F., Voisin, L.: Rodin: An Open Toolset for Modelling and Reasoning in Event-B. Technical Report, DE-PLOY Project (2009), http://deploy-eprints.ecs.soton.ac.uk/130/
10. Back, R.-J., Kurki-Suonio, R.: Distributed Cooperation with Action Systems. ACM Trans. Program. Lang. Syst. 10(4), 513–554 (1988)
11. Hallerstede, S.: Justifications for the Event-B Modelling Notation. In: Julliand, J., Kouchnarenko, O. (eds.) B 2007. LNCS, vol. 4355, pp. 49–63. Springer, Heidelberg (2006)
12. Salehi Fathabadi, A., Butler, M.: Applying Event-B Atomicity Decomposition to a Multi Media Protocol. In: de Boer, F.S., Bonsangue, M.M., Hallerstede, S., Leuschel, M. (eds.) FMCO 2009. LNCS, vol. 6286, pp. 89–104. Springer, Heidelberg (2010)
13. ESA Media Center, Space Science. Factsheet: Bepicolombo, http://www.esa.int/esaSC/120391_index_0_m.html
14. Zeyda, F., Cavalcanti, A.: Mechanised Translation of Control Law Diagrams into Circus. In: Leuschel, M., Wehrheim, H. (eds.) IFM 2009. LNCS, vol. 5423, pp. 151–166. Springer, Heidelberg (2009)
15. Hoare, C.A.R.: Communicating Sequential Processes. Prentice Hall, Englewood Cliffs (1985) ISBN 0-13-153289-8
16. Davies, J., Woodcock, J.: Using Z: Specification, Refinement and Proof. Prentice Hall International Series in Computer Science (1996) ISBN 0-13-948472-8
17. Butler, M.: csp2B: A Practical Approach to Combining CSP and B. Formal Aspects of Computing 12, 182–196 (2000) ISSN 0934-5043
18. Schneider, S., Treharne, H.: Verifying Controlled Components. In: Boiten, E.A., Derrick, J., Smith, G.P. (eds.) IFM 2004. LNCS, vol. 2999, pp. 87–107. Springer, Heidelberg (2004)
19. Iliasov, A.: Tutorial on the Flow plugin for Event-B. In: Workshop on B Dissemination [WOBD] Satellite event of SBMF, Natal, Brazil, November 8-9 (2010)
20. Schneider, S., Treharne, H., Wehrheim, H.: A CSP approach to control in event-B. In: Méry, D., Merz, S. (eds.) IFM 2010. LNCS, vol. 6396, pp. 260–274. Springer, Heidelberg (2010)
21. Said, M.Y., Butler, M., Snook, C.: Language and Tool Support for Class and State Machine Refinement in UML-B. In: Cavalcanti, A., Dams, D.R. (eds.) FM 2009. LNCS, vol. 5850, pp. 579–595. Springer, Heidelberg (2009)

A Theory of Skiplists with Applications to the Verification of Concurrent Datatypes*

Alejandro Sánchez[1] and César Sánchez[1,2]

[1] The IMDEA Software Institute, Madrid, Spain
[2] Spanish Council for Scientific Research (CSIC), Spain
{alejandro.sanchez,cesar.sanchez}@imdea.org

Abstract. This paper presents a theory of skiplists with a decidable satisfiability problem, and shows its applications to the verification of concurrent skiplist implementations. A skiplist is a data structure used to implement sets by maintaining several ordered singly-linked lists in memory, with a performance comparable to balanced binary trees. We define a theory capable of expressing the memory layout of a skiplist and show a decision procedure for the satisfiability problem of this theory. We illustrate the application of our decision procedure to the temporal verification of an implementation of concurrent lock-coupling skiplists. Concurrent lock-coupling skiplists are a particular version of skiplists where every node contains a lock at each possible level, reducing granularity of mutual exclusion sections.

The first contribution of this paper is the theory $\mathsf{TSL_K}$. $\mathsf{TSL_K}$ is a decidable theory capable of reasoning about list reachability, locks, ordered lists, and sublists of ordered lists. The second contribution is a proof that $\mathsf{TSL_K}$ enjoys a finite model property and thus it is decidable. Finally, we show how to reduce the satisfiability problem of quantifier-free $\mathsf{TSL_K}$ formulas to a combination of theories for which a many-sorted version of Nelson-Oppen can be applied.

1 Introduction

A skiplist [14] is a data structure that implements sets, maintaining several sorted singly-linked lists in memory. Skiplists are structured in multiple levels, where each level consists of a single linked list. The skiplist property establishes that the list at level $i+1$ is a sublist of the list at level i. Each node in a skiplist stores a value and at least the pointer corresponding to the lowest level list. Some nodes also contain pointers at higher levels, pointing to the next element present at that level. The advantage of skiplists is that they are simpler and more efficient to implement than search trees, and search is still (probabilistically) logarithmic.

---

* This work was funded in part by the EU project FET IST-231620 *HATS*, MICINN project TIN-2008-05624 *DOVES*, CAM project S2009TIC-1465 *PROMETIDOS*, and by the COST Action IC0901 *Rich ModelToolkit-An Infrastructure for Reliable Computer Systems*.

M. Bobaru et al. (Eds.): NFM 2011, LNCS 6617, pp. 343–358, 2011.
© Springer-Verlag Berlin Heidelberg 2011

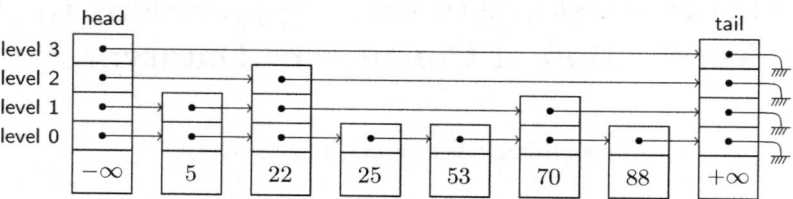

Fig. 1. A skiplist with 4 levels

Consider the skiplist shown in Fig. 1. Contrary to single-linked lists implementations, higher-level pointers allow to *skip* many elements during the search. A search is performed from left to right in a top down fashion, progressing as much as possible in a level before descending. For instance, in Fig. 1 a search for value 88 starts at level 3 of node *head*. From *head* the pointer at level 3 reaches *tail* with value $+\infty$, which is greater than 88. Hence the search algorithm moves down one level at *head* to level 2. The successor at level 2 contains value 22, which is smaller than 88, so the search continues at level 2 until a node containing a greater value is found. At that moment, the search moves down one further level again. The expected logarithmic search follows from the probability of any given node occurs at a certain level decreasing by 1/2 as a level increases (see [14] for an analysis of the running time of skiplists).

We are interested in the formal verification of implementations of skiplists, in particular in temporal verification (liveness and safety properties) of sequential and concurrent implementations. This verification activity requires to deal with unbounded mutable data. One popular approach to verification of heap programs is Separation Logic [17]. Skiplists, however, are problematic for separation-like approaches due to the aliasing and memory sharing between nodes at different levels. Based on the success of separation logic some researchers have extended this logic to deal with concurrent programs [23,7], but concurrent datatypes follow a programming style in which the activities of concurrent threads are not structured according to critical regions with memory footprints. In these approaches based on Separation Logic memory regions are implicitly declared (hidden in the separation conjunction), which makes the reasoning about unstructured concurrency more cumbersome.

Most of the work in formal verification of pointer programs follows program logics in the Hoare tradition, either using separation logic or with specialized logics to deal with the heap and pointer structures [9,24,3]. However, extending these logics to deal with concurrent programs is hard, and though some success has been accomplished it is still an open area of research, particularly for liveness.

Continuing our previous work [18] we follow a complementary approach. We start from temporal deductive verification in the style of Manna-Pnueli [11], in particular using general verification diagrams [5,19] to deal with concurrency. This style of reasoning allows a clean separation in a proof between the temporal part (why the interleavings of actions that a set of threads can perform satisfy a certain property) with the underlying data being manipulated. A verification diagram decomposes a formal proof into a finite collection of verification conditions (VC), each of which corresponds to the effect that a small step in the program has in the data. To automatize the process of checking the proof represented by a verification diagram it is necessary to use decision procedures for the kind of data structures manipulated. This paper studies the automatic verification of VCs for the case of skiplists.

Logics like [9, 24, 3] are very powerful to describe pointer structures, but they require the use of quantifiers to reach their expressive power. Hence, these logics preclude a combination a-la Nelson-Oppen [12] or BAPA [8] with other aspects of the program state. Instead, our solution starts from a quantifier-free theory of single-linked lists [16], and extends it in a non trivial way with order and sublists of ordered lists. The logic obtained can express skiplist-like properties without using quantifiers, allowing the combination with other theories. Proofs for an unbounded number of threads are achieved by parameterizing verification diagrams, splitting cases for interesting threads and producing a single verification condition to generalize the remaining cases. However, in this paper we mainly focus in the decision procedure. Since we want to verify concurrent lock-based implementations we extend the basic theory with locks, lock ownership, and sets of locks (and in general stores of locks). The decision procedure that we present here supports the manipulation of explicit regions, as in regional logic [2] equipped with *masked regions*, which enables reasoning about disjoint portions of the same memory cell. We use masked regions to "separate"different levels of the same skiplist node.

We call our theory $\mathsf{TSL_K}$, that allows to reason about skiplists of height at most K. To illustrate the use of this theory, we sketch the proof of termination of every invocation of an implementation of a lock-coupling concurrent skiplist.

The rest of the paper is structured as follows. Section 2 presents lock-coupling concurrent skiplists. Section 3 introduces $\mathsf{TSL_K}$. Section 4 shows that $\mathsf{TSL_K}$ is decidable by proving a finite model property theorem, and describes how to construct a more efficient decision procedure using the many-sorted Nelson-Oppen combination method. Finally, Section 5 concludes the paper. Some proofs are missing due to space limitation.

2 Fine-Grained Concurrent Lock-Coupling Skiplists

In this section we present a simple concurrent implementation of skiplists that uses lock-coupling [6] to acquire and release locks. This implementation can be seen as an extension of concurrent lock-coupling lists [6, 23] to multiple layers of pointers. This algorithm imposes a locking discipline, consisting of acquiring locks as the search progresses, and releasing a node's lock only after the lock of the next node in the search process has been acquired. A naïve implementation of this solution would equip each node with a single lock, allowing multiple threads to access simultaneously different nodes in the list, but protecting concurrent accesses to two different fields of the same node. The performance can

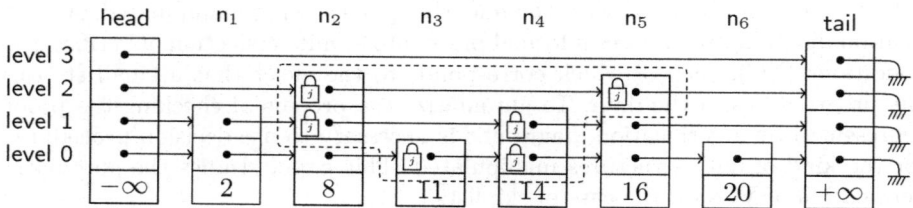

Fig. 2. A skiplist with the masked region given by the fields locked by thread j

be improved by carefully allowing multiple threads to simultaneously access the same node at different levels. We study here an implementation of this faster solution in which each node is equipped with a different lock at each level. At execution time a thread uses locks to protect the access to only some fields of a given node. A precise reasoning framework needs to capture those portions of the memory protected by a set of locks, which may include only *parts* of a node. Approaches based on strict separation (separation logic [17] or regional logic [2]) do not provide the fine grain needed to reason about individual fields of shared objects. Here, we introduce the concept of *masked regions* to describe regions and the fields within. A masked region consists of a set of pairs formed by a region (*Node* cell) and a field (a skiplist level): $\mathbf{mrgn} \triangleq 2^{Node \times \mathbb{N}}$ We call the field a mask, since it identifies which part of the object is relevant. For example, in Fig. 2 the region within dots represents the area of the memory that thread j is protecting. This portion of the memory is described by the masked region $\{(n_2, 2), (n_5, 2), (n_2, 1), (n_4, 1), (n_3, 0), (n_4, 0)\}$. As with regional logic, an empty set intersection denotes separation. In masked regions two memory nodes at different levels do not overlap. This notion is similar to data-groups [10].

Fig. 3(a) contains the pseudo-code declaration of the *Node* and *SkipList* classes. Throughout the paper we use //@ to denote ghost code added for verification purposes. Note that the structure is parametrized by a value K, which determines the maximum possible level of any node in the modeled skiplist. The fields *val* and *key* in the class *Node* contains the value and the key of the element used to order them. Then, we can store key-value pairs, or use the skiplist as a set of arbitrary elements as long as the key can be used to compare. The *next* array stores the pointers to the next nodes at each of the possible K different levels of the skiplist. Finally, the *lock* array keeps the locks, one for each level, protecting the access to the corresponding *next* field. The *SkipList* class contains two pointer fields: *head* and *tail* plus a ghost variable field r. Field *head* points to the first node of the skiplist, and *tail* to the last one. Variable r, only used for verification purposes, keeps the (masked) region represented by all nodes in the skiplist with all their levels. In this implementation, *head* and *tail* are sentinel nodes, with $key = -\infty$ and $key = +\infty$, respectively. For simplicity, these nodes are not eliminated during the execution and their *val* field remains unchanged.

```
class Node {                          class SkipList {
    Value val;                            Node* head;
    Key key;                              Node* tail;
    Array⟨Node*⟩(K) next;                 //@ mrgn r;
    Array⟨Node*⟩(K) lock;             }
}
```

(a) data structures

```
 1: procedure INSERT(SkipList sl, Value newval)
 2:     Vector⟨Node*⟩upd[0..K − 1]                    //@ mrgn m_r := ∅
 3:     lvl := randomLevel(K)
 4:     Node* pred := sl.head
 5:     pred.locks[K − 1].lock()                      //@ m_r := m_r ∪ {(pred, K − 1)}
 6:     Node* curr := pred.next[K − 1]
 7:     curr.locks[K − 1].lock()                      //@ m_r := m_r ∪ {(curr, K − 1)}
 8:     for i := K − 1 downto 0 do
 9:         if i < K − 1 then
10:             pred.locks[i].lock()                  //@ m_r := m_r ∪ {(pred, i)}
11:             if i ≥ lvl then
12:                 curr.locks[i + 1].unlock()        //@ m_r := m_r − {(curr, i + 1)}
13:                 pred.locks[i + 1].unlock()        //@ m_r := m_r − {(pred, i + 1)}
14:             end if
15:             curr := pred.next[i]
16:             curr.locks[i].lock()                  //@ m_r := m_r ∪ {(curr, i)}
17:         end if
18:         while curr.val < newval do
19:             pred.locks[i].unlock()                //@ m_r := m_r − {(pred, i)}
20:             pred := curr
21:             curr := pred.next[i]
22:             curr.locks[i].lock()                  //@ m_r := m_r ∪ {(curr, i)}
23:         end while
24:         upd[i] := pred
25:     end for
26:     Bool valueWasIn := (curr.val = newval)
27:     if valueWasIn then
28:         for i := 0 to lvl do
29:             upd[i].next[i].locks[i].unlock()      //@ m_r := m_r − {(upd[i].next[i], i)}
30:             upd[i].locks[i].unlock()              //@ m_r := m_r − {(upd[i], i)}
31:         end for
32:     else
33:         x := CreateNode(lvl, newval)
34:         for i := 0 to lvl do
35:             x.next[i] := upd[i].next[i]
36:             upd[i].next[i] := x                   //@ sl.r := sl.r ∪ {(x, i)}
37:             x.next[i].locks[i].unlock()           //@ m_r := m_r − {(x.next[i], i)}
38:             upd[i].locks[i].unlock()              //@ m_r := m_r − {(upd[i], i)}
39:         end for
40:     end if
41:     return ¬valueWasIn
42: end procedure
```

(b) insertion algorithm

Fig. 3. Data structure and insert algorithm for concurrent lock-coupling skiplist

Fig. 3(b) shows the implementation of the insertion algorithm. The algorithms for searching and removing are similar, and omitted due to space limitations. The ghost variable m_r stores a masked region containing all the nodes and fields currently locked by the running thread. The set operations $\cup$ and $-$ are used for the manipulation of the corresponding sets of pairs.

Let sl be a pointer to a skiplist (an instance of the class described in Fig. 3(a)). The following predicate captures whether sl points to a well-formed skiplist of height 4 or less:

$$SkipList_4(h, sl : SkipList) \ \hat{=} \ OList(h, sl, 0) \ \wedge \qquad (1)$$

$$\left(\begin{array}{l} h[sl].tail.next[0] = null \wedge h[sl].tail.next[1] = null \\ h[sl].tail.next[2] = null \wedge h[sl].tail.next[3] = null \end{array} \right) \wedge \qquad (2)$$

$$\left(\begin{array}{l} SubList(h, sl.head, sl.tail, 1, sl.head, sl.tail, 0) \ \wedge \\ SubList(h, sl.head, sl.tail, 2, sl.head, sl.tail, 1) \ \wedge \\ SubList(h, sl.head, sl.tail, 3, sl.head, sl.tail, 2) \end{array} \right) \qquad (3)$$

The predicate $OList$ in (1) describes that in heap h, the pointer sl is an ordered linked-lists when repeatedly following the pointers at level 0 starting at $head$. The predicate (2) indicates all levels are $null$ terminated, and (3) indicates that each level is in fact a sublist of its nearest lower level. Predicates of this kind also allow to express the effect of programs statements via first order transition relations. Consider the statement at line 36 in program $insert$ shown in Fig. 3(b) on a skiplist of height 4, taken by thread with id t. This transition corresponds to a new node x at level i being connected to the skiplist. If the memory layout from pointer sl is that of a skiplist before the statement at line 36 is executed, then it is also a skiplist after the execution:

$$SkipList_4(h, sl) \wedge \varphi_{aux} \wedge \rho_{36}^{[t]}(V, V') \rightarrow SkipList_4(h', sl')$$

The effect of the statement at line 36 is represented by the first-order transition relation $\rho_{36}^{[t]}$. To ensure this property, i is required to be a valid level, and the key of the nodes that will be pointing to x must be lower than the key of node x. Moreover, the masked region of locked nodes remains unchanged. Predicate φ_{aux} contains support invariants. For simplicity, we use $prev$ for $upd^{[t]}[i]$. Then, the full verification condition is:

$$SkipList_4(h, sl) \wedge \left(\begin{array}{l} x.key = newval \ \wedge \\ prev.key < newval \ \wedge \\ x.next[i].key > newval \ \wedge \\ prev.next[i] = x.next[i] \ \wedge \\ (x, i) \notin sl.r \wedge 0 \le i \le 3 \end{array} \right) \wedge \left(\begin{array}{l} at_{36}[t] \qquad \wedge \\ prev'.next[i] = x \ \wedge \\ at'_{37}[t] \qquad \wedge \\ h' = h \wedge sl = sl' \ \wedge \\ x' = x \qquad \ldots \end{array} \right) \rightarrow$$

$$SkipList_4(h', sl')$$

As usual, we use primed variables to describe the values of the variables after the transition is taken. Section 4 contains a full verification condition. This example illustrates that to be able to automatically prove VCs for the verification of skiplist manipulating algorithms, we require a theory that allows to reason about heaps, addresses, nodes, masked regions, ordered lists and sublists.

3   The Theory of Concurrent Skiplists of Height K: $\mathsf{TSL_K}$

We build a decision procedure to reason about skiplist of height K combining different theories, aiming to represent pointer data structures with a skiplist layout, masked regions and locks. We extend the Theory of Concurrent Linked Lists (TLL3) [18], a decidable theory that includes reachability of concurrent list-like structures in the following way:

- each node is equipped with a *key* field, used to reason about element's order.
- the reasoning about single level lists is extended to all the K levels.
- we extend the theory of regions with masked regions.
- lists are extended to ordered lists and sub-paths of ordered lists.

We begin with a brief description of the basic notation and concepts. A signature Σ is a triple (S, F, P) where S is a set of sorts, F a set of functions and P a set of predicates. If $\Sigma_1 = (S_1, F_1, P_1)$ and $\Sigma_2 = (S_2, F_2, P_2)$, we define $\Sigma_1 \cup \Sigma_2 = (S_1 \cup S_2, F_1 \cup F_2, P_1 \cup P_2)$. Similarly we say that $\Sigma_1 \subseteq \Sigma_2$ when $S_1 \subseteq S_2$, $F_1 \subseteq F_2$ and $P_1 \subseteq P_2$. If $t(\varphi)$ is a term (resp. formula), then we denote with $V_\sigma(t)$ (resp. $V_\sigma(\varphi)$) the set of variables of sort σ occurring in t (resp. φ).

A Σ-interpretation is a map from symbols in Σ to values. A Σ-structure is a Σ-interpretation over an empty set of variables. A Σ-formula over a set X of variables is satisfiable whenever it is true in some Σ-interpretation over X. Let Ω be a signature, $\mathcal{A}$ an Ω-interpretation over a set V of variables, $\Sigma \subseteq \Omega$ and $U \subseteq V$. $\mathcal{A}^{\Sigma,U}$ denotes the interpretation obtained from $\mathcal{A}$ restricting it to interpret only the symbols in Σ and the variables in U. We use $\mathcal{A}^\Sigma$ to denote $\mathcal{A}^{\Sigma,\emptyset}$. A Σ-theory is a pair $(\Sigma, \mathbf{A})$ where Σ is a signature and $\mathbf{A}$ is a class of Σ-structures. Given a theory $T = (\Sigma, \mathbf{A})$, a T-interpretation is a Σ-interpretation $\mathcal{A}$ such that $\mathcal{A}^\Sigma \in \mathbf{A}$. Given a Σ-theory T, a Σ-formula φ over a set of variables X is T-satisfiable if it is true on a T-interpretation over X. Formally, the theory of skiplists of height K is defined as $\mathsf{TSL_K} = (\Sigma_{\mathsf{TSL_K}}, \mathbf{TSLK})$, where

$$\Sigma_{\mathsf{TSL_K}} = \Sigma_{\mathsf{level_K}} \cup \Sigma_{\mathsf{ord}} \cup \Sigma_{\mathsf{thid}} \cup \Sigma_{\mathsf{cell}} \cup \Sigma_{\mathsf{mem}} \cup \Sigma_{\mathsf{reach}} \cup$$
$$\Sigma_{\mathsf{set}} \cup \Sigma_{\mathsf{setth}} \cup \Sigma_{\mathsf{mrgn}} \cup \Sigma_{\mathsf{bridge}}$$

The signature of $\mathsf{TSL_K}$ is shown in Fig. 4. **TSLK** is the class of $\Sigma_{\mathsf{TSL_K}}$-structures satisfying the conditions depicted in Fig. 5. The symbols of Σ_{set} and Σ_{setth} follow their standard interpretation over sets of addresses and thread identifiers resp.

Informally, sort addr represents addresses; elem the universe of elements that can be stored in the skiplist; ord the ordered keys used to preserve a strict order in the skiplist; thid thread identifiers; level_K the levels of a skiplist; cell models *cells* representing a node in a skiplist; mem models the heap, mapping addresses to cells or to *null*; path describes finite sequences of non-repeating addresses to

Signt	Sort	Functions	Predicates
$\Sigma_{\mathsf{level_K}}$	$\mathsf{level_K}$	$0, 1, \ldots, \mathsf{K} - 1 : \mathsf{level_K}$	$<: \mathsf{level_K} \times \mathsf{level_K}$
Σ_{ord}	ord	$-\infty, +\infty : \mathsf{ord}$	$\preceq : \mathsf{ord} \times \mathsf{ord}$
Σ_{thid}	thid	$\oslash : \mathsf{thid}$	
Σ_{cell}	cell elem ord addr thid	$error \quad : \mathsf{cell}$ $mkcell \quad : \mathsf{elem} \times \mathsf{ord} \times \mathsf{addr}^\mathsf{K} \times \mathsf{thid}^\mathsf{K} \to \mathsf{cell}$ $\_.data \quad : \mathsf{cell} \to \mathsf{elem}$ $\_.key \quad : \mathsf{cell} \to \mathsf{ord}$ $\_.next[\_] \quad : \mathsf{cell} \times \mathsf{level_K} \to \mathsf{addr}$ $\_.lockid[\_] \; : \mathsf{cell} \times \mathsf{level_K} \to \mathsf{thid}$ $\_.lock[\_] \quad : \mathsf{cell} \times \mathsf{level_K} \times \mathsf{thid} \to \mathsf{cell}$ $\_.unlock[\_] : \mathsf{cell} \times \mathsf{level_K} \to \mathsf{cell}$	
Σ_{mem}	mem addr cell	$null : \mathsf{addr}$ $\_[\_] \quad : \mathsf{mem} \times \mathsf{addr} \to \mathsf{cell}$ $upd \quad : \mathsf{mem} \times \mathsf{addr} \times \mathsf{cell} \to \mathsf{mem}$	
Σ_{reach}	mem addr path	$\epsilon \; : \mathsf{path}$ $[\_] : \mathsf{addr} \to \mathsf{path}$	$append : \mathsf{path} \times \mathsf{path} \times \mathsf{path}$ $reach_\mathsf{K} \; : \mathsf{mem} \times \mathsf{addr} \times \mathsf{addr}$ $\times \mathsf{level_K} \times \mathsf{path}$
Σ_{set}	addr set	$\emptyset \quad : \mathsf{set}$ $\{\_\} \quad : \mathsf{addr} \to \mathsf{set}$ $\cup, \cap, \setminus : \mathsf{set} \times \mathsf{set} \to \mathsf{set}$	$\in : \mathsf{addr} \times \mathsf{set}$ $\subseteq : \mathsf{set} \times \mathsf{set}$
Σ_{setth}	thid setth	$\emptyset_T \quad : \mathsf{setth}$ $\{\_\}_T \quad : \mathsf{thid} \to \mathsf{setth}$ $\cup_T, \cap_T, \setminus_T : \mathsf{setth} \times \mathsf{setth} \to \mathsf{setth}$	$\in_T : \mathsf{thid} \times \mathsf{setth}$ $\subseteq_T : \mathsf{setth} \times \mathsf{setth}$
Σ_{mrgn}	mrgn addr $\mathsf{level_K}$	$\mathbf{emp_{mr}} \quad : \mathsf{mrgn}$ $\langle\_,\_\rangle_{mr} \quad : \mathsf{addr} \times \mathsf{level_K} \to \mathsf{mrgn}$ $\cup_{mr}, \cap_{mr}, -_{mr} : \mathsf{mrgn} \times \mathsf{mrgn} \to \mathsf{mrgn}$	$\in_{mr} : \mathsf{addr} \times \mathsf{level_K} \times \mathsf{mrgn}$ $\subseteq_{mr} : \mathsf{mrgn} \times \mathsf{mrgn}$ $\#_{mr} : \mathsf{mrgn} \times \mathsf{mrgn}$
Σ_{bridge}	mem addr set path	$path2set \quad : \mathsf{path} \to \mathsf{set}$ $addr2set_\mathsf{K} : \mathsf{mem} \times \mathsf{addr} \times \mathsf{level_K} \to \mathsf{set}$ $getp_\mathsf{K} \quad : \mathsf{mem} \times \mathsf{addr} \times \mathsf{addr} \times \mathsf{level_K} \to \mathsf{path}$ $fstlock_\mathsf{K} \quad : \mathsf{mem} \times \mathsf{path} \times \mathsf{level_K} \to \mathsf{addr}$	$ordList : \mathsf{mem} \times \mathsf{path}$

Fig. 4. The signature of the $\mathsf{TSL_K}$ theory

model non-cyclic list paths; set models sets of addresses – also known as regions –, while setth models sets of thread identifiers and mrgn masked regions.

$\Sigma_{\mathsf{level_K}}$ contains symbols for level identifiers 0, 1, ..., $\mathsf{K} - 1$ and their conventional order. Σ_{ord} contains two special elements $-\infty$ and ∞ for the lowest and highest values in the order $\preceq$. Σ_{thid} only contains, besides $=$ and $\neq$ as for all the other theories, a special constant $\oslash$ to represent the absence of a thread identifier. Σ_{cell} contains the constructors and selectors for building and inspecting

Interpret. of sorts: addr, elem, thid, level$_K$, ord, cell, mem, path, set, setth and mrgn

Each sort σ in Σ_{TSL_K} is mapped to a non-empty set $\mathcal{A}_\sigma$ such that:
(a) $\mathcal{A}_{addr}$ and $\mathcal{A}_{elem}$ are discrete sets   (b) $\mathcal{A}_{thid}$ is a discrete set containing $\oslash$
(c) $\mathcal{A}_{level_K}$ is the finite collection $0,\ldots,K\text{-}1$ (d) $\mathcal{A}_{ord}$ is a total ordered set
(e) $\mathcal{A}_{cell} = \mathcal{A}_{elem} \times \mathcal{A}_{ord} \times \mathcal{A}_{addr}^K \times \mathcal{A}_{thid}^K$ (f) $\mathcal{A}_{mem} = \mathcal{A}_{cell}^{\mathcal{A}_{addr}}$
(g) $\mathcal{A}_{path}$ is the set of all finite sequences of   (h) $\mathcal{A}_{set}$ is the power-set of $\mathcal{A}_{addr}$
     (pairwise) distinct elements of $\mathcal{A}_{addr}$ (i) $\mathcal{A}_{setth}$ is the power-set of $\mathcal{A}_{thid}$
(j) $\mathcal{A}_{mrgn}$ is the power-set of $\mathcal{A}_{addr} \times \mathcal{A}_{level_K}$

Signature	Interpretation
Σ_{ord}	$x \preceq^{\mathcal{A}} y \wedge y \preceq^{\mathcal{A}} x \rightarrow x = y$ $x \preceq^{\mathcal{A}} y \vee y \preceq^{\mathcal{A}} x$ for any $x, y, z \in \mathcal{A}_{ord}$ $x \preceq^{\mathcal{A}} y \wedge y \preceq^{\mathcal{A}} z \rightarrow x \preceq^{\mathcal{A}} z$ $-\infty^{\mathcal{A}} \preceq^{\mathcal{A}} x \wedge x \preceq^{\mathcal{A}} +\infty^{\mathcal{A}}$
Σ_{cell}	$-\ mkcell^{\mathcal{A}}(e, k, \overrightarrow{a}, \overrightarrow{t}) = \langle e, k, \overrightarrow{a}, \overrightarrow{t} \rangle$ $-\ error^{\mathcal{A}}.next^{\mathcal{A}} = null^{\mathcal{A}}$ $-\ \langle e, k, \overrightarrow{a}, \overrightarrow{t} \rangle.data^{\mathcal{A}} = e$ $-\ \langle e, k, \overrightarrow{a}, \overrightarrow{t} \rangle.key^{\mathcal{A}} = k$ $-\ \langle e, k, \overrightarrow{a}, \overrightarrow{t} \rangle.next^{\mathcal{A}}[j] = a_j$ $-\ \langle e, k, \overrightarrow{a}, \overrightarrow{t} \rangle.lockid^{\mathcal{A}}[j] = t_j$ $-\ \langle e, k, \overrightarrow{a}, ...t_{j-1}, t_j, t_{j+1}...\rangle.lock^{\mathcal{A}}[j](t') = \langle e, k, \overrightarrow{a}, ...t_{j-1}, t', t_{j+1}...\rangle$ $-\ \langle e, k, \overrightarrow{a}, ...t_{j-1}, t_j, t_{j+1}...\rangle.unlock^{\mathcal{A}}[j] = \langle e, k, \overrightarrow{a}, ...t_{j-1}, \oslash, t_{j+1}...\rangle$ for each $e \in \mathcal{A}_{elem}$, $k \in \mathcal{A}_{ord}$, $t_0, \ldots, t_j, t_{j+1}, t_{j-1} \in \mathcal{A}_{thid}$, $t' \in \mathcal{A}_{thid}$, $\overrightarrow{a} \in \mathcal{A}_{addr}^K$, $\overrightarrow{t} \in \mathcal{A}_{thid}^K$ and $j \in \mathcal{A}_{level_K}$
Σ_{mem}	$m[a]^{\mathcal{A}} = m(a)$ $upd^{\mathcal{A}}(m, a, c) = m_{a \mapsto c}$ $m^{\mathcal{A}}(null^{\mathcal{A}}) = error^{\mathcal{A}}$ for each $m \in \mathcal{A}_{mem}$, $a \in \mathcal{A}_{addr}$ and $c \in \mathcal{A}_{cell}$

Σ_{reach}	$-\ \epsilon^{\mathcal{A}}$ is the empty sequence
	 $-\ [i]^{\mathcal{A}}$ is the sequence containing $i \in \mathcal{A}_{addr}$ as the only element
	 $-\ ([i_1 .. i_n], [j_1 .. j_m], [i_1 .. i_n, j_1 .. j_m]) \in append^{\mathcal{A}}$ iff $i_k \neq j_l$.
	 $-\ (m, a_{init}, a_{end}, l, p) \in reach_K{}^{\mathcal{A}}$ iff $a_{init} = a_{end}$ and $p = \epsilon$, or there exist
	    addresses $a_1, \ldots, a_n \in \mathcal{A}_{addr}$ such that:
	      (a) $p = [a_1 .. a_n]$ (c) $m(a_r).next^{\mathcal{A}}[l] = a_{r+1}$, for $r < n$
	      (b) $a_1 = a_{init}$ (d) $m(a_n).next^{\mathcal{A}}[l] = a_{end}$

Σ_{mrgn}	$-\ \mathbf{emp}_{mr}^{\mathcal{A}} = \emptyset$ $-\ r \cup_{mr}^{\mathcal{A}} s = r \cup s$ $-\ (a,j) \in_{mr}^{\mathcal{A}} r \leftrightarrow (a,j) \in r$ $-\ \langle a,j \rangle_{mr}^{\mathcal{A}} = \{(a,j)\}$ $-\ r \cap_{mr}^{\mathcal{A}} s = r \cap s$ $-\ r \subseteq_{mr}^{\mathcal{A}} s \leftrightarrow r \subseteq s$ $-\ r -_{mr}^{\mathcal{A}} s = r \setminus s$ $-\ r \#_{mr}^{\mathcal{A}} s \leftrightarrow r \cap_{mr}^{\mathcal{A}} s = \mathbf{emp}_{mr}^{\mathcal{A}}$ for each $a \in \mathcal{A}_{addr}$, $j \in \mathcal{A}_{level_K}$ and $r, s \in \mathcal{A}_{mrgn}$
Σ_{bridge}	$-\ path2set^{\mathcal{A}}(p) = \{a_1, \ldots, a_n\}$ for $p = [a_1, \ldots, a_n] \in \mathcal{A}_{path}$ $-\ addr2set_K{}^{\mathcal{A}}(m, a, l) = \{a' \mid \exists p \in \mathcal{A}_{path} \ . \ (m, a, a', l, p) \in reach_K\}$ $-\ getp_K{}^{\mathcal{A}}(m, a_{init}, a_{end}, l) = \begin{cases} p & \text{if } (m, a_{init}, a_{end}, l, p) \in reach_K{}^{\mathcal{A}} \\ \epsilon & \text{otherwise} \end{cases}$ for each $m \in \mathcal{A}_{mem}$, $p \in \mathcal{A}_{path}$, $l \in \mathcal{A}_{level_K}$ and $a_{init}, a_{end} \in \mathcal{A}_{addr}$ $-\ fstlock^{\mathcal{A}}(m, [a_1 .. a_n], l) = \begin{cases} a_k & \text{if there is } k \leq n \text{ such that} \\ & \text{for all } j < k, m[a_j].lockid[l] = \oslash \\ & \text{and } m[a_k].lockid[l] \neq \oslash \\ null & \text{otherwise} \end{cases}$ $-\ ordList^{\mathcal{A}}(m, p)$ iff $p = \epsilon$ or $p = [a]$ or $p = [a_1 .. a_n]$ with $n \geq 2$ and $m(a_i).key^{\mathcal{A}} \preceq m(a_{i+1}).key^{\mathcal{A}}$ for all $1 \leq i < n$, for any $m \in \mathcal{A}_{mem}$

Fig. 5. Characterization of a TSL_K-interpretation $\mathcal{A}$

cells, including *error* for incorrect dereferences. Σ_{mem} is the signature for heaps, with the usual memory access and single memory mutation functions. Σ_{set} and Σ_{setth} are theories of sets of addresses and thread ids resp. Σ_{mrgn} is the theory of masked regions. The signature Σ_{reach} contains predicates to check reachability of address using paths at different levels, while Σ_{bridge} contains auxiliary functions and predicates to manipulate and inspect paths and locks.

4   Decidability of $\mathsf{TSL_K}$

We show that $\mathsf{TSL_K}$ is decidable by proving that it enjoys the finite model property with respect to its sorts, and exhibiting upper bounds for the sizes of the domains of a small interpretation of a satisfiable formula.

Definition 1 (Finite Model Property). *Let Σ be a signature, $S_0 \subseteq S$ be a set of sorts, and T be a Σ-theory. T has the finite model property with respect to S_0 if for every T-satisfiable quantifier-free Σ-formula φ there exists a T-interpretation $\mathcal{A}$ satisfying φ such that for each sort $\sigma \in S_0$, $\mathcal{A}_\sigma$ is finite.*

The fact that $\mathsf{TSL_K}$ has the finite model property with respect to domains elem, addr, ord, $\mathsf{level_K}$ and thid, implies that $\mathsf{TSL_K}$ is decidable by enumerating all possible $\Sigma_{\mathsf{TSL_K}}$-structures up to a certain cardinality. We now define the set of normalized $\mathsf{TSL_K}$-literals.

Definition 2 ($\mathsf{TSL_K}$-normalized literals). *A TSL_K-literal is normalized if it is a flat literal of the form:*

$e_1 \neq e_2$	$a_1 \neq a_2$	$l_1 \neq l_2$
$a = null$	$c = error$	$c = rd(m, a)$
$k_1 \neq k_2$	$k_1 \preceq k_2$	$m_2 = upd(m_1, a, c)$
$c = mkcell(e, k, a_0, \ldots, a_{K-1}, t_0, \ldots, t_{K-1})$		
$s = \{a\}$	$s_1 = s_2 \cup s_3$	$s_1 = s_2 \setminus s_3$
$g = \{t\}_T$	$g_1 = g_2 \cup_T g_3$	$g_1 = g_2 \setminus_T g_3$
$r = \langle a, l \rangle_{\mathsf{mr}}$	$r_1 = r_2 \cup_{\mathsf{mr}} r_3$	$r_1 = r_2 -_{\mathsf{mr}} r_3$
$p_1 \neq p_2$	$p = [a]$	$p_1 = rev(p_2)$
$s = path2set(p)$	$append(p_1, p_2, p_3)$	$\neg append(p_1, p_2, p_3)$
$s = addr2set_K(m, a, l)$	$p = getp_K(m, a_1, a_2, l)$	
$t_1 \neq t_2$	$a = fstlock(m, p, l)$	$ordList(m, p)$

where e, e_1 and e_2 are elem-variables; a, a_0, a_1, $a_2, \ldots, a_{K-1}$ are addr-variables; c is a cell-variable; m, m_1 and m_2 are mem-variables; p, p_1, p_2 and p_3 are path-variables; s, s_1, s_2 and s_3 are set-variables; g, g_1, g_2 and g_3 are setth-variables; r, r_1, r_2 and r_3 are mrgn-variables; k, k_1 and k_2 are ord-variables; l, l_1 and l_2 are $\mathsf{level_K}$-variables and t, t_0, t_1, $t_2, \ldots, t_{K-1}$ are thid-variables.

Lemma 1. *Every TSL_K-formula is equivalent to a collection of conjunctions of normalized TSL_K-literals.*

Proof (sketch). First, transform a formula in disjunctive normal form. Then each conjunct can be normalized introducing auxiliary fresh variables when necessary.

The phase of normalizing a formula is commonly known [15] as the "variable abstraction phase". Note that normalized literals belong to just one theory.

Consider an arbitrary $\mathsf{TSL_K}$-interpretation $\mathcal{A}$ satisfying a conjunction of normalized $\mathsf{TSL_K}$-literals Γ. We show that if $\mathcal{A}$ consists of domains $\mathcal{A}_{\mathsf{elem}}$, $\mathcal{A}_{\mathsf{addr}}$, $\mathcal{A}_{\mathsf{thid}}$, $\mathcal{A}_{\mathsf{level_K}}$ and $\mathcal{A}_{\mathsf{ord}}$ then there are finite sets $\mathcal{B}_{\mathsf{elem}}$, $\mathcal{B}_{\mathsf{addr}}$, $\mathcal{B}_{\mathsf{thid}}$, $\mathcal{B}_{\mathsf{level_K}}$ and $\mathcal{B}_{\mathsf{ord}}$ with bounded cardinalities, where the finite bound on the sizes can be computed from Γ. Such sets can in turn be used to obtain a finite interpretation $\mathcal{B}$ satisfying Γ, since all the other sorts are bounded by the sizes of these sets.

Lemma 2 (Finite Model Property). *Let Γ be a conjunction of normalized $\mathsf{TSL_K}$-literals. Let $\overline{e} = |V_{\mathsf{elem}}(\Gamma)|$, $\overline{a} = |V_{\mathsf{addr}}(\Gamma)|$, $\overline{m} = |V_{\mathsf{mem}}(\Gamma)|$, $\overline{p} = |V_{\mathsf{path}}(\Gamma)|$, $\overline{t} = |V_{\mathsf{thid}}(\Gamma)|$ and $\overline{o} = |V_{\mathsf{ord}}(\Gamma)|$. Then the following are equivalent:*

1. *Γ is $\mathsf{TSL_K}$-satisfiable;*
2. *Γ is true in a $\mathsf{TSL_K}$ interpretation $\mathcal{B}$ such that*

$$|\mathcal{B}_{\mathsf{addr}}| \leq \overline{a} + 1 + \overline{m}\,\overline{a}\,K + \overline{p}^2 + \overline{p}^3 + (K+2)\overline{m}\,\overline{p} \qquad |\mathcal{B}_{\mathsf{elem}}| \leq \overline{e} + \overline{m}\,|\mathcal{B}_{\mathsf{addr}}|$$
$$|\mathcal{B}_{\mathsf{thid}}| \leq \overline{t} + K\overline{m}\,|\mathcal{B}_{\mathsf{addr}}| + 1 \qquad\qquad\qquad |\mathcal{B}_{\mathsf{ord}}| \leq \overline{o} + \overline{m}\,|\mathcal{B}_{\mathsf{addr}}|$$
$$|\mathcal{B}_{\mathsf{level_K}}| \leq K$$

Proof. $(2 \rightarrow 1)$ is immediate. $(1 \rightarrow 2)$ is proved on a case analysis over the set of normalized literals of $\mathsf{TSL_K}$. $\qquad\square$

4.1   A Combination-Based Decision Procedure for $\mathsf{TSL_K}$

Lemma 2 enables a brute force method to automatically check whether a set of normalized $\mathsf{TSL_K}$-literals is satisfiable. However, such a method is not efficient in practice. We describe now how to obtain a more efficient decision procedure for $\mathsf{TSL_K}$ applying a many-sorted variant [22] of the Nelson-Oppen combination method [12], by combining the decision procedures for the underlying theories. This combination method requires that the theories fulfill some conditions. First, each theory must have a decision procedure. Second, two theories can only share sorts (but not functions or predicates). Third, when two theories are combined, either both theories are stable infinite or one of them is polite with respect to the underlying sorts that it shares with the other. The stable infinite condition for a theory establishes that if a formula has a model then it has a model with infinite cardinality. In our case, some theories are not stable infinite. For example, $T_{\mathsf{level_K}}$ is not stably infinite, T_{ord}, and T_{thid} need not be stable infinite in same instances. The observation that the condition of stable infinity may be cumbersome in the combination of theories for data structures was already made in [16] where they suggest the condition of *politeness*:

Definition 3 (Politeness). *T is polite with respect to sorts $S : \{\sigma_1 \ldots \sigma_n\}$ whenever:*

(1) Let φ be a satisfiable formula in theory T, $\mathcal{A}$ be one model of φ and let $|\mathcal{A}_{\sigma_1}|, \ldots, |\mathcal{A}_{\sigma_n}|$ be the cardinalities of the domains of $\mathcal{A}$ for sorts in S. For every tuple of larger cardinalities $k_1 \geq |\mathcal{A}_{\sigma_1}|, \ldots, k_n \geq |\mathcal{A}_{\sigma_n}|$, there is a model $\mathcal{B}$ of φ with $|\mathcal{B}_{\sigma_i}| = k_i$.

(2) There is a computable function that for every formula φ returns an equivalent formula $(\exists \overline{v})\psi$ (where $\overline{v} = V_\psi \setminus V_\varphi$) such that, if ψ is satisfiable, then there is an interpretation $\mathcal{A}$ with $\mathcal{A}_\sigma = [V_\sigma(\psi)]^{\mathcal{A}}$ for each sort σ.

Condition *(1)* is called *smoothness*, and guarantees that interpretations can be enlarged as needed. Condition *(2)* is called *finite witnessability*, and gives a procedure to produce a model in which every element is represented by a variable. The Finite Model Property, Lemma 2 above, guarantees that every sub-theory of $\mathsf{TSL_K}$ is finite witnessable since one can add as many fresh variables as the bound for the corresponding sort in the lemma. The smoothness property can be shown for:

$$T_{\mathsf{cell}} \oplus T_{\mathsf{mem}} \oplus T_{\mathsf{path}} \oplus T_{\mathsf{set}} \oplus T_{\mathsf{setth}} \oplus T_{\mathsf{mrgn}}$$

with respect to sorts addr, $\mathsf{level_K}$, elem, ord and thid. Moreover, these theories can be combined because all of them are stably infinite. The following can also be combined: $T_{\mathsf{level_K}} \oplus T_{\mathsf{ord}} \oplus T_{\mathsf{thid}}$ because they do not share any sorts, so combination is trivial. The many-sorted Nelson-Oppen method allows to combine the first collection of theories with the second. Regarding the decision procedures for each individual theory, $T_{\mathsf{level_K}}$ is trivial since it is just a finite set of naturals with order. For T_{ord} we can adapt a decision procedure for dense orders as the reals [21], or other appropriate theory. For T_{cell} we can use a decision procedure for recursive data structures [13]. T_{mem} is the theory of arrays [1]. T_{set}, T_{setth} and T_{mrgn} are theories of (finite) sets for which there are many decision procedures [25, 8]. The remaining theories are T_{reach} and T_{bridge}. Following the approaches in [16, 18] we extend a decision procedure for the theory T_{path} of finite sequences of (non-repeated) addresses with the auxiliary functions and predicates shown in Fig. 6, and combine this theory to obtain:

$$T_{\mathsf{SLKBase}} = T_{\mathsf{addr}} \oplus T_{\mathsf{ord}} \oplus T_{\mathsf{thid}} \oplus T_{\mathsf{level_K}} \oplus T_{\mathsf{cell}} \oplus T_{\mathsf{mem}} \oplus T_{\mathsf{path}} \oplus T_{\mathsf{set}} \oplus T_{\mathsf{setth}} \oplus T_{\mathsf{mrgn}}$$

Using T_{path} all symbols in T_{reach} can be easily defined. The theory of finite sequences of addresses is defined by $T_{\mathsf{fseq}} = (\Sigma_{\mathsf{fseq}}, \mathsf{TGen})$, where $\Sigma_{\mathsf{fseq}} = (\{\mathsf{addr,fseq}\}, \{nil : \mathsf{fseq}, cons : \mathsf{addr} \times \mathsf{fseq} \to \mathsf{fseq}, hd : \mathsf{fseq} \to \mathsf{addr}, tl : \mathsf{fseq} \to \mathsf{fseq}\}, \emptyset)$ and TGen as the class of term-generated structures that satisfy the axioms of distinctness, uniqueness and generation of sequences using constructors, as well as acyclicity (see, for example [4]). Let Σ_{path} be Σ_{fseq} extended with the symbols of Fig. 6 and let *PATH* be the set of axioms of T_{fseq} including the ones in Fig. 6. Then, we can formally define $T_{\mathsf{path}} = (\Sigma_{\mathsf{path}}, \mathsf{ETGen})$ where ETGen is $\{\mathcal{A}^{\Sigma_{\mathsf{path}}} | \mathcal{A}^{\Sigma_{\mathsf{path}}} \models PATH \text{ and } \mathcal{A}^{\Sigma_{\mathsf{fseq}}} \in \mathsf{TGen}\}$. Next, we extend T_{SLKBase} with definitions for translating all missing functions and predicates from Σ_{reach} and Σ_{bridge} appearing in normalized $\mathsf{TSL_K}$-literals by definitions from T_{SLKBase}. Let *GAP* be the set of axioms that define ϵ, [_], *append*, $reach_K$, *path2set*, $getp_K$, *fstlock* and *ordList*. For instance: $ispath(p) \wedge ordPath(m, p) \leftrightarrow ordList(m, p)$. We now define $\widehat{\mathsf{TSL_K}} = (\Sigma_{\widehat{\mathsf{TSL_K}}}, \widehat{\mathsf{ETGen}})$ where $\Sigma_{\widehat{\mathsf{TSL_K}}}$ is $\Sigma_{T_{\mathsf{SLKBase}}} \cup \{$ *append*, $reach_K$, *path2set*, $getp_K$, *fstlock*, *ordList* $\}$ and $\widehat{\mathsf{ETGen}} := \{\mathcal{A}^{\Sigma_{\widehat{\mathsf{TSL_K}}}} | \mathcal{A}^{\Sigma_{\widehat{\mathsf{TSL_K}}}} \models GAP \text{ and } \mathcal{A}^{\Sigma_{T_{\mathsf{SLKBase}}}} \in \mathsf{ETGen}\}$.

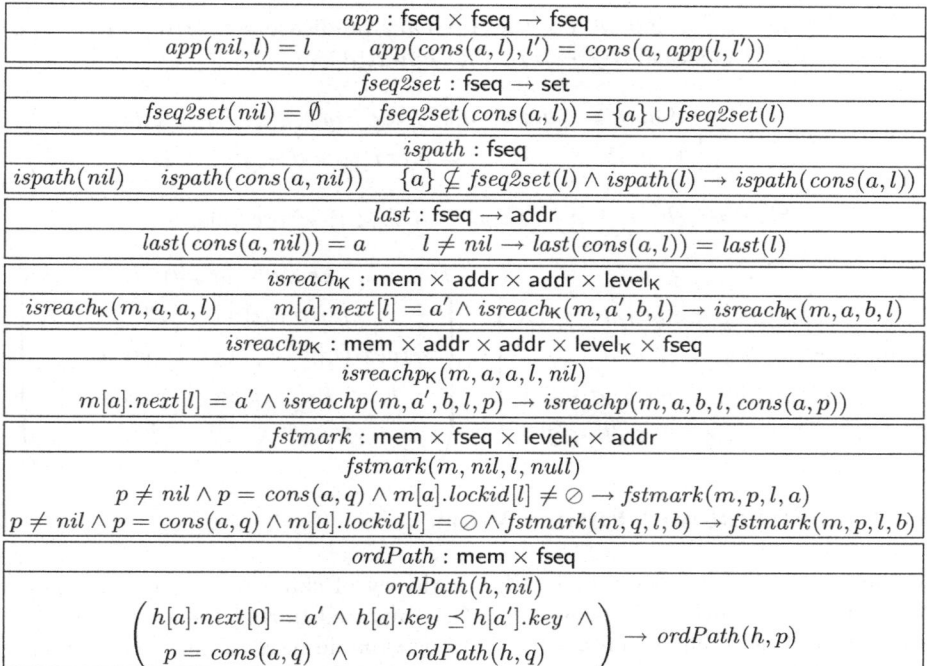

Fig. 6. Functions, predicates and axioms of T_{path}

Using the definitions of GAP it is easy to prove that if Γ is a set of normalized $\mathsf{TSL_K}$-literals, then Γ is $\mathsf{TSL_K}$-satisfiable iff Γ is $\widehat{\mathsf{TSL_K}}$-satisfiable. Therefore, $\widehat{\mathsf{TSL_K}}$ can be used in place of $\mathsf{TSL_K}$ for satisfiability checking. The reduction from $\widehat{\mathsf{TSL_K}}$ into T_{SLKBase} is performed in two steps. First, by the finite model theorem (Lemma 2), it is always possible to calculate an upper bound in the number of elements of sort addr, elem, thid, ord and level in a model (if there is one model), based on the input formula. Therefore, one can introduce one variable per element of each of these sorts and unfold all definitions in $PATH$ and GAP, by symbolic expansion, leading to terms in Σ_{fseq}, and thus, in T_{SLKBase}. This way, it is always possible to reduce a $\widehat{\mathsf{TSL_K}}$-satisfiability problem of normalized literals into a T_{SLKBase}-satisfiability problem. Hence, using a decision procedure for T_{SLKBase} we obtain a decision procedure for $\widehat{\mathsf{TSL_K}}$, and thus, for $\mathsf{TSL_K}$. Notice, for instance, that the predicate $subPath$: path × path for ordered lists can be defined using only $path2set$ as: $subPath(p_1, p_2) \hat{=} path2set(p_1) \subseteq path2set(p_2)$.

For space reasons, we do not provide complete specification and proofs of the temporal properties. However, in [18] is detailed an example of a termination proof over concurrent lists, which easily carries over to skiplists. For illustration purposes, we now show the full verification condition for the verification of the safety property $\Box\big(SkipList_4(h, sl)\big)$ when executing transition 36 of program *insert* by a thread with id t, from Section 2. For clarity, we again use *prev* as a short for $upd^{[t]}[i^{[t]}]$, and we use the auxiliary predicate $setnext(c, d, i, x)$ that makes the cell d identical to c except that $c.next[i] = x$.

$$setnext(c, d, i, x) \hat{=} \begin{pmatrix} d.data = c.data \land d.key = c.key \land d.lock[j] = c.lock[j] \land \\ (i \neq j) \rightarrow d.next[j] = c.next[j] \land d.next[i] = x \end{pmatrix}$$

The VC is $(SkipList_4(h, sl) \land \varphi \rightarrow SkipList_4(h', sl'))$ where φ is:

$$\begin{pmatrix} x^{[t]}.key = newval & \land \\ prev.key < newval & \land \\ x^{[t]}.next[i^{[t]}].key > newval & \land \\ prev.next[i^{[t]}] = x^{[t]}.next[i^{[t]}] & \land \\ (x^{[t]}, i^{[t]}) \notin sl.r \land 0 \leq i^{[t]} \leq 3 \end{pmatrix} \land \begin{pmatrix} at_{36}[t] \land at'_{37}[t] & \land \\ prev'.next[i^{[t]}] = x^{[t]} & \land \\ setnext(h[prev], newcell, i^{[t]}, x^{[t]}) \land \\ h' = upd(h, prev, newcell) & \land \\ sl = sl' \land x'^{[t]} = x^{[t]} & \land \end{pmatrix}$$

5 Conclusion and Future Work

In this paper we have presented TSL_K, a theory of skiplists of height at most K, useful for automatically prove the VCs generated during the verification of concurrent skiplist implementations. TSL_K is capable of reasoning about memory, cells, pointers, masked regions and reachability, enabling ordered lists and sublists, allowing the description of the skiplist property, and the representation of memory modifications introduced by the execution of program statements.

We showed that TSL_K is decidable by proving its finite model property, and exhibiting the minimal cardinality of a model if one such model exists. Moreover, we showed how to reduce the satisfiability problem of quantifier-free TSL_K formulas to a combination of theories using the many-sorted version of Nelson-Oppen, allowing the use of well studied decision procedures. The complexity of the decision problem for TSL_K is easily shown to be NP-complete since it properly extends TLL [16].

Current work includes the translation of formulas from T_{ord}, T_{level_K}, T_{set}, T_{setth} and T_{mrgn} into BAPA [8]. In BAPA, arithmetic, sets and cardinality aids in the definition of skiplists properties. Paths can be represented as finite sequences of addresses. We are studying how to replace the recursive functions from T_{reach} and Σ_{bridge} by canonical set and list abstractions [20], which would lead to a more efficient decision procedure, essentially encoding full TSL_K formulas into BAPA. The family of theories presented in the paper is limited to skiplists of a fixed maximum height. Typical skiplist implementations fix a maximum number of levels and this can be handled with TSL_K. Inserting more than than 2^{levels} elements into a skiplist may slow-down the search of a skiplist implementation but this issue affects performance and not correctness, which is the goal pursued in this paper. We are studying techniques to describe skiplists of arbitrary many levels. A promising approach consists of equipping the theory with a primitive predicate denoting that the skiplist property holds above and below a given level. Then the reasoning is restricted to the single level being modified. This approach, however, is still work in progress.

Furthermore, we are working on a direct implementation of our decision procedure, as well as its integration into existing solvers. Future work also includes the temporal verification of sequential and concurrent skiplists implementations, including one at the `java.concurrent` standard library. This can be accomplished by the design of verification diagrams that use the decision procedure presented in this paper.

References

1. Armando, A., Ranise, S., Rusinowitch, M.: A rewriting approach to satisfiability procedures. Information and Computation 183(2), 140–164 (2003)
2. Banerjee, A., Naumann, D.A., Rosenberg, S.: Regional logic for local reasoning about global invariants. In: Ryan, M. (ed.) ECOOP 2008. LNCS, vol. 5142, pp. 387–411. Springer, Heidelberg (2008)
3. Bouajjani, A., Dragoi, C., Enea, C., Sighireanu, M.: A logic-based framework for reasoning about composite data structures. In: Bravetti, M., Zavattaro, G. (eds.) CONCUR 2009. LNCS, vol. 5710, pp. 178–195. Springer, Heidelberg (2009)
4. Bradley, A.R., Manna, Z.: The Calculus of Computation. Springer, Heidelberg (2007)
5. Browne, A., Manna, Z., Sipma, H.B.: Generalized verification diagrams. In: Thiagarajan, P.S. (ed.) FSTTCS 1995. LNCS, vol. 1026, pp. 484–498. Springer, Heidelberg (1995)
6. Herlihy, M., Shavit, N.: The Art of Multiprocessor Programming. Morgran-Kaufmann, San Francisco (2008)
7. Hobor, A., Appel, A.W., Nardelli, F.Z.: Oracle semantics for concurrent separation logic. In: Gairing, M. (ed.) ESOP 2008. LNCS, vol. 4960, pp. 353–367. Springer, Heidelberg (2008)
8. Kuncak, V., Nguyen, H.H., Rinard, M.C.: An algorithm for deciding BAPA: Boolean algebra with presburger arithmetic. In: Nieuwenhuis, R. (ed.) CADE 2005. LNCS (LNAI), vol. 3632, pp. 260–277. Springer, Heidelberg (2005)
9. Lahiri, S.K., Qadeer, S.: Back to the future: revisiting precise program verification using smt solvers. In: Proc. of POPL 2008, pp. 171–182. ACM, New York (2008)
10. Leino, K.R.M.: Data groups: Specifying the modication of extended state. In: OOPSLA 1998, pp. 144–153. ACM, New York (1998)
11. Manna, Z., Pnueli, A.: Temporal Verification of Reactive Systems. Springer, Heidelberg (1995)
12. Nelson, G., Oppen, D.C.: Simplification by cooperating decision procedures. ACM Trans. Program. Lang. Syst. 1(2), 245–257 (1979)
13. Oppen, D.C.: Reasoning about recursively defined data structures. J. ACM 27(3), 403–411 (1980)
14. Pugh, W.: Skip lists: A probabilistic alternative to balanced trees. Commun. ACM 33(6), 668–676 (1990)
15. Ranise, S., Ringeissen, C., Zarba, C.G.: Combining data structures with nonstably infinite theories using many-sorted logic. In: Gramlich, B. (ed.) FroCos 2005. LNCS (LNAI), vol. 3717, pp. 48–64. Springer, Heidelberg (2005)

16. Ranise, S., Zarba, C.G.: A theory of singly-linked lists and its extensible decision procedure. In: Proc. of SEFM 2006. IEEE CS Press, Los Alamitos (2006)
17. Reynolds, J.C.: Separation logic: A logic for shared mutable data structures. In: Proc. of LICS 2002, pp. 55–74. IEEE CS Press, Los Alamitos (2002)
18. Sánchez, A., Sánchez, C.: Decision procedures for the temporal verification of concurrent lists. In: Dong, J.S., Zhu, H. (eds.) ICFEM 2010. LNCS, vol. 6447, pp. 74–89. Springer, Heidelberg (2010)
19. Sipma, H.B.: Diagram-Based Verification of Discrete, Real-Time and Hybrid Systems. Ph.D. thesis, Stanford University (1999)
20. Suter, P., Dotta, M., Kuncak, V.: Decision procedures for algebraic data types with abstractions. In: Proc. of POPL 2010, pp. 199–210. ACM, New York (2010)
21. Tarski, A.: A decision method for elementary algebra and geometry. University of California Press, Berkeley (1951)
22. Tinelli, C., Zarba, C.G.: Combining decision procedures for sorted theories. In: Alferes, J.J., Leite, J. (eds.) JELIA 2004. LNCS (LNAI), vol. 3229, pp. 641–653. Springer, Heidelberg (2004)
23. Vafeiadis, V.: Modular fine-grained concurrency verification. Ph.D. thesis, University of Cambridge (2007)
24. Yorsh, G., Rabinovich, A.M., Sagiv, M., Meyer, A., Bouajjani, A.: A logic of reachable patterns in linked data-structures. In: Aceto, L., Ingólfsdóttir, A. (eds.) FOSSACS 2006. LNCS, vol. 3921, pp. 94–110. Springer, Heidelberg (2006)
25. Zarba, C.G.: Combining sets with elements. In: Dershowitz, N. (ed.) Verification: Theory and Practice. LNCS, vol. 2772, pp. 762–782. Springer, Heidelberg (2004)

CORAL: Solving Complex Constraints for Symbolic PathFinder

Matheus Souza[1], Mateus Borges[1],
Marcelo d'Amorim[1], and Corina S. Păsăreanu[2]

[1] Federal University of Pernambuco, Recife, PE, Brazil
{mbas,mab,damorim}@cin.ufpe.br
[2] CMU SV/NASA Ames Research Center, Moffett Field, CA, USA
corina.s.pasareanu@nasa.gov

Abstract. Symbolic execution is a powerful automated technique for generating test cases. Its goal is to achieve high coverage of software. One major obstacle in adopting the technique in practice is its inability to handle complex mathematical constraints. To address the problem, we have integrated CORAL's heuristic solvers into NASA Ames' Symbolic PathFinder symbolic execution tool. CORAL's solvers have been designed to deal with mathematical constraints and their heuristics have been improved based on examples from the aerospace domain. This integration significantly broadens the application of Symbolic PathFinder at NASA and in industry.

1 Introduction

Systematic testing is widely accepted in academia and industry as a major approach to improve quality of general-purpose software. Perhaps less popularized is the role of testing as an economic viable technique to improve reliability of critical systems. In the aerospace domain, for instance, systematic testing has been used to reduce cost of bug finding, i.e., to increase application reliability. NASA, in particular, maintains open-source tools to assist systematic testing.

Symbolic execution [15] is an automated technique to generate test input data. The input to symbolic execution is a parameterized method m of the application under test and the output is a test suite that maximizes path coverage for m. Internally, a symbolic execution tool is organized in two components: the constraint generator and the constraint solver. The constraint generator builds constraints on the input parameters of m for achieving path coverage while the solver attempts to solve these constraints, i.e., to generate concrete assignments to input parameters. A major obstacle for techniques that build on constraint solvers, such as symbolic execution, is the inability to deal with complex constraints. In particular, constraints that build on undecidable theories, constraints that build on decidable theories but are very expensive to deterministically solve, and constraints that the solver cannot handle.

The goal of this work is to improve the solving of constraints that use floating-point variables and complex mathematical functions. Such constraints often

M. Bobaru et al. (Eds.): NFM 2011, LNCS 6617, pp. 359–374, 2011.

occur in the analysis of software from the aerospace domain; for example, consider software such as TSAFE [4,5] that helps air-traffic controllers in detecting and resolving short-term conflicts between aircrafts. This software estimates the location of an aircraft based on several factors including speed and direction and makes extensive use of floating-point variables and trigonometric functions. Good handling of complex constraints is fundamental for testing software of this kind using a symbolic execution tool such as NASA Ames' Symbolic Pathfinder [20].

Symbolic PathFinder (SPF) is a symbolic execution tool used at NASA and Fujitsu for testing complex applications. This paper reports the results of using the constraint solver CORAL to solve the complex mathematical constraints generated with SPF. CORAL uses meta-heuristic search, such as genetic algorithms [12] and particle-swarm optimization [14], to look for solutions to constraints that the SPF tool generates. The hypothesis is that *search* can be effective in solving such constraints not managed by traditional decision procedures. The principle of meta-heuristic search is to iteratively refine a set of solution candidates, initially chosen at random, for a fixed number of times. Informed fitness functions evaluate the quality of a candidate to solve a constraint in each generation round. A new generation of candidates is obtained with modifications to the best fit candidates. The search terminates after a determined number of iterations. In our case, it succeeds only when the best fit candidate is also a solution to the input constraint.

To deal with numeric constraints CORAL uses a specialized fitness function that conceptually measures the distance of a candidate solution to satisfying a particular constraint. To reduce the search space, it additionally tries to infer the units and ranges of variables from the functions where these variables are used. The design of CORAL has been influenced in part by the constraints that SPF generated from the analysis of several NASA applications. In particular, some rewriting rules have been added and the fitness function has been adjusted based on examples from the NASA domain.

This paper makes the following contributions:

- **New constraint solver:** We present CORAL a meta-heuristic constraint solver specialized to handle complex mathematical constraints;
- **Integration:** We report the integration of SPF and CORAL. This integration moves forward the limits of symbolic execution to manage a wider range of programs;
- **Evaluation:** We evaluate this integration on several examples from NASA and also compare the use of CORAL with other constraint solvers (with some support for real arithmetic) that have been previously integrated in SPF.

The rest of the paper is organized as follows. Section 2 briefly illustrates how symbolic execution works. Section 3 describes the Symbolic PathFinder tool. Section 4 describes CORAL. Section 5 evaluates the integration of CORAL in Symbolic PathFinder. Finally, Section 6 discusses related work and Section 7 gives our conclusions.

2 Symbolic Execution

Symbolic execution is a program analysis technique that executes a program with symbolic inputs as opposed to concrete inputs. It computes the effect of program execution on a symbolic state, which maps variables to symbolic expressions. When execution evaluates a branching instruction, the technique needs to decide which branching choice to select. In a regular execution the evaluation of a boolean expression is either true or false so only one branch of the conditional can be taken. In the case of a symbolic execution the evaluation of the boolean expression is a symbolic value, so both branches can be taken resulting in different paths through the program. Symbolic execution characterizes each path it explores with a path condition over the input variables $\overrightarrow{x}$. This condition is defined with a conjunction of boolean expressions $pc(\overrightarrow{x}) = \bigwedge b_i$. Each boolean expression b_i denotes a branching decision made during the execution of a distinct path in the program under test. Symbolic execution terminates when it explores all such paths corresponding to the different combinations of decisions. Note, however, that programs with loops and recursion can have an infinite number of paths. In those cases, symbolic execution needs to bound the number of paths it explores.

We illustrate symbolic execution using a simple example. Consider the fragment of code from Figure 1 (left) taken from a flight abort executive:

```
if(pressure < 640.0 ||
   pressure > 960.0) {
  abort();
} else { continue(); }
```

1.	$SYM < 640.0$
2.	$SYM >= 640.0 \wedge SYM > 960.0$
3.	$SYM >= 640.0 \wedge SYM <= 960.0$

Fig. 1. Abort example and corresponding path conditions

If the value of the *input* variable **pressure** is outside nominal values 640.0 and 960.0, then the mission is aborted, otherwise the mission is continued. Traditional testing of this code involves assigning some concrete values to the inputs and executing the code; for example, if the value of variable **pressure** is 460.0, testing will exercise only one path through the code, corresponding to the condition **pressure < 640.0** being true, resulting in an abort. In contrast, symbolic execution assigns a symbolic value to the input variable **pressure** and analyzes all the three possible paths through the code, corresponding to the three path conditions in Figure 1 (right). The path conditions correspond respectively to the cases where the first term of the disjunction ("||") is satisfied, the second term is satisfied, and none is satisfied. Note that due to the short-circuit operator, it is only possible to satisfy the second term of the condition negating the first. Solving these path conditions with a constraint solver gives the test inputs that achieve complete path coverage through the code.

3 Symbolic PathFinder

Symbolic PathFinder (SPF) is a symbolic execution tool for Java bytecode. SPF is used primarily for automated test case generation of code and also of Simulink/Stateflow and UML models, via a translation into bytecode [19]. SPF has been used at NASA (JSC Onboard Abort Executive, fault tolerant protocols, PadAbort-1 models, T-SAFE Java code), in industry (most notably at Fujitsu – 60K LOC), and in various research projects from academia. SPF is part of the Java PathFinder verification tool-set [8], a freely available open-source project. We describe here SPF's main features and how it builds complex mathematical constraints, which are then used with CORAL's heuristic solvers.

Features. The Java Pathfinder tool-set includes the JPF-core project, an explicit-state model checker for Java programs, and several extension projects, one of them being SPF (jpf-symbc Java project). The JPF-core implements an extensible custom Java Virtual Machine (VM), equipped with state storage and backtracking capabilities, different search strategies, as well as listeners for monitoring and influencing the search. By default, JPF-core executes the program based on the standard semantics of Java. SPF replaces this concrete execution semantics with a non-standard symbolic interpretation of bytecodes. It uses a custom bytecode instruction factory for that. More precisely, SPF uses the instruction factory class SymbolicInstructionFactory to build bytecode instructions that manipulate symbolic values and expressions. For example, the result of the symbolic interpretation of the bytecode IADD is to pop from the stack two symbolic integers sym_1 and sym_2 and to push the symbolic expression $sym_1 + sym_2$ back to the stack. SPF stores these symbolic values that symbolic execution computes in special "attributes" associated with the program data, i.e. variables, fields and stack operands.

The symbolic execution of conditional instructions (such as if statements) leads to the exploration of distinct program paths, corresponding to the boolean expression of the conditional evaluating to *true* or to *false*. SPF relies on the JPF-core framework to systematically explore the different choices of symbolic execution paths as well as thread interleavings. These choices are explored exhaustively (up to some bounds) using a mechanism of the JPF-core known as choice generators. The SPF implementation uses a specialized choice generator, the PCChoiceGenerator, for the construction of path conditions. Each generated choice is associated with a path condition encoding the condition or its negation, respectively. The path conditions are checked for satisfiability using off-the-shelf decision procedures or constraint solvers. If the path condition is satisfiable, the search continues; otherwise, the search backtracks (meaning that the path is unreachable).

Decision Procedures and Constraint Solvers. To check the feasibility of path conditions, SPF uses multiple decision procedures and constraint solvers through a generic interface. Currently, SPF supports the following solvers: CHOCO for integer/real constraints, CVC3 for linear constraints, and the interval arithmetic solver IASolver, as well as the SMT decision procedures CVC3 and

YICES. Both CHOCO and IASolver have support for handling constraints on reals and complex mathematical functions, however they both perform poorly in practice (in terms of correctness, speed and tool support). This paper reports the integration of a new constraint solver to SPF for handling complex mathematical constraints, namely CORAL.

Handling Math Functions. SPF uses JPF-core's native peers mechanism to model native libraries and any other program parts that cannot be analyzed directly with symbolic execution. Most notably, SPF incorporates native peers models for the methods in the `java.lang.Math` library; these models create symbolic expressions encoding the mathematical functions, that are left uninterpreted. Such use of native peers lifts the interpretation of Math functions from the concrete level to the abstract "model" level: whenever the symbolic execution reaches a call to a complex Math function, that call is intercepted by SPF and it is used as a symbolic operator to build a new symbolic expression. The path conditions containing such expressions are dispatched to an appropriate constraint solver that can handle complex Math constraints, such as CORAL.

SPF uses native peers for the following functions from the Java Math library: ACOS, ASIN, ATAN, ATAN2, COS, EXP, LOG, POW, ROUND, SIN, SQRT, TAN. For the rest of the Math functions, which are much simpler, we provide simple implementations that are interpreted directly by SPF.

```if (Math.pow(in,2.0)>16.0) {```
```   do1();```
```} else { do2(); }```
	1. $pow(in\_SYM, CONST\_2.0) < CONST\_16.0$
	2. $CONST\_16.0 == pow(in\_SYM, CONST\_2.0)$
	3. $pow(in\_SYM, CONST\_2.0) > CONST\_16.0$

Fig. 2. Example with Math function and corresponding path conditions

Figure 2 shows one example that uses the pow math function. Variable in stores the symbolic input in_SYM. The symbolic execution of this code produces the three path conditions to the right side of this figure. As mentioned before SPF does not directly interpret the call to the standard Java library function Math.pow. Instead, it constructs a symbolic expression pow(in_SYM,CONST_16.0) which is then used to build the symbolic constraints. When executing the if statement above, SPF creates a 3-choice split point related to the outcomes of the relational expression[1]. Each execution will explore one choice. As execution goes along, more boolean expression are added to the current path, building longer path constraints. The constraints are solved with an appropriate constraint solver; i.e., one that can handle such complex mathematical functions directly.

4   CORAL Heuristic Solvers

This section describes design and implementation of the CORAL heuristic constraint solvers. We first elaborate on the representation of the search space and

---

[1] The 3-way split reflects the three possible outcomes of the Java bytecode that compares two doubles, according to the Java semantics.

the search strategies used by CORAL. Then we illustrate the fitness function used, and finally, the optimizations.

4.1   Search Algorithms

Representation of Space and Search. Our characterization of a *candidate solution* is a map from symbolic variables to concrete values. A *population* corresponds to the set of candidates that are active in a given moment in the search. Our work follows an evolutionary approach to search. In this setting, the population evolves during the search according to some user-defined principle. Conceptually, each evolution step approximates the candidates to a solution. The search starts with a population obtained from the random assignment of values to variables and terminates after a fixed number of iterations or when it finds candidates with optimal fitness.

The CORAL infrastructure provides two different search strategies: random and Particle-Swarm Optimization (PSO). We discuss here PSO, the strategy that performed best in our experiments. Random search is described elsewhere [21,22]. PSO is a search algorithm, similar to the popular genetic algorithm search (GA), used in combinatorial optimization problems. Both PSO and GA use special operators to mutate candidates during the evolution process. While GA mimics biological evolution (e.g., with mutation and reproduction) PSO mimics movements of a group of animals in swarms. Although GA and PSO operate similarly with successive refinements of the population, they have different computational costs. At each iteration, GA needs to eliminate less fitted individuals, add new ones with crossover, and modify existing ones with mutation. The PSO algorithm updates the search state more efficiently: it uses efficient matrix arithmetic to update a fixed-size population. In PSO terminology candidate solutions are called *particles*. The particles collaborate to compute a solution (this is a central difference between GA and PSO). Each particle has a *position* in the search space and a contributing factor to the population, typically called *velocity*, which PSO uses to update the next position of each particle. The next position of a particle depends on its current position and velocity. The next velocity of a particle depends on the best position the swarm has seen from the start of the search (global) and the best position of that particle from the start (local). Details on design and implementation of these algorithms can be found elsewhere [12,14].

Fitness Functions. The role of a fitness function (a.k.a. objective function) is to drive the search towards (fitter) solutions. This function gives a score denoting the quality of an input candidate to solve the problem. Our solvers use a variation of the Stepwise Adaptive Weighting (SAW) fitness function that dynamically adjusts the importance of different sub-problems for solving the whole problem [9]. For constraint solving, the problem is to solve the entire path condition $pc(\overrightarrow{x}) = \bigwedge b_i$ and the sub-problem is to solve a clause b_i of the input path condition. The definition of SAW is as follows:

$$f(\overrightarrow{x}) = \sum_i w_i * g_i(\overrightarrow{x})$$

Function f is the weighted sum of $g_i(\overrightarrow{x})$, which denotes the score of candidate $\overrightarrow{x}$ to solve the clause b_i of the path condition. This score is given in the continuous interval $[0.0, 1.0]$ with higher values (respectively, low) indicating better (respectively, worse) fitness. The search goal is to maximize function f, i.e., to find inputs that produce maximal outcomes: high valuations of inputs on this function indicate fitter candidates. The search procedure dynamically increases the weight w_i associated to each clause b_i as that clause remains unsolved for longer than some specified number of times. The use of weights helps the search to positively differentiate candidate solutions that satisfy "difficult" clauses from solutions that satisfy many "easy" clauses. We note that a final solution is only relevant if it satisfies all clauses b_i.

SAW was originally created to solve SAT problems, i.e., propositional formula with boolean variables. We adjusted the definition to handle numeric variables. Recall that $g_i(\overrightarrow{x})$ denotes the score of $\overrightarrow{x}$ on b_i. Function g_i is defined as follows, where each clause b_i is a disjunction of terms $b_{i1} \vee \ldots \vee b_{im}$:

$$g_i(\overrightarrow{x}) = \max_{1<j<m} 1 - d(b_{ij}, \overrightarrow{x})$$

Note that the codomain of functions g_i and d are the same; the interval $[0.0, 1.0]$. Function d conceptually measures "how far" the candidate $\overrightarrow{x}$ is from a solution that satisfies the term b_{ij}. We want to maximize g_i and for that we need to minimize d, the distance to solution. For example, for the case where b_{ij} is an equality expression of the form $eq(e_1, e_2)$ we define the distance d as $norm(|e_2(\overrightarrow{x}) - e_1(\overrightarrow{x})|)$. The modulo of the difference denotes the distance between the evaluations of the expressions e_1 and e_2 on input $\overrightarrow{x}$. The function $norm$ normalizes the distance in the expected range. This function considers any input above some defined threshold t to return the upper bound 1 for the distance, otherwise it divides the input by t to obtain a value in the expected range. The evaluation of function d on a satisfying solution produces value 0. Definitions of the distance function d to other relational operators are similar.

Example. This example illustrates how the meta-heuristic search operates to find a solution to the constraint $sin(a) = -sin(b) \wedge sin(a) > 0$ using the fitness function we defined. Fig 3 illustrates the evolution of a fixed-size population of only two candidates. Each row details one candidate in a given iteration. Columns "it.", "(a, b)", "distance(weight)", and "fitness" show respectively the iteration number, the input-break assignment (candidate), the distance to satisfy a clause of the constraint with the current weight of the clause in parenthesis, and the fitness

it.	(a, b)	distance (weight)		fitness
		$sin(a)=-sin(b)$	$sin(a)>0$	
0	0.0000,0.0000	0.0 (1)	0.01 (1)	1.9900
	0.3927,5.7596	0.0011 (1)	0.0 (1)	1.9988
1	0.3927,6.2832	0.0038 (1)	0.0 (2)	2.9962
	0.3927,5.4978	0.0032 (1)	0.0 (2)	2.9968
2	0.5236,6.2832	0.0049 (2)	0.0 (2)	3.9900
	0.3927,5.7596	0.0011 (2)	0.0 (2)	3.9977
3	0.5236,5.2360	0.0036 (3)	0.0 (2)	4.9890
	0.3927,5.7596	0.0011 (3)	0.0 (2)	4.9965
4	0.0000,6.2832	0.0 (4)	0.01 (2)	5.9800
	0.5236,5.7596	0.0 (4)	0.0 (2)	6.0000

Fig. 3. Fitness-guided constraint solving

Table 1. Some rewriting rules of `CORAL`

1. $Math.Pow(E, E1) == Math.Pow(E, E2)$ $\Rightarrow$ $E1 = E2$
2. $Math.Pow(E1, E) == Math.Pow(E2, E)$ $\Rightarrow$ $E1 = E2$
3. $Math.Log(E) == c$ $\Rightarrow$ $E = POW(2, c)$
4. $Math.Log_{10}(E) == c$ $\Rightarrow$ $E = POW(10, c)$
5. $x_1 [+, -, *, /] x_2 == E$ $\Rightarrow$ $x_1 = E [-, +, /, *] x_2$
6. $x_1 + c * x_1 = E$ $\Rightarrow$ $x_1 = E/(1 + c)$

value of the candidate. The constraint is satisfied when the fitness equals the sum of the weights. Iteration 0 denotes the initial population. `CORAL` performs 4 iterations to find a solution. Note the increase in weight of the first clause (equality) relative to the second (inequality) as the search progresses.

Implementation. We have integrated `CORAL` in SPF by specializing SPF's generic decision procedure interface for `CORAL`; this involves encoding SPF's symbolic expressions into a format that is suitable for solving with `CORAL` and reading the solutions from `CORAL` back into SPF. `CORAL` currently uses the opt4j Java library [3] for implementing the search. The library essentially requires the user to define a fitness function and the representation of candidate solution, which, in our case, is a vector of integers and reals.

4.2 Optimizations

This section describes optimizations that `CORAL` uses.

Inference of Variable Domains. The quality of initial states is an important factor to determine overall search quality: a solution is obtained with a sequence of modifications on candidate inputs, starting from their initial assignments. `CORAL` tries to improve the quality of initial random assignments by inferring specific domains associated to each symbolic variable. The principle is that the search becomes more exhaustive when confined to a smaller space. For example, it infers the unit radian for variables that appear free within the context of sine and cosine expressions. For variables of this kind, `CORAL` starts the search assigning random values from a selection of values in the range $0 - 2\pi$. It also infers ranges which are explicit on the input constraint. For example, it will update the range $[lo_0, hi_0]$ associated to variable v to $[c, hi_0]$ if the constraint $v >= c$ is observed in the path condition and $c > lo_0$ holds, where c is a constant.

Elimination of Variables. Before passing a constraint to the search procedure, `CORAL` attempts to simplify the input formula. The approach it uses for that is to identify variables whose values can be fully determined by others. This is similar to a decision procedure for equality that partitions expressions in equivalence classes [17]. `CORAL` uses rewriting rules in attempt to isolate variables. Table 1 shows some of the rewriting rules it uses. Note that rule 2 is lossy, e.g., $E1=-2$ and $E2=2$. Rule 5 inverts the side of the arithmetic operation to isolate the variable. Rule 6 factors variable x_1 and inverts the side of the multiplication factor. Note that, considering fixed-precision arithmetic, the rules could lead to incorrect results. However, the search only terminates successfully if the optimal input satisfies the original constraint.

Table 2. Sample of constraints that `CORAL` handles

constraint	source
$(1.5 - x1 * (1 - x2)) == 0$	Beale
$(-13 + x1 + ((5 - x2) * x2 - 2) * x2) + (-29 + x1 + ((x2 + 1) * x2 - 14) * x2) == 0$	Freudenstein and Roth
$pow((1 - x1), 2) + 100 * (pow((x2 - x1 * x1), 2)) == 0$	Rosenbrock
$((pow(((x * (sin((((y * 0.017) - (z * 0.017)) + (((((((pow(w, 2.0))/((sin((t*0.017)))))/w)*0.0)/w)*-1.0)*x)/(((pow(x, 2.0))/((sin((t*0.017)))/(cos((t*0.017)))))/68443.0))))) - (w * 0.0)), 2.0) + (pow(((x*(cos((((y * 0.017) - (z * 0.017)) + (((((((pow(w, 2.0))/((sin((t * 0.017)))/(cos((t*0.017)))))/68443.0)*0.0)/w)*-1.0)*x)/(((pow(x, 2.0))/((sin((t*0.017)))/(cos((t*0.017)))))/68443.0))))) - (w * 1.0)), 2.0))) == 0.0$	TSAFE
$((exp(x) - exp((x * -1.0)))/(exp(x) + exp((x * -1.0)))) > (((exp(x) + exp((x * -1.0))) * 0.5)/((exp(x) - exp((x * -1.0))) * 0.5))$	PISCES
$x^{tan(y)} + z < x*atan(z) \wedge sin(y)+cos(y)+tan(y) >= x-z \wedge atan(x)+atan(y) > y$	manual

Evaluation of Boolean Expressions in Postfix Notation. In our context, evaluation refers to the operation that checks whether a candidate solution satisfies the input formula. Random(ized) search is very sensitive to evaluation time in general [12]. In principle, random search performs increasingly better as evaluation time decreases: more distinct inputs will be selected from an uniform distribution in the same allotted time. It is in our interest to improve evaluation time for a fair comparison with random solving and for more efficient solving. To that end the solver uses a postfix notation to evaluate path conditions on a given input. A postfix expression is scanned from left to right, therefore the operators can be applied efficiently to the operands located at the top of an operand stack. We use a fixed-size array of reals to implement such stack.

4.3   Sample Constraints

Table 2 gives a set of representative constraints that `CORAL` is able to solve. The first column shows the constraint and the second shows the source of the constraint. Some of these constraints are taken from the literature while others were generated by `CORAL` users and also by SPF from the analysis of NASA applications. Capitalized names indicate subjects from NASA. Note that most of the constraints are non-linear and use mathematical functions. The first 3 constraints are used elsewhere to evaluate the `FloPSy` constraint solver [18] (see also Section 6). The PISCES subject is discussed in Section 5.4. The manual constraints have been written by 3 users of `CORAL`. Note that solving equality constraints such as the first 4 in this table is challenging with random and heuristic search as they significantly reduce the solution space.

5   Evaluation

This section presents our evaluation of `CORAL`. Section 5.1 shows the setup of the various constraint solvers we used in our comparison. Section 5.2 compares the use of `CORAL` in SPF with other public solvers already integrated to SPF and also compares variations of `CORAL`. Section 5.3 discusses the impact of the number of search iterations set in CORAL on effectiveness and runtime. Finally, sections 5.4 and 5.5 discuss the analysis of the NASA PISCES library and the Java translation of the Apollo Lunar Autopilot Simulink model.

5.1   Setup

Solvers. The user can control the duration of a solving task in CORAL either by time or number of iterations. In our experiments we use number of iterations to obtain deterministic results. When not mentioned otherwise CORAL uses in each query request PSO as search strategy and 600 iterations (See Section 5.3). We consider the following solvers in our comparison: CORAL, CHOCO [1], CVC3 [2], and YICES [7]. All these solvers have been already integrated to SPF. We note that these solvers have different goals. For example, CVC3 and YICES are decision procedures for Satisfiability Modulo Theories (SMT). In particular, YICES can decide over linear real arithmetic and CVC3 can decide over rational linear arithmetic. But neither CVC3 nor YICES can handle complex mathematical functions directly. CHOCO, on the other hand, is a constraint-programming solver for the theories of integers and reals with support to mathematical functions.

The Wrapper Solver. In order to compare the different solvers, we developed a "wrapper solver" to encapsulate all the solvers considered in our evaluation. Similar to the basic solvers, such solver needs to implement a SPF-defined Java interface with operations for building the objects denoting the terms of a constraint and for calling the solver. We implemented the general solver for two reasons. First, it is possible that one of the solvers fails to solve a constraint that appears in a shallow exploration depth even though it could solve more elaborated constraints. With the wrapper solver, exploration will continue if at least one solver answers positively to a satisfiability check query. All solvers have the chance to answer each query generated with the symbolic execution. Second, the wrapper solver was useful to detect discrepancies between results that would often point to a bug in the SPF-solver integration or the solver itself.

5.2   Comparison with Other Solvers

Results for Decidable Constraints. We evaluated CORAL with all the other solvers for the symbolic execution of two set data-structures popularly used in testing: binary search tree and tree map. For these subjects, we used the implementation and test drivers available on the SPF codebase. The test drivers explore all sequences of method calls and inputs up to informed bounds. The symbolic execution of these data-structures generates constraints that only involve decidable theories. A decision procedure with support for linear integer arithmetic should be able to find solutions to all satisfiable path constraints. We observed that CORAL could solve as many constraints as any other solver in this experiment. The test driver was set to generate sequences up to bound 5. Solving decidable fragments is also important in this context as it is often the case that the input constraint mix decidable and undecidable parts.

Results for Constraints with Math Functions. We evaluated CORAL with 78 manually-written test cases including mathematical function expressions. In this setup, three users of CORAL first developed the constraints with the help of the Wolfram Alpha visualization tool [6] and then translated to Java. Although each constraint is satisfiable the translation to Java creates unsatisfiable paths. To note that Java models short-circuit boolean expression with control flow. In this setup we compared only CORAL and CHOCO since they provide support to math functions. Out of 678 queries CORAL solved 595 (87.7% of total). Of these, CHOCO did not solve 526. In addition, for no query CHOCO could solve and CORAL could not. CHOCO solved a total of 68 constraints (10.1% of total).

Results for Different Configurations of CORAL. In this experiment, we used all manually-written test cases. This includes complex constraints with and without math functions. The table from Figure 4 compares four instances of CORAL in this setup. We use a matrix to show how many constraints one solver could

	pso-opt	pso	ran-opt	ran	total
pso-opt	-	116	38	209	722
pso	36	-	50	118	642
ran-opt	10	102	-	179	694
ran	12	1	10	-	525
Total: # Queries=838, SOLVED=763					

Fig. 4. Different configurations of CORAL

solve that another could not. More specifically, each cell $A[i,j]$ of the square matrix A stores the number of constraints that solver i could solve and solver j could not. Last column and row show summaries. Last column shows the total of constraints the solver in that line could solve. Last row shows the total number of queries submitted to the solvers and the total number of queries solved. The label "pso" refers to CORAL using particle swarm optimization, while "ran" refers to use of random search. The label "-opt" indicates that the solver *enabled* optimization with the inference of variable domains and attempted to isolate variables that no other variables depend as discussed in Section 4. The random solvers use a bound of 360,000 iterations while the PSO solvers a bound of 600 corresponding to approximately the same time of search. (See Section 5.3.) We make the following observations:

- CORAL performed well even for cases where it was not designed for. It solved well the linear integer constraints generated from the symbolic execution of binary search and treemap. This result is important considering that symbolic execution of scientific applications builds constraints with both decidable and complex parts.
- CORAL performed significantly better than CHOCO for the queries including Math functions derived from constraints manually-written by CORAL developers. In particular, we found cases when CHOCO would report incorrect solutions (e.g. the constraint Math.sin(x)+Math.cos(y)==1). We note that we did not tweak any parameter of CHOCO. We used the configuration set in SPF.
- Figure 4 shows that the versions of CORAL with optimizations found more solutions on average. In addition, "pso-opt" found more solutions than "ran-opt". In some cases the optimized solver missed the solution of some

constraints that its non-optimized version finds. As discussed in Section 4.2 the optimizations can reduce not only the search space but also the solution space. Note also that the difference in total number of constraints solved between "pso-opt" and "ran-opt" is not huge. We observed that one affecting factor for this result is the relative high number of inequality constraints (e.g., $>=$) compared to that of equality constraints for which random search would conceptually have more difficulty to find solutions.

5.3   Impact of Number of Iterations on Precision and Runtime

This section discusses the impact of the number of iterations (using the PSO search) in runtime and precision (as measured by the number of solutions found) and present the method used to select a default value for the maximum number of iterations per query to the solve.

We considered manually-written and NASA's benchmarks in this experiment. We varied the number of iterations from 10 to 3000 and measured how many solutions the solver can find for each selection. The leftmost plot from Figure 5 relates number of iterations with numbers of solutions that CORAL finds for each assignment. The plot indicates that the ratio of increase varies in different rates. For the lower end of the range (say, less than 500 iterations) the increase is sharp; for larger values the increase is smoother and often unpredictable. For example, CORAL finds 1153 solutions when using 600 iterations and only 51 more when using 3000 iterations (which is 5x increase in number of iterations). The vertical line in the figure shows this point of "stabilization", which we use as default selection for the maximum number of iterations. It is perhaps worth mentioning that the plot is not increasing monotonically with the number of iterations. This occurs because the search algorithm in the opt4j library uses the number of iterations itself as a factor to regulate the perturbation of candidate solutions. That does not imply, however, that the search is non-deterministic for given seed and maximum number of iterations.

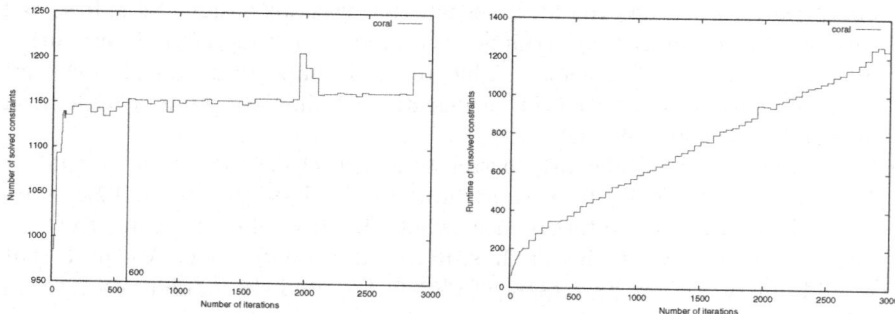

Fig. 5. Left plot relates number of iterations with number of solutions. Right plot relates number of iterations with runtime.

The rightmost plot shows average runtime in milliseconds for each assignment of number of iterations. For that, we used a machine with an Intel Core i7-920 processor (8M Cache, 2.66GHz), 8GB RAM, and running Ubuntu 10.04 32bits. In contrast to the previous experiment, we only considered unsolved constraints as they dominate runtime. In principle, the cost of a search iteration varies with the size of the constraint. In this setup, however, the size of the constraints does not vary significantly and the plot reveals an apparent linear relationship between number of iterations and average runtime.

Considering only the constraints that the solver could solve in the experiments from Section 5.2, CORAL took on average 60ms, CHOCO 3ms, CVC3 9ms, and YICES <1ms. As mentioned, this runtime difference can increase favorably to non-CORAL solvers when considering the constraints that CORAL cannot find solutions. Section 7 points to our plans to improve CORAL's runtime.

5.4   Analysis of the PISCES Library

We have applied SPF with the new CORAL solvers to the analysis of the PISCES (Platform Independent Software Components for the Exploration of Space) mathematical library. PISCES implements a collection of mathematical utility functions and it is used at NASA's Johnson Space Center for Web-based, collaborative development of computer programs for planning trajectories and trajectory-related aspects of spacecraft-mission design.

We have analyzed 20 methods in the library (version 2006), that perform complex mathematical computations such as hyperbolic (arc) sine, cosine, tangent, floating point reminder, factorial, as well as converting time and degrees into radians and back, etc. We were able to analyze all the methods with CORAL, and we discovered some problems, that were due to illegal arguments not properly caught in the code. Furthermore, we tested the implementations by performing checks of known mathematical properties of the PISCES functions.
For example, we checked the following:

```
public static void testHyperbolicTangent(double x) {
  double sinH = MathFunctions.sinh(x); /* hyperbolic sine */
  double cosH = MathFunctions.cosh(x); /* hyperbolic cosine */
  double tanH = MathFunctions.tanh(x); /* hyperbolic tangent */
  assert (tanH == sinH/cosH);
}
```

SPF with CORAL generates 6 path conditions, and it correctly determines that only 2 are feasible and that the assertion is not violated. If the assertion is changed to assert (tanH != sinH/cosH), SPF correctly finds two cases when the assertion is violated.

5.5   Analysis of the Apollo Lunar Autopilot

We have also applied SPF with CORAL to the analysis of the Apollo Lunar Autopilot, a Simulink model that was automatically translated to Java using the Vanderbilt tool-set [19]. This 2.6KLOC subject is deployed in a single package with 54 classes. (Numbers computed with the JavaNCSS tool [13].) The Simulink model was created by one of the engineers who worked on the Apollo Lunar Module digital autopilot design team to see how he would have done it using Simulink if it had been available in 1961. The model is available from MathWorks[2]. It contains both Simulink blocks and Stateflow diagrams and makes use of complex Math functions (e.g. `Math.sqrt`). The model could not be analyzed using `CHOCO` (or other constraint solvers that were previously in SPF), since these solvers could not handle the `sqrt` operation. In this experiment we set the bound on the length of a path condition to 50 and the bound on time to 2h. The bound on length makes the search to backtrack when it makes more than 50 consecutive

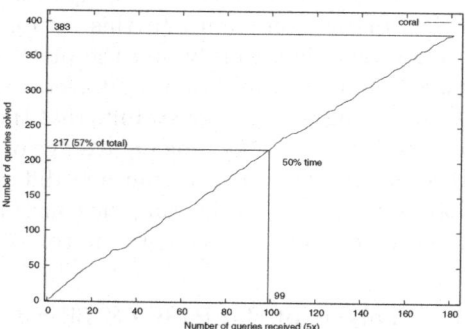

Fig. 6. Number of queries received vs. solved

branching choices. `CORAL` could solve 383 out of 905 queries generated (i.e., 42% of total) during the state-space exploration. Figure 6 summarizes the search. In one axis it shows the number of queries the constraint solver received (note the 5x scale) and the other shows the number of solutions found. The figure highlights the 1h data point. Note from the plot a small increase in saturation as time advances: in 50% of the time 57% (217 out of 383) of the total number of solutions are found. One reason for this is the increase of the path condition size (and cost of solving) with the increase of exploration depth. `CORAL` is sensitive to the path condition size in two ways. On the one hand as the path condition grows bigger the evaluation time also increases. On the other hand, a fitness function conceptually makes better judgments when more conjuncts appear in the path condition.

6 Related Work

Random-symbolic solving [11,21,22] has been recently proposed as an approach to solve constraints with undecidable fragments. The approach is to selectively randomize variables from the input constraint before passing a simplified version of it to a decision procedure. Empirical results show that such collaboration is very promising. We plan to investigate novel ways to promote collaboration between `CORAL` and decision procedures. For example, to first pass the input constraint to a decision procedure (with mathematical functions uninterpreted) and use solutions to seed the initial state of `CORAL`.

The constraint solver `FloPSy` [18] has been recently developed with similar purpose and approach as `CORAL`. `CORAL` and `FloPSy` use a similar notion of

---

[2] `http://www.mathworks.com/products/simulink/demos.html?file=/products/demos/shipping/simulink/aero_dap3dof.html`

distance in their fitness functions. Different from CORAL, FloPSy does not adjust the weights of constraint clauses in its fitness function as the search advances. As for the search, FloPSy uses a variation of the AVM method [16] and genetic algorithms. Another difference is that CORAL performs some optimizations (e.g., inference of domains and rewriting to eliminate variables) which are orthogonal to the search. (See Section 5.) FloPSy is used under the concolic execution of PEX [23], developed at Microsoft Research. CORAL has been customized specially for SPF; this could not be done readily with FloPSy.

Heuristic search has been previously proposed to improve random (concrete) testing [24,10] as opposed to symbolic testing. In the context of a concrete execution the fitness function operates directly over program elements. It measures how close execution is to discover a new program path using structural path coverage. One central distinction between the concrete and symbolic approaches is that, to evaluate fitness with concrete testing, one needs to execute the program to collect path coverage data while in the context of symbolic execution one needs to evaluate path conditions, which is an abstraction of the path.

7 Conclusions

This paper proposes the meta-heuristic solver CORAL for dealing with constraints involving mathematical functions and floating-point variables that symbolic execution can generate. The integration of CORAL with the NASA's Symbolic PathFinder tool (SPF) indicates that the approach is promising. The use of CORAL broadens the application of SPF at NASA and industry. CORAL is publicly available for use at the following address.

http://pan.cin.ufpe.br/coral

In future work, we plan to add incremental solving capability to CORAL (within the context of symbolic execution) and to investigate novel ways to collaborate with decision procedures. Finally, we plan to thoroughly evaluate CORAL in the context of constraints generated from the analysis of other NASA applications.

Acknowledgments. This work was partially supported by the National Institute of Science and Technology for Software Engineering (INES[3]), funded by CNPq and FACEPE, grants 573964/2008-4 and APQ-1037-1.03/08. Matheus Souza is supported by the CNPQ fellowship 118428/2010-1.

References

1. CHOCO web page, http://www.emn.fr/z-info/choco-solver/
2. CVC3 web page, http://www.cs.nyu.edu/acsys/cvc3/
3. Opt4J web page, http://opt4j.sourceforge.net/
4. TSAFE maryland, http://www.cs.umd.edu/~mvz/cmsc435-s09/

---

[3] www.ines.org.br

5. TSAFE mit, http://sdg.csail.mit.edu/TSAFE/downloads/
6. Wolfram Alpha web page, http://www.wolframalpha.com/
7. YICES web page, http://yices.csl.sri.com/
8. JPF project (2010), http://babelfish.arc.nasa.gov/trac/jpf
9. Back, T., Eiben, A.E., Vink, M.E.: A superior evolutionary algorithm for 3-SAT. In: Porto, V.W., Waagen, D. (eds.) EP 1998. LNCS, vol. 1447, pp. 125–136. Springer, Heidelberg (1998)
10. Baresi, L., Lanzi, P.L., Miraz, M.: Testful: An evolutionary test approach for java. In: ICST, pp. 185–194 (2010)
11. Godefroid, P., Klarlund, N., Sen, K.: DART: Directed Automated Random Testing. In: PLDI, pp. 213–223 (2005)
12. Goldberg, D.E.: Genetic Algorithms in Search, Optimization and Machine Learning. Addison-Wesley Longman Publishing Co., Inc., Boston (1989)
13. JavaNCSS website. JavaNCSS - A Source Measurement Suite for Java, http://www.kclee.de/clemens/java/javancss/
14. Kennedy, J., Eberhart, R.: Particle swarm optimization. IEEE Neural Networks, 1942–1948 (1995)
15. King, J.C.: Symbolic execution and program testing. Communications of ACM 19(7), 385–394 (1976)
16. Korel, B.: Automated software test data generation. IEEE Transactions on Software Engineering 16(8), 870–879 (1990)
17. Kroening, D., Strichman, O.: Decision Procedures – an Algorithmic Point of View. In: EATCS. Springer, Heidelberg (2008)
18. Lakhotia, K., Tillmann, N., Harman, M., de Halleux, J.: FloPSy - search-based floating point constraint solving for symbolic execution. In: Petrenko, A., Simão, A., Maldonado, J.C. (eds.) ICTSS 2010. LNCS, vol. 6435, pp. 142–157. Springer, Heidelberg (2010)
19. Pasareanu, C.S., Schumann, J., Mehlitz, P., Lowry, M., Karasai, G., Nine, H., Neema, S.: Model based analysis and test generation for flight software. In: Proceedings of SMC-IT (2009)
20. Păsăreanu, C.S., Mehlitz, P.C., Bushnell, D.H., Gundy-Burlet, K., Lowry, M., Person, S., Pape, M.: Combining unit-level symbolic execution and system-level concrete execution for testing nasa software. In: ISSTA, pp. 15–26 (2008)
21. Takaki, M., Cavalcanti, D., Gheyi, R., Iyoda, J., d'Amorim, M., Prudencio, R.: A comparative study of randomized constraint solvers for random-symbolic testing. In: NFM, pp. 56–65 (2009)
22. Takaki, M., Cavalcanti, D., Gheyi, R., Iyoda, J., d'Amorim, M., Prudencio, R.: Randomized constraint solvers: a comparative study. Innovations in Systems and Software Engineering (ISSE) 6(3), 243–253 (2010)
23. Tillmann, N., de Halleux, J.: Pex–white box test generation for .NET. In: Beckert, B., Hähnle, R. (eds.) TAP 2008. LNCS, vol. 4966, pp. 134–153. Springer, Heidelberg (2008)
24. Tonella, P.: Evolutionary testing of classes. In: ISSTA, pp. 119–128 (2004)

Automated Formal Verification of the *TTEthernet* Synchronization Quality

Wilfried Steiner[1] and Bruno Dutertre[2]

[1] TTTech Computertechnik AG, Chip IP Design
A-1040 Vienna, Austria
wilfried.steiner@tttech.com
[2] SRI International, Computer Science Laboratory
Menlo Park, CA 94025, USA
bruno@csl.sri.com

Abstract. Clock synchronization is the foundation of distributed real-time architectures such as the Timed-Triggered Architecture. Maintaining the local clocks synchronized is particularly important for fault tolerance, as it allows one to use simple and effective fault-tolerance algorithms that have been developed in the synchronous system model.

Clock synchronization algorithms have been extensively studied since the 1980s, and many fundamental results have been established. Traditionally, the correctness of a new clock synchronization algorithm is shown by reduction to these results. Until now, formal proofs of correctness all relied on interactive theorem provers such as PVS or Isabelle/HOL. In this paper, we present an automated proof of the *TTEthernet* clock-synchronization algorithm that is based on the SAL model checker.

1 Introduction

Distributed real-time systems are omnipresent in our daily lives and are becoming increasingly large and complex. It is becoming apparent that the correct development of such complex systems requires a sound architectural basis. The time-triggered architecture (TTA) [1] is intended to facilitate the development of fault-tolerant, real-time systems. TTA has been successfully adopted in industries that demand a high level of determinism, such as the avionics industry in which predictability of system operation is key. Upon others, *TTEthernet* (an implementation of the TTA) has been selected for the Orion Space Program [2]. The prime concept of TTA is a common perception of time in the devices that form the distributed system. These devices rely on local hardware clocks to build a common logical time base that is consistent across the system: any two logical clocks must read approximately equal values at any time during the system evolution. To maintain consistency, a clock synchronization algorithm must be used to compensate for the imperfection of the physical clocks. The maximal difference between two non-faulty logical clocks in the system is the *synchronization quality* or *precision* achieved by the algorithm.

M. Bobaru et al. (Eds.): NFM 2011, LNCS 6617, pp. 375–390, 2011.

Clock synchronization has been studied for decades. Fundamental results provide answer to basic questions such as how well can clocks be synchronized in a distributed system [3] or how to construct fault-tolerant clock synchronization algorithms (e.g., [4]). Applications of these results to specific implementations and industrial products is described in several publications (e.g., [5]). Even with recent technological improvements in hardware clocks (e.g., embedding atomic clocks on a chip), clock synchronization remains highly relevant to modern real-time distributed systems. Fault-tolerant synchronization algorithms are required to align the clocks initially and to tolerate clock failures.

In many systems, safety depends critically on correct clock synchronization. As a consequence, significant effort has been dedicated to developing rigorous correctness proofs of various clock-synchronization algorithms. Schneider has shown that these algorithms share very similar properties and has introduced a general proof scheme for establishing their correctness [6]. Formal proofs of clock-synchronization algorithms have been developed by Rushby et al. [7], Shankar [8], and Miner [9] using the EHDM theorem prover; other formal proofs used PVS, the successor of EHDM [10,11]. Both EHDM and PVS are interactive theorem provers that require human guidance and expertise. Recently, more automated proof methods have been investigated that attempt to reduce the need for human expertise, by leveraging advances in model checking technology and automated reasoning engines known as SMT solvers. For example, Barsotti et al. [12] combine Isabelle/HOL and the SMT solvers CVC3 and Yices to formally verify Schneider's generic scheme. Another example of combined PVS and SAL proof method has been presented by Pike [13]. In [14], we used a model-checking approach to verify the "compression master" functionality of the *TTEthernet* clock synchronization algorithm (see next section). This verification was almost automated except that it required us to provide a few auxiliary lemmas by hand. In this paper, we extend the latter work to the full *TTEthernet* clock-synchronization algorithm and to analyzing the synchronization quality achieved by this algorithm. Model-checking clock synchronization algorithms has been done before in [15]. However, these studies are limited to four fully connected nodes and symbolic representation of time with fixed timing parameters. In this paper we treat time as a continuous entity while leaving the parameters uninterpreted. Thus, our proofs are valid for all timing parameterizations of the *TTEthernet* clock synchronization protocol.

This paper continues in the following section with an informal presentation of the *TTEthernet* clock synchronization algorithm. In Section 3 we then give an overview of the proof method and discuss the formal model in detail. We present the results of the formal proofs as well as some example testcases in Section 4. Finally, we conclude in Section 5.

2   *TTEthernet* Clock Synchronization Algorithm

TTEthernet is an extension of the traditional Ethernet standard, with additional services that guarantee reliable, deterministic delivery of time-critical messages.

A *TTEthernet* network consists of end systems and switches. End systems are connected to switches with bi-directional communication links and switches may connect to each other. Each switch belongs to one and only one channel and in its simplest form a channel is formed by a single switch and the communication links to the end systems. For fault-tolerance reasons a *TTEthernet* network can implement redundant channels. An example network with two redundant channels is depicted in Figure 1.

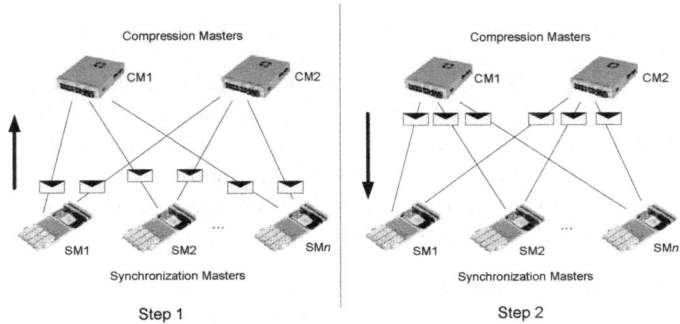

Fig. 1. Overview of the *TTEthernet* two step clock synchronization algorithm

2.1 Clock Synchronization Overview

End system and switches define physical components in the *TTEthernet* network and for the clock synchronization algorithm we use three different "roles": Synchronization Master (SM), Compression Master (CM), and the Synchronization Client (SC). For simplicity of discussion we assume a network consisting of five end systems and two channels, as depicted in Figure 1. Furthermore, end systems implement the SM role and the CMs are realized in the switches. SCs are only passively synchronizing to the timebase as maintained by the SMs and CMs and we exclude this role therefore from our discussion. In the clock synchronization algorithm SMs and CMs inform each other about their current state of their local clock by exchanging Protocol Control Frames (PCF).

In *TTEthernet* the clocks are synchronized in two steps. In the first step, the SMs send PCFs to the CMs. The CMs extract from the arrival points in time of the PCFs the current state of their local clocks and execute a first convergence function, the so-called compression function. The result of the convergence function is then delivered to the SMs in form of new PCFs (the "compressed" PCFs). In the second step the SMs collect the compressed PCFs from the CMs and execute a second convergence function. Our contribution in [14] has been restricted to showing the correctness of the implementation of the compression function, which we therefore assume in this paper.

TTEthernet requires an inconsistent-omission failure model for the CMs. This means that a faulty CM is able to arbitrarily accept and reject PCFs from the SMs and can also decide to which SMs it sends the compressed PCF and to which not. Babbling idiot failures of the CM are excluded by the design of the CM as self-checking pair. The SMs, on the other hand, may fail arbitrarily, and in particular, they may start to babble PCFs. The CMs implement a central guardian functionality that ensures that only one PCF per SM is used per resynchronization cycle. Though, in the worst case, we assume that the clock value provided by a faulty SM can be arbitrary.

2.2 First Step Convergence: Compression Master (CM)

The CMs collect the current states of the local clocks of the SMs. We denote these values by SM_clock_i, where $1 \leq i \leq |SM|$ and assume that the SM_clock_i values are sorted in increasing order. To minimize the impact of the faulty SMs *TTEthernet* uses a variant of the fault-tolerant median to calculate the new "compressed" clock. Following rules define the compressed clock depending on the number of SM_clock_i values received.

- one SM clock: $compressed\_clock = SM\_clock_1$
- two SM clocks: $compressed\_clock = \frac{SM\_clock_1 + SM\_clock_2}{2}$
- three SM clocks: $compressed\_clock = SM\_clock_2$
- four SM clocks: $compressed\_clock = \frac{SM\_clock_2 + SM\_clock_3}{2}$
- five SM clocks: $compressed\_clock = SM\_clock_3$
- more than five SM clocks: average of the $(k+1)^{th}$ largest and $(k+1)^{th}$ smallest clocks, where k is the number of faulty SMs to be tolerated.

The compressed clock is delivered back to the SMs in a new "compressed" PCF and the SMs are able to read the compressed clock value from the arrival point in time of the compressed PCF. In addition to the compressed clock value, the CMs also generate a membership vector $pcf\_membership\_new$. Each position in this vector is assigned to one and only one SM. The CMs will set the bit of a SM, if the respective SM i has provided a local clock value SM_clock_i and will clear the bit otherwise. The CMs transmit the $pcf\_membership\_new$ vector in the payload of the compressed PCF. The self-checking pair design of the CM guarantees that the compressed clock and the $pcf\_membership\_new$ vector are consistent. Hence, the design prevents a faulty CM to set an arbitrary number of bits in $pcf\_membership\_new$.

2.3 Second Step Convergence: Synchronization Master (SM)

In the second step of the clock synchronization algorithm, the SMs receive the compressed PCFs, extract the compressed clock values from them, and correct their local clocks. In the fault-free case each SM receives exactly one compressed PCF per CM from which it extracts the compressed clock values CM_clock_j, where $1 \leq j \leq |CM|$ and we assume the CM_clock_j values sorted in increasing order. Under the assumption of one CM per channel and up to three channels maximum, the convergence function has to cover following three cases:

- one CM clock: $SM\_clock = CM\_clock_1$
- two CM clocks: $SM\_clock = \frac{CM\_clock_1 + CM\_clock_2}{2}$
- three CM clocks: $SM\_clock = CM\_clock_2$

In the case of a faulty CM, a SM may receive at maximum one compressed PCF per CM (as the faulty CM may decide not to send its compressed PCF to some SMs). Furthermore, a SM will only use a compressed PCF in the convergence function discussed above if the *pcf_membership_new* field has at least *accept_threshold* of bits set. *accept_threshold* is calculated as follows:

1. *current_max* = maximum of bits set in the *pcf_membership_new* field of any compressed PCF
2. *accept_threshold* = *current_max* minus the allowed number of faulty SMs

The SM will discard a compressed PCF that has less than *accept_threshold* bits set in the *pcf_membership_new* field. This mechanism ensures that a SM excludes compressed PCFs that represent relative low numbers of SM clocks.

The *pcf_membership_new* vector is also used in other *TTEthernet* algorithms such as clique detection or startup as well as in network configurations that use more than one CM per channel. We do not discuss this functionality and configurations in this paper. For the analysis of the clock synchronization algorithm the description above is sufficient.

2.4   Clock Synchronization Example

Figure 2 gives and example scenario of the *TTEthernet* clock synchronization. The x-axis represents progress in time as alternating intervals of clock drift and re-synchronization using the two-steps approach. Note that these are logic steps and do not represent real time. Odd values on the x-axis represent the SM local clock values immediately before the synchronization, even values represent the values of the SM local clocks immediately after synchronization. The y-axis depicts the clock-time of the SMs. We will discuss the representation of the clock time in the next section.

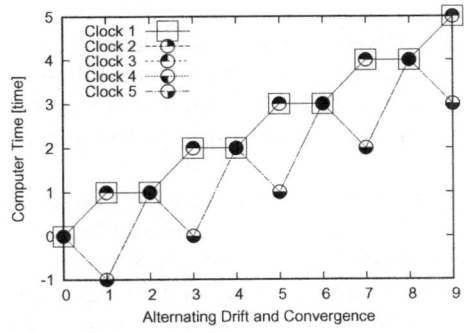

Fig. 2. Fault-free scenario of the *TTEthernet* clock synchronization algorithm

The example depicted in Figure 2 shows a fault-free execution trace of five SMs and two CMs. Initially, the SMs are perfectly synchronized. SMs 1, 2, and 3 have maximum positive drift and SMs 4, 5 have maximum negative drift. As there are no failures involved and when neglecting digitalization errors and transmission jitter on the network, the local clocks of the SMs become perfectly re-synchronized with each execution of the clock synchronization algorithm.

3   Automated Formal Verification Procedure

We give an overview of the proof method next. We then discuss the formal model in the SAL notation and the proof procedure.

3.1   Proof Method Overview

The *TTEthernet* clock synchronization algorithm has been formalized in SAL [16] as state-transition system of the form $\langle S, I, \rightarrow \rangle$. Here, S defines the set of system states σ_i, I the set of initial system states with $I \subseteq S$ and $\rightarrow$ the set of transitions between system states. Each system state σ maps the variables to particular values according their defined variable type. Furthermore, SAL supports structured modeling such that we can define the SM and CM functionality in encapsulated modules.

SAL provides several tools (symbolic, bounded, and bounded infinite-state model checking). While we experimented with all of them, we finally use the bounded infinite-state model checker `sal-inf-bmc` to prove the *TTEthernet* synchronization quality as well as to generate testcases. With `sal-inf-bmc` we can treat time as continuous entity and can use k-induction [17] as proof method. The proof of a property $\Box P$ by k-induction is a generalized form of regular induction and consists of following stages [18]:

- Base Case: Show that all the states reachable from I in no more than $k - 1$ steps satisfy P
- Induction Step: For all trajectories $\sigma_0 \rightarrow \ldots \rightarrow \sigma_k$ of length k, show that $\sigma_0 \models P \wedge \ldots \wedge \sigma_{k-1} \models P \Rightarrow \sigma_k \models P$

In our studies we have observed an interesting dependency between k and the synchronization quality: increasing k allowed to calculate the upper bound on the precision more tightly. This means there is a trade-off between the depth (k) of the proof and the quality of its result (calculated upper bound on the precision).

The SMs are modelled as state machines with two states representing the alternating drift and correction intervals. The example scenario in Figure 2 also gives an overview of our modelling method. As we are only interested in the maximum difference between any two non-faulty local clocks, we can abstract from the nominative length of the synchronization interval. All we need to model is the maximum difference to the nominative length that would result from a non-faulty clock. In many current industrial use cases the drift offset, i.e., the offset as a result of the imperfect physical local clocks, is the dominant part of this offset and we refer therefore to the offset as "drift offset". Note, although we use the term drift offset we implicitly also take into account network jitter, digitalization errors and similar error terms. We argue that these effects can be summarized by a sufficiently high value for what we call the drift offset. As we do not specify a particular value for the drift offset in our proofs, but only require an upper bound on it, the proofs are also valid for real systems rather than only for idealized models. We have been able to directly proof the value of the precision in certain *TTEthernet* networks by only specifying the functionality of the SM and the CM without any additional lemma or further modelling tricks. However, we see a significant performance gain if we use a lemma informing the model-checker that all SMs consistently change their state (from the drift interval to the correction interval and vice versa). For this lemma we use a simple system level abstraction as introduced in [18].

3.2 Formal Model

```
POSREAL: TYPE = {x: REAL | x>=0 };
max_drift: POSREAL; max_clock: REAL; max_SM:  NATURAL = 5; max_CM: NATURAL = 2;
```

The formal model[1] starts with some constants and types. `POSREAL` defines the positive real numbers. `max_drift` describes the absolute value of the maximum drift offset of a clock within one re-synchronization interval. `max_clock` describes the time horizon. Both, `max_drift` and `max_clock`, have no value assigned, hence, we leave them "uninterpreted". This means that they may have any value. `max_SM` defines the maximum number of SMs in the network. `max_CM` defines the number of redundant channels in the network. We define exactly one CM per channel.

We define dedicated types to denote the sets of nodes, SMs, channels etc. The formal model is then executed fully synchronously in alternating steps `send` and `sync` as denoted by the SM's state.

```
TYPE_drift: TYPE = REAL; TYPE_clock: TYPE = REAL;
TYPE_SM: TYPE = [1..max_SM]; TYPE_CM: TYPE = [1..max_CM];
TYPE_states: TYPE = {send, sync};
```

In the `send` state the SMs provide the values of their local clocks to the CMs which execute the first step convergence function and return the converged values back to the SMs. In the `sync` state, the SMs execute the second step convergence function and update their local clock accordingly. We discuss this process in more detail next based in the SM and CM implementation in SAL.

3.2.1 Synchronization Master Module. The synchronization master module `SM` is parameterized by `TYPE_SM`, to identify a particular SM by `id`.

---

[1] A more detailed report and the models can be found at
`http://sal-wiki.csl.sri.com`

```
SM[id:TYPE_SM]: MODULE = BEGIN
INPUT list_compressed_clock: ARRAY TYPE_CM OF TYPE_clock
OUTPUT state: TYPE_states, clock: ARRAY TYPE_CM OF TYPE_clock
LOCAL drift: TYPE_drift, interval_ctr: NATURAL
```

The SMs receive their input from the CMs. `list_compressed_clock` represents the first-step converged clock values, i.e., the compressed clock value. An SM will output its current state and the value of its local clock `clock`. `clock` is modeled as an array of size `TYPE_CM`, which allows us to model inconsistent faulty behavior of a faulty SM as discussed later on. In addition to the input and output we also define some local variables in for an SM. `drift` defines the drift offset for a given re-synchronization interval. `interval_ctr` counts the re-synchronization intervals; it is used to derive test traces. We initialize the model to a clean state.

```
INITIALIZATION  interval_ctr = 0; state = sync; clock = [[j:TYPE_CM] 0];
                drift IN {x: TYPE_drift | x=-max_drift OR x=max_drift};
```

We use the formal model for both testcase generation and formal proof of the synchronization quality. Depending on the purpose of the formal experiment `drift` can be set to a static value to pretty-print counterexamples or to an arbitrary value. In the case above, our aim is to generate a nice trace for which we initialize `drift` to take either the positive or the negative maximum drift offset. The model checker is free to chose either value once for the complete execution of the model. The SAL construct `IN` models this non-deterministic choice. It is interpreted as: let `drift` be an `x` which satisfies the condition as specified above. We use the `IN` construct at several positions in our model.

In case of the formal proof we want cover a more general case of clock drift, for which we have to define `drift` as a `DEFINITION`.

```
DEFINITION  %drift IN {x: TYPE_drift | x>=-max_drift AND x<=max_drift};
```

The "%" sign indicates a comment line in SAL. We use it here to emphasize that `drift` may either be initialized or defined, but not both. The definition of `drift` says that in every step of the model execution `drift` may take an arbitrary value in between the maximum negative and positive drift offset and this value may change with each step. We use this definition for the formal proofs.

```
[   state=sync -->
    state'=send;  interval_ctr'=interval_ctr + 1; clock'=[[j:TYPE_CM] clock[j] + drift];
[] state=send -->
    state'=sync;
    clock' IN {x: ARRAY TYPE_CM OF TYPE_clock |
      x[1]=x[2] AND average(list_compressed_clock[1], list_compressed_clock[2],x[1])}; ]
```

In the fault-free case there are only two transitions in the the state machine of an SM (expressed by guarded commands in the form `guard --> commands`). When the SMs are in the `sync` state their local clocks are closely synchronized. The next state will be `send` for which they increase the counter of the re-synchronization intervals, and select a new value for their local clocks. This new value is simply the sum of the current clock value and the drift offset as specified by `drift`. Our treatment of `clock` is different from the traditional correctness proofs which aim to show that clock-time simulates real-time with a certain accuracy. This is not necessary in our approach. We are only interested in the maximum difference of any two `clock` values of non-faulty components. Hence, we update `clock` only for the differences in the nominative length of the re-synchronization interval and can omit its actual length. In the `send` state the local clocks of the SMs are far apart and they process the compressed clock values received from the CMs to bring the local clocks back into agreement for the following `sync` state.

In a faulty-free system with two channels and one CM per channel, the SM applies the arithmetic average to the received compressed clock values. In the transition of the SM we specify that `clock` shall take a new value such that the `average` predicate is satisfied. The predicate is satisfied when the third parameter is the arithmetic mean of the first two parameters.

```
average(value1, value2, avg: TYPE_clock): BOOLEAN = avg=(value1+value2)/2
```

3.2.2   Compression Master Module. The CM is parameterized by TYPE_CM, such that `id` identifies a particular CM. It takes the clock values as input and returns the compressed clock value to the SMs as a result of the first step convergence. The CM uses the local variable `order` to sort the clock values as provided by the SMs.

```
CM[id:TYPE_CM]: MODULE = BEGIN
INPUT clocks_cm: ARRAY TYPE_SM OF TYPE_clock
OUTPUT compressed_clock: TYPE_clock
LOCAL order: ARRAY TYPE_SM OF TYPE_SM
```

We model the CM as a stateless process. Its only purpose is the calculation of the first step convergence function, the compression function. In a system with even number of SMs or more than five SMs the CM has to apply the averaging function as discussed previously.

```
compressed_clock IN {x: TYPE_clock | average(clocks_cm[order[2]], clocks_cm[order[3]], x)}
```

In a system with one, three, or five SMs, the compressed clock is simple the middle value, e.g., for five SMs (`compressed_clock=clocks_cm[order[3]];`). In both cases `order` determines the order of the clock values. We have introduced this method in [14] and summarize it here for completeness. We define `order` to be an array of SM identifiers that satisfies the `sort` predicate. `sort` is satisfied when `sorted_list` is an array in which the entries point to the elements of `unsorted_list` in increasing order.

```
order IN {x: ARRAY TYPE_SM OF TYPE_SM | sort(clocks_cm, x)};
sort(unsorted_list: ARRAY TYPE_SM OF TYPE_clock,
     sorted_list:ARRAY TYPE_SM OF TYPE_SM): BOOLEAN =
(FORALL (i:TYPE_SM): i<max_core =>
     unsorted_list[sorted_list[i]] <= unsorted_list[sorted_list[i+1]]) AND
(FORALL (i,j:TYPE_SM): sorted_list[i] = sorted_list[j] => i=j);
```

3.3 Automated Formal Proof

We are interested in verifying the precision in the system (property `distance`):

```
distance: LEMMA world |- G(FORALL (i,j: TYPE_SM):
 (list_states[i]=send AND list_clocks[i][1]>list_clocks[j][1] =>
  list_clocks[i][1]-list_clocks[j][1]) <= FACTOR*max_drift));
```

`distance` says that when the SMs are in the `send` state, which is just before the execution of the clock synchronization algorithm, the maximum difference of any two local clocks is bound by `FACTOR*max_drift` (when faulty SMs are present they have to be excluded). As introduced earlier, `max_drift` is the maximum drift offset of a correct clock in the system from real time within one synchronization inverval. The value of `FACTOR` has to be assigned by hand in the model. This value is typically the result of an informal analysis of the algorithm. E.g., in the case of faulty CM we suspect that the value is $(8/3)$. When `FACTOR` has not been determined upfront, we can even "search" for it by manually testing assignments for `FACTOR` until the model checker stops producing counter-examples and proves `distance` to be correct.

We invoke `sal-inf-bmc` using the following command, where `clocksync` is the model name, `--depth 3` specifies the analysis depth, and `-i` invokes k-induction:

```
> sal-inf-bmc clocksync distance --depth=3 -i
```

This direct proof works well for a low number of nodes and relatively benign failure modes. We can speed-up the verification time significantly with a simple abstraction method introduced in [18]. For this we define two abstract system states, `BIG` and `SMALL`.

```
abstractor: MODULE =
BIG = (FORALL (i:TYPE_SM): list_states[i]=send) AND
      (FORALL (i,j:TYPE_SM): list_clocks[i][1]>list_clocks[j][1] =>
           (list_clocks[i][1]-list_clocks[j][1] <= FACTOR * max_drift));
SMALL = (( ... <= FACTOR_small * max_drift));
```

In the system-level abstraction we formulate that all SMs are at the same time either in the `send` state or in the `sync` state and they are synchronously proceeding between the two states. Furthermore, we already define `FACTOR*max_drift` here, which makes the proof of `distance` later on trivial. We can proof that the system-level abstraction `abstract_invar` is correct and verify `distance` while using `abstract_invar` as lemma (option -l).

```
> sal-inf-bmc clocksync abstract_invar --depth=3 -i
> sal-inf-bmc clocksync distance -l abstract_invar --depth=3 -i
```

4   Fault-Injection Experiments and Results

In this section we show how to model failures and discuss the *TTEthernet* synchronization quality. We use a system model consisting of five SMs and two CMs. We want to show the synchronization quality of the *TTEthernet* clock synchronization algorithm under a single SM failure, a single CM failure, and under concurrent SM and CM failures.

4.1   Inconsistent Omission Faulty CM

In *TTEthernet* the failure mode of a CM is inconsistent omission faulty. Hence, a faulty CM may arbitrarily decide which clock values from the SMs to use and which to discard. It can also arbitrarily decide to which SMs it will send it's compressed clock value. However, even a faulty CM will correctly represent the set of SMs that is selected in the *pcf_membership_new* vector in its compressed PCF. We define CM with id 1 to be the faulty CM (FAULTY_CM: NATURAL=1).

A non-faulty CM receives clock values from all SMs. As in our network five SMs are present and they all are non-faulty (for now), a correct CM receives five SM values. According to the algorithm definition (Sec. 2.2) it selects the median value, i.e., the third value as its compressed clock value.

```
compressed_clock IN {x: TYPE_clock |
  IF id /= FAULTY_CM THEN x=clocks_cm[order[3]]
  ELSE x=clocks_cm[order[3]] OR
    average(clocks_cm[order[2]], clocks_cm[order[3]], x) OR
    average(clocks_cm[order[2]], clocks_cm[order[4]], x) OR
    average(clocks_cm[order[3]], clocks_cm[order[4]], x)
  ENDIF};
```

The inconsistent-omission faulty CM may accept only an arbitrary subset of the five SM clock values, but this choice is reflected in the number of bits it sets in the *pcf_membership_new* vector. As the correct CM will deliver its compressed PCF with a *pcf_membership_new* vector having five bits set, the *accept_threshold* will be four. Hence, in order that there is a chance at all that the compressed clock of the faulty CM is not excluded in the second convergence step in the SMs it may only discard one of the SMs clock values. There are four options for the CM to calculate the first step convergence function, as depicted in the SAL model above. In the first case the CM accepts all SM clocks, and the following three cases cover when it discards any one of the SM clocks.

In order to simulate the inconsistent-omission faulty transmission behavior of the faulty CM we define a second transition in the SM state machine for the send state.

```
[] state=send
   --> state'=sync; clock' = [[j:TYPE_CM] list_compressed_clock[CORRECT_CM]];
```

Hence, we map the inconsistent transmission failure of the CM to an non-deterministic choice in the SM: the SM is free to decide whether it received a clock value from the faulty CM or not.

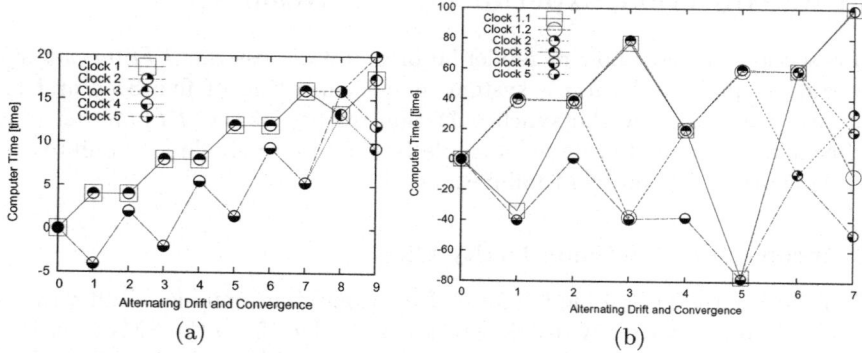

Fig. 3. Example scenarios of the *TTEthernet* clock synchronization algorithm with a faulty CM (left), and a faulty SM and CM (right)

Figure 3(a) gives an example trace of the algorithm execution in presence of a faulty CM. Again, the x-axis represents alternating intervals of drift and convergence and the y-axis the clock time. In contrast to Figure 2 we see the impact of the faulty CM resulting in a non-zero difference in the local clocks of the SMs after re-synchronization. We have formally verified the precision in this system setup to be $(8/3) \times drift\_offset$, by k-induction at depth three. FACTOR $= (8/3)$ has been calculated from an informal reasoning, which is also depicted in Figure 3(a): some SMs have fast clocks, some slow ones, and the faulty CM sends its compressed clock to only one of these groups. Consequently, only the SMs in one group correct their clocks towards the respective other one. In the figure we see that the faulty CM provides it clock only to the SMs with negative drift, which correct their clocks, while the SMs with positive drift do not correct their clock as they only receive the compressed clock values from the correct CM.

4.2   Arbitrarily Faulty SM

An arbitrarily faulty SM is free to fake its local clock values. We define the SM with id 1 to be faulty (FAULTY_SM: NATURAL=1). The communication of the local clock values is modelled by an array indexed by the CMs and the faulty SM may send different clock values to different CMs by assigning different values to different array entries. We model the arbitrary clock value by the failure term failure that simulates the faulty local clock values. failure can take any value for the faulty SM and is 0 for non-faulty SMs. Finally, we update the transition in the SM to reflect the failure (i.e., a change in the update of clock').

```
LOCAL failure: ARRAY TYPE_CM OF TYPE_drift
failure IN {x: ARRAY TYPE_CM OF TYPE_drift |
    IF id = FAULTY_SM THEN TRUE ELSE x[1]=0 AND x[2]=0 ENDIF};
clock'= [[j:TYPE_CM] clock[j] + drift + failure[j]];
```

The impact of a faulty SM only on the precision of the network is limited, even non-existent. Although the CMs can receive different clock values from the faulty SM and consequently derive different compressed clocks, they still will send their compressed clock values to all SMs. Hence, all non-faulty SMs receive the compressed clocks from the CMs consistently. The precision of the system with an arbitrarily faulty SM is, thus, the same as the precision in a fault-free system: $2 \times drift\_offset$.

4.3   Inconsistent Omission Faulty SM and CM

One particular failure combination of interest is when SM and CM are inconsistent-omission faulty. Hence, both may accept only a subset of clock values and send their clock value to only a subset of SMs or CMs. We can reuse the modelling of the faulty CM and have to introduce two additional transitions for the faulty SM. First, a faulty SM may decide to receive only the compressed clock from the faulty CM, and, secondly, the faulty SM may decide not to receive any compressed clock.

```
[] state=send AND id=FAULTY_SM
--> state'=sync; clock' = [[j:TYPE_CM] list_compressed_clock[FAULTY_CM]];
[] state=send AND id=FAULTY_SM
--> state'=sync;
```

For completeness, we note here that the remaining case of the faulty SM receiving only the compressed clock of the correct CM has been already covered by the transition below by modelling the faulty CM.

```
[] state=send
--> state'=sync; clock' = [[j:TYPE_CM] list_compressed_clock[CORRECT_CM]];
```

In addition to the model of the SM we also have to add additional cases to the calculation of the compressed clock in the CM. This can be done in a systematic way as depicted in the SAL source code below. There are two general cases, in the first case the faulty SM provides a clock value to the correct CM. This case is identical with the behavior of the faulty CM scenario discussed previously. In the second case the faulty SM does not provide a clock value to the correct CM. For this case we define the predicate **order_part** which is identical to the **order** predicate, except that it only orders the clock values from the correct SMs. The correct CM will calculate the compressed clock as the average from the second and third clock value. The faulty CM is free to use either option as discussed previously. In addition it may use the average as the correct CM or it may even decide to accept only three of the four correct SM clocks. In the latter case compressed clock from the faulty CM is either the second or the third correct SM clock.

```
compressed_clock IN {x: TYPE_clock |
  ( ... case as faulty CM only ... ) OR
(IF id /= FAULTY_CM THEN average(clocks_cm[order_part[2]], clocks_cm[order_part[3]],x)
 ELSE ( ... case as faulty CM only ... )  OR
      average(clocks_cm[order_part[2]], clocks_cm[order_part[3]],x) OR
      x=clocks_cm[order_part[2]] OR x=clocks_cm[order_part[3]] ENDIF) }
```

For a *TTEthernet* network with five SMs and two CMs and an inconsistent-omission faulty SM and CM we have verified the precision to be bound by $4 \times drift\_offset$.

4.4   Inconsistent Omission Faulty CM and Arbitrarily Faulty SM

The failure modelling of a network with inconsistent-omission fault CM and arbitrarily faulty SM is simply the combination of the individual failure models as introduced above. Figure 3(b) shows an example trace of the failure scenario. The main difference to the previous figures is that the clock of the faulty SM 1 is depicted as two clocks 1.1 and 1.2. This is, again, because the faulty SM may send different values to the different CMs, and, indeed, this scenario is shown in Figure 3(b). On the odd numbers on the x-axis the local clock readings just before the algorithm execution are depicted. Some clocks are fast and some are slow, and the faulty SM supports both groups. We also see that arbitrarily faulty behavior of the clock of SM 1, as the jumps from one extreme to the other. We have proven the precision to be $(12/3) \times drift\_offset$ in this configuration of five SMs and two CMs with faulty CM and SM. This number also confirms an informal argument of the worst-case scenario similar to the one discussed for a faulty CM only.

4.5   Summary of Verification Results

The verification times are summarized in Table 1. The precision Π for the scenarios follows from FACTOR as $\Pi = $ FACTOR $\times drift\_offset$. "distance" gives the verification times without system level abstraction. "abstraction" shows the verification times of the invariant for the system level abstraction, i.e., the verification that the abstraction is correct, and the last row depicts the verification times of "distance" when the abstraction is used as lemma.

Table 1. Verification results; FACTOR is a scalar, verification times are given in seconds

Property	No Faults	Faulty			
		CM	SM	CM/SM io	CM/SM a
FACTOR	2	(8/3)	2	(12/3)	(12/3)
distance	10.5	28.25	8.66	N/A	N/A
abstraction	0.5	0.58	0.49	85.3	44.43
distance+abst.	0.34	0.36	0.38	0.39	0.4

We clearly observe that the verification times decrease dramatically when the system-level abstraction is used. For difficult failure scenarios it is even essential to derive a formal proof. The arbitrarily faulty SM and inconsistent omission faulty CM faulty scenario terminates at depth five after eight-hundred seconds without counterexample and without proof. The inconsistent omission CM SM scenario returns the same result after sixteen-hundred seconds (indicated by N/A). Note that "N/A" in the table means that the inductive proof without the abstraction lemma was not possible. However, "distance" has been proven by k-induction using the abstraction lemma (as shown in the last row of the table), providing full coverage of our failure assumptions.

5   Conclusion

In this paper we have shown for the first time that fault-tolerant clock synchronization proofs can be fully automatized even in a model of continuous uninterpreted time. This is a significant advancement over the state-of-the-art which involves heavy-duty theorem provers or imposes significant modeling restrictions. We have shown that the precision in a *TTEthernet* network is between two and four times the drift offset (including network jitter and digitalization effects), depending on the failures to be tolerated. The only step requiring human interaction that one may argue being required in the synchronization verification is in the definition of the failure cases to model faulty components realistically. However, as we have discussed, the failure model can be constructed fairly systematically. For *TTEthernet* the failure cases are limited by design. For more complex protocols it can make sense to separately model check for the completeness of these cases. In our experiments we used a system of five Synchronization Masters and two Compression Masters. While this is a small system, industry trends indicate that mostly a core set of nodes for clock synchronization is used anyhow. Hence, the limitation to a small number of clocks does not impose an industrial shortcoming. On the other hand, given the fast verification times and low memory use of our approach, we will target larger systems in future work.

Acknowledgments

The research leading to these results has received funding from the European Community's Seventh Framework Programme (FP7/2007-2013) under grant agreement $n°236701$ (*CoMMiCS*). The second author was supported by NASA Cooperative Agreement NNX08AC59A. The authors would like to thank Günther Bauer for the informal proofs on the precision and feedback to this paper.

References

1. Kopetz, H., Bauer, G.: The Time-Triggered Architecture. Proceedings of the IEEE 91(1), 112–126 (2003)
2. Howard, C.E.: Orion avionics employ COTS technologies. In: Avionics Intelligence (June 2009)
3. Lundelius, J., Lynch, N.: An upper and lower bound for clock synchronization. Information and Control 62(2-3), 190–204 (1984)
4. Lamport, L., Melliar-Smith, P.M.: Byzantine clock synchronization. In: PODC 1984: Proceedings of the Third Annual ACM Symposium on Principles of Distributed Computing, pp. 68–74. ACM, New York (1984)

5. Kopetz, H.: TTP/C Protocol – Version 1.0. Vienna, Austria: TTTech Computertechnik AG (July 2002), http://www.ttagroup.org
6. Schneider, F.B.: Understanding protocols for byzantine clock synchronization. Cornell University, Ithaca, NY, USA, Tech. Rep. TR87–859 (1987)
7. Rushby, J., von Henke, F.: Formal verification of the interactive convergence clock synchronization algorithm. Computer Science Laboratory, SRI International, Menlo Park, CA, Tech. Rep. SRI-CSL-89-3R, (February 1989), http://www.csl.sri.com/papers/csl-89-3/ (revised online August 1991)
8. Shankar, N.: Mechanical verification of a generalized protocol for byzantine fault-tolerant clock synchronization. In: Vytopil, J. (ed.) FTRTFT 1992. LNCS, vol. 571, pp. 217–236. Springer, Heidelberg (1992)
9. Miner, P.S.: Verification of fault-tolerant clock synchronization systems. NASA, NASA Technical Paper 2249 (1993), http://ntrs.nasa.gov
10. Schwier, D., von Henke, F.: Mechanical verification of clock synchronization algorithms. In: Ravn, A.P., Rischel, H. (eds.) FTRTFT 1998. LNCS, vol. 1486, pp. 262–271. Springer, Heidelberg (1998)
11. Pfeifer, H., Schwier, D., von Henke, F.: Formal verification for time-triggered clock synchronization. In: Weinstock, C.B., Rushby, J. (eds.) Dependable Computing for Critical Applications, vol. 7, pp. 206–226 (January 1999)
12. Barsotti, D., Nieto, L., Tiu, A.: Verification of clock synchronization algorithms: experiments on a combination of deductive tools. Formal Aspects of Computing 19, 321–341 (2007)
13. Pike, L.: Modeling time-triggered protocols and verifying their real-time schedules. In: Proceedings of Formal Methods in Computer Aided Design (FMCAD 2007), pp. 231–238. IEEE, Los Alamitos (2007)
14. Steiner, W., Dutertre, B.: SMT-Based formal verification of a TTEthernet synchronization function. In: Kowalewski, S., Roveri, M. (eds.) FMICS 2010. LNCS, vol. 6371, pp. 148–163. Springer, Heidelberg (2010)
15. Malekpour, M.R.: Model checking a byzantine-fault-tolerant self-stabilizing protocol for distributed clock synchronization systems. NASA, Tech. Rep. NASA/TM-2007-215083 (2007)
16. de Moura, L., Owre, S., Rueß, H., Rushby, J., Shankar, N., Sorea, M., Tiwari, A.: Tool presentation: SAL2. In: Alur, R., Peled, D.A. (eds.) CAV 2004. LNCS, vol. 3114. Springer, Heidelberg (2004)
17. de Moura, L., Rueß, H., Sorea, M.: Bounded model checking and induction: From refutation to verification. In: Voronkov, A. (ed.) CAV 2003. LNCS, vol. 2725, pp. 14–26. Springer, Heidelberg (2003)
18. Dutertre, B., Sorea, M.: Modeling and verification of a fault-tolerant real-time startup protocol using calendar automata. In: Lakhnech, Y., Yovine, S. (eds.) FORMATS 2004 and FTRTFT 2004. LNCS, vol. 3253, pp. 199–214. Springer, Heidelberg (2004)

Extending the GWV Security Policy and Its Modular Application to a Separation Kernel

Sergey Tverdyshev

SYSGO AG, Germany
sergey.tverdyshev@sysgo.com

Abstract. Nowadays formal methods are required for high assurance security and safety systems. Formal methods allow a precise specification and a deep analysis of system designs. However, usage of formal methods in a certification process can be very expensive. In this context, we analyse the security policy proposed by Greve et al in the theorem prover Isabelle/HOL. We show how this policy with some extensions can be applied in a modular way, and hence, reduce the number of formal models and artifacts to certify. Thus, we show how the security policy for a separation kernel is derived from the security policy of the micro-kernel that forms the basis of the separation kernel. We apply our approach to an example derived from an industrial real-time operating system.

1 Introduction

Modern usage of software and hardware systems in safety and security critical areas requires certification. Usually certification depends on the application area, e.g. avionics [18, 19], railway [4], IT security [5]. Today formal methods are required for certification of high assurance security and safety systems, e.g. Common Criteria requires them in different depths for Evaluation Assurance Levels five (EAL5) and above with EAL7 being the highest level [5, 12]. Such a certification is an *extremely* expensive task [6, 3, 26, 16] and any industrial application has to keep these costs low.

This work is carried out as a part of the SeSaM and TECOM [23] projects. In our part we use formal methods to increase the level of trust as required by the Common Criteria. We apply formal methods to analyse information flow in an operating system. Keeping in mind certification costs we target creating reusable certification artifacts.

Related Work and Context. In this paper we analyse and apply the GWV security policy [11] which is well known and accepted in industry [10, 8, 9]. This policy models a separation kernel [20] which enforces partitioning between applications running on a single CPU system. The main benefit of a separation kernel is the control over direct communications between applications running in different partitions. Moreover, highly critical systems require absence of covert channels to ensure that no illicit information flow can take place. To prove that a separation kernel forbids such channels it has to possess the non-interference property [7]. The GWV policy satisfies this property [1]. There are several formalisations of the GWV policy: in ACL2 [11, 1], in PVS [21]. There are also applications of this policy to microprocessors and separation kernels [15]

M. Bobaru et al. (Eds.): NFM 2011, LNCS 6617, pp. 391–405, 2011.

as well as its usage for information flow analysis [2]. We extend these works, formalise the extended result in Isabelle/HOL, and show an interesting application.

Our case study is a separation kernel which is built on top of a micro-kernel [14]. The purpose of a separation kernel is to provide isolated execution environments, called partitions, for user applications. In our case study the separation functionality is based on the resource separation provided by the micro-kernel. Thus, security policies have to exist on both the micro- and the separation kernel levels. We apply the GWV policy on both components and formally show (via formal proofs) that the policy on the micro-kernel level implies the policy on the level of the separation kernel. Thus, the main contribution of this paper is the usage of the same security policy on two system components and a formal proof of the separation property for the separation kernel from the assumed separation property of the micro-kernel. We also show how the GWV policy can be applied in a modular way. The modular and reusable application of the security policy reduces the number of formal models, and hence, the number of artifacts to certify.

All our models are formalised in the theorem prover for higher-order logic Isabelle/HOL [17]. The largest part of this paper is directly synthesised from the formal theories, thus, the consistency between the paper and the formal theories is guaranteed. All our results are also available online [24].

The paper is organised as follows. In the next section we present the example which motivates this work. In Section 3 we describe the original the GWV policy and introduce and formalise clarifications proposed by Alves-Foss and Taylor [1]. Section 4 contains modifications to the GWV policy which are needed to apply it in a modular way. We apply the modified model on an abstract version of our motivating example in Section 5. Finally, we sum up the paper and present the future work.

2 Motivating Example: PikeOS

PikeOS is a real-time operating system for safety and security critical applications [13, 22]. PikeOS is certified for the DO-178B standard [18]. The PikeOS main usage lies in the avionic area (e.g. Airbus A350, A400M) which is well-known for requiring highly robust components. PikeOS is highly modular and runs on a variety of hardware platforms.

Architecturally PikeOS consists of two major components: a micro-kernel and a para-virtualisation layer (see Figure 1). The micro-kernel is very compact and provides the very basic functionality inspired by the ideas of Liedtke [14]. The para-virtualisation layer is implemented on the top of the micro-kernel and provides separated execution partitions for user applications. The para-virtualisation layer is a separation kernel. User applications run in the isolated partitions which can be "personalised" with APIs, e.g. POSIX, OSEK, Linux etc. Thus, the trusted base consists of only the micro-kernel, the separation kernel, the hardware, and some optional extensions which we don't cover in this paper.

In this paper we focus on the micro-kernel and the separation-kernel from the perspective of the access control and the separation of resources.

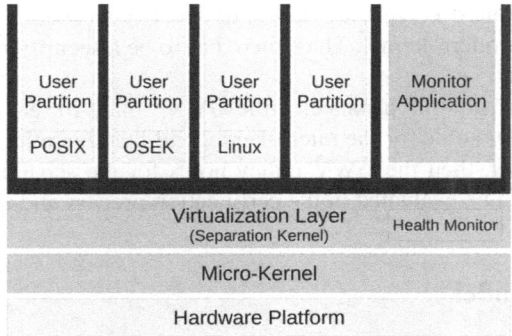

Fig. 1. Architecture of PikeOS

2.1 Micro-Kernel Layer

The micro-kernel runs on a hardware and controls all resources in the system. There are two major abstractions in the micro-kernel (MK): tasks and threads. A task is a passive entity and defined via an address space of the physical memory and can be considered as an object in the system. A thread is an active entity, i.e. a subject in a system which operates on the task state. A thread is always associated with a task.

Communication rules can be defined based on tasks and/or threads. For example, "a set of tasks can communicate with each other" means that the thread from a task from this set can access resources of other tasks in that set. Another example: "a thread can deny receiving any IPC message from a specific thread". The micro-kernel enforces the communication rules and controls access to all system resources.

2.2 Virtualisation Layer (Separation Kernel)

The separation kernel (SK) is implemented on the top of the micro-kernel. The main goal of this layer is to provide isolated partitions where user applications are executed. The isolation property guarantees that applications with different degree of trust (e.g. multiple independent levels of security MILS [25]) run on the same hardware without interference. Moreover, the separation kernel enforces a predefined communication policy between partitions. Thus, applications in different partitions cannot influence each other unless they are allowed to.

A partition consists of a set of tasks, a set of threads, and a set of communication ports. The communication ports are used to define uni-directed communication channels. Partitions can also communicate via shared memory which is implemented via a special built-in file system. User applications can access system resources under the supervision of the separation kernel.

2.3 The Motivation: Putting the Layers Together

The layered structure of PikeOS inspired the idea to use the GWV policy in a modular way. First, the PikeOS system integrator works directly with the separation kernel and she/he has to define a security policy between partitions. The separation kernel is built up on the top of the micro-kernel, and hence, its security mechanisms rely on the services provided by the micro-kernel. Thus, there has to be a security model on the level of the micro-kernel too.

In this paper we create one instance of the GWV policy for each kernel. We also show that if this policy holds for the micro-kernel and the separation kernel is defined in terms of the former, then the GWV policy holds for the separation kernel. Thus, only one policy has to be evaluated in the certification process and this is applied in a modular way.

3   The GWV Model

In this section we formalise the original GWV (Greve, Wilding, and Vanfleet) security policy [11] (Section 3.1 and Section 3.2) and its extension (Section 3.3) in the theorem prover Isabelle/HOL. Similarly to the original work we prove several corollaries which express some useful properties of the defined policy. This policy can also be referred to as a specification and be used in the certification of an implementation of the separation kernel [8, 10, 12]. The policy is defined abstractly without any system specific details. Thus, the same specification can be used to verify many implementations.

3.1   Definitions

We put all definitions for the original GWV model [11] into an Isabelle theory. This theory is parametrized and it will be used twice. In the Isabelle language such a theory is called "locale". Providing parameters to the locale instantiates all locale facts/theorems with the given ones. Additionally, refinement of a locale generates proof obligations for the assumptions from the locale which have to be met by the parameters. Definitions from the entire Section 3 and Section 4 are part of the *GWV* locale.

We start with a notion of the currently active partition which is returned by the function *current* for a given system state as input. Note, *'a* is a name for a polymorphic placeholder for a type named *a* and such types are input parameters when instantiating a locale; keyword **fixes** introduces a function name with its signature. Inside a locale such a function is fixed but it is a locale parameter for the outside world.

fixes *current* :: "*'sys_state_t* ⇒ *'partition_t*"

A partition has a number of assigned resources (i.e. objects) which in the GWV model are called *segments* and can be uniquely identified. Function *segs* for a given partition returns the set of associated segments.

fixes *segs* :: "*'partition_t* ⇒ *'segment_t set*"

A segment in a given system state has a value which can be accessed via the function *select*. It takes a system state and a segment name and returns the segment value (note that we employ curried notation).

fixes *select* :: "*'sys_state_t* ⇒ *'segment_t* ⇒ *'value_t*"

The GWV model contains an auxiliary function *equals* which tests whether segments from a given set have an equivalent value in two given system states. Note: the keyword *assume* will generate a proof obligation when refining a locale.

fixes *equals :: "'segment_t set ⇒ 'sys_state_t ⇒*
                  'sys_state_t ⇒ bool"
assumes *equals_def:*
"equals A sa sb = (∀ a ∈ A. select sa a = select sb a)"

A separation policy introduces communication rules between partitions. Such communication rules (e.g. allow/deny rules) are defined based on the use case of a system. These rules are then enforced by the implementation of a separation kernel. The GWV model specifies these rules via the function *dia* and the name stands for *direct interaction allowed*. For a given segment function *dia* returns the set of segments with which the segment is allowed to communicate.

fixes *dia :: "'segment_t ⇒ 'segment_t set"*

The function *Next* corresponds to one step in the model and computes the next state of the model based on a given state.

fixes *Next :: "'sys_state_t ⇒ 'sys_state_t"*

3.2   The GWV Security Policy

The GWV security policy is expressed as the assumption *separation* which has to be met by a concrete implementation. The policy is based on the interaction between segments and is stated about the model which progresses via the function *Next*. The assumption *separation* considers two arbitrary system states and claims that if

- the current partitions in these two states are the same,
- the value of some segment *a* is the same in both states, and
- values of segments which can interact with *a* are the same in both states, then

the value of the segment *a* after one step matches in both next states. Formally:

assumes *separation:*
"⟦current s = current t;
  select s a = select t a;
  equals ((dia a) ∩ (segs (current s))) s t
⟧ ⟹
select (Next s) a = select (Next t) a"

Thus, if the assumption *separation* is satisfied, then the only way partitions (on a single processor system) can communicate is via the function *dia.* We also prove several corollaries from *separation* which highlight several properties [24]. These are: 1. the corollary *exfiltration* states that computations in the current partition do not affect memory locations outside its access domain w.r.t. *dia*; 2. the corollary *infiltration* states that the data processing in the current partition is not affected by the data outside that partition unless a communication channel is defined; 3. the corollary *mediation* states that if segments of the current partition do not change, then an arbitrary memory cell does not change as well.

3.3   Clarifying the GWV Security Policy

Alves-Foss and Taylor [1] give several clarifications and propose changes that refine/restrict the GWV model. In this section we briefly describe them and apply them to the formalized GWV policy above.

Flow Based on Source Segment. The original GWV model is based on segment abstraction and ignores that these segments belong to a partition. Let us consider some partition A and three segments $seg1$, $seg2$, $seg3$ such that: (i) $seg1$ and $seg2$ belong to A (i.e. $\{seg1,\ seg2\} \subseteq segs\ A$), (ii) information can flow from $seg1$ to $seg3$ (i.e. $seg1 \in dia\ seg3$), (iii) information cannot flow from $seg2$ to $seg3$ (i.e. $seg2 \notin dia\ seg3$). Such a policy is too powerful for common hardware and operating systems because a subject in a partition could copy information from either $seg1$ or $seg2$ into a hardware register and then to $seg3$. To avoid this one has to tag information with the source name and this is not supported by modern micro-processors. Therefore, we weaken dia by allowing communication between segments of the same partition and we have now rather an "intra-partition communication" policy as suggested in [1].

> **assumes** $diaPartScope:$
> $"a \in segs\ p \implies segs\ p \subseteq dia\ a"$

Trustworthiness of Partitions. The original GWV model defines information flow only in terms of the information source. Thus, the following scenario is possible. If two partitions can read some segment $segS$ and this segment can influence another segment $segD$ (i.e. $segS \in dia\ segD$), then both partitions can write to this segment $segD$. In the case that one of these partitions is untrusted and should only read the segment $segS$, the original GWV policy cannot forbid writes to the segment $segD$. To avoid such write operations we add a restriction to dia by considering partition names. We define a function $diaStrong$ which for a given segment and a given partition name returns the set of segments which the given segment can influence.

> **fixes** $diaStrong :: "'segment\_t \Rightarrow 'partition\_t \Rightarrow$
> $'segment\_t\ set"$

We assume that $diaStrong$ restricts dia:

> **assumes** $diaStrongSubset:$
> $"(diaStrong\ a\ p) \subseteq (dia\ a)"$

We re-define $separation$ where we use $diaStrong$ instead of dia and prove that the new version implies the original version of $separation$ [24].

4   Extending the GWV Model with Subjects

Our goal is to instantiate the GWV model for two layers of the PikeOS system (Section 2.3) and to show that from the policy for the micro-kernel we can deduce the policy for the separation kernel. Note that there can be a confusion for the term partition, therefore, from now on whenever this term is used in the context of the GWV model we name it as the GWV-partition.

Let us consider two possible instantiations of the GWV model for PikeOS. First for the micro-kernel, we can instantiate the GWV-partitions with the PikeOS tasks and the GWV-segments with memory addresses. Second for the separation kernel, we can instantiate the GWV-partitions with the PikeOS partitions and the GWV-segments with memory addresses. In our context the first case is used as the base for the second one.

Our goal is to deduce the security policy for the separation kernel from the security policy for the micro-kernel.

If we try to deduce the security policy for the separation kernel from the security policy for the micro-kernel based on the so far presented GWV model, we run into the following problem: the separation theorem for separation kernel gives us that the current partition in two system runs is the same. However, to use the separation theorem for the micro-kernel (from the first instance of the original GWV model), we have to prove that the current task in these system runs is the same. This is impossible to prove because in the same current partition there can be different active tasks. To avoid it, we introduce a notion of *subject* into the GWV model. A subject is an active entity which operates on segments of a GWV-partition. Now, we add two functions to the Isabelle theory for GWV (i.e. to the GWV locale) and one consistency statement.

We define the currently active subject via the function *currentSubject* which returns it for a given system state as input.

fixes *currentSubject* :: "'*sys_state_t* ⇒ '*subject_t*"

Every subject is associated with a GWV-partition where it runs. Therefore, we add a function *subjectPart* which for a given system state and a subject returns the corresponding GWV-partition.

fixes *subjectPart* :: "'*sys_state_t* ⇒ '*subject_t* ⇒
                                        '*partition_t*"

We can easily express the consistency between the current partition and the current subject.

assumes *currentPart*:
  "*current s = subjectPart s (currentSubject s)*"

The modification in the assumption *separationII* is quite trivial: we add into its assumptions one more, i.e. the current subject in two system runs is the same. Formally:

assumes *separationII*:
  "⟦*current s = current t;*
   currentSubject s = currentSubject t;
   select s a = select t a;
   equals ((diaStrong a) ∩ (segs (current s))) s t ⟧ ⟹
   select (Next s) a = select (Next t) a"

We don't present here other adaptations of the GWV model because they are trivial and proofs for lemmas run without changes (see sources [24]).

5   A Modular Usage of the Modified GWV-Policy

In this section we instantiate the extended GWV model from Section 4. We create two instances: one for each layer in PikeOS (Section 2). First, we instantiate the GWV model with a model of the micro-kernel, where we target separation of tasks. Second, we instantiate the GWV model with a model of the separation kernel, where we target separation of partitions. In this section we also present a modular usage of the GWV policy by formally showing how the policy for the separation kernel is derived from the policy for the micro-kernel.

5.1 The Micro-Kernel Model

In this section we define a basic model for the micro-kernel. The major abstractions of the kernel are tasks (objects) and threads (subjects). First we introduce several basic types:

- $thread\_t$ – thread as the subject type
- $task\_t$ – task as the GWV-partition type, i.e. the object type
- $address\_t$ – physical address as the GWV-segment type (every task has a set of addresses it owns)
- $value\_t$ – value type for the data saved in the physical memory

We introduce components of the micro-kernel model. The state of the micro-kernel mk_state_t consists of the current thread and a set of tasks. Note that the Isabelle/HOL keyword **record** introduces a record type.

```
record mk_state_t =
  currentThreadMK :: "thread_t"
  tasks           :: "task_t set"
```

We define a task's address set via the function $taskAddrSet$ which returns the address set for a given task as input.

```
consts taskAddrSet :: "task_t ⇒ address_t set"
```

A thread is always assigned to a task. We model this relation via functions $threadTask$ and $taskThread$. Assumptions $threadTaskThread$ and $taskThreadTask$ specify the properties of those functions.

```
consts threadTask :: "thread_t ⇒ task_t"
consts taskThread :: "task_t ⇒ thread_t set"
axioms threadTaskThread:
  "th ∈ taskThread (threadTask th)"
axioms taskThreadTask:
  "∀ th ∈ taskThread tsk. threadTask th = tsk"
```

We define a set of all subjects in the system as the union of all threads for a given state s:

```
definition threads :: "mk_state_t ⇒ thread_t set" where
"threads s ≡ ⋃ tsk ∈ tasks s. taskThread tsk"
```

In the micro-kernel the current thread is one of the threads:

```
axioms currentThInThreads:
  "currentThreadMK s ∈ threads s"
```

The definitions above allow us to prove that for a given thread there is a unique task. Note: $\exists!$ denotes "exists uniquely". The proof is available on our website [24]
lemma *taskUniqueMK:*
  "th ∈ threads s ⟹ ∃! tsk ∈ tasks s. (taskThread tsk = th)"

The notion of the current thread (*currentSubject*, Section 4) allows us to define the notion of the current task (*current*, Section 3).
definition *currentTask ::* *"mk_state_t ⟹ task_t"* **where**
  "currentTask s ≡ threadTask s (currentThreadMK s)"

Retrieving a value for a given address is modelled as reading memory with function *readMem* which corresponds to *select* in the GWV model.
consts *readMem ::* *"mk_state_t ⟹ address_t ⟹ value_t"*

The security policy in the micro-kernel is defined on the level of memory addresses, i.e. which addresses can influence each other. For this purpose we introduce function *diaMK* which specifies the data flow policy for tasks. We don't give any specific definition because it depends on the use-case of the micro-kernel.
consts *diaMK ::* *"address_t ⟹ address_t set"*

In Section 3.3 we introduced a restriction to the original GWV policy which defines information flow in terms of the GWV-segment and the GWV-partition. We model this restriction as function *diaStrongMK* whose exact definition is not important for this paper but we assume that *diaStrongMK* restricts *diaMK*:
consts *diaStrongMK ::* *"address_t ⟹ task_t ⟹ address_t set"*
axioms *diaStrongSubsetTask:*
  "diaStrongMK a tsk ⊆ diaMK a"

The function *nextStepMK* represents the next-step function for the micro-kernel.
consts *nextStepMK ::* *"mk_state_t ⟹ mk_state_t"*

Finally, we instantiate the GWV model and call it *MicroKernel*. We instantiate all polymorphic types and fixed functions from Section 3.1 with the given ones (see parameters for *GWV* below). This instantiation inherits all assumptions from the *GWV* locale. Thus, herewith we assume that the instantiated separation property holds for the micro-kernel, i.e. the micro-kernel separates tasks. Note that instantiation of a locale inherits all assumptions and refinement of a locale generates proof obligations.

locale *MicroKernel =*
 GWV "λ x. threadTask (currentThreadMK x)"
                         — instantiation of *current*
     "currentThreadMK"   — instantiation of *currentSubject*
     "λ x. threadTask"   — instantiation of *subjectPart*
     "taskAddrSet"   — instantiation of *segs*
     "readMem"   — instantiation of *select*
     "diaMK"   — instantiation of *dia*

```
"nextStepMK"              — instantiation of Next
"λ A spa spb.
    ∀ addr ∈ A. readMem spa addr = readMem spb add"
                          — instantiation of equals
"diaStrongMK"             — instantiation of diaStrong
```

5.2 The Separation Kernel Model

In this section we define a model for the separation kernel. This model is built up on the top of the micro-kernel from the previous section. We wrap the definition of the separation kernel into a locale called *SeparationKernel*. We also present several properties to highlight how the micro-kernel model is related to the model of the separation kernel.

Subjects in the separation kernel are threads, thus, we reuse the subject type *thread_t*. A partition in the separation kernel consists of a set of tasks.

record *partition_t* =
  partTasks :: "task_t set"

We model the state of the separation kernel as a record consisting of the current thread and a set of partitions.

record *sk_state_t* =
  currentThreadSK :: "thread_t"
  parts :: "partition_t set"

We define the separation kernel on the top of the micro-kernel. First we need a way to relate the states of these two kernels. We introduce an abstraction function *absSkToMk* which constructs a state of the micro-kernel from a given state of the separation kernel by collecting tasks of all partitions into one set.

definition *absSkToMk* :: "sk_state_t ⇒ mk_state_t" **where**
"absSkToMk s ≡
  (| currentThreadMK = currentThreadSK s,
     tasks = ⋃ p ∈ parts s. partTasks p |)"

A thread in the separation kernel always belongs to a partition. We capture this property via function *threadPart* which takes a system state *s* and a thread *th* and returns the partition thread *th* belongs to. This function returns a unique partition such that (i) it is a partition in the system state *s* (ii) and there is a task in this partition where the thread *th* is running. Note: in this function we use the Isabelle/HOL operator *THE*; the expression *THE x. P x* defines a unique element satisfying *P* if one exists, otherwise the expression is typed but undefined (sometimes *THE* is called a strict version of the Hilbert's choice operator).

definition *threadPart* :: "sk_state_t ⇒ thread_t ⇒
                            partition_t" **where**
"threadPart s th ≡
  THE p. p ∈ parts s ∧
    (∃ !tsk ∈ partTasks p. tsk = threadTask (absSkToMk s) th)"

In a separation kernel there is the currently active partition which we define via the notion of the current thread.

definition currentPart :: "sk_state_t ⇒ partition_t" **where**
  "currentPart s ≡ threadPart s (currentThreadSK s)"

In the separation kernel a task is always assigned to a unique partition and we capture this fact by assumption taskUniqueSK

assumes taskUniqueSK:
  "⟦pA ∈ parts s; pB ∈ parts s; pA≠pB⟧ ⟹
      partTasks pA ∩ partTasks pB = ∅

In our architecture the separation kernel imports functionality of the micro-kernel. Thus, we define the transition function of the separation kernel model in terms of the micro-kernel model. We do the latter axiomatically to simplify definitions and proofs.

fixes nextStepSK :: "sk_state_t ⇒ sk_state_t"
assumes nextStepSKVianextStepMK:
" readMem (nextStepMK (absSkToMk s)) x =
  readMem (nextStepMK (absSkToMk t)) x
  ⟹
  readMem (absSkToMk (nextStepSK s)) x =
  readMem (absSkToMk (nextStepSK t)) x"

From the uniqueness of a thread we prove that the thread's task is always in the thread's partition (Lemma threadTaskInPart). This is the consistency between the mapping from threads to tasks and the mapping from threads to partitions.

lemma threadTaskInPart:
  "th ∈ threadsSK s ⟹
    threadPart s th ∈ parts s ∧
    threadTask th ∈ partTasks (threadPart s th)"

In PikeOS the security policy is defined over physical memory addresses. Therefore, we consider a GWV-segment as a memory address and associate a partition with memory addresses. For this purpose we introduce the function partAddrSet which defines the partition's address set by collecting addresses of all tasks of a partition into one set.

definition partAddrSet :: "partition_t ⇒ address_t set" **where**
  "partAddrSet p ≡ ⋃ tsk ∈ partTasks p. taskAddrSet tsk"

Since the security policy is to be defined on the level of memory addresses, we need a function which specifies the communication policy (the GWV-function dia, Section 3.1). Function diaSK defines such a communication policy between addresses of partitions. It relaxes the policy for tasks (Section 5.1) by allowing addresses in one partition to affect each other (Section 3.3). Note that {x. Q(x)} denotes a predicate set where every element satisfies predicate Q.

definition diaSK :: "address_t ⇒ address_t set" **where**
  "diaSK addr ≡ diaMK addr ∪
      {x. (∃ p. x ∈ partAddrSet p ∧ addr ∈ partAddrSet p)}"

At last, we define the stronger version of the communication policy (see `diaStrong`, Section 3.3) by the function `diaStrongSK`. It defines the communication policy between addresses of partitions and is based on the `diaStrongMK`.

definition `diaStrongSK :: "address_t ⇒ partition_t ⇒`
                              `address_t set"` **where**
```
"diaStrongSK addr p ≡
     (⋃ tsk ∈ partTasks p. diaStrongMK addr tsk) ∪
     {x. (∃ p. x ∈ partAddrSet p ∧ addr ∈ partAddrSet p)}"
```

5.3   The Separation Kernel Security Policy

In Isabelle/HOL one can refine a locale with a desired model such that (i) all assumptions of the locale become proof obligations and one has to prove them (ii) then one enjoys all lemmata/definitions/theorems etc. based on this locale. We employ the Isabelle/HOL keyword **sublocale** to state that the locale `SeparationKernel` is a refinement of the locale `GWV` (note this is the second usage of the generic model `GWV`, the first one is at the end of Section 5.1). This refinement starts a proof for all assumptions of the `GWV`. In this proof we can use the information of the model `SeparationKernel` from the section above.

sublocale `SeparationKernel` ⊆
```
    GWV  "currentPart"              — instantiation of current
         "currentThreadSK"         — instantiation of currentSubject
         "threadPart"              — instantiation of subjectPart
         "partAddrSet"             — instantiation of segs
         "(λ st add. readMem (absSkToMk st) add)"
                                   — instantiation of select
         "diaSK"                   — instantiation of dia
         "nextStepSK"              — instantiation of Next
         "(λ A sta stb. ∀ addr ∈ A.
           ((readMem (absSkToMk sta) add) =
            (readMem (absSkToMk stb) add)))"
                                   — instantiation of equals
         "diaStrongSK"             — instantiation of diaStrong
```

The refinement proof consists of proofs for assumptions from the `GWV` locale: `equals_def` from Section 3.1, `diaPartScope` and `diaStrongSubset` from Section 3.3 as well as `separation` and `currentPart` from Section 4. This proof is rather lengthy and therefore we don't present it here in the full length (yet it can be found on our web page [24]). The main goal of the proof is to show that the assumption `separation` holds for the separation kernel. The instantiation of the GWV model with the micro-kernel (Section 5.1) gives us the fact that `separation` holds for the micro-kernel. Applying the former fact to the main goal is *the modular usage of the GWV policy*. To apply it, we show how the separation kernel is related to the micro-kernel, i.e. show that the separation kernel is correctly defined on the top of the micro kernel (e.g. we use lemma `threadTaskInPar` as one of the crucial facts) and that definition does not violate the policy.

5.4   An Alternative Way to Instantiate the GWV Model with PikeOS

We can define the security policy of the separation kernel on the level of the tasks. Thus, we hide physical memory addresses at this level and consider the GWV segments as the tasks. This alternative instantiation follows the pattern presented above. One important change is that the *dia* function has to be defined for tasks, i.e. for a given task it will return a set of tasks the task is allowed to communicate. The proof is barely changed. The complete formal description of the alternative can also be found on our web page [24].

6   Summary

Formal models can be a great help to understand how a system works and to provide additional assurance in the system behavior. This is also recognized by the industry and is reflected in different standards, e.g. DO-178C [19], Common Criteria [5]. We use a formal model proposed by Greve et al and we are the first to present its usage in a modular way and applied the results to a separation kernel: we have one uniform specification for all layers and we apply a uniform instantiation mechanism, and thus, only one policy as a certification artifact.

We propose extensions to the GWV model and formal proofs to illustrate a formal modular usage of the modified GWV security policy. This paper is mainly generated directly from the theorem prover Isabelle/HOL, thus, usage of definitions, lemmas, and proofs can be replayed independently. This is also an important fact for the certification process because, for instance, the "Common Criteria" require that a certifier has to be able to re-run tests/proofs for the system.

The motivating example comes from analysis of PikeOS which is an operating system for safety and security critical applications developed at SYSGO AG [22]. We formally proved the that the separation kernel indeed separates based on the assumption that the micro-kernel works correctly. Our results are quite generic and can be reused for similar designs. The next step is to apply the developed formal models on PikeOS to produce artifacts for certification for high EALs of the "Common Criteria".

References

1. Alves-Foss, J., Taylor, C.: An analysis of the GWV security policy. In: ACL2 Workshop (2004)
2. Amtoft, T., Hatcliff, J., Rodrguez, E., Robby, H.J., Greve, D.: Specification and checking of software contracts for conditional information flow. In: Design and Verification of Microprocessor Systems for High-Assurance Applications, pp. 341–380 (2010)
3. Bill Hart, G.H.S.: SDR security threats in an open source world (2009), http://www.sdrforum.org/pages/sdr04/3.5%20Security%20%20Dillinger/3.5-%3%20Hart.pdf
4. CENELEC: DIN EN50128:2001: Railway applications. Communications, signalling and processing systems. Software for railway control and protection systems (2001)
5. Common Criteria Sponsoring Organizations: Common criteria for information technology security evaluation. version 3.1, revision 1 (September 2006), http://www.commoncriteriaportal.org/thecc.html

6. Ganssle, J.: Code: Getting it Right. A new OS has been proven to be correct using mathematical proofs. The cost: astronomical (2009), http://www.embedded.com/design/220900551

7. Goguen, J.A., Meseguer, J.: Security policies and security models. In: IEEE Symposium on Security and Privacy (1982)

8. Green Hills Software: INTEGRITY-178B Separation kernel security target (2008), http://www.niap-ccevs.org/cc-scheme/st/vid10119/

9. Greve, D.: Information security modeling and analysis. In: Design and Verification of Microprocessor Systems for High-Assurance Applications, pp. 249–300 (2010)

10. Greve, D., Richards, R., Wilding, M.: A Summary of Intrinsic Partitioning Verification. In: Proceedings of the Fifth International Workshop on the ACL2 Theorem Prover and Its Applications, ACL2 (2004)

11. Greve, D., Wilding, M., Vanfleet, W.M.: A separation kernel formal security policy. In: Fourth International Workshop on the ACL2 Prover and its Applications, ACL2-2003 (2003), http://www.cs.utexas.edu/users/moore/acl2/books/books/workshops/2003/gr%eve-wilding-vanfleet/security-policy.pdf.gz

12. Information Assurance Directorate: U.S. Government Protection Profile for Separation Kernels in Environments Requiring High Robustness. Version 1.03 (SKPP) (June 2007), http://www.niap-ccevs.org/cc-scheme/pp/pp_skpp_hr_v1.03/

13. Kaiser, R., Wagner, S.: Evolution of the PikeOS microkernel. In: Kuz, I., Petters, S.M. (eds.) MIKES: 1st International Workshop on Microkernels for Embedded Systems (2007), http://ertos.nicta.com.au/publications/papers/Kuz_Petters_07.pdf

14. Liedtke, J.: On micro-kernel construction. In: Proceedings of the 15th ACM Symposium on Operating Systems Principles, pp. 237–250. ACM Press, New York (1995)

15. Miller, S.P.: Will this be formal? In: Mohamed, O.A., Muñoz, C., Tahar, S. (eds.) TPHOLs 2008. LNCS, vol. 5170, pp. 6–11. Springer, Heidelberg (2008)

16. NICTA: L4.verified: Numbers (2009), http://ertos.nicta.com.au/research/14.verified/numbers.pml

17. Paulson, L.C.: Isabelle: a generic theorem prover. LNCS, vol. 828. Springer, New York (1994)

18. RTCA SC-167 / EUROCAE WG-12: DO-178B: Software Considerations in Airborne Systems and Equipment Certification. Radio Technical Commission for Aeronautics (RTCA), Inc., 1828 L St. NW., Suite 805, Washington, D.C. 20036 (December 1992)

19. RTCA SC-205/EUROCAE WG-71: Discussion and development site for Software Considerations in Airborne Systems: Discussion forum for DO-178C (2009), http://forum.pr.erau.edu/SCAS/

20. Rushby, J.: Design and verification of secure systems. In: Eighth ACM Symposium on Operating System Principles, pp. 12–21 (1981)

21. Rushby, J.: A Separation Kernel Formal Security Policy in PVS. SRI International (2004), http://www.sri.com

22. SYSGO AG: PikeOS RTOS technology embedded system software for safety critical real-time systems (2008), http://www.sysgo.com

23. TECOM Consortium: TECOM Project: Trusted Embedded Computing (2008), http://www.tecom-project.eu

24. Tverdyshev, S.: Formalisation and Modular Usage of GWV Security Policy in Isabelle/HOL: Source files (2010), ftp://ftp.sysgo.com/FormalMethods/Modular-GWV-Policy/

25. Vanfleet, W.M., Luke, J.A., Beckwith, R.W., Taylor, C., Calloni, B., Uchenick, G.: MILS: Architecture for high-assurance embedded computing. Crosstalk (August 2005), http://www.stsc.hill.af.mil/crosstalk/2005/08/0508Vanfleet_etal.html

26. Wind River: New Capability for the Warfighter Multilevel Secure Systems Based on a MILS Architecture (2009), http://ftp.windriver.speedera.net/ftp.windriver/2009-rc-presentations/arlington/breakouts/Wind_River_Presents_MILS.pdf

Combining Partial-Order Reduction and Symbolic Model Checking to Verify LTL Properties[*]

José Vander Meulen[1] and Charles Pecheur[2]

[1] Université catholique de Louvain
jose.vandermeulen@uclouvain.be
[2] Université catholique de Louvain
charles.pecheur@uclouvain.be

Abstract. BDD-based symbolic techniques and partial-order reduction (POR) are two fruitful approaches to deal with the combinatorial explosion of model checking. Unfortunately, past experience has shown that BDD-based techniques do not work well for loosely-synchronized models, whereas POR methods allow explicit-state model checkers to deal with large concurrent models. This paper presents an algorithm that combines symbolic model checking and POR to verify linear temporal logic properties without the next operator (LTL_X), which performs better on models featuring asynchronous processes. Our algorithm adapts and combines three methods: Clarke et al.'s tableau-based symbolic LTL model checking, Iwashita et al.'s forward symbolic CTL model checking and Lerda et al.'s ImProviso symbolic reachability with POR. We present our approach, outline the proof of its correctness, and present a prototypal implementation and an evaluation on two examples.

1 Introduction

Two common approaches are commonly exploited to fight the combinatorial state-space explosion in model-checking, with different perspectives: partial-order reduction methods (POR) explore a reduced state space in a property-preserving way [1,2] while symbolic techniques use efficient structures such as binary decision diagrams (BDDs) to concisely encode and compute large state spaces [3]. In their basic form, symbolic approaches tend to perform poorly on asynchronous models where concurrent interleavings are the main source of explosion, and explicit-state model-checkers with POR such as Spin [4] have been the preferred approach for such models.

This paper presents an approach that integrates POR in BDD-based model checking for LTL_X to provide an efficient and scalable symbolic verification solution for models featuring asynchronous processes. Our approach proceeds as follows:

---

[*] This work is supported by project MoVES under the Interuniversity Attraction Poles Programme — Belgian State — Belgian Science Policy.

M. Bobaru et al. (Eds.): NFM 2011, LNCS 6617, pp. 406–421, 2011.
© Springer-Verlag Berlin Heidelberg 2011

1. We start from the tableau-based reduction of LTL verification to fair-CTL of Clarke et al. [5], which results in looking for fair executions in the product P of the model and a tableau-based encoding of the (negated) property.
2. We construct P_r, a property-preserving partial-order reduction of P, using an adaptation of Lerda et al.'s ImProviso algorithm [6]. We also implemented the algorithm of Alur et al. [7] for comparison purposes.
3. Finally, we check within P_r whether P contains a fair cycle using the forward traversal approach of Iwashita et al. [8]. We also implemented the classical backward as a basis for comparison, though experimental results show the forward approach to be more efficient than the backward approach.

We have implemented this new approach in a prototype and obtained experimental results that show a significant performance gain with respect to symbolic techniques without POR.

The main contributions of this paper are the global symbolic verification algorithm for checking LTL_X properties which adapts and combines tableau-based LTL, fair-cycle detection and partial-order reduction, a proof of correctness of the global algorithm, a prototype implementation, and an experimental evaluation on two models.

The remainder of the paper is structured as follows. Section 2 establishes basic definitions and notations and presents the tableau-based reduction of LTL to fair-CTL and the forward traversal approach. Section 3 presents partial-order reduction and its application to symbolic model checking in ImProviso. In Section 4, we present our new approach for LTL model-checking with POR and detail our adapation of the ImProviso algorithm. Section 5 presents our implementation and reports experimental results. Section 6 reviews related work. Finally, Section 7 gives conclusions as well as directions for future work.

2   Symbolic LTL Model Checking

2.1   Transitions Systems

We represent the behavior of a system as a *transition system*, with labelled transitions and propositions interpreted over states. In the rest of this paper, we assume a set AP of *atomic propositions* and a set A of *actions*[1]. Without loss of generality, the set AP can be restricted to the propositions that appear in the property to be verified on the system. A *fair transition system* is a transition system enriched with a set of *fairness constraints*, each constraint consisting of a set of states.

Definition 1 (Transition System). *Given a set of actions A and a set of atomic propositions AP, a transition system (over A and AP) is a structure $M = (S, R, I, L)$ where S is a finite set of states, $I \subseteq S$ are initial states, $R \subseteq S \times A \times S$ is a transition relation, and $L : S \to 2^{AP}$ is an interpretation function over states.*

---

[1] Often called *transitions* in the literature, notably in [9]. For clarity, we only call *transitions* specific transition instances $s \xrightarrow{a} s'$.

Definition 2 (Fair Transition System). *A* fair transition system *is a structure* $M = (S, R, I, L, F)$ *where* (S, R, I, L) *is a transition system and* $F \subseteq 2^S$ *is a set of fairness constraints.*

We write $s \xrightarrow{a} s'$ for $(s, a, s') \in R$. An action a is *enabled* in a state s iff there is a state s' such that $s \xrightarrow{a} s'$. We write enabled(s, R) for the set of enabled actions of R in s. When the context is clear, we write enabled(s) instead of enabled(s, R). We assume that R is total (i.e. enabled$(s) \neq \emptyset$ for all $s \in S$). The set of all *paths* of M is defined as $\mathrm{tr}(M) = \{s_0 \xrightarrow{a_0} s_1 \xrightarrow{a_1} \ldots \mid s_0 \in I \wedge \forall i \in \mathbb{N} \cdot s_i \xrightarrow{a_i} s_{i+1}\}$. A path π is said to be *fair* if and only if for every $F_i \in F$, $\inf(\pi) \cap F_i \neq \emptyset$, where $\inf(\pi)$ is the set of states that appear infinitely often in π. The set of all fair paths, or fair traces, of M is defined as $\mathrm{ftr}(M) = \{\pi \mid \pi \in \mathrm{tr}(M) \wedge \forall F_i \in F \cdot \inf(\pi) \cap F_i \neq \emptyset\}$.

We write $M \sqsubseteq M'$ iff M is a *sub-transition system* of M', in the following sense:

Definition 3 (Inclusion of fair transition systems). *Let* $M = (S, R, I, L, F)$, $M' = (S', R', I', L', F')$ *be two fair transition systems. M is a sub-transition system of M', denoted $M \sqsubseteq M'$, if and only if $S \subseteq S'$, $R \subseteq R'$, $I \subseteq I'$, $L(s) = L'(s)$ for $s \in S$, and $\forall F'_i \in F' \cdot \exists F_i \in F \cdot F_i \subseteq F'_i$.*

We can see that if $M \sqsubseteq M'$, each fair path of M is a fair path of M'.

Lemma 1. *if* $M \sqsubseteq M'$ *then* $\mathrm{ftr}(M) \subseteq \mathrm{ftr}(M')$.

2.2   From LTL to Fair-CTL

This section outlines the algorithm, introduced in [5], to verify LTL properties using BDD-based symbolic model checking.

We consider the verification of properties expressed in LTL$_X$, linear propositional temporal logic without the next operator. LTL formulæ are interpreted over each (infinite) execution path of the model. We denote the classical temporal operators as **F**, **G** and **U**. Informally, let π be a execution path, **G** f (globally f) says that f will hold in all future states of π, **F**f (finally f) says that f will hold in some future state of π, f **U** g (f until g) says that g will hold in some future state of π and, at every preceding state of π, f will hold. We will reason for the most part in terms of *(un)satisfiability* of the negation of the desired property $\neg f$. We write $(M, s) \models \mathbf{E}g$ to express that there exists a path from state s in M that satisfies a formula g.

Given a transition system M and an LTL property f, the *tableau* of $\neg f$ is constructed. The tableau of a formula g is a fair transition system $T = (S_T, R_T, I_T, L_T, F_T)$ over the singleton alphabet $A = \{\bot\}$ and the set AP of propositions which appear in g. Each state of the tableau is a set of formulae derived from g, which characterizes the sub-formulae of g that are satisfied on fair traces from that state. Initial states are those that entail g, and the fairness constraints ensure that all eventualities occurring in g are fulfilled. The fair traces of the tableau correspond to the traces that satisfy g. See [5] for details.

The tableau of $\neg f$ is then composed with the initial system M to produce a new fair transition system P. If P contains deadlocks, we remove from S_P all the states which lead necessarily to deadlocks and restrict R_p to the remaining states.

Definition 4 (Product of M and T). *Given a system $M = (S, R, I, L)$ and a tableau $T = (S_T, R_T, I_T, L_T, F_T)$, the product of M and T, denoted $M \times T$, is a fair transition system $P = (S_P, R_P, I_P, L_P, F_P)$ where:*

- $S_P = \{(s_t, s) \in S_T \times S \mid L_T(s_t) = L(s)\}$
- $R_P = \{((s_t, s), a, (s_t', s')) \mid R_T(s_t, \bot, s_t') \wedge R(s, a, s')\}$
- $I_P = S_P \cap (I_T \times I)$
- $L_P((s_t, s)) = L_T(s_t) = L(s)$
- $F_P = \{\{(s_t, s) \in S_P \mid s_t \in F_T^i\} \mid F_T^i \in F_T\}$

It is shown in [5] that M contains a path which satisfies $\neg f$ iff there is an infinite fair path in P that starts from an initial state (i_t, i). Furthermore, the existence of fair traces is captured by the fair CTL formula $\mathbf{E}_F\mathbf{G}$ *true*, to be read as "there exists a fair path such that globally true". The interest is that fair-CTL formulae can be verified with BDD-based symbolic model checking.

Theorem 1. *Let T be the tableau of $\neg f$ and P be the product of M and T. Given a state $i \in I$, $(M, i) \models \mathbf{E}\neg f$ if and only if there is a state (i_t, i) in I_P such that $(P, (i_t, i)) \models \mathbf{E}_F\mathbf{G}$ true.*

2.3   Forward Symbolic Model-Checking

In [8], Iwashita et al. present a model-checking algorithm for a fragment of fair-CTL based on forward state traversal. In the following sections, we enrich this algorithm with partial-order reduction to efficiently check the unsatisfiability of the $\mathbf{E}_F\mathbf{G}$ *true* formula derived from tableau-based LTL model-checking.

The semantic of a CTL formula f is defined as a relation $s \models f$ over states $s \in S$. We define the *language* of f as $\mathcal{L}(f) = \{s \in S \mid s \models f\}$. In the sequel we assimilate a temporal logic formula f to the set of states $\mathcal{L}(f)$ that it denotes, for the sake of simplifying the notations.

Given a model M, a formula f and initial conditions i, conventional BDD-based symbolic model-checking can be described as evaluating $\mathcal{L}(f)$ over the sub-formulæ of f in a bottom-up manner, and checking whether $\mathcal{L}(i) \subseteq \mathcal{L}(f)$. The evaluation of (future) CTL operators in f results in a backward state-space traversal of the model. $\mathcal{L}(i) \subseteq \mathcal{L}(f)$ can be expressed as checking whether $i \implies f$, or equivalently, checking unsatisfiability of $i \wedge \neg f$ in M.

The forward exploration from [8] works by transforming a property $h \wedge op(g)$ into $op'(h) \wedge g$, where a future, backward-traversal CTL operator op in the right term is transformed into a past, forward-traversal operator op' in the left term. It is shown in [8] that these formulae are *equisatisfiable in M*, in the sense that there exists a state in M which satisfies the transformed formula iff there exists a state in M which satisfies the original formula.

In general, h is then a past-CTL formula. The following (past-temporal) operations over formulæ are defined[2]:

$$\text{FwdUntil}(h, g) = \mu Z.[h \vee \text{post}(Z \wedge g)]$$
$$\text{FairEH}(h) = \nu Z.[h \wedge \text{post}(\textstyle\bigwedge_{F_i \in F} \text{FwdUntil}(F_i, Z) \wedge Z)]$$

where $\text{post}(X) = \{s' \in S \,|\, \exists s \in X, a \in A \cdot s \xrightarrow{a} s'\}$ is the post-image of X. $\text{FwdUntil}(h, g)$ computes states s that can be reached from h within g (except for s itself), and $\text{FairEH}(h)$ computes states reachable from a fair cycle all within h[3]. On this basis, it is established that $h \wedge \mathbf{E}_F \mathbf{G} \, g$ is equisatisfiable in M to $\text{FairEH}(\text{FwdUntil}(h, g) \wedge g)$.

In particular, for $h = i$ and $g = \mathit{true}$ this reduces to $\text{FairEH}(\text{FwdUntil}(i, \mathit{true}))$, where $\text{FwdUntil}(i, \mathit{true})$ exactly computes the reachable state space of M, which we denote $\mathit{Reachable}(M)$. We thus obtain the following fact.

Theorem 2

$$\exists i \in I \cdot (M, i) \models \mathbf{E}_F \mathbf{G} \, \mathit{true} \quad \mathit{iff} \quad \exists s \in S \cdot (M, s) \models \text{FairEH}(\mathit{Reachable}(M))$$

In essence, this theorem captures the fact that the fair-CTL model-checking problem resulting from the tableau-based reduction of LTL can be decomposed into two distinct parts, the computation of the reachable state space and the search for a fair cycle. Besides, the POR theory shows that only a subset of the reachable state space needs to be computed to see whether a property is satisfied or not. The following sections will demonstrate that different methods can be used to compute the (reduced) reachable state space, and also that different methods can be used to perform the fair-cycle detection.

3   Partial-Order Reduction

The goal of partial-order reduction methods (POR) is to reduce the number of states explored by model-checking, by avoiding the exploration of different equivalent interleavings of concurrent transitions [10,2,9].

Partial-order reduction is based on the notions of *visibility* of actions and *independence* between actions. An action a is *invisible* if and only if it does not affect atomic propositions, i.e. if $L(s) = L(s')$ for any $s \xrightarrow{a} s'$ (and *visible* otherwise). Two actions are *independent* if they do not disable one another and executing them in either order results in the same state. Intuitively, if two independent actions a and b are invisible with respect to the property f that one wants to verify, then it does not matter whether a is executed before or after b, because they lead

---

[2] The notation $\mu Z.\tau(Z)$ (resp. $\nu Z.\tau(Z)$) denotes the *least fixed point* (resp. *greater fixed point*) of the *predicate transformer* τ. For more details, we refer the reader to [5].

[3] Both FwdUntil and FairEH can be expressed in the past version of fair-CTL: $\text{FairEH}(h)$ corresponds to $\mathbf{E}_F \mathbf{G} \, h$ and $\text{FwdUntil}(h, f)$ corresponds to $h \vee \mathbf{EX} \, \mathbf{E}[f \, \mathbf{U} \, (h \wedge f)]$, where the direction of temporal operators is reversed.

to the same state and do not affect the truth of f. Partial-order reduction consists in identifying such situations and restricting the exploration to either of these two alternatives. Given a transition system $M = (S, R, I, L)$, POR amounts to exploring a reduced model $M_R = (S_R, R_R, I, L_R)$ with $S_R \subseteq S$, $R_R \subseteq R$, and $L_R = \{(s_r, A) \in L \mid s_r \in S_R\}$. In practice, classical POR algorithms [2,9] execute a modified depth-first search (DFS). At each state s, an adequate subset ample(s) of the actions enabled in s are explored. To ensure that this reduction is adequate, that is, that verification results on the reduced model hold for the full model, ample(s) must respect the following set of conditions as set forth in [9,10]:

C_0 ample(s) = $\emptyset$ if and only if enabled(s) = $\emptyset$.

C_1 Along every path in the full state graph that starts at s, an action $a \notin$ ample(s) that is dependent on an action in ample(s) cannot be executed without an action in ample(s) occurring first.

C_2 If ample(s) $\neq$ enabled(s), then all actions in ample(s) are invisible.

C_3 A cycle is not allowed if it contains a state in which some action is enabled, but is never included in ample(s) on the cycle.

Conditions C_0, C_1, C_2 and C_3 are sufficient to guarantee that the reduced model preserves properties expressed in LTL_X, but does not preserve properties expressed in LTL [9]:

Theorem 3. *Given M a transition system, f a LTL_X property, if M_R is a POR reduction of M using an ample(s) that satisfies conditions C_0–C_3, then $(M, i) \models \mathbf{E}f$ iff $(M_R, i) \models \mathbf{E}f$.*

Conditions C_1 and C_3 depend on the whole state graph. C_1 is not directly exploitable in a verification algorithm. Instead, one uses sufficient conditions, typically derived from the structure of the model description, to safely decide where reduction can be performed. Contrary to C_1, C_3 can be checked on the reduced graph, though in a nontrivial way. However, a stronger condition can be used. A sufficient condition for C_3 is that at least one state along each cycle is fully expanded.

3.1 Process Model

In the sequel, we assume a process-oriented modeling language. We define a *safe process model* as an extension of a transition system which distinguishes disjoint subsets of local actions A_i, that are suitable candidates for partial-order reduction. Typically, such actions will correspond to local transitions of different processes p_i in a concurrent program.

Definition 5 (Safe Process Model). *Given a transition system $M = (S, R, I, L)$, a process model for M consists of a finite set of disjoint sets of local actions $A_0, A_1, \ldots, A_{m-1}$ with $A_i \subseteq A$. The local transitions are defined as $R_i = R \cap (S \times A_i \times S)$. A process model is safe with respect to M iff all its local transitions are safe, that is, for all $a \in A_i$, a is invisible, and for all $s \in S$, ample(s) = enabled(s, R_i) satisfies condition C_1.*

Note that this definition guarantees that ample(s) = enabled(s, R_i) respects conditions C_1 and C_2, but not C_3, which is ensured dynamically by detecting cycles within the reduction algorithm.

3.2   Partial-Order Reduction with BDDs

In this section we discuss two algorithms which implement a symbolic version of the POR method presented in Section 3. Both approaches can be used to compute a reduced reachable state space.

In [6], Lerda et al. propose ImProviso, a BDD-based symbolic version of the Two-Phase POR algorithm for computing a reduced state space. The Two-Phase algorithm was first presented by Nalumasu and Gopalakrishnan in [11]. ImProviso alternates between two distinct phases: Phase-1 and Phase-2. Phase-1 expands only safe transitions considering each process at a time, in a fixed order. As long as a process offers safe transitions, those transitions alone are executed, otherwise the algorithm moves on to the next process. Phase-2 performs a full expansion of the final states reached in Phase-1, then Phase-1 is recursively applied to the reached states.

In [7], Alur et al. propose another approach based on a modified *breadth-first search* (BFS) algorithm which respects conditions C_0–C_3, using BDD techniques. It produces a reduced graph by expanding at each step a subset of the transition relation.

Both approaches perform a BFS instead of a DFS. Hence, it is much harder to detect cycles. To tackle this problem, both algorithms over-approximate the cycles. The over-approximation guarantees that all cycles are correctly identified, but possibly needlessly decreases the number of states where the reduction can be applied.

Although Alur's method and ImProviso are similar, they differ in the following ways:

- In Alur's method, a single subset of the whole transition relation is computed at each step. In ImProviso, for each process a transition relation which contains only safe actions is precomputed. These transition relations are used during Phase-1. We contend that this leads to better performance because each Phase-1 step is computed with much smaller BDDs.
- The Two-Phase approach reduces the over-approximation by limiting cycle detection to the current execution of Phase-1.

4   LTL Model Checking with Partial-Order Reduction

In this section we bring together the computation of the reachable state space by means of POR and the fair-cycle detection. Given a transition system $M = (S, R, I, L)$ with a safe process model $A_1, \ldots, A_n$ and a LTL$_X$ property f, our algorithm verifies whether M satisfies f by building a tableau T for $\neg f$ and checking the absence of accepting traces in $P = (S_P, R_P, I_P, L_P, F_P)$, the product of M and T.

This check is performed symbolically. We first compute a reduced state space of P, and then we check whether P contains a fair cycle within the reduced state space. In this section, we use a variant of the ImProviso to compute the reduced state space. We also use the forward model checking to perform the fair-cycle detection. In other words, this check is performed symbolically, by checking the emptiness of the following formula using BDDs: FairEH(ReachablePOR(P)). In Section 5, we compare different methods to compute the reduced graph, as well as to look for fair cycles.

4.1 Computation of the Reachable States

A key new element is the algorithm ReachablePOR which constructs a reduced reachable state space of P. It is given in Figure 1, and is based on the ImProviso algorithm of [6]. In order to apply partial-order reduction on the product system P, we lift the process model from M to P and pre-compute, for each safe action set A_i, the BDD of the partial transition relation $R_{P,i} = R_P \cap (S_P \times A_i \times S_P)$.

```
 1   global R_P
 2   global R_{P,i}[0..m-1]
 3
 4   global frontier // current frontier
 5   global visited  // visited states
 6
 7   procedure ReachablePOR(P_I)
 8     frontier,visited := I_P, I_P
 9     while (frontier ≠ {}) {
10       phase1()
11       phase2()
12     }
13   }
14
15   function deadStates(R, X) {
16     return X \ dom R
17   }
18
19   procedure phase2() {
20     local image := post(R_P, frontier)
21     frontier := image\visited
22     visited :=  visited ∪ image
23   }
24
25
26   procedure phase1() {
27     local cycleApprox := {}
28     local stack := frontier
29
30     foreach (i in 0,···,m-1) {
31       local image :=
32         post(R_{P,i}[i], frontier)
33       local dead :=
34         deadStates(R_{P,i}[i], frontier)
35
36       while ((image\stack) ≠ {}) {
37         stack := stack ∪ image
38         cycleApprox := cycleApprox ∪
39           (image ∩ stack)
40         frontier := image\stack
41         image := post(R_{P,i}[i], frontier)
42         dead := dead ∪
43           deadStates(R_{P,i}[i], frontier)
44       }
45
46       frontier := frontier ∪ dead
47     }
48     frontier := frontier ∪ cycleApprox
49     visited := visited ∪ stack
50   }
```

Fig. 1. ReachablePOR algorithm

ReachablePOR performs the two phases alternatively until no states to visit remain. The global variable **frontier** contains the current frontier, that is, the set of states which have been reached but not expanded yet. The global variable **visited** contains all the reached states. The first phase (**phase1**) performs partial expansion of the safe transitions of each process. The outer loop (lines 30–47) iterates over the processes. The inner loop (lines 36–44) expands all safe transitions of the current process, until no more new states can be found. The following invariants hold at line 36:

- The `stack` variable contains all the states which have already been reached during the current run of `phase1`.
- The `cycleApprox` contains all the states already in `stack` which have been reached again in a consecutive iteration. Those states over-approximate the set of states closing a cycle; they are added back to the current frontier when moving to Phase-2 (line 48).
- The `dead` variable contains all the reached states with no enabled transitions for the current process, as computed by `deadStates`. Those states are added back to the frontier when moving to the next process (line 46).

The second phase (`phase2`) performs a single-step full expansion of the states of the current frontier.

ReachablePOR differs from ImProviso in the following ways:

1. ReachablePOR explores a product system P. The ample sets, captured in $R_{P,i}$, depend only on the model M, while the cycle condition is checked on the product P. In ImProviso, there is no tableau and everything is computed on the original model M.

2. When a presumed cycle is detected on state s in Phase-1, ImProviso will expand s during the expansion of the next process, whereas ReachablePOR will postpone expansion of s to the next Phase-2. When a product P is reduced, we have noticed that this modification tends to improve both the number of visited states and the verification time.

3. ReachablePOR keeps track of states that have no transition with the current process (lines 34 and 43) and passes them to the next processes. If this computation was not done, we could have missed some states during the BFS. So, we could have violated the condition $C0$. The need for this computation was apparently not addressed in [6].

4. ImProviso performs an additional outermost loop in Phase-1, to expand any additional safe actions that have been enabled by the previous round over all processes, such as receiving a message on a channel where it has been previously sent. This is not needed in ReachablePOR because by construction our notion of safe action does not allow this kind of situation. It would easily be added back if it were to become useful.

ReachablePOR(P) explores a reduced transition system $P_R = (S_R, R_R, I_P, L_R, F_R)$, where $L_R = \{(s_r, A) \in L_P \mid s_r \in S_R\}$, and F_R is the restriction of F_P to S_R, i.e. $F_R = \{F_i \cap S_R \mid F_i \in F_P\}$. It returns the explored states $S_R = $ ReachablePOR(P) as the final value of `visited`. By construction, $P_R \sqsubseteq P$.

4.2   Fair-Cycle Detection

The reduced state space S_R is used to search for infinite fair paths by computing FairEH(S_R). From the definition of FairEH, it is clear that FairEH(S_R) only explores states within S_R. Note, however, that FairEH uses the full transition relation R_P rather than the reduced transition relation R_R implicitly explored by ReachablePOR. FairEH(S_R) thus explores an *induced* fair transition system $P_I = (S_R, R_P \cap (S_R \times A \times S_R), I_P, L_P, F_P)$. By construction $P_R \sqsubseteq P_I \sqsubseteq P$ and thus, by Lemma 1, $\mathrm{ftr}(P_R) \subseteq \mathrm{ftr}(P_I) \subseteq \mathrm{ftr}(P)$. Note that S_R is evidently equal to Reachable(P_I). Hence evaluating FairEH(ReachablePOR(P)) in P amounts to evaluating FairEH(Reachable(P_I)) in P_I and we have the following lemma:

Lemma 2. *Given* $P = (S_P, R_P, I_P, L_P, F_P)$, $S_R = \mathrm{ReachablePOR}(P)$, $P_I = (S_R, R_P \cap (S_R \times A \times S_R), I_P, L_P, F_P)$ *and* $s_P \in S_P$, $(P, s_P) \models$ FairEH(ReachablePOR(P)) *if and only if* $(P_I, s_P) \models$ FairEH(Reachable(P_I)). *When* $(P_I, s_P) \models$ FairEH(Reachable(P_I)), $s_P \in S_R$.

4.3   Correctness

To demonstrate the correctness of our approach, we have to prove that, given a property f and a model M, f holds in M iff FairEH(ReachablePOR(P)) returns an empty set, where P is the product of M and the tableau for $\neg f$. Conversely, we will prove that there is a path from an initial state i in M on which $\neg f$ holds, written $(M, i) \models \mathbf{E}\neg f$, iff there is a state in P satisfying FairEH(ReachablePOR(P)).

Before getting to this main result, we need to address two technical issues. First, the following two lemmas establish that the preservation of properties when reducing M to M_R is carried over when reducing P to P_R based on the transitions of M, as performed in ReachablePOR.

Lemma 3. *Given a product system* $P = M \times T$ *and* P_R *the reduced transition system explored by* $\mathrm{ReachablePOR}(P)$, *there exists a reduced transition system* M_R *such that* $P_R = M_R \times T$ *and* M_R *is a property-preserving reduction of* M, *i.e.* $(M, i) \models \mathbf{E}\neg f$ *iff* $(M_R, i) \models \mathbf{E}\neg f$.

Proof. We follow the same reasoning as Theorem 4.2 in [1], which we only outline here. In [1], given a transition system G and a LTL property f, a Büchi automaton B which accepts the language $\mathcal{L}(\neg f)$ is constructed[4]. It is shown that $G \models \mathbf{A}f$ if and only if the intersection (i.e. product) A of G and B is empty, or equivalently if A does not contain any cycle, reachable from some initial state, that contains some accepting state. A reduced version A' of A is constructed by choosing at each step of the DFS a valid ample set. The conditions C_1 and C_2 are checked on G alone, while C_0 and C_3 are checked on the whole product. It is shown that A' corresponds to a product of a reduced system G_R and B such that G_R is a property-preserving reduction of G, i.e. $(G, i) \models \mathbf{E}\neg f$ iff $(G_R, i) \models \mathbf{E}\neg f$.

The ReachablePOR procedure follows the same process. It constructs a reduced version P_R of P by choosing at each step a valid ample set, i.e $\mathrm{ample}((s_t, s)) = \{a \mid (s_t, s) \xrightarrow{a} (s'_t, s') \land a \in \mathrm{ample}(s)\}$. By following the same

---

[4] Although Theorem 4.2 in [1] considers only deterministic transition systems, both the theorem and its proof remain valid with non-deterministic transition systems. The proof remains exactly the same.

reasoning as in [1] we can conclude that P_R is the product of a reduced system M_R and T such that M_R is a property-preserving reduction of M, i.e. $(M, i) \models \mathbf{E} \neg f$ iff $(M_R, i) \models \mathbf{E} \neg f$.   □

Together with Theorem 1, the following lemma follows directly.

Lemma 4. $(P, i_P) \models \mathbf{E}_F \mathbf{G}$ *true if and only if* $(P_R, i_P) \models \mathbf{E}_F \mathbf{G}$ *true.*

Secondly, the following lemma establishes that P_I, which corresponds to the system explored by FairEH, preserves the properties of P_R, which corresponds to the system explored by ReachablePOR.

Lemma 5. $(P_R, i_P) \models \mathbf{E}_F \mathbf{G}$ *true if and only if* $(P_I, i_P) \models \mathbf{E}_F \mathbf{G}$ *true.*

Proof. We know that $\mathrm{ftr}(P_R) \subseteq \mathrm{ftr}(P_I) \subseteq \mathrm{ftr}(P)$. Therefore, any fair path of P_R is also a fair path of P_I. Conversely, any fair path of P_I is also a fair path of P and therefore there exists a corresponding fair path in P_R by Lemma 4.   □

We now get to the main result.

Theorem 4. *Given a model M, a property f and the product P of M and the tableau of* $\neg f$*, there exists a state* $i \in I$ *such that* $(M, i) \models \mathbf{E} \neg f$ *iff there exists a state* $s_P \in S_P$ *such that* $(P, s_P) \models \mathrm{FairEH}(\mathrm{ReachablePOR}(P))$.

Proof. Let P_R and P_I be defined as previously. We have successively:

$$\exists i \in I \cdot (M, i) \models \mathbf{E} \neg f$$
$$\Leftrightarrow \exists i_P \in I_P \cdot (P, i_P) \models \mathbf{E}_F \mathbf{G} \text{ true} \qquad \text{(Theorem 1)}$$
$$\Leftrightarrow \exists i'_P \in I_P \cdot (P_R, i'_P) \models \mathbf{E}_F \mathbf{G} \text{ true} \qquad \text{(Lemma 4)}$$
$$\Leftrightarrow \exists i''_P \in I_P \cdot (P_I, i''_P) \models \mathbf{E}_F \mathbf{G} \text{ true} \qquad \text{(Lemma 5)}$$
$$\Leftrightarrow \exists s_P \in S_R \cdot (P_I, s_P) \models \mathrm{FairEH}(\mathrm{Reachable}(P_I)) \qquad \text{(Theorem 2)}$$
$$\Leftrightarrow \exists s_P \in S_P \cdot (P, s_P) \models \mathrm{FairEH}(\mathrm{ReachablePOR}(P)) \qquad \text{(Lemma 2)}$$
□

Given any algorithm which constructs a valid POR-reduced reachable state set Reduced(M) of a transition system M, we can use that algorithm instead of ReachablePOR in our approach, checking the emptiness of FairEH(Reduced(P)). In the same way, other algorithms can be used to detect fair cycles, for instance the classical backward CTL model-checking algorithm can be used. Actually, these approaches are valid, and the demonstration of Section 4.3 remains the same.

5   Evaluation

We extended the Milestones model checker presented in [12] to support the method presented in this paper. Milestones is available under the GNU General Public License at http://lvl.info.ucl.ac.be/Tools/Milestones. Milestones allows us to describe concurrent systems and to verify LTL_X properties. It defines a language for describing transition systems. The design of the language

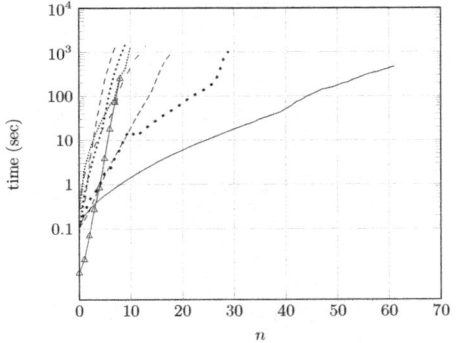
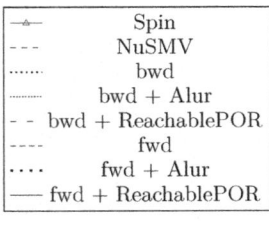

Fig. 2. Verification times for the Producer-Consumer property P3

has been influenced by the NuSMV language [13] and by the synchronization by rendez-vous mechanism. Milestones detects fair cycles either with the classical backward fair CTL model-checking algorithm [9] (hereafter denoted as *bwd*), or with the forward approach described in Section 2.3 (denoted as *fwd*). The reachable state space can be generated using the ReachablePOR approach, as well as Alur's method mentioned in Section 3.2, or without any POR reduction. Together these offer $2 \times 3 = 6$ different modes of operation.

In order to assess the effectiveness and scalability of the approach proposed in this paper, we discuss two models which were translated both into the language of Milestones, NuSMV, and Spin. This section presents the models and the results we obtained. All the tests have been run on a 2.16 GHz Intel Core 2 Duo with 2 GB of RAM. We compare the verification performance between all six different modes. We also compare to NuSMV, which performs bwd without POR, and Spin which performs explicit model checking [4].

The first model is a variant of a producer-consumer system where all producers and consumers contribute on the production of every single item. The model is composed of $2 \times m$ processes: m producers and m consumers. Each producer and each consumer has two local transitions. The producers and consumers communicate together via a bounded buffer composed of eight slots. Each producer works locally on a piece p, then it waits until all producers terminate their task. Then, p is added to the buffer, and the producers start processing the next piece. When the consumers remove p from the buffer, they work locally on it. When all the consumers have terminated their local work, another piece can be removed from the buffer. The size of the reachable state space grows exponentially by a factor of 40 at each step of m.

Five properties have been analyzed on this model. For instance, P_3 states that at any time the producers will eventually add a piece into the buffer (satisfied), and P_4 states that the buffer will never overflow (unsatisfied). Figure 2 compares the times for the verification of the property P_3. Similar results have been obtained for the other four properties.

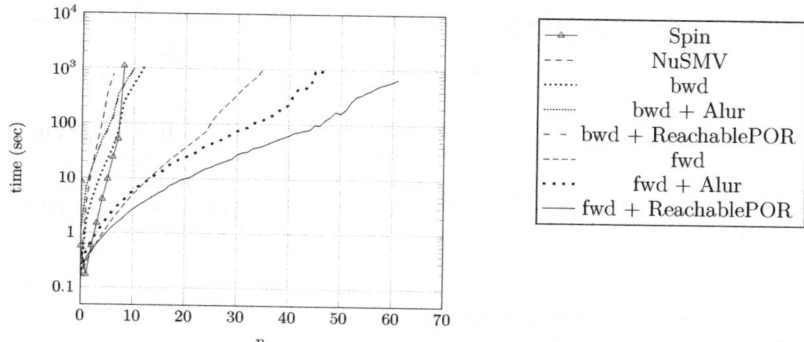

Fig. 3. Verification times for the Turntable property T_3

The second model is a turntable model, described in [14]. The turntable system consists of a round turntable, an input place, an output place, n drills and a testing device. The turntable transports products in sequence between the different tools, where they are drilled and tested. The turntable has $n + 3$ slots that each can hold a single product. The original model had only one drill; we extended it to represent an arbitrary number of drills. The size of the reachable state space grows exponentially by a factor of 7 at each step of n.

We have verified six properties on this system: four properties that the system satisfies, and two properties which are not fulfilled. For instance, the property T_3 states that if in the future there will be a piece which is not well drilled, the alarm will necessarily resonate. Here is the translation of this property in LTL: **G [F a piece is not well drilled $\implies$ F an alarm is raised]**. Figure 3 compares the times for the verification of the property T_3.

Table 1 compares the state space computed by the three forward methods (without POR, with Alur's method and with ReachablePOR), in terms of number of BDD nodes, number of states and computation time. It is quite interesting to note that while POR substantially decreases the number of reached states, the number of BDD nodes is increased (likely due to breaking some symmetry in the full state space). However, it still results in substantial speed improvements. We also notice that the state spaces produced by the Alur's method and the ReachablePOR method have approximately the same size.

6 Related Work

Besides the approaches of Alur [7] and Improviso [6] on which this work is based (as presented in Section 3.2), several other approaches have been proposed that combine symbolic model checking and POR to verify different classes of properties.

This paper builds on our previous work combining POR and the forward state traversal approach to verify CTL_X properties[5] [12]. It remains to evaluate the

---

[5] CTL_X is the subset of the CTL logic without the next operator.

Table 1. BDD size (in # nodes), state space size (in # states) and computation time (in seconds) for the P reachable state space computed either by the forward method without POR, or the forward method and Alur's approach, or the ReachablePOR method. "–" indicates that the computation did not end within 1000 seconds.

# drills	# nodes			# states			time (sec)		
	Fwd	Fwd + Alur	ReachPOR	Fwd	Fwd + Alur	ReachPOR	Fwd	Fwd + Alur	ReachPOR
1	197	318	282	86488	27408	26668	.11	.28	.17
2	442	880	744	521944	39188	38044	.17	.45	.23
4	975	2444	2021	$2.49 \times 10^{+7}$	62804	60796	.35	.90	.34
8	2040	5832	4717	$5.98 \times 10^{+10}$	111012	106300	1.08	2.59	.65
16	4168	17165	14126	$3.45 \times 10^{+17}$	207120	197308	5.13	9.34	1.49
32	8421	57467	47218	$1.15 \times 10^{+31}$	394208	379324	37.09	57.68	4.35
40	–	91214	75104	–	495668	470332	–	173.84	6.37
47	–	125194	103300	–	580484	549964	–	971.60	8.59
50	–	–	105844	–	–	584092	–	–	9.76
61	–	–	146912	–	–	709228	–	–	14.9

compared merits of the two approaches for properties that can be expressed in both LTL_X and CTL_X. For conventional BDD-based model checking, experiments in [5] have found that, in the absence of POR, CTL verification tends to be faster.

In [15], we present another LTL_X model-checking algorithm which combines the Two-Phase algorithm and SAT-based bounded model checking (BMC). On the property P_2 of the producer-consumer system of Section 5, the BMC algorithm of [15] takes approximately 68 minutes to find a counter-example of length 1,017, while the algorithm presented here takes only 314 milliseconds to show the violation.

In [16], Abdulla et al. present a general method for combining POR and symbolic model checking. Their method can check safety properties either by backward or forward reachability analysis. So as to perform the reduction, they employ the notion of commutativity in one direction, a weakening of the dependency relation which is usually used to perform POR. This approach deals both with backward and forward analysis but for reachability only, while we are able to check LTL_X properties but using only forward analysis.

In [17], Kurshan et al. perform partial-order reduction at compile time. The method applies static analysis techniques to discover local cycles and produce a reduced model, which can be verified using standard symbolic model checking. It could be interesting to investigate whether this kind of analysis could help in ensuring the cycle condition C_3 in our approach.

In [18], Holzmann performs a reduction of individual processes by *merging local transitions*. Then, the processes are put in parallel to be explicitly verified. Merging local transitions can be seen as a special application of partial reduction method. It avoids to create intermediate states between local transitions. Actually, all the transitions which are merged will be considered as safe transitions by our approach. Those transitions will be explored dynamically during phase1, i.e. when POR is applied. By contrast, the Holzmann algorithm removes them statically at compile time. We notice that our approach might visit more than once a state which will be removed by the Holzmann algorithm.

7   Conclusion

In this paper, we presented an improved BDD-based model-checking algorithm for verifying LTL_X properties on asynchronous models. Our approach combines the tableau-based reduction of LTL model-checking to fair-CTL from [5], forward state-traversal of fair-CTL formulæ from [8] used to detect fair cycles, and a symbolic partial-order reduction based on ImProviso [6] to reduce the forward state traversal.

We implemented the new algorithm in our existing model checker and observed on two case studies that our approach achieves a significant improvement in comparison to the tableau-based approach of [5] without POR, in both its backward and forward versions. It remains to confirm those results on a larger range of case studies and to compare with other methods and tools.

The reduced state set computed by ReachablePOR could as well be used in other BDD-based model-checking circumstances: as a filter during fixpoint computations in classical backward model-checking, or even to restrict the BDD of the transition relation before standard, non-POR techniques are applied. It would be interesting to compare the benefits of the reduction in the different approaches. For the latter case, however, the size of the BDD representing the transition relation of M_R could become unmanageable due to the loss of some symmetry.

References

1. Peled, D.: Combining partial order reductions with on-the-fly model-checking. Formal Methods in System Design 8(1), 39–64 (1996)
2. Godefroid, P.: Partial-Order Methods for the Verification of Concurrent Systems. LNCS, vol. 1032. Springer, Heidelberg (1996)
3. Burch, J.R., Clarke, E.M., McMillan, K.L., Dill, D.L., Hwang, J.: Symbolic model checking: 10^{20} states and beyond. Information and Computation 98(2), 142–170 (1992)
4. Holzmann, G.J.: The model checker SPIN. IEEE Transactions on Software Engineering 23(5) (1997)
5. Clarke, E.M., Grumberg, O., Hamaguchi, K.: Another look at LTL model checking. Form. Methods Syst. Des. 10(1), 47–71 (1997)
6. Lerda, F., Sinha, N., Theobald, M.: Symbolic model checking of software. In: Cook, B., Stoller, S., Visser, W. (eds.) Electronic Notes in Theoretical Computer Science, vol. 89. Elsevier, Amsterdam (2003)
7. Alur, R., Brayton, R.K., Henzinger, T.A., Qadeer, S., Rajamani, S.K.: Partial-order reduction in symbolic state space exploration. In: Grumberg, O. (ed.) CAV 1997. LNCS, vol. 1254, pp. 340–351. Springer, Heidelberg (1997)
8. Iwashita, H., Nakata, T., Hirose, F.: CTL model checking based on forward state traversal. In: ICCAD 1996: Proceedings of the 1996 IEEE/ACM International Conference on Computer-aided Design, pp. 82–87. IEEE Computer Society, Washington, DC (1996)
9. Clarke, E.M., Grumberg, O., Peled, D.: Model Checking. MIT Press, Cambridge (1999)

10. Gerth, R., Kuiper, R., Peled, D., Penczek, W.: A partial order approach to branching time logic model checking. Information and Computation 150(2), 132–152 (1999)
11. Nalumasu, R., Gopalakrishnan, G.: A new partial order reduction algorithm for concurrent system verification. In: CHDL 1997: Proceedings of the IFIP TC10 WG10.5 International Conference on Hardware Description Languages and their Applications: Specification, Modelling, Verification and Synthesis of Microelectronic Systems, pp. 305–314. Chapman & Hall, Ltd., London (1997)
12. Vander Meulen, J., Pecheur, C.: Efficient symbolic model checking for process algebras. In: Cofer, D., Fantechi, A. (eds.) FMICS 2008. LNCS, vol. 5596, pp. 69–84. Springer, Heidelberg (2009)
13. Cimatti, A., Clarke, E., Giunchiglia, F., Roveri, M.: NUSMV: A new symbolic model verifier. In: Halbwachs, N., Peled, D.A. (eds.) CAV 1999. LNCS, vol. 1633, pp. 495–499. Springer, Heidelberg (1999)
14. Bortnik, E.M., Trčka, N., Wijs, A., Luttik, B., van de Mortel-Fronczak, J.M., Baeten, J.C.M., Fokkink, W., Rooda, J.E.: Analyzing a χ model of a turntable system using spin, cadp and uppaal. J. Log. Algebr. Program. 65(2), 51–104 (2005)
15. Vander Meulen, J., Pecheur, C.: Combining partial order reduction with bounded model checking. In: Communicating Process Architectures 2009 - WoTUG-32. Concurrent Systems Engineering Series, vol. 67, pp. 29–48. IOS Press, Amsterdam (2009)
16. Abdulla, P.A., Jonsson, B., Kindahl, M., Peled, D.: A general approach to partial order reductions in symbolic verification (extended abstract). In: Y. Vardi, M. (ed.) CAV 1998. LNCS, vol. 1427, pp. 379–390. Springer, Heidelberg (1998)
17. Kurshan, R.P., Levin, V., Minea, M., Peled, D., Yenigün, H.: Static partial order reduction. In: Steffen, B. (ed.) TACAS 1998. LNCS, vol. 1384, pp. 345–357. Springer, Heidelberg (1998)
18. Holzmann, G.J.: The engineering of a model checker: the GNU i-protocol case study revisited. In: Dams, D.R., Gerth, R., Leue, S., Massink, M. (eds.) SPIN 1999. LNCS, vol. 1680, pp. 232–244. Springer, Heidelberg (1999)

Towards Informed Swarm Verification

Anton Wijs*

Eindhoven University of Technology, 5612 AZ Eindhoven, The Netherlands
A.J.Wijs@tue.nl

Abstract. In this paper, we propose a new method to perform large scale grid model checking. A manager distributes the workload over many embarrassingly parallel jobs. Only little communication is needed between a worker and the manager, and only once the worker is ready for more work. The novelty here is that the individual jobs together form a so-called cumulatively exhaustive set, meaning that even though each job explores only a part of the state space, together, the tasks explore all states reachable from the initial state.

Keywords: parallel model checking, state space exploration.

1 Introduction

In (explicit-state) Model checking (MC), the truth-value of a logical statement about a system specification, i.e. design, (or directly software code) is checked by exploring all its potential behaviour, implicitly described by that specification, as a directed graph, or state space. A flawed specification includes undesired behaviour, which is represented by a trace through the corresponding state space. With MC, we can find such a trace, and report it to the developers. To show flaw (bug) absence, full exploration of the state space is crucial. However, in order to explore a state space at once, it needs to be stored in the computer's main memory, and often, state spaces are too large, possibly including billions of states. A secondary point of concern was raised in [22,23]: as the amount of available main memory gets bigger, it becomes technically possible to explore large state spaces using existing sequential, i.e. single-processor, techniques, but the time needed to do so is practically too long. Therefore new techniques are needed, which can exploit multi-core processors and grid architectures.

We envision an 'MC@Home', similar to SETI@Home [34], where machines in a network or grid can contribute to solving a computationally demanding problem. In many application areas, this is very effective. BOINC [8] has about 585,000 computers processing around 2.7 petaFLOPS, topping the current fastest super-computer (IBM Roadrunner with 1.026 PFLOPS). However, flexible grid MC does not exist yet; current *distributed* MC methods, in which multiple machines are employed for a single MC task, need lots of synchronisation between the

---

* Supported by the Netherlands Organisation for Scientific Research (NWO) project 612.063.816 *Efficient Multi-Core Model Checking*.

M. Bobaru et al. (Eds.): NFM 2011, LNCS 6617, pp. 422–437, 2011.

computers ('workers'), which is a serious bottleneck. MC is both computationally expensive, and cannot obviously be distributed over so-called *embarrassingly parallel* [14] processes, i.e. processes which do not synchronise with each other. In this paper, we propose a method to divide a state space reachability task into multiple smaller, embarrassingly parallel, subtasks. The penalty for doing so is that some parts of the state space may be explored multiple times, but, as noted by [22], this is probably unavoidable, and not that important, if enough processing power is available. What sets the method which we present in this paper apart from previous ones is that we distribute the work over a so-called *cumulatively exhaustive set* (CES) of search instructions, where each individual instruction yields a strictly non-exhaustive search, in which a strict subset of the set of reachable states is explored, hence less time and memory is needed, while it is also guaranteed that the searches yielded by all instructions *together* search the whole state space. This is novel, since partial (or non-exhaustive) searches, such as random walk [40] and beam search [28,38,41] are typically very useful to detect bugs quickly, but cannot provide a guarantee of bug-absence. In our case, if all searches instructed by the (finitely-sized) CES cannot uncover a bug, we can conclude that the state space is bug-free. We believe that a suitable method for large scale grid MC must be efficient both memory-wise and time-wise; distributed MC techniques tend to scale very well memory-wise, but disappointingly time-wise, while papers on *multi-core* MC, in which multiple cores on a single processor are employed for a single MC task, tend to focus entirely on speedup, while assuming that the state space fits entirely in the main memory of a single machine. Therefore, we wish to focus more on the combination of time and memory improvements. Our approach is built on the observation that state space explosion is often due to the fact that a system specification is defined as a set of processes in parallel composition, while those processes in isolation do not yield large state spaces. The only serious requirement which systems must meet for our method to be applicable at the moment, is that at least one process in the specification yields finite, i.e. cycle-free, behaviour; we can enforce this by performing a bounded analysis on a process yielding infinite behaviour, but then, we do not know a priori whether the swarm will be cumulatively exhaustive.

The structure of the paper is as follows: in the next Section, related work is discussed. In Section 3, preliminary notions are explained. Then, Section 4 contains a discussion on directed search techniques. After that, in Section 5, we explain the basics of our method for system with independent parallel processes. A more challenging setup with synchronising parallel processes, together with our algorithms, are presented in Section 6. Then, experimental results are given in Section 7, and finally, conclusions and future work appear in Section 8.

2 Related Work

Concerning the state space explosion problem, over the years, many techniques have been developed to make explicit-state MC tasks less demanding. Prominent examples are reduction techniques like partial order reduction [31], and directed

MC [12], which covers the whole range of state space exploration algorithms. Some of these use heuristics to find bugs quickly, but if these are inaccurate, or bugs are absent, they have no effect on the time and memory requirements.

In distributed algorithms such as e.g. in [3,4,5,6,10,15,27,32], multiple workers in a cluster or grid work together to perform an MC task. This has the advantage that more memory is available; in practice, though, the techniques do not scale as well as desired. Since the workers need to synchronise data quite frequently, for very large state spaces, the time spent on synchronisation tends to be longer than the time spent on the actual task. Furthermore, if one of the workers is considerably slower than the others or fails entirely, this has a direct effect on the whole process. Another development is multi-core MC. Since a few years, Moore's Law no longer holds, meaning that the speed of new processors does not double every two years anymore. Instead, new computers are equipped with a growing number of processor cores. For e.g. MC, this means that in order to speedup the computations, the available algorithms must be adapted. In multi-core MC, we can exploit that the workers share memory. Major achievements are reported in e.g. [1,20,21,26]. [26] demonstrates a significant speedup in a multi-core breadth-first search (BFS) using a lock-free hash table. However, papers on multi-core MC tend to focus on reducing the time requirements, and it is assumed that the entire state space fits in the main memory of a single machine.

A major step towards efficient grid MC was made with Swarm Verification (SV) [22,23,24] and Parallel Randomized State Space Search [11,35], which involve embarrassingly parallel explorations. They require little synchronisation, and have been very successful in finding bugs in large state spaces quickly. Bug absence, though, still takes as much time and memory to detect than a traditional, sequential search, since the individual workers are unaware of each other's work, and each worker is not bounded to a specific part of the state space. The method we propose is based on SV, and since each worker uses particular information about the specification to guide the search, we call it *informed* SV (ISV), relating it to informed search techniques in directed MC. Similar ideas appear in related work: in [27], it is proposed to distribute work based on the behaviour of a single process. The workers are not embarrassingly parallel, though. A technique to restrict analysis of a program based on a given trace of events is presented in [16]. It has similarities with ISV, but also many differences; their technique performs slicing on a deterministic C program, and is not designed for parallel MC, whereas ISV distributes work to analyse concurrent behaviour of multiple processes based on the behaviour of a subsystem. Finally, a similar approach appears in [36], but there, it is applied on symbolic execution trees to generate test cases for software testing. Unlike ISV, they distribute the work based on a (shallow) bounded analysis of the whole system behaviour.

3   Preliminaries

Labelled Transition Systems Labelled transition systems (LTSs) capture the operational behaviour of concurrent systems. An LTS consists of transitions $s \xrightarrow{\ell} s'$, meaning that being in a state s, an action ℓ can be executed, after which a state s' is reached. In model checking, a system specification, written in a modelling language, has a corresponding LTS, defined by the structural operational semantics of that language.

Definition 1. *A labelled transition system (LTS) is a tuple $\mathcal{M} = (\mathcal{S}, \mathcal{A}, \mathcal{T}, s_{in})$, where $\mathcal{S}$ is a set of states, $\mathcal{A}$ a set of actions or transition labels, $\mathcal{T}$ a transition relation, and s_{in} the initial state. A transition $(s, \ell, s') \in \mathcal{T}$ is denoted by $s \xrightarrow{\ell} s'$.*

A sequence of labels $\sigma = \langle \ell_1, \ell_2, \ldots, \ell_n \rangle$, with $n > 0$, describes a sequence of events relating to a trace in an LTS, starting at s_{in}, with matching labels, i.e. it maps to traces in the LTS with $s_0, \ldots, s_n \in \mathcal{S}$, $\ell_1, \ldots, \ell_n \in \mathcal{A}$, with $s_0 = s_{in}$, such that $s_0 \xrightarrow{\ell_1} s_1 \xrightarrow{\ell_2} \cdots \xrightarrow{\ell_n} s_n$. Note that σ maps to a single trace iff the LTS is *label-deterministic*, i.e. that for all $s \in \mathcal{S}$, if there exist $s \xrightarrow{\ell'} s'$ and $s \xrightarrow{\ell''} s''$ with $s' \neq s''$, then also $\ell' \neq \ell''$. If the LTS is not label-deterministic, then σ may describe a set of traces. In this paper, we assume that LTSs are label-deterministic, but this is strictly not required. The set of enabled transitions restricted to a set of labels $A \in \mathcal{A}$ in state s of LTS $\mathcal{M}$ is defined as $en_{\mathcal{M}}(s, A) = \{t \in \mathcal{T} \mid \exists s' \in \mathcal{S}, \ell \in A. \, t = s \xrightarrow{\ell} s'\}$. Whenever $en_{\mathcal{M}}(s, \mathcal{A}) = \emptyset$, we call s a *deadlock* state. For $T \subseteq \mathcal{T}$, we define $nxt(T) = \{s \in \mathcal{S} \mid \exists s' \xrightarrow{\ell} s \in T\}$. This means that $nxt(en_{\mathcal{M}}(s, \mathcal{A}))$ is the set of immediate successors of s.

System specifications often consist of a finite number of process specifications in parallel composition. Then, the process specifications describe the potential behaviour of individual system components. The potential behaviour of all these processes concurrently then constitutes the LTS of the system as a whole. What modelling language is being used to specify these systems is unimportant here; we only assume that the process specifications can be mapped to process LTSs, and that the processes can interact using *synchronisation* actions.

Next, we will highlight how a system LTS can be derived from a given set of process LTSs and a so-called *synchronisation function*. System behaviour can be described by a finite set Π of $n > 0$ process LTSs $\mathcal{M}_i = (\mathcal{S}_i, \mathcal{A}_i, \mathcal{T}_i, s_{in,i})$, for $1 \leq i \leq n$, together with a partial function $\mathfrak{C} : \mathcal{A}^s \times \mathcal{A}^s \to \mathcal{A}^f$, with $\mathcal{A}^s = \bigcup_{1 \leq i \leq n} \mathcal{A}_i$ and $\mathcal{A}^f$ a set of actions representing successful synchronisation, describing the potential synchronisation behaviour of the system, i.e. it defines which actions $\ell, \ell' \in \bigcup_{1 \leq i \leq n} \mathcal{A}_i$ can synchronise with each other, resulting in an action $\ell'' \in \mathcal{A}^f$. We write $\mathfrak{C}(\{\ell, \ell'\}) = \ell''$, to indicate that the order of ℓ and ℓ' does not matter.[1] Furthermore, we assume that each action ℓ is always only involved in at most one synchronisation rule, i.e. for each ℓ, there are no two distinct ℓ', ℓ'' such that both $\mathfrak{C}(\{\ell, \ell'\})$ and $\mathfrak{C}(\{\ell, \ell''\})$ are defined. Definition 2 describes how to construct a system LTS from a finite set Π of process LTSs.

Definition 2. *Given a set Π of $n > 0$ process LTSs $\mathcal{M}_i = (\mathcal{S}_i, \mathcal{A}_i, \mathcal{T}_i, s_{in,i})$, for $1 \leq i \leq n$, and synchronisation function $\mathfrak{C} : \mathcal{A}^s \times \mathcal{A}^s \to \mathcal{A}^f$, with $\mathcal{A}^s =$*

---

[1] In practice, synchronisation rules can also be defined for more than two parties, resulting in *broadcasting* rules. In this paper, we restrict synchronisation to two parties. Note, however, that the definitions can be extended to support broadcasting.

$\bigcup_{1 \leq i \leq n} \mathcal{A}_i$ and $\mathcal{A}^f$ a set of actions representing successful synchronisation, we construct a system LTS $\mathcal{M} = (\mathcal{S}, \mathcal{A}, \mathcal{T}, s_{in})$ as follows:

- $s_{in} = (s_{in,1}, \ldots, s_{in,n})$;
- Let $z_1 = (s_1, \ldots, s_i, \ldots, s_j, \ldots, s_n) \in \mathcal{S}$ with $i \neq j$.
  - If for some $\mathcal{M}_i \in \Pi$, $s_i \xrightarrow{\ell} s_i'$ with $\ell \in \mathcal{A}_i$, and there does not exist $\ell' \in \mathcal{A}^s$ such that $\mathfrak{C}(\{\ell, \ell'\}) = \ell''$, for some $\ell'' \in \mathcal{A}^f$, then $z_2 = (s_1, \ldots, s_i', \ldots, s_j, \ldots, s_n) \in \mathcal{S}$. In this case, $\ell \in \mathcal{A}$ and $z_1 \xrightarrow{\ell} z_2 \in \mathcal{T}$;
  - If for some $\mathcal{M}_i \in \Pi$, $s_i \xrightarrow{\ell} s_i'$ with $\ell \in \mathcal{A}_i$, and for some $\mathcal{M}_j \in \Pi$ $(i \neq j)$, $s_j \xrightarrow{\ell'} s_j'$ with $\ell' \in \mathcal{A}_j$, and $\mathfrak{C}(\{\ell, \ell'\}) = \ell''$, for some $\ell'' \in \mathcal{A}^f$, then $z_2 = (s_1, \ldots, s_i', \ldots, s_j', \ldots, s_n) \in \mathcal{S}$. In this case, $\ell'' \in \mathcal{A}$ and $z_1 \xrightarrow{\ell''} z_2 \in \mathcal{T}$.

4   Directed LTS Search Techniques

The two most basic LTS exploration algorithms available in model checkers are BFS and depth-first search (DFS). They differ in the order in which they consider states for exploration. In BFS, states are explored in order of their distance from s_{in}. DFS gives priority to searching at increasing depth instead of exploring all states at a certain depth before continuing. If at any point in the search, the selected state has no successors, or they have all been visited before, then DFS will backtrack to the parent of this state, and explore the next state from the parent's set of successors, according to the ordering function. BFS and DFS are typical *blind* searches, since they do not take additional information about the system under verification into account. In contrast to this are the *informed* searches which do use such information. Examples of informed searches are *Uniform-cost search* [33], also known as *Dijkstra's search* [9], and A^* [19].

All searches, both blind and informed, mentioned so far are examples of *exhaustive* searches, i.e. in the absence of deadlocks, they will explore all states reachable from s_{in}. Another class of searches is formed by the *non-exhaustive* searches. These searches prune the LTS on-the-fly, completely ignoring those parts which are deemed uninteresting according to some heuristics. At the cost of losing completeness, these searches can find deadlocks in very large LTSs fast, since they can have drastically lower memory and time requirements, compared to exhaustive searches. A blind search in this category is random walk, or simulation, in which successor states are chosen randomly. An informed example is *beam search*, which is basically a BFS-like search, where in each iteration, i.e. depth, only up to $\beta \in \mathbb{N}$, which is given a priori, states are selected for exploration. For the selection procedure, various functions can be used; in classic beam search, a (state-based) function as in A^* is used, in *priority* beam search [38,39,41], a selection procedure based on transition labels is employed, while *highway search* [13] uses random selection, and is therefore a blind variant of beam search.

For a search L, let us define its scope $\text{REACH}_\mathcal{M}(L)$ in a given LTS $\mathcal{M}$ as the set of states in $\mathcal{M}$ that it will explore. For all exhaustive L, we have $\text{REACH}_\mathcal{M}(L) = \mathcal{S}$, while for all non-exhaustive L, $\text{REACH}_\mathcal{M}(L) \subset \mathcal{S}$. Let us consider two searches L_1 and L_2, such that $\text{REACH}_\mathcal{M}(L_1) \cup \text{REACH}_\mathcal{M}(L_2) = \mathcal{S}$. Then, we propose to call $\{L_1, L_2\}$ *cumulatively exhaustive* on $\mathcal{M}$. Such *cumulatively exhaustive sets* (CESs) are very interesting for SV; the elements can be run independently in parallel, and once they are all finished, the full LTS will have been explored.

Some existing searches lead to CESs. *Iterative deepening* [25] uses depth-bounded DFS in several iterations, each time relaxing the bound. Each iteration can be seen as an individual search, subsequent searches having increasing scopes. Iterative searches form a class, which includes e.g. IDA* [25]. Another class leading to CESs consists of random searches like random walk and highway search. However, all these are not suitable for grid computing. Iterative searches form CESs containing an exhaustive search. If $\mathcal{M}$ is bug-free, then eventually this search is performed, which is particularly inefficient. With random searches, there is no guarantee that after n searches, all reachable states are visited. If $n \to \infty$, the set will eventually be cumulatively exhaustive, but performing the searches may take forever. Moreover, the probabilities to visit states in a random walk are not uniformly distributed, but depend on the graph structure [30].

We want to derive CESs with a bounded number of non-exhaustive elements from a system under verification. Preferably, all scopes have equal size, to ensure load-balancing, but this is not necessary (in SV, workers do not synchronise, hence load-balancing is less important [22,23]). To achieve this, we have developed a search called *informed swarm search* (ISS) which accepts a guiding function, and we have a method to compute a set of guiding functions $f_0, f_1, \ldots, f_n$ given a system specification, such that $\{\text{ISS}(f_0), \text{ISS}(f_1), \ldots\}$ is a CES. The guiding functions actually relate to traces through the LTS of a subsystem π of the system under verification, which are derived from an LTS exploration of π. Such an LTS can in practice be much smaller than the system LTS. For now, we require that π yields finite behaviour, i.e. that its LTS is cycle-free.

ISS only selects those transitions for exploration which either do not stem from π, or which correspond with the current position in the given trace through the LTS of π. This is computationally inexpensive, since it only entails label comparison. The underlying assumption is that labels can be uniquely mapped to process LTSs; given a label, we know from which process it stems. If a given set of LTSs does not meet this requirement, some label renaming can fix this.

5   Systems with Independent Processes

In this section, we will explain our method for constructing CESs for very basic specifications which consist of completely independent processes in parallel composition. We are aware of the fact that such specifications may practically not be very interesting. However, they are very useful for our explanation.

Figure 1 presents two LTSs of a beverage machine, and a snack machine, respectively. Both are able to dispense goods when one of their buttons is pressed.

There is no interaction between the machines. We have $\Pi = \{\mathcal{M}_b, \mathcal{M}_s\}$, with $\mathcal{M}_b$ and $\mathcal{M}_s$ the LTSs of the beverage machine and the snack machine, respectively.

If we perform a DFS through $\mathcal{M}_b$, and we record the encountered traces whenever backtracking is required, we get the following set: $\{\langle push\_button(1),$ $get\_coffee\rangle, \langle push\_button(2), get\_tea\rangle\}$. Note that all reachable states of $\mathcal{M}_b$ have been visited. We use these two traces as guiding principles for two different searches through $\mathcal{M}_{bs}$, which is the LTS obtained by placing the two LTSs in parallel composition. Algorithm 1 presents the pseudo-code of our ISS, which accepts a trace σ and a set of transition labels $\mathcal{A}^{ex}$ to guide the search. In our example, each ISS, with one of the two traces and $\mathcal{A}^{ex} = \mathcal{A}_b$ as input, focusses on specific behaviour of the beverage machine within the bigger system context.

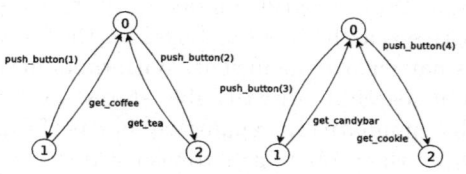

Fig. 2 shows which states will be visited in $\mathcal{M}_{bs}$ if we perform an ISS based on $\langle push\_button(1),$ $get\_coffee\rangle$. Alg. 1 explains how this is done. Initially, we put s_{in} in *Open*. Then, we add all successors

Fig. 1. Two LTSs of a beverage and a snack machine

reached via a transition with a label not in $\mathcal{A}^{ex}$ in *Next*, and add all successors reached via a transition labelled $\sigma(i)$ in *Step*. Here, $\sigma(i)$ returns the $(i+1)^{th}$ element in the trace σ; if i is bigger than or equal to the trace length, we say that $\sigma(i) = \bot$, where $\bot$ represents 'undefined', and $\{\bot\}$ is equivalent to $\emptyset$. For now, please ignore the next step concerning $\mathcal{F}_i$; it has to do with feedback to the manager, and will be explained later. Finally, s_{in} is added to *Closed*, i.e. the set of explored states, and the states in *Next* which have not been explored constitute the new *Open*. Then, the whole process is repeated. This continues until *Open* $= \emptyset$. Then, the contents of *Step* is moved to *Open*, and the ISS moves on to the next step in σ. In this way, the ISS explores all traces $\gamma = \langle \alpha_0, \sigma(0), \alpha_1, \ldots, \alpha_n, \sigma(n), \alpha_{n+1}\rangle$, with n the length of σ and the α_i traces

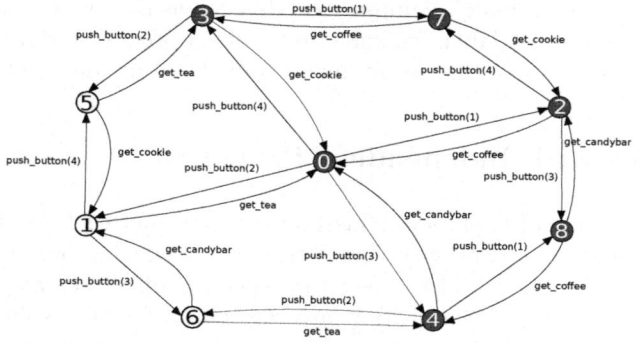

Fig. 2. A search through $\mathcal{M}_{bs}$ with $\mathcal{M}_b$ restricted to $\langle push\_button(1), get\_coffee\rangle$

Algorithm 1. Bfs-based Informed Swarm Search

Require: Implicit description of $\mathcal{M}$, exclusion action set $\mathcal{A}^{ex}$, swarm trace σ
Ensure: $\mathcal{M}$ restricted to σ is explored
$\quad i \leftarrow 0$
$\quad Open \leftarrow s_{in}; \ Closed, Next, Step, \mathcal{F}_i \leftarrow \emptyset$
$\quad \textbf{while } Open \neq \emptyset \vee Step \neq \emptyset \textbf{ do}$
$\quad\quad \textbf{if } Open = \emptyset \textbf{ then}$
$\quad\quad\quad i \leftarrow i + 1$
$\quad\quad\quad Open \leftarrow Step \setminus Closed; \ Step, \mathcal{F}_i \leftarrow \emptyset$
$\quad\quad \textbf{end if}$
$\quad\quad \textbf{for all } s \in Open \textbf{ do}$
$\quad\quad\quad Next \leftarrow Next \cup nxt(en_{\mathcal{M}}(s, \mathcal{A} \setminus \mathcal{A}^{ex}))$
$\quad\quad\quad Step \leftarrow Step \cup nxt(en_{\mathcal{M}}(s, \{\sigma(i)\}))$
$\quad\quad\quad \mathcal{F}_i \leftarrow \mathcal{F}_i \cup \{\ell \mid \exists s' \in \mathcal{S}.(s \xrightarrow{\ell} s') \in en_{\mathcal{M}}(s, \mathcal{A}^{ex})\}$
$\quad\quad \textbf{end for}$
$\quad\quad Closed \leftarrow Closed \cup Open$
$\quad\quad Open \leftarrow Next \setminus Closed; \ Next \leftarrow \emptyset$
$\quad \textbf{end while}$

---

containing only labels from $\mathcal{A} \setminus \mathcal{A}_b$. If we perform an Iss for every trace through $\mathcal{M}_b$, we will visit all reachable states in $\mathcal{M}_{bs}$. Figure 2 shows what the Iss with $\langle push\_button(1), get\_coffee \rangle$ explores; out of the 9 states, 6 are explored, meaning that 33% of $\mathcal{M}_{bs}$ could be ignored. The Iss using $\langle push\_button(2), get\_tea \rangle$ also explores 6 states, namely 0, 3 and 4 (the states reachable via behaviour from $\mathcal{M}_s$), and 1, 5, and 6. In this way, some states are explored multiple times, but we have a Ces of non-exhaustive searches through $\mathcal{M}_{bs}$.

6 Systems with Synchronising Processes

Next, we consider parallel processes which synchronise. In such a setting, things get slightly more complicated. Before we continue with an example, let us first formally define a subsystem and the Lts it yields.

Definition 3. *Given a set Π of $n > 0$ process Ltss $\mathcal{M}_i = (\mathcal{S}_i, \mathcal{A}_i, \mathcal{T}_i, s_{in}i)$, for $1 \leq i \leq n$, and a synchronisation function $\mathfrak{C} : \mathcal{A}^s \times \mathcal{A}^s \to \mathcal{A}^f$, we call a subset $\pi \subseteq \Pi$ of Ltss a subsystem of Π. We can derive a synchronisation function $\mathfrak{C}_\pi : \mathcal{A}_\pi^s \times \mathcal{A}_\pi^s \to \mathcal{A}_\pi^f$ from $\mathfrak{C}$ as follows: $\mathcal{A}_\pi^s = \bigcup_{\mathcal{M}_i \in \pi} \mathcal{A}_i$, and for all $\ell, \ell' \in \mathcal{A}_\pi^s$, if $\mathfrak{C}(\{\ell, \ell'\}) = \ell''$, for some $\ell'' \in \mathcal{A}^f$, we define $\mathfrak{C}_\pi(\{\ell, \ell'\}) = \ell''$ and $\ell'' \in \mathcal{A}_\pi^f$.*

Note that the Lts of a subsystem π, which can be obtained with Definition 2, describes an over-approximation of the potential behaviour of π within the bigger context of Π. This is because in π, it is implicitly assumed that all synchronisation with external processes (which are in Π, but not in π) can happen whenever a process in π can take part in it. For the Iss, we have to choose $\mathcal{A}^{ex}$ more carefully, and we need to post-process traces σ yielded from π; we must take synchronisation between π and the larger context into account. For this, we define a relabelling function $R : \mathcal{A}_\pi \to \mathcal{A}^f$ as follows: $R(\ell) = \ell$ if there exists no ℓ' such that $\mathfrak{C}(\{\ell, \ell'\})$ is defined, and $R(\ell) = \ell''$ if there exists an ℓ' such that $\mathfrak{C}(\{\ell, \ell'\}) = \ell''$. Then, we say that $\mathcal{A}^{ex} = \{R(\ell) \mid \ell \in \mathcal{A}_\pi\}$ and $\sigma'(i) = R(\sigma(i))$ for all defined $\sigma(i)$, such that $\mathcal{A}^{ex}$ and σ' are applicable in the system Lts, because actions from π which are forced to synchronise are relabelled to the results of those synchronisations. In this case, σ' relates to a single trace iff $\mathfrak{C}$ is injective.

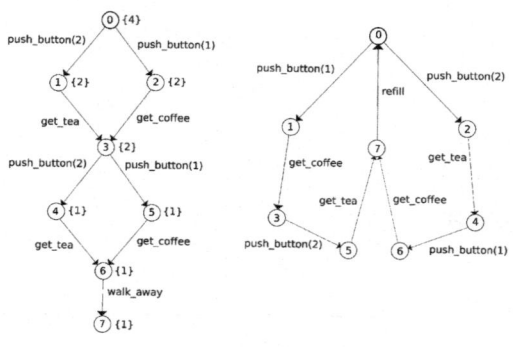

Fig. 3 shows two LTSs: one of a modified beverage machine $\mathcal{M}_b$, and one of a user $\mathcal{M}_u$ (for now, please ignore the numbers between curly brackets). For the parallel composition $\mathcal{M}_{ub}$, we define: $\mathfrak{C}(\{push\_button, push\_button\}) = button\_pushed^2$, $\mathfrak{C}(\{get\_coffee, get\_coffee\}) = take\_coffee$, $\mathfrak{C}(\{get\_tea, get\_tea\}) = take\_tea$.

Fig. 3. LTSs of a user of a beverage machine, and a beverage machine, respectively

First, if $\pi = \{\mathcal{M}_u\}$, observe that a DFS through $\mathcal{M}_u$ does not give us the full set of traces through $\mathcal{M}_u$; if we consider the transition ordering from left to right, a DFS will first provide trace $\langle push\_button(2), get\_tea, push\_button(2), get\_tea, walk\_ away \rangle$. Then, it will backtrack to state 3, and continue via 5 to 6.

Since 6 has already been explored, the DFS produces trace $\langle push\_button(2), get\_tea, push\_button (1), get\_coffee \rangle$ and backtracks to 0. Note that this new trace does not finish with $walk\_away$. Continuing from 0, the search will finally produce the trace $\langle push\_button(1), get\_coffee \rangle$.

Figure 4 presents $\mathcal{M}_{ub}$. If we use these three traces (after relabelling with R) as guiding functions as in Alg. 1, and define $\mathcal{A}^{ex}$ as mentioned earlier, none of the searches will visit (the marked) state 5! The reason for this is that although in $\mathcal{M}_u$, multiple

---

Algorithm 2. Trace-counting DFS

Require: Implicit description of cycle-free $\mathcal{M}$
Ensure: $\mathcal{M}$ and $tc : \mathcal{S} \to \mathbb{N}$ are constructed

$Closed \leftarrow \emptyset$
$tc(s_{in}) \leftarrow dfs(s_{in})$
$dfs(s) =$
  if $s \notin Closed$ **then**
    $tc(s) \leftarrow 0$
    for all $s' \in nxt(en_{\mathcal{M}}(s, \mathcal{A}))$ **do**
      $tc(s) \leftarrow tc(s) + dfs(s')$
    end for
    if $nxt(en_{\mathcal{M}}(s, \mathcal{A})) = \emptyset$ **then**
      $tc(s) \leftarrow 1$
    end if
    $Closed \leftarrow Closed \cup \{s\}$
  end if
  return $tc(s)$

---

traces may lead to the same state, in $\mathcal{M}_{ub}$, the corresponding traces may not. This is due to synchronisation. Since the different traces in $\mathcal{M}_u$ synchronise with different traces in $\mathcal{M}_b$ which do not lead to the same state, also in $\mathcal{M}_{ub}$, the resulting traces will lead to different states. One solution is to fully explore $\mathcal{M}_u$ with a DFS without a $Closed$ set. However, this is very inefficient, since the complete reachable graph from a state s needs to be explored n times, if n different traces reach s. Instead, we opt for constructing a *weighted* LTS, where each state is assigned a value indicating the number of traces that can be explored from

---

[2] When transition labels have parameters, they can synchronise iff they have the same parameter values. Then, the resulting transition also has these parameter values.

that state. In Figure 3, these numbers are displayed between curly brackets. The advantage of this is that we can avoid re-exploration of states, and it is in fact possible to uniquely identify traces through $\mathcal{M}_u$ by a trace ID $\in \mathbb{N}$.

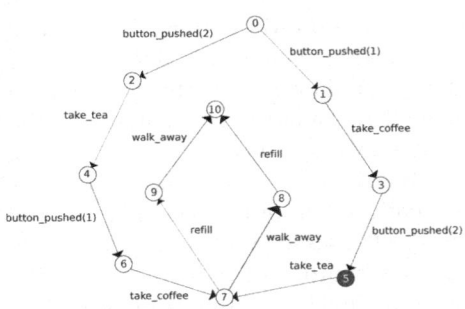

Fig. 4. The LTS of a beverage machine and a user in concurrency

Alg. 2 presents our trace counting search (in which *Closed* is a global variable), which not only explores a full LTS, but also constructs a function $tc : \mathcal{S} \to \mathbb{N}$ indicating the number of traces one can follow from a state. Deadlock states get weight 1, and other states get a weight equal to the sum of the weights of their immediate successors. In DFS, a state s is placed in *Closed* once all states reachable from s have been explored, hence at that moment, we know the final weight of s. This allows us to reuse weight information whenever we visit an explored state. Note that $tc(s_{in})$ equals the number of possible traces through the LTS. Alg. 3 shows how to reconstruct a trace, given its ID between 0 and $tc(s_{in})$. It is important here that for each state, its successor states are always ordered in the same way. In the weighted LTS, each trace from s_{in} to a state s represents a range of trace IDs from *lower*, the maintained lower-bound, to *lower* + $tc(s)$. Starting at s_{in}, the algorithm narrows the matching range down to the exact ID. At each state, it explores the transition with a matching ID interval to the next state, and continues like this until a deadlock state is reached.

The method works as follows: first, an explicit weighted LTS of a subsystem π is constructed. Whenever a worker is ready for work, he contacts a manager, who then selects a trace ID from the given set of IDs, and constructs the associated trace σ using the weighted LTS. This trace, after relabelling with R, is used by the worker to guide his ISS. Next, we discuss the $\mathcal{F}_i$ in Alg. 1. For each $\sigma(i)$, set $\mathcal{F}_i$ is constructed to hold all labels from $\mathcal{A}_\pi$

Algorithm 3. Trace Reconstruction

Require: Cycle-free $\mathcal{M}$, $tc : \mathcal{S} \to \mathbb{N}$, ID $\in \mathbb{N}$
Ensure: Trace with given ID is constructed in σ
  $i, lower \leftarrow 0$
  $crt \leftarrow s_{in}$
  for all $(crt \xrightarrow{\ell} s) \in en_{\mathcal{M}}(crt, \mathcal{A})$ **do**
    if $lower + tc(s) >$ ID **then**
      $crt \leftarrow s$
      $\sigma(i) \leftarrow \ell; i \leftarrow i + 1$
    else
      $lower \leftarrow lower + tc(s)$
    end if
  end for

---

after relabelling which are encountered in $\mathcal{M}$ while searching for $\sigma(i)$. Since $\mathcal{M}_\pi$ is an over-approximation of the potential behaviour of π in $\mathcal{M}$,[3] the set of trace IDs is likely to contain many false positives, i.e. behaviour of π which cannot be fully followed in $\mathcal{M}$.

---

[3] Note that only 2 of the 4 traces through $\mathcal{M}_u$ can be followed completely in $\mathcal{M}_{ub}$ by a swarm search.

These $\mathcal{F}_i$ provide invaluable feedback, allowing the manager to prune non-executable traces from the work-list. This is essential to on-the-fly reduce the number of ISSs drastically. The manager performs this pruning by traversing the weighted LTS, similar to Algorithm 3, and removing all ranges of trace IDs corresponding to traces which are known to be false positives from a maintained trace set. E.g. if a worker discovers that after an action a of π, the only action of π which can be performed is b, then the manager will first follow a in the weighted LTS of π from s_{in} to a state s, and remove all ID ranges corresponding to the immediate successors of s which are not reached via a transition labelled b from the trace set. The manager will also remove the ID of the trace followed by that worker, if the worker was able to process it entirely. From that moment on, the manager will select trace IDs from the pruned set. This allows for embarrassingly parallel workers, and the manager can dynamically process feedback and provide new work in the form of traces. Furthermore, only little communication is needed, and if a worker fails due to a technical issue, the work can easily be redone.

Note that trace-counting DFS, like DFS, runs in $O(|\mathcal{S}| + |\mathcal{T}|)$, and the trace reconstruction search and the ID pruning algorithm run in $O(n + (n * b))$, with n the length of the largest trace through the weighted LTS, and b the maximum number of successors of a state in the LTS. Finally, the complexity of ISS depends on the LTS structure; it is less than $O(|\mathcal{S}| + |\mathcal{T}|)$, as it is non-exhaustive, but also not linear in the length of the longest trace through the LTS, like e.g. beam search, as it prunes less aggressively (not every BFS-level has the same size).

7   Experiments

All proposed algorithms are implemented as an extension of LTSMIN [7]. The advantage of using this toolset as a starting point is that it has interfaces with multiple popular model checkers, such as DIVINE [2] and MCRL2 [17]. However, DIVINE is based on Kripke structures, where the states instead of the transitions are labelled. Future work, therefore, is to develop a state-based ISV.

We have two `bash` scripts for performing real multi-core ISVs and simulating ISVs, in case not enough processors are available. We do not yet support communication between workers and the manager over a network. The functionality of the manager is implemented in new tools to perform the pruning on the current trace ID set, and select new trace IDs from the remaining set. All intermediary information is maintained on disk; Initially, the user has to create a subsystem specification based on the system specification, which is used for trace-counting, leading to a weighted LTS on disk. The selection tool can select available trace IDs from that LTS, and write explicit traces into individual files. Relabelling can also be applied, given the synchronisation rules. The IDs are currently selected such that they are evenly distributed over the ID range. The trace files are accepted by the ISS in LTSMIN, applied on the system specification. Finally, the written feedback can be applied on a file containing the current trace ID set.

Table 1. Results for two protocol specifications. ISV n indicates an ISV with n workers. # π-*traces*: estimated # Isss needed. (1 for single Bfs)# *Isss*: actual # Isss needed. *max. states*: largest # states explored by an Iss, or # states explored by Bfs. *max. time*: longest running time of an Iss, or running time of Bfs.

case	search	results			
		# π-*traces*	# Isss	*max. states*	*max. time*
	Bfs	1	1	13,246,976	19,477 s
DRM (1nnc, 3ttp)	ISV 10	$1.31 * 10^{13}$	7,070	70,211	177 s
	ISV 100	$1.31 * 10^{13}$	9,900	70,211	175 s
	Bfs	1	1	137,935,402	105,020 s
1394 (3 link ent.)	ISV 10	$3.01 * 10^{9}$	1,160	236,823	524 s
	ISV 100	$3.01 * 10^{9}$	1,400	236,823	521 s

We performed a number of experiments using μCRL [18] specifications of a DRM protocol [37] and the Link Layer Protocol of the IEEE-1394 Serial Bus (Firewire) [29] with three parallel link protocol entities. In the first case, we performed trace-counting on the two iPod devices in parallel composition. In the second case, we isolated one of the link protocol entities for the trace-counting, and bounded its infinite behaviour, only allowing the traversal through cyclic behaviour up to 2 times. The experiments were performed on a machine with two dual-core AMD OPTERON (tm) processors 885 2.6 GHz, 126 GB RAM, running RED HAT 4.3.2-7. Creating the weighted LTss took no more than a few minutes; the DRM weighted LTS contained 962 states, the 1394 weighted LTS 73 states. We simulated the ISVs, executing Isss in sequence. This influences the outcome slightly, since we process feedback each time n new Isss have been performed; we do this to approach a real ISV, where, on the one hand, feedback can be processed as it becomes available, but on the other hand, many Isss can be launched in parallel at the same time, hence when a certain amount of feedback has been processed. Therefore, updating the remaining set of traces after each individual Iss is not really fair. The need to simulate was the main reason that we have not looked at larger LTss yet. Table 1 presents the results. For smaller instances, we validated that the swarm was cumulatively exhaustive, by writing the full state vectors of states to disk. Observe that initially, the first analyses produced a large over-approximation of the number of Isss needed (see "#π-*traces*"). The quality of the estimation has an effect on the ISV efficiency, even with feedback. This also has an effect on the difference in efficiency when changing n (the number of parallel workers); when the over-approximation is large, many Isss may be launched which can not process the given trace entirely, since it is a false positive. As n is increased, the probability of launching such Isss gets higher, as more Isss are launched before any new feedback is given. On the other hand, increasing n still reduces the overall execution time, because there is more parallellism.

The ISVs take more time[4] compared to a Bfs. This seems to indicate that the method is not interesting. However, keep in mind that already the first few Isss

---

[4] Note that the overall execution time takes at most *max. time* $* (\#\text{Isss}/n)$ seconds.

reach great depths, and they go into several directions. Therefore, even though we did no bug-hunting, in many cases, it is to be expected that like SV, our ISV will find bugs much quicker than a single search. This could perhaps even be improved by using a DFS-based version of Alg. 1. We plan to do an empirical comparison between such a DFS-based ISV and SV. For full exploration, ISV does not provide a speedup, but this was not intended; instead, observe that the maximum number of states explored in an Iss is much smaller than the LTS size (in the DRM case about $\frac{1}{2}$% of the overall size, in the 1394 case about $\frac{1}{6}$%). A better trade-off could be realised by guiding each Iss with a set of subsystem traces; for the DRM case, an Iss following 10 traces would still explore no more than 5% of the LTS (probably less depending on the amount of redundant work which could now be avoided), while the number of Isss could be decreased by an order of magnitude. This makes ISV applicable in clusters and grids with large numbers of processors, each having access to e.g. 2 GB memory, even if exploring the whole LTS at once would require much more memory.

8   Conclusions

In this paper, we have proposed a new approach to parallel MC, aimed at large scale grid computing. Part of the system behaviour is analysed in isolation, yielding a set of possible traces, each representing a search through the full LTS. These searches are embarrassingly parallel, since only a trace, the set of actions of the subsystem, and the specification are needed for input. Once a search is completed, feedback is sent to the manager, giving him information on the validity of the remaining traces, which is invaluable, since the set of traces is an over-approximation of the possible traces of the subsystem in the bigger context of the full system. We believe that our method is fully compatible with existing techniques. E.g. one can imagine having multiple multi-core machines available; the ISV method can then be used to distribute the work over these machines, but each machine individually can perform the work using a multi-core search. Also reduction techniques like partial order reduction should be compatible with ISV. For good results, we expect it to be important that both during the analysis of the subsystem and the full system, the same reduction techniques are applied.

For future work, we plan to test ISV more thoroughly to see if it scales to real-life problems, and make the tools more mature. As the size of the subsystem has an effect on the work distribution, it is interesting to investigate what an ideal subsystem relative to a system would be. We also wish to generalise ISV, such that subsystems yielding infinite behaviour can be analysed, and to improve the trace set approximation with e.g. static analysis. One could also construct the trace set according to the MC task, e.g. taking the property to check into account. We plan to investigate different strategies for trace selection. A good strategy sends the workers into very different directions. Finally, [16,36] provide good pointers to develop a state-based ISV.

References

1. Barnat, J., Brim, L., Ročkai, P.: Scalable multi-core LTL model-checking. In: Bošnački, D., Edelkamp, S. (eds.) SPIN 2007. LNCS, vol. 4595, pp. 187–203. Springer, Heidelberg (2007)
2. Barnat, J., Brim, L., Ročkai, P.: DiVinE multi-core – A parallel LTL model-checker. In: Cha, S(S.), Choi, J.-Y., Kim, M., Lee, I., Viswanathan, M. (eds.) ATVA 2008. LNCS, vol. 5311, pp. 234–239. Springer, Heidelberg (2008)
3. Barnat, J., Brim, L., Stříbrná, J.: Distributed LTL model-checking in SPIN. In: Dwyer, M.B. (ed.) SPIN 2001. LNCS, vol. 2057, pp. 200–216. Springer, Heidelberg (2001)
4. Behrmann, G., Hune, T., Vaandrager, F.: Distributing Timed Model Checking - How the Search Order Matters. In: Emerson, E.A., Sistla, A.P. (eds.) CAV 2000. LNCS, vol. 1855, pp. 216–231. Springer, Heidelberg (2000)
5. Blom, S.C.C., Calamé, J.R., Lisser, B., Orzan, S., Pang, J., van de Pol, J.C., Torabi Dashti, M., Wijs, A.J.: Distributed Analysis with μCRL: A Compendium of Case Studies. In: Grumberg, O., Huth, M. (eds.) TACAS 2007. LNCS, vol. 4424, pp. 683–689. Springer, Heidelberg (2007)
6. Blom, S.C.C., Lisser, B., van de Pol, J.C., Weber, M.: A Database Approach to Distributed State Space Generation. In: Haverkort, B., Černá, I. (eds.) PDMC 2007. ENTCS, vol. 198, pp. 17–32. Elsevier, Amsterdam (2007)
7. Blom, S.C.C., van de Pol, J., Weber, M.: LTSMIN: distributed and symbolic reachability. In: Touili, T., Cook, B., Jackson, P. (eds.) CAV 2010. LNCS, vol. 6174, pp. 354–359. Springer, Heidelberg (2010)
8. BOINC: Visited on 18 February (2011), http://boinc.berkeley.edu
9. Dijkstra, E.W.: A note on two problems in connection with graphs. Numerische Mathematik 1, 269–271 (1959)
10. Dill, D.: The Murphi Verification System. In: Alur, R., Henzinger, T.A. (eds.) CAV 1996. LNCS, vol. 1102, pp. 390–393. Springer, Heidelberg (1996)
11. Dwyer, M.B., Elbaum, S.G., Person, S., Purandare, R.: Parallel Randomized State-space Search. In: 29th Int. Conference on Software Engineering, pp. 3–12. IEEE Press, New York (2007)
12. Edelkamp, S., Leue, S., Lluch-Lafuente, A.: Directed explicit-state model checking in the validation of communication protocols. STTT 5(2), 247–267 (2004)
13. Engels, T.A.N., Groote, J.F., van Weerdenburg, M.J., Willemse, T.A.C.: Search Algorithms for Automated Validation. JLAP 78(4), 274–287 (2009)
14. Foster, I.: Designing and Building Parallel Programs. Addison-Wesley, Reading (1995)
15. Garavel, H., Mateescu, R., Bergamini, D., Curic, A., Descoubes, N., Joubert, C., Smarandache-Sturm, I., Stragier, G.: DISTRIBUTOR and BCG_MERGE: Tools for distributed explicit state space generation. In: Hermanns, H., Palsberg, J. (eds.) TACAS 2006. LNCS, vol. 3920, pp. 445–449. Springer, Heidelberg (2006)
16. Groce, A., Joshi, R.: Exploiting traces in static program analysis: better model checking through printfs. STTT 10(2), 131–144 (2008)
17. Groote, J.F., Keiren, J., Mathijssen, A., Ploeger, B., Stappers, F., Tankink, C., Usenko, Y.S., Weerdenburg, M.J.: The MCRL2 Toolset. In: 1st Int. Workshop on Academic Software Development Tools and Techniques 2008, pp. 5-1/10 (2008)
18. Groote, J.F., Ponse, A.: The Syntax and Semantics of μCRL. In: Algebra of Communicating Processes 1994. Workshops in Computing, pp. 26–62. Springer, Heidelberg (1995)

19. Hart, P.E., Nilsson, N.J., Raphael, B.: A Formal Basis for the Heuristic Determination of Minimum Cost Paths. IEEE Trans. on Systems, Science and Cybernetics 2, 100–107 (1968)

20. Holzmann, G.J.: A Stack-Slicing Algorithm for Multi-Core Model Checking. In: Haverkort, B., Černá, I. (eds.) PDMC 2007. ENTCS, vol. 198, pp. 3–16. Elsevier, Amsterdam (2007)

21. Holzmann, G.J., Bošnački, D.: The Design of a Multicore Extension of the SPIN Model Checker. IEEE Trans. On Software Engineering 33(10), 659–674 (2007)

22. Holzmann, G.J., Joshi, R., Groce, A.: Swarm Verification. In: 23rd IEEE/ACM Int. Conference on Automated Software Engineering, pp. 1–6. IEEE Press, New York (2008)

23. Holzmann, G.J., Joshi, R., Groce, A.: Tackling large verification problems with the swarm tool. In: Havelund, K., Majumdar, R. (eds.) SPIN 2008. LNCS, vol. 5156, pp. 134–143. Springer, Heidelberg (2008)

24. Holzmann, G.J., Joshi, R., Groce, A.: Swarm Verification Techniques. IEEE Trans. On Software Engineering (2010) (to appear)

25. Korf, R.E.: Depth-First Iterative-Deepening: An Optimal Admissible Tree Search. Artificial Intelligence 27(1), 97–109 (1985)

26. Laarman, A., van de Pol, J.C., Weber, M.: Boosting Multi-Core Reachability Performance with Shared Hash Tables. In: Int. Conference on Formal Methods in Computer-Aided Design (2010)

27. Lerda, F., Sisto, R.: Distributed-Memory Model Checking with SPIN. In: Dams, D.R., Gerth, R., Leue, S., Massink, M. (eds.) SPIN 1999. LNCS, vol. 1680, pp. 22–39. Springer, Heidelberg (1999)

28. Lowerre, B.T.: The HARPY speech recognition system. PhD thesis, Carnegie-Mellon University (1976)

29. Luttik, S.P.: Description and Formal Specification of the Link Layer of P1394. Technical Report SEN-R 9706, CWI (1997)

30. Pelánek, R., Hanžl, T., Černá, I., Brim, L.: Enhancing Random Walk State Space Exploration. In: 10th Int. Workshop on Formal Methods for Industrial Critical Systems. ACM SIGSOFT, pp. 98–105 (2005)

31. Peled, D., Pratt, V., Holzmann, G.J. (eds.): Partial Order Methods in Verification. Series in Discrete Mathematics and Theoretical Computer Science 29 (1996)

32. Romein, J.W., Plaat, A., Bal, H.E., Schaeffer, J.: Transposition Table Driven Work Scheduling in Distributed Search. In: 16th National Conference on Artificial Intelligence, pp. 725–731. AAAI Press, Menlo Park (1999)

33. Russell, S., Norvig, P.: Artificial intelligence: A modern approach. Prentice-Hall, New Jersey (1995)

34. SETI@home, http://setiathome.berkeley.edu (Visited on 18 February 2011)

35. Sivaraj, H., Gopalakrishnan, G.: Random Walk Based Heuristic Algorithms for Distributed Memory Model Checking. ENTCS, vol. 89, pp. 51–67 (2003)

36. Staats, M., Păsăreanu, C.: Parallel Symbolic Execution for Structural Test Generation. In: 19th Int. Conference on Software Testing and Analysis, pp. 183–194. ACM, New York (2010)

37. Torabi Dashti, M., Krishnan Nair, S., Jonker, H.L.: Nuovo DRM Paradiso: Towards a Verified Fair DRM Scheme. Fundamenta Informaticae 89(4), 393–417 (2008)

38. Torabi Dashti, M., Wijs, A.J.: Pruning state spaces with extended beam search. In: Namjoshi, K.S., Yoneda, T., Higashino, T., Okamura, Y. (eds.) ATVA 2007. LNCS, vol. 4762, pp. 543–552. Springer, Heidelberg (2007)

39. Valente, J.M.S., Alves, R.A.F.S.: Filtered and recovering beam search algorithms for the early/tardy scheduling problem with no idle time. Computers & Industrial Engineering 48(2), 363–375 (2005)
40. West, C.H.: Protocol Validation by Random State Exploration. In: 8th Int. Conference on Protocol Specification, Testing and Verification, pp. 233–242. North-Holland, Amsterdam (1986)
41. Wijs, A.J.: What to Do Next?: Analysing and Optimising System Behaviour in Time. PhD thesis, Vrije Universiteit Amsterdam (2007)

Scaling Up with Event-B: A Case Study[*]

Faqing Yang and Jean-Pierre Jacquot

LORIA – DEDALE Team – Nancy Université
Vandoeuvre-Lès-Nancy, France
firstname.lastname@loria.fr

Abstract. Ability to scale up from toy examples to real life problems is a crucial issue for formal methods. Formalizing a algorithm used in vehicle automation (platooning control) in a certification perspective, we had the opportunity to study the scaling up when going from a (toy) model in 1D to a (more realistic) model in 2D. The formalism, Event-B, belongs to the family of mathematical state based methods. Increase was quantitative: 3 times more events and 4 times more proofs; and qualitative: trigonometric functions and integrals are used. Edition and verification of the specification scale up well. The crucial part of the work was the adaptation of the mathematical and physical model through standard heuristics. The validation of temporal properties and behaviors do not scale up so well. Analysis of the difficulties suggests improvements in both tool support and formalism.

1 Introduction

This paper relates our experience with the specification of a realistic algorithm for the control of autonomous vehicles. The problem to solve has interesting characteristics:

- the development should lead to a certified product (a component for a car moving in the public space),
- the physical and mathematical model uses common mathematical notions such as trigonometric, kinematics, integrals, and so on,
- there is an existing empirical solution.

The problem is known as *platooning:* autonomous vehicles that move as virtual trains. All vehicles follow the virtual track defined by the first vehicle while keeping to a minimum the distance between them. The issue is then to guarantee that the control algorithm is safe, i.e., that vehicles can never collide.

We chose Event-B because the concepts of proof obligation and formal refinement are well fitted for the task of guaranteeing an implementation. However, several points needed to be assessed. Could a non-functional property such as non-collision be specified in Event-B? Is the support environment, particularly the provers, strong enough for such a problem? Can we model adequately a system which contains continuous functions, real numbers, or geometric relationships, in a framework which is based on discrete sets and integers?

---

[*] Work partially supported by ANR under project ANR-06-SETI-017 TACOS (http://tacos.loria.fr), and by Pôle de Compétitivité Alsace/Franche-Comté under CRISTAL project (http://www.projet-cristal.net).

M. Bobaru et al. (Eds.): NFM 2011, LNCS 6617, pp. 438–452, 2011.

A first specification was written on a simplified version of the problem. The model considered only a linear track (1D) and the control was only aimed at keeping some ideal distance while avoiding collisions. While simplistic, this model was important on three respects: it allowed us to identified the "hard" parts, it prototyped the properties of interest, and it provided us with a neat structure for the development.

The next specification considered the platooning problem from a realistic point of view. Vehicles are now moving on a plane, the leader of the platoon is not constrained to a predefined track, the properties of interest are now the non-collision and the distance form the virtual track drawn by the leading vehicle.

The changes between a model in 1D to a model in 2D do not seem that big. Instead of one value, the control law must now compute two values: linear acceleration and derivative of the curvature. Furthermore, the system is assumed to stay within the boundaries which guarantee that the lateral and longitudinal controls can be modeled and computed independently.

However, working on a plane introduces notions such as trigonometric functions, curvature, and so on. While this means a modest increase in complexity for a mathematically literate person, those new concepts introduce genuine difficulties for the specifier; e.g, how to model a sine function when co-domains are restricted to integers?

The paper discusses some scaling-up issues when going from a 1D model to a 2D model. We could solve some, most importantly the consistency proofs and the adaptation of the mathematical model. Other issues can be solved but at a high cost, proving global temporal properties is among them. Last, some are yet beyond our reach because the tools cannot deal with the complexity; animation falls in this case for instance.

The paper is structured as follows: Section 2 introduces the notation and semantics for Event-B; Section 3 describes the platoon problem and the model used; Sections 4, 5, 6, 7 and 8 discuss the different aspects of scaling up with Event-B: mathematics, specification structure, temporal properties, tools and process; finally, Section 9 concludes.

2 Event-B Language

Event-B [18,2] is an evolution of the classic B method [1]. Designed for modeling the environment where a piece of software developed with the B-method must execute, Event-B proved to be a good formalism for specifying and reasoning about systems such as concurrent systems, reactive systems, or complex algorithms. Event-B is a state based specification technique. It embodies a process: formal refinement.

Formal Model. A formal model consists of a *state* and *events*. A state is a set of variables constrained by invariants. Values associated to variables are either symbols, integers, or set-theoretic constructions upon those (powersets, relations, functions, etc.). Invariants are expressed as formulae in first-order predicate calculus. Events are guarded generalized substitutions. Guards are formulae on the state and substitutions apply simultaneously on a subset of state's variables.

The semantics of a model is given by a few rules. Substitutions use the weakest precondition calculus of Dijkstra [8]; events must keep the invariant; when several guards are true, the choice of the event to fire is non-deterministic; there must exist a computable initial state. The intuitive behavior of a specification is easy to explain: first, the *INITIALISATION* event is fired, then a cycle begins where: all guards are evaluated, one event is picked among those with a "true" guard, and its substitutions are executed. The cycle ends when no guard is true, which means either that the system has reached a terminal state, or that the system is deadlocked. Infinite cycles are also possible, which could be the correct behavior or indicate a reachability problem for a terminal state.

The formal semantic rules are implemented as *proof obligations*. To *verify* a model, that is, to assess its consistency, we need to discharge all the proof obligations.

Refinement. Event-B allows one to express that a model is a refinement of another, more abstract model. Refinement consists in introducing new variables. An abstraction invariant relates the new variables to the abstract variables. Events from the abstract model can be kept untouched in the refinement, or can be rewritten using the new variables. New events can be introduced too. In practice, it is often useful to think in term of *reification* of variables and of *decomposition* of an event into several smaller ones.

The semantics of refinement is given by proof obligations. Proving a refinement correct amounts to prove that concrete events maintain the invariant of the abstract model, the abstraction invariant, and do not prevent abstract events to be triggered.

The syntactic structure of the language was designed so that the proof obligations can be easily generated and broken into small formulae. The Rodin platform [20] provides the practical framework to carry out modeling in Event-B. It seamlessly integrates modeling and proving, and provides mechanisms for extension and configuration so that it can be tailored to different application domains or development methods.

3 The Platooning Problem

3.1 Platoons

Research on urban mobility systems based on fleets of small electric vehicles stresses the importance of a new moving mode: platooning. A platoon is defined as a convoy of autonomous vehicles which follow exactly the same path and which are spaced at very close distance one from the other.

In this work, we consider platoons formed by a *leader* vehicle and *followers*. Leaders and followers have different control laws. We specify only the follower control law. Its aim is to keep as close as possible to the preceding vehicle while following a virtual ideal track without colliding. We use a model of vehicle where the control can be decomposed into longitudinal (distance with preceding vehicle) and lateral laws. We assume operating conditions such that the two controls can be set independently [7].

There are numerous strategies to form and maintain platoons, characterized by their degree of centralization and the volume of communication. We specify a minimal strategy: no central control and no communication between vehicles other than perception, i.e., a vehicle can sense a few information from the preceding vehicle (distance, speed, etc.). The virtual track is set by the leader. The control is local to each vehicle, based on current state and perceptions. This strategy may not be the most efficient but it is very robust. In particular, it can be used as a fall-back in case of failure in a system using more sophisticated algorithms. Hence the need to guarantee its correctness.

Within this problem setting, platoons can be considered as situated multi-agent systems (MAS) which evolve following the Influence/Reaction model [10,9]. Development of the specification follows a stepwise refinement process based on this model: (i) driving systems perceive, (ii) decisions are taken, and (iii) physical vehicles move.

3.2   Research Goal and System Hypotheses

We aim at modeling formally a pragmatic strategy known as *Daviet-Parent algorithm* [7] in order to prove that implementations enjoy certain properties [12] such as: (i) the model is sound bound-wise, (ii) no collision occurs between the vehicles, (iii) no unhooking occurs, and (iv) no oscillation occurs.

Presently, we focus on two essential safety properties: no collision within a platoon occurs[1] and the soundness is maintained.

Our model is based on the following system hypotheses:

- we consider a set of $N(\geq 2)$ vehicles forming a linear platoon,
- motion of vehicles is limited by fixed bounds on velocity, acceleration, curvature and derivative of curvature,
- we consider forward-only motions on a non self-intersecting track,
- we suppose that the frequency of the control algorithm is the same for all vehicles, so they can be modeled as synchronized,
- sensors are perfect and their accuracy is such that the velocity of the previous vehicle can be precisely known,
- actuators of the engine are perfect.

The hypotheses are strong but not that far from reality considering (1) we are not modeling fault, fault-tolerance, or such matters, (2) near perfect abstract sensors or actuators can be built from merging results of several concrete ones.

3.3   State of Vehicles

In the 1D model, vehicles move on a linear track, equivalent to a rail. The state of the ith vehicle at time t is the pair $(xpos_i(t), speed_i(t))$, where $xpos_i$ represents position on the track and $speed_i$ represents the velocity. Control consists in setting of an acceleration to modulate speed. The behavior law is represented by (1) extracted from [24], where $MaxSpeed$ is the maximum velocity, $accel_i$ is the acceleration, and Δt is the time increment:

$$\begin{cases} n\_speed = speed_i(t) + accel_i(t).\Delta t \\ xpos_i(t+\Delta t) = \begin{cases} xpos_i(t) + MaxSpeed.\Delta t & \text{if } n\_speed > MaxSpeed \\ xpos_i(t) - \frac{speed_i(t)^2}{2.accel_i(t)} & \text{if } n\_speed < 0 \\ \left(\begin{array}{c} xpos_i(t) + speed_i(t).\Delta t \\ + \frac{accel_i(t).\Delta t^2}{2} \end{array} \right) & \text{otherwise} \end{cases} \\ speed_i(t+\Delta t) = \begin{cases} MaxSpeed & \text{if } n\_speed > MaxSpeed \\ 0 & \text{if } n\_speed < 0 \\ n\_speed & \text{otherwise .} \end{cases} \end{cases} \quad (1)$$

---

[1] Collisions between platoons or between a vehicle and an obstacle should of course be considered in a real system. First kind should be taken care by the control law of leaders, second kind is dealt with by lower level emergency systems. Both are outside the scope of this work.

The acceleration $accel_i$ is chosen according to the current state of the ith vehicle and the values sensed on the preceding vehicle.

In the 2D model, vehicles move on a plane. The vehicle state η must model its position, represented by cartesian coordinates (x, y), and its attitude, represented by the orientation θ of vehicle's axis with respect to x-axis. The behavior law now contains a velocity v and a trajectory's curvature κ which are controlled by application of a linear acceleration a and a derivative of the curvature χ. When a control (a, χ) is applied to a state $(x_0, y_0, \theta_0, v_0, \kappa_0)$ at time t for a period Δt, the new state at time $t + \Delta t$ becomes:

$$
\begin{cases}
x = x_0 + \cos\theta_0 F_C(\Delta t, v_0, \kappa_0, a, \chi) - \sin\theta_0 F_S(\Delta t, v_0, \kappa_0, a, \chi) \\
y = y_0 + \cos\theta_0 F_S(\Delta t, v_0, \kappa_0, a, \chi) - \sin\theta_0 F_C(\Delta t, v_0, \kappa_0, a, \chi) \\
\theta = \theta_0 + v_0\kappa_0\Delta t + (a\kappa_0 + v_0\chi)\frac{\Delta t^2}{2} + a\chi\frac{\Delta t^3}{3} \\
v = v_0 + a\Delta t \\
\kappa = \kappa_0 + \chi\Delta t
\end{cases}
\tag{2}
$$

where $F_C(\Delta t, v_0, \kappa_0, a, \chi) = \int_0^{\Delta t}(v_0 + at)\cos(v_0\kappa_0 t + (a\kappa_0 + v_0\chi)t^2/2 + a\chi t^3/3)dt$
and $F_S(\Delta t, v_0, \kappa_0, a, \chi) = \int_0^{\Delta t}(v_0 + at)\sin(v_0\kappa_0 t + (a\kappa_0 + v_0\chi)t^2/2 + a\chi t^3/3)dt$.

At this mathematical level, the increase in complexity is noticeable but not dramatic: most of it boils down to the expansion of standard geometric formulae.

4 Scaling Up with Event-B: Mathematics

4.1 1D Model Adaptation

The formulae in the 1D model are basic arithmetic expressions; they contain no special mathematical functions. The values of $xpos_i$, $speed_i$ and $accel_i$ can easily be modeled as integer numbers. It suffices to choose a system of units small enough to reach the accuracy needed in practice. Hence, they can be expressed straight away in Event-B.

4.2 2D Model Adaptation

By contrast, a simple look at the 2D model shows that we need to transform the model so it can be expressed in Event-B. The most obvious "problems" are the sine and cosine functions (meaningless on Integers) and the integral in F_C and F_S expressions.

The Discretization Issue. The heart of the difficulty lies in the discretization of continuous kinematic values such as position, speed or acceleration. The question is then: Why not use continuous values? We are not ready to answer positively for two reasons.

First reason is practical. Current provers within the B world consider only integer numbers. Even with these "simple" numbers, proofs are often complex and intricate. It is not clear that provers doing a good job with real numbers will be available soon.

Second reason is deeper. Software systems are inherently discrete. Because of numerous latencies in the autonomous car (sensing data, computing controls, driving actuators), the control system will operate at a rather slow frequency. So, the actual system will run as if time is discrete.

The B formal method aims at producing code which is proven to maintain functional invariants. So, we need to introduce the discretization at some point. Our position is that we must introduce this fundamental feature early in the models: as soon as we need "continuous" values in the specification. Of course, we must then develop techniques and strategies to take care of this feature.

Although reasonably simple from a mathematician point of view, the 2D model cannot be translated directly in Event-B. We need to "refine" it. We used three heuristics.

(1) Free Physical Units. In Event-B, the easiest representation of continuous kinematic values such as position, velocity, acceleration is integer numbers. By keeping the physical units unspecified but homogeneous (e.g., the unit of velocity is equal to unit of distance divided by unit of time.), we can adapt the representation to the desired accuracy of the computations. Distances can be millimeters as well as meters, and times can be milliseconds as well as seconds.

(2) Approximate Mathematical Functions. The restriction to Integers of the ranges of sine or cosine is a three value set: not very interesting. To solve this problem, we introduce a special dimensionless constant μ and we consider $\mu\cos\theta$ and $\mu\sin\theta$ instead of $\cos\theta$ and $\sin\theta$. We do the same with F_C and F_S and consider μF_C and μF_S. By choosing a μ with a big value, expressions can be reasonably coded with integers.

Event-B provers know about standard rules of arithmetic but ignore trigonometric or general calculus rules. In order to use the provers, we use Taylor series and identities to transform expressions into arithmetic approximations.

Last, the vehicle state $\eta = (x, y, \theta, v, \kappa)$ is represented by a 6-tuple $(x, y, \gamma^\theta, \sigma^\theta, v, \kappa)$. The values of x, y, v and κ are integers; the units must be taken small enough to obtain a good accuracy. The values of γ^θ and σ^θ are also integers which respectively represent $\mu\cos\theta$ and $\mu\sin\theta$. We define a carrier set POINT to denote the set of all possible vehicle states. The approximate, but accurate, model of 2D platooning is then:

$$\begin{cases} x = x_0 + (\gamma_0^\theta \tilde{F}_C(\Delta t, v_0, \kappa_0, a, \chi) - \sigma_0^\theta \tilde{F}_S(\Delta t, v_0, \kappa_0, a, \chi))/\mu^2 \\ y = y_0 + (\gamma_0^\theta \tilde{F}_S(\Delta t, v_0, \kappa_0, a, \chi) + \sigma_0^\theta \tilde{F}_C(\Delta t, v_0, \kappa_0, a, \chi))/\mu^2 \\ \gamma^\theta = (\mu_C \gamma_0^\theta - \mu_S \sigma_0^\theta)/\mu \\ \sigma^\theta = (\mu_C \sigma_0^\theta - \mu_S \gamma_0^\theta)/\mu \\ v = v_0 + a\Delta t \\ \kappa = \kappa_0 + \chi\Delta t \end{cases} \qquad (3)$$

where $\mu_C = \mu - \beta^2/(2\mu)$, $\mu_S = \beta - \beta^3/(6\mu^2)$ with $\beta = v_0\kappa_0\Delta t + (a\kappa_0 + v_0\chi)\Delta t^2/2 + a\chi\Delta t^3/3$, $\tilde{F}_C(\Delta t, v_0, \kappa_0, a, \chi)$ and $\tilde{F}_S(\Delta t, v_0, \kappa_0, a, \chi)$ expanded as Taylor series.

Event-B translation is straightforward but yields overly long expressions.

(3) Check and Rewrite Mathematical Formulae for Provability. Many properties of formulae on real numbers can be safely assumed when we restrict their use to integer numbers, but not all. Consider the true equality with real numbers: $a * (b/c) = (a * b)/c$. Its equivalent with natural numbers is $a * (b \div c) = (a * b) \div c$. Unfortunately, this equality is not true anymore (hint: $\div$ denotes the integer quotient). So any proof which relies on the equality cannot be discharged anymore.

In the initial 1D platooning model, we found two reviewed goals that were instances of the above example. The "obvious" formula below, straight translation of the 1D mathematical model, introduces non provable proof obligations.

$$xpos0+MAX\_SPEED-(((MAX\_SPEED-speed0)*(MAX\_SPEED-speed0))/(2*accel0))$$

We proved the model by avoiding the problem with the equivalent expression:

$$xpos0 + ((MAX\_SPEED-speed0)*(MAX\_SPEED-speed0))/(2*accel0) +$$
$$speed0*((MAX\_SPEED-speed0)/accel0) +$$
$$MAX\_SPEED*(1-(MAX\_SPEED-speed0)/accel0)$$

5 Scaling Up with Event-B: Specification Structure

An important question when we started the 2D modeling was: can we keep the same development structure as for 1D modeling? We had two reasons. First, a great deal of effort had been put into it so it is intelligible, consistent with the general MAS model, then easy to validate. Second, we can expect proofs structures (and even whole proofs) too to be similar if developments are similar.

We have been able to keep the exact same structure of the development. The same refinements with the same rationales are present in both specifications. In fact, we used the development of 1D specification as a "road-map" for development of the new model.

The structure of the specification [12] consists of an abstract machine Platoon and four refinements. Each development introduces a clearly identified concept. Platoon sets the "vocabulary" and the safety property of interest. Platoon_1 splits the platoon's movement into each vehicles' movements. Platoon_2 implements the physical reaction laws. Platoon_3 introduces the decision step. Platoon_4 introduces the perception step and implements the decision laws.

5.1 Decomposition of Events

During the refinement, the number of events increased much more steeply. A simple pattern explains this explosion.

Both specifications implement the reaction laws in machine Platoon_2 and the decision laws in machine Platoon_4. We need to decompose the abstract events move1 and move in machine Platoon_1, and the abstract events decide1 and decide in machine Platoon_3 into more concrete ones. Let us consider the decomposition of event move.

The mathematical model (1) in 1D indicates that three cases must be considered when computing a new state. This is due to the fact that *speed* is bounded. In Event-B, conditional definitions are expressed by the use of guards. This means that the move event must be decomposed into three events, one for each situation (*speed* reaching lower bound, reaching upper bound, or within bounds.)

All events where *speed* is a parameter are decomposed following the analysis exemplified in Table 1.

In the 2D model, we have to consider two bounded parameters: speed and curvature.

$$\begin{cases} n\_speed_i = speed_i + accel_i.\Delta t \\ n\_\kappa_i \quad = \kappa_i + \chi_i.\Delta t \ . \end{cases} \tag{4}$$

The analysis must then take into account the combination of three cases for n_speed_i and three cases for n_κ_i. So, events are refined into nine following the pattern of Table 2.

Table 1. Decomposition of move event in the 1D model

n_speed_i	< 0	$\in 0..MAX\_SPEED$	$> MAX\_SPEED$
move	move_reduce	move_normal	move_max

Table 2. Decomposition of move event in the 2D model

$n\_speed_i \setminus n\_\kappa_i$	$< -MAX\_\kappa$	$\in -MAX\_\kappa..MAX\_\kappa$	$> MAX\_\kappa$
< 0	move_vmin_κmin	move_vmin_κ	move_vmin_κmax
$\in 0..MAX\_SPEED$	move_v_κmin	move_v_κ	move_v_κmax
$> MAX\_SPEED$	move_vmax_κmin	move_vmax_κ	move_vmax_κmax

5.2 Statistics of the Specifications

The multiplication of events depicted above happened a few times. It should be noted that other refinement strategies for the abstract event move could have been chosen. For instance, we could have kept the refined event unique, but at the expense of very complex guards. Our trade-off lengthens the specification text and increases the number of proof obligations but each proof is much simpler.

Table 3 shows the increase in complexity when passing from the 1D initial model, to the 1D revised model using the technique presented in Sect. 4.2 and augmented with deadlock-freeness, to the 2D model.

Introducing a safety property such as deadlock-freeness has little impact on complexity. While the number of variables roughly doubled when going 2D, all other measures varied by a four-fold increase. Interestingly, the ratio between manually and automatically discharged proof obligations increases just a little: most of the new proof obligations are simple ones. The most important increase is the number of theorems. They are used to ease the proofs by introducing only once standard mathematical properties.

Table 3. Statistics of the specifications

	1D initial model	1D revised model	2D model
Sets	0	0	1
Constants	15	15	50
Axioms	27	27	86
Variables (last refinement)	10	10	16
Invariants	16	17	29
Events (last refinement)	15	15	39
Guards (last refinement)	81	81	354
Theorems	1	4	46
Variants	3	3	3
Automatic POs	187	196	743
Manual POs	29	36	177
Reviewed POs	4	1	0
Undischarged POs	0	0	0
Total POs	220	233	920

This is a consequence of the introduction of more complex arithmetic expressions in the 2D model.

6 Scaling Up with Event-B: Temporal Properties

Temporal properties can be classified into two categories. Safety properties specify that nothing bad will happen. Liveness properties specify that something good will eventually happen [11]. Our model must guarantee a safety property: deadlock-freeness.

This safety property was introduced after a long period of perplexity where we were baffled by observing collisions in our programmed simulations of the exact same model that was verified, i.e., where we had proven that firing any event kept a strictly positive distance between vehicles as invariant [12]. The mystery was lifted when we realized that if a moving vehicle cannot react, i.e., its control system is deadlocked, then it will likely collide with something.

Deadlock-freeness is not well integrated into the Event-B framework. As for many other temporal properties, we can use "tricks". For deadlock-freeness, we build the disjunction of the guards of all events other than *INITIALISATION* and we prove it is a theorem. So, we can be sure that one event at least can always be fired.

The trick works well on small models but does not scale up. In the 1D specification, the deadlock-freeness theorem in machine Platoon_2 is a formula with around 42 lines (7 events times 6 lines per guard). It would be around 390 lines long in the last refinement of the 2D specification (39 events times 10 lines per guard). Two problems arise then. One concerns the management of a proof of a disjunction of 39 cases. Rodin proof explorer is well thought out and we are quite confident that, with time and patience, we could discharge the proof. The second concerns the construction of the formula. Right now, we must rely on a manual cut and paste procedure. Needless to say, the probability of introducing a non obviously detectable error is too high for the result to be trusted. Clearly, we need a tool to build automatically this formula.

7 Scaling Up with Event-B: Tools

Formal methods depend heavily on automated support. They require long, intricate, and generally tedious chains of reasoning to discharge or establish properties, even trivial ones. This is the nature of formal proof systems. Effective tools are not a "nice addition" to a formal method, but a key factor for its deployment. Event-B is supported by Rodin, a framework which integrates gracefully tools to edit, verify and validate models.

7.1 Edition and Verification

The increase in size shown in Table 3 did not pose any problems to Rodin editors, either the native structural editor or pluggable text-editors such as Camille [5]. Likewise, visualization tools such as the LaTeX generator or the pretty-printer were up to the task.

Provers were also able to deal with the increase in complexity. In fact, the general strategy of breaking a verification proof into several smaller proof obligations as used by B spreads the increase in complexity on much more proof obligations, but each one remains reasonably simple. Some proofs were more complex because the model uses more complex formulae, not because it is bigger.

7.2   Animation Plug-In for Validation

Animation is a technique to *execute* specifications. Thus, we can play, experiment and observe the behavior of models. Several tools support the technique [4,13,26,21,23]. The principle of animating an Event-B specification is a simple three-step process:

1. the user gives values to the constants and carrier sets in the contexts,
2. the *INITIALISATION* event is fired to set the system in its initial state,
3. the animator enters a loop:
   (a) compute the guard of all events, enable those for which the guard is true. When events are parameterized, pick one value, if any, which makes the guard true,
   (b) the specifier fires one of the enabled event; the substitutions are computed,
   (c) check if the invariants still hold [optional].

The computation of the invariant is superfluous when the animated specification has been fully proven. However, it is a very valuable feature when animation is used on unproven specifications, in particular to check potential candidate for invariant formulae.

It may sound strange to use anecdotal observations in a context where mathematical proofs are pivotal to the method. However, we firmly believe that such semi-formal activities are useful, and even sometimes necessary, for three reasons.

The first reason relates to the notion of *validation*. Proofs show that a particular text is logically consistent and that the last model in a sequence of refinements is a correct concretization of the initial model. However, proofs do not tell if a model is an adequate description of the desired behavior of the system. Animation exhibits behavior.

The second reason concerns temporal properties. Not all properties can be expressed in Event-B. Animation can then be used to "test" the specification for certain properties. We can set up scenarios and look if the system goes only through safe states. Like tests, animation does not prove correctness but shows errors.

The last reason is that animation is a good, practical, tool to get deep insights on complex specifications. Actually, animation was mainly invented for this reason [3].

We used intensively animation to understand the collision problem in the 1D platooning specification. In particular, animation helped us to understand which values lead to deadlocks. From those, we could abstract to general configurations, and then relate to parts of the deadlock-freeness theorem.

We also found out, in another work [16], that animation could be a reasonably cheap way to get a correct refinement of an abstract fact into a complex behavior. Animation helped to define the guards and the explicit coding of causal order required by Event-B.

7.3   Breaking Animators

The positive experience with 1D induced us to use animation early in the 2D model development. Unfortunately, animators failed us even on the first refinement.

We tried with two different animators, Brama[2] and ProB [14]. In both cases, the notion of POINT (3) was the first visible obstruction. We needed to code it, either crudely as integers with Brama or more abstractedly as symbols with ProB. Either way, a list should be provided by hand. This is not realistic and even meaningful.

One way to get rid of POINTs is to refine them as their coordinates. Each coordinate is an integer function which could be individually managed but six of them create too complex a space. Brama seems to enumerate values, ProB uses more sophisticated constraint solving techniques; both strategies fail on the 6-dimensions space.

This can be explained by two important features of Event-B: non-determinism and definition of values by their properties. Animators are then more oriented toward "picking" values rather than "computing" values. This orientation is fine most of the time, but it should not be exclusive. Sometimes, even in abstract specifications, we know some expressions are deterministic computations: kinematic functions are a good example. In those cases, we would appreciate to be able to tell this to the animator.

In the transformations we developed to make specifications "animatable" [17], we have some ad-hoc heuristics which force a computation. Their major drawback is to make the text of the guards and the substitutions much more complex. While in principle they are applicable, we have not yet tried them in the 2D specification. We lack the automated editors required by the number of expressions to transform.

8   Scaling Up with Event-B: Process

Like for any complex artifact, we need a precise and definite process to build *good* formal specifications. A *good* formal specification should have the following properties: (a) it is logically consistent, (b) it has proven functional properties, (c) it meets non-functional properties, and (d) it is a reasonable model of the problem.

Formal refinement is the keystone around which the B-method is designed. Its embodiment into the language and the support tools allows one to develop pieces of software where an implementation is proven against its specification. Refinements break down the verification process into discharging many, but small, proof obligations. So issues (a) and (b) are well taken care of. Event-B uses the same strategy. Here, models are complexified while retaining the same functional properties.

To deal with the issues (c) and (d), i.e., the validation of the specification, we have defined an extended refinement process depicted in Fig. 1.

The idea is to associate validation activities to formal refinement steps. A development step is then composed of four activities:

1. refinement of the physical/mathematical model. The mathematical expressions are refined so that they can be translated into Event-B and lead to provable properties. Discretization is studied at this stage.
2. formal refinement of the Event-B specification with all proof obligations discharged.
3. animation of the specification. The specification is transformed in order to be animatable [17], scenarios are elaborated, and behavior is observed with the animator.

---

[2] http://www.brama.fr

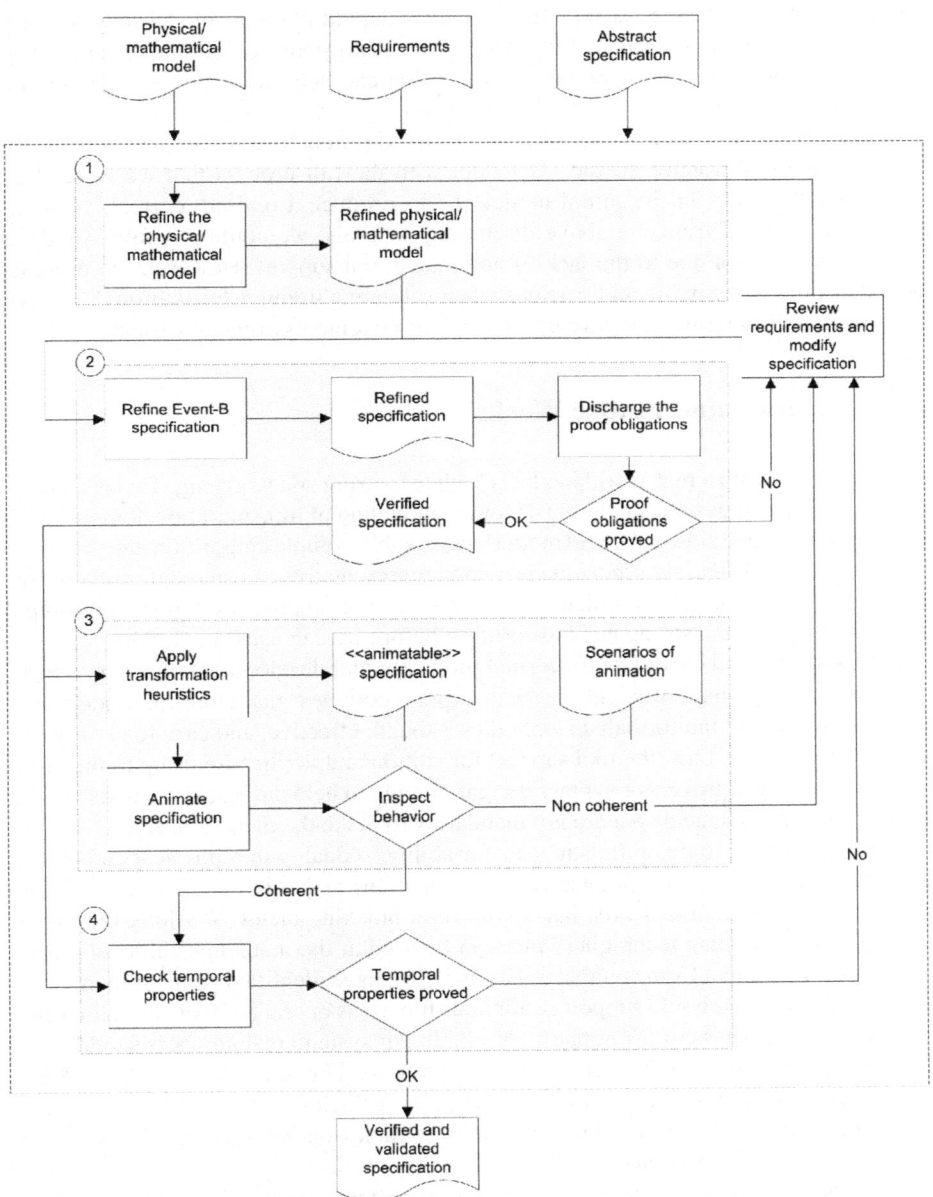

Fig. 1. A step of development process

4. analysis and refinement of temporal properties. This activity leads to the introduction of variants and theorems for temporal properties and their proof.

The order of the activities is important. A mis-adapted physical model may lead to non provable, although correct, formulae. There is no point to validate an unproven specification as it may be inconsistent. Animation can help get better insights on the temporal constraints which need to be formalized.

The process was successfully tested with the 1D model. Actually, it was through replaying the construction of the 1D model with the full process that we were able to identify the flaws in the initial model and to produce a revised, correct, version. As can be inferred from the above discussion on tools, we could not fully validate the 2D specification due to the lack of automated tool support. However, the process helped us the ask the pertinent questions when validating through "walkthrough" of the specification. Furthermore, it gave us ideas on improvements of the environment.

9   Conclusion and Future Works

Our experience with a real-world model is both reassuring and worrying. This is consistent with our findings on using Event-B for the modeling of transportation domain [15]. On the very positive side: we could model a reasonably complex algorithm and prove its correctness. *Daviet-Parent algorithm* is a good representative of a class of problems of great practical importance: problems for which we have empirical solutions, prototype implementation, and a strong need to certify it before we can use it in practice.

Event-B is a good candidate for formal modeling and development of real systems. First, the language has sufficient power to express complex mathematical models and algorithms. Second, the formalism embodies a sound, effective, and easy to use refinement based process. Last, the tool support for edition and verification is up to the task. Of course, stronger provers or syntactic sugar-coating to help navigate long texts would be welcome improvements but are not mandatory to make the method usable.

A second reason to be optimistic was our ability to deal with a physical and mathematical model which incorporates complex functions and continuous time. This has required some sweat and efforts, but was never a blocking factor. We think that there are a few "conditioning techniques" that can be used at the mathematical level to put continuous model in a form suitable to Event-B. We have identified some of them.

Whether Event-B should support continuous functions or real numbers is an interesting question. The answer may not be clear-cut. In our system, real numbers would have eased the writing and maybe some consistency proofs. However, they would not help much improving an implementation because actual vehicles will operate on a discrete time. The control software of actual prototype vehicles operates around 20 Hz. That makes for quite a discrete time.

The worrying issue lies with the checking of temporal properties, either formally through proofs, or pragmatically through animation.

On the formal side, the current situation is not adequate. Only "coarse" properties can be expressed and even then, awkwardly. This is clearly an area where research is needed. A way to overcome this limitation is to associate B or Event-B with another formalism which supports temporal modelling. CSP||B proposed in [22,25] is a candidate. Experiments [6,19] conducted on the specification of the platoon problem indicate a good potential. However, two big issues are still opened: the automation of the proof of consistency between CSP and B parts, and the refinement divergence between CSP and B. Whether CSP||B will scale up soon to realistic models is not yet clear.

On the pragmatic side, we need better animators. Such tools have a very important property: they act on the specification itself. We can be confident that observing animations is observing the model's behavior. It is not clear that the failures to use animators that we have identified can be overcome soon. We are currently working on the idea of translating the specification into an executable language like C or MATLAB. Such translators could be used as a "fall-back" when standard animation fails. They will not exhibit the real model's behavior but a reasonably close version.

Acknowledgments. We are indebted to our colleagues from MAIA and TRIO groups at LORIA with special thanks to A. Scheuer for his development of the 2D model.

References

1. Abrial, J.R.: The B Book. Cambridge University Press, Cambridge (1996)
2. Abrial, J.R.: Modeling in Event-B: System and Software Engineering. Cambridge University Press, Cambridge (2010)
3. Balzer, R.M., Goldman, N.M., Wile, D.S.: Operational specification as the basis for rapid prototyping. SIGSOFT Softw. Eng. Notes 7(5), 3–16 (1982)
4. Bendisposto, J., Leuschel, M., Ligot, O., Samia, M.: La validation de modèles Event-B avec le plug-in ProB pour RODIN. Technique et Science Informatiques 27(8), 1065–1084 (2008)
5. Bendisposto, J., Fritz, F., Leuschel, M.: Developing Camille, a Text Editor for Rodin. In: Proc. Workshop on Tool Building in Formal Methods. colocated with ABZ Conference, Orford, Canada (2010)
6. Colin, S., Lanoix, A., Kouchnarenko, O., Souquières, J.: Using CSP||B Components: Application to a Platoon of Vehicles. In: Cofer, D., Fantechi, A. (eds.) FMICS 2008. LNCS, vol. 5596, pp. 103–118. Springer, Heidelberg (2009)
7. Daviet, P., Parent, M.: Longitudinal and Lateral Servoing of Vehicles in a Platoon. In: Proceeding of the IEEE Intelligent Vehicles Symposium, pp. 41–46 (1996)
8. Dijkstra, E.W.: A Discipline of Programming. Prentice Hall, Englewood Cliffs (1976)
9. Ferber, J.: Multi-Agent Systems: An Introduction to Distributed Artificial Intelligence. Addison-Wesley Professional, Reading (1999)
10. Ferber, J., Muller, J.P.: Influences and Reaction: a Model of Situated Multiagent Systems. In: 2nd Int. Conf. on Multi-agent Systems, pp. 72–79 (1996)
11. Lamport, L.: Proving the Correctness of Multiprocess Programs. IEEE Transactions on Software Engineering 3(2), 125–143 (1977)
12. Lanoix, A.: Event-B Specification of a Situated Multi-Agent System: Study of a Platoon of Vehicles. In: 2nd IFIP/IEEE International Symposium on Theoretical Aspects of Software Engineering (TASE), pp. 297–304. IEEE Computer Society, Los Alamitos (2008)
13. Leuschel, M., Adhianto, L., Butler, M., Ferreira, C., Mikhailov, L.: Animation and Model Checking of CSP and B using Prolog Technology. In: Proceedings of the ACM Sigplan Workshop on Verification and Computational Logic VCL 2001, pp. 97–109 (2001)
14. Leuschel, M., Butler, M.: ProB: An Automated Analysis Toolset for the B Method. STTT 10(2), 185–203 (2008)

15. Mashkoor, A., Jacquot, J.P.: Domain Engineering with Event-B: Some Lessons We Learned. In: 18th International Requirements Engineering Conference - RE 2010, pp. 252–261. IEEE, Sydney (2010)

16. Mashkoor, A., Jacquot, J.P., Souquières, J.: B événementiel pour la modélisation du domaine: application au transport. In: Approches Formelles dans l'Assistance au Développement de Logiciels (AFADL 2009), France, Toulouse, p. 19 (2009), http://hal.inria.fr/inria-00326355/en/

17. Mashkoor, A., Jacquot, J.P., Souquières, J.: Transformation Heuristics for Formal Requirements Validation by Animation. In: 2nd International Workshop on the Certification of Safety-Critical Software Controlled Systems - SafeCert 2009, Royaume-Uni York (2009), http://hal.inria.fr/inria-00374082/en/

18. Metayer, C., Voisin, L.: The Event-B Mathematical Language (October 2007)

19. Nguyen, H.N., Jacquot, J.P.: A Tool for Checking CSP∥B Specifications. In: Proc. Workshop on Tool Building in Formal Methods. colocated with ABZ Conference, Orford, Canada (2010)

20. RODIN: Rigorous Open Development Environment for Complex Systems. website (August 2007), http://rodin-b-sharp.sourceforge.net

21. Schmid, R., Ryser, J., Berner, S., Glinz, M., Reutemann, R., Fahr, E.: A Survey of Simulation Tools for Requirements Engineering. Tech. Rep. 2000.06, University of Zurich (2000)

22. Schneider, S., Treharne, H.: Communicating B machines. In: Bert, D., Bowen, J.P., Henson, M.C., Robinson, K. (eds.) B 2002 and ZB 2002. LNCS, vol. 2272, pp. 416–435. Springer, Heidelberg (2002)

23. Siddiqi, J.I., Morrey, I.C., Roast, C.R., Ozcan, M.B.: Towards quality requirements via animated formal specifications. Ann. Softw. Eng. 3, 131–155 (1997)

24. Simonin, O., Lanoix, A., Colin, S., Scheuer, A., Charpillet, F.: Generic Expression in B of the Influence/Reaction Model: Specifying and Verifying Situated Multi-Agent Systems. INRIA Research Report 6304, INRIA (September 2007), http://hal.inria.fr/inria-00173876/en/

25. Treharne, H.: Combining Control Executives and Software Specifications. Ph.D. thesis, University of London (2000)

26. Van, H.T., van Lamsweerde, A., Massonet, P., Ponsard, C.: Goal-Oriented Requirements Animation. In: RE 2004: Proceedings of the Requirements Engineering Conference, 12th IEEE International, pp. 218–228. IEEE Computer Society, Washington, DC (2004)

D-Finder 2: Towards Efficient Correctness of Incremental Design

Saddek Bensalem[1], Andreas Griesmayer[1], Axel Legay[3], Thanh-Hung Nguyen[1], Joseph Sifakis[1], and Rongjie Yan[1,2]

[1] Verimag Laboratory, Université Joseph Fourier Grenoble, CNRS
[2] State Key Laboratory of Computer Science, Institute of Software, CAS, Beijing
[3] INRIA/IRISA, Rennes

Abstract. *D-Finder 2* is a new tool for deadlock detection in concurrent systems based on effective invariant computation to approximate the effects of interactions among modules. It is part of the BIP framework, which provides various tools centered on a component-based language for incremental design. The presented tool shares its theoretical roots with a previous implementation, but was completely rewritten to take advantage of a new version of BIP and various new results on the theory of invariant computation. The improvements are demonstrated by comparison with previous work and reports on new results on a practical case study.

1 Context

Language. *D-Finder 2* is part of a framework of tools that share a common language, *BIP*, to describe component-based systems [1]. The language is based on *atomic components* and *connectors* to describe their interactions. Components can also be hierarchically organized to build new components. An atomic component is a transition system $B = (L, P, \mathcal{T})$, where $L = \{l_1, l_2, \ldots, l_k\}$ is a set of control locations, P is a set of ports, and $\mathcal{T} \subseteq L \times P \times L$ is a set of transitions. A component additionally can contain data and use C code for actions and conditions on the transitions to manipulate this data. Figure 1 shows a graphical representation of two atomic components B_1 and B_2. We use cycles for locations and arrows for transitions. Every transition is labeled by a port to synchronize with ports of other components to create interactions. In the example, the ports *trigger* and *tick* of the two components are synchronized, which means the corresponding transitions have to be executed concurrently, and are only available if the guards in both components are fulfilled. The transition *rel* can be taken whenever a component is in the *fire* location. We can give only a very brief description of *BIP* here, please refer to [1] for more details.

Verification. Previous work [3,4] introduced an efficient verification method for the models above. Key to this method is the approximation of the reachable states by compositional invariant computation based on (1) *component invariants* Φ_i that capture the constraints on local data of a component B_i, and (2)

M. Bobaru et al. (Eds.): NFM 2011, LNCS 6617, pp. 453–458, 2011.

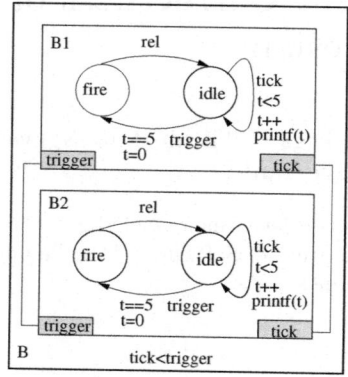

Fig. 1. A BIP model

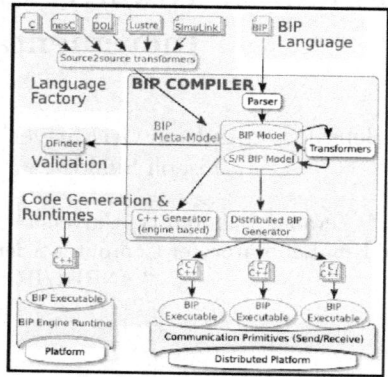

Fig. 2. BIP tools and work-flow

the *interaction invariant* Ψ, which captures constraints on the global state space induced by the synchronization. More formally, we have the following rule:

$$\frac{\{B_i < \Phi_i >\}_i, \ \Psi \in II(\|_\gamma\{B_i\}_i, \{\Phi_i\}_i), \ (\bigwedge_i \Phi_i) \wedge \Psi \Rightarrow \Phi}{\|_\gamma\{B_i\}_i < \Phi >}$$

The rule states that if all components B_i fulfill their respective *component invariants* Φ_i, the composition of all components $II(\|_\gamma\{B_i\}_i, \{\Phi_i\}_i)$ with the interactions γ fulfills an *interaction invariant* Ψ, and if furthermore the conjunction of the invariants $(\bigwedge_i \Phi_i) \wedge \Psi$ implies a predicate on the global system Φ, then also the global system $\|_\gamma\{B_i\}_i$ itself fulfills Φ. In this paper we concentrate on global deadlock-freedom. Indeed, it suffices to prove the invariance of the predicate $\neg\mathcal{DIS}$, where $\mathcal{DIS}$ is the set of states of the system from which all the interactions are disabled.

Tool Chain. The design flow between *BIP* and *D-Finder* is sketched in Figure 2. The framework allows to (1) start from scratch and describe a composite system with the BIP language, or (2) to use the *Language Factory* to translate existing models described in languages such as C, DOL [15] or Simulink [13] into the BIP framework. These models then are used for validation, verification, model to model transformation and eventually generation of C++ code for simulation or deployment. *D-Finder* plays a central role in this process to verify the initial models as well as ensuring correctness after transformation steps.

2   D-Finder 2

Recently, *BIP* has been updated and enriched with new features to improve the modeling process for building hierarchical models and add new interactions in an incremental manner. Furthermore, since the tool presentation in [4], new, more efficient techniques for computing Ψ were introduced in [6,2]. To show the results of unifying those recent developments, this paper presents *D-Finder 2*, the

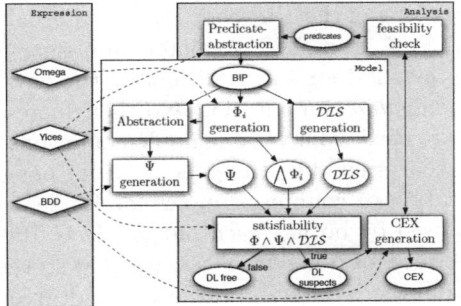

```
#> dfinder -f p1000.bip --incr_file incr_15.incr
   --method=pm --analysis=dl
# overall analysis                                 :
#   compute II using incremental pm                :
#     Eliminate Variables Abstraction(Phil...      :
#       Compute CI for Philosopher                 : 0:01
#     Eliminate Variables Abstraction(Phil...      : 0:02
...
#     get common locations                         : 0:03
#     compute BBCs[0]                               : 0:01
...
#     integrate for increment[1]                    : 0:00
...
#     dual compuatation                            : 0:00
#     concretization                               : 0:02
#   compute II using incremental pm                : 0:41
#   incremental DIS                                : 0:24
    Found 1 deadlocks:
# overall analysis                                 : 1:07
```

Fig. 3. Structure of the D-Finder tool   **Fig. 4.** Call from the command line

second edition of the *D-Finder* tool-set. The tool has been entirely rewritten and new techniques for computing invariants have been implemented in a modular manner.

2.1 Computing Interaction Invariants in an Incremental Manner: the Theory behind *D-Finder 2*

D-Finder 2 implements new efficient techniques for computing Ψ that were recently introduced in [6,2]. Those techniques build on the new concept of *Boolean Behavioral Constraints* (BBCs) that allow to relate the communication between different components with their internal transitions and hence model a unified invariant of the model. Solutions of BBCs can be used to symbolically compute a strong interaction invariant. There are two different techniques that exploit BBCs: (1) a symbolic computation based on a *Fixed-Point iteration* (FP), and (2) a symbolic algorithm to solve the BBCs using so called *Positive Mapping* (PM). Both methods allow an efficient implementation for computing interaction invariants using BDDs and show their strengths for different topologies of the model to check. The main advantage of the two aforementioned techniques is that they allow to exploit the component based design of *BIP* and compute interaction invariants incrementally. In the *Incremental Fixed Point* (IFP) and *Incremental Positive Mapping* (IPM) methods, *D-Finder 2* partitions the model into subsystems (also called increments). The internal interactions in these subsystems are used to compute "partial" interaction invariants. Relations between different increments are considered in a second step and used to integrate the intermediate results to the final Ψ. Computing the global interaction invariant from smaller intermediate results allows to reduce the size of the data structures involved in the computation.

2.2 Implementation Details

D-Finder 2 was developed with modularity and extensibility in mind. The tool is written in Java and uses external tools and native code via the Java Native Interface (JNI) for computations. Fig. 3 gives an overview of the main modules of the tool. The *Model* block handles the parsing of the *BIP* code into an internal model and provides the means to compute Φ, Ψ, and $\mathcal{DIS}$. Available implementations comprise the methods from the previous tool and additionally the new algorithms using *fixed-point* and *positive mapping* computation and their incremental versions. The results from *Model* are used from various implementations of the *Analysis* block, which perform further steps like the generation of possible deadlocks and, most recently, generation of counterexamples for Boolean systems (CEX). The *Expression* block is used by both *Model* and *Analysis*. Its main purpose is to provide an uniform interface for different back-ends that store the actual expressions. The abstraction from the back-ends allows a high degree of flexibility for implementing the algorithms. The most general implementation is a wrapper to the actual parse tree representation of the expressions (using the Eclipse Modeling Framework, EMF) and uses external tools for computations. For algorithms on Boolean variables, like computation of Ψ, a more succinct implementation with BDDs as back-end is used, while large systems that incorporate non-Boolean data require to directly create and maintain input files for an SMT solver on disk. These different versions of expressions can be used interchangeably in many contexts, with the respective tools being called transparently for actual computations. The *Expression* block also provides methods to translate between representation and manage the scopes of variables. Fig. 3 shows the use of external tools for models with non-Boolean data; for models with only Boolean variables, BDDs are used in all computation steps.

The three main blocks are complemented by a common configuration module that reads settings from default values, configuration files and command line and provides the means to instantiate the proper modules for an example to check. The used tools for SMT solving (Yices, [16]) and variable quantification (omega library, [12] are accessed using wrappers, giving rise for easy extension and replacement. Similarly, the used BDD-Manager (JavaBDD, [11]) provides a Java implementation, but has the option to use native BDD managers on supported machines. Currently we use CUDD [14] on Linux and OS X. This flexibility provides the means to develop and maintain new and experimental algorithms in the tool while leaving the main behavior intact, which is currently done, e.g., for experimental modules to perform predicate abstraction to create Boolean systems, check the reachability of deadlocks to remove false positives, and to construct error traces to understand the causes of reachable deadlocks, all of which were not present in the previous tool.

2.3   Availability of the Tool and Example of Use

The *D-Finder 2* and *BIP* tools, along with the examples discussed in this paper, can be freely downloaded from [9] and [7] respectively. An excerpt of a call to *D-Finder 2* for the case of dining philosophers is given in Figure 4. The example shows verification of a problem size of 1000 Philosophers partitioned into 20 increments. The first step is the computation of an abstraction to remove the variables for Ψ computation (using the post conditions from Φ computation to split the states), followed by the local computations (BBC) for each of the

Table 1. Comparison of D-Finder versions. Times in min, timeout one hour.

System Information				D-Finder 1	D-Finder 2			
scale	comps	locs	intrs	Enum	PM	FP	IPM	IFP
Dining Philosopher								
100 philos	200	600	500	0:06	0:09	-	0:03	0:21
500 philos	1000	3000	2500	1:51	3:32	-	0:22	3:09
1000 philos	2000	6000	5000	7:08	14:57	-	0:50	19:05
1500 philos	3000	9000	7500	19:30	34:23	-	1:34	-
3000 philos	6000	18000	15000	-	-	-	4:57	-
Gas Station								
300 pumps	3301	12902	12000	33:02	36:01	11:32	2:03	4:18
400 pumps	4401	17202	16000	-	-	21:40	3:41	10:30
500 pumps	5501	21502	20000	-	-	-	5:48	20:05
ATM System								
2 atms	6	48	38	0:59	0:05	0:02	0:02	0:02
20 atms	42	444	362	-	1:12	1:00	0:43	1:13
50 atms	102	1104	902	-	7:14	8:00	1:57	11:22
100 atms	202	2204	1802	-	-	-	4:60	-
200 atms	402	4404	3602	-	-	-	17:07	-

Table 2. Verification times for the Dala robot

module	comps	locs	intrs	vars	time	
					D-Finder 1	D-Finder 2
RFLEX	56	308	227	35	9:39	3:07
NDD	27	152	117	27	8:16	1:15
SICK	43	213	202	29	1:22	1:04
Aspect	29	160	117	21	0:39	0:21
Antenna	20	97	73	23	0:14	0:13
Combined	198	926	724	132	-	5:05

increments and their integration. Computation of the dual and mapping to the concrete values finishes the computation of Ψ, which is used to directly compute the intersection with $\mathcal{DIS}$. Finally, the tool successfully reports one deadlock.

Large examples from real world applications may require manual assumptions on the components to rule out false positives. *D-Finder 2* supports these additional inputs on the component and global level. To support organization of the required models, specifications, and output files, the tool supports so called example configuration files, which allow to collect the required files in own directories. The examples and case studies on the web site are organized in this way. The Web page of *BIP* [7] gives more information to introduce the language as well as details on usage and case studies. The web page of *D-Finder 2* [9] comes up with illustrations of the use of the tool as well as many other case studies of huge size. The sites also reference a series of publications that give more details on the theory implemented in the tool.

3 Experimental Results

We compare the performance of the original version of *D-Finder* with the new version of the tool presented in this paper on some case studies (see [9] for more experiments). Experiments where conducted with a 32Bit Linux on Xeon 2.67GHz We started by considering verification of deadlock properties for the classical case studies of Dining Philosopher, the Gas Station [10], for which we assume that every pump has 10 customers, and the Automatic Teller Machine (ATM) [8]. The results are given in Table 1, where *scale* is the parameter of the example, *comps* the number of components, *locs* the number of control locations, and *intrs* the total number of interactions. The experiments where performed on Mac-Book Pro laptops with CUDD as back end for BDD computations. We see that especially the incremental versions of the new Ψ computation methods led to major improvements in run time compared to the original version of the tool from [4].

To demonstrate the application of *D-Finder 2* to industrial problems we want to refer to a case study on the application of *BIP* to autonomous robots *Dala* [5].

The case study uses software written in *Genom*, a tool to design modular real time software architectures, which were then translated to *BIP*. The translated modules implement features of the robot like movement (RFLEX), navigation (NDD) and self localization using a laser range finder (SICK), and themselves consist of more internal components, for more details see [5]. The case study shows how to use *BIP* code generation tools and the *BIP engine* to create C code from the model, which runs at the functional level of the robot to guarantee coordination of the various modules in a correct manner. The previous tool was not able to verify this model. And while prototypes were able to show the deadlock-freedom of single modules in the past, only *D-Finder 2* allowed us recently to verify the combination of all five main components. (Results are reported in Table 2). This use of the work flow of *D-Finder 2* and *BIP* is a major change with respect to other methodologies to design autonomous robots. Indeed, most of other existing works propose functional levels that are designed manually, without any formal guarantee of correctness.

References

1. Basu, A., Bozga, M., Sifakis, J.: Modeling heterogeneous real-time components in BIP. In: SEFM, Washington, DC, USA, pp. 3–12. IEEE, Los Alamitos (2006)
2. Bensalem, S., Bogza, M., Legay, A., Nguyen, T.-H., Sifakis, J., Yan, R.: Incremental component-based construction and verification using invariants. In: FMCAD (2010)
3. Bensalem, S., Bozga, M., Nguyen, T.-H., Sifakis, J.: Compositional verification for component-based systems and application. In: Cha, S(S.), Choi, J.-Y., Kim, M., Lee, I., Viswanathan, M. (eds.) ATVA 2008. LNCS, vol. 5311, pp. 64–79. Springer, Heidelberg (2008)
4. Bensalem, S., Bozga, M., Nguyen, T.-H., Sifakis, J.: D-finder: A tool for compositional deadlock detection and verification. In: Bouajjani, A., Maler, O. (eds.) CAV 2009. LNCS, vol. 5643, pp. 614–619. Springer, Heidelberg (2009)
5. Bensalem, S., de Silva, L., Gallien, M., Ingrand, F., Yan, R.: Rock solid software: A verifiable and correct by construction controller for rover and spacecraft functional layers. In: ISAIRAS, pp. 859–866 (2010)
6. Bensalem, S., Legay, A., Nguyen, T.-H., Sifakis, J., Yan, R.: Incremental invariant generation for compositional design. In: TASE, pp. 157–167 (2010)
7. BIP tool page, http://www-verimag.imag.fr/BIP-Tools,93.html
8. Chaudron, M.R.V., Eskenazi, E.M., Fioukov, A.V., Hammer, D.K.: A framework for formal component-based software architecting. In: SVCS (2001)
9. DFinder tool page, http://www-verimag.imag.fr/dfinder/
10. Heimbold, D., Luckham, D.: Debugging Ada tasking programs. IEEE Softw. 2(2), 47–57 (1985)
11. JavaBDD tool page, http://javabdd.sourceforge.net/
12. Omega library tool page, http://www.cs.umd.edu/projects/omega/
13. Simulink, http://www.mathworks.com/products/simulink/
14. Somenzi, F.: CUDD tool page, http://vlsi.colorado.edu/~fabio/CUDD/
15. Thiele, L., Bacivarov, I., Haid, W., Huang, K.: Mapping applications to tiled multiprocessor embedded systems. In: ACSD, pp. 29–40. IEEE, Los Alamitos (2007)
16. Yices tool page, http://yices.csl.sri.com/

Infer: An Automatic Program Verifier for Memory Safety of C Programs

Cristiano Calcagno and Dino Distefano

Monoidics Ltd, UK

Abstract. Infer[1] is a new automatic program verification tool aimed at proving memory safety of C programs. It attempts to build a compositional proof of the program at hand by composing proofs of its constituent modules (functions/procedures). Bugs are extracted from failures of proof attempts. We describe the main features of Infer and some of the main ideas behind it.

1 Introduction

Proving memory safety has been traditionally a core challenge in program verification and static analysis due to the high complexity of reasoning about pointer manipulations and the heap. Recent years have seen an increasing interest of the scientific community for developing reasoning and analysis techniques for the heap. One of the several advances is separation logic, a formalism for reasoning about mutable data structures and pointers [14], and based on that, techniques for automatic verification. Infer is a commercial program analyzer aimed at the verification of memory safety of C code. Infer combines many recent advances in automatic verification with separation logic. Some of the features are:

- It performs *deep-heap analysis* (a.k.a. shape analysis) in presence of dynamic memory allocation. Infer's abstract domain can precisely reason about complex dynamic allocated data structures such as singly/doubly linked lists, being them circular or non-circular, nested with other lists, etc.
- It is *sound* w.r.t. the underlying model of separation logic. Infer synthesizes sound Hoare triples which imply memory safety w.r.t. that model.
- It is *scalable*. Infer implements a compositional inter-procedural analysis and has been applied to several large software projects containing up to several millions of lines of code (e.g. the Linux kernel).
- It is completely *automatic*: the user is not required to add any annotations or modify the original source code. Moreover, for large software projects, Infer exploits the information in the project's build to perform the analysis.
- It can analyze *incomplete code*. Infer can be applied to a piece of code in isolation, independently from the context where the code will be used.

---

[1] Special thanks to Dean Armitage, Tim Lownie, and John Lownie from Monoidics USA, Richard Retting and Bill Marjerison from Monoidics Japan, and Hongseok Yang, for their invaluable contributions to Infer.

M. Bobaru et al. (Eds.): NFM 2011, LNCS 6617, pp. 459–465, 2011.

Being an automatic program verifier, when run, Infer attempts to build a proof of memory safety of the program. The outcomes of the attempt can be several:

- Hoare triples of some procedures are found. Then, because of soundness, one can conclude that those procedures will not make erroneous use of memory in any of their executions. If a triple for the top level procedure (e.g., main) is found, one can conclude the memory safety of the entire program.
- The proof attempt fails for some procedures. Infer extracts from this failed attempt the possible reasons which prevented it to establish memory safety. These findings are then returned to the user in the form of a bug report.
- For some procedures the proof attempt fails due to internal limitations of Infer (e.g., expressivity of the abstract domain, or excessive over-approximation). In this case nothing can be concluded for those procedures.

Infer's theoretical foundations are mainly based on [5], but also include techniques from [8,2,4,11]. This paper focusses on the tool mainly from a user perspective. We refer the interested reader to the above articles for the underlying theory.

2   Procedure-Local Bugs

In this section we illustrate Infer's concept of *procedure-local* bugs by example, focusing on memory leaks in the context of incomplete code (e.g. without main) and inter-procedural reasoning. The usual notion that all memory allocated must be eventually freed does not apply in the context of incomplete code. The question then arises of what is a memory leak in this context and how to assign blame. We will apply the following general principle: when a new object is allocated during the execution of a procedure, it is the procedure's responsibility to either deallocate the object or make it available to its callers; there is no such obligation for objects received from the caller.

Consider the function alloc0() in Figure 1. It allocates an integer cell and stores its address into i when the flag b is true, or sets i to zero when b is false. This function by itself does not cause memory leaks, because it returns the newly allocated cell to the caller using the reference parameter i. It is the caller's responsibility to make good use of the returned cell. example1() shows a first example of procedure-local bug. The first call to alloc0() sets the local variable i to zero. However, after the second call, i will point to a newly allocated integer cell. This cell is then leaked since it is not freed before example1() completes. Infer blames example1() for leaking the cell pointed to by i. The bug is fixed in example2() where i is returned to the caller. It becomes the caller's responsibility to manage the cell, perhaps by freeing it, perhaps by making it available to its own caller. This passing of responsibility carries on, up the call chain, as long as source code is available. In the extreme case, when the whole program with a main function is available, we recover the usual (global) notion of memory leak. Even in that case, it is important to blame the appropriate procedure. A more subtle leak is present in example3(), which is a slight modification of example2(). As in the previous case, after the second call to alloc0(), i points

```
void alloc0(int **i, int b) {        int *example3() {
  if (b) *i = malloc(sizeof (int));    int *i;
  else *i = 0;                         alloc0(&i, 0); // ok
}                                      alloc0(&i, 1); // ok, malloc
                                       alloc0(&i, 1); // leak: i overwritten
void example1() {                      return i; }
  int *i;
  alloc0(&i, 0); //ok                int *global;
  alloc0(&i, 1); // memory leak
}                                    int *example4() {
                                       int *i;
int *example2() {                      alloc0(&i, 0); // ok
  int *i;                              alloc0(&i, 1); // ok, malloc
  alloc0(&i, 0); // ok                 global = i;
  alloc0(&i, 1); // ok, malloc         alloc0(&i, 1); // ok, i in global
  return i; // no memory leak }        return i; }
```

Fig. 1. Examples of procedure-local bugs

to a newly allocated cell. However, the third call to `alloc0()` creates a second cell and makes the first one unreachable and therefore leaked. The problem is fixed in `example4()` where the first cell is stored in a global variable, and the second one passed to the caller. Hence, `example4()` does not leak any memory.

Specifications. Infer automatically discovers specs for the functions that can be proven to be memory safe (in this case, those which do not leak memory). The spec discovered for `example2()` is[2] $\{emp\}$ `example2()` $\{ret \mapsto -\}$ meaning that the return value points to an allocated cell which did not exist in the (empty) precondition. The spec discovered for `example4()` is

$$\{\&global \mapsto -\} \; example4() \; \{\exists x. \&global \mapsto x \; * \; x \mapsto - \; * \; ret \mapsto -\}$$

meaning that variable `global` in the postcondition contains the address x of some memory cell, and the return value points to a separate cell. Notice that a call to `example4()` could produce a leak if `global` were overwritten. In that case, since the leaked cell exists before the call, Infer would blame the caller.

3   Infer

3.1   Bi-abduction and Compositional Analysis

The theoretical notion used by Infer to automatically synthesize specifications is *bi-abductive inference* [5]. It consists in solving the following extension of entailment problem: $H * X \vdash H' * Y$. Here H and H' are given formulae in separation

---

[2] We use the standard separation logic notation: $x \mapsto y$ describes a single heap-allocated cell at address x whose content is y; we write "$-$" to indicate *some* value; *emp* represents the empty heap; the $*$ operator separates allocated cells.

logic and X (anti-frame) and Y (frame) needs to be deduced. The usefulness of bi-abductive inference derives from the fact that it allows to discover the part of the heap missing (the anti-frame in the above entailment) for using a specification within a particular calling context of another procedure. For example, consider the specification

$$\{x \mapsto -\} \ \texttt{void use\_cell(int *x)} \{emp\}$$

and the function `void g(int *x, int y) { y=0; use_cell(x);}` . Infer uses bi-abductive inference for comparing the calling heap of `use_cell` within g against the spec's preconditon. In doing so, Infer will understand by means of the resulting anti-frame that g's precondition must require x to be allocated. The construction of g's specification is therefore compositional (that is: the specs of a procedure are determined by the specs of the procedures it calls). The preconditions of specs computed with bi-abductive inference approximate the procedures' footprint (the parts of memory that a procedure uses). One consequence of this feature when combined with the principle of local reasoning is that these specs can be used independently of the calling context in the program. Moreover, these specs can be used as procedure summaries for implementing an inter-procedural shape analysis. Such analysis can be seen as the attempt to build proofs for Hoare triples of a program. The triples are constructed by symbolically executing the program and by *composing* triples of procedures in the program in a bottom-up fashion according to the call graph. For mutually-recursive procedures an iterative fixed-point computation is performed. This bottom-up analysis provides Infer with the ability to analyze *incomplete code* (a useful features since in practice the entire program is not always available).

An immediate consequence of Infer's compositional nature is its great ability to scale. Procedures are analyzed in isolation, and therefore, when analyzing large programs, only small parts of the source code needs to be loaded into memory. Infer has a low memory requirement even when analyzing programs composed by millions of lines of code. Moreover, the analysis results can be reused: Infer implements an *incremental analysis*, and in successive runs of the analysis of the same program, only the modified procedures need to be re-analyzed. The results of previous analyses for unchanged procedures are still valid.

3.2   Infer's Architecture

Figure 2 shows Infer's basic architecture. Infer refers to a collection of related source code files to be analyzed together as a 'project'. When analyzing a project, a number of intermediate files are created and stored in the "Project Results Directory" for use between different phases of the verification.

The Infer system consists of three main components: InferCapture, InferAnalyze, and InferPrint. These implement the three phases of the verification.

Capture Phase. Like ordinary compilation, the source code is parsed and converted to an internal representation necessary for analysis and verification.

Analysis Phase. This phase performs the actual static and verification analysis, based on the internal representation produced during the Capture Phase.

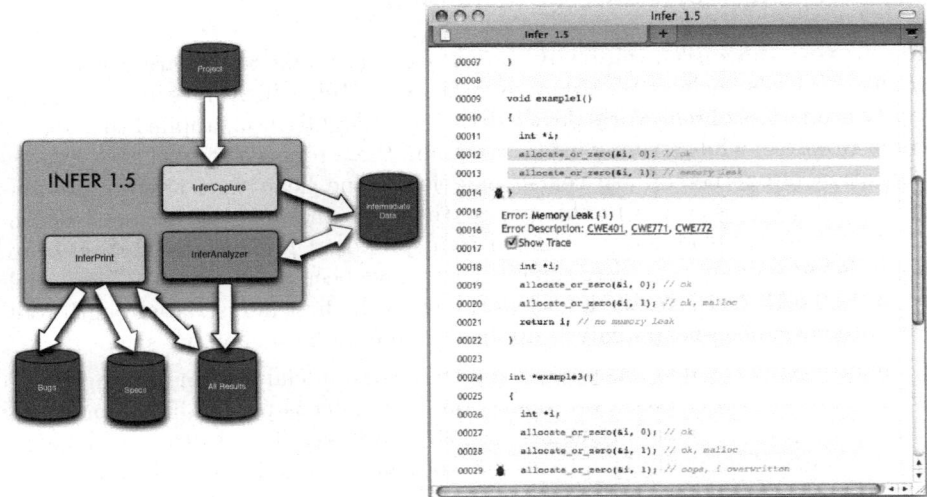

Fig. 2. Left: Infer's architecture. Right: screenshot of error trace high-lighted in yellow.

Results Post-Processing Phase. For the set of analyzed procedures where bugs have been identified, a list of bugs can be output in CSV or XML format. In addition, the list of specifications obtained for the set of procedures can be generated in either text or graphical format for use with external tools.

3.3   Implementation

Infer is available as a command line tool, in the cloud, or as an Eclipse plug-in. **Command-line.** Infer can be used in a terminal as a command-line tool. This facilitates the integration in an existing tool chain. The output are lists of errors and procedure specifications in several formats (CSV, XML, dotty, SVG). **In the cloud.** Alternatively, Infer comes with a GUI which can be used with any web-browser. In this version, the core back-end and the GUI are hosted in servers in the cloud. Users can login into their account, upload projects and run the analysis from anywhere. The GUI visualizes errors and procedure specifications in a user-friendly way (see the screenshot on the right of Fig. 2). Moreover, statistics comparing results of different runs of the analysis on the same projects are also visualized. This makes it easy for programmers or managers to track improvements of the reliability of their software during the development process. **Eclipse plug-in.** Infer can also be used within Eclipse with a special plug-in. The developer can benefit from a complete integrated environment which goes from editing, to compilation, to verification. Since Infer can analyze incomplete code, the developer can constantly check his code before committing to the central repository of his organization and avoid critical errors at very early stages.

4   Related Work and Conclusions

Recent years have seen impressive advances in automatic software verification thanks to tools such as SLAM [1] and BLAST [10], which have been used to verify properties of real-world device drivers and ASTREE [3] applied to avionics code. However, while striking in their domain, these tools either eschew dynamic allocation altogether or use coarse models for the heap that assume pointer safety. Instead, as shown in this paper, these are the areas of major strength of Infer. Several academic tools have been proposed for automatic deep-heap analysis but only on rare cases some of these have been applied to real industrial code [12,9,6,13,7,11]. To our knowledge, Infer is the first industrial-strength tool for automatic deep-heap analysis applicable to C programs of any size.

Conclusions. This paper has presented Infer, a commercial tool for proving memory safety of C code. Based on separation logic, Infer is precise in the presence of deep-heap updates and dynamic memory allocation. Thanks to the compositional nature of its analysis, Infer is able to scale to large industrial code.

References

1. Ball, T., Majumdar, R., Millstein, T., Rajamani, S.: Automatic predicate abstraction of C programs. In: PLDI, pp. 203–213 (2001)
2. Berdine, J., Calcagno, C., Cook, B., Distefano, D., O'Hearn, P.W., Wies, T., Yang, H.: Shape analysis for composite data structures. In: Damm, W., Hermanns, H. (eds.) CAV 2007. LNCS, vol. 4590, pp. 178–192. Springer, Heidelberg (2007)
3. Blanchet, B., Cousot, P., Cousot, R., Feret, J., Mauborgne, L., Miné, A., Monniaux, D., Rival, X.: A static analyzer for large safety-critical software. In: PLDI (2003)
4. Calcagno, C., Distefano, D., O'Hearn, P.W., Yang, H.: Footprint analysis: A shape analysis that discovers preconditions. In: Riis Nielson, H., Filé, G. (eds.) SAS 2007. LNCS, vol. 4634, pp. 402–418. Springer, Heidelberg (2007)
5. Calcagno, C., Distefano, D., O'Hearn, P.W., Yang, H.: Compositional shape analysis by means of bi-abduction. In: POPL, pp. 289–300. ACM, New York (2009)
6. Chang, B., Rival, X., Necula, G.: Shape analysis with structural invariant checkers. In: Riis Nielson, H., Filé, G. (eds.) SAS 2007. LNCS, vol. 4634, pp. 384–401. Springer, Heidelberg (2007)
7. Chin, W.-N., David, C., Nguyen, H., Qin, S.: Enhancing modular OO verification with separation logic. In: Proceedings of POPL, pp. 87–99. ACM, New York (2008)
8. Distefano, D., O'Hearn, P., Yang, H.: A local shape analysis based on separation logic. In: Hermanns, H. (ed.) TACAS 2006. LNCS, vol. 3920, pp. 287–302. Springer, Heidelberg (2006)
9. Hackett, B., Rugina, R.: Region-based shape analysis with tracked locations. In: POPL, pp. 310–323 (2005)
10. Henzinger, T.A., Jhala, R., Majumdar, R., McMillan, K.L.: Abstractions from proofs. In: POPL (2004)
11. Yang, H., Lee, O., Berdine, J., Calcagno, C., Cook, B., Distefano, D., O'Hearn, P.: Scalable shape analysis for systems code. In: Gupta, A., Malik, S. (eds.) CAV 2008. LNCS, vol. 5123, pp. 385–398. Springer, Heidelberg (2008)
12. Lev-Ami, T., Sagiv, M.: TVLA: A system for implementing static analyses. In: SAS 2000. LNCS, vol. 1824, pp. 280–302. Springer, Heidelberg (2000)

13. Marron, M., Hermenegildo, M., Kapur, D., Stefanovic, D.: Efficient context-sensitive shape analysis with graph based heap models. In: Hendren, L. (ed.) CC 2008. LNCS, vol. 4959, pp. 245–259. Springer, Heidelberg (2008)
14. Reynolds, J.C.: Separation logic: A logic for shared mutable data structures. In: LICS (2002)

Model Construction and Priority Synthesis for Simple Interaction Systems

Chih-Hong Cheng[1], Saddek Bensalem[2], Barbara Jobstmann[2],
Rongjie Yan[3], Alois Knoll[1], and Harald Ruess[4]

[1] Department of Informatics, Technischen Universität München, Germany
[2] Verimag Laboratory, Grenoble, France
[3] State Key Laboratory of Computer Science, Institute of Software, CAS, China
[4] Fortiss GmbH, Munich, Germany

Abstract. VISSBIP is a software tool for visualizing and automatically orchestrating component-based systems consisting of a set of components and their possible interactions. The graphical interface of VISSBIP allows the user to interactively construct BIP models [3], from which executable code (C/C++) is generated. The main contribution of VISSBIP is an analysis and synthesis engine for orchestrating components. Given a set of BIP components together with their possible interactions and a safety property, the VISSBIP synthesis engine restricts the set of possible interactions in order to rule out unsafe states. The synthesis engine of VISSBIP is based on automata-based (game-theoretic) notions. It checks if the system satisfies a given safety property. If the check fails, the tool automatically generates additional constraints on the interactions that ensure the desired property. The generated constraints define priorities between interactions and are therefore well-suited for conflict resolution between components.

1 Introduction

We present VISSBIP[1], an open-source tool to construct, analyze, and synthesize component-based systems. Component-based systems can be modeled using three ingredients: (a) *Behaviors*, which define for each basic component a finite set of labeled transitions (i.e., an automaton), (b) *Interactions*, which define synchronizations between two or more transitions of different components, and (c) *Priorities*, which are used to choose between possible interactions [3].

In the BIP framework [3], the user writes a model using a programming language based on the Behavior-Interaction-Priority principle. Using the BIP toolset, this model can be compiled to run on a dedicated hardware platforms. The core of the execution is the *BIP engine*, which decides which interactions are executed and ensures that the execution follows the semantics.

---

[1] VISSBIP is a shortcut for **Vi**sualization and **S**ynthesis of **S**imple **BIP** models. It is available at http://www6.in.tum.de/~chengch/vissbip

M. Bobaru et al. (Eds.): NFM 2011, LNCS 6617, pp. 466–471, 2011.
© Springer-Verlag Berlin Heidelberg 2011

VissBIP is a tool for constructing and visualizing BIP models. Its graphical interface allows the user to model hierarchical systems. The analysis and synthesis engine can currently only interpret non-hierarchical model, which we call *simple*. In BIP, a system is built by constructing a set of basic components and composing them using interactions and priorities. The interactions and priorities are used to ensure global properties of the systems. For instance, a commonly seen problem is mutual exclusion, i.e., two components should avoid being in two dedicated states at the same time. Intuitively, we can enforce this property by requiring that interactions that exit one of the dedicated states have higher priority than interactions that enter the states. Adding interactions or priorities to ensure a desired behavior of the overall systems is often a non-trivial task. VissBIP supports this step by automatically adding a set of priorities that enforce a desired safety property of the composed systems. We call this technique **priority synthesis**. We concentrate on adding priorities because (1) priorities preserve already established safety properties as well as deadlock-freedom of the system, and (2) priorities can be implemented efficiently by allowing the components to coordinate temporarily [6].

2   Visualizing Simple Interaction Systems

The user can construct a system using the drag-and-drop functionality of VissBIP's graphical user interface shown in Figure 1. BIP objects (components, places, properties, and edges) can be simply dragged from the menu on the left to the drawing window on the right.

We use the system shown in Figure 1 to illustrate how a system is represented. The system consisting of two components (`Process1` and `Process2`) depict as boxes. Each component has two places (`high` and `low`) and a local variable (`var1` and `var2`, respectively). A place (also called location) is represented by a circle. A green circle indicates that this place is an initial location of a behavioral component. E.g., place `v1` is marked as initial in `Process1`. Squares denotes variables definitions and their initialization within a component. E.g., `var1` and `var2` are both initialized to 1. Edges between two locations represent *transitions*. Each transition is of the format `{precondition} port-name {postcondition}`. E.g., the transition of `Process1` from place `low` to `high` is labeled with port name `a` and upon its execution the value of `var1` is increased by 1. For simplicity we use **port-name bindings** to construct interactions between components, i.e., that transitions using the same port name are automatically grouped to a single interaction and are executed jointly[2]. In the following, we refer to an interaction by its port name. Finally, additional squares outside of any component, are used to define system properties such as priorities over interactions and winning conditions (for synthesis or verification). In particular, we use the keyword `PRIORITY` to state priorities. E.g., the statement `Process2.d < Process1.b` means that whenever interactions `b` and `d` are available, the BIP engine always

---

[2] It is possible to pass data through an interaction. The user specifies the data flow associated to an interaction in the same way she describes priorities (see below).

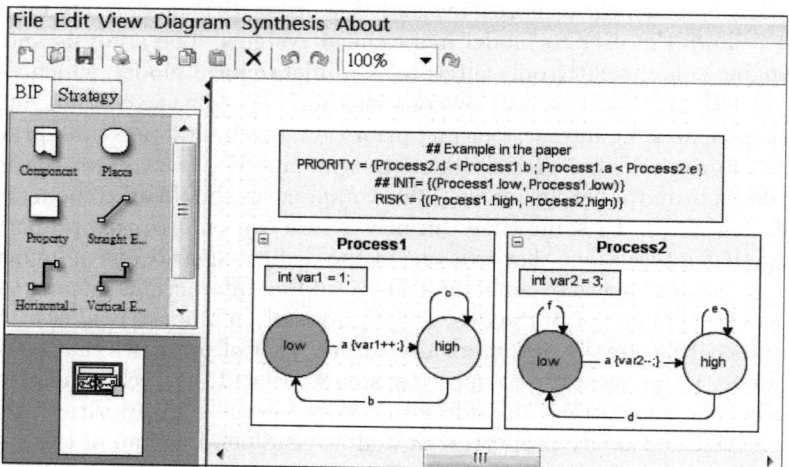

Fig. 1. Model construction using VISSBIP

executes b. The keyword RISK is used to state risk conditions. E.g., the condition RISK = {(Process1.high, Process2.high)} states that the combined location pair (Process1.high, Process2.high) is never reached. Apart from the stated conditions, we also implicitly require that the system is deadlock-free, i.e., at anytime, at least one interaction is enabled. When only deadlock avoidance is required, the keyword DEADLOCK can be used instead. Lines started with ## are comments.

3   Priority Synthesis

We define priority synthesis as an automatic method to introduce a set of new priorities over interactions on a BIP system such that the augmented system satisfies the specified property. A priority is *static* if it does not contain evaluations over ports or locations in components as a precondition to make the priority active. We focus on synthesizing static priorities. We consider safety (co-reachability) properties, i.e., the property specifies a set of risk states, and the system should never reach any of them. For the rest of this section, we first show the results of priority synthesis under VISSBIP. Then, we give some details about the underlying algorithm and the implementation.

3.1   Safety Synthesis by Adding Global Priorities: Examples

The user can invoke the synthesis engine on a system like to one shown in Figure 1. The engine responds in one of the following three ways: (1) It reports that no additional priorities are required. (2) It returns a set priority constraints that ensure the stated property. (3) It states that no solution based on priorities can be found by the engine.

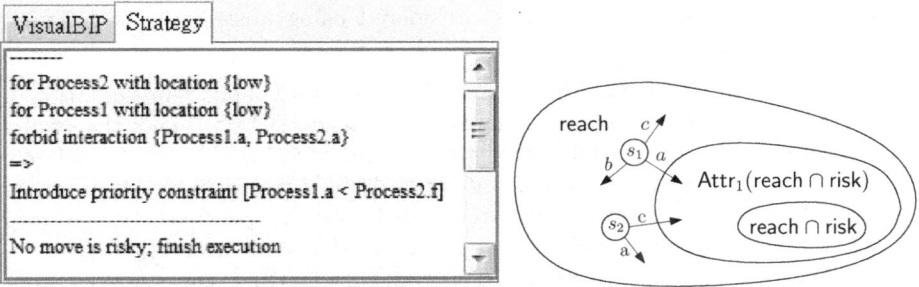

Fig. 2. The automatically synthesized priority for the model in Figure 1 (left), and an example for priority resolution (right)

Figure 2 shows the *Strategy* panel of VISSBIP, which displays the results obtained by invoking the synthesis engine on the example of Figure 1. Recall, that in the example, we stated that the combined location pair (Process1.high, Process2.high) is never reached. The engine reports that the priority constraint Process1.a < Process2.f should be added. Note that if the system is in state (Process1.low, Process2.low), then the interaction Process1.a (which is a joint action from Process 1 and Process 2) would immediately leads to a risk state (a state satisfying the risk condition). This can be avoided by executing Process2.f first. The new priority ensures that interaction f is executed forever, which is also deadlock-free.

3.2 Safety Synthesis by Adding Global Priorities: Algorithmic Issues

The algorithm of priority synthesis is based on concepts in games [5]. A game is a graph partitioned into player-0 and player-1 vertices. We refer to Player 0 as the system (which is controllable) and Player 1 as the environment (which is uncontrollable). In controllable vertices, the system can choose among the set of available transitions during execution. Conceptually, a play between two players is proceeded as follows:

1. A player-1 vertex is a product of (i) locations of behavioral components and (ii) evaluations of their variables. Since the values of the variables are not relevant in our example, we omit them for simplicity. E.g., in Figure 1, (Process1.low, Process2.low) is a player-1 vertex.
2. From a player-1 vertex the game moves to a player-0 vertex that represents all the available interactions. E.g., from location (Process1.low, Process2.low), the game proceeds to state labeled with the interactions a and f.
3. Then, Player 0 responds by selecting one interaction and updates the location (i.e., to a new vertex of player-1). Note that its selection is constrained by the pairing of interactions as well as the priority specified in the system.

Admittedly now in our formulation player-1 is deterministic (thus it can be viewed as 1-player game, or automaton). Nevertheless, a non-deterministic player-1 is introduced when data is considered using abstraction techniques. The algorithmic flow for priority synthesis is as follows.

- (GAME CONSTRUCTION) First, VISSBIP creates a symbolic representation of the game specified above using the BDD package JDD [2]. Then, the engine compute the set of reachable risk states by intersecting the set of reachable states reach with the set of risk states risk derived from the specification. In order to obtain a good initial variable ordering, we use heuristics to keep variables from component that participate in the same interaction close.
- (GAME SOLVING AND RISK-EDGE GENERATION) Once the symbolic representation of the game is constructed, VISSBIP solves the safety game by symbolically computing the *risk attractor* $\mathsf{Attr}_1(\mathsf{reach} \cap \mathsf{risk})$, which is the set of states from which player-1 can force to move to a risk state, regardless of moves done by Player 0 [5]. If all the reachable moves of Player 0 avoid the risk attractor, then the model running on the BIP engine is guaranteed to be safe. Otherwise, we derive a set of *risk edges*, which are all the edges leading to the risk attractor. E.g., Figure 2 shows that from (Process1.low, Process2.low) interaction a corresponds to a risk edge. We compute the set of risk edges T_{risk} with the following formula: $(\mathsf{reach} \setminus \mathsf{Attr}_1(\mathsf{reach} \cap \mathsf{risk})) \cap T_0^\# \cap \mathsf{Attr}_1'(\mathsf{reach} \cap \mathsf{risk})$, where $\mathsf{Attr}_1'(\mathsf{reach} \cap \mathsf{risk})$ is the primed version of $\mathsf{Attr}_1(\mathsf{reach} \cap \mathsf{risk})$.
- (RISK-EDGE INTERPRETATION) We aim to introduce priority constraints to prevent the BIP engine from selecting transitions in T_{risk}. This can be done by examining interactions that are also available at the locations from which a risk edge can be taken. E.g., in our example (Figure 1), the engine examines all alternative interactions at state (Process1.low, Process2.low). Since Process2.f can also be selected, the engine generates the priority constraint Process1.a < Process2.f to avoid using Process1.a.
  To avoid enumerating all the risk edges, VISSBIP aims to rule out risk-edges in a symbolic fashion. More precisely, VISSBIP proceeds on cubes of the risk-edges, which are sets of edges that can be represented by a conjunction over the state variables or their negation. For each cube, the engine generates a set of candidate priorities that can be used to avoid these risk edges.
- (PRIORITY GENERATION) When the engine has collected the set of priorities for each cube in T_{risk}, these priorities are as priority fixes having preconditions (E.g., in Figure 2 if on state s_2 we should use priority $c < a$ to escape from the attractor). We are interested in adding static priorities, which are independent of the actual state. VISSBIP offers the user to select between two incomplete algorithms to obtain static priorities.
  - *Priority resolution using SAT solvers*: From the set of candidate priorities obtained for each cube, the engine needs to select a set of non-conflicting priorities. E.g., consider the example in Figure 2, for state s_1 VISSBIP generates a candidate set $\{(a < b), (a < c)\}$ and for s_2 it gives $\{(c < a)\}$. The engine should not report $\{(a < c), (c < a)\}$ as a static priority fix, as it does not satisfy the strict partial order. Finding a set

of priorities satisfying each cube while obeying the strict partial order can be done using SAT solvers. In VISSBIP we use SAT4J [1].

- *Fixed-size priority selection*: For this scheme, the engine examines all possible subsets of the collected priorities with the size bounded by a user-specified number, starting from the smallest subset. Since in general a small number of modification of the system is desirable, this is a natural approach. Furthermore, it gives appealing results on our examples.

4   Evaluation and Summary

In this paper, we present a tool for constructing simple BIP systems together with a technique called priority synthesis, which automatically adds priorities over interactions to maintain system safety and deadlock-freedom.

We have evaluated the tool on some examples, e.g., traditional dining philosophers problems, and problems related to critical region control. On these examples, VISSBIP enables to generate small yet interesting priority fixes. Due to the limitation concerning the number of components to be placed in a single canvas, examples under investigations are admittedly not very big, but they can be solved within reasonable time. E.g., the dining philosophers problem with size 12 (i.e., a total of 24 interacting components) is solved within 1 seconds using the SAT-resolution method and within 3 seconds using constant-depth fixing[3]. As the fix is automatically generated without human intervention, we treat this as a promising step towards computer-aided synthesis in BIP. Algorithms in VISSBIP can be viewed as new features for the future D-Finder tool [4], which focuses on deadlock finding and verification over safety properties. Nevertheless, the front-end GUI targeted for the ease of model construction is also important.

Lastly, as priority synthesis is essence a method of *synthesizing simple component glues for conflict resolution*, under suitable modifications, our technique is applicable for multicore/manycore systems working on task models for resource protection, which is our next step.

References

1. SAT4J: a SAT-solver based on Java, http://www.sat4j.org
2. The JDD project, http://javaddlib.sourceforge.net/jdd/
3. Basu, A., Bozga, M., Sifakis, J.: Modeling heterogeneous real-time components in bip. In: SEFM 2006, pp. 3–12. IEEE, Los Alamitos (2006)
4. Bensalem, S., Bozga, M., Nguyen, T., Sifakis, J.: D-finder: A tool for compositional deadlock detection and verification. In: Bouajjani, A., Maler, O. (eds.) CAV 2009. LNCS, vol. 5643, pp. 614–619. Springer, Heidelberg (2009)
5. Grädel, E., Thomas, W., Wilke, T.: Automata, Logics, and Infinite Games. LNCS, vol. 2500. Springer, Heidelberg (2002)
6. Graf, S., Peled, D., Quinton, S.: Achieving distributed control through model checking. In: Touili, T., Cook, B., Jackson, P. (eds.) CAV 2010. LNCS, vol. 6174, pp. 396–409. Springer, Heidelberg (2010)

---

[3] The result is evaluated under the Java Virtual Machine with parameter -Xms1024m -Xmx2048m. The system runs under Linux with an Intel 2.8 Ghz CPU.

OpenJML: JML for Java 7 by Extending OpenJDK

David R. Cok

GrammaTech, Inc., Ithaca NY, 14850, USA
dcok@grammatech.com

Abstract. The Java Modeling Language is a widely used specification language for Java. However, the tool support has not kept pace with advances in the Java language. This paper describes OpenJML, an implementation of JML tools built by extending the OpenJDK Java tool set. OpenJDK has a readily extendible architecture, though its details could be revised to further facilitate extension. The result is a suite of JML tools for Java 7 that provides static analysis, specification documentation, and runtime checking, an API that is used for other tools, uses Eclipse as an IDE, and can be extended for further research. In addition, OpenJML can leverage the community effort devoted to OpenJDK.

Keywords: OpenJML, JML, specification, verification, OpenJDK, ESC/Java2.

1 The Java Modeling Language

The Java Modeling Language [16] was first proposed in 1999 [14] as a specification language for Java programs; it is accompanied by a tool suite that can process those specifications for a variety of purposes. The language expresses specifications in a traditional style: pre-, frame and post-conditions for methods, and invariants for objects and classes. Examples can be found in the reference manual [16] and in tutorials (e.g. [15]). Part of the goal of JML is to be easy to read and write, by drawing as much as possible from the syntax and semantics of its host programming language and thereby to be familiar to Java programmers. The specifications are often intermingled with the code, in the same .java file; the specifications can also be written in separate files, for situations in which the source code is not available or not writable by the specification author.

JML is now widely used[1] as a basis for research and education about formal methods using Java. Tools supporting JML have been applied to verification of industrial software [2,6,8]. Maintaining current and robust tool support is essential to further exploration, education, and use of formal methods in Java.

As a specification language, JML enables many different, complementary capabilities:

---

[1] cf. the list of groups on http://www.eecs.ucf.edu/~leavens/JML/index.shtml

M. Bobaru et al. (Eds.): NFM 2011, LNCS 6617, pp. 472–479, 2011.

- verifying that specifications and implementations logically agree (software verification and bug finding) [8]
- inferring specifications (e.g. loop invariants, pre- and post-conditions) from an implementation [13] or live use [11]
- generating run-time checks, compiled into programs [5]
- augmenting documentation with specification information [3]
- generating test oracles from specifications [21]
- as a guide to generating test cases that exercise relevant execution paths

OpenJML extends OpenJDK to create JML tools for Java 7. The tools are available at http://jmlspecs.sourceforge.net. Its source code is available at http://jmlspecs.svn.sourceforge.net/viewvc/jmlspecs.

2   Precursors to OpenJML

JML was first used in an early extended static checker (ESC/Java [17]) and was implemented in a set of tools called JML2 [3]. The second generation of ESC/Java, ESC/Java2 [8], was made current with Java 1.4 and with the definition of JML. Further research produced tools for runtime checking, documentation generation (jmldoc), test generation, and integration with Eclipse [3].

However, the JML2 tools were based on hand-crafted compilers; the effort of maintaining those Java compilers overwhelmed the volunteer and academic resources as Java evolved. A new approach was needed, one that built on an existing compiler in a way that leveraged further developments in that compiler, yielded compact command-line tools, but allowed easy integration with a Java IDE environment, and was readily maintainable and extensible.

The JML community considered two alternatives: the Eclipse JDT [10] and, later, OpenJDK [19], with discussion on relevant mailing lists[2]. Both are full-fledged, well-supported compilers; neither is designed for extension; the Eclipse compiler is well-integrated into an IDE; the OpenJDK compiler is more compact, functions well stand-alone, but has no natural IDE. The JML3 project to extend the Eclipse environment with pure plug-ins was abandoned: too much integration with the Eclipse internals was required. A later attempt to extend the Eclipse JDT directly, JML4 [4], found that the internals were complicated, progress was slow and resource intensive, and the result would be difficult to extend by a wide group of researchers. The JML4 work has transformed into the JIR/JMLEclipse projects, still based on the Eclipse JDT but with a different emphasis.

OpenJDK offered an alternative; the OpenJML project described here found the OpenJDK compiler, while not designed for extension, to be much more extensible in a less invasive way, than the current Eclipse JDT. Command-line tools to parse and type-check JML constructs along with the Java source were readily produced; verification checking made use of back-end SMT solvers (experiments [7] were performed with Yices, CVC3, and Simplify). IDE integration

---

[2] jmlspecs-reloaded@sourceforge.net, particularly from Sept. 2007 on, and white papers on http://sourceforge.net/apps/trac/jmlspecs/wiki/DevelopmentNotes

is accomplished with conventional Eclipse plug-ins. The principal drawback is that in an Eclipse-integrated system, the Eclipse compiler is used (as is) for Java compilation and the OpenJML/OpenJDK compiler is used as a back-end tool for handling JML and verification tasks.

OpenJML is a reimplementation, on a completely new code-base, of tools such as the JML2 tools for JML on Java 1.4 (including jmldoc and jmlrac) and ESC/Java2 and ESC/Java for static checking.

3   OpenJDK

History. OpenJDK [19] is an open-source version of the Java toolkit, announced by Sun in 2006. Since then, numerous groups have collaborated to establish a firm open foundation for OpenJDK, including porting it to other architectures. Researchers have also used the OpenJDK to create other applications or to experiment with language extensions[3].

Extending the OpenJDK Architecture. The goal of the OpenJML project is to produce a usable and extensible set of JML tools while being minimally invasive to the OpenJDK implementation. The architecture of OpenJDK is quite amenable to extension, with the difficulties lying in the details rather than the overall architecture. The compilation process makes use of a chain of components: lexical analysis, parsing, symbol table maintenance, annotation processing, name resolution, type-checking, semantic use checks, flow checks, AST desugaring, and code generation. Each component is registered in a *compilation context*; the context creates instances of or references to various tools as needed. The tools are called in succession, passing partially-processed ASTs from tool to tool.

The behavior of the compiler is readily altered by registering replacement, JML-aware components in the compilation context before a compilation begins. The new components need to be derived classes of the old components, as no Java interfaces are defined. In some cases, such as the scanner and parser, the replacement component has significant new functionality; in others, the replacement component simply inserts a new phase. For example, the desugaring component is altered simply to invoke static analysis rather than continuing on to code generation; alternately, a component that adds runtime assertion checks can be inserted between desugaring and code generation.

Some of the complications in extending OpenJDK were these:

- The lexical scanner and parser are hand-coded, not generated by code-generation tools. This complicates altering the parsing phase. However, the parser design is mostly a top-down parser with limited look-ahead, so overriding relevant methods in the parser can accomplish most of the needed extension. JML text is wholly contained in Java comments, so overriding the comment handling methods (which are already present to process javadoc comments) makes it straightforward to parse the JML specifications.

---

[3] cf. the projects listed on http://openjdk.java.net/

- The JML expression language includes nearly all of the Java expression language, leaving out only expressions with side-effects, but including method and class declaration. However, JML also adds some new syntax and many reserved tokens with special meaning (e.g. \result for the return value of a method) that must be recognized at the leaves of the AST.
- JML specifications can occur in more than one file and in the absence of any Java source code. Merging the specifications and connecting the specifications with the correct Java construct requires some amount of type resolution before the ASTs can be completely built.
- Declarations that occur in specifications are not in scope in Java code; but both Java and JML specifications are in scope in specifications. Name resolution must be sensitive to context while still using a common symbol table.
- The set of AST nodes must be extended to include JML constructs. This refactoring was more complicated than it might be because Java Enums are used to indicate the kind of AST node and Enums are not extensible.
- Lack of interfaces for tools, AST nodes, and visitor classes.

Overall, the tools code of OpenJDK (as of the current base build) contains 683 source files. OpenJML modified just 41 of them, with only 9 requiring significant change; one-third required only visibility changes, with others needing minor changes or corrections. The code contains cautions that most classes do not constitute a public API and may change; however, merges to new builds have generally been smooth and not resulted in significant rework. OpenJML is currently based on OpenJDK build 116 (November 2010); merges are performed periodically to remain current with the upcoming release of Java 1.7.

4   OpenJML

The result of the tool development work described above is the OpenJML command-line tool, with the following functionality.

- the ability to parse and type-check current JML, producing internal ASTs. The definition of JML is currently under review, simplifying its syntax and clarifying its semantics; corresponding changes in OpenJML are in progress.
- the ability to perform static verification checks using backend SMT solvers
- the ability to explore counterexamples (models) provided by the solver [7]
- integration with the features of Java 1.5 and 1.6, including generics and annotations, with Java 1.7 features in progress; the main work is to represent the semantics of the features in the verification logic, rather than, as was previously the case, compiler development
- partial implementation of JML-aware documentation generation
- a proof of concept implementation of runtime assertion checking
- independently, JMLUnitNG [21] has used OpenJML to create a test generation tool, using OpenJML's API to access the parsed specifications

In addition, the previously developed Eclipse plug-in was modernized and integrated with OpenJML, providing an IDE that permits working with JML within Eclipse's Java development environment, with this functionality:

- the ability to parse and type-check JML (either embedded in .java files or in standalone specification files), showing any errors or warnings as Eclipse problems, but with a custom icon and problem type;
- the ability to check JML specifications against the Java code — verification conditions are produced from the internal ASTs and submitted to a back-end SMT solver, and any proof failures are shown as Eclipse problems;
- the ability to use files with runtime checks along with Eclipse-compiled files;
- the ability to explore specifications and counterexamples within the GUI;
- functionality integrated as Eclipse menus, commands, and editor windows.

Exploring Counterexamples from Static Checking. The Eclipse GUI enables exploring counterexamples produced by failed static checking much more effectively than previous JML tools. Tools such as ESC/Java and ESC/Java2 created verification conditions, shipped them to a back-end solver, which produced counterexample information that was essentially a dump of the prover state and was notoriously difficult to debug. The Eclipse GUI for OpenJML interprets the counterexample information and relates it directly to the program as seen in the Eclipse editor windows.

This capability allows implementing two particularly useful bits of functionality. First, the counterexample information can be interpreted to determine the control flow through the program that the particular counterexample represents. This control flow can then be highlighted with suitable color coding, making it clear which sequence of computations violates which verification condition.

Second, either from the counterexample or from subsequent interactions with the prover, the value of any subexpression along the flow of control can be determined. Thus the user can interrogate, through the GUI, the value of any expression, in order to understand the precise sequence of computations that lead to the invalid verification condition.

Note that this interaction is completely static. One can obtain similar information from a dynamic debugger. But one must proceed step-by-step, forward through the program. Most debuggers do not permit backtracking. Furthermore, the code module under study must be executable, is only explorable for the particular conditions in which it is called, and does not execute specifications.

In contrast, using "static debugging", the user can explore backwards and forwards at will, always obtaining program variable values in the context in which they occur (that is, with the values they have at that point in the program execution). Assertions and other specifications that are not executable (but are part of the logical representation) are also explorable. The program snippet need not be executable.

5   Related Tools

JML is sufficiently widely used in education, application, and research that there are other tools that support and use the language. Some of the publicly available tools with continuing development are these:

- Key [2] – This substantial project researched software verification workflow and produced corresponding tools. The project began by targeting OCL as a specification language, but has changed to using JML for Java programs. The project uses its own parser for both Java and JML, creating proof obligations; these verification conditions are then proved interactively (with semi-automated assistance) using the Why and Coq proof systems. Its target has been JavaCard rather than full Java.
- JMLEclipse [4] – JMLEclipse is a partial integration of JML functionality into the Eclipse IDE, by modifying the Eclipse JDT.
- Mobius [18] – The Mobius project combined a large number of tools in an integrated verification environment, including existing JML-aware tools and a companion Byte-code Modeling Language (BML).
- Why [22] – The Why tool integrates both back-end solvers and programming language-specific verification condition generators (e.g. Krakatoa for Java) into its custom IDE, creating a verification workbench. Krakatoa uses a custom compiler and has been applied particularly to JavaCard applications.
- Spec# [1], CodeContracts [12] – These tools target C# (and .NET), rather than Java; the analog to JML is the Spec# specification language, which is integrated with the Visual Studio tools for C#. Spec# is available in a research mode, but has not been commercialized. Instead, Microsoft has released the CodeContracts system as part of .NET.

Pluggable type-checkers [9] are another mechanism for adding static analysis to Java. These type checkers use Java's annotation mechanism (with the JSR308 extensions expected in Java 1.8) to implement a variety of type checks, such as non-null types, readonly types, and types that check for interned values. It would be possible to implement static verification-like checks through this mechanism as well. The principal drawback is the extra complication and messy syntax of writing annotation expressions as annotation arguments (and as character strings) [20]. Nevertheless the type annotation mechanism is well-engineered and is slated to become part of Java, so it is worth exploring its integration with JML and with static checking of verification conditions.

6 Availability and Use

The OpenJML tool suite is available at `http://jmlspecs.sourceforge.net`, with the Eclipse update site at `http://jmlspecs.sourceforge.net/openjml-updatesite`. The parsing and type-checking aspects of OpenJML have been stable for more than a year; that portion has been used in the JmlUnitNG tool [21]. The translations to verification conditions and to compiled run-time checks are still in alpha release stage. Producing output in BoogiePL and in SMT-LIBv2 is in progress. A draft user guide is also available at the above web site.

To date, I am aware of OpenJML being used in publications (e.g. [21][7]), in an MSc and a PhD thesis, in a released tool, and in teaching.

7   Future Work

The future tasks for OpenJML development are these (as of 14 February 2011):

- adapt OpenJML to recent evolutions in the JML language
- complete the SMT-LIBv2 interface so that OpenJML's static checking can use any (conforming) SMT solver
- complete capabilities that are underway: verification generation, runtime assertion checking, documentation generation, integration with test generators (e.g. JMLUnitNG)
- integrate specification inference to simplify the task of writing specifications
- implement specification refactoring capabilities within the Eclipse plug-in
- complete a BoogiePL interface
- integration with Java 1.7 features and, eventually, with JSR 308 (type annotations in Java)

More important is applying the tool to further application and research. In particular, the field needs

- substantial verification case studies of a variety of realistic sets of code,
- review and research of the constructs needed to specify mid- and high-level design features, not just the absence of low-level errors,
- integration of specification concepts developed for easier reasoning about frame conditions, ownership, memory separation, and concurrency,
- specification inference to reduce the burden of user-written specifications,
- and continued development of OpenJML as a basis for building other tools.

References

1. Barnett, M., Leino, K.R.M., Schulte, W.: The Spec# programming system: An overview. In: Barthe, G., Burdy, L., Huisman, M., Lanet, J.-L., Muntean, T. (eds.) CASSIS 2004. LNCS, vol. 3362, pp. 49–69. Springer, Heidelberg (2005)
2. Beckert, B., Hähnle, R., Schmitt, P.H. (eds.): Verification of Object-Oriented Software. LNCS (LNAI), vol. 4334. Springer, Heidelberg (2007)
3. Burdy, L., Cheon, Y., Cok, D.R., Ernst, M.D., Kiniry, J.R., Leavens, G.T., Leino, K.R.M., Poll, E.: An overview of JML tools and applications. In: Arts, T., Fokkink, W. (eds.) Eighth International Workshop on Formal Methods for Industrial Critical Systems (FMICS 2003). ENTCS, vol. 80, pp. 73–89. Elsevier, Amsterdam (2003)
4. Chalin, P., Robby, James, P., Lee, J., Karabotsos, G.: Towards an industrial grade IVE for Java and next generation research platform for JML. STTT 12, 429–446 (2010) 10.1007/s10009-010-0164-8
5. Cheon, Y., Leavens, G.T.: A runtime assertion checker for the Java Modeling Language (JML). In: Arabnia, H.R., Mun, Y. (eds.) Proceedings of the International Conference on Software Engineering Research and Practice (SERP 2002), Las Vegas, Nevada, USA, June 24-27, pp. 322–328. CSREA Press (2002)
6. Cochran, D., Kiniry, J.R.: Votail: A Formally Specified and Verified Ballot Counting System for Irish PR-STV Elections. In: FoVeOOS 2010 (2010)

7. Cok, D.: Improved usability and performance of SMT solvers for debugging specifications. STTT 12, 467–481 (2010)
8. Cok, D.R., Kiniry, J.R.: ESC/Java2: Uniting eSC/Java and JML. In: Barthe, G., Burdy, L., Huisman, M., Lanet, J.-L., Muntean, T. (eds.) CASSIS 2004. LNCS, vol. 3362, pp. 108–128. Springer, Heidelberg (2005)
9. Dietl, W., Dietzel, S., Ernst, M.D., Muşlu, K., Schiller, T.: Building and using pluggable type-checkers. In: Proceedings of the 33rd International Conference on Software Engineering, ICSE 2011, Waikiki, Hawaii, USA, May 25–27 (2011)
10. http://www.eclipse.org
11. Ernst, M.D., Perkins, J.H., Guo, P.J., McCamant, S., Pacheco, C., Tschantz, M.S., Xiao, C.: The Daikon system for dynamic detection of likely invariants. Science of Computer Programming 69(1–3), 35–45 (2007)
12. Fähndrich, M., Barnett, M., Logozzo, F.: Embedded contract languages. In: Proceedings of the 2010 ACM Symposium on Applied Computing, SAC 2010, pp. 2103–2110. ACM, New York (2010)
13. Flanagan, C., M. Leino, K.R.: Houdini, an annotation assistant for eSC/Java. In: Oliveira, J.N., Zave, P. (eds.) FME 2001. LNCS, vol. 2021, pp. 500–517. Springer, Heidelberg (2001)
14. Leavens, G.T., Baker, A.L., Ruby, C.: JML: A notation for detailed design. In: Kilov, H., Rumpe, B., Simmonds, I. (eds.) Behavioral Specifications of Businesses and Systems, pp. 175–188. Kluwer Academic Publishers, Boston (1999)
15. Leavens, G.T., Kiniry, J.R., Poll, E.: A JML tutorial: Modular specification and verification of functional behavior for Java. In: Damm, W., Hermanns, H. (eds.) CAV 2007. LNCS, vol. 4590, Springer, Heidelberg (2007)
16. Gary, T.: Leavens, Erik Poll, Curtis Clifton, Yoonsik Cheon, Clyde Ruby, David R. Cok, Peter Müller, Joseph Kiniry, Patrice Chalin, and Daniel M. Zimmerman. JML reference manual (September 2009), http://www.jmlspecs.org
17. Rustan, K., Leino, M.: Greg Nelson, and James B. Saxe. ESC/Java user's manual. Technical note, Compaq Systems Research Center (October 2000)
18. http://mobius.ucd.ie
19. http://www.openjdk.org
20. Taylor, K.B., Rieken, J., Leavens, G.T.: Adapting the Java Modeling Language (JML) for Java 5 annotations. Technical Report 08-06, Department of Computer Science, Iowa State University, 226 Atanasoff Hall, Ames, Iowa 50011 (April 2008)
21. Zimmerman, D.M., Nagmoti, R.: JMLUnit: The Next Generation. In: Beckert, B., Marché, C. (eds.) FoVeOOS 2010. LNCS, vol. 6528, pp. 183–197. Springer, Heidelberg (2011), http://formalmethods.insttech.washington.edu/software/jmlunitng/
22. Why web site, http://why.lri.fr

jSMTLIB: Tutorial, Validation and Adapter Tools for SMT-LIBv2

David R. Cok

GrammaTech, Inc., Ithaca NY 14850, USA
dcok@grammatech.com

Abstract. The SMT-LIB standard defines an input format and response requirements for Satisfiability-Modulo-Theories automated reasoning tools. The standard has been an incentive to improving and comparing the increasing supply of SMT solvers. It could also be more widely used in applications, providing a uniform interface and portability across different SMT tools. This tool paper describes a tutorial and accompanying software package, *jSMTLIB*, that will help users of SMT solvers understand and apply the newly revised SMT-LIB format; the tutorial also describes fine points of the SMT-LIB format which, along with a compliance suite, will be useful to SMT implementors. Finally, the tool suite includes adapters that allow using some older solvers, such as Simplify, as SMT-LIB compliant tools.

Keywords: SMT solvers, SMT-LIB, validation, software verification, automated reasoning, jSMTLIB, OpenJML.

1 SMT Solvers and SMT-LIB

Automated reasoning engines for Satisfiability-Modulo-Theories (SMT) logics have improved in capability and increased in number in recent years. This trend is partly driven by challenging applications in software verification and model checking. The progress is made visible through public competitions, spurring practical advances, theoretical research, and collegial rivalry. The SMT-COMP competition has been held annually since 2005 in conjunction with well-known conferences. Competing solvers vie to solve the most problems in the shortest time. At the 2010 SMT workshop [8], 10 different provers demonstrated their capability and performance in open competition, with 8 other tool development groups participating in 2008 or 2009.

But such a competition needs to be able to express problems in a format that is readable by all participants — hence the need for the SMT-LIB language. This standard form for stating SMT problems was proposed in 2003 [13]; a significant revision as version 2.0 [3] has just been released. Version 2.0 is significantly different from previous versions. It simplified the syntax, removed the distinction between terms and formulas, introduced a command language, and added simple parameterization and polymorphism. Version 2.0 was developed with considerable input from SMT implementors and users; intentionally, it is much more

M. Bobaru et al. (Eds.): NFM 2011, LNCS 6617, pp. 480–486, 2011.

useful for representing the semantics of software than was the previous version and for interacting with front-end, possibly interactive, tools. In the remainder of this document, SMT-LIB refers to this new version.

This standard language for SMT solvers enables an application that states problems in SMT-LIB to (eventually) use any of a wide set of conforming SMT provers as its back-end constraint solver. The transition from one solver to another is not seamless: solvers may specialize in one kind of problem or another, and the way a problem is stated may still affect different solvers differently. Nevertheless, not having to change the software in order to interface with a new solver promotes flexibility and experimentation. A position paper by Barrett and Conway [1] identified gaps between users and implementors as a significant impediment to more wide-spread application, and therefore improvement, of SMT solvers. The goal of the tools and tutorial described in this paper is to narrow that gap by providing information and readily available tools to those implementing applications that use SMT solvers or those simply wishing to experiment with and use SMT solvers.

2   Tools and Materials

While implementors of SMT solvers are quite aware of the SMT competition and SMT-LIB, potential new users of the SMT-LIB format and of SMT solvers may not be. Thus, in support of the overall SMT-LIB endeavor and to encourage use of SMT solvers in appropriate applications, the following tools and materials have been created and made available. These items complement related material described in Section 5.

- a tutorial introduction to SMT-LIB, targeting users of SMT solvers, but also useful to implementors
- a Java parser and type-checker for SMT-LIB command scripts
- a wrapper for the tools that operates as a network client and server
- adapters that translate SMT-LIB command scripts into the input languages of various existing solvers
- a Java API for programmatic interaction with SMT solvers
- an Eclipse plug-in for editing and executing SMT-LIB scripts
- a validation suite of command scripts to test SMT-LIB compliance
- extensibility
- a user guide to the listed software tools

The SMT-LIB Tutorial. The first contribution is a tutorial overview of the SMT-LIB standard. The tutorial provides students and application developers with an introduction to SMT solvers and to using the SMT-LIB format. SMT solver developers will be very familiar with the details of SMT solver implementations, but not necessarily with the SMT-LIB interface; this tutorial gives them an overview and a discussion of fine points. The tutorial includes the following:

- quick start examples that enable an impatient reader to learn by example and experimentation;

- an informal but detailed description of the syntax, sort and expression languages, and command language, with examples;
- an overview of the different logics and theories that SMT-LIB provides;
- a description of available tools for SMT-LIB;
- a survey of recent SMT solvers.

The tutorial is not intended to be an introduction to first-order logic or SMT solver implementations, though some important concepts are presented briefly to put the SMT-LIB format in context. It is also not a formal definition of SMT-LIB; rather it is an informal, but correct and detailed description of the capabilities and use of the language. Just as occurs with software development, the writing of a tutorial by someone not an author of the original standard was effective in ferreting out misstatements, ambiguities and omissions in the first versions of the SMT-LIB v.2 definition.[1]

The first version of the tutorial[2], incorporating a round of review and correction of typos, is available from http://www.grammatech.com/resources/smt/SMTLIBTutorial.pdf. Other resources described in this paper are also available at http://www.grammatech.com/resources/smt.

jSMTLIB: An SMT-LIB Type-checker. A Java parser and type-checker for SMT-LIB was written as a tool, called jSMTLIB, that validates SMT-LIB scripts and constitutes an alternate implementation of the standard. This exercise was also effective in uncovering unspecified aspects of the language.[3]

The tool is packaged as an executable jar file, and operates on SMT-LIB command scripts. In type-checking mode, it will report any syntax errors, mis-sorted (mis-typed) expressions, and errors in the use of the command language; the tool does not do any satisfiability checking. It is also the front-end to the adapters and the basis for the API described below. The tool accepts the standard concrete syntax defined by SMT-LIB. However, the tool is readily extensible; alternate parsers and printers can be written for alternate syntax, and translation tools are easily created.

An alpha version of the jSMTLIB software package is available from http://www.grammatech.com/resources/smt/jSMTLIB.tar.

A Network Service. The command script validator takes its input from files. A simple modification allows the input to come through a network port. This enables jSMTLIB to act as a network service for checking SMT-LIB input; a companion piece of software is a client. Though there are no immediate plans to do so, this functionality could provide a publicly available network service for experimenting with SMT-LIB.

---

[1] As acknowledged in the preface of [4]. A few examples are correcting inconsistencies with the definition, clarification of unspecified behavior, setting restrictions on the order of commands, and clarifying the allowed escape characters in strings.

[2] As of 13 February 2011.

[3] Some, but not all, of these have been discussed on the smt-lib@cs.nyu.edu or smt-api@cs.nyu.edu mailing lists. Others have been resolved privately and are being included in the next (as of 12/21/2010) revision of the standard.

jSMTLIB: SMT Solver Adapters. The SMT-LIB v.2 language is relatively new and few, if any, current SMT solvers completely adhere to it as yet. Solvers under active development will likely eventually be compliant. However, adjusting to a new input language is not the highest priority for some development groups, and some tools, such as Simplify [9], are no longer under development but still frequently used. Thus adapters that convert SMT-LIB to the input language of some existing but non-compliant tools are useful.

The SMT-LIB parser produces, internally, an abstract syntax tree of the parsed SMT-LIB command script. It is a straightforward matter to translate the parsed tree into the input syntax of existing SMT tools. What is not so straightforward, but still possible, is to adjust to other differing aspects of SMT tools: some tools require all function symbols and sorts to be declared, others do not; some distinguish terms and formulas, others do not; some implement a sorted first-order logic, others use unsorted first-order logic.

The tool contains preliminary implementations of these adapters on the Windows OS: Simplify 1.5.4 [9], CVC3 2.2 [5], Yices 1.0.28 [10], and Z3 2.11 [7].

jSMTLIB: A Java API. The jSMTLIB software package can also be used as a Java API and library for a tool that wishes to link in its parsing, printing, and type-checking functionality; such a tool would thereby have access to the back-end interfaces to compliant and, through the adapters described above, non-compliant SMT solvers. The OpenJML [6] project is using this capability to connect SMT solvers performing static analysis and verification proofs to OpenJML's implementation of the Java Modeling Language using OpenJDK (cf. http://openjdk.java.net/).

The SMT-LIB standard explicitly did not standardize the abstract syntax of the language, only one instance of a concrete syntax. Nevertheless, the abstract syntax proved to be a good abstract interface for an API for SMT-LIB. ASTs are built using the abstract interface; a parser for the concrete syntax uses object factories to produce concrete AST node instances. Visitors over the AST implement type-checkers, printers (to any concrete syntax) or translators (for specific SMT solvers). The design enables easy implementation of other concrete syntaxes and other extensions of the standard.

Note that the API allows direct linking into the jSMTLIB library; the library still communicates with solvers through text-based (SMT-LIB-based) inter-process communication channels. This is not a generic API directly into each solver's implementation.

An Eclipse Plug-in for jSMTLIB. The jSMTLIB library can also be connected to the Eclipse (www.eclipse.org) IDE through a conventional Eclipse plug-in. The overall functionality is the same as the command-line version of jSMTLIB, but the IDE environment may be more accessible to students or for small-scale experimentation with SMT-LIB and SMT solvers.

The plug-in provides this functionality:

- a GUI text editor customized to SMT-LIB with syntax coloring, allowing creation and editing of files containing SMT-LIB command scripts, automatically associated with the .smt2 filename suffix
- syntax and type-checking with errors reported by Eclipse problem markers
- Eclipse menu items to perform actions on SMT-LIB files, interfaces for setting preferences, and the ability to associate keyboard combinations with menu commands
- the ability to send scripts to a choice of backend solvers for evaluation, receiving the responses within the Eclipse environment
- (in later versions) the ability to explore proofs and counterexamples using GUI interactions
- (in later versions) the ability to refactor SMT-LIB scripts

The plug-in (alpha release) is available from a typical Eclipse plug-in site: http://www.grammatech.com/resources/smt/jSMTLIB-UpdateSite .

Validating SMTLIB-compliant Solvers. Part of the development of the jSMTLIB library and application included the creation of a test suite. The test suite also constitutes a compliance test suite for solvers that seek to be SMT-LIB compliant. The test suite does not test the actual solving capabilities of the solvers — that is the task of the SMT-COMP competitions. Rather the tests determine whether all of the standard commands and expressions are properly accepted and produce the appropriate responses.

The results of compliance tests applied to SMT solvers are not given here, both because space is limited and because those tools are under active development. None of the discrepancies found affect the core utility of SMT solvers — determining satisfiability—but, they do affect their usability in a setting that expects an SMT-LIB-compliant interface. This compliance suite should be viewed and used as an aid to more compliant SMT solver interfaces.

Until some nuances of the SMT-LIB definition are resolved, documented, and can be implemented in the validation suite, the suite is available by email from the author, rather than by download.

Extending jSMTLIB. The jSMTLIB software library is designed to be extensible. Java naturally provides extension in one dimension by inheritance; the library uses Java reflection to provide extension in multiple dimensions. The library can readily be extended to provide

- parsers and printers for new concrete syntaxes,
- additional commands,
- additional solver adapters,
- additional logics and theories (some aspects of SMT-LIB logics and theories must currently be built-in),
- and, in a future version, additional kinds of expressions.

Extending jSMTLIB will allow a researcher or developer to provide additional functionality that is still integrated with the rest of SMT-LIB; it also allows easy experimentation with features proposed for future versions of SMT-LIB.

3   Availability

The materials described in this paper are available from http://www.grammatech. com/resources/smt. During the two months since the first announcement, there have been emails giving feedback and expressing appreciation. In addition an MSc thesis using jSMTLIB has been started.

4   Future Work

The tools and documents described here are under active development. E-mail feedback on the documents has already been incorporated into revisions or is slated for later major additions. The pace and direction of future work will be driven by feedback, interest and needs of users. The primary items currently envisioned are these (as of 15 February 2011):

- for the tutorial:
  addition of examples and case studies;
  more information on the logics and theories for new users;
  maintaining currency with the SMT-LIB language as it evolves;
  expansion of the discussion of current SMT solvers and related tools
- the jSMTLIB tool, library, and user guide:
  completion and refactoring of some language features (e.g. par definitions in theories);
  complete the API and validate it with use cases in a user guide;
  complete and add to the set of adapters, expanding to Linux implementations as well;
  fill out the user guide
- the Eclipse plugin:
  adding additional editor options, including word-completion, word-wrapping and pretty printing;
  customizing and expanding built-in choices such as syntax coloring;
  additional navigation short-cuts;
  exploring proofs and counterexamples through the GUI;
  refactoring capabilities
- the validation suite:
  expand the range of tests;
  document the performance of many current solvers

5   Related Tools and Materials

The formal definition of SMT-LIB is the technical report by Barrett, Stump and Tinelli [4]. The most recent version is dated December 21, 2010. The most recent SMT-COMP, its results, and the participating solvers are described on http://www.smtcomp.org/2010/, which also has links to the competitions of previous years. Barrett et al. [2] describes the results of the 2007 competition; journal publications about the more recent competitions will appear.

The tools described in this paper are implemented for Java. Some tools have also been announced for other languages: a parser and lexer implemented in C99 [11], an parser in OCaml [14], and one in Haskell [12].

References

1. Barrett, C., Conway, C.: Leveraging SMT: Using SMT Solvers to Improve Verification; Using Verification to Improve SMT Solvers. Technical report, Department of Computer Science, New York University (2010)
2. Barrett, C., Deters, M., Oliveras, A., Stump, A.: Design and results of the 3rd annual satisfiability modulo theories competition (SMT-COMP 2007). International Journal on Artificial Intelligence Tools (IJAIT) 17(4), 569–606 (2008)
3. Barrett, C., Stump, A., Tinelli, C.: The SMT-LIB Standard: Version 2.0. In: Gupta, A., Kroening, D. (eds.) Proceedings of the 8th International Workshop on Satisfiability Modulo Theories, Edinburgh, England (2010)
4. Barrett, C., Stump, A., Tinelli, C.: The SMT-LIB Standard: Version 2.0. Technical report, Department of Computer Science, The University of Iowa (2010)
5. Barrett, C.W., Tinelli, C.: CVC3. In: Damm, W., Hermanns, H. (eds.) CAV 2007. LNCS, vol. 4590, pp. 298–302. Springer, Heidelberg (2007)
6. Cok, D.: Improved usability and performance of SMT solvers for debugging specifications. International Journal on Software Tools for Technology Transfer (STTT) 12, 467–481 (2010); 10.1007/s10009-010-0138-x
7. de Moura, L., Bjørner, N.S.: Z3: An Efficient SMT Solver. In: Ramakrishnan, C.R., Rehof, J. (eds.) TACAS 2008. LNCS, vol. 4963, pp. 337–340. Springer, Heidelberg (2008)
8. Deters, M.: http://www.smtcomp.org/2010
9. Detlefs, D., Nelson, G., Saxe, J.B.: Simplify: a theorem prover for program checking. Journal of the ACM 52(3), 365–473 (2005)
10. Dutertre, B., De Moura, L.: The yices SMT solver. Technical report (2006)
11. Griggio, A.: https://es.fbk.eu/people/griggio/misc/smtlib2parser.html
12. Hawkins, T.: http://hackage.haskell.org/package/smt-lib
13. Ranise, S., Tinelli, C.: The SMT-LIB Format: An Initial Proposal. In: Proceedings of the 1st Workshop on Pragmatics of Decision Procedures in Automated Reasoning, Miami Beach, USA (2003)
14. Tinelli, C.: http://www.smt-lib.org

opaal: A Lattice Model Checker

Andreas Engelbredt Dalsgaard, René Rydhof Hansen, Kenneth Yrke Jørgensen,
Kim Gulstrand Larsen, Mads Chr. Olesen, Petur Olsen, and Jiří Srba

Department of Computer Science, Aalborg University,
Selma Lagerlöfs Vej 300, DK-9220 Aalborg East, Denmark
{andrease,rrh,kyrke,kgl,mchro,petur,srba}@cs.aau.dk

Abstract. We present a new open source model checker, opaal, for automatic verification of models using lattice automata. Lattice automata
allow the users to incorporate abstractions of a model into the model
itself. This provides an efficient verification procedure, while giving the
user fine-grained control of the level of abstraction by using a method
similar to Counter-Example Guided Abstraction Refinement. The opaal
engine supports a subset of the UPPAAL timed automata language extended with lattice features. We report on the status of the first public
release of opaal, and demonstrate how opaal can be used for efficient
verification on examples from domains such as database programs, lossy
communication protocols and cache analysis.

1 Introduction

Common to almost all applications of model checking is the notion of an underlying concrete system with a very large—or sometimes even infinite—concrete
state space. In order to enable model checking of such systems, it is necessary to
construct an abstract model of the concrete system, where some system features
are only modelled approximately and system features that are irrelevant for a
given verification purpose are "abstracted away".

The opaal model checker described in this paper allows for such abstractions
to be integrated in the model through user-defined lattices. Models are formalised
by *lattice automata*: synchronising extended finite state machines which may
include lattices as variable types. The lattice elements are ordered by the amount
of behaviour they induce on the system, that is, larger lattice elements introduce
more behaviour. We call this the *monotonicity property*. The addition of explicit
lattices makes it possible to apply some of the advanced concepts and expressive
power of abstract interpretation directly in the models.

Lattice automata, as implemented in opaal, are a subclass of well-structured
transition systems [1]. The tool can exploit the ordering relation to reduce the
explored state space by not re-exploring a state if its behaviour is *covered* by
an already explored state. In addition to the ordering relation, lattices have a
join operator that joins two lattice elements by computing their least upper
bound, thereby potentially overapproximating the behaviour, with the gain of a
reduced state space. Model checking the overapproximated model can however

M. Bobaru et al. (Eds.): NFM 2011, LNCS 6617, pp. 487–493, 2011.

be inconclusive. We introduce the notion of a *joining strategy* affording the user more control over the overapproximation, by specifying which lattice elements are joinable. This allows for a form of user-directed CEGAR (Counter-Example Guided Abstraction Refinement) [2,3]. The CEGAR approach can easily be automated by the user, by exploiting application-specific knowledge to derive more fine-grained joining strategies given a spurious error trace. Thus providing, for some systems and properties, efficient model checking and conclusive answers at the same time.

The `opaal` model checker is released under an open source license, and can be freely downloaded from our webpage: `www.opaal-modelchecker.com`. The tool is available both in a GUI and CLI version, shown in Fig. 1. The UPPAAL [4] GUI is used for creation of models.

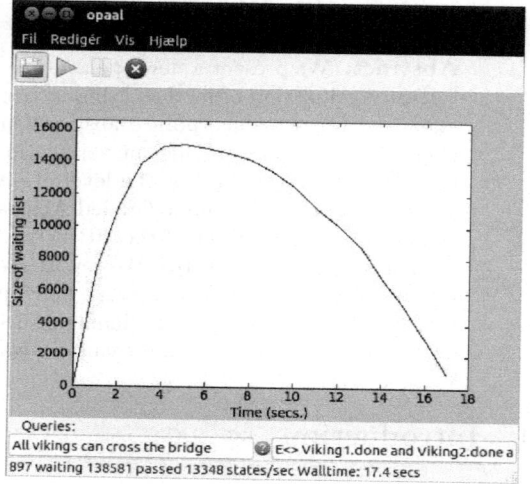

The `opaal` tool is implemented in Python and is a stand-alone model checking engine. Models are specified using the UPPAAL XML format, extended with some specialised lattice features. Using an interpreted language has the advantage that it is easy to develop and integrate new lattice implementations in the core model checking algorithm. Our experiments indicate that although `opaal` uses an interpreted language, it is still sufficiently fast to be useful.

Users can create new lattices by implementing simple Python class interfaces. The new classes can then be used directly in the

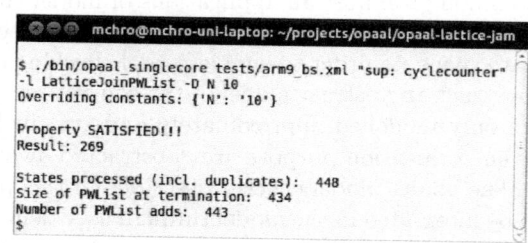

Fig. 1. `opaal` GUI and CLI

model (including all user-defined methods). Joining strategies are defined as Python functions.

An overview of the `opaal` architecture is given in Fig. 2, showing the five main components of `opaal`. The "Successor Generator" is responsible for generating a transition function for the transition system based on the semantics of UPPAAL automata. The transition function is combined with one or more lattice implementations from the "Lattice Library".

The "Successor Generator" exposes an interface that the "Reachability Checker" can use to perform the actual verification. During this process a

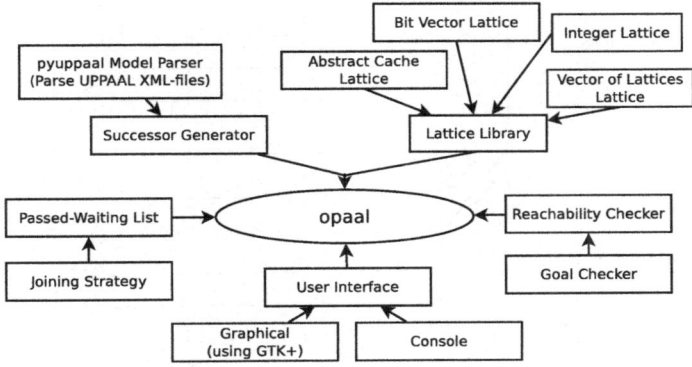

Fig. 2. Overview of opaal's architecture

"Passed-Waiting List" is used to save explored and to-be explored states; it employs a user-provided "Joining Strategy" on the lattice elements of states, before they are added to the list.

2   Examples

In this section we present a few examples to demonstrate the wide applicability of opaal. The tool currently has a number of readily available lattices that are used to abstract the real data in our examples.

2.1   Database Programs

In recent work by Olsen et al. [5], the authors propose using present-absent sets for the verification of database programs. The key idea is that many behavioural properties may be verified by only keeping track of a few representative data values.

This idea can be naturally described as a lattice tracking the definite present- and absent-ness of database elements. In the model, this is implemented using a bit-vector lattice. For the experiment we adopt a model from [5], where users can login, work, and logout. The model has been updated to fit within the lattice framework, as shown in Fig. 3(a). In the code in Fig. 3(b), the construct **extern** is used on line 3 to import a lattice from the library. Subsequently two lattice variables, pLogin and aLogin, are defined at line 4 and 5, both vectors of size N_USERS. The lattice variables are used in the transitions of the graphical model, where e.g. a special method "num0s()" is used to count the number of 0's in the bitvector. The definition of a lattice type in Fig. 3(c) is just an ordinary Python class with at least two methods: join and the ordering.

We can verify that two users of the system cannot work at the same time using explicit exploration, or by exploiting the lattice ordering to do cover checks, see Fig. 4.

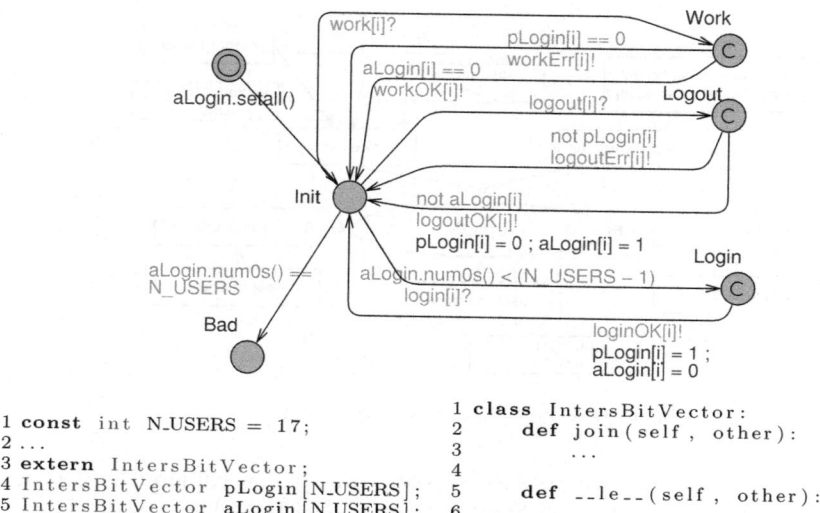

```
1 const int N_USERS = 17;
2 ...
3 extern IntersBitVector;
4 IntersBitVector pLogin[N_USERS];
5 IntersBitVector aLogin[N_USERS];
```

```
1 class IntersBitVector:
2     def join(self, other):
3         ...
4
5     def __le__(self, other):
6         ...
```

Fig. 3. (a) Database model (b) Lattice variables (c) Lattice library (in Python)

Number of users	explicit exploration	cover check
2	224 (<1s)	56 (<1s)
3	2352 (2s)	336 (<1s)
4	21952 (28s)	1792 (2s)
5	192080 (8:22m)	8960 (9s)
6	-	43008 (48s)
7	-	200704 (4:38m)

Fig. 4. Explored states and time for the property "no two users work at the same time"

Another property to check is that the database cannot become full. For this property we can exploit a CEGAR approach: A naïve joining strategy will give inconclusive results, but refining the joining strategy not to join two states if the resulting state has a full database, leads to conclusive results while still preserving a significant speedup, see Fig. 5.

2.2 Asynchronous Lossy Communication Protocol: Leader Election

Communication protocols where messages are asynchronously passed via an unreliable (lossy and duplicating) medium can be modelled as a lattice automaton. As long as we are interested in safety properties, such a communication can be modelled as a set of already sent messages called *pool*. Initially the set *pool* is empty. Once a message it sent, it is added to the set *pool* and it remains

Number of users	explicit exploration	joining (naïve strategy)	joining (refined strategy)
8	6312 (15s)	(Inconclusive) 51 (<1s)	787 (1s)
9	14228 (56s)	(Inconclusive) 57 (<1s)	1238 (2s)
10	31614 (4:19m)	(Inconclusive) 63 (<1s)	976 (2s)
11	69478 (21:35m)	(Inconclusive) 69 (<1s)	1036 (2s)
12	-	(Inconclusive) 75 (<1s)	1707 (3s)
16	-	(Inconclusive) 99 (<1s)	25900 (4:18m)
17	-	(Inconclusive) 105 (<1s)	66490 (25:01m)

Fig. 5. Explored states and time for the property "database cannot become full"

Number of agents	explicit exploration	cover check	joining
5	840 (5s)	37 (<1s)	17 (<1s)
6	5760 (5:20m)	58 (<1s)	23 (<1s)
7	45360 (671:02m)	86 (1s)	30 (<1s)
15	-	682 (4:21m)	122 (2s)
25	-	2927 (283:16m)	327 (12s)
50	-	-	1277 (4:19m)
100	-	-	5052 (98:45m)

Fig. 6. Explored states and time for the leader election protocol

there forever (duplication). As the protocol parties are not forced to read any message from *pool* and we ask about safety properties, lossiness is covered by the definition too.

It is obvious that 2^{pool}, i.e. the set of all subsets of *pool*, together with the subset ordering is a complete lattice. As long as the set of messages is finite and all parties in the protocol behave in the way that their steps are conditioned only on the presence of a message in the pool and not on its absence, the system will satisfy the monotonicity property and we can apply our model checker.

We have modelled the asynchronous leader election protocol [6] in opaal. Here we have N agents with their unique identifications $0, 1, \ldots, N-1$ and they select a leader with the highest id. Experimental data, for the property that only the agent with the highest id can become leader, are provided in Fig. 6. The cover check column refers to using only the monotonicity property to reduce the explored state-space. We can see that while being exact (no overapproximation), the speed-up is considerable. Moreover, using the join strategy provides even more significant speed-up while still providing conclusive answers.

2.3 Cache Analysis

To ensure safe scheduling of real-time systems, the estimation of Worst-Case Execution Time (WCET) of each task in a given system is necessary [7]. One major part of determining WCETs for modern processors is accounting for the effects of the memory cache. Efficient abstractions exist for analysing some types of caches [8], which we have implemented as a lattice. By recasting the cache analysis into our framework we gain the ability to give WCET guarantees, and gradually refine those guarantees by being more and more concrete with respect to the data-flow of the program.

On a simple program (binary search in array of size 100) and a simple cache we get the same WCET using all approaches. The complete state space has 5726 states (computed in 6s), cover update reduces this to 4043 states (3s), while join only needs to store 3944 states (3s). On more complex examples join will start to give overapproximated guarantees, which can be further refined.

2.4 Timed Automata

It is well-known that the theory of *zones* of timed automata (see e.g. [9,10]) is a finite-state abstraction of clock values with a lattice structure. A zone-lattice is currently being developed for use in opaal, but has not matured to a point where meaningful experiments can be made yet.

3 Conclusion

We presented a new model checker, opaal, for lattice automata and provided a number of applications. The expressiveness of the formalism, derived from well-structured transition systems, promises broad applicability of the tool. Our initial experiments indicate that careful abstraction using the techniques implemented in opaal lead to efficient verification.

We plan on extending the foundations of opaal to additional formalisms such as Petri nets, as well as on improving the performance of the tool by rewriting core parts in a compiled language. Of course, additional lattices and areas of application are also to be investigated.

References

1. Finkel, A., Schnoebelen, P.: Well-structured transition systems everywhere! Theoretical Computer Science 256(1-2), 63–92 (2001)
2. Henzinger, T.A., Jhala, R., Majumdar, R., Sutre, G.: Lazy abstraction. In: POPL 2002, pp. 58–70. ACM, New York (2002)
3. Ball, T., Rajamani, S.: The SLAM toolkit. In: Berry, G., Comon, H., Finkel, A. (eds.) CAV 2001. LNCS, vol. 2102, pp. 260–264. Springer, Heidelberg (2001)
4. Behrmann, G., David, A., Larsen, K.G.: A tutorial on UPPAAL. In: Bernardo, M., Corradini, F. (eds.) SFM-RT 2004. LNCS, vol. 3185, pp. 200–236. Springer, Heidelberg (2004)
5. Olsen, P., Larsen, K.G., Skou, A.: Present and absent sets: Abstraction for testing of reactive systems with databases. In: Sixth Workshop on Model-Based Testing, Paphos, Cyprus (2010)
6. Garcia-Molina, H.: Elections in a distributed computing system. IEEE Trans. Comput. 31(1), 48–59 (1982)
7. Wilhelm, R., Engblom, J., Ermedahl, A., Holsti, N., Thesing, S., Whalley, D., Bernat, G., Ferdinand, C., Heckmann, R., Mitra, T., Mueller, F., Puaut, I., Puschner, P.P., Staschulat, J., Stenstrm, P.: The Worst-Case Execution Time Problem - Overview of Methods and Survey of Tools. Trans. on Embedded Comp. Sys. 7(3), 1–53 (2008)

8. Alt, M., Ferdinand, C., Martin, F., Wilhelm, R.: Cache Behavior Prediction by Abstract Interpretation. In: Cousot, R., Schmidt, D.A. (eds.) SAS 1996. LNCS, vol. 1145, pp. 52–66. Springer, Heidelberg (1996)
9. Henzinger, T., Nicollin, X., Sifakis, J., Yovine, S.: Symbolic model checking for real-time systems. Information and Computation 111(2), 193–244 (1994)
10. Bengtsson, J., Yi, W.: Timed automata: Semantics, algorithms and tools. In: Desel, J., Reisig, W., Rozenberg, G. (eds.) Lectures on Concurrency and Petri Nets. LNCS, vol. 3098, pp. 87–124. Springer, Heidelberg (2004)

A Tabular Expression Toolbox for Matlab/Simulink

Colin Eles* and Mark Lawford*

McMaster Centre for Software Certification
McMaster University, Hamilton, Ontario, Canada L8S 4K1
{elesc,lawford}@mcmaster.ca

Abstract. Tabular expressions have been successfully used in developing safety critical systems, however insufficient tool support has hampered their wider adoption. To address this shortfall we have developed the Tabular Expression Toolbox for Matlab/Simulink[1]. An intuitive user interface allows users to easily create, modify and check the completeness and disjointness of tabular expressions using the ATP PVS or SMT solver CVC3. The tabular expressions are translated to m-functions allowing their seamless use with Matlab's simulation and code generation.

1 Introduction

Model based design (MBD) has gained increased industrial acceptance, but successful commercial tools such as Matlab/Simulink lack formal semantics and notations that would directly support formal methods. On the other hand formal (and semi-formal) methods have not provided support tools that integrate with existing industrial software development practices and typically overburden developers with the complexity of their formal notation and user interfaces. For example, tabular expressions provide a formal method of specifying mathematical functions that are readable by domain experts, but inadequate tool support has hampered industrial adoption of tabular expressions. To address these problems we have designed a tabular expression toolbox for Matlab/Simulink. We support some of the most common types of tables and have designed the toolbox to allow easy extension to other table types.

The rest of the paper is organized as follows. In Section 2 we discuss some background on work done on formalizing Simulink and existing table tools. Section 3 presents some preliminary information on tabular expressions, and Matlab/Simulink. In Section 4 we provide details of the development and use of the toolbox. A case study using the toolbox is briefly presented in section 5.

* Supported by the Ontario Research Fund, and the National Science and Engineering Research Council of Canada.
[1] Toolbox is available from
http://www.mathworks.com/matlabcentral/
fileexchange/28812-tabular-expression-toolbox

M. Bobaru et al. (Eds.): NFM 2011, LNCS 6617, pp. 494–499, 2011.

2   Related Work

In the area of formalizing Simulink models Roy and Shankar [2] describe a tool that provides a richer type system for Simulink diagrams. Whalen *et al.* [3] have discussed an integrated approach for formally analyzing model based designs using a combination of commercial and custom tools. Tiwari [10] has proposed a method for formally analyzing Simulink and Stateflow models using push down automata.

Examples of table tools for software engineering include an Eclipse IDE plugin for designing tabular expressions using the OMDoc language to represent tables and PVS for verification purposes [4]. The SCR* toolset [5] supports a wide variety of tables and formal analysis techniques but is only available under strict licensing terms. Our work differs from these tools by providing an open source toolset for formally checking a Simulink tabular expression block, or set of tabular expression blocks from within Matlab. We do not attempt to verify the entire model. The Toolbox is the first attempt to integrate tabular methods with a widely available commercial MBD framework.

3   Preliminaries

3.1   Tabular Expression

To specify software, designers often need to describe what should be done for different equivalence classes of inputs. It has been shown that tables provide a formal yet convenient way to specify these functions [6]. We believe that tabular expressions allow for easier readability of documentation, and facilitate inspection of completeness and consistency of specified functionality. Below, a simple example is used to explain tabular expressions. For a detailed discussion of tabular expressions semantics, we refer the reader to Jin and Parnas [7].

In Fig. 1 a formal logical specification of a function appears on the left and its semantically equivalent two dimensional tabular expression as displayed by the toolbox appears on the right. For the one-dimensional function table in (1), we require the Boolean conditions in x, y (the c_i's) in the table's *predicate grid* to be complete (2) and disjoint (3). We can then return the unique value of the expression e_i from the *output grid* when c_i is true. Disjointness ensures that the specified function is deterministic, and completeness guarantees that we have considered all possible inputs, both critical properties for safety applications. Although the graphical layout of tables lends itself to ease of visual inspection of these properties, we would prefer to automate the checking of these obligations, to avoid human errors.

$$f(x, y) \stackrel{\mathrm{df}}{=} \left\{ \begin{array}{|c|c|c|c|} \hline c_1 & c_2 & \dots & c_n \\ \hline e_1 & e_2 & \dots & e_n \\ \hline \end{array} \right. \tag{1}$$

$$\textbf{disjointness} \stackrel{\mathrm{df}}{=} i \neq j \rightarrow (c_i \wedge c_j \leftrightarrow \bot) \tag{2}$$

$$\textbf{completeness} \stackrel{\mathrm{df}}{=} (c_1 \vee c_2 \vee \dots \vee c_n) \leftrightarrow \top \tag{3}$$

$$f(x,y) \stackrel{\mathrm{df}}{=} \begin{cases} x + y & \text{if } x > 1 \wedge y < 0 \\ x - y & \text{if } x \leq 1 \wedge y < 0 \\ x & \text{if } x > 1 \wedge y = 0 \\ xy & \text{if } x \leq 1 \wedge y = 0 \\ y & \text{if } x > 1 \wedge y > 0 \\ x/y & \text{if } x \leq 1 \wedge y > 0 \end{cases}$$

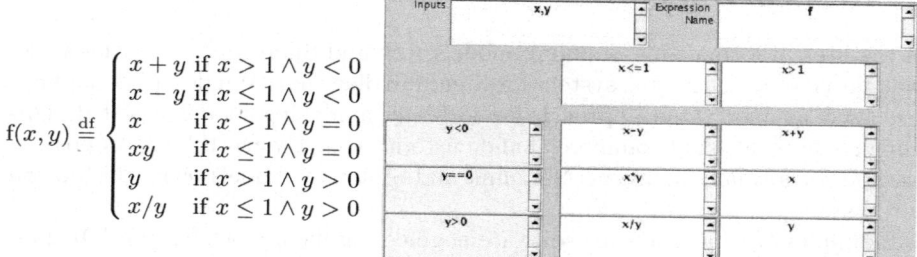

Fig. 1. A logical description of a function and its tabular expression

3.2 Matlab/Simulink

The Simulink modeling language is considered the *de facto* Model Based Design platform for industrial MBD applications [1]. A commonly cited problem with Matlab/Simulink is its lack of formal semantics. We believe that by considering a smaller "safe" subset of the Matlab language, we will be able to convincingly argue that the semantics of Matlab tables are consistent with target languages[1,3].

4 Toolbox

We desired an intuitive user interface that facilitated tabular expression creation and editing, code generation, verification, and graphical counter-example generation. Rather than building all of components for such a tool from the ground up, we have leveraged the power of existing tools, namely Matlab/Simulink, PVS and CVC3. The toolbox combines these tools, integrating them into the Simulink workflow while hiding the detailed verification steps from the end user.

The tool currently supports one and two dimensional normal function tables and also allows for sub-grids in one dimension. The toolbox generates embedded Matlab code for each table which can be saved to a Simulink block or to an M-file. Thus once created, a table can be immediately executed, integrated with other Matlab scripts and functions, or used to generate code for the target platform.

We support predicate subtyping on inputs/outputs of tables, as both CVC3 and PVS have support for predicate subtyping. The complexity of conditional and output expressions is only limited by that of the embedded Matlab language, as well as the capabilities of the backend verification languages. As complexity of expressions increases the verification time generally increases and the chance of finding a counter-example decreases.

4.1 Model

Our model of tabular expressions is similar to that presented by Jin and Parnas [7]. We identify two different constituents which are related to form the table; one or more predicate grids and an output grid. We can evaluate a grid by locating the root predicate grids, determine the true predicate cell, if the cell

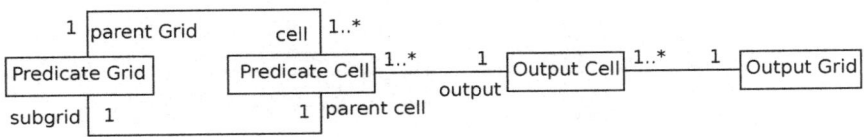

Fig. 2. Class Diagram

has a sub-grid we then evaluate that recursively, otherwise we select the output cell. A class diagram of this model is presented in Fig. 2, where the numbers on the relationships represent the multiplicity of the classes.

4.2   Table Toolbox GUI

The current version of the toolbox has been developed using the Matlab GUI API. The Tool has been integrated into Matlab/Simulink so that users do not have to leave the primary development environment to use the tool. In Fig. 3 a screenshot of the current version of the tool shows a table that has failed the disjointness check and the m-function generated from the table has been executed and viewed to assist in debugging. The table can be fixed by changing the "||" to "&&" in the second predicate. Colour coding is used to differentiate the columns that overlap for the counter example generated by CVC3. Both c_1 and c_2 are one colour, since they are true, while c_3 is a different colour to indicate it is false.

4.3   Verification and Validation

For V & V of functions we have chosen to use the theorem prover PVS [8], as well as the SMT solver CVC3 [9]. These tools offer a diverse approaches to verifying the disjointness (2) and completeness (3) conditions.

PVS has built-in support for tabular expressions, so the toolbox only needs to generate PVS for the table and then PVS generates proof obligations for (2) and (3). We make use of the random-testing functionality of PVS to attempt to find counterexamples to unprovable obligations.

We use CVC3 for the same purpose as PVS, to ensure that tables are disjoint and complete. As CVC3 does not directly support tabular expressions, we must generate the obligations (2) and (3) for CVC3 in the form of queries which are pushed onto the proof stack. CVC3 will output a counter-model of a query if it is shown to be invalid.

Strategy. We have found that by utilizing tools based on different technologies and theoretical models, we can solve a greater variety of problems. While the automatic proof strategies are often adequate for simple tables, PVS allows users to manually control proofs when required for more complicated obligations. The cost of this flexibility is a larger overhead and greater time required to prove. CVC3 as an SMT solver does not allow for guided proofs but in our experience is

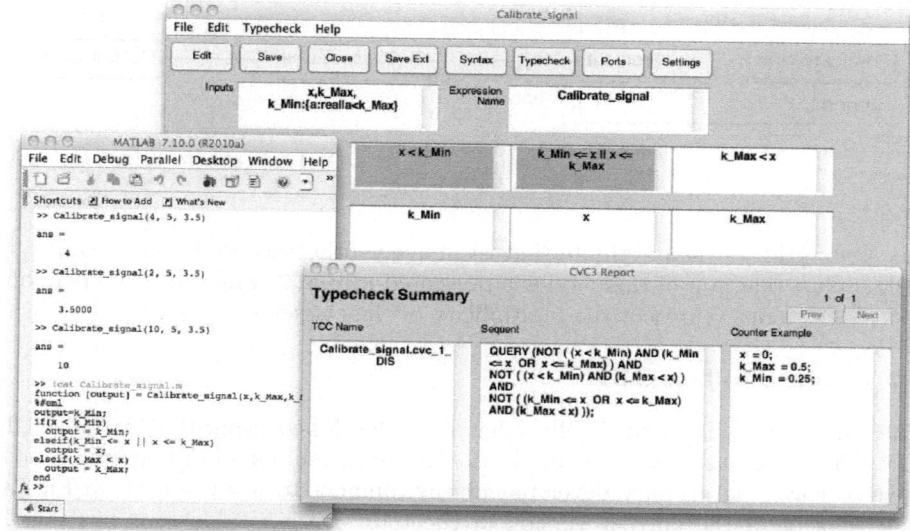

Fig. 3. Screenshot of Tabular Expression Toolbox

very fast and guarantees generation of counter-example via its constraint solving when it is applicable. For tables involving nonlinear expressions, PVS generally gives better results as CVC3 if incomplete for nonlinear constraints.

When both tools are applicable, having two diverse tools to check the tables has the potentially benefit of mitigating against an error in one of the tools. In regulated industries this has the effect of lowering the level of rigour required for qualification of a V & V tool.

5 Case Study

The table tool was used to model a power estimation module based upon requirements for the shutdown system of a nuclear power plant. This system, previously described in Wassying and Lawford [6], used tabular expressions in word documents to document the software requirements. Based upon the requirements document, the example module was implemented in Simulink by an undergraduate student.The module design used 42 different tabular expressions for the power estimation module. Upon typechecking the implemented tabular blocks in the model it was discovered that two contained typographical errors which affected the blocks functionality but would not produce syntax errors or compilation errors. Fig. 3 is an example of one of the detected errors. Errors of this nature are very easy to fix if detected immediately, and become much more difficult and expensive to detect and correct later in the development life cycle.

6 Conclusion and Future Work

Early case studies and feedback have shown that the Tabular Expression Toolbox can be a great asset in developing requirements, designs and implementations in a MBD software development process. By leveraging the power and diversity of tools such as PVS and CVC3 we achieve a greater level of assurance of table correctness. We have managed to hide the formal verification process "under the hood" leaving the designer to concentrate on the design of their system rather than having to learn two new, diverse formal system. Future goals of this project involve additional case studies; consideration of inter-block typing issues, including leveraging existing tools [2]; as well as a more detailed investigation of the consistency of the semantics of Matlab and the analysis tools employed.

References

1. Scaife, N., Sofronis, C., Caspi, P., Tripakis, S., Maraninchi, F.: Defining and trans-lating a "safe" subset of simulink/stateflow into lustre. In: EMSOFT 2004, pp. 259–268. ACM, New York (2004)
2. Roy, P., Shankar, N.: SimCheck: An expressive type system for Simulink. In: NASA Formal Methods Symposium, pp. 149–160 (2010)
3. Whalen, M., Cofer, D., Miller, S., Krogh, B.H., Storm, W.: Integration of formal analysis into a model-based software development process. In: Leue, S., Merino, P. (eds.) FMICS 2007. LNCS, vol. 4916, pp. 68–84. Springer, Heidelberg (2008)
4. Peters, D.K., Lawford, M., Trancon y Widemann, B.: An IDE for software devel-opment using tabular expressions. In: CASCON 2007, pp. 248–251. ACM, Toronto (2007)
5. Bharadwaj, R., Heitmeyer, C.: Developing high assurance avionics systems with the SCR requirements method. In: Proceedings of the 19th on Digital Avionics Systems Conferences, DASC, vol. 1, pp. 1D1/1–1D1/8 (2000)
6. Wassyng, A., Lawford, M.: Lessons learned from a successful implementation of formal methods in an industrial project. In: Araki, K., Gnesi, S., Mandrioli, D. (eds.) FME 2003. LNCS, vol. 2805, pp. 133–153. Springer, Heidelberg (2003)
7. Jin, Y., Parnas, D.L.: Defining the meaning of tabular mathematical expressions. Science of Computer Programming 75(11), 980–1000 (2010)
8. Owre, S., Rushby, J.M., Shankar, N.: PVS: A prototype verification system. In: Kapur, D. (ed.) CADE 1992. LNCS (LNAI), vol. 607, pp. 748–752. Springer, Hei-delberg (1992)
9. Barrett, C., Tinelli, C.: CVC3. In: Damm, W., Hermanns, H. (eds.) CAV 2007. LNCS, vol. 4590, pp. 298–302. Springer, Heidelberg (2007)
10. Tiwari, A.: Formal semantics and analysis methods for Simulink Stateflow models. Technical report, SRI International (2002), http://www.csl.sri.com/~tiwari/stateflow.html

LLVM2CSP: Extracting CSP Models from Concurrent Programs

Moritz Kleine, Björn Bartels, Thomas Göthel, Steffen Helke, and Dirk Prenzel

Technische Universität Berlin
Department of Software Engineering and Theoretical Computer Science
Berlin, Germany
{mkleine,bbartels,tgoethel,helke,prenzel}@cs.tu-berlin.de

Abstract. In this paper, we present the *llvm2csp* tool which extracts CSP models from the LLVM compiler intermediate representation of concurrent programs. The generation of CSP models is controlled by user annotations and designed to create models of different levels of abstraction for subsequent analysis with standard CSP tools.

1 Introduction

Communicating Sequential Processes (CSP) [9] is a mature formalism supporting the design and verification of safety critical concurrent systems. However, verifying that a concurrent C/C++ implementation refines such a CSP design remains a major obstacle. In [4] we present our approach to the verification of concurrent C/C++ programs using CSP. The approach is based on the automated extraction of CSP models from the Low Level Virtual Machine (LLVM) [5] compiler intermediate representation (IR) of programs. The models are tailored for subsequent analysis with established CSP tools such FDR [9], ProB [6] or the CSP Prover [2]. Conformance of the LLVM IR of a program and the generated CSP model can be established using the approach presented in [1]. In this paper, the *llvm2csp* tool is described, which extracts CSP models from concurrent C/C++ programs. It is realized as an LLVM compiler back-end that outputs CSP models encoded in the machine-readable dialects CSP_M (for automated refinement checking with FDR or animation and LTL model checking with ProB) and CSP_{TP} (for interactive theorem proving with the CSP Prover).

Background: The LLVM compiler infrastructure [5] provides a modular framework that is designed to be extended by user-defined compilation passes and custom compiler back-ends. It also offers a diverse set of predefined analyses and optimizations that can be used out of the box. The heart of the compiler infrastructure project is its intermediate representation (IR). It is a typed assembler-like language using SSA form for all scalar register values. The LLVM framework provides gcc-based front-ends for a variety of programming languages, including C and C++.

M. Bobaru et al. (Eds.): NFM 2011, LNCS 6617, pp. 500–505, 2011.

CSP [9] is a process calculus supporting the specification and verification of concurrent systems. It is based on *events* and *processes*. Processes are computations observable only by the sets of events that they may accept or refuse at any point in their evolution. Events are commonly regarded as abstractions of arbitrary actions. In the context of *llvm2csp*, events model arbitrary user-defined observation points (e.g. "a" or "debug.5") of a system or communication of a processing unit with the memory of the computing system (when observing a variable).

Related Work: SVA [10] defines a simple proprietary input language to express concurrent shared variable programs and a compiler to transform such programs into CSP$_M$ for subsequent analysis with FDR. Any feature present in the input language is translated, only type widths are reduced to rather small sets. SVA targets the analysis of concurrent algorithms on a rather low level. In [11], Scuglik and Sveda present an approach go generate CSP models from UML diagrams or while-languages. Java2CSP [12] inputs Java Bytecode and generates CSP models for analysis with FDR. This tool attempts to avoid the state space explosion problem by some built-in abstractions. Our approach also suffers from the state space explosion problem on the CSP$_M$ level when used with FDR or ProB. However, *llvm2csp* also supports generation of CSP$_{TP}$ models for interactive theorem proving and the use of annotations (ghost code) allowing the user to define his own abstractions. Using the approach described in [1] soundness of the abstractions can then be proved with the help of the operational semantics of LLVM.

2   Assembling CSP Models of Concurrent Programs

The *llvm2csp* tool is implemented as an LLVM tool, providing an LLVM backend and using additional compiler passes and analyses. It compiles concurrent C/C++ programs into CSP models. The tool outputs CSP models that come in different flavors. The first is a CSP$_M$ model particularly suited for subsequent animation and verification with ProB. Then there are two CSP$_M$ models optimized for use with FDR. The tool also outputs CSP$_{TP}$ models for verification with the CSP Prover. In each of the models LLVM instructions are translated into terminating sequential processes. Memory is divided into private and shared memory and modeled as separate processes. Composing these processes with a domain-specific scheduling process results in a CSP model reflecting the semantics of the program (with respect to its operational LLVM semantics [1]).

Figure 1 shows that models generated by *llvm2csp* consist of three parts: an application-specific part, describing the behavior of threads; a domain-specific part, which encapsulates low-level software concepts such as scheduling; and a platform-specific part, which is the hardware model. The terminating sequential processes extracted from the instructions of a program, form the application-specific part of the final model. The application-specific part, and parameters for the domain-specific and the platform-specific parts are generated by our *llvm2csp* tool, while the fixed fraction of the other parts is modeled manually.

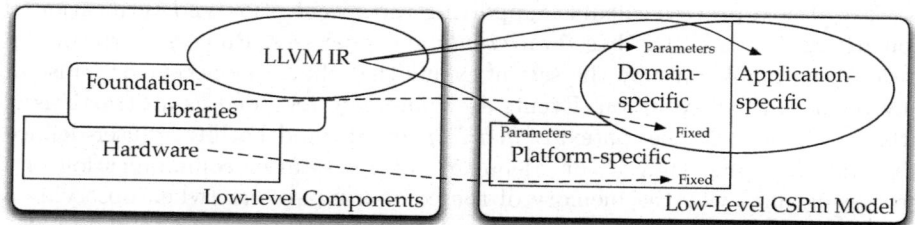

Fig. 1. Extracting CSP models from LLVM IR requires manual modeling of platform- and domain-specific parts

Parameters of these two manually modeled parts are typing information for the channels and the set of thread identifiers, for example. Automated generation is symbolized by solid and manual modeling by dashed arcs in Figure 1. While the domain-spefic part is configurable, the platform-specific part is not configurable yet and limited to a single processor machine for now.

2.1   Optimizing Models for FDR

The structure of a CSP_M model determines how efficiently it can be handled by FDR. It is, for example, important to synchronize processes as early as possible and to hide events as low in the process structure as possible. Furthermore, renaming and copying processes is to be used in preference to the parameterizing of a process definition (see [9] for details). Since the domain structure heavily influences the performance gain that can be achieved using the rules above, it is made configurable to match the application's needs. *llvm2csp* supports two predefined domain specific parts modeling a nondeterministic preemptive scheduler. In the first version, threads are modeled as a union process while the second version models threads as independent processes.

The combined union process contains the instructions of the program, which are composed with a scheduler and communicate with processes modeling the system's memory. This way, the threads are implicitly defined by the union process. It is generated as follows: First, the instructions are embedded into the scheduler process, which executes them one after another, possibly handling preemption in between. The resulting process is then synchronized with the threads' private memory processes. Finally, it is synchronized with the shared memory.

In the case of multiple independent processes, the threads are first synchronized with their private memory, then copied and finally composed with the scheduler and the shared memory. This latter model is structured as follows: First, the threads' instructions are embedded into a sequential process possibly handling preemption between any two instructions. This raw thread process is then synchronized with its private memory, resulting in the basic thread process. The private memory holds all variables not meant to be manipulated by other threads. These processes can then be copied if there are multiple threads using the same code-base. Each thread is then synchronized with the scheduler process. Finally, it is synchronized with the shared memory.

The two differently structured models result in semantically equivalent processes. Which of the models performs best, depends on the number of instructions, the size of the private memory and the usage of shared memory. Experience shows that highly interacting threads are best put together in the former way, while the latter way suits independent threads better.

2.2   The *llvm2csp* User Interface

The *llvm2csp* user interface consists of annotations defined in header files and command line switches of the tool. The annotations are realized as two ghost functions [4] and a dedicated ghost constant (a configuration string named *CSP_USERSCRIPT*). Since we advocate the idea that specification and verification should reside in the same development artifacts, this ghost constant is supposed to hold the specification and verification information (e.g. the FDR assertions). The value of this ghost constant is always included in the generated CSP_M script. Although not being a native feature of C/C++, *llvm2csp* deals with concurrent programs. Since concurrency does not show up in the IR of a program, the ghost functions are used to annotate concurrency related aspects of a program (which must match the predefined platform- and domain-specific parts, of course).

The *llvm2csp_static* ghost function is used to set static values, e.g. the entry points of threads, the size of the model's integer or a maximum number of stackframes for a function. Once set, these values are fixed at runtime. Calls to this function are eliminated by the compiler after interpreting the function's parameters.

The *llvm2csp_dynamic* ghost function is used to include dynamic parts in the final model. Call sites of this function are only relevant if they reside within functions that are included in the generated model. Call sites are replaced with the elements computed from the parameters of this ghost function. This ghost function allows custom CSP_M code snippets to be inserted into the final model (to control the domain-specific part of the model by inserting user-defined events or to enable and disable preemption) or to replace modules of the program with their respective specifications, for example. Replacing parts of the program with user-defined processes allows the user to define abstractions attacking the state space explosion problem to simplify the generated models while maintaining the desired properties. For example, conditionals can be abstracted to nondeterministic (internal) choice and functions or blocks of code can be replaced by user-defined processes. Soundness of the abstractions cannot be proved within *llvm2csp* itself but is to be done using the approach presented in [1].

Context-bounded verification [8] is supported using the command line option *numswitches*. This option defines the upper bound of context switches and thus reduces the state-space of the generated model. The output format is also chosen on the command line.

Examples of how to use these ghost functions and the ghost constant are given on the *llvm2csp* demo-website `https://group.swt.tu-berlin.de/llvm2csp`.

2.3 Functions and Variables

Models generated by *llvm2csp* contain only those functions of the program that are actually called in the translated code (if not annotated otherwise). The set of functions that may get called in the code that is visible to the compiler is computed by a custom analysis. Function pointers are also supported. This analysis works with different entries for all threads and includes a cautious approximation on what function pointers may get exchanged between two locations. Recursive functions are supported up to a user-defined recursion depth, which can be set using a ghost function.

The compiler reduces the set of global variables to those that are actually used by the program. These are then treated as shared variables. They can be referenced and accessed by any thread. Shared variables, which are neither referenced by any translated instruction nor pointed to by any other shared variable, are not included in the generated model.

Local variables are treated as shared variables, because, in general, pointers to local variables can be exposed to other threads. Since many instances of a local variable can coexist, a number of them is allocated, depending on the number of threads and the allowed number of calls to the function they belong to. Any other value (e.g. constants) are inlined.

2.4 Types and Arithmetics

Booleans, bytes, integers of any size, pointers to any data type and also function pointers are all mapped to the model's integer type. Owing to the state-space explosion problem, *llvm2csp* uses reduced ranges of data-types. Its initial size is set by ghost functions. If the address space of the shared variables in the model or the total number of functions translated is too big for the initial size of the model's integer, its upper bound is adjusted accordingly. Complex data types such as classes, structs or arrays, containing any other primitive or complex types recursively are also supported.

Arithmetic is supported on the data-types mentioned above. Since the final model uses reduced ranges, arithmetic operations can produce overflows that are different from those that would occur in the original program. As described in [4], these situations are handled by so-called error code events, using the following subprocess: `error_code -> STOP` where `error_code` is a fresh event introduced to signal the incident. Using error codes ensures that no false positives occur during refinement checking and maintains the soundness of our approach.

There is currently no support to execute globals constructors before any threads start in their entry functions. The current workaround is that one thread calls all global constructors while preemption is disabled. Exception handling is not supported. Neither are floating point types, floating point arithmetic and dynamic memory allocation.

3   Conclusions

To date, *llvm2csp* is a quite generic tool supporting the generation of detailed or abstract CSP models optimized for use with FDR, ProB or the CSP Prover. Models generated for use with FDR or ProB may suffer from state space explosion. Even simple programs (just a few dozen lines of code) require severe abstractions if used with FDR. Sometimes restructuring the model or using FDR's built-in compression functions solves the problem. However, many interesting properties like deadlock-freedom are provable on a rather abstract view of a system. Thus, it is advisable to always start with generating very abstract models when using the *llvm2csp* tool. Another result obtained so far is that the modular structure of the generated models can be exploited to replace the generated models of lower-level components with their specifications. This way, compositionality of CSP can be exploited. The tool was used to generate the CSP model of an operating system scheduler [3]. It is currently being used for a case study on verifying the core components and basic applications of the BOSS operating system [7].

References

1. Bartels, B., Glesner, S.: Formal Modeling and Verification of Low-Level Software Programs. In: QSIC 2010. IEEE Computer Society, Los Alamitos (2010)
2. Isobe, Y., Roggenbach, M.: A generic theorem prover of CSP refinement. In: Halbwachs, N., Zuck, L.D. (eds.) TACAS 2005. LNCS, vol. 3440, pp. 108–123. Springer, Heidelberg (2005)
3. Kleine, M., Bartels, B., Göthel, T., Glesner, S.: Verifying the Implementation of an Operating System Scheduler. In: TASE 2009, pp. 285–286. IEEE Computer Society Press, Los Alamitos (2009)
4. Kleine, M., Helke, S.: Low Level Code Verification Based on CSP Models. In: Oliveira, M.V.M., Woodcock, J. (eds.) SBMF 2009. LNCS, vol. 5902, pp. 266–281. Springer, Heidelberg (2009)
5. Lattner, C., Adve, V.: LLVM: A Compilation Framework for Lifelong Program Analysis & Transformation. In: CGO 2004, pp. 75–85. IEEE Computer Society Press, Los Alamitos (2004)
6. Leuschel, M., Fontaine, M.: Probing the Depths of CSP-M: A new FDR-compliant Validation Tool. In: Liu, S., Araki, K. (eds.) ICFEM 2008. LNCS, vol. 5256, pp. 278–297. Springer, Heidelberg (2008)
7. Montenegro, S., Briess, K., Kayal, H.: Dependable Software (BOSS) for the BEESAT Pico Satellite. In: DASIA 2006, ESTEC (2006)
8. Qadeer, S.: The case for context-bounded verification of concurrent programs. In: Havelund, K., Majumdar, R. (eds.) SPIN 2008. LNCS, vol. 5156, pp. 3–6. Springer, Heidelberg (2008)
9. Roscoe, A.W.: The Theory and Practice of Concurrency. Prentice Hall, Englewood Cliffs (2005)
10. Roscoe, A.W., Hopkins, D.: SVA, a tool for analysing shared-variable programms. In: AVoCS 2007, pp. 177–183 (2007)
11. Scuglik, F., Sveda, M.: Automatically Generated CSP Specifications. Journal of Universal Computer Science 9(11), 1277–1295 (2003)
12. Shi, H.: Java2CSP: A System for Verifying Concurrent Java Programs. In: FM-TOOLS, pp. 111–115 (2000)

Multi-Core LTSmin:
Marrying Modularity and Scalability

Alfons Laarman, Jaco van de Pol, and Michael Weber

Formal Methods and Tools, University of Twente, The Netherlands
{a.w.laarman,vdpol,michaelw}@cs.utwente.nl

Abstract. The LTSMIN toolset provides multiple generation and on-the-fly analysis algorithms for large graphs (*state spaces*), typically generated from concise behavioral specifications (*models*) of systems. LTSMIN supports a variety of input languages, but its key feature is modularity: language frontends, optimization layers, and algorithmic backends are completely decoupled, without sacrificing performance. To complement our existing symbolic and distributed model checking algorithms, we added a multi-core backend for checking safety properties, with several new features to improve efficiency and memory usage: low-overhead load balancing, incremental hashing and scalable state compression.

1   LTSmin in a Nutshell

The LTSMIN[1] toolset serves as a testbed for our research in the design of model checking tools which sacrifice neither modularity and composability nor performance. Previously, we described general features of LTSMIN [4]: its wide support for input languages through reuse of existing implementations (MCRL, NIPSVM, DVE, MAPLE and GNA, ETF), which can be combined with algorithms for checking safety properties: *enumerative*, *distributed* and BDD-based *symbolic* reachability analysis, several language-independent on-the-fly optimizations (*local transition caching, regrouping*) [2], as well as off-line state-space minimization algorithms.

The unifying concept in LTSMIN is an **I**nterface based on a **P**artitioned **N**ext-**S**tate function. PINS connects language frontends, on-the-fly optimizations, and algorithmic backends. In Sec. 2, we describe how our new multi-core (MC) backend utilizes PINS for parallel shared-memory reachability [6] for all supported thread-safe language frontends (DVE, NIPSVM, ETF).

Our MC backend provides several new contributions in the area of high- performance model checking: multi-core load balancing (Sec. 2.1), *incremental hashing* (Sec. 3), and scalable *state compression* (Sec. 4). The latter reduces memory requirements drastically, but can also improve running time of the MC tool. This is remarkable, as compression techniques generally trade space off for computational overhead.

---

[1] http://fmt.cs.utwente.nl/tools/ltsmin/, current version: 1.6, open-source.

M. Bobaru et al. (Eds.): NFM 2011, LNCS 6617, pp. 506–511, 2011.

2   LTSmin Multi-Core Architecture

PINS carefully exposes just enough structure of the models to enable high-performance algorithms and optimizations, while remaining abstract to the specific modeling language. For the purpose of this exposition, we limit the PINS description to the state and transition representation, their dependency matrices, and the next-state function. Further details can be found elsewhere [2].

In LTSMIN, states are generally represented by fixed-length vectors of N slots: $\langle s_1, \ldots, s_N \rangle \in S$. The transition relation $\rightarrow \subseteq S \times S$ is partitioned disjunctively into K *transition groups* $(\rightarrow_1, \ldots, \rightarrow_K), \rightarrow_i \subseteq \rightarrow$. Language modules provide these subrelations. We exploit that often, a transition group depends not on the full state vector, but only on a small subset of all slots, which can be statically approximated. Hence, a $K \times N$ binary *dependency matrix* records which slots are needed per group (*read matrix*, D^R), and another records which slots are modified (*write matrix*, D^W). A value $D_{i,j}^R = 0$ indicates that all transitions in group $\rightarrow_i$ are independent of slot j, hence its value s_j can be arbitrary. A value $D_{i,j}^W = 0$ indicates that slot j will not be modified by any transition in group $\rightarrow_i$. The dependency matrices are utilized by our multi-core tool via incremental hashing and state compression.

Our multi-core backend is implemented using the pthreads library. The same reachability algorithm [6] is started in multiple threads (*workers*) that share a state storage holding the set of states already visited by the search algorithm (*closed set*). The main operation is the FINDORPUT(s) function, which (atomically) reports if state s is already present in the set and otherwise inserts it. We have shown that this architecture is at least as efficient as a widely used approach based on static (hash-based) partitioning [6], despite being simpler.

2.1   Multi-Core Load Balancing

To provide all processors with some initial work, *static load balancing* (SLB) can be used. E.g., we could (sequentially) explore a sufficiently large prefix of the state space, and partition it over all workers. In parallel, each worker then explores all states reachable from its initial partition until no unvisited states are left. This simple scheme is surprisingly effective for many models, but predictably, for some inputs it leads to bad work distribution, or starvation of workers. Therefore, we tailored a *synchronous random polling* (SRP) load balancing algorithm [9] to our multi-core setting by using atomic reads and writes on shared data.

The number of explored transitions are used as measure for the work load, since it gives a close estimation of the number of actual computations (or rather, memory accesses) performed by a worker. Our measurements show that SRP provides almost perfect work distribution (less than 1% deviation from average) with negligible overhead compared to SLB. Together with shared state storage we obtain linear scalability for the LTSMIN multi-core backend, which currently outperforms both SPIN [5] and DiVinE [1] on the BEEM benchmark set [8].

2.2   Example Use Cases

LTSMIN tool names are composed of a prefix for the language frontend and a suffix for the algorithmic backend: `<language><algorithm>`. For example, the ETF frontend in combination with the multi-core backend is named `etf2lts-mc`. *Multi-Core Reachability Analysis* using ETF can be launched with:

```
etf2lts-mc --threads=4 -s22 --lb=srp leader-7-14.etf
```

The command performs multi-core reachability with four workers (`--threads=4`) and the SRP load balancer (`--lb=srp`, default as described in `--help`). The hash table size is fixed to 2^{22} states (`-s`). This parameter needs to be chosen carefully to fit the model size or the available memory of the machine, because of our hash table design decisions [6]. Slow language frontends like NIPSVM and MCRL can optionally enable transition caching (`-c`) to speed up state generation. Caching is implemented efficiently using the dependency matrix [2].

The following command searches for deadlocks (`-d`):

```
etf2lts-mc -s22 --strategy=bfs -d --trace=trace.gcf leader-7-14.etf
```

A parallel (pseudo) breadth-first search (`bfs`) generally finds a short counter example, which is stored in file `trace.gcf` and can be analyzed in detail, for example by conversion into *comma-separated value* format (only recording differences between subsequent state vectors), and loading into a spreadsheet:

```
ltsmin-tracepp --diff trace.gcf trace.csv
```

3   Incremental State Hashing

Hash tables are a common implementation choice to represent the closed set of a search. Hence, the previously mentioned FINDORPUT(s) operation calculates the hash value of a given state s. For large state vectors and small *transition delays* (the time needed to calculate the effects of a transition on a state), hash calculations can easily take up to 50% of the overall run time (e.g., for C-compiled DVE2 models), even when using optimized hash functions. Given the observation that for most transitions $s \rightarrow s'$, the difference between s and s' are small (often in the order of 1–4 slots), *incremental hashing* has been investigated [7]. We have added an alternative scheme to LTSMIN, which is based on Zobrist hashing [10] commonly used in games like computer chess. We believe this is the first time that Zobrist's approach has been used in the context of model checking.

Zobrist hashing incrementally composes a hash value from a matrix Z of random numbers. Each random number is bound to a fixed configuration of the game, for example, pawn at H3. When the numbers are combined using the XOR ($\oplus$) operation, the hash value can be updated incrementally between different game configurations. For example, if a pawn P moves from H3 to H4, we manipulate the hash value h as follows: $h' := (h \oplus Z[P][H3]) \oplus Z[P][H4]$. Algebraic properties of $\oplus$ guarantee that a hash is unique for a configuration, independently of the path through which the configuration was reached.

The number of possible configurations of our models (slot values) is usually not known up front or too large to generate random numbers for. Therefore, we only generate a fixed amount of L numbers per state slot and map each slot value to one of them using the *modulo* operation (the Z matrix is of size $L \times N$).

Alg. 1 shows how PINS can be used to update only those slots of a state s', which (potentially) changed with respect to its predecessor s. Based on initial experimenting, we concluded that $L = 2^6$ is sufficient to yield a hash distribution at least as good as standard hash functions.[2] The size of the Zobrist matrix Z is insignificant ($4L \times N$ bytes).

The following command launches a multi-core state space exploration (reachability) with the DVE2 frontend using Zobrist hashing with $L = 2^6$ (option -z6), and a hash table of size 2^{18} (option -s18):

> **Input** : transition $s \to_i s'$
> **Input** : hash value h of s
> **Output**: hash value h' of s'
> $s = \langle s_1, \dots, s_N \rangle$
> $s' = \langle s'_1, \dots, s'_N \rangle$
> $h' \leftarrow h$
> **for** $j \in \{j \mid D^W_{i,j} = 1\}$ **do**
> $\quad h' \leftarrow h' \oplus Z[j][s_j \bmod L]$
> $\quad h' \leftarrow h' \oplus Z[j][s'_j \bmod L]$

Algorithm 1. Calculating a hash h' for successor s' of state s with hash h, using Zobrist and PINS.

```
dve22lts-mc -s18 -z6 firewire_tree.4.dve
```

While the availability of large amounts of RAM in recent years shifted the "model checking bottleneck" towards processing time (we would run out of patience before running out of memory), with our improved multi-core algorithms we can easily surpass 10 million states/sec with 16 cores, sometimes claiming memory at a rate of 1 GB/sec. This causes memory to be the bottleneck again.

4   Multi-Core State Compression

We improve the memory efficiency of our tools by introducing a multi-core version of *tree compression* [3]. The following command uses it:

```
dve22lts-mc --state=tree --threads=16 firewire_tree.5.dve
```

Compared to a hash table (--state=table, default), memory usage for the closed set drops from 14 GB to 96 MB, while the run-time decreases as well, from 5.4 sec to 3.3 sec! The model, firewire_tree.5.dve, is an extreme case because of its long state vectors of 443 integers. In Sec. 5, we show that tree compression also performs well for 250 other models from the BEEM database.

The tree structure used for compression is a binary tree of *indexed sets* which map pairs of integers to indices, starting at the fringe of the tree with the slots of a state vector [3]. To provide the necessary stable indexing efficiently, we inject all indexed sets I_k into a single table [6] by appending the set number k to the lookup key. In addition to our earlier work, the tree structure is now updated incrementally using the PINS dependency matrix.

---

[2] Results available at: http://fmt.ewi.utwente.nl/tools/ltsmin/nfm-2011/

Reducing Open Set Memory. In the above case of `firewire_tree.5.dve`, the *open set* becomes the new memory hot-spot, using 200 MB. Hence, we can also opt to only store (32-bit) references to state vectors in the open set, at the expense of extra lookup operations:

```
dve22lts-mc --state=tree --threads=16 --ref firewire_tree.5.dve
```

This reduces the memory footprint of the open set from 200 MB to about 250 KB. Alternatively, depth-first search could be used, which often succeeds with a smaller open set than BFS:

```
dve22lts-mc --state=tree --strategy=dfs firewire_tree.5.dve
```

5   Experiments

We performed benchmarks on a 16-core AMD Opteron 8356 with 64 GB RAM. All models of the BEEM database [8] were used with command lines illustrated in the previous section. The hash table size was fixed for all tools to avoid resizing.

Tab. 1 shows an example of the effects of tree compression, Zobrist and references on the run-time and the memory usage of the different algorithms. The memory totals represent the space occupied by states on the open set and closed set (tree or hash table). Zobrist is not implemented for the tree structure.

Table 1. All possible combinations of the use cases for model `firewire_link.5`

		Cores:	1			16		
		Options:	none	--ref	-z6	none	--ref	-z6
Total time [sec]	bfs	table	5.4	5.7	4.7	0.3	0.3	0.3
		tree	4.8	4.4	–	0.2	0.2	–
	dfs	table	5.7	5.7	4.8	0.4	0.4	0.3
		tree	4.1	4.4	–	0.2	0.2	–
Total mem. [GB]	bfs	table	12.6	12.5	12.6	12.6	12.5	12.6
		tree	0.9	0.7	–	0.9	0.7	–
	dfs	table	12.5	12.5	12.5	12.5	12.5	12.5
		tree	0.7	0.7	–	0.7	0.7	–

Analysis revealed that the compression factors of tree compression and SPIN's COLLAPSE are primarily (linearly) dependent on the state length [3]. Fig. 1 shows *absolute* compression factors as values for all BEEM models that fitted into memory (250 out of 300). We established a maximum line for both compression techniques. On average, tree compression is about four times as effective as COLLAPSE.

Fig. 2 compares the performance of our MC backend with other tools. We translated BEEM models to PROMELA for SPIN; only those 100 models with similar state counts were used (less than 20% difference). Despite slower sequential performance due to the (larger) PINS state format, LTSMIN ultimately scales better than DiVinE and SPIN. Tree compression results in only 20% run-time overhead (aggregated) compared to the fastest hash table-based method.

Future Work. In the lab, we have working versions of LTSMIN that support full LTL model checking, partial-order reduction and multi-core *swarmed* LTL. All of these features are implemented as additional PINS layers and search strategies, building on the current infrastructure.

Conclusions. Several use cases and experiments show how LTSMIN can be applied to solve verification problems. Multi-core runs with Zobrist hashing can

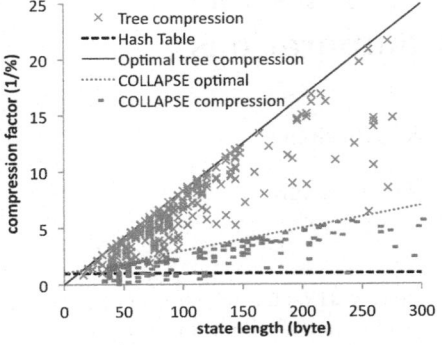

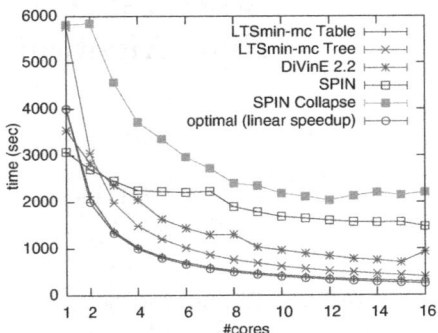

Fig. 1. Tree/COLLAPSE compression for 250 models

Fig. 2. Aggregate run-times for all tools

solve problems quickly provided that enough memory is available, while tree compression and state references can solve problems with large state vectors or on machines with little memory.

References

1. Barnat, J., Ročkai, P.: Shared hash tables in parallel model checking. Elec. Notes in Theor. Comp. Sc. 198(1), 79–91 (2008); Proc. of the 6th International Workshop on Parallel and Distributed Methods in verifiCation (PDMC 2007)
2. Blom, S., van de Pol, J., Weber, M.: Bridging the gap between enumerative and symbolic model checkers. Tech. Rep. TR-CTIT-09-30, Centre for Telematics and Information Technology, University of Twente, Enschede (2009)
3. Blom, S., Lisser, B., van de Pol, J., Weber, M.: A database approach to distributed state space generation. In: Sixth Intl. Workshop on Par. and Distr. Methods in verifiCation, PDMC, pp. 17–32. CTIT, Enschede (2007)
4. Blom, S., van de Pol, J., Weber, M.: LTSmin: Distributed and symbolic reachability. In: Touili, T., Cook, B., Jackson, P. (eds.) CAV 2010. LNCS, vol. 6174, pp. 354–359. Springer, Heidelberg (2010)
5. Holzmann, G.J., Bošnacki, D.: The design of a multicore extension of the SPIN model checker. IEEE Trans. Softw. Eng. 33(10), 659–674 (2007)
6. Laarman, A.W., van de Pol, J.C., Weber, M.: Boosting multi-core reachability performance with shared hash tables. In: Sharygina, N., Bloem, R. (eds.) Proceedings of the 10th International Conference on Formal Methods in Computer-Aided Design, Lugano, Swiss. IEEE Computer Society, USA (2010)
7. Nguyen, V.Y., Ruys, T.C.: Incremental hashing for SPIN. In: Havelund, K., Majumdar, R., Palsberg, J. (eds.) SPIN 2008. LNCS, vol. 5156, pp. 232–249. Springer, Heidelberg (2008)
8. Pelánek, R.: BEEM: Benchmarks for explicit model checkers. In: Bošnacki, D., Edelkamp, S. (eds.) SPIN 2007. LNCS, vol. 4595, pp. 263–267. Springer, Heidelberg (2007)
9. Sanders, P.: Load Balancing Algorithms for Parallel Depth First Search. Ph.D. thesis, University of Karlsruhe (1997)
10. Zobrist, A.L.: A new hashing method with application for game playing. Tech. Rep. 88, Computer Sciences Department, University of Wisconsin (1969)

GiNaCRA: A C++ Library for Real Algebraic Computations

Ulrich Loup and Erika Ábrahám

RWTH Aachen University, Germany

Abstract. We present the growing C++ library GiNaCRA, which provides efficient and easy-to-integrate data structures and methods for *real algebra*. It is based on the C++ library GiNaC, supporting the symbolic representation and manipulation of polynomials. In contrast to other similar tools, our *open source* library aids *exact*, real algebraic computations based on an appropriate *data type representing real zeros* of polynomials. The only non-standard library GiNaCRA depends on is GiNaC, which makes the installation and usage of our library simple. Our long-term goal is to integrate decision procedures for real algebra within the Satisfiability-Modulo-Theories (SMT) context and thereby provide tool support for many applied formal methods.

GiNaCRA – GiNaC Real Algebra package
http://ginacra.sourceforge.net/

1 Introduction

Formal methods for simulation, analysis, and synthesis have been making great progress during the last decades. The success of new methods in these fields often depends on efficient solvers for specific, well-established problems. For instance, there is a growing interest in Satisfiability-Modulo-Theories (SMT) solvers, implementing decision procedures for first-order logics over various theories [9]. The demand on these solvers is also growing; in particular, there is a need for more expressive logics. One example is the highly expressive but still decidable first-order logic over the reals with addition and multiplication, called *real algebra*. Whereas SMT-solvers for the linear fragment of this logic are very successful nowadays, even in industrial contexts, full real algebra still has not crossed this border. Nevertheless, several decision procedures were developed since the 1950s, which are currently operational in some computer algebra systems.

Although those computer algebra systems are frequently used and well-suited to solve a wide range of problems, they have some common drawbacks. First of all, many of them are either not *free*, or depend on non-free software. Furthermore, most of the free systems are not *open source*, restricting the extensibility and modification of the underlying algorithms. Another disadvantage is

M. Bobaru et al. (Eds.): NFM 2011, LNCS 6617, pp. 512–517, 2011.

that many computer algebra systems do not allow an *easy integration* of their functionalities into other external programs: Firstly, because they only offer graphical or textual user interfaces rather than a programming interface. Secondly, their *output* is usually a string displayed on screen, complicating its reuse in further external, exact computations.

Our long-term goal is to integrate an efficient decision procedure for real algebra into an SMT-solver. However, in view of the drawbacks mentioned above, current implementations of these decision procedures as, for example, the cylindrical algebraic decomposition (CAD) method, the virtual substitution method, or methods using Gröbner bases are not suited for an SMT-integration.

In this paper we introduce our open source C++ library GiNaCRA, which is free of the above-mentioned drawbacks. Besides the standard C++ library, GiNaCRA is based on a single non-standard library GiNaC [1]. GiNaCRA is under active and continuous development, and it already aids some functionalities not yet supported by any other C++ library. For example, GiNaCRA provides data types for *real algebraic numbers* as well as arithmetic and relational operations on them. In addition, an algorithm finding the common real roots of a set of univariate polynomials with rational coefficients is available. These features are useful not only for SMT-solving, but in a variety of other domains for computations with real algebraic constraints. Support for finding common real roots of multivariate polynomials is being implemented at the present time. In the near future, GiNaCRA will be also capable of computing realizable sign conditions of a set of multivariate polynomials [10, Algorithm 13.1].

Related Work. To our knowledge, there is currently no open source C++ implementation of real algebraic numbers, able to perform exact arithmetic operations from scratch. Nevertheless, very close to at least providing real-root computations are Libreduce, a C++ library of the Reduce computer algebra system, CoCoALib [8], a C++ library of the CoCoA computer algebra system, Givaro [2], and SYNAPS [3]. Singular [12], KANT [4], and PARI/GP [15] are examples of software packages for arithmetic and algebraic computations supporting algebraic numbers, but no computations with real algebraic ones. Maple [5], MATLAB [6], and Mathematica [7] are prominent examples of quite a number of further computer algebra systems providing computations with polynomials. There are also several programs implementing decision procedures for real algebra. QEPCAD [11] is a C++ implementation of the CAD method. Another example is the Redlog package [13] of the computer algebra system Reduce, offering an optimized combination of the virtual substitution and the CAD method.

Most of GiNaCRA's algorithms are based on the textbook [10], a comprehensive guide to real algebra comprising many methods of practical importance.

2   Real Algebra

Real algebra denotes the first-order logic over the reals with addition, multiplication, and the order relation. A real algebraic *formula* φ is a possibly quantified

Boolean combination of real algebraic *constraints c*. Each constraint, in turn, is an equality or inequality of polynomials in one or more variables x. The syntax can be formalized by the following abstract grammar:

$$
\begin{array}{rcl}
p & ::= & 0 \quad | \quad 1 \quad | \quad x \quad | \quad (p+p) \quad | \quad (p \cdot p) \\
c & ::= & p = p \quad | \quad p < p \\
\varphi & ::= & c \quad | \quad (\neg\varphi) \quad | \quad (\varphi \wedge \varphi) \quad | \quad (\exists x \varphi)
\end{array}
$$

Real algebra belongs to the mathematical theory of algebraic geometry, which is, to a great extent, based on Hilbert's Nullstellensatz. This field of research is the source of highly topical algorithms for solving the satisfiability problem of real algebra, that is, the question if we can assign real values to the variables of a quantifier-free real algebraic formula such that the formula evaluates to true. Solving this problem involves a lot of sophisticated computations, but they ultimately depend on a central problem: *finding real roots* of a univariate polynomial with rational coefficients. Note that there is no restriction to the degree of the polynomial. In particular, this problem can not be solved by applying a solution formula as in the quadratic case. In addition, it requires a representation of the root itself, a real algebraic number.

3 Features of GiNaCRA

GiNaCRA is a C++ library providing a collection of basic and advanced methods for real algebraic computations. It supports different representations of real algebraic numbers (order, sign, and interval representation [14, p. 327]). The different representations can be transformed into each other, such that for each computation the most suitable one can be chosen. A wrapping class for all representation types is under current development, providing an easy way of computing with real algebraic number objects and at the same time the ability to switch to an appropriate representation efficiently. In addition, a numerical representation of a real algebraic number can be computed.

The following list comprises some more features of GiNaCRA.

Open Source: GiNaCRA is meant to be accessible by everyone: researchers, students, industrial and commercial developers. It shall be possible to enhance GiNaCRA by everyone as well as to use GiNaCRA in other non-proprietary projects. Therefore, GiNaCRA is licensed under the GNU Lesser General Public License version 3 (LGPLv3).

Standalone: This library depends solely on one non-standard library GiNaC, which is licensed under the GNU General Public License version 2 (GPLv2). In particular, GiNaCRA does not depend on any closed-source software. GiNaC is currently available in many Linux distributions innately, what makes it very easy to install GiNaCRA, once downloaded.

Object-oriented: Being written in C++, GiNaCRA is class-based and object-oriented. It offers classes as types for real algebraic numbers, univariate polynomials with rational coefficients, multivariate polynomials, multivariate monomials, open intervals with rational endpoints, and many more.

This increases maintainability, extensibility, and reduces error-proneness in programs using this library.

Powerful interface: We provide a clear interface for an uncomplicated and efficient communication between a C++ application and GiNaCRA's real algebraic functionality. In contrast, several state-of-the-art computer-algebra systems for real algebraic computations, such as Reduce, MATLAB or Maple, can not be integrated so smoothly. In addition to the library, a simple console application for testing purposes is included in the GiNaCRA package.

Reliable: GiNaCRA comes with an extensive CppUnit test suite, currently containing nearly 100 test cases for the various functions of the library. There are separate test classes defining test cases for any GiNaCRA class, providing the opportunity to enhance the testing framework easily, for example, in case of a bug being found. Moreover, we make use of prevailing, well-tested C++ implementations, like the C++ standard library, wherever possible.

SMT-compliant: This library was specially designed to enable the implementation of an SMT-solver for real algebra. For this purpose, existing techniques for solving real algebraic constraint systems have to be adapted to the SMT-framework. This particularly means incremental solving algorithms for real algebra, being capable of (re)storing a state during search. This functionality can be realized by means of GiNaCRA's data structures, and is a content of current developments.

4 How to Use GiNaCRA

We give three simple examples on how GiNaCRA can be used inside a C++ program.

Example 1: Enumerating and Refining Real Roots

The following example program computes the real roots of the polynomial $x^5 - 39x^4 + 574x^3 - 3954x^2 + 12673x - 15015 = -(3-x)(5-x)(7-x)(11-x)(13-x)$.

```
1   #include <iostream>
    using namespace std;
    #include <ginacra/ginacra.h>
    using namespace GiNaC;
5
    int main(int argc, char **argv)
    {
      symbol x("x");
9     ex e(-15015 + 12673*x - 3954*pow(x,2) + 574*pow(x,3) - 39*pow(x,4) + pow(x,5));
      RationalUnivariatePolynomial p( e, x );
      list<IntervalRepresentation> roots = IntervalRepresentation::realRoots( p );
      cout << p << "_has_" <<
13          roots.size() << "_real_roots:" << endl;
      for(list<IntervalRepresentation>::const_iterator
          root = roots.begin();
          root != roots.end();
17        ++root)
        cout << "__" << *root << endl;
      cout << "List_of_refined_intervals_for_the_roots:" << endl;
      for(list<IntervalRepresentation>::iterator
21        root = roots.begin();
          root != roots.end();
          ++root)
      {
25      for(register unsigned i = 0; i < 10; ++i)
          root->refine();
        cout << "__" << root->Order() << ":_" << root->Interval() << endl;
      }
29    return 0;
    }
```

Line 9 contains the definition of the input polynomial as a `GiNaC` expression e. In line 10, a univariate polynomial p with rational coefficients is constructed from e and the `GiNaC` symbol x. After computing the real roots of p as a list of real algebraic numbers in line 11, the program outputs them in the first loop (lines 14 to 18). The second loop (lines 20 to 28) iterates through the roots again, but this time we change the iterated objects by calling the method refine () ten times, thereby gaining tighter bounds on the roots. refine () is implemented by a divide and conquer algorithm, using Sturm's theorem for real root counting. The refinement is done automatically whenever necessary for the interval representation in arithmetic or relational operations.

The listing below shows the compiler call followed by the call and the output of the example program. In the first loop, the output displays each real root as a triplet, consisting of the polynomial, an interval that contains exactly this single root, and the position of the root with respect to the order <. In the second loop, the polynomials are omitted in the output.

```
> g++ −lginac −lginacra −o example1 example1.cpp
> ./example1
−15015+574*x^3−3954*x^2+12673*x−39*x^4+x^5(x) has 5 real roots:
  {−15015+574*x^3−3954*x^2+12673*x−39*x^4+x^5(x):  ]0, 1877/512[  : 1}
  {−15015+574*x^3−3954*x^2+12673*x−39*x^4+x^5(x):  ]1877/512, 5631/1024[  : 2}
  {−15015+574*x^3−3954*x^2+12673*x−39*x^4+x^5(x):  ]5631/1024, 1877/256[  : 3}
  {−15015+574*x^3−3954*x^2+12673*x−39*x^4+x^5(x):  ]5631/512, 13139/1024[  : 4}
  {−15015+574*x^3−3954*x^2+12673*x−39*x^4+x^5(x):  ]13139/1024, 1877/128[  : 5}
List of refined intervals for the roots:
  1:  ]1571049/524288, 786463/262144[
  2:  ]5242461/1048576, 2622169/524288[
  3:  ]3669535/524288, 7340947/1048576[
  4:  ]11534165/1048576, 5768021/524288[
  5:  ]6815387/524288, 13632651/1048576[
```

Example 2: Computing Common Real Roots of Two Polynomials

This example addresses the computation of the common real roots of $x^5 - 39x^4 + 574x^3 - 3954x^2 + 12673x - 15015$ and $-x^4 + 26x^3 - 236x^2 + 886x - 1155$ where $-x^4 + 26x^3 - 236x^2 + 886x - 1155 = -(3-x)(5-x)(7-x)(11-x)$ is a factor of the polynomial of Example 1. Since the same headers are needed as in Example 1, we show only the important snippet from the main method here.

```
   ex e1(−15015 + 12673*x − 3954*pow(x,2) + 574*pow(x,3) − 39*pow(x,4) + pow(x,5));
2  ex e2(−1155 + 886*x − 236*pow(x,2) + 26*pow(x,3) − pow(x,4));
   list<RationalUnivariatePolynomial> l = list<RationalUnivariatePolynomial>();
   l.push_back(RationalUnivariatePolynomial( e1, x ));
6  l.push_back(RationalUnivariatePolynomial( e2, x ));
   list<IntervalRepresentation> roots = IntervalRepresentation::commonRealRoots( l );
   cout << l.front() << "_and_" << l.back() << endl << "have_" <<
        roots.size() << "_common_real_roots:" << endl;
   for(list<IntervalRepresentation>::const_iterator
10      root = roots.begin();
        root != roots.end();
        ++root)
     cout << "__" << *root << endl;
```

A list containing the two polynomials is constructed in lines 1 to 5. In line 6, the method commonRealRoots is called with this list as input. The method returns a list of real algebraic numbers, which is displayed as follows:

```
−15015−39*x^4+x^5−3954*x^2+574*x^3+12673*x(x) and −1155−x^4−236*x^2+26*x^3+886*x(x)
have 4 common real roots:
  {1155+x^4+236*x^2−26*x^3−886*x(x):  ]0, 289/64[  : 1}
  {1155+x^4+236*x^2−26*x^3−886*x(x):  ]289/64, 867/128[  : 2}
  {1155+x^4+236*x^2−26*x^3−886*x(x):  ]867/128, 289/32[  : 3}
  {1155+x^4+236*x^2−26*x^3−886*x(x):  ]289/32, 289/16[  : 4}
```

Example 3: Real Algebraic Number Arithmetic

Finally, we give a short code snippet illustrating arithmetic and relational operations on real algebraic numbers in GiNaCRA. For the sake of simplicity, we take the zeros of $x^2 - 2$ and $x - 1$ as example numbers.

```
   RationalUnivariatePolynomial p1( pow(x,2)-2, x );
 2 RationalUnivariatePolynomial p2( x-1, x );
   list<IntervalRepresentation> sqrt2s = IntervalRepresentation::realRoots( p1 );
   cout << "Interval_representation_of_sqrt(2):_" << sqrt2s.back() << endl;
   list<IntervalRepresentation> one = IntervalRepresentation::realRoots( p2 );
 6 cout << "Interval_representation_of_1:_" << one.front() << endl << endl;
   cout << "sqrt(2)+(-sqrt(2))_=_" << sqrt2s.back()+sqrt2s.front() << endl;
   IntervalRepresentation minustwo = sqrt2s.back()*sqrt2s.front();
   cout << "sqrt(2)*-sqrt(2)_=_" << minustwo << endl;
10 IntervalRepresentation two = one.front()+one.front();
   cout << "1+1_=_" << two << endl;
   cout << "(sqrt(2)*-sqrt(2))_==_-(1+1)?_" << ((minustwo==-two)?"Yes!":"No!") << endl;
```

This program generates the following output:

```
Interval representation of sqrt(2): {-2+x^2(x): ]0, 2.236067977499789696964[ : 2}
Interval representation of 1: {-1+x(x): ]0, 1.414213562373095048[ : 1}

sqrt(2) + (-sqrt(2)) = {x^4-8*x^2(x): ]0, 0[ : 2}
sqrt(2) * -sqrt(2) = {16+x^4-8*x^2(x): ]-6.708203932499369089, -1/17[ : 1}
1 + 1 = {-2+x(x): ]1/3, 2.828427124746190975[ : 1}
(sqrt(2) * -sqrt(2)) == -(1 + 1)? Yes!
```

These arithmetic operations each produce a new interval representation by computing a new polynomial, based on the original ones, and a new interval, whose bounds depend on the original bounds. The original intervals are then refined until the new interval isolates exactly one real root of the new polynomial.

References

1. http://www.ginac.de/
2. http://ljk.imag.fr/CASYS/LOGICIELS/givaro/
3. http://www-sop.inria.fr/galaad/logiciels/synaps/
4. http://www.math.tu-berlin.de/~kant/kash.html
5. http://www.maplesoft.com/
6. http://www.mathworks.de/
7. http://www.wolfram.com/
8. Abbott, J., Bigatti, A.: CoCoALib: a c++ library for doing Computations in Commutative Algebra, http://cocoa.dima.unige.it/cocoalib/
9. Barrett, C., Stump, A., Tinelli, C.: The Satisfiability Modulo Theories Library, SMT-LIB (2010), http://www.SMT-LIB.org
10. Basu, S., Pollack, R., Roy, M.: Algorithms in real algebraic geometry, 2nd edn., vol. 10. Springer, Heidelberg (2010)
11. Brown, C.W.: QEPCAD B: a program for computing with semi-algebraic sets using CADs. SIGSAM Bulletin 37(4), 97–108 (2003)
12. Decker, W., Greuel, G.M., Pfister, G., Schönemann, H.: Singular 3-1-2 — A computer algebra system for polynomial computations (2010), http://www.singular.uni-kl.de
13. Dolzmann, A., Sturm, T.: REDLOG: Computer algebra meets computer logic. SIGSAM Bulletin 31(2), 2–9 (1997)
14. Mishra, B.: Algorithmic Algebra. Texts and Monographs in Computer Science. Springer, New York (1993)
15. The PARI Group, Bordeaux: PARI/GP, version 2.3.5 (2008), http://pari.math.u-bordeaux.fr/

Kopitiam: Modular Incremental Interactive Full Functional Static Verification of Java Code

Hannes Mehnert

IT University of Copenhagen, 2300 København S, Denmark
hame@itu.dk

Abstract. We are developing Kopitiam, a tool to interactively prove full functional correctness of Java programs using separation logic by interacting with the interactive theorem prover Coq. Kopitiam is an Eclipse plugin, enabling seamless integration into the workflow of a developer. Kopitiam enables a user to develop proofs side-by-side with Java programs in Eclipse.

1 Introduction

It is challenging to reason about object-oriented programs, because these contain implicit side effects, shared mutable data and aliasing. Reasoning with Hoare logic always has to consider the complete heap, which does not preserve the abstractions of the programming language. Separation logic [18] extends Hoare logic to allow modular local reasoning about programs with shared mutable state.

Coq [4] is an interactive theorem prover based on the calculus of constructions with inductive definitions. Kopitiam generates proof obligations from specifications written in Java, which the user needs to discharge by providing Coq proof scripts. A proof script is a sequence of tactics.

The contribution is Kopitiam, a tool combining the following verification properties:

- **Modular.** Extensions of a verified Java library can rely on the specification of the library, without reverifying the library.
- **Incremental.** While parts of the code can be verified and proven, other parts might remain unverified, and development of proofs and code can be interleaved, as in Code Contracts [10].
- **Interactive.** Automated proof systems like jStar [8] are limited in what they can prove. We use an interactive approach where the user discharges the proof obligations using provided tactics, thus Kopitiam does not limit what a user can prove.
- **Full functional.** Given a complete, precise formal specification the proof shows that the implementation adheres to its specification.
- **Static.** The complete verification is done at compile time, without execution of the program. Other code verification approaches, like design by contract [15], may depend on run time checks. Especially in mission critical systems, compile time verification is indispensable, since a failing run time check would be disastrous.

M. Bobaru et al. (Eds.): NFM 2011, LNCS 6617, pp. 518–524, 2011.

The structure of the paper is: we give an overview of Kopitiam in Section 2, demonstrate a detailed example in Section 3, relate Kopitiam to similar tools in Section 4, and in Section 5 conclude and present future work.

2   Overview of Kopitiam

Kopitiam provides an environment that is familiar to both Java programmers and Coq users. Coq developers use Proof General (based on Emacs) or CoqIDE (a self-hosted user interface). Many Java programmers use an IDE for development, the major Java IDEs are Eclipse and IntelliJ. To integrate seamlessly into the normal development workflow we develop Kopitiam as a plugin for Eclipse, so a developer does not have to switch tools to prove her code correct. We base Kopitiam on Eclipse because it is open source, popular and easily extendible via plugins. While an Eclipse integration for Coq [6] already exists, Kopitiam provides a stronger integration of Java code and Coq proofs. This is achieved by a single intermediate representation for both code and proofs. A change to either code or proof directly changes this intermediate representation.

```
public int fac (int n) {                    Lemma fac_valid :
    Coq.requires("ege n 0");                  I-G {{spec_p fac_spec ()}}
    Coq.ensures("ret .-. fac_Z n");                Fac.fac_body
    int x;                                         {{spec_qret fac_spec () "x"}}.
    if (n > 0)                              Proof.
        x = n * fac(n - 1);                     unfold_valid.
    else                                        forward. forward.
        x = 1;                                  call_rule (TClass "FacC") ().
    return x;                                   - substitution. unentail. intuition.
}                                               - reflexivity. substitution.
}                                               forward.
```

Fig. 1. Java and Coq editor side-by-side; closeup of Coq editor in Fig 2

In Figure 1 Kopitiam is shown. It consists of a standard Eclipse Java editor on the left and a specially developed Coq proof editor on the right. The content of the Java editor is the method `fac`, a recursive implementation of the factorial function. The Java code contains a call to `Coq.requires` and a call to `Coq.ensures`, whose arguments are the pre- and postcondition of the method. The right side shows the Coq lemma `fac_valid`, stating that factorial fulfills its specification, together with parts of the proof script (full code in Section 3). Due to the single intermediate language, Kopitiam reflects every change to the content of one editor to the other editor, e.g. a change to the specification on the Java side changes the Coq proof obligation.

Kopitiam consists of a Java parser, with semantic analysis, a transformer to SimpleJava (presented in Section 2.2), a Coq parser, and communication to Coq via standard input and output. All these parts are expressible in a functional way, so we chose Scala [16] as the implementation language of Kopitiam. Scala is a type-safe functional object-oriented language supporting pattern matching. It compiles to Java bytecode, allowing for seamless integration with Eclipse (every

Scala object is a Java object and vice versa). Kopitiam is open source under the Simplified BSD License and available at https://github.com/hannesm/Kopitiam.

2.1   Coq Editor and Goal Viewer

To develop proofs, Kopitiam provides a Coq editor and a goal viewer, shown in Figure 2. The Coq code on the left side states the lemma `fac_step`: for all n, n greater than 0 implies that n * fac(n - 1) equals fac(n) (lines 1-3). All except the last 2 lines of the Coq code that have been processed by Coq (highlighted in blue in Kopitiam, the unprocessed ones are black). The goal viewer on the right side shows the current state of proof assumptions, proof obligations and subgoals. The current state is after doing induction over n and discharging the base case using the `intuition` tactic. The remaining proof obligation is the induction step.

As in other Coq user interfaces, there are buttons (not shown) to step forward and backward through the proof.

```
Lemma fac_step : forall n,
    n > 0 ->
    (n * (fac (n - 1))) = fac n.
Proof.
    induction n.
    - intuition.
    - simpl; intuition.
Qed.
```

> ⬚ Outline / 🌐 Goal Viewer ✕
>
> n : nat
> IHn : n > 0 -> n * fac (n - 1) = fac n
>
> S n > 0 -> S n * fac (S n - 1) = fac (S n)
> other subgoals

Fig. 2. Coq editor and goal viewer of Kopitiam, closeup of Figure 1

If Coq signals an error while processing, this error is highlighted in Kopitiam. Figure 3 shows on the left side the erroneous Coq proof script next to Eclipse's corresponding problems tab. Errors are indicated by red wiggly lines, similar to the way programming errors are displayed in Eclipse.

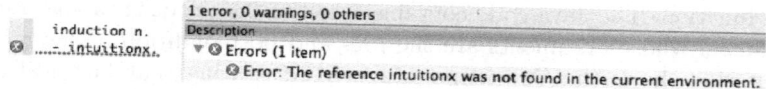

> induction n.
> ❌ ....intuitionx.

1 error, 0 warnings, 0 others
Description
▼ ❌ Errors (1 item)
❌ Error: The reference intuitionx was not found in the current environment.

Fig. 3. Coq proof script containing an error and Eclipse's problems tab

2.2   The SimpleJava Programming Language

We formalized SimpleJava, a subset of Java, and implemented it using a shallow embedding in Coq (details in an upcoming paper by Bengtson, Birkedal, Jensen and Sieczkowski). SimpleJava syntax is a prefix (S-expression) notation of Java's abstract syntax tree. Dynamic method dispatch is the core ingredient of object oriented programming, and supported by SimpleJava. A SimpleJava

```
class FacC {              (cif (egt (var_expr "n") 0)  Fixpoint fac n :=
  int fac (int n) {         (cseq                        match n with
    Coq.requires("ege n 0");  (ccall "x" "this" "fac"      | S n => (S n) * fac n
    Coq.ensures               ((eminus                     | 0 => 1
       ("ret ·=· facZ n");       (var_expr "n") 1))        end.
    int x;                    (TClass "FacC"))           Definition facZ :=
    if (n > 0)              (cassign "x"                  fun (n:Z) =>
      x = n * fac(n - 1);    (etimes                       match ((n ?= 0)%Z) with
    else x = 1;               (var_expr "n")               | Lt => 0
    return x;                 (var_expr "x"))))            | _ =>
  }                        (cassign "x" 1))                 Z_of_nat(fac(Zabs_nat n))
}                                                         end.
```

Fig. 4. Java code   **Fig. 5.** SimpleJava code   **Fig. 6.** Coq definitions

program consists of classes and interfaces. An interface contains a set of method signatures and a set of interfaces, that it inherits from; a class consists of a set of implemented interfaces, a set of fields, and a set of method implementations. A method body consists of a sequence of statements (allocation, conditional, loop, call, field read, field write and assignment) followed by a single return statement. Automatic transformation of unstructured returns to a single return would impose method-global control flow changes; and the SimpleJava code would distract the Java programmer while proving.

3   Example Verification of Factorial

An example program is the factorial, shown in Figure 4. Figure 5 shows the SimpleJava code, automatically translated by Kopitiam. A call (lines 3-6) consists of the return value binding (x), the receiver (this), the method (fac), the argument list and the receiver class (TClass "FacC").

In Figure 6 the fixpoint fac is defined, which is the common factorial function on natural numbers. Our Java code uses integers, so we additionally need facZ, which extends the domain of fac to integers.

The specification of a program consists of specifications for all classes and interfaces. An example specification of method fac is shown in Figure 7, whose code is automatically generated by Kopitiam from the Java code (Figure 4). The precondition (line 3 of both Figures) requires that the parameter n must be equal or greater (ege) than 0. The postcondition (lines 4-5 of both Figures) ensures that the returned value (ret) is equal to facZ n. The bottom block of Figure 7 defines Spec, which connects the specification fac_s to the actual program, class FacC, method fac.

Figure 8 shows the hand-written proof that the Java implementation of factorial satisfies its specification. The proof uses the forward tactic [2]. This extracts the first Hoare triple; the resulting proof obligation (Hoare triple) is the original precondition combined with the extracted postcondition, the remaining statement sequence, the original postcondition. If the extracted precondition cannot be discharged trivially, the user has to do it. After applying forward twice (line 4, for cif and cseq), the proof obligation for the call is discharged by the call_rule tactic (line 5).

```
Definition fac_s :=
  Build_spec unit (fun _ =>
    (ege "n" 0,
    ((("ret":expr) .=.
      (facZ ("n":expr))):asn))).


Definition Spec := TM.add
  (TClass "FacC")
  (SM.add "fac" ("n" :: nil, fac_s))
  (SM.empty _)
  (TM.empty _).
```

Fig. 7. Specification

```
Lemma fac_valid : |=G {{spec_p Fac_spec.fac_spec ()}}
  Fac.fac_body {{spec_qret Fac_spec.fac_spec () "x"}}.
Proof.
  unfold_valid. forward. forward.
  call_rule (TClass "FacC") ().
  - substitution. unentail. intuition.
  - reflexivity. substitution.
  forward. unentail. intuition. subst. simpl.
    rewrite Fac_spec.facZ_step; [reflexivity | omega].
  forward. unentail. intuition. subst.
    destruct (Z_dec (val_to_int k) 0).
    assert False; [|intuition]. destruct s; intuition.
    rewrite e. intuition.
  Existential 1:=().
Qed.
```

Fig. 8. Coq proof script for factorial

4   Related Work

Several currently available proof tools are compared in Table 1. Only Krakatoa [11], jStar [8] and Kopitiam target Java. Krakatoa uses Why, which uses a simple While language where mutable variables cannot be aliased. The automated proof system jStar targets Jimple [19], a Java intermediate language built from Java bytecode. Kopitiam directly translates from a subset of Java source code to SimpleJava.

Different code contracts [15] implementations focus on C# (Code Contracts [10]) and Java (JML [5]). Code contract implementations translate some non-trivial specifications to run time checks, while we focus on static verification. The integration of code contracts in an IDE is beneficial, as the developer can incrementally develop code and proofs in the same environment. An example for an industrial grade IDE with code contracts is the KeY tool [1], based on UML and OCL. Code Contracts [10] do not focus on full functional correctness, while some JML tools such as Mobius [3] do. In contrast to those tools, we use separation logic, thus a user does not need to specify frame conditions.

Dafny [14] is a proof tool for an imperative programming language supporting generics and algebraic data types, but not subtyping. Dafny is well integrated into Microsoft Visual Studio and also allows incremental proofs. It provides a multi-sorted first-order logic as specification logic.

Ynot [7] uses a shallow embedding in Coq for a higher-order imperative programming language without inheritance. Thus to verify code with Ynot the program has to be reimplemented in the Ynot tool.

Table 1. Comparison of verification tools

Name	T	Language	Specification logic	Automation
Krakatoa	sta	Java; While	multi-sorted FOL	several provers
Ynot	sta	higher-order imp	separation logic	Coq tactics
jStar	sta	Java; Jimple	separation logic	user proof rules, SMT
Spec#	dyn	C#	C#/Java	run time assertions
Dafny	inc	imp + generics	Boogie	Z3 (SMT-solver)
Kopitiam	inc	Java; SimpleJava	separation logic	Coq tactics

The jStar [8] tool is fully automated and does a proof search on available proof rules, which are extensible by the user. A user can introduce unsound proof rules, since these are treated as axioms and are not verified. Moreover it is difficult to guide the proof search in jStar, since the order of rules matters. Both Ynot and Kopitiam use the proof assistant Coq, in which proof rules have to be proven before usage.

5   Conclusion and Future Work

We are developing Kopitiam, an Eclipse plugin for interactive full functional static verification of Java code using separation logic. Our implementation is complete enough to prove correctness of factorial and in-place reversal of linked lists. We currently do not handle the complete Java language, e.g. unstructured returns and `switch` statements. Class to class inheritance is also not supported. Kopitiam does not support more advanced Java features like generics and exceptions.

We plan to integrate more automation: We will provide context aware suggestions, a technique widely used in Eclipse for code completion, for specifications, whose syntax we also plan to improve. We will provide separation logic lemmas and tactics for Coq, allowing the user to focus on the non-trivial proof obligations. We also want the user to discharge separation logic proof obligations instead of exposing the Coq layer.

We are also working on more and larger case studies ranging from simple object-oriented code (`Cell` and `ReCell` from [17]), to the composite pattern and other verification challenges [20], to real-world data structures like Linked Lists with Views [12] and Snapshottable Trees [9], to the C5 collection library [13], the extensive case study of our research project.

Acknowledgement. We want to thank Peter Sestoft, Jesper Bengtson, Joe Kiniry and the anonymous reviewers for their valuable feedback.

References

1. Ahrendt, W., Baar, T., Beckert, B., Bubel, R., Giese, M., Hähnle, R., Menzel, W., Mostowski, W., Roth, A., Schlager, S., Schmitt, P.H.: The KeY tool. Software and System Modeling 4 (2005)
2. Appel, A.W.: Tactics for separation logic. INRIA Rocquencourt and Princeton University, Early Draft (2006)
3. Barthe, G., Crégut, P., Grégoire, B., Jensen, T., Pichardie, D.: The MOBIUS proof carrying code infrastructure. In: Boer, F.S., Bonsangue, M.M., Graf, S., Roever, W.P. (eds.) Formal Methods for Components and Objects. Springer, Heidelberg (2008)
4. Bertot, Y., Castéran, P.: Interactive Theorem Proving and Program Development. In: Coq'Art: the Calculus of Inductive Constructions. Springer, Heidelberg (2004)

5. Burdy, L., Cheon, Y., Cok, D.R., Ernst, M.D., Kiniry, J.R., Leavens, G.T., Leino, K.R.M., Poll, E.: An overview of JML tools and applications. International Journal on Software Tools for Technology Transfer 7 (June 2005), also published by Springer http://dx.doi.org/10.1007/s10009-004-0167-4

6. Charles, J., Kiniry, J.R.: A lightweight theorem prover interface for Eclipse. UITP at TPHOL 2008 (2008)

7. Chlipala, A., Malecha, G., Morrisett, G., Shinnar, A., Wisnesky, R.: Effective interactive proofs for higher-order imperative programs. In: ACM Proc. of ICFP 2009 (August 2009)

8. Distefano, D., Parkinson, M.J.: jStar: towards practical verification for Java. In: ACM Proc. of OOPSLA 2008 (October 2008)

9. Driscoll, J., Sarnak, N., Sleator, D., Tarjan, R.: Making data structures persistent. In: ACM Proc. of STOC 1986 (November 1986)

10. Fähndrich, M., Barnett, M., Logozzo, F.: Embedded contract languages. In: ACM Proc. of SAC 2010 (March 2010)

11. Filliâtre, J.-C., Marché, C.: The Why/Krakatoa/Caduceus platform for deductive program verification. In: Damm, W., Hermanns, H. (eds.) CAV 2007. LNCS, vol. 4590, pp. 173–177. Springer, Heidelberg (2007)

12. Jensen, J.B., Birkedal, L., Sestoft, P.: Modular verification of linked lists with views via separation logic. In: Proc. of FTfJP 2010 (May 2010)

13. Kokholm, N., Sestoft, P.: The C5 generic collection library for C# and CLI. Tech. Rep. ITU-TR-2006-76, IT University of Copenhagen (2006)

14. Leino, K.R.M.: Dafny: An automatic program verifier for functional correctness. In: Proc. of LPAR-16 (March 2010)

15. Meyer, B.: Design by contract. In: Advances in Object-Oriented Software Engineering (1991)

16. Odersky, M., et al.: An overview of the Scala programming language. Tech. Rep. IC/2004/64, EPFL Lausanne, Switzerland (2004)

17. Parkinson, M.J., Bierman, G.: Separation logic, abstraction and inheritance. In: ACM Proc. of POPL 2008 (January 2008)

18. Reynolds, J.C.: Separation logic: A logic for shared mutable data structures. In: IEEE Proc. of 17th Symp. on Logic in CS (November 2002)

19. Vallée-Rai, R., Hendren, L.J.: Jimple: Simplifying Java bytecode for analyses and transformations. Tech. Rep. 4, McGill University (1998)

20. Weide, B., Sitaraman, M., Harton, H., Adcock, B., Bucci, P., Bronish, D., Heym, W., Kirschenbaum, J., Frazier, D.: Incremental benchmarks for software verification tools and techniques. In: Shankar, N., Woodcock, J. (eds.) VSTTE 2008. LNCS, vol. 5295, pp. 84–98. Springer, Heidelberg (2008)

Milestones: A Model Checker Combining Symbolic Model Checking and Partial Order Reduction*

José Vander Meulen and Charles Pecheur

Université catholique de Louvain
{jose.vandermeulen,charles.pecheur}@uclouvain.be

Abstract. Symbolic techniques and partial order reduction (POR) are two fruitful approaches to deal with the combinatorial explosion of model checking. Unfortunately, past experience has shown that symbolic techniques do not work well for loosely-synchronized models, whereas, by applying POR methods, explicit-state model checkers are able to deal with large concurrent models. This paper presents the Milestones model checker which combines symbolic techniques and POR. Its goal is to verify temporal properties on concurrent systems. On such a system, Milestones allows to check the absence of deadlock, LTL properties, and CTL properties. In order to compare our approach to others, Milestones is able to translate a model into an equivalent Spin model [7] or NuSMV model [4]. We briefly present the theoretical foundation on which Milestones is based on. Then, we present the Milestones model checker, and an evaluation based on an example.

1 Introduction

Two common approaches are commonly exploited to fight the combinatorial state-space explosion problem in model checking. On one hand, the *partial-order reduction* methods (POR) explore a reduced state space in a property-preserving way [10,6]. On the other hand, symbolic techniques use functional representations of the state space to tackle the state-space explosion problem. Two different approaches to symbolic model-checking have been broadly considered: the BDD-based approach uses *binary decision diagrams* (BDDs) to concisely encode and compute state spaces [3], while the bounded model-checking (BMC) approach translates the original problem into a SAT problem. In their basic form, symbolic approaches tend to perform poorly on asynchronous models where concurrent interleavings are the main source of explosion, and explicit-state model-checkers with POR have been the preferred approach for such models.

This paper presents the Milestones model checker, which combines POR techniques and symbolic methods. Milestones defines a language for describing transition systems. CTL properties (as well as absence of deadlock) can be checked

---

* This work is supported by project MoVES under the Interuniversity Attraction Poles Programme — Belgian State — Belgian Science Policy.

M. Bobaru et al. (Eds.): NFM 2011, LNCS 6617, pp. 525–531, 2011.

by combining BDD-based approach and POR [11]. LTL properties can be verified by combining POR either with the BDD-based approach or with the BMC approach [12]. In order to evaluate our approaches, Milestones can translate its model into a Promela model [7] or into a NuSMV model [4]. In order to make the comparison as fair as possible, the resulting state machines are (almost) exactly the same as those generated by Milestones. In the case of NuSMV, the generated BDDs are the same as well. Milestones is available under the GNU General Public License at http://lvl.info.ucl.ac.be/Tools/Milestones.

2 Model Checking

When applying model-checking to verify a concurrent system, the size of the combined state space can grow exponentially in the number of processes, due to all the different interleavings among the executions of all the processes. Different approaches were developed to tackle this problem, among which the partial order methods (POR) and symbolic model checking.

The goal of partial-order reduction is to reduce the number of states explored by model-checking, by not exploring different equivalent interleavings of concurrent events. Naturally, these methods are best suited for strongly asynchronous programs. Interleavings which are required to be preserved may depend on the property to be checked. Intuitively, if two concurrent transitions α and β do not interfere with each other and do not affect the property f that one wants to verify, then it does not matter whether α is executed before or after β, and the exploration can be restricted to either of these two alternatives.

Symbolic model-checking, based on Binary Decision Diagrams (BDD), allows to reason on set of states rather than individual states. This technique made it possible to verify systems with a very large number of states [3]. However for large models, the size of the BDD structures themselves can become intractable. In contrast, bounded model-checking characterizes an error execution path of length k as a propositional formula, and searches for solutions to that formula with a SAT solver. BMC is limited by the need to fix the bound k but takes only polynomial space with respect to the model.

3 The Milestones Symbolic Model-Checker

Our tool, Milestones, is a symbolic model-checker which takes as input a model of a concurrent system annotated with temporal logic properties and produces as output the truth value of those properties. It also generates statistical data such as verification time, memory usage, BDD construction time, etc. Because collecting this information can significantly influence the verification time, the data generation can be switched off.

3.1 Modeling Language

Milestones defines a language for describing transition systems. The design of the language has been influenced by the NuSMV language [4] but supports

```
 1   SYNC                                    34      CASE [0]
 2      ADD;                                 35      [exit] p3 == 0:
 3      ERR;                                 36         pc := 3;
 4                                           37      [REQremoveTrue] p3==1 & tr==1:
 5      exit;                                38         pc := 1;
 6      REQremoveTrue;                       39      [REQremoveFalse] p3==1 & tr==0:
 7      REQremoveFalse;                      40         pc := 1;
 8      ...                                  41      END // CASE [0]
 9   END //SYNC                              42
10                                           43      CASE [1]
11   VARIABLE                                44         ...
12      INTEGER r:1:=0;                      45      END //CASE [1]
13      INTEGER tr:1:=0;                     46   END //LOCAL MC4
14      ...                                  47
15   END //VARIABLES                         48   GLOBAL
16                                           49      VAR mc1: MC1;
17   LOCAL MC4                               50      VAR mc2: MC2[3];
18      SYNC                                 51      VAR mc3: MC3;
19         exit;                             52      VAR mc4: MC4;
20         REQremoveTrue;                    53      VAR mc5: MC5;
21         REQremoveFalse;                   54
22         ...                               55      CASE
23      END //SYNC                           56         [exit] true : p3 := 0;
24                                           57            p0 := 1;
25      ACTION                               58         ...
26         tau;                              59      END //CASE
27      END //ACTION                         60   END//GLOBAL
28                                           61
29      VARIABLE                             62   LTL
30         INTEGER pc:4:=0;                  63      F (p0 == 1);
31         INTEGER mr:4;                     64   END //LTL
32         ...
33      END //VARIABLES
```

Fig. 1. A Milestone model of a turntable system

synchronization by rendez-vous. Figure 1 shows parts of the Milestones model of the turntable system discussed in Section 4.

A model of a concurrent system declares a set of integer variables. Each variable is declared as follows **INTEGER** *name*: n [:= **expr**] where *name* is its name, n is the number of bits which are used to encode it, and **expr** is an expression which represents its initial value. The expression **expr** is optional. If it is not mentioned the initial value can be any values which are representable with **n** bits. A model declares a set of global variables (line 11), a set of shared actions (line 1) and a set of processes (line 17). A process p declares a set of local variables (line 29), a set of local actions (line 25) and the set of shared actions which p is synchronized on (line 18). Each process has a distinguished local program counter variable pc (line 30). For each value of pc, the behavior of a process is defined by means of a list of *action-labelled guarded commands* of the form $[\alpha]$ c : u, where α is an action, c is a condition on variables and u is an assignment updating some variables (line 36). *Shared* actions are used to define synchronization between the processes. A *shared* action occurs simultaneously in all the processes that share it, and only when all enable it. Properties can be expressed in CTL as well as LTL (line 62).

Milestones consists of a set command-line tools; it does not provide any graphical interface.

3.2 BDD-Based CTL Verification

Milestones allows to check whether a model verifies a CTL property. The check can be performed with or without POR reduction (commands `checkCTLWithPOR` and `checkCTL`). Without POR, the classical *backward* CTL model checking algorithm of [3] is applied, as in NuSMV. With POR, Milestones uses the FwdUntilPOR approach which was first presented in [11]. In order to perform the verification, a *forward* CTL model checking approach of Iwashita et al. [8] is combined with a symbolic POR forward exploration derived from Lerda et al.'s Improviso [9]. Contrary to the backward algorithm which does not apply POR methods, the forward approach is only applicable for a subset of CTL. Both methods can be combined together to check all the possible CTL formulæ.

3.3 BDD-Based LTL Verification

Milestones is able to verify LTL properties, with or without POR reduction (commands `checkLTLWithPOR` and `checkLTL`), using the symbolic *tableau-based* LTL model checking algorithm of [5]. This method results in looking for fair executions in the product P of the model and a tableau-based encoding of the (negated) property. With POR, we construct P_r, a property-preserving partial-order reduction of P, using an adaptation of Lerda et al.'s ImProviso algorithm [9]. Finally, we check within P_r whether P contains a fair cycle using the forward traversal approach of Iwashita et al. [8].

3.4 SAT-Based LTL Verification

LTL properties can also be checked by means of the *Bounded Model Checking* (BMC) approach, either with or without POR (command `checkLTLWithSBTP` and `checkLTLWithBMC`). Without POR, the algorithm of Biere et al. is executed [1]. With POR, we use the Stuttering Bounded Two-Phase algorithm (SBTP) first described in [12]. This algorithm merges a variant of Improviso [9] with the BMC procedure. In short, from a model and the negation of a property f, the BMC method constructs a propositional formula which represents a finite unfolding of the transition relation and $\neg f$. Our method proceeds in the same way, but instead of using the entire transition relation during the unfolding of the model, we only use a safe subset based on POR considerations.

Intuitively, SBTP alternately executes two phases. For each process of the model under verification, the first phase of SBTP unfolds some fixed number m of safe deterministic transitions. If less than m deterministic transitions are allowed, an idle transition which does not modify anything is performed instead. In the second phase, a full expansion occurs, even if there are safe deterministic transitions remaining. This avoids cycles of partial expansions, thus ensuring a property-preserving reduction. Because the generated propositional formula contains only few disjunctions, its satisfiability verification generates little back-tracking, making it well-suited for modern DPLL-based SAT-solvers.

3.5 Exporting to Promela and NuSMV

Milestones can translate a model which is defined in its own language to NuSMV [4] (command `translateIntoNuSMV`) or to Promela, the language used by Spin [7] (command `translateIntoPromela`). We think it is important to compare Milestones to NuSMV and Spin, respectively the most prominent symbolic and explicit-state model-checkers.

The generated NuSMV model defines exactly the same state machine as the Milestones model. Because BDD variable ordering can considerably influence the size of BDDs, and so the performance of the algorithm, a file which represents this order is generated[1]. This file can be used by NuSMV to construct its BDDs. Together these allow a close and fair comparison between Milestones and NuSMV.

The generated Promela model also defines almost the same state machine as the Milestones model, except for one more state which is necessary to correctly initialize the variables. We thus have good support for fair comparison between Spin and Milestones as well.

The accuracy of the translation was confirmed by comparing number of states and BDD nodes (in the NuSMV case), as well as by detailed comparison on small examples.

4 Evaluation

To evaluate the performance of Milestones, we used it to verify a turntable model which was first described in [2]. We also verified this model with NuSMV and Spin.

The turntable system consists of a round turntable, n drills and a testing device. The turntable transports products between the drills, the testing device and input and output positions. The drills bore holes in the products. After being drilled, the products are delivered to the tester, where the depth of the holes is measured, since it is possible that drilling went wrong. The turntable has $n + 3$ slots that each can hold a single product.

In [11], thirteen CTL properties have been checked on this model, For instance, the $p11$ property states that *each piece will be removed from the turntable after it is tested*. It is shown that a turntable with 40 drills can be checked in approximatively 40 seconds with the classical backward approach and in 4 seconds when the POR reduction is applied.

Six LTL properties, three of which are invalid, have been verified on the turntable model. For instance, the property T_3 states that *if in the future there is a piece which is not well drilled then the alarm will necessarily resonate*. Within a max time of 16 minutes, we are able to check T_3 on a turntable model composed of 61 drills with the Milestones model checker (with POR), 20 drills with the Spin model checker, 6 drills with the NuSMV model checker.

---

[1] For more details about the Milestones variable ordering, we refer the reader to [11].

In essence, the length of the failure traces influences greatly the performance of the BMC algorithms. It turns out that the turntable model features failure traces that are too long for BMC approaches. On a variant of the producer-customer model which is composed of n producers and n consumers, SBTP achieves an improvement in comparison to the classical bounded model checking algorithm [12]. For $n = 3$ (resp. $n = 7$), the reachable state space of this system is approximatively equal to 3×10^6 states (resp. 10^{14}). When $n = 3$, the classical BMC approach verifies such a system in 11,679 seconds, and it takes more than 8 hours to verify a bigger model. By contrast, when $n = 7$, SBTP checks such a system in 77 seconds.

5   Conclusion

In this paper, we introduced the Milestones model checker. It merges POR methods and symbolic approaches to provide automatic verification of CTL and LTL properties on asynchronous models. The CTL properties are checked by means of a BDD-based approach, and the LTL properties can be verified either by the BDD-based technique or by the bounded model checking approach.

We show on a realistic-sized case study that our methods achieve an improvement in comparison to the classical algorithms. Although it is usually considered that symbolic model checking is inadequate for asynchronous systems, our results show that with appropriate optimization this approach might in fact be quite effective to tackle the state space explosion problem.

Although Milestones is able to check temporal properties, it needs to be extended by adding generation of counter-examples for failed properties. The heuristic used to determine safe transitions, i.e. transitions which can be exploited to perform POR, is quite simple. Instead of defining a set of processes, we could define a hierarchy of processes, and exploit it to discover more safe transitions.

References

1. Biere, A., Cimatti, A., Clarke, E.M., Strichman, O., Zhu, Y.: Bounded model checking. Advances in Computers 58, 118–149 (2003)
2. Bortnik, E.M., Trčka, N., Wijs, A., Luttik, B., van de Mortel-Fronczak, J.M., Baeten, J.C.M., Fokkink, W., Rooda, J.E.: Analyzing a χ model of a turntable system using spin, cadp and uppaal. J. Log. Algebr. Program 65(2), 51–104 (2005)
3. Burch, J.R., Clarke, E.M., McMillan, K.L., Dill, D.L., Hwang, J.: Symbolic model checking: 10^{20} states and beyond. Information and Computation 98(2), 142–170 (1992)
4. Cimatti, A., Clarke, E., Giunchiglia, F., Roveri, M.: NUSMV: A new symbolic model verifier. In: Halbwachs, N., Peled, D.A. (eds.) CAV 1999. LNCS, vol. 1633, pp. 495–499. Springer, Heidelberg (1999)
5. Clarke, E.M., Grumberg, O., Hamaguchi, K.: Another look at LTL model checking. Form. Methods Syst. Des. 10(1), 47–71 (1997)
6. Godefroid, P.: Partial-Order Methods for the Verification of Concurrent Systems. LNCS, vol. 1032. Springer, Heidelberg (1996)

7. Holzmann, G.J.: The model checker SPIN. IEEE Transactions on Software Engineering 23(5) (1997)
8. Iwashita, H., Nakata, T., Hirose, F.: CTL model checking based on forward state traversal. In: ICCAD 1996: Proceedings of the 1996 IEEE/ACM International Conference on Computer-aided Design, pp. 82–87. IEEE Computer Society, Washington, DC (1996)
9. Lerda, F., Sinha, N., Theobald, M.: Symbolic model checking of software. In: Cook, B., Stoller, S., Visser, W. (eds.) Electronic Notes in Theoretical Computer Science, vol. 89. Elsevier, Amsterdam (2003)
10. Peled, D.: Combining partial order reductions with on-the-fly model-checking. Formal Methods in System Design 8(1), 39–64 (1996)
11. Vander Meulen, J., Pecheur, C.: Efficient symbolic model checking for process algebras. In: Cofer, D., Fantechi, A. (eds.) FMICS 2008. LNCS, vol. 5596, pp. 69–84. Springer, Heidelberg (2009)
12. Vander Meulen, J., Pecheur, C.: Combining partial order reduction with bounded model checking. In: Communicating Process Architectures 2009 - WoTUG-32. Concurrent Systems Engineering Series, vol. 67, pp. 29–48. IOS Press, Amsterdam (2009)

Author Index